Advanced Mathematics

Precalculus with Discrete Mathematics and Data Analysis

Solution Key

McDougal Littell/Houghton Mifflin
Evanston, Illinois
Boston   Dallas   Phoenix

Contents

Class Exercises 1-1 · Pages 4–5

1. $CD = \sqrt{(8-0)^2 + (6-0)^2} = \sqrt{100} = 10$; $M = \left(\dfrac{0+8}{2}, \dfrac{0+6}{2}\right) = (4,3)$

2. $CD = \sqrt{(6-4)^2 + (6-2)^2} = \sqrt{20} = 2\sqrt{5}$; $M = \left(\dfrac{4+6}{2}, \dfrac{2+6}{2}\right) = (5,4)$

3. $CD = \sqrt{(3-(-3))^2 + (-2-4)^2} = \sqrt{72} = 6\sqrt{2}$; $M = \left(\dfrac{-3+3}{2}, \dfrac{4-2}{2}\right) = (0,1)$

4. $CD = \sqrt{(7-7)^2 + (-1-(-9))^2} = \sqrt{64} = 8$; $M = \left(\dfrac{7+7}{2}, \dfrac{-9-1}{2}\right) = (7,-5)$

5. $M = \left(\dfrac{2+6}{2}, \dfrac{3+7}{2}\right) = (4,5)$; midpoint of $\overline{AM} = \left(\dfrac{2+4}{2}, \dfrac{3+5}{2}\right) = (3,4)$;

midpoint of $\overline{MB} = \left(\dfrac{4+6}{2}, \dfrac{5+7}{2}\right) = (5,6)$

6. a. $2(3) + 3(3) = 15$; yes **b.** $2(9) + 3(-1) = 15$; yes **c.** $2(2.5) + 3(3.5) = 15.5$; no
d. $2(-10.5) + 3(12) = 15$; yes

7. a. Answers may vary. Examples: $(-3,4), (0,4), (2,4)$ **b.** $y = 4$

8. a. Answers may vary. Examples: $(8,0), (8,-1), \left(8,\frac{3}{2}\right)$ **b.** $x = 8$

9. $4x - 3 \cdot 0 = 18$, $x = 4.5$; x-axis, $(4.5, 0)$; $4 \cdot 0 - 3y = 18$, $y = -6$; y-axis, $(0, -6)$

10. a. $(2,3)$ **b.** Adding, $3x = 6$ and $x = 2$; $2 + y = 5$ and $y = 3$; $(2,3)$
c. The answers are the same.

Written Exercises 1-1 · Pages 5–7

1. $CD = \sqrt{(7-1)^2 + (8-0)^2} = \sqrt{36+64} = 10$; $M = \left(\dfrac{1+7}{2}, \dfrac{0+8}{2}\right) = (4,4)$

2. $CD = \sqrt{(15-3)^2 + (12-3)^2} = \sqrt{144+81} = 15$; $M = \left(\dfrac{3+15}{2}, \dfrac{3+12}{2}\right) = \left(9, \dfrac{15}{2}\right)$

3. $CD = \sqrt{(7+8)^2 + (5+3)^2} = \sqrt{225+64} = 17$; $M = \left(\dfrac{-8+7}{2}, \dfrac{-3+5}{2}\right) = \left(-\dfrac{1}{2}, 1\right)$

4. $CD = \sqrt{(4+2)^2 + (9+1)^2} = \sqrt{36+100} = 2\sqrt{34}$; $M = \left(\dfrac{-2+4}{2}, \dfrac{-1+9}{2}\right) = (1,4)$

5. $CD = \sqrt{\left(-2 - \dfrac{1}{2}\right)^2 + \left(-\dfrac{3}{2} - \dfrac{9}{2}\right)^2} = \sqrt{\dfrac{25}{4} + 36} = \sqrt{\dfrac{169}{4}} = \dfrac{13}{2}$;

$M = \left(\dfrac{\frac{1}{2} - 2}{2}, \dfrac{\frac{9}{2} - \frac{3}{2}}{2}\right) = \left(-\dfrac{3}{4}, \dfrac{3}{2}\right)$

6. $CD = \sqrt{\left(-\dfrac{5}{2} - \dfrac{7}{2}\right)^2 + \left(\dfrac{7}{2} + 1\right)^2} = \sqrt{36 + \dfrac{81}{4}} = \dfrac{15}{2}$;

$M = \left(\dfrac{\frac{7}{2} - \frac{5}{2}}{2}, \dfrac{-1 + \frac{7}{2}}{2}\right) = \left(\dfrac{1}{2}, \dfrac{5}{4}\right)$

7. $CD = \sqrt{(4.8 - 4.8)^2 + (-2.8 - 2.2)^2} = \sqrt{0 + 25} = 5$;

$M = \left(\dfrac{4.8 + 4.8}{2}, \dfrac{2.2 - 2.8}{2}\right) = (4.8, -0.3)$

8. $CD = \sqrt{(-2.3 - 1.7)^2 + (5.7 - 5.7)^2} = \sqrt{16 + 0} = 4$;

$M = \left(\dfrac{1.7 - 2.3}{2}, \dfrac{5.7 + 5.7}{2}\right) = (-0.3, 5.7)$

9. a. $3(9) - 2(6) = 15$; yes   **b.** $3(8) - 2(4) = 16$; no

   c. $3\left(-\dfrac{4}{3}\right) - 2\left(-\dfrac{19}{2}\right) = 15$; yes   **d.** $3(3.4) - 2(-3.2) = 16.6$; no

   e. $3(-9) - 2(-22) = 17$; no

10. a. $-5(-1.2) + 4(3.0) = 18$; yes   **b.** $-5(3) + 4\left(-\dfrac{3}{4}\right) = -18$; no

   c. $-5(-18) + 4(24) = 186$; no   **d.** $-5(-6) + 4(-3) = 18$; yes

   e. $-5(3.6) + 4(9) = 18$; yes

11. $A = \dfrac{1}{2}bh = \dfrac{1}{2}(2)(3) = 3$ (sq. units)   **12.** $A = \dfrac{1}{2}bh = \dfrac{1}{2}(6)(8) = 24$ (sq. units)

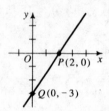

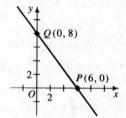

13. $(5, 3)$; $x = 5$ and $y = 3$ **14.** $(-2, -1)$; $x = -2$ and $y = -1$

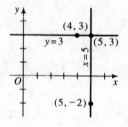

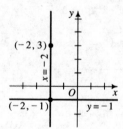

15.
$$\begin{array}{ll} 3x - 5y = 9 & 3x - 5y = 9 \\ x + y = 3 & 3x + 3y = 9 \\ & \overline{-8y = 0}; \; y = 0; \; x + 0 = 3; \; x = 3; \; (3, 0) \end{array}$$

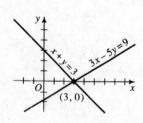

Ex. 15

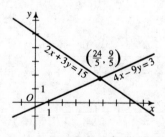

Ex. 16

16. $2x + 3y = 15$ $4x + 6y = 30$
     $4x - 9y = 3$ $\underline{4x - 9y = 3}$
                      $15y = 27; y = \dfrac{9}{5}; 2x + 3\left(\dfrac{9}{5}\right) = 15; x = \dfrac{24}{5}; \left(\dfrac{24}{5}, \dfrac{9}{5}\right)$

17. $x - 3y = 4$ $5x - 15y = 20$
    $5x + y = -8$ $\underline{5x + y = -8}$
                    $-16y = 28; y = -\dfrac{7}{4}; x - 3\left(-\dfrac{7}{4}\right) = 4; x = -\dfrac{5}{4}; \left(-\dfrac{5}{4}, -\dfrac{7}{4}\right)$

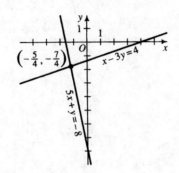

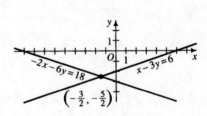

Ex. 17   Ex. 18

18. $-2x - 6y = 18$ $-2x - 6y = 18$
      $x - 3y = 6$ $\underline{-2x + 6y = -12}$
                    $-12y = 30; y = -\dfrac{5}{2}; x - 3\left(-\dfrac{5}{2}\right) = 6;$

$x = -\dfrac{3}{2}; \left(-\dfrac{3}{2}, -\dfrac{5}{2}\right)$

19. $AB = \sqrt{(3 - 1)^2 + (5 - 7)^2} = \sqrt{4 + 4} = 2\sqrt{2}; BC = \sqrt{(4 - 3)^2 + (-1 - 5)^2} = \sqrt{1 + 36} = \sqrt{37}; CD = \sqrt{(2 - 4)^2 + (1 + 1)^2} = \sqrt{4 + 4} = 2\sqrt{2}; AD = \sqrt{(2 - 1)^2 + (1 - 7)^2} = \sqrt{1 + 36} = \sqrt{37}; AB = CD$ and $AD = BC$, so $ABCD$ is a parallelogram.

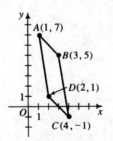

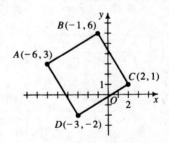

Ex. 19   Ex. 20

20. $AB = \sqrt{(-1 + 6)^2 + (6 - 3)^2} = \sqrt{25 + 9} = \sqrt{34}; BC = \sqrt{(2 + 1)^2 + (1 - 6)^2} = \sqrt{9 + 25} = \sqrt{34}; CD = \sqrt{(-3 - 2)^2 + (-2 - 1)^2} = \sqrt{25 + 9} = \sqrt{34}; AD = \sqrt{(-3 + 6)^2 + (-2 - 3)^2} = \sqrt{9 + 25} = \sqrt{34}; AB = BC = CD = AD$, so $ABCD$ is a rhombus. $AC = \sqrt{(2 + 6)^2 + (1 - 3)^2} = \sqrt{64 + 4} = 2\sqrt{17}; BD = \sqrt{(-3 + 1)^2 + (-2 - 6)^2} = \sqrt{4 + 64} = 2\sqrt{17}; AC = BD$, so $ABCD$ is a rectangle. Thus, $ABCD$ is a square.

21. $\overline{AC}$ has midpoint $\left(\dfrac{5+1}{2}, \dfrac{1-3}{2}\right) = (3, -1)$; $\overline{BD}$ has midpoint $\left(\dfrac{7-1}{2}, \dfrac{-1-1}{2}\right) = (3, -1)$.

Since diagonals $\overline{AC}$ and $\overline{BD}$ have the same midpoint, the diagonals bisect each other. Thus, $ABCD$ is a parallelogram.

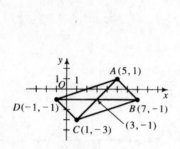

Ex. 21   **Ex. 22**

22. $\overline{AC}$ has midpoint $\left(\dfrac{2+9}{2}, \dfrac{0+1}{2}\right) = \left(\dfrac{11}{2}, \dfrac{1}{2}\right)$; $\overline{BD}$ has midpoint $\left(\dfrac{4+7}{2}, \dfrac{-6+7}{2}\right) =$ $\left(\dfrac{11}{2}, \dfrac{1}{2}\right)$. Since $\overline{AC}$ and $\overline{BD}$ intersect at their midpoints, they bisect each other. Thus, $ABCD$ is a parallelogram.

23. $AB = \sqrt{(1+3)^2 + (11-3)^2} = \sqrt{16+64} = 4\sqrt{5}$; $BC = \sqrt{(3-1)^2 + (15-11)^2} =$ $\sqrt{4+16} = 2\sqrt{5}$; $AC = \sqrt{(3+3)^2 + (15-3)^2} = \sqrt{36+144} = 6\sqrt{5}$; $AB + BC =$ $4\sqrt{5} + 2\sqrt{5} = 6\sqrt{5} = AC$, so B is on $\overline{AC}$.

24. $AB = \sqrt{(-1+3)^2 + (4-7)^2} = \sqrt{4+9} = \sqrt{13}$; $BC = \sqrt{(3+1)^2 + (-2-4)^2} =$ $\sqrt{16+36} = 2\sqrt{13}$; $AC = \sqrt{(3+3)^2 + (-2-7)^2} = \sqrt{36+81} = 3\sqrt{13}$; $AB + BC =$ $\sqrt{13} + 2\sqrt{13} = 3\sqrt{13} = AC$, so B is on $\overline{AC}$.

25. a. $AP = \sqrt{(9-4)^2 + (2-2)^2} = \sqrt{25+0} = 5$;
$BP = \sqrt{(1-4)^2 + (6-2)^2} = \sqrt{9+16} = 5$; $AP = BP$

b. $\sqrt{(9-2)^2 + (2-k)^2} = \sqrt{(1-2)^2 + (6-k)^2}$;
$49 + 4 - 4k + k^2 = 1 + 36 - 12k + k^2$; $8k = -16$; $k = -2$

26. a. $AP = \sqrt{(-5-1)^2 + (-3-4)^2} = \sqrt{36+49} = \sqrt{85}$;
$BP = \sqrt{(-1-1)^2 + (-5-4)^2} = \sqrt{4+81} = \sqrt{85}$; $AP = BP$

b. $\sqrt{(-5-3)^2 + (-3-k)^2} = \sqrt{(-1-3)^2 + (-5-k)^2}$;
$64 + 9 + 6k + k^2 = 16 + 25 + 10k + k^2$; $32 = 4k$; $8 = k$

27. If $P = (a, 0)$, $\sqrt{(a+3)^2 + (0-5)^2} = 13$; $a^2 + 6a + 9 + 25 = 169$;
$a^2 + 6a - 135 = 0$; $(a+15)(a-9) = 0$; $a = -15$ or $a = 9$; $(-15, 0)$ or $(9, 0)$

28. If $Q = (0, b)$, $\sqrt{(0-6)^2 + (b-1)^2} = 2\sqrt{10}$; $36 + b^2 - 2b + 1 = 40$;
$b^2 - 2b - 3 = 0$; $(b-3)(b+1) = 0$; $b = 3$ or $b = -1$; $(0, 3)$ or $(0, -1)$

29. $x + 3y = 19$
$\underline{x - 2y = 4}$

$5y = 15$; $y = 3$; $x + 3(3) = 19$; $x = 10$. Thus lines $x + 3y = 19$ and $x - 2y = 4$ intersect at $(10, 3)$. $2x - 5y = 5$; $2(10) - 5(3) \stackrel{?}{=} 5$; $20 - 15 = 5$, so $(10, 3)$ is on all three lines. The lines intersect in one point.

30. $3x + 2y = 4$

$\underline{5x - 2y = 0}$

$8x \quad\quad = 4; x = \dfrac{1}{2}; 3\left(\dfrac{1}{2}\right) + 2y = 4; y = \dfrac{5}{4}$. Thus lines $3x + 2y = 4$ and

$5x - 2y = 0$ intersect at $\left(\dfrac{1}{2}, \dfrac{5}{4}\right)$. $4x + 3y = 3; 4\left(\dfrac{1}{2}\right) + 3\left(\dfrac{5}{4}\right) \overset{?}{=} 3; 5\dfrac{3}{4} \neq 3$; the lines

do not intersect in one point.

31. a. $(BC)^2 + (AC)^2 = (\sqrt{(-2 - 6)^2 + (-1 - 3)^2})^2 + (\sqrt{(-2 + 6)^2 + (-1 - 7)^2})^2 =$

$(64 + 16) + (16 + 64) = 160; (AB)^2 =$

$(\sqrt{(6 + 6)^2 + (3 - 7)^2})^2 = 144 + 16 =$

$160; (BC)^2 + (AC)^2 = (AB)^2$, so $\angle C$ is a

right angle.

b. $M = \left(\dfrac{-6 + 6}{2}, \dfrac{7 + 3}{2}\right) = (0,5); CM =$

$\sqrt{(0 + 2)^2 + (5 + 1)^2} = \sqrt{4 + 36} = 2\sqrt{10};$

from part (a), $(AB)^2 = 160$, so $AB = \sqrt{160} =$

$4\sqrt{10}$; thus, $CM = 2\sqrt{10} = \dfrac{1}{2}(4\sqrt{10}) = \dfrac{1}{2}AB.$

32. $AB = \sqrt{(5 + 13)^2 + (17 - 2)^2} = \sqrt{324 + 225} = \sqrt{549};$

$BC = \sqrt{(22 - 5)^2 + (-4 - 17)^2} = \sqrt{289 + 441} = \sqrt{730};$

$AC = \sqrt{(22 + 13)^2 + (-4 - 2)^2} = \sqrt{1225 + 36} = \sqrt{1261};$

$s = \dfrac{\sqrt{549} + \sqrt{730} + \sqrt{1261}}{2}$; area $= \sqrt{s(s - a)(s - b)(s - c)} = 316.5$ sq. units

33. Let $A = (a,b)$, $B = (c,d)$, and $C = (e,f)$. Then $\left(\dfrac{a + c}{2}, \dfrac{b + d}{2}\right) = (7,3);$

$\left(\dfrac{c + e}{2}, \dfrac{d + f}{2}\right) = (10,9)$, and $\left(\dfrac{a + e}{2}, \dfrac{b + f}{2}\right) = (5,5)$. Thus: (1) $a + c = 14,$

(2) $b + d = 6$, (3) $c + e = 20$, (4) $d + f = 18$, (5) $a + e = 10$, and (6) $b + f = 10.$

(5) $-$ (3) gives $a - c = -10$ and since (1) $a + c = 14$, $2a = 4$, and $a = 2$. Hence

$c = 12$ and $e = 8$. (6) $-$ (4) gives $b - d = -8$ and since (2) $b + d = 6$, $2b = -2$ and

$b = -1$. Hence $d = 7$ and $f = 11$. $A = (2,-1)$, $B = (12,7)$, and $C = (8,11).$

34. $\overline{AB}$ is the hypotenuse of right triangle ABC. $AC = |x_2 - x_1|$ and $BC = |y_2 - y_1|$. By

the Pythagorean Theorem, $(AB)^2 = (AC)^2 + (BC)^2 =$

$|x_2 - x_1|^2 + |y_2 - y_1|^2 = (x_2 - x_1)^2 + (y_2 - y_1)^2;$

since distances are nonnegative, $AB = \sqrt{(x_2 - x_1)^2 + (y_2 - y_1)^2}.$

35. a. $\overleftrightarrow{BC}$ has equation $x = x_2$, so the x-coordinate of P is x_2. P is the midpoint of $\overline{BC}$, so

if $P = (x_2,k)$, $|y_2 - k| = |k - y_1|$; $y_2 - k = k - y_1$; $2k = y_1 + y_2$; $k = \dfrac{y_1 + y_2}{2};$

$P = \left(x_2, \dfrac{y_1 + y_2}{2}\right).$

b. $\overleftrightarrow{AC}$ has equation $y = y_1$, so let $Q = (j,y_1)$. Since Q is the midpoint of $\overline{AC}$,

$|x_2 - j| = |j - x_1|$; $x_2 - j = j - x_1$; $2j = x_1 + x_2$; $j = \dfrac{x_1 + x_2}{2}$; $Q = \left(\dfrac{x_1 + x_2}{2}, y_1\right).$

c. M and P are midpoints, so $\overline{MP} \parallel \overline{AC}$; thus, since $\overline{AC}$ is horizontal, $\overline{MP}$ is horizontal

and M and P have the same y-coordinate.

(Continued)

d. M and Q are midpoints, so $\overline{MQ} \parallel \overline{BC}$; thus, since $\overline{BC}$ is vertical, $\overline{MQ}$ is vertical and M and Q have the same x-coordinate.

e. Let $M = (x_3, y_3)$. From parts (b) and (d), $x_3 = \dfrac{x_1 + x_2}{2}$. From parts (a) and (c),

$y_3 = \dfrac{y_1 + y_2}{2}$. Thus, $M = \left(\dfrac{x_1 + x_2}{2}, \dfrac{y_1 + y_2}{2} \right)$.

36. Given $A(-3,1)$, $B(1,4)$, and $C(4,3)$, let $X(a,b)$ be the fourth vertex. The diagonals of a parallelogram bisect each other. *Case 1:* $\overline{AX}$ bisects $\overline{BC}$. $\left(\dfrac{-3+a}{2}, \dfrac{1+b}{2} \right) =$ $\left(\dfrac{1+4}{2}, \dfrac{4+3}{2} \right)$; $-3 + a = 5$ and $1 + b = 7$; $X = (8,6)$. *Case 2:* $\overline{BX}$ bisects $\overline{AC}$. $\left(\dfrac{1+a}{2}, \dfrac{4+b}{2} \right) = \left(\dfrac{-3+4}{2}, \dfrac{1+3}{2} \right)$; $1 + a = 1$ and $4 + b = 4$; $X = (0,0)$. *Case 3:* $\overline{CX}$ bisects $\overline{AB}$. $\left(\dfrac{4+a}{2}, \dfrac{3+b}{2} \right) = \left(\dfrac{-3+1}{2}, \dfrac{1+4}{2} \right)$; $4 + a = -2$ and $3 + b = 5$; $X = (-6,2)$. There are three possible answers.

Activity 1-2 · page 10

a. They are complementary; they are congruent. b. They are complementary.

c. They are perpendicular. d. $\dfrac{a}{b}$; $-\dfrac{b}{a}$

Class Exercises 1-2 · page 11

1. $\dfrac{7-0}{5-0} = \dfrac{7}{5}$ **2.** $\dfrac{6-0}{3-2} = 6$ **3.** $\dfrac{0-3}{1+2} = -1$ **4.** $\dfrac{-3+6}{8-2} = \dfrac{1}{2}$ **5.** $\dfrac{2-2}{8+8} = 0$

6. Since $x_2 - x_1 = 0$, the line has no slope. **7.** slope, 3; y-intercept, 4

8. slope, $\dfrac{3}{5}$; y-intercept, -3 **9.** slope, $-\dfrac{4}{3}$; y-intercept, 3 **10.** slope, 0; y-intercept, -2

11. The denominator $x_2 - x_1$ in the slope formula is equal to zero. Division by zero is undefined.

12. $\dfrac{1}{2}$ **13. a.** $-\dfrac{3}{2}$ **b.** $\dfrac{2}{3}$

14. slope of $\overline{AB} = \dfrac{9-5}{-1+4} = \dfrac{4}{3}$; slope of $\overline{BC} = \dfrac{3-9}{7+1} = -\dfrac{3}{4}$; slope of $\overline{CD} = \dfrac{-1-3}{4-7} = \dfrac{4}{3}$;

slope of $\overline{AD} = \dfrac{-1-5}{4+4} = -\dfrac{3}{4}$

a. slope of $\overline{AB} =$ slope of $\overline{CD}$, and slope of $\overline{BC} =$ slope of $\overline{AD}$, so $\overline{AB} \parallel \overline{CD}$ and $\overline{BC} \parallel \overline{AD}$.

b. $\dfrac{4}{3}\left(-\dfrac{3}{4} \right) = -1$; since adjacent sides have slopes $\dfrac{4}{3}$ and $-\dfrac{3}{4}$, adjacent sides are

perpendicular.

Written Exercises 1-2 · pages 11–13

1. $\dfrac{5-2}{9-4} = \dfrac{3}{5}$ **2.** $\dfrac{0-4}{12-0} = -\dfrac{1}{3}$ **3.** $\dfrac{-6+2}{2+4} = -\dfrac{2}{3}$ **4.** $\dfrac{-2-6}{2+2} = -2$ **5.** $\dfrac{5-5}{-7-8} = 0$

6. Since $x_2 - x_1 = 0$, the line has no slope. **7.** $\dfrac{-\frac{1}{3}-2}{\frac{1}{4}+\frac{1}{3}} = \dfrac{-\frac{7}{3}}{\frac{7}{12}} = -4$

8. $\dfrac{1-1.5}{0.5-0.25} = \dfrac{-0.5}{0.25} = -2$ **9.** $\dfrac{a-b}{b-a} = \dfrac{-(b-a)}{b-a} = -1$ if $a \neq b$

10. $\dfrac{\frac{b}{a}-\frac{a}{b}}{b-a} = \dfrac{b^2-a^2}{ab} \cdot \dfrac{1}{b-a} = \dfrac{(b+a)(b-a)}{ab(b-a)} = \dfrac{b+a}{ab}$ if $a \neq b$ **11.** slope, 3; y-intercept, 5

12. slope, -5; y-intercept, 4 **13.** $y = 2x - 4$; slope, 2; y-intercept, -4

14. $y = -\dfrac{1}{3}x + \dfrac{7}{9}$; slope, $-\dfrac{1}{3}$; y-intercept, $\dfrac{7}{9}$ **15.** $y = \dfrac{11}{3}x$; slope, $\dfrac{11}{3}$; y-intercept, 0

16. slope, 0; y-intercept, 5

17. a. $m = \dfrac{5}{2}$ **b.** $m = \dfrac{15}{6} = \dfrac{5}{2}$ **c.** $m = -\dfrac{4}{10} = -\dfrac{2}{5}$ The lines in (a) and (b) are parallel;

the line in (c) is perpendicular to the other two.

18. a. $m = \dfrac{5}{3}$ **b.** $m = -\dfrac{3}{5}$ **c.** $m = -\dfrac{6}{10} = -\dfrac{3}{5}$ The lines in (b) and (c) are parallel;

the line in (a) is perpendicular to the other two.

19. $m_1 = \dfrac{2+3}{7-2} = 1$; $m_2 = \dfrac{2-7}{2+3} = -1$; $m_1 \cdot m_2 = 1(-1) = -1$, so the lines are

perpendicular.

20. $m_1 = \dfrac{-2-3}{5-2} = -\dfrac{5}{3}$; l_2 has equation $y = \dfrac{3}{5}x - 3$, so $m_2 = \dfrac{3}{5}$; $m_1 \cdot m_2 = -\dfrac{5}{3}\left(\dfrac{3}{5}\right) =$

-1, so the lines are perpendicular.

21. $m_1 = \dfrac{8-k}{6-4} = \dfrac{8-k}{2}$ and $m_2 = \dfrac{8-4}{0+1} = 4$ **a.** $m_1 = m_2$; $\dfrac{8-k}{2} = 4$; $8 - k = 8$; $k = 0$

b. $m_1 = -\dfrac{1}{m_2}$; $\dfrac{8-k}{2} = \dfrac{-1}{4}$; $8 - k = -\dfrac{1}{2}$; $k = 8\dfrac{1}{2}$

22. $m_1 = \dfrac{10-h}{5-3} = \dfrac{10-h}{2}$; $m_2 = 3$ **a.** $m_1 = m_2$; $\dfrac{10-h}{2} = 3$; $10 - h = 6$; $h = 4$

b. $m_1 = -\dfrac{1}{m_2}$; $\dfrac{10-h}{2} = -\dfrac{1}{3}$; $10 - h = -\dfrac{2}{3}$; $h = 10\dfrac{2}{3}$

23. a. slope of $\overline{AB} = \dfrac{4+6}{2+4} = \dfrac{5}{3}$; slope of $\overline{BC} = \dfrac{6-4}{8-2} = \dfrac{1}{3}$; slope of $\overline{CD} = \dfrac{-4-6}{2-8} = \dfrac{5}{3}$;

slope of $\overline{AD} = \dfrac{-4+6}{2+4} = \dfrac{1}{3}$; the slopes of opposite sides are equal, so $ABCD$ is a

parallelogram. **b.** midpoint of $\overline{AC} = \left(\dfrac{-4+8}{2}, \dfrac{-6+6}{2}\right) = (2,0)$;

midpoint of $\overline{BD} = \left(\dfrac{2+2}{2}, \dfrac{4-4}{2}\right) = (2,0)$

24. a. There are several ways to show that $ABCD$ is a parallelogram. *Example*: midpoint of

$\overline{AC} = \left(\dfrac{-4 + 4}{2}, \dfrac{1 + 9}{2}\right) = (0, 5)$; midpoint of $\overline{BD} = \left(\dfrac{2 - 2}{2}, \dfrac{3 + 7}{2}\right) = (0, 5)$; thus, the

diagonals bisect each other and $ABCD$ is a parallelogram. Slope of $\overline{AC} = \dfrac{9 - 1}{4 + 4} = 1$;

slope of $\overline{BD} = \dfrac{7 - 3}{-2 - 2} = -1$; $1(-1) = -1$, so $\overline{AC} \perp \overline{BD}$. **b.** rhombus

25. Given $P(-4, -5)$, $Q(-3, 0)$, $R(0, 2)$, and $S(5, 1)$, slope of $\overline{PS} = \dfrac{1 + 5}{5 + 4} = \dfrac{2}{3}$ and slope of

$\overline{QR} = \dfrac{2 - 0}{0 + 3} = \dfrac{2}{3}$; $\overline{PS} \parallel \overline{QR}$. Slope of $\overline{PQ} = \dfrac{0 + 5}{-3 + 4} = 5$ and slope of $\overline{RS} = \dfrac{1 - 2}{5 - 0} = -\dfrac{1}{5}$;

$\overline{PQ} \nparallel \overline{RS}$; $PQRS$ has exactly two parallel sides, so it is a trapezoid.
$PQ = \sqrt{(-3 + 4)^2 + (0 + 5)^2} = \sqrt{26}$; $RS = \sqrt{(5 - 0)^2 + (1 - 2)^2} = \sqrt{26}$;
$PQ = RS$, so trap. $PQRS$ is isosceles.

26. Given $W(-1, -1)$, $X(9, 4)$, $Y(20, 6)$, and $Z(10, 1)$, first show that $WXYZ$ is a

parallelogram. *Example*: midpoint of $\overline{WY} = \left(\dfrac{-1 + 20}{2}, \dfrac{-1 + 6}{2}\right) = \left(\dfrac{19}{2}, \dfrac{5}{2}\right)$; midpoint

of $\overline{XZ} = \left(\dfrac{9 + 10}{2}, \dfrac{4 + 1}{2}\right) = \left(\dfrac{19}{2}, \dfrac{5}{2}\right)$; since the diagonals bisect each other, $WXYZ$ is a

parallelogram. Slope of $\overline{WY} = \dfrac{6 + 1}{20 + 1} = \dfrac{1}{3}$; slope of $\overline{XZ} = \dfrac{1 - 4}{10 - 9} = -3$; $\dfrac{1}{3}(-3) = -1$,

so diagonals $\overline{WY}$ and $\overline{XZ}$ are perpendicular. Thus, $WXYZ$ is a rhombus. $d_1 = WY = $
$\sqrt{(20 + 1)^2 + (6 + 1)^2} = 7\sqrt{10}$; $d_2 = XZ = \sqrt{(10 - 9)^2 + (1 - 4)^2} = \sqrt{10}$; area $=$

$\dfrac{1}{2}d_1 d_2 = \dfrac{1}{2}(7\sqrt{10})(\sqrt{10}) = 35$ sq. units

27. Since (x_1, y_1) and (x_2, y_2) are points on $y = Mx + k$, $y_2 = Mx_2 + k$ and $y_1 = Mx_1 + k$.

Subtract to get $y_2 - y_1 = Mx_2 - Mx_1$; $y_2 - y_1 = M(x_2 - x_1)$; $M = \dfrac{y_2 - y_1}{x_2 - x_1}$. (Note that

$x_2 \neq x_1$ since a vert. line cannot have eq. $y = Mx + k$.)

28. a. Since $y = m_1 x + b_1$ and $y = m_2 x + b_2$, $m_1 x + b_1 = m_2 x + b_2$; $(m_1 - m_2)x =$

$b_2 - b_1$; $x = \dfrac{b_2 - b_1}{m_1 - m_2}$.

b. It is undefined; either the lines coincide, or they do not intersect and are parallel.

29. a. slope of $l_1 = \dfrac{BC}{AC}$; slope of $l_2 = \dfrac{EF}{DF}$

b. slope of $l_1 = $ slope of l_2 (Given); $\dfrac{BC}{AC} = \dfrac{EF}{DF}$ (Def. of slope); $\angle BCA \cong \angle EFD$ (All rt. $\angle$s

are $\cong$.); (1) $\triangle ABC \sim \triangle DEF$ (SAS Similarity); (2) $\angle BAC \cong \angle EDF$ (Corr. $\angle$s of $\sim$ $\triangle$s
are $\cong$.); (3) $l_1 \parallel l_2$ (If 2 lines are cut by a trans. so that corr. $\angle$s are $\cong$, the lines are $\parallel$.)

c. $l_1 \parallel l_2$ (Given); (1) $\angle BAC \cong \angle EDF$ (If 2 $\parallel$ lines are cut by a trans., corr. $\angle$s are $\cong$.);

$\angle BCA \cong \angle EFD$ (All rt. $\angle$s are $\cong$.); (2) $\triangle ABC \sim \triangle DEF$ (AA Similarity); $\dfrac{BC}{AC} = \dfrac{EF}{DF}$

(Corr. sides of $\sim$ $\triangle$s are proportional.); (3) slope of $l_1 = $ slope of l_2 (Def. of slope)

30. a. l_1 has slope $m_1 = \dfrac{BA}{OA}$; l_2 has slope $m_2 = -\dfrac{OC}{DC}$.

b. $m_1 = -\dfrac{1}{m_2}$ (Given); $\dfrac{BA}{OA} = -\dfrac{1}{-\dfrac{OC}{DC}} = \dfrac{DC}{OC}$ (Subst. prop.); $\dfrac{BA}{DC} = \dfrac{OA}{OC}$ (A prop. of

proportions); $\angle C \cong \angle A$ (All rt. $\angle$s are $\cong$.); (1) $\triangle OAB \sim \triangle OCD$ (SAS Similarity);
(2) $\angle 1 \cong \angle 2$ (Corr. $\angle$s of $\sim$ $\triangle$s are $\cong$.); $m\angle 1 + m\angle 3 = 90$ (x- and y-axes intersect at
rt. $\angle$s.); $m\angle 2 + m\angle 3 = 90$ (Subst. prop.); $l_1 \perp l_2$ (Lines that form a right angle
are $\perp$.)

c. $l_1 \perp l_2$ (Given); $m\angle DOB = 90$ (Def. of $\perp$ lines and rt. angle); $m\angle 2 + m\angle 3 = 90$
(Subst. prop.); $m\angle 1 + m\angle 3 = 90$ (x- and y-axes intersect at rt. $\angle$s.); $m\angle 2 + m\angle 3 =$
$m\angle 1 + m\angle 3$ (Subst. prop.); (1) $\angle 2 \cong \angle 1$ (Subtr. prop. of =.); $\angle C \cong \angle A$ (All rt. $\angle$s

are $\cong$.); (2) $\triangle OAB \sim \triangle OCD$ (AA Similarity); $\dfrac{BA}{OA} = \dfrac{DC}{OC}$ (Corr. sides of $\sim$ $\triangle$s are

prop.); $\dfrac{BA}{OA} = -\dfrac{1}{-\dfrac{OC}{DC}}$; $\dfrac{BA}{OA}\left(-\dfrac{OC}{DC}\right) = -1$; slope of $l_1 \cdot$ slope of $l_2 = -1$ (Def. of slope;

subst. prop.)

31. a. $OA = \sqrt{(1-0)^2 + (m_1 - 0)^2} = \sqrt{(m_1)^2 + 1}$; $OB = \sqrt{(1-0)^2 + (m_2 - 0)^2} =$
$\sqrt{(m_2)^2 + 1}$; $AB = \sqrt{(1-1)^2 + (m_1 - m_2)^2} = \sqrt{(m_1 - m_2)^2}$.

b. If $(OA)^2 + (OB)^2 = (AB)^2$, then $(m_1)^2 + 1 + (m_2)^2 + 1 = (m_1 - m_2)^2$;
$(m_1)^2 + (m_2)^2 + 2 = (m_1)^2 - 2m_1 m_2 + (m_2)^2$; $2 = -2m_1 m_2$; $m_1 \cdot m_2 = -1$.

c. For the converse, suppose that $m_1 \cdot m_2 = -1$. Then $(AB)^2 = (m_1 - m_2)^2 =$
$(m_1)^2 - 2m_1 m_2 + (m_2)^2 = (m_1)^2 - 2(-1) + (m_2)^2 = (m_1)^2 + 2 + (m_2)^2 =$
$[(m_1)^2 + 1] + [(m_2)^2 + 1] = (OA)^2 + (OB)^2$. Thus, if $m_1 \cdot m_2 = -1$,
$(OA)^2 + (OB)^2 = (AB)^2$.

32. Let $C = (e,f)$ and $D = (g,h)$. Since $\overline{AB} \perp \overline{AD}$, $\dfrac{8-0}{6-0} \cdot \dfrac{h-0}{g-0} = -1$; $\dfrac{4h}{3g} = -1$;

$h = -\dfrac{3}{4}g$; thus, $D = \left(g, -\dfrac{3}{4}g\right)$. $AB = AD$; $\sqrt{(6-0)^2 + (8-0)^2} =$

$\sqrt{(g-0)^2 + \left(-\dfrac{3}{4}g - 0\right)^2}$; $36 + 64 = g^2 + \dfrac{9}{16}g^2$; $100 = \dfrac{25}{16}g^2$; $g^2 = 64$; $g = \pm 8$;

$h = -\dfrac{3}{4}g$, so $h = \pm 6$ and $D = (8, -6)$ or $D = (-8, 6)$. *Case 1:* If $D = (8, -6)$, the

midpoint M of $\overline{BD} = \left(\dfrac{8+6}{2}, \dfrac{-6+8}{2}\right) = (7, 1)$; the diagonals bisect each other, so the

midpoint of $\overline{AC} = (7, 1) = \left(\dfrac{0+e}{2}, \dfrac{0+f}{2}\right)$; $e = 14$ and $f = 2$; $C = (14, 2)$. *Case 2:* If

$D = (-8, 6)$, $M = \left(\dfrac{-8+6}{2}, \dfrac{6+8}{2}\right) = (-1, 7) = \left(\dfrac{0+e}{2}, \dfrac{0+f}{2}\right)$; $e = -2$ and $f = 14$;

$C = (-2, 14)$.

33. Let $R = (a, b)$ and $S = (c, d)$. Since $\overline{PQ} \perp \overline{PS}$, $\dfrac{5 + 1}{-6 + 4} \cdot \dfrac{d + 1}{c + 4} = -1$; $-3 = -\dfrac{c + 4}{d + 1}$;

$3d + 3 = c + 4$; $c = 3d - 1$; thus, $S = (3d - 1, d)$. $PQ = 2(QR) = 2(PS)$;
$\sqrt{(-6 + 4)^2 + (5 + 1)^2} = 2\sqrt{(3d - 1 + 4)^2 + (d + 1)^2}$; $4 + 36 = 4[(3d + 3)^2 + (d + 1)^2]$; $10 = 9d^2 + 18d + 9 + d^2 + 2d + 1$; $0 = 10d^2 + 20d = 10d(d + 2)$; $d = 0$
or $d = -2$; thus, $S = (-1, 0)$ or $S = (-7, -2)$. *Case 1*: If $S = (-1, 0)$, the midpoint M of

$\overline{QS} = \left(\dfrac{-1 - 6}{2}, \dfrac{0 + 5}{2}\right) = \left(-\dfrac{7}{2}, \dfrac{5}{2}\right)$; the diagonals bisect each other, so the midpoint of

$\overline{PR} = \left(-\dfrac{7}{2}, \dfrac{5}{2}\right) = \left(\dfrac{a - 4}{2}, \dfrac{b - 1}{2}\right)$; $a = -3$ and $b = 6$; $R = (-3, 6)$. *Case 2*: If $S =$

$(-7, -2)$, $M = \left(\dfrac{-7 - 6}{2}, \dfrac{-2 + 5}{2}\right) = \left(-\dfrac{13}{2}, \dfrac{3}{2}\right) = \left(\dfrac{a - 4}{2}, \dfrac{b - 1}{2}\right)$; $a = -9$ and $b = 4$;

$R = (-9, 4)$.

Calculator Exercises • page 13

1. slope of $\overline{AB} = \dfrac{4.4 - 2.3}{-0.4 + 1.8} = 1.5$; slope of $\overline{AC} = \dfrac{-0.5 - 2.3}{2.4 + 1.8} = -0.\overline{6} = -\dfrac{2}{3}$; $1.5\left(-\dfrac{2}{3}\right) =$

-1, so $\overline{AB} \perp \overline{AC}$; thus, $\triangle ABC$ is a right triangle with hyp. $\overline{BC}$. $AB = \sqrt{(-0.4 + 1.8)^2 + (4.4 - 2.3)^2} = \sqrt{6.37}$; $AC = \sqrt{(2.4 + 1.8)^2 + (-0.5 - 2.3)^2} = \sqrt{25.48}$; area $= \dfrac{1}{2}(AB)(AC) = \dfrac{1}{2}\sqrt{(6.37)(25.48)} = 6.37$ sq. units

2. The midpoint M of $\overline{BC} = \left(\dfrac{-0.4 + 2.4}{2}, \dfrac{4.4 - 0.5}{2}\right) = (1, 1.95)$; $MB = MC$. $MA = \sqrt{(-1.8 - 1)^2 + (2.3 - 1.95)^2} = \sqrt{7.9625}$; $MB = \sqrt{(-0.4 - 1)^2 + (4.4 - 1.95)^2} = \sqrt{7.9625}$; $MA = MB = MC$.

Class Exercises 1-3 • page 16

1. $y = \dfrac{5}{3}x - 2$ **2.** $\dfrac{y - 6}{x + 4} = 3$

3. $y = 4x - 3$ has slope 4, so the perpendicular line has slope $-\dfrac{1}{4}$; $\dfrac{y - 2}{x - 1} = -\dfrac{1}{4}$

4. $\dfrac{x}{-1} + \dfrac{y}{6} = 1$, or $-x + \dfrac{y}{6} = 1$ **5.** Slope is $\dfrac{-1 - 3}{2 - 8} = \dfrac{2}{3}$; $\dfrac{y + 1}{x - 2} = \dfrac{2}{3}$ or $\dfrac{y - 3}{x - 8} = \dfrac{2}{3}$

6. a. $y = mx$ **b.** $\dfrac{y}{x} = m$ **c.** No intercept form

Written Exercises 1-3 • pages 16–18

1. $y = -2x + 8$ or $2x + y = 8$ **2.** $y = \dfrac{3}{5}x$ or $3x - 5y = 0$

3. $\dfrac{x}{-2} + \dfrac{y}{4} = 1$ or $2x - y = -4$ **4.** The slope is $-\dfrac{5}{4}$; $y = -\dfrac{5}{4}x - 6$ or $5x + 4y = -24$

5. $m = \dfrac{8 - 4}{5 + 1} = \dfrac{2}{3}$; $\dfrac{y - 4}{x + 1} = \dfrac{2}{3}$ or $\dfrac{y - 8}{x - 5} = \dfrac{2}{3}$ or $2x - 3y = -14$

6. $m = \dfrac{1 - 5}{6 - 0} = -\dfrac{2}{3}$; $\dfrac{y - 5}{x} = -\dfrac{2}{3}$ or $\dfrac{y - 1}{x - 6} = -\dfrac{2}{3}$ or $2x + 3y = 15$

7. $m = 0$; $y = 0x - 7$; $y = -7$ **8.** The line has no slope; $x = 5$

9. The line is vertical; $x = 2$ **10.** The line is horizontal; $y = -3$

11. $0.3x - 1.2y = 6.4$ has slope $\dfrac{0.3}{1.2} = 0.25$; $y = 0.25x + 1.8$ or $0.25x - y = -1.8$

12. $m = \dfrac{7 - 1}{5 - 1} = \dfrac{3}{2}$; $\dfrac{y - 4}{x + 2} = \dfrac{3}{2}$ or $3x - 2y = -14$

13. Line $y = 7 - 2x$ has slope -2, so $m = \dfrac{1}{2}$; $\dfrac{y + 2}{x - 8} = \dfrac{1}{2}$ or $x - 2y = 12$

14. Line $x - 3y = 9$ has slope $\dfrac{1}{3}$, so $m = -3$; $\dfrac{y - 0}{x - 0} = -3$; $\dfrac{y}{x} = -3$; $y = -3x$ or

$3x + y = 0$

15. Methods may vary. The segment has slope $\dfrac{5 - 3}{-4 - 0} = -\dfrac{1}{2}$, so $m = 2$;

midpoint $M = \left(\dfrac{0 - 4}{2}, \dfrac{3 + 5}{2}\right) = (-2, 4)$; $\dfrac{y - 4}{x + 2} = 2$ or $2x - y = -8$

16. Methods may vary. The segment has slope $\dfrac{-4 - 4}{4 - 2} = -4$, so $m = \dfrac{1}{4}$;

midpoint $M = \left(\dfrac{2 + 4}{2}, \dfrac{4 - 4}{2}\right) = (3, 0)$; $\dfrac{y - 0}{x - 3} = \dfrac{1}{4}$ or $x - 4y = 3$

17. *Method 1:* $M = \left(\dfrac{2 + 8}{2}, \dfrac{0 + 4}{2}\right) = (5, 2)$; slope of $\overline{AB} = \dfrac{4 - 0}{8 - 2} = \dfrac{2}{3}$; slope of $\overline{PM} =$

$\dfrac{2 - 5}{5 - 3} = -\dfrac{3}{2}$; $\dfrac{2}{3}\left(-\dfrac{3}{2}\right) = -1$, so $\overline{AB} \perp \overline{PM}$. *Method 2:* $PA = \sqrt{(3 - 2)^2 + (5 - 0)^2} =$

$\sqrt{26}$; $PB = \sqrt{(3 - 8)^2 + (5 - 4)^2} = \sqrt{26}$; $PA = PB$

18. *Method 1:* $M = \left(\dfrac{0 + 2}{2}, \dfrac{7 - 1}{2}\right) = (1, 3)$; slope of $\overline{AB} = \dfrac{-1 - 7}{2 - 0} = -4$; slope of $\overline{PM} =$

$\dfrac{4 - 3}{5 - 1} = \dfrac{1}{4}$; $-4\left(\dfrac{1}{4}\right) = -1$, so $\overline{AB} \perp \overline{PM}$. *Method 2:* $PA = \sqrt{(5 - 0)^2 + (4 - 7)^2} =$

$\sqrt{34}$; $PB = \sqrt{(5 - 2)^2 + (4 + 1)^2} = \sqrt{34}$; $PA = PB$

19. a. Let M = the midpoint of $\overline{PQ} = \left(\dfrac{4 - 2}{2}, \dfrac{-1 + 7}{2}\right) = (1, 3)$; the median contains M and

R; $m = \dfrac{9 - 3}{9 - 1} = \dfrac{3}{4}$; $\dfrac{y - 9}{x - 9} = \dfrac{3}{4}$ or $3x - 4y = -9$

b. $\overline{PQ}$ has slope $\dfrac{7 + 1}{-2 - 4} = -\dfrac{4}{3}$, so the altitude from R has slope $\dfrac{3}{4}$; $\dfrac{y - 9}{x - 9} = \dfrac{3}{4}$ or

$3x - 4y = -9$ **c.** yes; yes

20. a. $DF = \sqrt{(5 + 2)^2 + (6 - 5)^2} = \sqrt{49 + 1} = 5\sqrt{2}$; $EF = \sqrt{(5 - 6)^2 + (6 + 1)^2} =$

$\sqrt{1 + 49} = 5\sqrt{2}$; since $DF = EF$, $\triangle DEF$ is isosceles.

b. The bisector of $\angle F$ bisects $\overline{DE}$ and is $\perp$ to $\overline{DE}$. *Sol. 1:* $\overline{DE}$ has midpoint $M =$

$\left(\dfrac{-2 + 6}{2}, \dfrac{5 - 1}{2}\right) = (2, 2)$; $\overleftrightarrow{MF}$ has slope $\dfrac{6 - 2}{5 - 2} = \dfrac{4}{3}$; $\dfrac{y - 6}{x - 5} = \dfrac{4}{3}$ or $4x - 3y = 2$.

Sol. 2: $\overline{DE}$ has slope $\dfrac{-1 - 5}{6 + 2} = -\dfrac{3}{4}$, so $m = \dfrac{4}{3}$; $\dfrac{y - 6}{x - 5} = \dfrac{4}{3}$ or $4x - 3y = 2$.

21. a. $\overline{AB}$ has midpoint $M_1 = (-1, 3)$, $\overline{BC}$ has midpoint $M_2 = (4, 5)$, and $\overline{AC}$ has midpoint

$M_3 = (3, 1)$. Since $\overleftrightarrow{CM_1}$ is horizontal, $\overleftrightarrow{CM_1}$ has eq. $y = 3$. Slope of $\overleftrightarrow{AM_2} = \dfrac{5+1}{4+2} = 1$,

so $\overleftrightarrow{AM_2}$ has eq. $\dfrac{y+1}{x+2} = 1$, or $x - y = -1$. Slope of $\overleftrightarrow{BM_3} = \dfrac{7-1}{0-3} = -2$, so $\overleftrightarrow{BM_3}$ has

eq. $\dfrac{y-7}{x-0} = -2$, or $y = -2x + 7$.

b. From $\overleftrightarrow{CM_1}$, $y = 3$; from $\overleftrightarrow{AM_2}$, $x - 3 = -1$; $x = 2$; $G = (2, 3)$ lies on $\overleftrightarrow{CM_1}$ and $\overleftrightarrow{AM_2}$; $\overleftrightarrow{BM_3}$ has eq. $y = -2x + 7$; $3 = -2(2) + 7$; thus G lies on each median.

22. a. $\overline{PQ}$ has midpoint $(4, 1)$ and slope $-\dfrac{3}{4}$, so its $\perp$ bis. has eq. $\dfrac{y-1}{x-4} = \dfrac{4}{3}$, or

① $4x - 3y = 13$. $\overline{QR}$ has midpoint $\left(\dfrac{15}{2}, \dfrac{3}{2}\right)$ and slope -7, so its $\perp$ bis. has

eq. $\dfrac{y - \frac{3}{2}}{x - \frac{15}{2}} = \dfrac{1}{7}$, or ② $x - 7y = -3$. $\overline{PR}$ has midpoint $\left(\dfrac{7}{2}, \dfrac{9}{2}\right)$ and slope $\dfrac{1}{7}$, so its

$\perp$ bis. has eq. $\dfrac{y - \frac{9}{2}}{x - \frac{7}{2}} = -7$, or ③ $7x + y = 29$.

b. From ②, $x = 7y - 3$. From ③, $7x + y = 29$; $7(7y - 3) + y = 29$; $y = 1$; $x = 7(1) - 3 = 4$; ② and ③ contain $(4, 1)$; from ①, $4x - 3y = 13$; $4(4) - 3(1) = 13$; $C(4, 1)$ lies on each $\perp$ bisector. **c.** an isosceles right triangle

23. a. $\overline{KL}$ has slope 0, so the altitude to $\overline{KL}$ contains $M(6, 12)$ and has no slope; ① $x = 6$. $\overline{LM}$ has slope -1, so the altitude to $\overline{LM}$ has slope 1 and contains $K(0, 0)$; ② $y = x$.

$\overline{KM}$ has slope 2, so the altitude to $\overline{KM}$ has slope $-\dfrac{1}{2}$ and contains $L(18, 0)$;

$\dfrac{y-0}{x-18} = -\dfrac{1}{2}$, or ③ $x + 2y = 18$.

b. From ①, $x = 6$. From ②, $y = 6$. $6 + 2(6) = 18$, so $O(6, 6)$ lies on each altitude.

24. $\overline{RS}$ has midpoint $M_1(13, 2)$ and slope 0; $\overline{ST}$ has midpoint $M_2\left(\dfrac{35}{2}, 11\right)$ and slope $-\dfrac{6}{5}$; $\overline{RT}$ has midpoint $M_3\left(\dfrac{11}{2}, 11\right)$ and slope 2.

a. $\overleftrightarrow{TM_1}$ has eq. $\dfrac{y-20}{x-10} = -6$, or ① $6x + y = 80$. $\overleftrightarrow{RM_2}$ has eq. $\dfrac{y-2}{x-1} = \dfrac{6}{11}$, or

② $6x - 11y = -16$. $\overleftrightarrow{SM_3}$ has eq. $\dfrac{y-2}{x-25} = -\dfrac{6}{13}$, or ③ $6x + 13y = 176$.

① $-$ ② gives $12y = 96$; $y = 8$. From ①, $6x + 8 = 80$; $x = 12$. $G(12, 8)$ satisfies the eq. of each median.

b. The $\perp$ bis. of $\overline{RS}$ contains M_1 and has no slope; ① $x = 13$. The $\perp$ bis. of $\overline{ST}$ contains

M_2 and has slope $\dfrac{5}{6}$; $\dfrac{y-11}{x - \frac{35}{2}} = \dfrac{5}{6}$, or ② $10x - 12y = 43$. The $\perp$ bis. of $\overline{RT}$ contains

M_3 and has slope $-\dfrac{1}{2}$; $\dfrac{y-11}{x - \frac{11}{2}} = -\dfrac{1}{2}$, or ③ $2x + 4y = 55$. From ①, $x = 13$.

From ③, $2(13) + 4y = 55$; $y = \dfrac{29}{4}$. $C\left(13, \dfrac{29}{4}\right)$ satisfies the eq. of each $\perp$ bis.

c. The altitude to $\overline{RS}$ contains T and has no slope; ① $x = 10$. The altitude to $\overline{ST}$

contains R and has slope $\dfrac{5}{6}$; $\dfrac{y-2}{x-1} = \dfrac{5}{6}$, or ② $5x - 6y = -7$. The altitude to $\overline{RT}$

contains S and has slope $-\dfrac{1}{2}$; $\dfrac{y-2}{x-25} = -\dfrac{1}{2}$, or ③ $x + 2y = 29$. From ①, $x = 10$.

From ③, $10 + 2y = 29$; $y = \dfrac{19}{2}$. $O\left(10, \dfrac{19}{2}\right)$ satisfies the eq. of each altitude.

d. $G = (12, 8)$, $C = \left(13, \dfrac{29}{4}\right)$, $O = \left(10, \dfrac{19}{2}\right)$; $\overleftrightarrow{GC}$ has slope $\dfrac{\frac{29}{4} - 8}{13 - 12} = -\dfrac{3}{4}$ and $\overleftrightarrow{GO}$ has

slope $\dfrac{8 - \frac{19}{2}}{12 - 10} = -\dfrac{3}{4}$, so G, C, and O lie on a line.

25. When $m = -1$, $\overleftrightarrow{PQ}$ has eq. $\dfrac{y-1}{x-2} = -1$, or $x + y = 3$; $P = (0, 3)$ and $Q = (3, 0)$, so

$A = \dfrac{1}{2}(3)(3) = 4.5$ sq. units. When $m = -2$, $\overleftrightarrow{PQ}$ has eq. $\dfrac{y-1}{x-2} = -2$, or $2x + y = 5$;

$P = (0, 5)$ and $Q = \left(\dfrac{5}{2}, 0\right)$, so $A = \dfrac{1}{2}\left(\dfrac{5}{2}\right)(5) = 6.25$ sq. units. When $m = -3$, $\overleftrightarrow{PQ}$ has

eq. $\dfrac{y-1}{x-2} = -3$, or $3x + y = 7$; $P = (0, 7)$ and $Q = \left(\dfrac{7}{3}, 0\right)$, so $A = \dfrac{1}{2}\left(\dfrac{7}{3}\right)(7) = $

$8.1\overline{6}$ sq. units. The triangle at the left ($m = -1$) has the least area.

26. If $m = -1$, $\overleftrightarrow{PQ}$ has eq. $\dfrac{y-6}{x-2} = -1$, or $x + y = 8$; $P = (0, 8)$; $Q = (8, 0)$;

$A = \dfrac{1}{2}(8)(8) = 32$ sq. units. If $m = -2$, $\overleftrightarrow{PQ}$ has eq. $\dfrac{y-6}{x-2} = -2$, or $2x + y = 10$;

$P = (0, 10)$; $Q = (5, 0)$; $A = \dfrac{1}{2}(5)(10) = 25$ sq. units. If $m = -3$, $\overleftrightarrow{PQ}$ has

eq. $\dfrac{y-6}{x-2} = -3$, or $3x + y = 12$; $P = (0, 12)$, $Q = (4, 0)$; $A = \dfrac{1}{2}(4)(12) = 24$ sq. units.

If $m = -4$, $\overleftrightarrow{PQ}$ has eq. $\dfrac{y-6}{x-2} = -4$, or $4x + y = 14$; $P = (0, 14)$, $Q = \left(\dfrac{7}{2}, 0\right)$;

$A = \dfrac{1}{2}\left(\dfrac{7}{2}\right)(14) = 24.5$ sq. units. The triangle of least area occurs when $m = -3$.

27. a. $BC = \sqrt{(9-0)^2 + (-2-1)^2} = 3\sqrt{10}$; $m = \dfrac{-2-1}{9-0} = -\dfrac{1}{3}$; $\dfrac{y-1}{x-0} = -\dfrac{1}{3}$, or

$x + 3y = 3$.

b. $\overline{BC}$ has slope $-\dfrac{1}{3}$, so altitude has slope 3; $\dfrac{y-5}{x-8} = 3$, or $3x - y = 19$.

c. From part (a), $x = 3 - 3y$. From part (b), $3(3 - 3y) - y = 19$; $y = -1$;
$x = 3 - 3(-1) = 6$; $D(6, -1)$.

d. $AD = \sqrt{(8-6)^2 + (5+1)^2} = 2\sqrt{10}$

e. $A = \dfrac{1}{2}(BC)(AD) = \dfrac{1}{2} \cdot 3\sqrt{10} \cdot 2\sqrt{10} = 30$ sq. units

28. Line $4x - 3y = -4$ has slope $\dfrac{4}{3}$, so a line $\perp$ to it has slope $-\dfrac{3}{4}$; the $\perp$ through $A(9,5)$

has eq. $\dfrac{y - 5}{x - 9} = -\dfrac{3}{4}$, or $3x + 4y = 47$. Find the intersection pt. B:

$$4x - 3y = -4 \qquad 12x - 9y = -12$$
$$3x + 4y = 47 \qquad \underline{12x + 16y = 188}$$
$$-25y = -200;\ y = 8;\ 4x - 3(8) = -4;\ x = 5;\ B = (5,8);$$
$$AB = \sqrt{(9 - 5)^2 + (5 - 8)^2} = 5$$

29. a. Line j through (x_1, y_1) with slope $\dfrac{b}{a}$ has eq. $\dfrac{y - y_1}{x - x_1} = \dfrac{b}{a}$, or $bx - ay = bx_1 - ay_1$.

b.
$$ax + by = c \qquad\qquad abx + b^2y = bc$$
$$bx + ay = bx_1 - ay_1 \qquad \underline{abx - a^2y = abx_1 - a^2y_1}$$
$$(b^2 + a^2)y = bc - abx_1 + a^2y_1;\ y = \dfrac{a^2y_1 - abx_1 + bc}{a^2 + b^2};$$

$$a^2x + aby = ac$$
$$\underline{b^2x - aby = b^2x_1 - aby_1}$$
$$(a^2 + b^2)x = ac + b^2x_1 - aby_1;\ x = \dfrac{b^2x_1 - aby_1 + ac}{a^2 + b^2};$$

$$Q = \left(\dfrac{b^2x_1 - aby_1 + ac}{a^2 + b^2},\ \dfrac{a^2y_1 - abx_1 + bc}{a^2 + b^2} \right)$$

c.
$$PQ = \sqrt{\left(\dfrac{b^2x_1 - aby_1 + ac}{a^2 + b^2} - x_1 \right)^2 + \left(\dfrac{a^2y_1 - abx_1 + bc}{a^2 + b^2} - y_1 \right)^2} =$$

$$\sqrt{\left(\dfrac{-a^2x_1 - aby_1 + ac}{a^2 + b^2} \right)^2 + \left(\dfrac{-b^2y_1 - abx_1 + bc}{a^2 + b^2} \right)^2} =$$

$$\sqrt{\dfrac{[-a(ax_1 + by_1 - c)]^2 + [-b(by_1 + ax_1 - c)]^2}{(a^2 + b^2)^2}} = \sqrt{\dfrac{(a^2 + b^2)(ax_1 + by_1 - c)^2}{(a^2 + b^2)^2}} =$$

$$\dfrac{|ax_1 + by_1 - c|}{\sqrt{a^2 + b^2}}$$

d. $P(x_1, y_1) = (9,5);\ PQ = \dfrac{|ax_1 + by_1 - c|}{\sqrt{a^2 + b^2}} = \dfrac{|4 \cdot 9 + (-3)5 - (-4)|}{\sqrt{4^2 + (-3)^2}} = \dfrac{|25|}{\sqrt{25}} = 5$

30. a. $d_1 = \dfrac{|12 \cdot 7 + (-5)9 - 0|}{\sqrt{12^2 + (-5)^2}} = \dfrac{|39|}{\sqrt{169}} = 3;\ d_2 = \dfrac{|3 \cdot 7 + (-4)9 - 0|}{\sqrt{3^2 + (-4)^2}} = \dfrac{|-15|}{\sqrt{25}} = 3$

b. Any point equidistant from the sides of an angle is on the angle bisector. The bisector

contains points $(0,0)$ and $(7,9)$; its slope is $\dfrac{9}{7}$, so its equation is $\dfrac{y - 0}{x - 0} = \dfrac{9}{7}$, or

$9x - 7y = 0;\ m_1 = \dfrac{12}{5},\ m_2 = \dfrac{3}{4}$, and $\dfrac{1}{2}\left(\dfrac{12}{5} + \dfrac{3}{4} \right) = \dfrac{1}{2}\left(\dfrac{63}{20} \right) = \dfrac{63}{40} \neq \dfrac{9}{7}$.

31. a. Using the formula of Ex. 29, the given equation states that the distance from P to
$\overleftrightarrow{OD}$ equals the distance from P to $\overleftrightarrow{OE}$. Any point equidistant from the sides of an
angle lies on the angle bisector.

b. $\dfrac{|4x - 3y - (-6)|}{\sqrt{4^2 + (-3)^2}} = \dfrac{|3x - 4y - (-4)|}{\sqrt{3^2 + (-4)^2}};\ \dfrac{|4x - 3y + 6|}{5} = \dfrac{|3x - 4y + 4|}{5};$

$4x - 3y + 6 = 3x - 4y + 4$ or $4x - 3y + 6 = -(3x - 4y + 4)$; the lines have
equations $x + y = -2$ and $7x - 7y = -10$

Activity 1-4 · page 19

a. 10.001; 10.002; 10.003; 10.004

c. The slope of the graph is close to zero.

b.

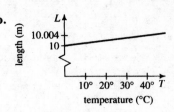

Class Exercises 1-4 · pages 21–22

1. a. $f(3) = 5(3) - 10 = 5; f(-3) = 5(-3) - 10 = -25$

b. $0 = 5x - 10; x = 2$; zero of f is 2.

c. Intersects vertical axis at $(0, -10)$;
intersects horizontal axis at $(2, 0)$.

d. At the intersection with the horizontal axis

2. functions in (a) and (b)

3. $t = \dfrac{-s + 8}{2}; t(s) = -\dfrac{s}{2} + 4$

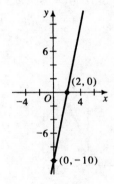

Ex. 1c

4. a.

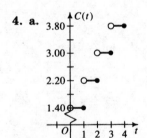

b.

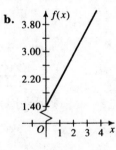

5. $f(a) - f(b) = (ma + k) - (mb + k) =$
$ma - mb = m(a - b)$. Subtract 69.6
from 79.6 to get 10 and multiply this
result by m (which is 50) to get 500.
$f(12.8) - f(12.3) = (0.5)(50) = 25$

Written Exercises 1-4 · pages 22–25

1. a. $f(2) = \dfrac{3}{4}(2) - \dfrac{1}{2} = 1; f(-2) = \dfrac{3}{4}(-2) - \dfrac{1}{2} = -2$

b. $0 = \dfrac{3}{4}x - \dfrac{1}{2}; x = \dfrac{2}{3}$; zero is $\dfrac{2}{3}$.

2. a. $C(0) = 20 - \dfrac{5}{8}(0) = 20; C(16) = 20 - \dfrac{5}{8}(16) = 10$

b. $0 = 20 - \dfrac{5}{8}x; x = 32$; zero is 32.

3. No; $f(2) + f(6) = 3 \cdot 2 - 7 + 3 \cdot 6 - 7 = 10; f(8) = 3 \cdot 8 - 7 = 17; f(2) + f(6) \neq f(8)$

4. No; $h(4.5) - h(3.5) = -4.5 - (-2.5) = -2; h(1) = \dfrac{5}{2}; h(4.5) - h(3.5) \neq h(1)$

5. a. Each has slope 0. See diagram.

   b. $g(x) = 0x + 2$;

      $h(x) = 0x - 1$

6. a. -0.5 **b.** No; $0 = -0.5$ has no solutions.

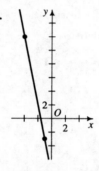

7. a. 1.5 **b.** $0 = 1.5x - 2$; $x = \dfrac{4}{3}$; zero is $\dfrac{4}{3}$.

   c. $0 = 1.5x - 2$; $x = \dfrac{4}{3}$; x-int. $= \dfrac{4}{3}$; $y = 1.5(0) - 2$; $y = -2$, y-int. $= -2$

8. a. 80 **b.** $C(0) = 5.2$; at $(0, 5.2)$

9. $m = -2$; $f(6) = 0$; $\dfrac{f(n) - 0}{n - 6} = -2$; $f(n) = -2n + 12$

10. $S(3) = 0$; $S(0) = -2$; $m = \dfrac{2}{3}$; $S(x) = \dfrac{2}{3}x - 2$

11. a.

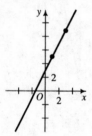

12. a.

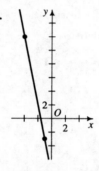

b. Using $(1, 5)$ and $(3, 9)$,

$$m = \frac{9 - 5}{3 - 1} = 2; \quad \frac{y - 5}{x - 1} = 2,$$

or $y = f(x) = 2x + 3$

b. Using $(-1, -3)$ and $(-4, 12)$,

$$m = \frac{12 + 3}{-4 + 1} = -5; \quad \frac{y + 3}{x + 1} = -5,$$

or $y = g(x) = -5x - 8$

13. a. $m = \dfrac{3.2 - 4}{5 - 0} = -0.16$; $D(t) = -0.16t + 4$

   b. $0 = -0.16t + 4$; $t = 25$; about 25 min

14. a.

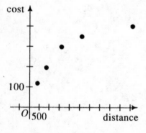

b. No; a linear function appears to be a good model for the first three points. However, it appears that there is no good linear approximation for the entire data set.

15. a. $C(m) = 0.15m + 280$ **b.** 0.15

16. a. $C(t) = 1.5t + 24{,}000$ **b.** $R(t) = 6.5t$ **c.** See below.

    d. $6.5t = 1.5t + 24{,}000$; $t = 4800$; $R(t) = C(t) = 6.5(4800) = 31{,}200$; break-even pt. is
    $(4800, 31{,}200)$.

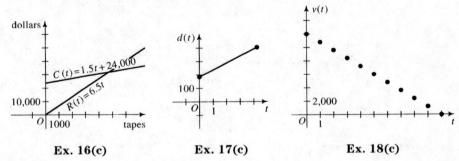

      Ex. 16(c)   **Ex. 17(c)**   **Ex. 18(c)**

17. a. $\dfrac{220}{4.25} \approx 51.8$ mi/h; $d(t) = 51.8t + 185$ **b.** the real numbers t such that $0 \le t \le 4\dfrac{1}{4}$

    c. See above; slope is 51.8. **d.** They are equal.

18. a. $v(t) = 12{,}000 - 1200t$ **b.** The set of integers from 0 to 10, inclusive

    c. See above; slope is -1200. **d.** They are equal.

19. Answers may vary. Examples: We walked up the gentle slope of the grassy hillside; the
ramp had a slope of 0.2. Both usages refer to grade or steepness. The common
meaning can refer to the ground itself and is relative, ranging from gentle to steep. The
math term refers to a line or something modeled by a line and has a precise numerical
value. A vertical line has no mathematical slope, but in everyday usage a vertical face is
a very steep gradient.

20. a.   **b.**   **21. a.**   **b.**

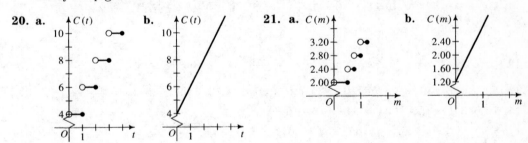

    In part (b), $C(t) = 2t + 4$.   In part (b), $C(m) = 1.6m + 1.2$.
    Other answers are possible.   Other answers are possible.

22. a. The numbers of sides increase by equal amounts and the angle measures do not.

    b. The set of integers ≥ 3 **c.** The sum of the meas. of the n $\angle$s is $(n - 2)180$, so

$$A(n) = \frac{180n - 360}{n} = 180 - \frac{360}{n}.$$

23. $m = \dfrac{90 - 65}{78 - 47} = \dfrac{25}{31}$; $\dfrac{f(x) - 90}{x - 78} = \dfrac{25}{31}$; $f(x) = \dfrac{25}{31}x + \dfrac{840}{31}$

24. a. If he were driving at a constant speed

b. $m = \dfrac{46{,}081 - 45{,}973}{\dfrac{1}{2} - \dfrac{7}{8}} = -288;\ \dfrac{f(x) - 46{,}081}{x - \dfrac{1}{2}} = -288;$

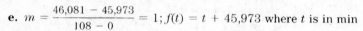

$f(x) = -288x + 46{,}225$

c. The y-coordinate of that pt. gives the odometer reading when the fuel tank is empty.

d. No; his car travels 288 mi/tank, or 36 mi/$\frac{1}{8}$ tank.

e. $m = \dfrac{46{,}081 - 45{,}973}{108 - 0} = 1;\ f(t) = t + 45{,}973$ where t is in min

Class Exercises 1-5 · page 28

1. $7 + 2i$ **2.** $-1 - 5i$ **3.** $9 - i^2 = 9 + 1 = 10$ **4.** $27 - 6i - i^2 = 28 - 6i$

5. $16 - i^2 = 17$ **6.** $9 + 25 = 34$ **7.** $2 + 1 = 3$ **8.** $a^2 + b^2$

9. $4 = 2x$ and $-5 = y;\ x = 2$ **10.** $1; 1; 1; 1;$ equal to 1

11. a. $\sqrt{4} \cdot \sqrt{9} = 2 \cdot 3 = 6 = \sqrt{36} = \sqrt{4 \cdot 9}$

    b. $\sqrt{-4} \cdot \sqrt{9} = 2i \cdot 3 = 6i = \sqrt{-36} = \sqrt{-4 \cdot 9}$

    c. $\sqrt{-4} \cdot \sqrt{-9} = 2i \cdot 3i = -6;\ \sqrt{-4(-9)} = \sqrt{36} = 6;\ -6 \neq 6$ **d.** No

Written Exercises 1-5 · pages 28–29

1. $2i + 4i + i = 7i$ **2.** $7i - 3i + 6i = 10i$ **3.** $i \cdot 3i = 3i^2 = -3$

4. $i\sqrt{2} \cdot i\sqrt{5} = i^2\sqrt{10} = -\sqrt{10}$ **5.** $\dfrac{2i\sqrt{3}}{i\sqrt{3}} = 2$ **6.** $\dfrac{5i}{5i\sqrt{2}} = \dfrac{\sqrt{2}}{2}$ **7.** $-2 + 5i$

8. $1 - 10i$ **9.** $(12 + 20i) - (4 - 12i) = 8 + 32i$

10. $\left(\dfrac{7}{6} - \dfrac{1}{3}i\right) + \left(\dfrac{10}{3} - \dfrac{10}{3}i\right) = \dfrac{9}{2} - \dfrac{11}{3}i$ **11.** $36 - i^2 = 36 + 1 = 37$

12. $49 - 9i^2 = 49 + 9 = 58$ **13.** $25 + 5 = 30$ **14.** $3 + 32 = 35$

15. $16 - 34i - 15i^2 = 31 - 34i$ **16.** $-5 + 17i - 6i^2 = 1 + 17i$

17. $16 - 2(20i) + 25i^2 = (16 - 25) - 40i = -9 - 40i$

18. $16 + 2(28i) + 49i^2 = (16 - 49) + 56i = -33 + 56i$

19. $\dfrac{1}{2 + 5i} \cdot \dfrac{2 - 5i}{2 - 5i} = \dfrac{2 - 5i}{4 + 25} = \dfrac{2}{29} - \dfrac{5}{29}i$ **20.** $\dfrac{1}{4 - 3i} \cdot \dfrac{4 + 3i}{4 + 3i} = \dfrac{4 + 3i}{16 + 9} = \dfrac{4}{25} + \dfrac{3}{25}i$

21. $\dfrac{5 + i}{5 - i} \cdot \dfrac{5 + i}{5 + i} = \dfrac{25 + 2(5i) + i^2}{25 + 1} = \dfrac{24 + 10i}{26} = \dfrac{12}{13} + \dfrac{5}{13}i$

22. $\dfrac{3 - 2i}{3 + 2i} \cdot \dfrac{3 - 2i}{3 - 2i} = \dfrac{9 - 2(6i) + (4i^2)}{9 + 4} = \dfrac{5 - 12i}{13} = \dfrac{5}{13} - \dfrac{12}{13}i$

23. $\dfrac{3 + i\sqrt{2}}{7 - i\sqrt{2}} \cdot \dfrac{7 + i\sqrt{2}}{7 + i\sqrt{2}} = \dfrac{21 + 10i\sqrt{2} + 2i^2}{49 + 2} = \dfrac{19}{51} + \dfrac{10\sqrt{2}}{51}i$

24. $\dfrac{2 + i\sqrt{5}}{3 - i\sqrt{5}} \cdot \dfrac{3 + i\sqrt{5}}{3 + i\sqrt{5}} = \dfrac{6 + 5i\sqrt{5} + (-5)}{9 + 5} = \dfrac{1 + 5i\sqrt{5}}{14} = \dfrac{1}{14} + \dfrac{5\sqrt{5}}{14}i$

25. $\dfrac{5}{i} \cdot \dfrac{-i}{-i} = \dfrac{-5i}{-(-1)} = -5i$ **26.** $i + 2i^2 = -2 + i$

27. $i + i^2 + i(i^2) + (i^2)^2 + (i^2)^2 i = i + (-1) - i + 1 + 1 \cdot i = i$

28. $(i^2)^{23} + (i^2)^{23} \cdot i = (-1)^{23} + (-1)^{23} \cdot i = -1 - i$ **29.** $i^{-3} = \dfrac{1}{i^3} = \dfrac{i}{i^4} = i$

30. $i^{-6} = \dfrac{1}{i^6} = \dfrac{i^2}{i^8} = i^2 = -1$ **31.** $i^{-35} = \dfrac{1}{i^{35}} = \dfrac{i}{i^{36}} = \dfrac{i}{(i^2)^{18}} = \dfrac{i}{(-1)^{18}} = \dfrac{i}{1} = i$

32. $(i^n)^4 = (i^4)^n = [(i^2)^2]^n = [(-1)^2]^n = 1^n = 1$

33. $2x + y = 1$ and $3 - 5x = -7$; $10 = 5x$, so $x = 2$; $2(2) + y = 1$; $y = -3$

34. $3x - 4y = 0$ and $6x + 2y = 5$; $6x - 8y = 0$ and $6x + 2y = 5$; $10y = 5$, so $y = \dfrac{1}{2}$;

$3x - 4\left(\dfrac{1}{2}\right) = 0$; $x = \dfrac{2}{3}$

35. $a - bi$ is the conjugate of $a + bi$; $(a + bi) + (a - bi) = 2a + (b - b)i = 2a$; a is a real number, so $2a$ is a real number.

36. $a - bi$ is the conjugate of $a + bi$; $(a + bi)(a - bi) = a^2 - b^2 i^2 = a^2 + b^2$; since a and b are real numbers, a^2 and b^2 are nonnegative real numbers; thus, $a^2 + b^2$ is a nonnegative real number.

37. a. Calculate 72^2 and compare the result with 6241.

b. Show that $(3 - i)^2 = 9 - 6i + i^2 = 8 - 6i$.

38. $(4 - 3i)^2 = 16 - 24i + 9i^2 = 7 - 24i$

39. $\left[\dfrac{\sqrt{2}}{2}(1 + i)\right]^2 = \left(\dfrac{\sqrt{2}}{2}\right)^2 (1 + i)^2 = \dfrac{2}{4}(1 + 2i + i^2) = \dfrac{1}{2}(2i) = i$

40. $(a + bi)^2 = 3 + 4i$; $a^2 - b^2 + 2abi = 3 + 4i$; (1) $a^2 - b^2 = 3$ and (2) $2ab = 4$; from

(2): $b = \dfrac{2}{a}$; substitute in (1): $a^2 - \dfrac{4}{a^2} = 3$; $a^4 - 3a^2 - 4 = 0$; $(a^2 - 4)(a^2 + 1) = 0$;

$a = \pm 2$; when $a = 2$, $b = \dfrac{2}{2} = 1$; when $a = -2$, $b = \dfrac{2}{-2} = -1$, roots are $2 + i$ and

$-2 - i$.

41. If $z = \bar{z}$, $a + bi = a - bi$; $bi = -bi$; $2bi = 0$; $b = 0$; thus $z = a + 0i = a$, and z is a real number.

42. If $z_1 = a + bi$ and $z_2 = c + di$, $\overline{z_1 + z_2} = \overline{(a + bi) + (c + di)} = \overline{(a + c) + (b + d)i} = (a + c) - (b + d)i = (a - bi) + (c - di) = \overline{a + bi} + \overline{c + di} = \overline{z_1} + \overline{z_2}$

43. If $z_1 = a + bi$ and $z_2 = c + di$, $\overline{z_1 z_2} = \overline{(a + bi)(c + di)} = \overline{ac + adi + bci + bdi^2} = \overline{(ac - bd) + (ad + bc)i} = (ac - bd) - (ad + bc)i$; $\overline{z_1} \cdot \overline{z_2} = (a - bi)(c - di) = ac - adi - bci + bdi^2 = (ac - bd) - (ad + bc)i$; $\overline{z_1 z_2} = \overline{z_1} \cdot \overline{z_2}$

44. If $z_1 = a + bi$ and $z_2 = c + di \neq 0$, $\overline{\left(\dfrac{z_1}{z_2}\right)} = \overline{\left(\dfrac{a + bi}{c + di}\right)} = \overline{\left(\dfrac{a + bi}{c + di} \cdot \dfrac{c - di}{c - di}\right)} =$

$\overline{\left(\dfrac{(ac + bd) + (bc - ad)i}{c^2 + d^2}\right)} = \dfrac{ac + bd}{c^2 + d^2} - \dfrac{bc - ad}{c^2 + d^2}i$; $\dfrac{\overline{z_1}}{\overline{z_2}} = \dfrac{a - bi}{c - di} = \dfrac{a - bi}{c - di} \cdot \dfrac{c + di}{c + di} =$

$\dfrac{ac + bd - (bc - ad)i}{c^2 + d^2} = \dfrac{ac + bd}{c^2 + d^2} - \dfrac{bc - ad}{c^2 + d^2}i$; $\overline{\left(\dfrac{z_1}{z_2}\right)} = \dfrac{\overline{z_1}}{\overline{z_2}}$

45. $\overline{z^2} = \overline{z \cdot z} = \bar{z} \cdot \bar{z}$ (from Ex. 43) $= (\bar{z})^2$

46. a. $\overline{z^3} = \overline{z^2 \cdot z} = \overline{z^2} \cdot \bar{z}$ (from Ex. 43) $= (\bar{z})^2 \cdot \bar{z}$ (from Ex. 45) $= (\bar{z})^3$

b. $\overline{z^n} = (\bar{z})^n$, where n is a natural number

Calculator Exercise · page 29

$(6 + 7i)^8 = [(6 + 7i)^2]^4 = [(6^2 - 7^2) + (2 \cdot 6 \cdot 7)i]^4 = (-13 + 84i)^4 = [(-13 + 84i)^2]^2 = [(-13)^2 - (84)^2 + 2(-13)(84)i]^2 = (-6887 - 2184i)^2 = (-6887)^2 - (-2184)^2 + 2(-6887)(-2184)i = 42,660,913 + 30,082,416i$

Computer Exercise · page 29

```
10 PRINT "COMPUTE (A + BI)↑N, ";
20 PRINT "WHERE I = SQR(-1)."
30 PRINT
40 PRINT "INPUT A POSITIVE INTEGER N ";
50 INPUT N
60 IF (N < 1 OR INT(N) < N) THEN 40
70 PRINT "INPUT REAL NUMBERS A, B ";
80 INPUT A,B
90 LET X = A
100 LET Y = B
110 IF N = 1 THEN 180
120 FOR I = 2 TO N
130 LET S = A * X - B * Y
140 LET T = A * Y + B * X
150 LET X = S
160 LET Y = T
170 NEXT I
180 PRINT
190 PRINT "RESULT: ";X;" + ";Y;" * I"
200 END
```

Class Exercises 1-6 · page 34

1. $\left(\dfrac{1}{2} \cdot 8\right)^2 = 16$ **2.** $\left[\dfrac{1}{2}(-10)\right]^2 = 25$ **3.** $\left(\dfrac{1}{2} \cdot 7\right)^2 = \dfrac{49}{4}$ **4.** $\left(\dfrac{1}{2} \cdot 2a\right)^2 = a^2$

5–10. Answers may vary.   **5.** completing the square   **6.** factoring   **7.** factoring
8. quadratic formula   **9.** quadratic formula   **10.** completing the square

11. When $b^2 - 4ac < 0$, then $\sqrt{b^2 - 4ac} = i\sqrt{-(b^2 - 4ac)}$; roots are $\dfrac{-b \pm i\sqrt{-(b^2 - 4ac)}}{2a}$,

which are imaginary numbers.

12. When $b^2 - 4ac = 0$, the two roots of the equation are given by $\dfrac{-b \pm 0}{2a}$, or $-\dfrac{b}{2a}$. If a

and b are real, then $-\dfrac{b}{2a}$ is real. Thus, there is one real solution, $-\dfrac{b}{2a}$, which is a

double root. Note that the original equation $ax^2 + bx + c = 0$ can be factored as

$\left(x + \dfrac{b}{2a}\right)\left(x + \dfrac{b}{2a}\right)$ in this case.

13. If $b^2 - 4ac$ is the square of an integer k, then the roots are $\dfrac{-b \pm k}{2a}$. Since a, b, and k are

all integers, the roots are rational.

14. No; the coefficient of x is $-\sqrt{5}$, which is not an integer.

15. Yes; the only restriction on a, b, and c is that $a \neq 0$. The same laws of algebra that apply to the quadratic formula with rational coefficients apply with irrational (or even imaginary) coefficients.

16. Dividing by $x + 4$ causes the root -4 to be lost.

17. $(2x - 5)(x + 6) - 7(x + 6) = 0$; $(2x - 5 - 7)(x + 6) = 0$; $2x - 12 = 0$ or $x + 6 = 0$; $x = 6$ or $x = -6$

18. $(3x - 5)(4x - 1) - (4x - 1) = 0$; $(3x - 5 - 1)(4x - 1) = 0$; $3x - 6 = 0$ or

$4x - 1 = 0$; $x = 2$ or $x = \dfrac{1}{4}$

19. Squaring introduces the extraneous root -2, which does not satisfy the original equation.

Written Exercises 1-6 · pages 35–36

1. $(3x - 7)(x + 1) = 0$; $x = \dfrac{7}{3}$ or $x = -1$

2. $4(x^2 - 2x - 8) = 0$; $4(x - 4)(x + 2) = 0$; $x = 4$ or $x = -2$

3. $(2x - 3)(x + 4) = 6$; $2x^2 + 5x - 18 = 0$; $(2x + 9)(x - 2) = 0$; $x = -\dfrac{9}{2}$ or $x = 2$

4. $(3y - 2)(y + 4) = 24$; $3y^2 + 10y - 32 = 0$; $(3y + 16)(y - 2) = 0$; $y = -\dfrac{16}{3}$ or $y = 2$

5. $x^2 - 10x = 1575$; $x^2 - 10x + 25 = 1600$; $(x - 5)^2 = 1600$; $x - 5 = \pm 40$; $x = 5 \pm 40$; $x = 45$ or $x = -35$

6. $x^2 - 6x = 391$; $x^2 - 6x + 9 = 400$; $(x - 3)^2 = 400$; $x - 3 = \pm 20$; $x = 3 \pm 20$; $x = 23$ or $x = -17$

7. $2z^2 - 16z = 1768$; $z^2 - 8z = 884$; $z^2 - 8z + 16 = 900$; $(z - 4)^2 = 900$; $z - 4 = \pm 30$; $z = 4 \pm 30$; $z = 34$ or $z = -26$

8. $x^2 - 8x = 20$; $x^2 - 8x + 16 = 36$; $(x - 4)^2 = 36$; $x - 4 = \pm 6$; $x = 4 \pm 6$; $x = 10$ or $x = -2$

9. $x^2 + 6x = -10$; $x^2 + 6x + 9 = 9 - 10$; $(x + 3)^2 = -1$; $x + 3 = \pm i$; $x = -3 \pm i$

10. $y^2 + 10y = -35$; $y^2 + 10y + 25 = 25 - 35$; $(y + 5)^2 = -10$; $y + 5 = \pm i\sqrt{10}$; $y = -5 \pm i\sqrt{10}$

11. $5x^2 + 2x - 1 = 0$; $x = \dfrac{-2 \pm \sqrt{2^2 - 4(5)(-1)}}{2 \cdot 5} = \dfrac{-2 \pm \sqrt{24}}{10} = \dfrac{-2 \pm 2\sqrt{6}}{10} = \dfrac{-1 \pm \sqrt{6}}{5}$

12. $4x^2 - 4x - 17 = 0$; $x = \dfrac{-(-4) \pm \sqrt{(-4)^2 - 4(4)(-17)}}{2 \cdot 4} = \dfrac{4 \pm \sqrt{288}}{8} = \dfrac{1 \pm 3\sqrt{2}}{2}$

13. $t^2 - 4t + 5 = 0$; $t = \dfrac{-(-4) \pm \sqrt{(-4)^2 - 4(1)(5)}}{2 \cdot 1} = \dfrac{4 \pm \sqrt{-4}}{2} = \dfrac{4 \pm 2i}{2} = 2 \pm i$

14. $5u^2 - 5u + 2 = 0$; $u = \dfrac{-(-5) \pm \sqrt{(-5)^2 - 4(5)(2)}}{2 \cdot 5} = \dfrac{5 \pm \sqrt{-15}}{10} = \dfrac{5 \pm i\sqrt{15}}{10}$

15. $4v - 16 = v^2 - 6v$; $v^2 - 10v + 16 = 0$;

$v = \dfrac{-(-10) \pm \sqrt{(-10)^2 - 4(1)(16)}}{2 \cdot 1} = \dfrac{10 \pm \sqrt{36}}{2} = \dfrac{10 \pm 6}{2}$; $v = \dfrac{16}{2} = 8$ or $v = \dfrac{4}{2} = 2$

16. $4z - 12 = 3z^2$; $3z^2 - 4z + 12 = 0$;

$$z = \frac{-(-4) \pm \sqrt{(-4)^2 - 4(3)(12)}}{2 \cdot 3} = \frac{4 \pm \sqrt{-128}}{6} = \frac{4 \pm 8i\sqrt{2}}{6} = \frac{2 \pm 4i\sqrt{2}}{3}$$

17. $8x^2 + 10x - 7 = 0$; $b^2 - 4ac = 10^2 - 4(8)(-7) = 324 = 18^2$; $(4x + 7)(2x - 1) = 0$;

$$x = -\frac{7}{4} \text{ or } x = \frac{1}{2}$$

18. $15t^2 - 4t + 1 = 0$; $b^2 - 4ac = (-4)^2 - 4(15)(1) = -44$;

$$t = \frac{-(-4) \pm \sqrt{-44}}{2 \cdot 15} = \frac{4 \pm 2i\sqrt{11}}{30} = \frac{2 \pm i\sqrt{11}}{15}$$

19. $(3x - 2)^2 = 121$; $3x - 2 = \pm 11$; $3x = 2 \pm 11$; $x = \dfrac{2 \pm 11}{3}$; $x = \dfrac{13}{3} \text{ or } x = -3$

20. $(4y + 4)^2 = -16$; $4y + 4 = \pm 4i$; $y = \dfrac{-4 \pm 4i}{4} = -1 \pm i$

21. $(4x + 7)(x - 1) - 2(x - 1) = 0$; $(4x + 7 - 2)(x - 1) = 0$; $4x + 5 = 0 \text{ or } x - 1 = 0$;

$$x = -\frac{5}{4} \text{ or } x = 1$$

22. $(2x + 1)(4x - 3) - 3(4x - 3)^2 = 0$; $(4x - 3)(2x + 1 - 3(4x - 3)) = 0$;

$$(4x - 3)(-10x + 10) = 0; x = \frac{3}{4} \text{ or } x = 1$$

23. $2w(4w - 1) - w(1 - 4w) = 0$; $(4w - 1)(2w + w) = 0$; $w = \dfrac{1}{4} \text{ or } w = 0$

24. $3(2x - 3)^2 - 4x(3 - 2x) = 0$; $(2x - 3)(3(2x - 3) + 4x) = 0$; $(2x - 3)(10x - 9) = 0$;

$$x = \frac{3}{2} \text{ or } x = \frac{9}{10}$$

25. $(x + 3)^2 + (x - 3)^2 = 18 - 6x$; $2x^2 + 6x = 0$; $2x(x + 3) = 0$; $x = 0 \text{ or } x = -3$
(reject); $x = 0$

26. $r(r + 1) - r(r - 1) = 2$; $r^2 + r - r^2 + r = 2$; $2r = 2$; $r = 1$ (reject); no solution

27. $3(t^2 + 1) = t(t + 2) + 5 \cdot 3$; $3t^2 + 3 = t^2 + 2t + 15$; $2t^2 - 2t - 12 = 0$;
$t^2 - t - 6 = 0$; $(t - 3)(t + 2) = 0$; $t = 3 \text{ or } t = -2$ (reject); $t = 3$

28. $\dfrac{1}{x - 3} = 3 - \dfrac{4}{x - 3}$; $\dfrac{5}{x - 3} = 3$; $5 = 3x - 9$; $x = \dfrac{14}{3}$

29. $(2\sqrt{x})^2 = (x - 8)^2$; $4x = x^2 - 16x + 64$; $x^2 - 20x + 64 = 0$; $(x - 16)(x - 4) = 0$;
$x = 16 \text{ or } x = 4$ (reject); $x = 16$

30. $(\sqrt{2x + 5})^2 = (x + 1)^2$; $2x + 5 = x^2 + 2x + 1$; $x^2 = 4$; $x = \pm 2$; reject $x = -2$; $x = 2$

31. $\dfrac{x}{x + 3} = \dfrac{8}{2x}$; $2x^2 = 8x + 24$; $x^2 - 4x - 12 = 0$; $(x - 6)(x + 2) = 0$; $x = 6 \text{ or } x = -2$

(reject); $x = 6$

32. $\dfrac{x + 6}{2x + 6} = \dfrac{6x}{7x + 1}$; $7x^2 + 43x + 6 = 12x^2 + 36x$; $5x^2 - 7x - 6 = 0$;

$$(5x + 3)(x - 2) = 0; x = -\frac{3}{5} \text{ (reject) or } x = 2; x = 2$$

33. $x(x + 7) = 14(20); x^2 + 7x - 280 = 0; x = \dfrac{-7 \pm \sqrt{49 - 4(-280)}}{2}; x = \dfrac{-7 \pm \sqrt{1169}}{2};$

$x = \dfrac{-7 - \sqrt{1169}}{2}$ (reject) or $x = \dfrac{-7 + \sqrt{1169}}{2}$

34. $x \cdot 5x = 9(x + 14); 5x^2 - 9x - 126 = 0; (5x + 21)(x - 6) = 0; x = -\dfrac{21}{5}$ (reject) or

$x = 6; x = 6$

35. a. $8^2 - 4(4)k = 64 - 16k$ **b.** when $64 - 16k = 0$, or $k = 4$

c. when $64 - 16k > 0$, or $k < 4$ **d.** when $64 - 16k < 0$, or $k > 4$

e. Examples of values for which $64 - 16k$ is a perfect square: $k = 0, k = 3, k = \dfrac{15}{4}$,

$k = -5$

36. a. $8^2 - 4(5)(k) = 64 - 20k$ **b.** when $64 - 20k = 0$, or $k = \dfrac{16}{5}$

c. when $64 - 20k > 0$, or $k < \dfrac{16}{5}$ **d.** when $64 - 20k < 0$, or $k > \dfrac{16}{5}$

e. Examples of values for which $64 - 20k$ is a perfect square: $k = 0, k = 3, k = 2.4$,

$k = -4$

37. $x = \dfrac{5 \pm \sqrt{(-5)^2 - 4 \cdot \sqrt{2} \cdot \sqrt{8}}}{2\sqrt{2}} = \dfrac{5 \pm \sqrt{9}}{2\sqrt{2}} = \dfrac{5 \pm 3}{2\sqrt{2}}; x = \dfrac{8}{2\sqrt{2}} = 2\sqrt{2}$ or $x = \dfrac{2}{2\sqrt{2}} = \dfrac{\sqrt{2}}{2}$

38. $x = \dfrac{2\sqrt{5} \pm \sqrt{(-2\sqrt{5})^2 - 4(4)(-1)}}{2 \cdot 4} = \dfrac{2\sqrt{5} \pm \sqrt{36}}{8} = \dfrac{2\sqrt{5} \pm 6}{8} = \dfrac{\sqrt{5} \pm 3}{4}$

39. $x^2 - 6ix - 9 = 0; x = \dfrac{6i \pm \sqrt{(-6i)^2 - 4(1)(-9)}}{2 \cdot 1} = \dfrac{6i \pm \sqrt{-36 + 36}}{2} = 3i$

40. $ix^2 - 3x - 2i = 0; x = \dfrac{3 \pm \sqrt{(-3)^2 - 4i(-2i)}}{2i} = \dfrac{3 \pm \sqrt{1}}{2i}; x = \dfrac{4}{2i} = \dfrac{2}{i} = \dfrac{2}{i} \cdot \dfrac{-i}{-i} = -2i$ or

$x = \dfrac{2}{2i} = \dfrac{1}{i} = \dfrac{1}{i} \cdot \dfrac{-i}{-i} = -i$

41. $(\sqrt{2x + 5})^2 = (2\sqrt{2x} + 1)^2; 2x + 5 = 8x + 4\sqrt{2x} + 1; -6x + 4 = 4\sqrt{2x};$

$-3x + 2 = 2\sqrt{2x}; 9x^2 - 12x + 4 = 8x; 9x^2 - 20x + 4 = 0; (9x - 2)(x - 2) = 0$

$x = \dfrac{2}{9}$ or $x = 2$ (reject); $x = \dfrac{2}{9}$

42. $(\sqrt{y - 3})^2 = (1 - \sqrt{2y - 4})^2; y - 3 = 1 - 2\sqrt{2y - 4} + 2y - 4; 2\sqrt{2y - 4} = y;$

$4(2y - 4) = y^2; y^2 - 8y + 16 = 0; (y - 4)^2 = 0; y = 4$ (reject); no solution

43. (41) 0.22;
(42) No solution
No extraneous solutions are encountered with this method.

44. $ax^2 + bx + c = 0; x^2 + \dfrac{b}{a}x + \dfrac{c}{a} = 0; x^2 + \dfrac{b}{a}x + \left(\dfrac{b}{2a}\right)^2 = -\dfrac{c}{a} + \left(\dfrac{b}{2a}\right)^2;$

$\left(x + \dfrac{b}{2a}\right)^2 = \dfrac{b^2 - 4ac}{4a^2}; x + \dfrac{b}{2a} = \pm\sqrt{\dfrac{b^2 - 4ac}{4a^2}};$

$x = -\dfrac{b}{2a} \pm \dfrac{\sqrt{b^2 - 4ac}}{2a} = \dfrac{-b \pm \sqrt{b^2 - 4ac}}{2a}$ if $a \neq 0$

45. a. $b^2 - 4ac = 1^2 - 4a(-6) = 1 + 24a$; we are to find the 5 smallest positive integral
values of a such that $1 + 24a$ is a perfect square.   By trial and error: when $a = 1$,
$1 + 24a = 25 = 5^2$; when $a = 2$, $1 + 24a = 49 = 7^2$; when $a = 5$, $1 + 24a = 121 = 11^2$; when $a = 7$, $1 + 24a = 169 = 13^2$; when $a = 12$, $1 + 24a = 289 = 17^2$

   b. Examples: $53^2 = 2809$; $2808 \div 24 = 117$, thus, we have $a = 117$.   Also,
$59^2 = 3481$; $3480 \div 24 = 145$; thus, we have $a = 145$.

Computer Exercise · page 36

```
10 PRINT "COMPUTE ROOTS OF AX↑2 + BX + C = 0"
15 PRINT
20 PRINT "INPUT A, B, C ";
30 INPUT A,B,C
40 PRINT
50 IF A = 0 THEN 20
60 LET D = B * B - 4 * A * C
70 LET R1 = -B/(2 * A)
80 IF D = 0 THEN 150
90 IF D > 0 THEN 180
100 LET R2 = SQR(-D)/(2 * ABS(A))
110 PRINT "IMAGINARY ROOTS: "
120 PRINT R1; "+";R2; "*I AND "
130 PRINT R1; "-";R2; "*I"
140 GOTO 220
150 PRINT "DOUBLE ROOT: "
160 PRINT R1
170 GOTO 220
180 PRINT "REAL ROOTS: "
190 LET R2 = SQR(D)/(2 * A)
200 PRINT R1 + R2;" AND ";
210 PRINT R1 - R2
220 END
```

Class Exercises 1-7 · pages 40–41

1. a. since $a > 0$, upward   **b.** $b^2 - 4ac = (-2)^2 - 4(1)(3) < 0$; no points

2. a. since $a < 0$, downward   **b.** $b^2 - 4ac = 144 - 4(-2)(-18) = 0$; one point

3. a. since $a < 0$, downward   **b.** $b^2 - 4ac = 0^2 - 4(-1)(9) > 0$; two points

4. $(5, 4)$ **5.** $(-2, -3)$ **6.** $(0, -10)$

7. The zeros of a quadratic function $f(x)$, if real, are the x-intercepts of the graph of $f(x)$.
If $f(x)$ has only one (real) zero, then the graph has only one x-intercept.   If $f(x)$ has
imaginary zeros, then the graph has no x-intercepts.

Written Exercises 1-7 · pages 41–43

1. a.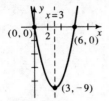

b.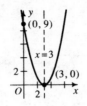

c.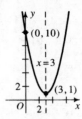

2. a.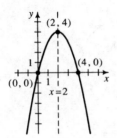

b.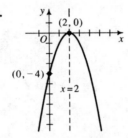

c.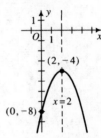

3.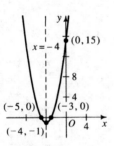

4.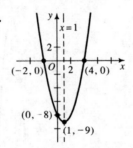

5.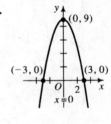

6.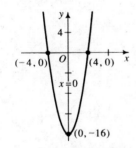

7.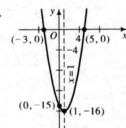

8.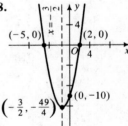

9.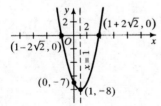

10.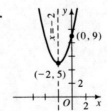

26

11.

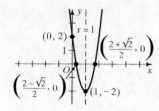

12.

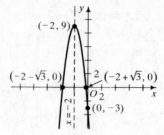

13.

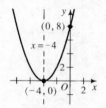

14.

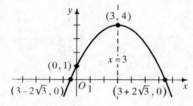

15.

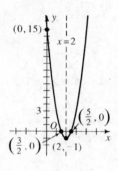

16.

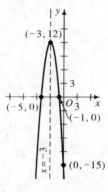

17.

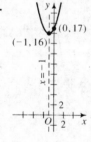

18.

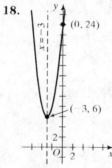

19.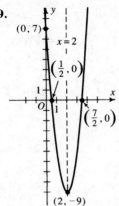

20.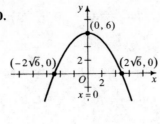

21. $y = 4 - 2x$ and $y = x^2 - 6x + 8$; $4 - 2x = x^2 - 6x + 8$; $x^2 - 4x + 4 = 0$;
$(x - 2)^2 = 0$; $x = 2$ and $y = 4 - 2(2) = 0$; one point: $(2, 0)$

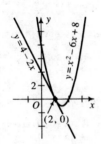

Ex. 21

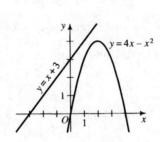

Ex. 22

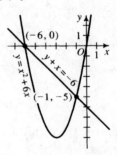

Ex. 23

22. $y = x + 3$ and $y = 4x - x^2$; $x + 3 = 4x - x^2$;
$x^2 - 3x + 3 = 0$; $b^2 - 4ac = (-3)^2 - 4(1)(3) < 0$;
there is no real root and thus no points of intersection.

23. $y = -x - 6$ and $y = x^2 + 6x$; $-x - 6 = x^2 + 6x$;
$x^2 + 7x + 6 = 0$; $(x + 6)(x + 1) = 0$; $x = -6$ and
$y = 6 - 6 = 0$, or $x = -1$ and $y = 1 - 6 = -5$;
two points: $(-6, 0)$ and $(-1, -5)$

24. $y = -2x + 10$ and $y = 9 - x^2$; $-2x + 10 = 9 - x^2$;
$x^2 - 2x + 1 = 0$; $(x - 1)^2 = 0$; $x = 1$ and $y = -2(1) + 10 = 8$;
one point: $(1, 8)$

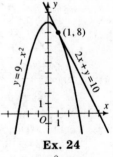

Ex. 24

25. To show that $y = 2x - 5$ intersects the graph in one point, solve $2x - 5 = x^2 - 4$.
$2x - 5 = x^2 - 4$; $x^2 - 2x + 1 = 0$; $(x - 1)^2 = 0$; $x = 1$. When $x = 1$, $y = 2(1) - 5$.
Thus, the line and the parabola intersect only at the point $(1, -3)$.

26. A value of -4 will give $y = x - 4$. To show that the line $y = x - 4$ intersects the
parabola $y = x^2 + 5x$ exactly once, solve $x - 4 = x^2 + 5x$; $x - 4 = x^2 + 5x$;
$x^2 + 4x + 4 = 0$; $(x + 2)^2 = 0$; $x = -2$. When $x = -2$, $y = -2 - 4 = -6$. Thus, the
line and the parabola intersect only at $(-2, -6)$.

27. Since the x-intercepts are 2 and -1, $f(x) = a(x - 2)(x + 1)$; substituting $(0, 6)$,
$6 = a(0 - 2)(0 + 1)$; $6 = -2a$; $a = -3$; $f(x) = -3(x - 2)(x + 1)$;
$f(x) = -3x^2 + 3x + 6$.

28. Since the x-intercepts are 5 and 1, $f(x) = a(x - 5)(x - 1)$; substituting $(0, 1)$,

$$1 = a(0 - 5)(0 - 1) = 5a; \ a = \frac{1}{5}; f(x) = \frac{1}{5}(x - 5)(x - 1); f(x) = \frac{1}{5}x^2 - \frac{6}{5}x + 1.$$

29. Since the vertex is $(4, 8)$, $f(x) = a(x - 4)^2 + 8$; substituting $(0, 0)$,

$$0 = a(0 - 4)^2 + 8 = 16a + 8; \ a = -\frac{1}{2}; f(x) = -\frac{1}{2}(x - 4)^2 + 8; f(x) = -\frac{1}{2}x^2 + 4x.$$

30. Since the vertex is $(3, -8)$, $f(x) = a(x - 3)^2 - 8$; substituting $(0, 0)$,

$$0 = a(0 - 3)^2 - 8 = 9a - 8; \ a = \frac{8}{9}; f(x) = \frac{8}{9}(x - 3)^2 - 8; f(x) = \frac{8}{9}x^2 - \frac{16}{3}x.$$

31. Since the vertex is $(3, -5)$, $f(x) = a(x - 3)^2 - 5$; substituting $(1, 2)$,

$$2 = a(1 - 3)^2 - 5 = 4a - 5; \ 7 = 4a; \ a = \frac{7}{4}; f(x) = \frac{7}{4}(x - 3)^2 - 5.$$

32. Since the vertex is $(-1, 6)$, $g(x) = a(x + 1)^2 + 6$; substituting $(-3, 4)$,

$$4 = a(-3 + 1)^2 + 6 = 4a + 6; \ -2 = 4a; \ a = -\frac{1}{2}; g(x) = -\frac{1}{2}(x + 1)^2 + 6.$$

33. Since the parabola is tangent to a horizontal line at $(4, 0)$, $(4, 0)$ is the vertex; thus,

$$f(x) = a(x - 4)^2 + 0 = a(x - 4)^2; \text{substituting } (0, 6), \ 6 = a(0 - 4)^2 = 16a; \ a = \frac{3}{8};$$

$$f(x) = \frac{3}{8}(x - 4)^2.$$

34. By symmetry, the vertex is $(1, 8)$, so $f(x) = a(x - 1)^2 + 8$; substituting $(3, 0)$ gives
$0 = a(3 - 1)^2 + 8$; $a = -2$; $f(x) = -2(x - 1)^2 + 8$; $f(x) = -2x^2 + 4x + 6$.

35.

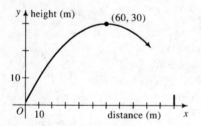

36.

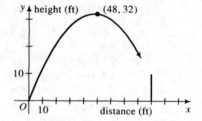

a. Assume that the ball's path begins at the origin. The vertex is $(60, 30)$, so $y = a(x - 60)^2 + 30$; also

$$0 = a(0 - 60)^2 + 30; \ a = -\frac{29}{3600};$$

$$y = -\frac{29}{3600}(x - 60)^2 + 30.$$

b. When $x = 110$, $y \approx 9.2$; yes, the ball will go over the 4 m fence. (Note: If the ball's path begins above the ground, say 1 m

high, then $f(x) = -\dfrac{29}{3600}(x - 60)^2 + 30$;

$f(110) > 9$, so the ball still clears the fence.)

a. The vertex is $(48, 32)$, so $y = a(x - 48)^2 + 32$; substituting $(0, 0)$ gives $0 = a(0 - 48)^2 + 32$;

$$a = -\frac{1}{72};$$

$$y = -\frac{1}{72}(x - 48)^2 + 32.$$

b. When $x = 90$, $y = 7.5$; since the

ball is $7\dfrac{1}{2}$ ft above the ground

when it reaches the goal post, the kicker will not make the field goal.

37. a. $f(1 + k) = 2(1 + k)^2 - 4(1 + k) + 7 = 2 + 4k + 2k^2 - 4 - 4k + 7 = 2k^2 + 5$;
$f(1 - k) = 2(1 - k)^2 - 4(1 - k) + 7 = 2 - 4k + 2k^2 - 4 + 4k + 7 = 2k^2 + 5$

b. Since $f(1 + k) = f(1 - k)$, if a point $(1 + k, f(1 + k))$ is on the parabola, so is $(1 - k, f(1 + k))$; the line $x = 1$ is the vertical line halfway between each pair of points on the parabola, so $x = 1$ is the axis of symmetry.

38. a. $h(3) = 2(3)^2 - 12(3) = -18$; $h(3 + k) = 2(3 + k)^2 - 12(3 + k) = 18 + 12k + 2k^2 - 36 - 12k = -18 + 2k^2$; for all real k, $2k^2 \geq 0$, so $-18 + 2k^2 \geq -18$; $h(3 + k) \geq h(3)$. The parabola $h(x) = 2x^2 - 12x$ has vertex $(3, -18)$ and opens upward; thus, $(3, -18)$ is a minimum point; the y-coordinate of every point on $h(x)$ is at least -18.

b. $f(4) = 9 + 8(4) - 4^2 = 25$; $f(4 + k) = 9 + 8(4 + k) - (4 + k)^2 = 9 + 32 + 8k - 16 - 8k - k^2 = 25 - k^2$; for all real k, $k^2 \geq 0$, so $25 - k^2 \leq 25$; $f(4 + k) \leq f(4)$. The parabola $f(x) = 9 + 8x - x^2$ has vertex $(4, 25)$ and opens downward; thus, $(4, 25)$ is a maximum; the y-coordinate of every point on $f(x) \leq 25$.

39. The roots of $ax^2 + bx + c = 0$ are $\dfrac{-b \pm \sqrt{b^2 - 4ac}}{2a}$; average $= \dfrac{1}{2}\left(\dfrac{-2b}{2a}\right) = -\dfrac{b}{2a}$

40. $y = ax^2 + bx + c$; $y - c = a\left(x^2 + \dfrac{b}{a}x\right)$; $y - c + \dfrac{b^2}{4a} = a\left(x^2 + \dfrac{b}{a}x + \dfrac{b^2}{4a^2}\right)$;

$y + \dfrac{b^2 - 4ac}{4a} = a\left(x + \dfrac{b}{2a}\right)^2$; $y = a\left(x + \dfrac{b}{2a}\right)^2 - \dfrac{b^2 - 4ac}{4a} = a\left(x + \dfrac{b}{2a}\right)^2 + \dfrac{4ac - b^2}{4a}$;

$y = a(x - h)^2 + k$, so $h = -\dfrac{b}{2a}$ and $k = \dfrac{4ac - b^2}{4a}$.

41. a. $y = x^2$ and a small portion near $(1, 1)$ are at the right; $(0.95, 0.9)$ and $(1.05, 1.1)$ are on the graph and near $(1, 1)$; slope $\approx \dfrac{1.1 - 0.9}{1.05 - 0.95} = 2$

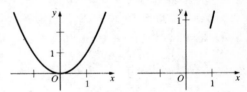

b. $y - 1 = 2(x - 1)$; To show that $y - 1 = 2(x - 1)$ intersects the graph of $y = x^2$ in one point, solve $2(x - 1) + 1 = x^2$; $x^2 - 2x + 1 = 0$; $(x - 1)^2 = 0$; $x = 1$. When $x = 1$, $y = 1$. Thus, the line and the parabola intersect only at the point $(1, 1)$.

42. Establish a coordinate system with the origin at the highest point of the bridge; the parabola has eq. $y = a(x - 0)^2 + 0 = ax^2$; $(6, -6)$ and $(-6, -6)$ are points on the graph, so $-6 = 36a$ and $a = -\dfrac{1}{6}$; $y = -\dfrac{1}{6}x^2$; the trailer is 9 m wide, so consider $x = \pm\dfrac{1}{2}(9) = \pm 4.5$; $y = -\dfrac{1}{6}(4.5)^2 = -3.375$; $6 - 3.375 = 2.625$. At a point 4.5 m to the right of center of arc, there is only 2.625 m of clearance. The trailer will not fit.

Class Exercises 1-8 · page 45

1. Quadratic　　**2.** Linear　　**3.** Quadratic　　**4.** Other

5. As the flight speed of parakeets increases from 12 mi/h, the energy consumption decreases, and the birds achieve peak efficiency at approximately 22 mi/h. As speed increases from 22 mi/h, efficiency decreases. Parakeets appear to fly most efficiently in the middle of their speed range.

Written Exercises 1-8 · pages 45–48

1. (1) $f(0) = 0a + 0b + c = 5$ From (1), $c = 5$; substituting in (2) and (3) gives
   (2) $f(1) = a + b + c = 10$ (4) $a + b = 5$ and (5) $4a + 2b = 14$, or $2a + b = 7$;
   (3) $f(2) = 4a + 2b + c = 19$ from (4) and (5), $a = 2$ and $b = 3$;
   $$f(x) = 2x^2 + 3x + 5$$

2. (1) $f(1) = a + b + c = 4$
   (2) $f(2) = 4a + 2b + c = 12$ (2) $-$ (1): $3a + b = 8$; $b = 8 - 3a$
   (3) $f(4) = 16a + 4b + c = 46$ (3) $-$ (1): $15a + 3b = 42$
   $15a + 3(8 - 3a) = 42$; $a = 3$; $b = 8 - 3(3) = -1$; $3 - 1 + c = 5$; $c = 3$;
   $f(x) = 3x^2 - x + 2$

3. (1) $f(0) = 0a + 0b + c = 6$ From (1), $c = 6$; substituting in (2) and (3) gives
   (2) $f(2) = 4a + 2b + c = 18$ (4) $4a + 2b = 12$, or $2a + b = 6$ and
   (3) $f(4) = 16a + 4b + c = 34$ (5) $16a + 4b = 28$, or $4a + b = 7$;
   from (4) and (5), $2a = 1$, so $a = \dfrac{1}{2}$; $2\left(\dfrac{1}{2}\right) + b = 6$; $b = 5$; $f(x) = \dfrac{1}{2}x^2 + 5x + 6$

4. (1) $f(1) = a + b + c = 10.5$
   (2) $f(2) = 4a + b + c = 13$ (2) $-$ (1): $3a + b = 2.5$; $b = 2.5 - 3a$
   (3) $f(5) = 25a + 5b + c = 32.5$ (3) $-$ (1): $24a + 4b = 22$
   $24a + 4(2.5 - 3a) = 22$; $a = 1$; $b = 2.5 - 3(1) = -0.5$; $1 + (-0.5) + c = 10.5$;
   $c = 10$; $f(x) = x^2 - 0.5x + 10$

5. **a.** $0.00625(8^2) + 0.2625(8) + 2 = 4.50$
   $0.00625(10^2) + 0.2625(10) + 2 = 5.25$
   $0.00625(16^2) + 0.2625(16) + 2 = 7.80$

   b. Yes; $0.00625(18^2) + 0.2625(18) + 2 = 8.75$

   c. No; $0.00625(4^2) + 0.2625(4) + 2 = 3.15$; $3.15 is too much for a 4 in. pizza.

6. **a.** $f(20) = 20 + \dfrac{20^2}{20} = 40$; $f(30) = 30 + \dfrac{30^2}{20} = 75$; $75 - 40 = 35$; 35 ft

   b. $f(50) = 50 + \dfrac{50^2}{20} = 175$; $f(60) = 60 + \dfrac{60^2}{20} = 240$; $240 - 175 = 65$ (ft)

   c. No; it assumes a linear function rather than a quadratic function.

7. **a.** (1) $D(20) = 400a + 20b + c = 26$
   (2) $D(40) = 1600a + 40b + c = 34$ (2) $-$ (1): $1200a + 20b = 8$
   (3) $D(50) = 2500a + 50b + c = 32$ (3) $-$ (2): $900a + 10b = -2$
   $\underline{\begin{array}{l} 600a + 10b = 4 \\ 900a + 10b = -2 \end{array}}$

   $300a = -6$; $a = -0.02$; $900(-0.02) + 10b = -2$; $b = 1.6$; from (1),
   $400(-0.02) + 20(1.6) + c = 26$; $c = 2$; $D(s) = -0.02s^2 + 1.6s + 2$

   b. $D(65) = -0.02(65^2) + 1.6(65) + 2 = 21.5$ (mi)

   c. Solve $16 = -0.02s^2 + 1.6s + 2$; $s^2 - 80s + 700 = 0$; $(s - 70)(s - 10) = 0$; if you
   drive at any speed from 10 mi/h to 70 mi/h, you will reach the gas station; thus, the
   speed limit of 55 mi/h is the maximum driving speed.

8. a. Answers may vary. Example: the increased crowding of seeds

   b. (1) $f(3) = 9a + 3b + c = 70$

       (2) $f(4) = 16a + 4b + c = 85$ (2) $-$ (1): $7a + b = 15$

       (3) $f(5) = 25a + 5b + c = 80$ (3) $-$ (2): $\underline{9a + b = -5}$

                                     $-2a \quad = 20; a = -10$

       $7(-10) + b = 15; b = 85; 9(-10) + 3(85) + c = 70; c = -95;$

       $f(x) = -10x^2 + 85x - 95$

   c. The maximum occurs when $x = -\dfrac{b}{2a} = -\dfrac{85}{2(-10)} = 4\dfrac{1}{4}$ bags.

9. a. 6 in. gives $\dfrac{9\pi}{8} \approx 3.53$ in.^{2}/dollar; 8 in. gives $\dfrac{16\pi}{12} \approx 4.19$ in.^{2}/dollar; 10 in. gives

    $\dfrac{25\pi}{20} \approx 3.93$ in.^{2}/dollar; 8 in. gives you the most cheesecake per dollar.

   b. (1) $f(6) = 36a + 6b + c = 8$

       (2) $f(8) = 64a + 8b + c = 12$ (2) $-$ (1): $28a + 2b = 4$

       (3) $f(10) = 100a + 10b + c = 20$ (3) $-$ (2): $\underline{36a + 2b = 8}$

                                     $8a \quad = 4; a = \frac{1}{2}$

       $28\left(\dfrac{1}{2}\right) + 2b = 4; b = -5; 36\left(\dfrac{1}{2}\right) + 6(-5) + c = 8; c = 20;$

       $f(x) = \dfrac{1}{2}x^2 - 5x + 20$

10. a. (1) $P(2) = 4a + 2b + c = 176$

       (2) $P(4) = 16a + 4b + c = 224$ (2) $-$ (1): $12a + 2b = 48$

       (3) $P(5) = 25a + 5b + c = 200$ (3) $-$ (2): $9a + b = -24; b = -9a - 24$

       $12a + 2b = 48; 6a + b = 24; 6a + (-9a - 24) = 24; a = -16;$

       $b = -9(-16) - 24 = 120; 4(-16) + 2(120) + c = 176; c = 0;$

       $P(I) = -16I^2 + 120I$

   b. $P(1) = 104; P(2) = 176; P(3) = 216; P(4) = 224; P(5) = 200;$

       a setting of 4 amperes yields the most power; the 5 settings yield the following

       watts/amp: $104, \dfrac{176}{2} = 88; \dfrac{216}{3} = 72; \dfrac{224}{4} = 56; \dfrac{200}{5} = 40;$ a setting of 1 ampere

       yields the most watts of power per ampere, 104 watts/amp.

11. a. $h(t) = -4.9t^2 + 14t + 30$

   b. Maximum pt. for $h(t)$ occurs at $t = -\dfrac{14}{2(-4.9)} = \dfrac{10}{7}$ s later.

   c. $0 = -4.9t^2 + 14t + 30; (-7t + 30)(0.7t + 1) = 0; t = \dfrac{30}{7}$ or $t = -\dfrac{10}{7}$ (reject);

    $\dfrac{30}{7}$ s later

12. $h(t) = -4.9t^2 + h_0; 0 = -4.9(3.5)^2 + h_0; h_0 = 60.025;$ about 60 m

13. The time for the diver to fall from her maximum height to the water is found by solving

    $0 = -4.9t^2 + 4.225; t = \dfrac{13}{14}$; total time in the air is $\dfrac{13}{14} + \dfrac{1}{2} = \dfrac{10}{7}$ s.

14. If h_w is the distance from the top of the well to the surface of the water, then the time for the stone to drop is given by $0 = -4.9t_1{}^2 + h_w$, or $t_1 = \sqrt{\dfrac{h_w}{4.9}}$. The time for the sound to travel back up the well is given by $t_2 = \dfrac{h_w}{343}$. We are given

$$t_1 + t_2 = 4 = \sqrt{\frac{h_w}{4.9}} + \frac{h_w}{343}.$$

$4 - \dfrac{h_w}{343} = \sqrt{\dfrac{h_w}{4.9}}$; $16 - \dfrac{8h_w}{343} + \dfrac{(h_w)^2}{117649} = \dfrac{h_w}{4.9}$; $0.0000085h_w{}^2 - 0.2274h_w + 16 = 0$;

Solving by using the quadratic formula gives $h_w \approx 71$ m.

15. a. $R(x) = (24 - x)(1000 + 100x) = -100x^2 + 1400x + 24{,}000$

b. Maximum for $R(x)$ occurs at $x = \dfrac{-1400}{2(-100)} = 7$, so the maximizing price is $24 - 7$, or \$17.

16. $f(x) = (540 - 2x)(150 + x) = -2x^2 + 240x + 81{,}000$; since the plane has a maximum capacity of 200 passengers, and there is a minimum of 150 passengers, the domain is the set of integers from 0 to 50, inclusive. The range is 81,000 to 88,000, inclusive. The maximum value occurs when $x = 50$. (Note that the maximum pt. of the quadratic function is outside the domain of x.) $f(50) = 88{,}000 = $ max. value.

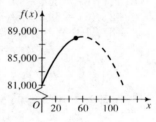

17. a. The radius of the opening will be $r = \dfrac{D}{2} - t$, so the carrying capacity is

$$f(t) = \pi r^2 h = \pi L\left(\frac{D}{2} - t\right)^2.$$

b. $f\left(\dfrac{1}{4}D\right) = \pi L\left(\dfrac{D}{2} - \dfrac{D}{4}\right)^2 = \pi L\left(\dfrac{D}{4}\right)^2 = \dfrac{1}{16}\pi L D^2$;

$\dfrac{1}{4} \cdot f(0) = \dfrac{1}{4} \cdot \pi L\left(\dfrac{D}{2} - 0\right)^2 = \dfrac{1}{4}\pi L \cdot \dfrac{D^2}{4} =$

$\dfrac{1}{16}\pi L D^2$; $f\left(\dfrac{1}{4}D\right) = \dfrac{1}{4} \cdot f(0)$. The carrying capacity of the pipe with mineral deposits is one fourth that of the original pipe. In other words, when the radius is halved, the carrying capacity is quartered.

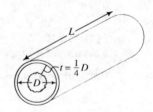

Chapter Test · pages 50–51

1. a. $AB = \sqrt{(-4 + 2)^2 + (2 + 6)^2} = \sqrt{4 + 64} = 2\sqrt{17}$

b. $M = \left(\dfrac{-2 - 4}{2}, \dfrac{-6 + 2}{2}\right) = (-3, -2)$

2. $2(4) + a(-2) = 14$; $-6 = 2a$; $-3 = a$

3. $6x - y = -4$, so $y = 6x + 4$; $2x + 3y = 2$;

$2x + 3(6x + 4) = 2$; $20x = -10$; $x = -\dfrac{1}{2}$;

$y = 6\left(-\dfrac{1}{2}\right) + 4 = 1$; $\left(-\dfrac{1}{2}, 1\right)$; see graph at right.

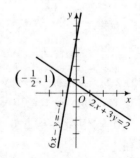

Ex. 3

4. $4x - 2y = 7;\ 2y = 4x - 7;\ y = 2x - \dfrac{7}{2};$ slope is 2 and y-intercept is $-\dfrac{7}{2}$.

5. a. $y = -\dfrac{2}{3}x + \dfrac{1}{3};\ m = -\dfrac{2}{3}$ **b.** $m = \dfrac{3}{2}$ **c.** $y = \dfrac{6x - 10}{4} = \dfrac{3}{2}x - \dfrac{5}{2};\ m = \dfrac{3}{2}.$ The lines
in (b) and (c) are parallel; the line in (a) is perpendicular to the lines in (b) and (c).

6. $m = \dfrac{1 + 2}{-3 - 6} = -\dfrac{1}{3};\ y = -\dfrac{1}{3}x + k;\ -2 = -\dfrac{1}{3}(6) + k;\ k = 0;\ y = -\dfrac{1}{3}x$

7. $4x + 3y = -2$ has slope $-\dfrac{4}{3};$ thus, an eq. is $\dfrac{y - 5}{x - 5} = -\dfrac{4}{3},$ or $4x + 3y = 35.$

8. $\dfrac{x}{-3} + \dfrac{y}{-5} = 1,$ or $5x + 3y = -15$ **9.** $x = 4$

10. a. Find the slope m and the midpoint $M(a, b)$ of the line segment. Write the eq. of line with
slope $-\dfrac{1}{m}$ and containing M: $\dfrac{y - b}{x - a} = -\dfrac{1}{m}.$

b. The midpoint $= \left(\dfrac{7 + 1}{2}, \dfrac{0 + 8}{2}\right) = (4, 4);$ the segment has slope $\dfrac{8 - 0}{1 - 7} = -\dfrac{4}{3},$ so its
$\perp$ bis. has slope $\dfrac{3}{4};\ \dfrac{y - 4}{x - 4} = \dfrac{3}{4},$ or $3x - 4y = -4.$

11. a. $m = \dfrac{7 - 0}{63 - 0} = \dfrac{1}{9};\ \dfrac{f(t) - 0}{t - 0} = \dfrac{1}{9};\ f(t) = \dfrac{1}{9}t$

b. $23 = \dfrac{1}{9}t;\ t = 207;\ 207\ \text{min} = 3\ \text{h}\ 27\ \text{min}$

12. $5i\sqrt{2} - 2i\sqrt{2} = 3i\sqrt{2}$ **13.** $4 + 12i + 9i^2 = -5 + 12i$

14. $\dfrac{1}{2 + 3i} \cdot \dfrac{2 - 3i}{2 - 3i} = \dfrac{2 - 3i}{4 - 9i^2} = \dfrac{2}{13} - \dfrac{3}{13}i$

15. $\dfrac{\sqrt{3} + i}{\sqrt{3} - i} \cdot \dfrac{\sqrt{3} + i}{\sqrt{3} + i} = \dfrac{3 + 2\sqrt{3}i + i^2}{3 - i^2} = \dfrac{2 + 2\sqrt{3}i}{4} = \dfrac{1}{2} + \dfrac{\sqrt{3}}{2}i$

16. $12 + 8i - 5 + 5i = 7 + 13i$ **17.** $i^{16} \cdot i = (i^4)^4 \cdot i = 1 \cdot i = i$

18. a. $7x^2 - 5x - 2 = 0;\ (7x + 2)(x - 1) = 0;\ x = -\dfrac{2}{7}$ or $x = 1$

b. $x^2 - 4x + 4 = 9 + 4;\ (x - 2)^2 = 13;\ x - 2 = \pm\sqrt{13};\ x = 2 \pm \sqrt{13}$

c. $x^2 + 2x = -2;\ x^2 + 2x + 1 = -1;\ (x + 1)^2 = -1;\ x + 1 = \pm i;\ x = -1 \pm i$

19. $b^2 - 4ac = (-2)^2 - 4(3)(2) = -20;\ x = \dfrac{2 \pm \sqrt{-20}}{2(3)} = \dfrac{2 \pm 2i\sqrt{5}}{6} = \dfrac{1 \pm i\sqrt{5}}{3}$

20. a.

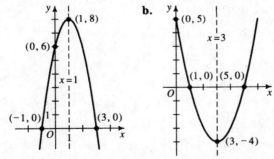

b.

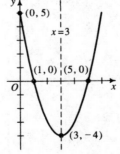

21.

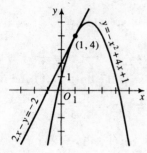

$y = 2x + 2;\ y = -x^2 + 4x + 1;\ 2x + 2 = -x^2 + 4x + 1;$
$x^2 - 2x + 1 = 0;\ (x - 1)^2 = 0;\ x = 1;\ y = 2(1) + 2 = 4;$
$(1, 4)$

22. a. (1) $f(3) = 9a + 3b + c = 29$

(2) $f(4) = 16a + 4b + c = 39$ (2) − (1): $7a + b = 10$

(3) $f(5) = 25a + 5b + c = 55$ (3) − (2): $\underline{9a + b = 16}$

$ 2a = 6;\ a = 3$

$7(3) + b = 10;\ b = -11;\ 9(3) + 3(-11) + c = 29;\ c = 35;\ f(x) = 3x^2 - 11x + 35$

b. $f(7) = 3(49) - 11(7) + 35 = 105;\ \$1.05$

Class Exercises 2-1 · page 55

1. Quadratic; $5x^2$; 5; 2   2. Quartic; $5x^4$; 5; 4   3. Cubic; $-4x^3$; -4; 3

4. Quintic; $-2x^5$; -2; 5   5. $f(3) = 2(3)^2 + 5 = 23$

6. $f(3i) = 2(3i)^2 + 5 = 2(-9) + 5 = -13$ 7. $f(3n) = 2(3n)^2 + 5 = 18n^2 + 5$

8. $f(n + 3) = 2(n + 3)^2 + 5 = 2(n^2 + 6n + 9) + 5 = 2n^2 + 12n + 23$

9. $(x + 5)(3x - 9) = 0$; $x = -5, 3$

10. $(x^2 - 16)(x^2 + 25) = 0$; $x^2 = 16$ or $x^2 = -25$; $x = \pm\sqrt{16} = \pm 4$ or $x = \pm\sqrt{-25} = \pm 5i$

11. $(x - 3)^2 = 0$; $x = 3$

12. $ax^2 + bx + c = 0$; by the quadratic formula, $x = \dfrac{-b \pm \sqrt{b^2 - 4ac}}{2a}$

13.
$$\begin{array}{r|rrrr} 3 & 2 & -9 & 0 & 27 \\ & & 6 & -9 & -27 \\ \hline & 2 & -3 & -9 & \boxed{0} = P(3) \end{array} \qquad \begin{array}{r|rrrr} -3 & 2 & -9 & 0 & 27 \\ & & -6 & 45 & -135 \\ \hline & 2 & -15 & 45 & \boxed{-108} = P(-3) \end{array}$$

Written Exercises 2-1 · pages 56–57

1. yes; zero: $\dfrac{17}{3}$ 2. yes; $(x - 4)(x - 2) = 0$; zeros: $4, 2$ 3. yes; no zeros

4. no; $x - \dfrac{1}{x} = 0$; $x^2 - 1 = (x + 1)(x - 1) = 0$; zeros: $1, -1$

5. no; $\dfrac{x^2 - 3x - 4}{x^2 + 1} = 0$; $(x - 4)(x + 1) = 0$; zeros: $4, -1$

6. yes; $\dfrac{x^2 + 2}{2} = 0$; $x^2 = -2$; zeros: $\pm i\sqrt{2}$

7. yes; $x(x^2 - 9) = x(x + 3)(x - 3) = 0$; zeros: $0, -3, 3$

8. yes; $(1 - 3x)(1 - 2x) = 0$; zeros: $\dfrac{1}{3}, \dfrac{1}{2}$

9. yes; $px^2 + qx + r = 0$; zeros: $\dfrac{-q \pm \sqrt{q^2 - 4pr}}{2p}$ if $p \neq 0$ (Note that zero is $-\dfrac{r}{q}$ if $p = 0$

and $q \neq 0$ and there is no zero if $p = 0$ and $q = 0$.)

10. yes; zeros: $7, \pm i\sqrt{7}$ 11. yes; $x(x^2 + 2x + 1) = x(x + 1)^2 = 0$; zeros: $0, -1$

12. yes; $x^2(2x^2 - x - 1) = x^2(2x + 1)(x - 1) = 0$; zeros: $0, -\dfrac{1}{2}, 1$

13. a. $\dfrac{x^3 + 2x^2 - x - 2}{x + 2} = \dfrac{x^2(x + 2) - (x + 2)}{x + 2} = \dfrac{(x^2 - 1)(x + 2)}{x + 2} = x^2 - 1$

   b. The form of a polynomial is $a_nx^n + a_{n-1}x^{n-1} + \cdots + a_1x + a_0$ where n is a nonnegative integer. The first is not of the form; the second is.

   c. The first expression is undefined for $x = -2$. The second is never undefined.

14. a. $f(x)$ is undefined when $x - 2 = 0$; $x = 2$.

   b. $\dfrac{2x^3 - 3x^2 - 8x + 12}{x - 2} = 0$; $\dfrac{(x^2 - 4)(2x - 3)}{x - 2} = 0$; $\dfrac{(x + 2)(x - 2)(2x - 3)}{x - 2} = 0$; $f(x)$ is

   undefined for $x = 2$; zeros: $-2, \dfrac{3}{2}$

15. a. $2(-1)^2 - 5(-1) + 6 = 13$ **b.** $2(2i)^2 - 5(2i) + 6 = -8 - 10i + 6 = -2 - 10i$

   c. $2(1 + i)^2 - 5(1 + i) + 6 = 2(2i) - 5 - 5i + 6 = 1 - i$

   d. $2(3a)^2 - 5(3a) + 6 = 18a^2 - 15a + 6$

16. a. $8(2\sqrt{3}) - 4(2\sqrt{3})^2 = 16\sqrt{3} - 48$

   b. $8(1 - \sqrt{2}) - 4(1 - \sqrt{2})^2 = 8 - 8\sqrt{2} - 12 + 8\sqrt{2} = -4$

   c. $8(1 + 2i) - 4(1 + 2i)^2 = 8 + 16i + 12 - 16i = 20$

   d. $8\left(\dfrac{2}{x}\right) - 4\left(\dfrac{2}{x}\right)^2 = \dfrac{16}{x} - \dfrac{16}{x^2}$, or $\dfrac{16x - 16}{x^2}$

17. a. $\left(-\dfrac{\sqrt{2}}{3}\right)^3 - 9\left(-\dfrac{\sqrt{2}}{3}\right) = -\dfrac{2\sqrt{2}}{27} + \dfrac{9\sqrt{2}}{3} = \dfrac{-2\sqrt{2} + 81\sqrt{2}}{27} = \dfrac{79\sqrt{2}}{27}$

   b. $(i\sqrt{3})^3 - 9(i\sqrt{3}) = -3i\sqrt{3} - 9i\sqrt{3} = -12i\sqrt{3}$

   c. $\left(\dfrac{x}{3}\right)^3 - 9\left(\dfrac{x}{3}\right) = \dfrac{x^3}{27} - 3x$, or $\dfrac{x^3 - 81x}{27}$

   d. $(x - 3)^3 - 9(x - 3) = (x^3 - 9x^2 + 27x - 27) - 9x + 27 = x^3 - 9x^2 + 18x$

18. a. $(4 - \sqrt{2})^2[(4 - \sqrt{2})^2 + 16] = (18 - 8\sqrt{2})[(18 - 8\sqrt{2}) + 16] =$
   $(18 - 8\sqrt{2})(34 - 8\sqrt{2}) = 740 - 416\sqrt{2}$

   b. $(1 + i)^2[(1 + i)^2 + 16] = 2i[2i + 16] = -4 + 32i$

   c. $\left(\dfrac{p}{q}\right)^2\left[\left(\dfrac{p}{q}\right)^2 + 16\right] = \dfrac{p^2}{q^2}\left(\dfrac{p^2 + 16q^2}{q^2}\right) = \dfrac{p^4 + 16p^2q^2}{q^4}$, or $\dfrac{p^4}{q^4} + \dfrac{16p^2}{q^2}$

   d. $(x^2)^2[(x^2)^2 + 16] = x^4(x^4 + 16) = x^8 + 16x^4$

19. a.

$$\begin{array}{r|rrrr}
2 & 4 & -5 & 7 & -9 \\
  &   & 8 & 6 & 26 \\
\hline
  & 4 & 3 & 13 & \boxed{17} = P(2)
\end{array}$$

b.

$$\begin{array}{r|rrrr}
3 & 4 & -5 & 7 & -9 \\
  &   & 12 & 21 & 84 \\
\hline
  & 4 & 7 & 28 & \boxed{75} = P(3)
\end{array}$$

c.

$$\begin{array}{r|rrrr}
-2 & 4 & -5 & 7 & -9 \\
   &   & -8 & 26 & -66 \\
\hline
   & 4 & -13 & 33 & \boxed{-75} = P(-2)
\end{array}$$

d.

$$\begin{array}{r|rrrr}
-3 & 4 & -5 & 7 & -9 \\
   &   & -12 & 51 & -174 \\
\hline
   & 4 & -17 & 58 & \boxed{-183} = P(-3)
\end{array}$$

20. a.

$$\begin{array}{r|rrrrr}
3 & 1 & 0 & -3 & 7 & 8 \\
  &   & 3 & 9 & 18 & 75 \\
\hline
  & 1 & 3 & 6 & 25 & \boxed{83} = P(3)
\end{array}$$

b.

$$\begin{array}{r|rrrrr}
-3 & 1 & 0 & -3 & 7 & 8 \\
   &   & -3 & 9 & -18 & 33 \\
\hline
   & 1 & -3 & 6 & -11 & \boxed{41} = P(-3)
\end{array}$$

c.

$$\begin{array}{r|rrrrr}
-1 & 1 & 0 & -3 & 7 & 8 \\
   &   & -1 & 1 & 2 & -9 \\
\hline
   & 1 & -1 & -2 & 9 & \boxed{-1} = P(-1)
\end{array}$$

d.

$$\begin{array}{r|rrrrr}
-2 & 1 & 0 & -3 & 7 & 8 \\
   &   & -2 & 4 & -2 & -10 \\
\hline
   & 1 & -2 & 1 & 5 & \boxed{-2} = P(-2)
\end{array}$$

21. a.

$$\begin{array}{r|rrrr}
\frac{1}{3} & 3 & -7 & 2 & 3 \\
            &   & 1 & -2 & 0 \\
\hline
            & 3 & -6 & 0 & \boxed{3} = P\left(\frac{1}{3}\right)
\end{array}$$

b.

$$\begin{array}{r|rrrr}
-\frac{2}{3} & 3 & -7 & 2 & 3 \\
             &   & -2 & 6 & -\frac{16}{3} \\
\hline
             & 3 & -9 & 8 & \boxed{-\frac{7}{3}} = P\left(-\frac{2}{3}\right)
\end{array}$$

22. a.

$$\begin{array}{r|rrrrr}
-\frac{1}{4} & 4 & -3 & 0 & 7 & -2 \\
             &   & -1 & 1 & -\frac{1}{4} & -\frac{27}{16} \\
\hline
             & 4 & -4 & 1 & \frac{27}{4} & \boxed{-\frac{59}{16}} = P\left(-\frac{1}{4}\right)
\end{array}$$

b.

$$\begin{array}{r|rrrrr}
\frac{3}{4} & 4 & -3 & 0 & 7 & -2 \\
            &   & 3 & 0 & 0 & \frac{21}{4} \\
\hline
            & 4 & 0 & 0 & 7 & \boxed{\frac{13}{4}} = P\left(\frac{3}{4}\right)
\end{array}$$

23. $f(4) = 0; 0 = 3(4^3) + k(4) - 2 = 190 + 4k; k = -\dfrac{190}{4} = -\dfrac{95}{2}$, or -47.5

24. $f(2i) = 0; 0 = (2i)^4 + (2i)^2 + a = 16(1) + 4(-1) + a = 12 + a; a = -12$

25. $P(x) = -2x^2 + 6 = -2(x^2 - 3); -2(x^2 - 3) = 0; x^2 = 3; x = \pm\sqrt{3}$; zeros: $\pm\sqrt{3}$

26. $P(x) = 2x^3 + ax^2 - 5x + 7; 21 = 2(8) + a(4) - 5(2) + 7 = 13 + 4a; 8 = 4a; 2 = a;$
$P(x) = 2x^3 + 2x^2 - 5x + 7; P(3) = 2(3^3) + 2(3^2) - 5(3) + 7 = 64$

27. a. $2(x - 1)(x - 4) = 0$; zeros: $1, 4$ **b.** $f(x) = a(x - 2)(x - 3)$ for any real $a \neq 0$
   c. $g(x) = ax(x - 2)(x - 3)$ for any real $a \neq 0$

28. a. $3(x + 1)(x - 2) = 0$; zeros: $-1, 2$ **b.** $f(x) = a(x - 3)(x + 4)$ for any real $a \neq 0$
   c. $g(x) = a(x + 3)(x - 3)(x + 4)(x - 4)$ for any real $a \neq 0$

29. a. $f(9.2) - f(8.2) = [7(9.2) + 2] - [7(8.2) + 2] = 66.4 - 59.4 = 7$
   b. $f(x + 1) - f(x) = [7(x + 1) + 2] - [7x + 2] = (7x + 9) - (7x + 2) = 7$

30. a. $g(6.25) - g(4.25) = [3 - 8(6.25)] - [3 - 8(4.25)] = (-47) - (-31) = -16$
   b. $g(x + 2) - g(x) = [3 - 8(x + 2)] - [3 - 8x] = (-8x - 13) - (-8x + 3) = -16$

31. For all x and all nonzero h, $\dfrac{f(x + h) - f(x)}{h} = \dfrac{[m(x + h) + k] - (mx + k)}{h} =$

$\dfrac{mx + mh + k - mx - k}{h} = \dfrac{mh}{h} = m$, which is the slope of the graph of f; the value

depends only on m, which is a constant, so the slope of a nonvertical line is constant.

32. For all x and all nonzero h, $\dfrac{f(x + h) - f(x)}{h} = \dfrac{(x + h)^2 - x^2}{h} = \dfrac{x^2 + 2xh + h^2 - x^2}{h} =$

$\dfrac{h(2x + h)}{h} = 2x + h$, which is dependent on the values of x and h.

33. The degree of the product = the sum of the degrees of the factors. If the polynomial 0 were to have degree 0, then the product of the polynomial 0 and a polynomial of positive degree n would have degree $0 + n = n$. Since the product would be the polynomial 0, its degree must also be 0. To avoid this contradiction, we say that the polynomial 0 has no degree.

34. a. Answers may vary. Examples: As x increases by 1, the differences increase by 2, or $f(x + 1) - f(x) = 2x + 3$.

b.

(x)	0	1	2	3	4	5
$g(x)$	-1	-2	1	8	19	34
diff.		-1	3	7	11	15

Answers may vary. Example: As x increases by 1, the differences increase by 4, or $g(x + 1) - g(x) = 4x - 1$.

c. Let $h(x) = ax^2 + bx + c$.

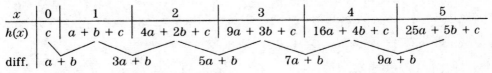

x	0	1	2	3	4	5
$h(x)$	c	$a + b + c$	$4a + 2b + c$	$9a + 3b + c$	$16a + 4b + c$	$25a + 5b + c$
diff.		$a + b$	$3a + b$	$5a + b$	$7a + b$	$9a + b$

Answers may vary. Example: As x increases by 1, the differences increase by $2a$, or $h(x + 1) - h(x) = (2x + 1)a + b$.

35. Let $k(x) = ax^3 + bx^2 + cx + d$.

x	0	1	2	3	4
$k(x)$	d	$a+b+c+d$	$8a+4b+2c+d$	$27a+9b+3c+d$	$64a+16b+4c+d$

diff. $a+b+c$ $7a+3b+c$ $19a+5b+c$ $37a+7b+c$

second
diff. $6a+2b$ $12a+2b$ $18a+2b$

Answers may vary. Example: As x increases by 1, the second differences, or the differences of the differences, increase by $6a$, or
$$k(x+1) - k(x) = (3x^2 + 3x + 1)a + (2x+1)b + c.$$

Computer Exercises · page 58

1. Times may vary. The second program is faster.

2. This program is faster than the first, but slower than the second.

Class Exercises 2-2 · page 60

1. a. Remainder is $P(1) = (1)^{15} + 3(1)^{10} + 2 = 6$.

  b. Remainder is $P(-1) = (-1)^{15} + 3(-1)^{10} + 2 = 4$.

2. a. Dividend: $x^4 - 8x^2 + 5x - 1$; divisor: $x + 3$

  b. Quotient: $x^3 - 3x^2 + x + 2$; remainder: -7 **c.** -7

3. a. Dividend: $2x^3 - 5x^2 + 8x - 4$; divisor: $x - \dfrac{1}{2}$

  b. Quotient: $2x^2 - 4x + 6$; remainder: -1

    The value of the polynomial when $x = \dfrac{1}{2}$ is -1.

4.
```
-1 | 1   3   1  -1
   |    -1  -2   1
     1   2  -1  |0 = P(-1)
```
Since $P(-1) = 0$, $x + 1$ is a factor of $P(x)$. The other factors can be found by the quadratic formula to find the zeros of the quotient, $x^2 + 2x - 1$. If the zeros are a and b, the other factors are $x - a$ and $x - b$.

5. $x - a$ is a factor of $P(x)$ if and only if the division of $P(x)$ by $x - a$ results in a remainder of 0. Since the remainder theorem tells us that the remainder is $P(a)$, $x - a$ is a factor of $P(x)$ if and only if $P(a) = 0$.

Written Exercises 2-2 · pages 61–62

1. a. $1^5 - 2 \cdot 1^3 + 1^2 - 4 = -4 = R$ **b.** $(-1)^5 - 2(-1)^3 + (-1)^2 - 4 = -2 = R$

  c. $2^5 - 2 \cdot 2^3 + 2^2 - 4 = 16 = R$ **d.** $(-2)^5 - 2(-2)^3 + (-2)^2 - 4 = -16 = R$

2. a. $2^3 - 3 \cdot 2^2 + 5 = 1 = R$ **b.** $(-2)^3 - 3(-2)^2 + 5 = -15 = R$

  c. $3^3 - 3 \cdot 3^2 + 5 = 5 = R$ **d.** $(-3)^3 - 3(-3)^2 + 5 = -49 = R$

3.
```
1 | 1  -2   5   1
  |     1  -1   4
    1  -1   4  |5
```
quotient: $x^2 - x + 4$; rem.: 5

4.
```
-2 | 2   1   3    7
   |    -4   6  -18
     2  -3   9  |-11
```
quotient: $2x^2 - 3x + 9$; rem.: -11

5.
$$
\begin{array}{r|rrrrr}
-1 & 1 & -2 & 0 & 5 & 2 \\
   &   & -1 & 3 & -3 & -2 \\
\hline
   & 1 & -3 & 3 & 2 & \boxed{0} \\
\end{array}
$$
quotient: $x^3 - 3x^2 + 3x + 2$; rem.: 0

6.
$$
\begin{array}{r|rrrrr}
1 & 2 & -3 & 4 & -5 & 2 \\
  &   & 2 & -1 & 3 & -2 \\
\hline
  & 2 & -1 & 3 & -2 & \boxed{0} \\
\end{array}
$$
quotient: $2x^3 - x^2 + 3x - 2$; rem.: 0

7.
$$
\begin{array}{r|rrrrrr}
3 & 1 & 0 & 1 & 0 & 1 & 0 \\
  &   & 3 & 9 & 30 & 90 & 273 \\
\hline
  & 1 & 3 & 10 & 30 & 91 & \boxed{273} \\
\end{array}
$$
quotient: $x^4 + 3x^3 + 10x^2 + 30x + 91$; rem.: 273

8.
$$
\begin{array}{r|rrrrr}
-2 & -3 & 0 & 1 & 0 & 0 \\
   &   & 6 & -12 & 22 & -44 \\
\hline
   & -3 & 6 & -11 & 22 & \boxed{-44} \\
\end{array}
$$
quotient: $-3x^3 + 6x^2 - 11x + 22$; rem.: -44

9.
$$
\begin{array}{r}
3x^2 - 8x + 21 \\
x^2 + 2x \overline{)\, 3x^4 - 2x^3 + 5x^2 + x + 1} \\
\underline{3x^4 + 6x^3} \\
-8x^3 + 5x^2 \\
\underline{-8x^3 - 16x^2} \\
21x^2 + x \\
\underline{21x^2 + 42x} \\
-41x + 1 \\
\end{array}
$$
quotient: $3x^2 - 8x + 21$
remainder: $-41x + 1$

10.
$$
\begin{array}{r}
x^3 - 2x^2 + 3x - 1 \\
x^2 + 2x + 1 \overline{)\, x^5 + 0x^4 + 0x^3 + 3x^2 + 0x + 4} \\
\underline{x^5 + 2x^4 + x^3} \\
-2x^4 - x^3 + 3x^2 \\
\underline{-2x^4 - 4x^3 - 2x^2} \\
3x^3 + 5x^2 + 0x \\
\underline{3x^3 + 6x^2 + 3x} \\
-x^2 - 3x + 4 \\
\underline{-x^2 - 2x - 1} \\
-x + 5 \\
\end{array}
$$
quotient: $x^3 - 2x^2 + 3x - 1$
remainder: $-x + 5$

11. $(1)^{100} - 4(1)^{99} + 3 = 1 - 4 + 3 = 0$, so $x - 1$ is a factor.
$(-1)^{100} - 4(-1)^{99} + 3 = 1 + 4 + 3 = 8$, so $x + 1$ is not a factor.

12. $(2)^{20} - 4(2)^{18} + 3(2) - 6 = 2^{20} - 2^{20} + 6 - 6 = 0$, so $x - 2$ is a factor.
$(-2)^{20} - 4(-2)^{18} + 3(-2) - 6 = 2^{20} - 2^{20} - 6 - 6 = -12$, so $x + 2$ is not a factor.

13. a. $(1)^3 - 5(1)^2 + 3(1) + 9 = 8$; no   **b.** $(-3)^3 - 5(-3)^2 + 3(-3) + 9 = -72$; no
c. $(3)^3 - 5(3)^2 + 3(3) + 9 = 0$; yes

14. a. $(-2)^4 - 3(-2)^3 + 5(-2) - 2 = 28$; no   **b.** $(2)^4 - 3(2)^3 + 5(2) - 2 = 0$; yes
c. $(-4)^4 - 3(-4)^3 + 5(-4) - 2 = 426$; no

15. If $P(x) = x^n - a^n$ and n is a positive integer, then $P(a) = (a)^n - a^n = 0$. By the factor theorem, $x - a$ is a factor of $P(x)$.

16. If $P(x) = x^n + a^n$ and n is odd, then $P(-a) = (-a)^n + a^n = -a^n + a^n = 0$. By the factor theorem, $x + a$ is a factor of $P(x)$.

17. $P(x) = (x^2 - x + 4)(2x + 1) + 3 = 2x^3 - x^2 + 7x + 7$

18. $P(x) = (x^3 + 2x + 2)(3x - 4) - 1 = 3x^4 - 4x^3 + 6x^2 - 2x - 9$

19. $3 \mid$ $\begin{array}{rrrr} 2 & -5 & -4 & 3 \end{array}$ $2x^3 - 5x^2 - 4x + 3 = (x - 3)(2x^2 + x - 1) =$

$\underline{\begin{array}{rrr} 6 & 3 & -3 \end{array}}$ $(x - 3)(2x - 1)(x + 1)$; the other roots are $\dfrac{1}{2}$ and -1.

$\begin{array}{rrr|r} 2 & 1 & -1 & 0 \end{array}$

20. $-2 \mid$ $\begin{array}{rrrr} 6 & 11 & -4 & -4 \end{array}$ $6x^3 + 11x^2 - 4x - 4 = (x + 2)(6x^2 - x - 2) =$

$\underline{\begin{array}{rrr} -12 & 2 & 4 \end{array}}$ $(x + 2)(3x - 2)(2x + 1)$; the other roots are $\dfrac{2}{3}$ and $-\dfrac{1}{2}$.

$\begin{array}{rrr|r} 6 & -1 & -2 & 0 \end{array}$

21. $1 \mid$ $\begin{array}{rrrrr} 2 & -9 & 2 & 9 & -4 \end{array}$ $-1 \mid$ $\begin{array}{rrrr} 2 & -7 & -5 & 4 \end{array}$

$\underline{\begin{array}{rrrr} 2 & -7 & -5 & 4 \end{array}}$ $\underline{\begin{array}{rrr} -2 & 9 & -4 \end{array}}$

$\begin{array}{rrrr|r} 2 & -7 & -5 & 4 & 0 \end{array}$ $\begin{array}{rrr|r} 2 & -9 & 4 & 0 \end{array}$

$2x^4 - 9x^3 + 2x^2 + 9x - 4 = (x - 1)(x + 1)(2x^2 - 9x + 4) =$

$(x - 1)(x + 1)(2x - 1)(x - 4)$; the other roots are $\dfrac{1}{2}$ and 4.

22. $-2 \mid$ $\begin{array}{rrrrr} 4 & -4 & -25 & 1 & 6 \end{array}$ $3 \mid$ $\begin{array}{rrrr} 4 & -12 & -1 & 3 \end{array}$

$\underline{\begin{array}{rrrr} -8 & 24 & 2 & -6 \end{array}}$ $\underline{\begin{array}{rrr} 12 & 0 & -3 \end{array}}$

$\begin{array}{rrrr|r} 4 & -12 & -1 & 3 & 0 \end{array}$ $\begin{array}{rrr|r} 4 & 0 & -1 & 0 \end{array}$

$4x^4 - 4x^3 - 25x^2 + x + 6 = (x + 2)(x - 3)(4x^2 - 1) =$

$(x + 2)(x - 3)(2x - 1)(2x + 1)$; the other roots are $\dfrac{1}{2}$ and $-\dfrac{1}{2}$.

23. $-4 \mid$ $\begin{array}{rrrrr} 1 & 3 & -3 & 3 & -4 \end{array}$ $1 \mid$ $\begin{array}{rrrr} 1 & -1 & 1 & -1 \end{array}$

$\underline{\begin{array}{rrrr} -4 & 4 & -4 & 4 \end{array}}$ $\underline{\begin{array}{rrr} 1 & 0 & 1 \end{array}}$

$\begin{array}{rrrr|r} 1 & -1 & 1 & -1 & 0 \end{array}$ $\begin{array}{rrr|r} 1 & 0 & 1 & 0 \end{array}$

$x^4 + 3x^3 - 3x^2 + 3x - 4 = (x + 4)(x - 1)(x^2 + 1) = (x + 4)(x - 1)(x - i)(x + i)$; the other roots are i and $-i$.

24. $-1 \mid$ $\begin{array}{rrrrr} 1 & -2 & 1 & 0 & -4 \end{array}$ $2 \mid$ $\begin{array}{rrrr} 1 & -3 & 4 & -4 \end{array}$

$\underline{\begin{array}{rrrr} -1 & 3 & -4 & 4 \end{array}}$ $\underline{\begin{array}{rrr} 2 & -2 & 4 \end{array}}$

$\begin{array}{rrrr|r} 1 & -3 & 4 & -4 & 0 \end{array}$ $\begin{array}{rrr|r} 1 & -1 & 2 & 0 \end{array}$

$x^4 - 2x^3 + x^2 - 4 = (x + 1)(x - 2)(x^2 - x + 2)$; $x = \dfrac{1 \pm \sqrt{1 - 4(1)(2)}}{2 \cdot 1} = \dfrac{1 \pm i\sqrt{7}}{2}$;

the other roots are $\dfrac{1}{2} + \dfrac{\sqrt{7}}{2}i$ and $\dfrac{1}{2} - \dfrac{\sqrt{7}}{2}i$.

25. If $P(x) = x^2(a - b) + a^2(b - x) + b^2(x - a)$, then $P(a) =$
$a^2(a - b) + a^2(b - a) + b^2(a - a) = a^2(a - b + b - a) + 0 = 0$; thus, by the factor theorem, $x - a$ is a factor of $P(x)$.

26. If $P(x) = (x - b)^3 + (b - c)^3 + (c - x)^3$, then $P(c) = (c - b)^3 + (b - c)^3 + (c - c)^3 =$
$(c - b)^3 + (-1)^3(c - b)^3 + 0 = (c - b)^3 - (c - b)^3 = 0$; thus, by the factor theorem, $x - c$ is a factor of $P(x)$.

Activity 1 · page 62

a.

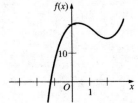

b.

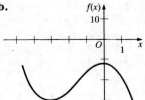

As x increases without bound, $f(x)$ increases without bound. As x decreases without bound, $f(x)$ decreases without bound.

As x increases without bound, $f(x)$ decreases without bound. As x decreases without bound, $f(x)$ increases without bound.

c. Answers will vary. The graphs will look like one or the other of those above.

d. The graph of a cubic function looks like a "sideways S." If the leading coefficient is positive, the graph will behave like the one in part **a**. If the leading coefficient is negative, the graph will behave like the one in part **b**.

Activity 2 · page 63

a.

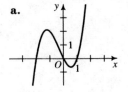

b.

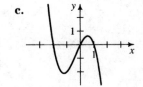

c.

d.

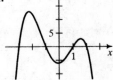

Activity 3 · page 64

a.

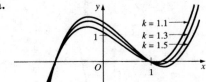

b. The graph will be tangent to the x-axis at the root associated with the squared factor. If $x > -1$, then $(x + 1)(x - 1)^2 \geq 0$. This means that the graph cannot cross the x-axis at the x-intercept.

The distance between the x-intercepts k and 1 approaches 0.

Class Exercises 2-3 · page 65

1. Let a be the leading coefficient of $P(x)$. If $P(x)$ is of odd degree, then the graph of $y = P(x)$ will rise from left to right for $a > 0$, and will fall from left to right for $a < 0$. If $P(x)$ is of even degree, then the graph of $y = P(x)$ will open up if $a > 0$, and will open down if $a < 0$.

2. Suppose $P(x) = Q(x)(x - c)^3$. The closer x is to c, the closer $(x - c)^3$ is to 0. If x is close enough to c, then $(x - c)^3$ will be close enough to 0 to make the product $Q(x)(x - c)^3$ very close to 0 as well. This means that for values of x very near c, the point $(x, P(x))$ is very close to the x-axis, causing the graph to flatten out around the x-intercept $x = c$.

3. Fig. (1): 2 distinct real roots; 3 is a double root.
   Fig. (2): 3 distinct real roots; 3 is a double root.
   Fig. (3): 3 distinct real roots; 0 and 3 are double roots.
   Fig. (4): 1 distinct real root; 2 is a triple root.
   Fig. (5): 1 distinct real root; 2 is a triple root.
   Fig. (6): 2 distinct real roots; 1 is a triple root.

4. a. 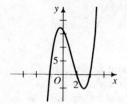 **b.** 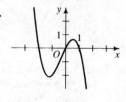 **c.**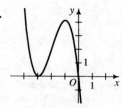

d. 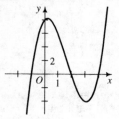 **e.** 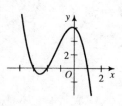 **f.**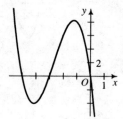

5. Answers may vary. Examples are given.   **a.** $y = (x - 1)(x - 4)^2$

   b. $y = -x(x + 1)(x - 3)$ **c.** $y = -(x - 1)^2$ **d.** $y = x(x - 2)^3$

Written Exercises 2-3 · pages 66–68

1. 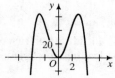 **2.** 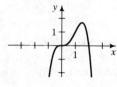 **3.**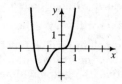

4. 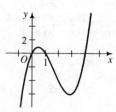 **5.** 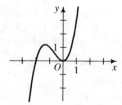 **6.**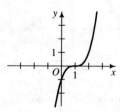

7. 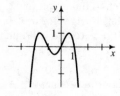 **8.** 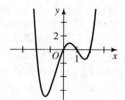 **9.**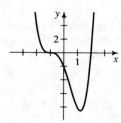

10.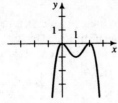

11.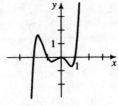

12.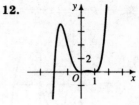

13. $f(x) = x(x + 2)(x - 2)$

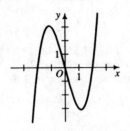

14. $f(x) = x(x + 1)(x - 5)$

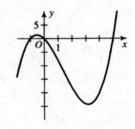

15. $f(x) = x^2(x + 1)(x - 1)$

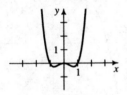

16. $f(x) = x^2(x - 1)^2$

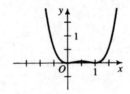

17. $f(x) = (x + 1)(x - 1)^3$

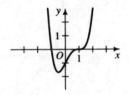

18. $f(x) = (2x + 1)(2x - 1)(x - 3)^2$

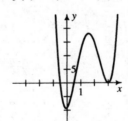

19. a.

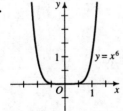

b. $(-1, 1), (0, 0), (1, 1)$

20. a.

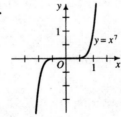

b. $(-1, -1), (0, 0), (1, 1)$

21. $y = -(x + 3)(x + 1)(x - 1)$

22. $y = x(x + 2)^2$

23. $y = (x + 3)^2(x + 1)(x - 1)$

24. $y = -x^2(x - 3)^2$

25. a. $P(x) = -x(x + 4)(x - 4)$ **b.** $P(2) = 16(2) - 2^3 = 24$;

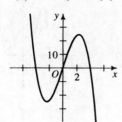

$P(2.1) = 16(2.1) - (2.1)^3 = 24.339; P(2) < P(2.1)$,
so the highest point does not occur at $(2, 24)$. By
trial-and-error, a local maximum occurs at $x \approx 2.3$.

c. $P(-x) = 16(-x) - (-x)^3 = -16x + x^3 = -(16x - x^3) = -P(x)$

26. a. $P(x) = x^2(x + 2)(x - 2)$ **b.** $P(1) = 1^4 - 4(1^2) = -3$;

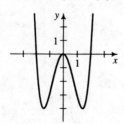

$P(1.1) = (1.1)^4 - 4(1.1)^2 = -3.3759; P(1.1) < P(1)$,
so the lowest point does not occur at $(1, -3)$. By
trial-and-error, a local minimum occurs at $x \approx 1.4$.

c. $P(-x) = (-x)^4 - 4(-x)^2 = x^4 - 4x^2 = P(x)$

27. $y = x^3 - 4x$ and $y = -3x; x^3 - 4x = -3x; x^3 - x = 0;$
$x(x + 1)(x - 1) = 0; x = 0$ and $y = -3(0) = 0$, or $x = -1$
and $y = -3(-1) = 3$, or $x = 1$ and $y = -3(1) = -3; (0, 0),$
$(-1, 3), (1, -3)$

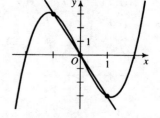

28. $y = -x(x - 2)^2$ and $y = -x; -x(x - 2)^2 = -x;$
$x - x(x - 2)^2 = 0; x[1 - (x - 2)^2] = 0; x = 0$ or
$1 = (x - 2)^2; x = 0$ or $x = 2 \pm 1; x = 0$ and $y = -0 = 0,$
or $x = 1$ and $y = -1$, or $x = 3$ and $y = -3; (0, 0), (1, -1),$
$(3, -3)$

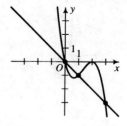

29. Let $y = k(x + 4)(x - 2)$; substituting $(0, 3)$, $3 = k(4)(-2) = -8k; k = -\dfrac{3}{8}$;

$y = -\dfrac{3}{8}(x + 4)(x - 2)$

30. Let $y = kx(x - 1)(x - 2)$; substituting $(-1, -6)$, $-6 = k(-1)(-2)(-3) = -6k; k = 1$;
$y = x(x - 1)(x - 2)$

31. Let $y = kx(x + 2)^2$; substituting $(1, -3)$, $-3 = k(1)(1 + 2)^2 = 9k; k = -\dfrac{1}{3}$;

$y = -\dfrac{1}{3}x(x + 2)^2$

32. Let $y = k(x + 1)^2(x - 1)(x - 2)$; substituting $(0, -3)$, $-3 = k(1^2)(-1)(-2) = 2k$;

$k = -\dfrac{3}{2}; y = -\dfrac{3}{2}(x + 1)^2(x - 1)(x - 2)$

33, 34. Graphs will vary. Examples are given.

33. a.
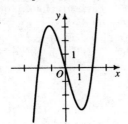

b.
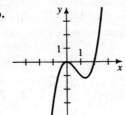

c.
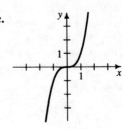

d. Impossible

34. a.
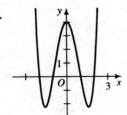

b.
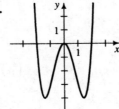

c.
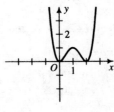

d.
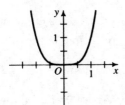

e.
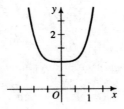

35. $0, 1, 2, 3, 4, 5, 6$ **36.** $1, 2, 3, 4, 5$

37. Since the graph contains $(3, 0)$ and is tangent at $(0, 0)$, let $y = kx^2(x - 3)$; substituting $(1, 4)$, $4 = k(1^2)(1 - 3) = -2k$; $k = -2$; $y = -2x^2(x - 3)$

38. Since the equation has double roots of -1 and 2, let $y = k(x + 1)^2(x - 2)^2$; substituting $(0, -2)$, $-2 = k(1^2)(-2)^2 = 4k$; $k = -\dfrac{1}{2}$; $y = -\dfrac{1}{2}(x + 1)^2(x - 2)^2$

39. Since the function has zeros -3, -1, and 2, let $P(x) = k(x + 3)(x + 1)(x - 2)$; substituting $(0, 6)$, $6 = k(3)(1)(-2) = -6k$; $k = -1$; $P(x) = -(x + 3)(x + 1)(x - 2)$

40. Since $P(x) > 0$ only when $x > 4$, the graph of $P(x)$ is tangent at $(0, 0)$ and contains $(4, 0)$; let $P(x) = kx^2(x - 4)$; substituting $(2, -4)$, $-4 = k(2^2)(2 - 4) = -8k$; $k = \dfrac{1}{2}$; $P(x) = \dfrac{1}{2}x^2(x - 4)$

Class Exercises 2-4 • pages 70–71

1. a. minimum   **b.** $x = \dfrac{1 + 7}{2} = 4$ **2. a.** maximum   **b.** $x = 2$

3. a. minimum   **b.** $x = -\dfrac{-6}{2(2)} = \dfrac{3}{2}$ **4. a.** maximum   **b.** $x = -\dfrac{-4}{2(-3)} = -\dfrac{2}{3}$

5. No, for example, $f(x) = x^3$ has no local maximum or minimum.

6. A cubic function always takes on values greater than its local maximum and less than its local minimum.   The adjective "local" is used to emphasize that a maximum (or minimum) value is such only on a bounded subset of the function's domain.   By contrast, a quadratic function either has a maximum value greater than any of its other values or has a minimum value less than any of its other values.   Thus, the word "local" is not necessary when discussing a maximum or minimum value of a quadratic function.

7. Change lines 20, 40, and 80 as follows:

```
20 FOR X = 3 TO 5 STEP 0.01
40 IF V > = M THEN 70
80 PRINT "MINIMUM VALUE BETWEEN 3 AND 5 IS APPROXIMATELY"; M; "
   AT X = "; X1
```

Written Exercises 2-4 · pages 71–74

Quadratic functions

1. a. The enclosure has width x and length $120 - 2x$; $A(x) = x(120 - 2x)$; set of real numbers x such that $0 < x < 60$

b. Since the zeros are 0 and 60, the maximum value occurs at $x = \dfrac{0 + 60}{2} = 30$.

2. $80 = 2x + 2l$; $40 = x + l$; $l = 40 - x$; $A(x) = x(40 - x)$; since the zeros are 0 and 40, the maximum value occurs at $x = \dfrac{0 + 40}{2} = 20$; $A(20) = 20(20) = 400 \text{ cm}^2$.

3. If each short piece has length x, then each of the longer pieces has length

$\dfrac{1}{2}(102 - 3x) = 51 - 1.5x$; $A(x) = x(51 - 1.5x)$; since the zeros are 0 and 34, the

maximum occurs at $x = \dfrac{0 + 34}{2} = 17$; $A(17) = 17(51 - 1.5(17)) = 433.5 \text{ m}^2$.

4. If one square enclosure is made, its sides are $\dfrac{200}{4} = 50$ m long and its area is

$50^2 = 2500 \text{ m}^2$. If two square enclosures are made and one has sides of length x meters, then the other has sides of length $(200 - 4x) \div 4$, or $50 - x$ meters; $A(x) = x^2 + (50 - x)^2 = 2x^2 - 100x + 2500 = 2(x^2 - 50x + 1250)$; the minimum

occurs when $x = \dfrac{50}{2(1)} = 25$; $A(25) = 2(25^2 - 50(25) + 1250) = 1250$; $0 \le x \le 50$, and $A(0) = A(50) = 2500$. Thus, the plan that gives the least area is when $x = 25$, that is, when two square enclosures each with sides 25 m long are built. The plan that gives the greatest area is when $x = 0$ or $x = 50$, that is, when one square enclosure with sides 50 m long is built.

5. $P(x) = (x - x_1)^2 + (x - x_2)^2 = x^2 - 2x_1 x + x_1^2 + x^2 - 2x_2 x + x_2^2 = 2x^2 - (2x_1 + 2x_2)x + (x_1^2 + x_2^2)$; the graph of $P(x)$ is a parabola opening upward, so

its minimum point occurs when $x = -\dfrac{-(2x_1 + 2x_2)}{2(2)} = \dfrac{2(x_1 + x_2)}{4} = \dfrac{x_1 + x_2}{2}$.

6. $P(x) = (x - x_1)^2 + (x - x_2)^2 + (x - x_3)^2 = (x^2 - 2x_1x + x_1{}^2) +$
$(x^2 - 2x_2x + x_2{}^2) + (x^2 - 2x_3x + x_3{}^2) = 3x^2 - (2x_1 + 2x_2 + 2x_3)x +$
$(x_1{}^2 + x_2{}^2 + x_3{}^2)$; the graph of $P(x)$ is a parabola opening upward, so its minimum
point occurs when $x = -\dfrac{-(2x_1 + 2x_2 + 2x_3)}{2(3)} = \dfrac{2(x_1 + x_2 + x_3)}{2(3)} = \dfrac{x_1 + x_2 + x_3}{3}$.

7. If n measurements $x_1, x_2, x_3, \ldots, x_n$ are made of a quantity whose true measure x is
not known, then the sum of the squares of the errors is a minimum if the value
$\dfrac{x_1 + x_2 + x_3 + \cdots + x_n}{n}$ (that is, the arithmetic average of the n measurements) is
assigned to x.

8. If a rectangle has perimeter P and width x, then its length is $\dfrac{1}{2}(P - 2x) = \dfrac{1}{2}P - x$ and
its area is $A(x) = x\left(\dfrac{1}{2}P - x\right)$; the graph of $A(x)$ is a parabola that opens downward
and has zeros 0 and $\dfrac{1}{2}P$. The maximum value occurs when $x = \dfrac{1}{2}\left(0 + \dfrac{1}{2}P\right) = \dfrac{1}{4}P$; if
the width is $\dfrac{1}{4}P$, then the length is $\dfrac{1}{2}P - \dfrac{1}{4}P = \dfrac{1}{4}P$ and the rectangle is a square.

9. a. The ball hits the ground when $h(t) = 0$ and $t > 0$; $30t - 5t^2 = 0$; $5t(6 - t) = 0$;
reject $t = 0$; when $t = 6$ s.   **b.** The set of real nos. t such that $0 \le t \le 6$

 c. The maximum height occurs when $t = \dfrac{1}{2}(0 + 6) = 3$; $h(3) = 30(3) - 5(3)^2 = 45$;
the ball goes up 45 m.

10. a. The graph of $h(t) = 30 + 25t - 5t^2 = 5(6 - t)(1 + t)$ is a parabola that opens
downward and has zeros 6 and -1. The ball hits the ground when $h(t) = 0$; $t > 0$, so
reject $t = -1$; thus, $t = 6$ and the ball hits the ground 6 s after it is thrown.

 b. The set of real nos. t such that $0 \le t \le 6$

 c. The maximum height occurs when $t = \dfrac{1}{2}(-1 + 6) = \dfrac{5}{2}$;
$h\left(\dfrac{5}{2}\right) = 5\left(6 - \dfrac{5}{2}\right)\left(1 + \dfrac{5}{2}\right) = 61.25$; the ball goes up 61.25 m.

11. Let x be the number of 40¢ increases. Then $1.6 + 0.4x =$ the price in dollars and
$80{,}000 - 10{,}000x =$ the circulation. Thus, the income is
$I(x) = (1.6 + 0.4x)(80{,}000 - 10{,}000x) = (0.4)(10{,}000)(4 + x)(8 - x) =$
$4000(4 + x)(8 - x)$; the function $I(x)$ has its maximum value when
$x = \dfrac{1}{2}(-4 + 8) = 2$; the price should be $1.6 + 0.4(2) = 2.4$, or \$2.40.

12. Let $d =$ the number of days. Then $400 + 20d =$ the number of crates and
$60 - 2d =$ the price in dollars per crate. Thus, the income is
$I(d) = (400 + 20d)(60 - 2d) = 20(2)(20 + d)(30 - d)$; the function $I(d)$ has its
maximum value when $d = \dfrac{1}{2}(-20 + 30) = 5$; the grower should ship in 5 days.

13. Let the rectangle have dimensions x and y, as shown. $BD = \sqrt{15^2 - 9^2} = 12$; using similar right triangles, we have $\dfrac{x}{12} = \dfrac{9 - \frac{y}{2}}{9}$;

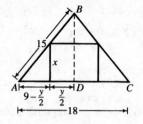

$9x = 108 - 6y; x = 12 - \dfrac{2}{3}y$;

area $= A(y) = y\left(12 - \dfrac{2}{3}y\right)$; the function $A(y)$ has its maximum value when $y = \dfrac{1}{2}(0 + 18) = 9$;

then $x = 12 - \dfrac{2}{3}(9) = 6$; the rectangle with the greatest area has width 9 and height 6.

14. Given $P(k, 0)$, $(PA)^2 + (PB)^2 + (PC)^2 = [(k - 0)^2 + (0 - 5)^2] + [(k - 3)^2 + (0 - 7)^2] + [(k - 6)^2 + (0 - 2)^2] = 3k^2 - 18k + 123 = 3(k^2 - 6k + 41)$; the minimum value occurs when $k = -\dfrac{-6}{2 \cdot 1} = 3$; $P = (3, 0)$.

Cubic functions

1. a. The box has length $14 - 2x$, width $8 - 2x$, and height x;
$V(x) = lwh = x(8 - 2x)(14 - 2x)$.

b. The set of real nos. x such that $0 < x < 4$

c. Use the program on page 70 with these changes:

```
20 FOR X = 0 TO 4 STEP 0.01
30 LET V = X * (8 - 2 * X) * (14 - 2 * X)
```

This gives $x \approx 1.64$; max. volume is approx. 82.98 cm³.

2. a. The box has length $10 - x$, width $10 - 2x$, and height x;
$V(x) = lwh = x(10 - x)(10 - 2x)$.

b. The set of real nos. x such that $0 < x < 5$

c. Use the program on page 70 with these changes:

```
20 FOR X = 0 TO 5 STEP 0.01
30 LET V = X * (10 - X) * (10 - 2 * X)
```

This gives $x \approx 2.11$; max. volume is approx. 96.22 cm³.

3. a. The height of the box is $\dfrac{20 - 8x}{2} = 10 - 4x$;

$V(x) = Bh = x^2(10 - 4x)$.

b. The set of real nos. x such that $0 < x < 2.5$

c. Use the program on page 70 with these changes:

```
20 FOR X = 0 TO 2.5 STEP 0.01
30 LET V = X * X * (10 - 4 * X)
```

This gives $x \approx 1.67$; max. volume is approx. 9.26 ft³.

4. a. The height of the prism is $\dfrac{30 - 6x}{2} = 15 - 3x$. The base of the prism is an

equilateral triangle with area $\dfrac{1}{2}x\left(\dfrac{\sqrt{3}}{2}x\right) = \dfrac{\sqrt{3}}{4}x^2$. $V(x) = Bh = \dfrac{\sqrt{3}}{4}x^2(15 - 3x)$.

b. The set of real nos. x such that $0 < x < 5$.

c. Use the program on page 70 with these changes:

```
20 FOR X = 0 TO 5 STEP 0.01
30 LET V = (SQR(3)/4) * X * X * (15 - 3 * X)
```

This gives $x \approx 3.33$; max. volume is approx. 24.06 in.^{3}.

5. a. The height of the cylinder is $\dfrac{50 - 2x}{2} = 25 - x$. The base of the cylinder has

circumference x, so $2\pi r = x$; $r = \dfrac{x}{2\pi}$. $V(x) = \pi r^2 h = \pi\left(\dfrac{x}{2\pi}\right)^2(25 - x) =$

$\dfrac{1}{4\pi}x^2(25 - x)$; the domain is the set of real nos. x such that $0 < x < 25$.

b. Use the program on page 70 with these changes:

```
20 FOR X = 0 TO 25 STEP 0.01
30 LET V = X * X * (25 - X)/(4 * 3.14159)
```

This gives $x \approx 16.67$; max. volume is approx. 184.21 cm.^{3}.

6. a. The radius of the cylinder is x and the height is $\dfrac{12 - 2x}{2} = 6 - x$.

$V(x) = \pi r^2 h = \pi x^2(6 - x)$; the domain is the set of real nos. x such that $0 < x < 6$.

b. Use the program on page 70 with these changes:

```
20 FOR X = 0 TO 6 STEP 0.01
30 LET V = 3.14159 * X * X * (6 - X)
```

This gives $x \approx 4.00$; max. volume is approx. 100.53 in.^{3}.

7. a. The radius of the cylinder is $\sqrt{25 - x^2}$ and the height is $2x$.

$V(x) = \pi r^2 h = \pi(25 - x^2)(2x) = 2\pi x(25 - x^2)$; the domain is the set of real nos. x such that $0 < x < 5$.

b. Use the program on page 70 with these changes:

```
20 FOR X = 0 TO 5 STEP 0.01
30 LET V = 2 * 3.14159 * X * (25 - X * X)
```

This gives $x \approx 2.89$; max. volume is approx. 302.30 cubic units.

8. a. The cone has radius $\sqrt{36 - x^2}$ and height $x + 6$. $V(x) = \dfrac{1}{3}\pi r^2 h =$

$\dfrac{1}{3}\pi(36 - x^2)(x + 6)$; the domain is the set of real nos. x such that $0 < x < 6$.

b. Use the program on page 70 with these changes:

```
20 FOR X = 0 TO 6 STEP 0.01
30 LET V = (3.14159/3) * (36 - X↑2) * (X + 6)
```

This gives $x \approx 2.00$; max. volume is approx. 268.08 cubic units.

9. Using similar right triangles, we have $\dfrac{5-r}{h} = \dfrac{5}{10}$; $h = 10 - 2r$; $V = \pi r^2 h$, so

$V(r) = \pi r^2 (10 - 2r)$. Use the program on page 70 with these changes:

```
20 FOR X = 0 TO 5 STEP 0.01
30 LET V = 3.14159 * X↑2 * (10 - 2 * X)
65 LET X2 = 10 - 2 * X
80 PRINT "MAXIMUM VOLUME IS APPROXIMATELY"; M;" WHEN R = "; X1;
   "AND H = "; X2
```

This gives $r \approx 3.33$, $h \approx 3.33$, and the max. volume is approx. 116.36 cubic units.

Calculator Exercises • page 74

Let x be the length of the piece used to form the circle; then $40 - x$ is the length used to

form the square; $x = 2\pi r$ and $r = \dfrac{x}{2\pi}$; each side of the square has length

$$\frac{1}{4}(40 - x) = 10 - \frac{1}{4}x;\; A(x) = \pi\left(\frac{x}{2\pi}\right)^2 + \left(10 - \frac{1}{4}x\right)^2 = \frac{x^2}{4\pi} + \left(10 - \frac{1}{4}x\right)^2.$$

1. $A(x) = \dfrac{x^2}{4\pi} + 100 - 5x + \dfrac{x^2}{16} = \left(\dfrac{1}{4\pi} + \dfrac{1}{16}\right)x^2 - 5x + 100$; the minimum value occurs

when $x = \dfrac{5}{2\left(\dfrac{1}{4\pi} + \dfrac{1}{16}\right)} = \dfrac{40\pi}{\pi + 4}$; the required lengths are $\dfrac{40\pi}{\pi + 4} \approx 17.596034$ and

$\left(40 - \dfrac{40\pi}{\pi + 4}\right) \approx 22.403966$; use about 17.6 cm for the circle and about 22.4 cm for

the square.

2. $0 \le x \le 40$; $A(0) = 100$ and $A(40) = \dfrac{400}{\pi} \approx 127.32395$; the maximum occurs when

$x = 40$, that is, when the entire length of wire is used to form a circle.

Class Exercises 2-5 • page 77

1. $P(1) = -3$ and $P(2) = 5$, so there is a real root between 1 and 2.

2. $P(1) = -2$ and $P(2) = 11$, so there is a real root between 1 and 2.

3. $P(-1) = 3$ and $P(0) = -3$; also, $P(2) = -9$ and $P(3) = 15$; there is one real root between -1 and 0 and another real root between 2 and 3.

4. $P(-5) = -3$, $P(-4) = 13$, $P(-1) = 1$, $P(0) = -3$, and $P(1) = 3$; there is a real root between -5 and -4, between -1 and 0, and between 0 and 1.

5. Answers may vary. The roots are approximately -2.2, 0.4, and 2.8.

6. There is no change in sign at a double root.

Written Exercises 2-5 · pages 78–79

1–6. The initial graph is shown.   Use the ZOOM or TRACE feature to find the real roots to the nearest tenth.

1.

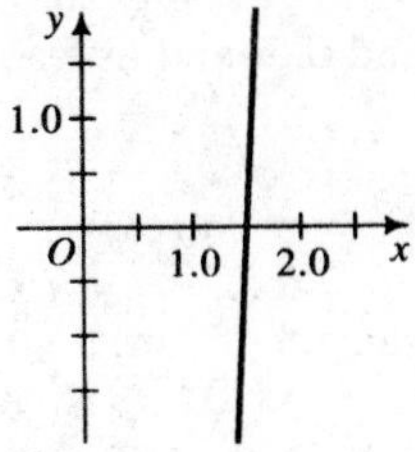

$x = 1.5$

2.

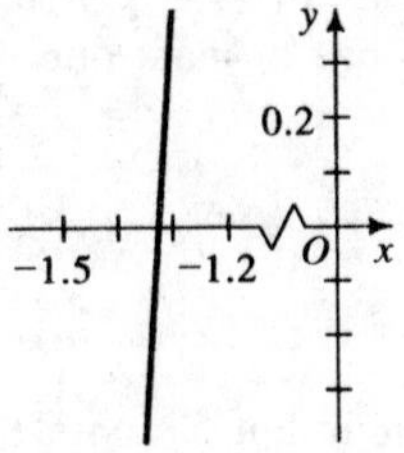

$x \approx -1.3$

3.

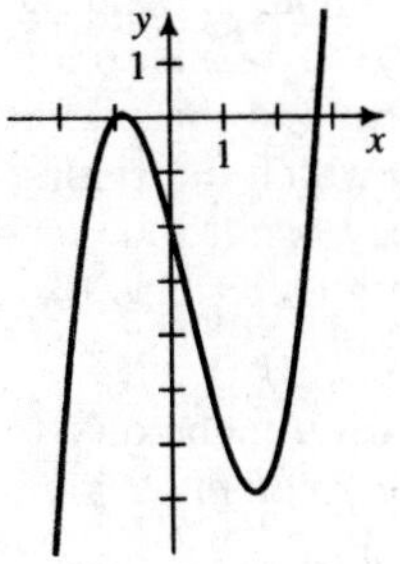

$x = -1$
$x \approx -0.7$
$x \approx 2.7$

4.

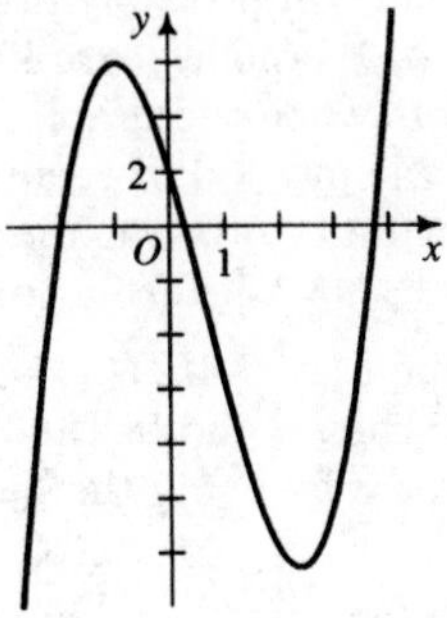

$x = -2$
$x \approx 0.3$
$x \approx 3.7$

5.

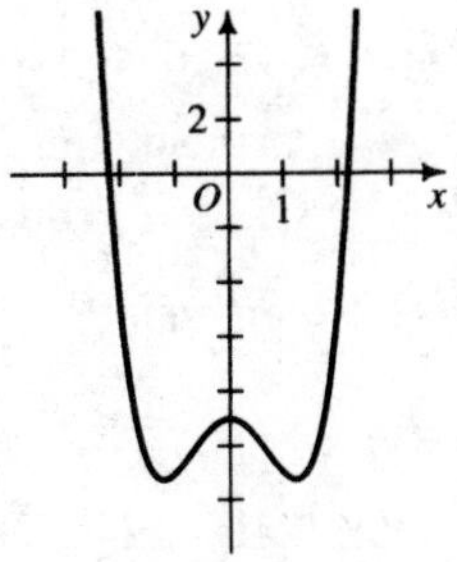

$x \approx \pm 2.2$

6.

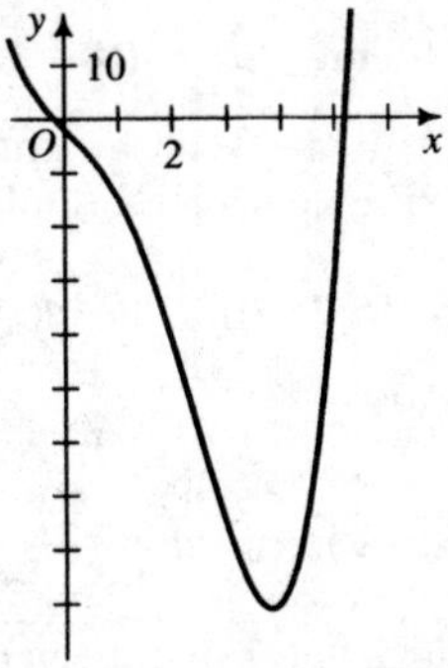

$x \approx -0.2$
$x \approx 5.2$

7. Running a program like the one on page 76 shows that $P(5) = -22$ and $P(6) = 11$.   By changing line 20 you can show that $x \approx 5.7$.

8. Running a program like the one on p. 76 shows that $P(-1) = -2$, $P(0) = 2$, $P(1) = 0$, $P(2) = -2$, and $P(3) = 2$.   Thus, 1 is a root, another is between -1 and 0, and a third is between 2 and 3.   By changing line 20 you can show that the second and third roots are approximately -0.7 and 2.7.

9. Running a program like the one on p. 76 shows that $P(0) = 0$, $P(1) = -3.1$, and $P(2) = 9.2$.   Thus, 0 is a root and there is another root between 1 and 2.   By changing line 20 you can show that this root is approximately 1.6.

10. Running a program like the one on p. 76 shows that $P(-2) = 2.98$, $P(-1) = -3.97$, $P(2) = -5.02$, and $P(3) = 0.43$.   Thus, there is one root between -2 and -1, and another root between 2 and 3.   By changing line 20 you can show that these roots are approximately -1.7 and 3.0.

11.
$$\begin{array}{r|rrrr} -2 & 1 & -1 & -5 & 2 \\ & & -2 & 6 & -2 \\ \hline & 1 & -3 & 1 & 0 \end{array}$$

The quotient is $x^2 - 3x + 1$. The zeros of the quotient are
$$x = \frac{3 \pm \sqrt{9 - 4(1)(1)}}{2(1)} = \frac{3 \pm \sqrt{5}}{2}; x \approx 0.38 \text{ or } x \approx 2.62.$$

12. If $P(x) = 9x^4 - 8x^2 + 1$, then $P(0) = 1$, $P(0.5) = -0.4375$, and $P(1) = 2$. By the location principle, $P(x) = 0$ has at least one real root between 0 and 0.5, and at least one real root between 0.5 and 1.

13. Since $P(x) = 42x^2 - 13x + 1 = (6x - 1)(7x - 1)$, the roots of $P(x) = 0$ are
$x = \dfrac{1}{6} \approx 0.167$ and $x = \dfrac{1}{7} \approx 0.143$. Because these roots are so close together, a step size of 0.1 will not reveal them, but a step size of 0.01 will.

14. Answers will vary. Students will probably prefer the method of Example 1. Among the limitations of the method of Example 1 are knowing how far along the x-axis to look for roots and determining roots when the graph is tangent—or nearly so—to the x-axis. Among the limitations of the method of Example 2 are knowing in which interval(s) to look for roots and the possibility of skipping over roots because they occur too close together (see Written Ex. 13) or because a root has even multiplicity (see Class Ex. 6).

15. The smaller box has volume $lwh = 4(2)(1) = 8$; $16 = (4 + x)(2 + x)(1 + x)$; $x^3 + 7x^2 + 14x - 8 = 0$; the second box is larger than the one with dimensions 4, 2, and 1, so $x > 0$; solving for x by using the method of Example 1 or Example 2 gives $x \approx 0.46$.

16. The shaved block has dimensions $3 - 2x$, $5 - 2x$, and $4 - 2x$ cm and volume
$\dfrac{1}{2} lwh = \dfrac{1}{2}(3)(5)(4) = 30$ cm^3. $30 = (3 - 2x)(5 - 2x)(4 - 2x)$;
$4x^3 - 24x^2 + 47x - 15 = 0$; x cm is the amount that is shaved, so $x > 0$; solving for x by using the method of Example 1 or Example 2 gives $x \approx 0.39$.

17. $P(x)$ contains $(a, P(a))$ and $(b, P(b))$. The slope of $\overleftrightarrow{AB}$ is $\dfrac{P(b) - P(a)}{b - a}$, so an equation of
$\overleftrightarrow{AB}$ is $y - P(a) = \dfrac{P(b) - P(a)}{b - a}(x - a)$. At the x-intercept, $y = 0$;
$0 - P(a) = \dfrac{P(b) - P(a)}{b - a}(x - a)$; $x - a = -\dfrac{P(a) \cdot (b - a)}{P(b) - P(a)}$; $x = a - \dfrac{P(a) \cdot (b - a)}{P(b) - P(a)}$.

18. a. Let d_1 and d_2 be the horizontal distances of the intersection of the ladders from building 1 and from building 2, respectively. The shorter ladder is $\sqrt{15^2 - d^2}$ high on building 1; the longer ladder is $\sqrt{20^2 - d^2}$ high on building 2. By similar triangles, $\dfrac{d_1}{c} = \dfrac{d}{\sqrt{400 - d^2}}$ and $\dfrac{d_2}{c} = \dfrac{d}{\sqrt{225 - d^2}}$. Therefore,

$\dfrac{d}{\sqrt{400 - d^2}} + \dfrac{d}{\sqrt{225 - d^2}} = \dfrac{d_1}{c} + \dfrac{d_2}{c} = \dfrac{d_1 + d_2}{c} = \dfrac{d}{c}$; thus,

$\dfrac{1}{\sqrt{400 - d^2}} + \dfrac{1}{\sqrt{225 - d^2}} = \dfrac{1}{c}$.

b. If $d = 12$, then $\dfrac{1}{\sqrt{400 - 12^2}} + \dfrac{1}{\sqrt{225 - 12^2}} = \dfrac{1}{c}$; $\dfrac{1}{c} = \dfrac{1}{\sqrt{256}} + \dfrac{1}{\sqrt{81}} = \dfrac{1}{16} + \dfrac{1}{9} = \dfrac{25}{144}$;

$c = \dfrac{144}{25} = 5.76$.

c. `10 FOR D = 0 TO 15`
`20 LET Y = 1/SQR(400 - D↑2) + 1/SQR(225 - D↑2) - 1/8`
`30 PRINT D, Y`
`40 NEXT D`
`50 END`

The computer print-out indicates that the value of y changes from negative to positive between 5 and 6. We change line 10 to: `10 FOR D = 5 TO 6 STEP 0.1`; the print-out now indicates a change between 5.9 and 6.0. Again, changing line 10 to: `10 FOR D = 5.9 TO 6.0 STEP 0.01`, we find that the print-out indicates a change between 5.95 and 5.96. Since 5.95 gives a y-value closer to 0, we choose 5.95 as the solution to the nearest hundredth.

Computer Exercise • pages 79–80

```
10 PRINT "ENTER THE DEGREE OF THE POLYNOMIAL P(X)."
20 INPUT N
30 DIM P(N)
40 FOR D = N TO 0 STEP -1
50 PRINT "ENTER THE COEFFICIENT OF THE TERM HAVING DEGREE ";D;"."
60 INPUT P(D)
70 NEXT D
80 PRINT "ENTER TWO NUMBERS A AND B SUCH THAT P(A) AND P(B) HAVE
   OPPOSITE SIGNS."
90 INPUT A,B
100 LET M = (A + B)/2
110 LET PA = P(N): LET PB = P(N): LET PM = P(N)
120 FOR D = N - 1 TO 0 STEP -1
130 LET PA = PA * A + P(D): LET PB = PB * B + P(D):
    LET PM = PM * M + P(D)
140 NEXT D
150 IF ABS (PM) < 0.01 THEN 200
160 IF PA * PM < 0 THEN LET B = M
170 IF PB * PM < 0 THEN LET A = M
180 PRINT "A ZERO OF P(X) IS BETWEEN ";A;" AND ";B;"."
190 GOTO 100
200 PRINT "THE ZERO IS APPROXIMATELY ";M;"."
210 END
```

To get greater accuracy in the approximation of the root, change the "0.01" in line 150 to a smaller positive number.

Class Exercises 2-6 • page 83

1. $x^2(x + 4) - 9(x + 4) = 0$; $(x^2 - 9)(x + 4) = 0$; $(x + 3)(x - 3)(x + 4) = 0$; $x = -3, 3, -4$

2. a. Let $y = x^2$. Then $x^4 - 5x^2 + 4 = 0$ becomes $y^2 - 5y + 4 = 0$, which is quadratic in y.

  b. Solve $y^2 - 5y + 4 = 0$ by factoring. Then substitute x^2 for y and solve for x.

3. a. Yes; substitute y for $2x - 1$. **b.** Yes; substitute y for x^3.

4. a. First factor out x. Then solve the remaining quadratic by factoring or by using the quadratic formula.

b. The method of part (a) can be used only on cubic equations with no constant term.

5. a. $\pm 4,\ \pm 2,\ \pm 1,\ \pm\dfrac{1}{2}$　**b.** $\pm 5,\ \pm 1,\ \pm\dfrac{5}{2},\ \pm\dfrac{1}{2},\ \pm\dfrac{5}{3},\ \pm\dfrac{1}{3},\ \pm\dfrac{5}{6},\ \pm\dfrac{1}{6}$

6. No; by the location principle there must be at least one real root between $x = 1$ and $x = 2$, but the root may be irrational. (For example, $P(x) = x^2 + x - 3 = 0$ has $P(1) = -1$, $P(2) = 3$, and one root $\dfrac{-1 + \sqrt{13}}{2} \approx 1.3$.)

7. a. $P(-2) = -1 < 0$ and $P(-1) = 1 > 0$, so by the location principle there must be at least one real root between $x = -1$ and $x = -2$.

b. The only possible rational roots are ± 1, so this root must be irrational.

c. $P(x) = x^3 + 2x^2 + x + 1$ is greater than 1 for all positive numbers x.

Written Exercises 2-6 • pages 83–85

1. b; $x^4 - 4x^2 - 12 = 0$; $(x^2 - 6)(x^2 + 2) = 0$; $x = \pm\sqrt{6}$ or $\pm i\sqrt{2}$

2. a; $x^2(x + 6) - 4(x + 6) = 0$; $(x^2 - 4)(x + 6) = 0$; $x = \pm 2$ or -6

3. a; $x^2(3x - 16) - 4(3x - 16) = 0$; $(x^2 - 4)(3x - 16) = 0$; $x = \pm 2$ or $\dfrac{16}{3}$

4. b; $(x^2 - 8)(x^2 + 1) = 0$; $x = \pm 2\sqrt{2}$ or $\pm i$.

5. b; $2x^4 + 7x^2 - 15 = 0$; $(2x^2 - 3)(x^2 + 5) = 0$; $x = \pm\sqrt{\dfrac{3}{2}}$ or $\pm i\sqrt{5}$; $x = \pm\dfrac{\sqrt{6}}{2}$ or $\pm i\sqrt{5}$

6. a; $x^2(2x - 1) - 1(2x - 1) = 0$; $(x^2 - 1)(2x - 1) = 0$; $x = \pm 1$ or $\dfrac{1}{2}$

7. a; $x^3 + 2x^2 - 6x - 12 = 0$; $x^2(x + 2) - 6(x + 2) = 0$; $(x^2 - 6)(x + 2) = 0$; $x = \pm\sqrt{6}$ or -2

8. a; $2x^3 - 3x^2 + 8x - 12 = 0$; $x^2(2x - 3) + 4(2x - 3) = 0$; $(x^2 + 4)(2x - 3) = 0$; $x = \pm 2i$ or $\dfrac{3}{2}$

9. a; $10x^3 - 6x^2 + 5x - 3 = 0$; $2x^2(5x - 3) + 1(5x - 3) = 0$; $(2x^2 + 1)(5x - 3) = 0$; $x = \pm\dfrac{\sqrt{2}}{2}i$ or $\dfrac{3}{5}$

10. b; $(2x^2 - 5)(x^2 + 4) = 0$; $x = \pm\sqrt{\dfrac{5}{2}}$ or $\pm 2i$; $x = \pm\dfrac{\sqrt{10}}{2}$ or $\pm 2i$

11. Let $y = x^3$; then $y^2 - 7y - 8 = 0$, which is quadratic in y. $(y - 8)(y + 1) = 0$; $y = 8$ or $y = -1$; $x^3 = 8$ or $x^3 = -1$; $x^3 - 8 = (x - 2)(x^2 + 2x + 4) = 0$ or $x^3 + 1 = (x + 1)(x^2 - x + 1) = 0$; $x = 2$ or $x = \dfrac{-2 \pm\sqrt{4 - 4(1)(4)}}{2(1)} = -1 \pm i\sqrt{3}$, or $x = -1$, or $x = \dfrac{1 \pm\sqrt{1 - 4(1)(1)}}{2(1)} = \dfrac{1}{2} \pm\dfrac{\sqrt{3}}{2}i$.

12. a. $y^2 - y - 1 = 0$; $y = \dfrac{1 \pm\sqrt{1 - 4(1)(-1)}}{2(1)} = \dfrac{1 \pm\sqrt{5}}{2}$; $y \approx 1.62$ or -0.62

b. $x^2 \approx 1.62$ or $x^2 \approx -0.62$; $x \approx \pm\sqrt{1.62} \approx \pm 1.27$ or $x \approx \pm i\sqrt{0.62} \approx \pm 0.79i$

13. $P(1) = 1 - 1 - 1 + 1 = 0$

$$\underline{1|}\ \begin{array}{rrrr} 1 & -1 & -1 & 1 \\ & 1 & 0 & -1 \\ \hline 1 & 0 & -1 & \underline{|0} \end{array}$$

$P(x) = (x - 1)(x^2 - 1) =$
$(x - 1)^2(x + 1); x = 1, -1$

14. $P(1) = 1 + 2 - 1 - 2 = 0$

$$\underline{1|}\ \begin{array}{rrrr} 1 & 2 & -1 & -2 \\ & 1 & 3 & 2 \\ \hline 1 & 3 & 2 & \underline{|0} \end{array}$$

$P(x) = (x - 1)(x^2 + 3x + 2) =$
$(x - 1)(x + 1)(x + 2); x = 1, -1, -2$

15. $P(1) = P(-1) = 1 - 10 + 9 = 0$

$$\begin{array}{r} \underline{1|} \\ \\ \underline{-1|} \\ \\ \\ \end{array} \begin{array}{rrrrr} 1 & 0 & -10 & 0 & 9 \\ & 1 & 1 & -9 & -9 \\ \hline 1 & 1 & -9 & -9 & \underline{|0} \\ & -1 & 0 & 9 & \\ \hline 1 & 0 & -9 & \underline{|0} & \end{array}$$

$P(x) = (x - 1)(x + 1)(x^2 - 9) =$
$(x - 1)(x + 1)(x - 3)(x + 3); x = \pm 1, \pm 3$

16. $P(1) = 2 - 9 + 3 + 4 = 0$

$$\underline{1|}\ \begin{array}{rrrr} 2 & -9 & 3 & 4 \\ & 2 & -7 & -4 \\ \hline 2 & -7 & -4 & \underline{|0} \end{array}$$

$P(x) = (x - 1)(2x^2 - 7x - 4) =$

$(x - 1)(2x + 1)(x - 4); x = 1, -\dfrac{1}{2}, 4$

17. $P(-1) = -3 - 4 + 5 + 2 = 0$

$$\underline{-1|}\ \begin{array}{rrrr} 3 & -4 & -5 & 2 \\ & -3 & 7 & -2 \\ \hline 3 & -7 & 2 & \underline{|0} \end{array}$$

$P(x) = (x + 1)(3x^2 - 7x + 2) = (x + 1)(3x - 1)(x - 2);$

$x = -1, \dfrac{1}{3}, 2$

18. $P(-1) = 3 - 2 - 9 + 12 - 4 = 0; P(2) = 48 + 16 - 36 - 24 - 4 = 0$

$$\begin{array}{r} \underline{-1|} \\ \\ \underline{2|} \\ \\ \\ \end{array} \begin{array}{rrrrr} 3 & 2 & -9 & -12 & -4 \\ & -3 & 1 & 8 & 4 \\ \hline 3 & -1 & -8 & -4 & \underline{|0} \\ & 6 & 10 & 4 & \\ \hline 3 & 5 & 2 & \underline{|0} & \end{array}$$

$P(x) = (x + 1)(x - 2)(3x^2 + 5x + 2) =$
$(x + 1)(x - 2)(3x + 2)(x + 1) =$

$(x + 1)^2(x - 2)(3x + 2); x = -1, 2, -\dfrac{2}{3}$

19. $P(-1) = 1 - 2 - 2 + 6 - 3 = 0$

$$\underline{-1|}\ \begin{array}{rrrrr} 1 & 2 & -2 & -6 & -3 \\ & -1 & -1 & 3 & 3 \\ \hline 1 & 1 & -3 & -3 & \underline{|0} \end{array}$$

$P(x) = (x + 1)(x^3 + x^2 - 3x - 3) =$
$(x + 1)[x^2(x + 1) - 3(x + 1)] =$
$(x + 1)[(x^2 - 3)(x + 1)] = (x + 1)^2(x^2 - 3);$
$x = -1, \pm\sqrt{3}$

20. Possible rat. roots: $\pm 1, \pm 2, \pm 3, \pm 4, \pm 6, \pm 12, \pm\dfrac{1}{3}, \pm\dfrac{2}{3}, \pm\dfrac{4}{3}$.

$P(0) = 12$ and $P(1) = -22$, so try $x = \dfrac{1}{3}$.

$$\underline{\tfrac{1}{3}|}\ \begin{array}{rrrr} 3 & -1 & -36 & 12 \\ & 1 & 0 & -12 \\ \hline 3 & 0 & -36 & \underline{|0} \end{array}$$

$P(x) = \left(x - \dfrac{1}{3}\right)(3x^2 - 36) = 3\left(x - \dfrac{1}{3}\right)(x^2 - 12);$

$x = \dfrac{1}{3}, \pm 2\sqrt{3}$

21. $P(-2) = -16 + 14 + 2 = 0$

$$\begin{array}{r|rrrr} -2 & 2 & 0 & -7 & 2 \\ & & -4 & 8 & -2 \\ \hline & 2 & -4 & 1 & \underline{|0} \end{array}$$

$P(x) = (x + 2)(2x^2 - 4x + 1); x = -2$ or

$$x = \frac{4 \pm \sqrt{16 - 4(2)(1)}}{2 \cdot 2} = \frac{2 \pm \sqrt{2}}{2}$$

22. Synthetic division shows that 2 and $-\dfrac{1}{2}$ are roots. Also,

$$P(x) = (x - 2)\left(x + \frac{1}{2}\right)(2x^2 + 2x - 2) =$$

$$2(x - 2)\left(x + \frac{1}{2}\right)(x^2 + x - 1); x = 2 \text{ or } x = -\frac{1}{2} \text{ or}$$

$$x = \frac{-1 \pm \sqrt{1 - 4(1)(-1)}}{2 \cdot 1} = \frac{-1 \pm \sqrt{5}}{2}$$

$$\begin{array}{r|rrrrr} 2 & 2 & -1 & -7 & 1 & 2 \\ & & 4 & 6 & -2 & -2 \\ \cline{2-6} -\dfrac{1}{2} & 2 & 3 & -1 & -1 & \underline{|0} \\ & & -1 & -1 & 1 & \\ \cline{2-5} & 2 & 2 & -2 & \underline{|0} & \end{array}$$

23. $f(x) = x^2(x + 2) - 9(x + 2) = (x^2 - 9)(x + 2) = (x + 3)(x - 3)(x + 2)$

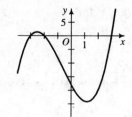   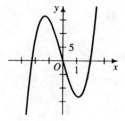   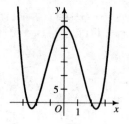

        Ex. 23   **Ex. 24**   **Ex. 25**

24. $g(x) = x(4x^2 + x - 18) = x(4x + 9)(x - 2)$

25. $h(x) = (x^2 - 7)(x^2 - 4) = (x + \sqrt{7})(x - \sqrt{7})(x + 2)(x - 2)$

26. $k(x) = -(x^4 - 10x^2 + 24) = -(x^2 - 6)(x^2 - 4) =$
    $-(x + \sqrt{6})(x - \sqrt{6})(x + 2)(x - 2)$

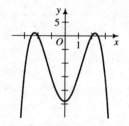   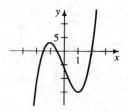   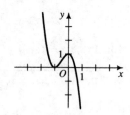

        Ex. 26   **Ex. 27**   **Ex. 28**

27. $m(x) = (x - 2)(4x^2 + 8x + 3) = (x - 2)(2x + 3)(2x + 1)$

28. $n(x) = -(2x^3 + 3x^2 - 1) = -(x + 1)(2x^2 + x - 1) = -(x + 1)(2x - 1)(x + 1) =$
    $-(x + 1)^2(2x - 1)$

29. The only possible rational roots are ±1 and ±3. Test these values:
    $P(1) = 1 + 1 - 3 = -1; P(-1) = -1 + 1 - 3 = -3; P(3) = 27 + 9 - 3 = 33;$
    $P(-3) = -27 + 9 - 3 = -21.$ None of these values is a root, so there are no rational
    roots. $P(1) = -1$ and $P(2) = 8 + 4 - 3 = 9; P(x)$ changes sign between $x = 1$ and
    $x = 2$, so there is an irrational root between $x = 1$ and $x = 2$.

30. $3x^3 - 4x^2 + 5x - 2$ is negative for all negative values of x, so there are no negative roots. Thus, the only possible rational roots are 1, 2, $\dfrac{1}{3}$, and $\dfrac{2}{3}$. Test these values:

$P(1) = 2$, $P(2) = 16$, $P\left(\dfrac{1}{3}\right) = -\dfrac{2}{3}$, and $P\left(\dfrac{2}{3}\right) = \dfrac{4}{9}$. None of these is a root, so there are no rational roots. $P(0) = -2$ and $P(1) = 2$; $P(x)$ changes sign between $x = 0$ and $x = 1$, so there is an irrational root between $x = 0$ and $x = 1$.

31. $y = x^3 - 3x$ and $y = 2$; $x^3 - 3x = 2$; $x^3 - 3x - 2 = 0$; the possible rat. roots are ± 1 and ± 2; synthetic division gives $P(-1) = P(2) = 0$; $(x + 1)^2(x - 2) = 0$; the graphs intersect at $(2, 2)$ and are tangent at $(-1, 2)$.

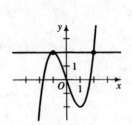
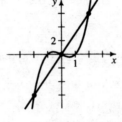
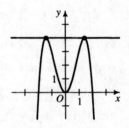

Ex. 31   Ex. 32   Ex.33

32. $y = x^3 - x$ and $y = 3x$; $x^3 - x = 3x$; $x^3 - 4x = 0$; $x(x^2 - 4) = 0$;
$x(x + 2)(x - 2) = 0$; $x = 0$ and $y = 3(0) = 0$, or $x = -2$ and $y = 3(-2) = -6$, or $x = 2$ and $y = 3(2) = 6$; the graphs intersect at $(0, 0)$, $(-2, -6)$, and $(2, 6)$.

33. $4x^2 - x^4 = 4$; $x^4 - 4x^2 + 4 = 0$; $(x^2 - 2)(x^2 - 2) = 0$; $x = \pm\sqrt{2}$; the graphs are tangent at $(\sqrt{2}, 4)$ and $(-\sqrt{2}, 4)$.

34. $x^4 - 6x^2 = -9$; $(x^2 - 3)^2 = 0$; $x = \pm\sqrt{3}$; the graphs are tangent at $(\sqrt{3}, -9)$ and $(-\sqrt{3}, -9)$.

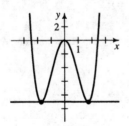
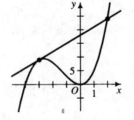
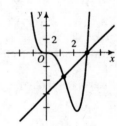

Ex. 34   Ex. 35   Ex. 36

35. $x^3 + 4x^2 = 3x + 18$; $x^3 + 4x^2 - 3x - 18 = 0$; the possible rat. roots are ± 1, ± 2, ± 3, ± 6, ± 9, and ± 18; synthetic division gives $P(2) = P(-3) = 0$; $(x - 2)(x + 3)^2 = 0$; $x = 2$ and $y = 3(2) + 18 = 24$, or $x = -3$ and $y = 3(-3) + 18 = 9$; the graphs intersect at $(2, 24)$ and are tangent at $(-3, 9)$.

36. $x^4 - 3x^3 = 2x - 6$; $x^4 - 3x^3 - 2x + 6 = 0$; $x^3(x - 3) - 2(x - 3) = 0$;
$(x^3 - 2)(x - 3) = 0$; $x = \sqrt[3]{2}$ or $x = 3$; the graphs intersect at $(3, 0)$ and $(\sqrt[3]{2}, 2\sqrt[3]{2} - 6)$.

37. Let $x = \sqrt[3]{12}$; then $x^3 = 12$, or $x^3 - 12 = 0$. The possible rational roots are ± 1, ± 2, ± 3, ± 4, ± 6, and ± 12. We need check only positive roots: $1^3 - 12 = -11 \neq 0$, $2^3 - 12 = -4 \neq 0$, $3^3 - 12 = 15 \neq 0$; 4^3, 6^3 and 12^3 are all greater than 12, so no rational root exists; however, since $P(2) < 0$ and $P(3) > 0$, there is an irrational root between 2 and 3.

38. Let $x = \sqrt[5]{100}$; then $x^5 = 100$, or $x^5 - 100 = 0$. The possible rational roots are ± 1, ± 2, ± 4, ± 5, ± 10, ± 20, ± 25, ± 50, and ± 100. We need check only positive roots: $1^5 - 100 = -99 \neq 0$; $2^5 - 100 = -68 \neq 0$; $4^5 - 100 = 924 \neq 0$; 5^5, 10^5, 20^5, 25^5, 50^5, and 100^5 are all greater than 100, so no rational root exists; however, since $P(2) < 0$ and $P(4) > 0$, there is an irrational root between 2 and 4.

39. a. $V = \dfrac{1}{3}\pi r^2 h$ and $r^2 + h^2 = 9$; thus, $r^2 = 9 - h^2$ and $V(h) = \dfrac{1}{3}\pi(9 - h^2)h$

   b. $V(h) = \dfrac{10}{3}\pi$; $\dfrac{1}{3}\pi(9 - h^2)h = \dfrac{10}{3}\pi$; $9h - h^3 = 10$; $h^3 - 9h + 10 = 0$; possible rat. roots are ± 1, ± 2, ± 5, and ± 10; $V(2) = 0$, and $h^3 - 9h + 10 =$ $(h - 2)(h^2 + 2h - 5)$; $h = 2$ or $h = \dfrac{-2 \pm \sqrt{4 - 4(1)(-5)}}{2 \cdot 1} = -1 \pm \sqrt{6}$; reject the negative root; $h = 2$ or $h = -1 + \sqrt{6}$.

40. a. $V = \pi r^2 h$, $h = 2x$, and $r^2 + x^2 = 16$; thus, $r^2 = 16 - x^2$ and $V = \pi(16 - x^2)h = \pi(16 - x^2)(2x) = 2\pi x(16 - x^2)$.

   b. $V(x) = 42\pi$; $2\pi x(16 - x^2) = 42\pi$; $16x - x^3 = 21$; $x^3 - 16x + 21 = 0$; possible rat. roots are ± 1, ± 3, ± 7, and ± 21; $3^3 - 16(3) + 21 = 0$, so $x^3 - 16x + 21 =$ $(x - 3)(x^2 + 3x - 7)$; $x = 3$ or $x = \dfrac{-3 \pm \sqrt{9 - 4(1)(-7)}}{2 \cdot 1} = \dfrac{-3 \pm \sqrt{37}}{2}$; reject the negative root; $x = 3$ or $x = \dfrac{-3 + \sqrt{37}}{2}$.

41. $(n - 2)((n + 3) - 2)((n + 9) - 2) = \dfrac{1}{2}n(n + 3)(n + 9)$; $2(n - 2)(n + 1)(n + 7) =$ $n(n + 3)(n + 9)$; $2n^3 + 12n^2 - 18n - 28 = n^3 + 12n^2 + 27n$; $n^3 - 45n - 28 = 0$; $(n - 7)(n^2 + 7n + 4) = 0$; n must be an integer, so $n = 7$; the dimensions of the original block are 7 cm by 10 cm by 16 cm.

Class Exercises 2-7 · page 89

1. $\sqrt{3} - 2i$ **2.** $-3 - \sqrt{2}$ and $2i$

3. a. sum $= -\dfrac{5}{2}$; product $= -\dfrac{3}{2}$ **b.** sum $= -\dfrac{-2}{4} = \dfrac{1}{2}$; product $= -\dfrac{-6}{4} = \dfrac{3}{2}$

   c. sum $= -\dfrac{2}{1} = -2$; product $= \dfrac{0}{1} = 0$ **d.** sum $= -\dfrac{0}{1} = 0$; product $= -\dfrac{32}{1} = -32$

4. As the graphs on p. 62 show, the graph of a cubic function is shaped like a "sideways S"; therefore, the graph of a cubic eq. must cross the x-axis at least once.

5. As the graphs at the top of p. 64 show, the graph of a quartic function is shaped like a "W" or an "M"; such a graph may lie entirely above or entirely below the x-axis, in which case, there are no real roots.

Written Exercises 2-7 · pages 89–91

1. False by Theorem 4 (See also Class Ex. 4.)

2. False, for example, $x^2 = -1$ has only imaginary roots.

3. False; if $1 + i$ is a root, then $1 - i$ must also be a root and the equation cannot be quartic.

4. True; $(x + 3)(x - 4)(x - (1 - \sqrt{2}))(x^2 + 1) = 0$ is such an equation. (We do not know that the coefficients are rational.)

5. False; if the graph of $P(x)$ is tangent at $x = k$, then $(x - k)$ must be a double root; hence, the degree of the equation would be at least 6, not 3.

6. True by Theorem 2   **7.** True by Theorem 3 with $a = 0$

8. False; for example, $x^3 + 1 = 0$ has 2 imaginary roots and 1 real root.

9. sum $= -\dfrac{-3}{4} = \dfrac{3}{4}$; product $= \dfrac{6}{4} = \dfrac{3}{2}$ **10.** sum $= -\dfrac{-9}{6} = \dfrac{3}{2}$; product $= -\dfrac{0}{6} = 0$

11. sum $= -\dfrac{5}{3}$; product $= -\dfrac{-2}{3} = \dfrac{2}{3}$ **12.** sum $= -\dfrac{0}{1} = 0$; product $= \dfrac{-5}{1} = -5$

13–16. Answers may vary. Examples are given.

13. sum $= (1 + i) + (1 - i) = 2$; product $= (1 + i)(1 - i) = 2$; $x^2 - 2x + 2 = 0$

14. sum $= (4 + \sqrt{3}) + (4 - \sqrt{3}) = 8$; product $= (4 + \sqrt{3})(4 - \sqrt{3}) = 13$;
$x^2 - 8x + 13 = 0$

15. sum $= (3 + \sqrt{2}) + (3 - \sqrt{2}) = 6$; product $= (3 + \sqrt{2})(3 - \sqrt{2}) = 7$;
$x^2 - 6x + 7 = 0$

16. sum $= \dfrac{1 + i\sqrt{2}}{3} + \dfrac{1 - i\sqrt{2}}{3} = \dfrac{2}{3}$; product $= \left(\dfrac{1 + i\sqrt{2}}{3}\right)\left(\dfrac{1 - i\sqrt{2}}{3}\right) = \dfrac{3}{9} = \dfrac{1}{3}$;

$x^2 - \dfrac{2}{3}x + \dfrac{1}{3} = 0$; $3x^2 - 2x + 1 = 0$

17. Since $2 + i\sqrt{5}$ is a root, $2 - i\sqrt{5}$ is also a root. Let r be the third root. Since the quadratic term is 0, $2 + i\sqrt{5} + 2 - i\sqrt{5} + r = 0$; $4 + r = 0$; $r = -4$.

18. Since $3 - i\sqrt{7}$ is a root, $3 + i\sqrt{7}$ is also a root. Let r and s be the other two roots. Since the cubic term is 0, $3 + i\sqrt{7} + 3 - i\sqrt{7} + r + s = 0$ and $r + s = -6$; since the constant term is 0, $(3 + i\sqrt{7})(3 - i\sqrt{7})(r)(s) = 0$ and so $r = 0$ or $s = 0$; $r = 0$ and $s = -6$, or $s = 0$ and $r = -6$. The other roots are 0 and -6.

19–24. Answers may vary. Examples are given.

19. A quadratic eq. with roots $4 + i$ and $4 - i$ is $x^2 - 8x + 17 = 0$;
$(x - 2)(x^2 - 8x + 17) = 0$; $x^3 - 10x^2 + 33x - 34 = 0$.

20. A quadratic eq. with roots $7 - i$ and $7 + i$ is $x^2 - 14x + 50 = 0$;
$(x - 3)(x^2 - 14x + 50) = 0$; $x^3 - 17x^2 + 92x - 150 = 0$.

21. $\dfrac{4 + i\sqrt{3}}{2} + \dfrac{4 - i\sqrt{3}}{2} = 4$; $\dfrac{4 + i\sqrt{3}}{2} \cdot \dfrac{4 - i\sqrt{3}}{2} = \dfrac{19}{4}$; a quadratic eq. with these roots is

$x^2 - 4x + \dfrac{19}{4} = 0$, or $4x^2 - 16x + 19 = 0$; $(x + 1)(4x^2 - 16x + 19) = 0$;

$4x^3 - 12x^2 + 3x + 19 = 0$

22. $(x + i\sqrt{2})(x - i\sqrt{2})(x - 5) = 0$; $(x^2 + 2)(x - 5) = 0$; $x^3 - 5x^2 + 2x - 10 = 0$

23. A quadratic eq. with roots $5 \pm i\sqrt{3}$ is $x^2 - 10x + 28 = 0$; a quadratic eq. with roots $\pm i$ is $x^2 + 1 = 0$; $(x^2 + 1)(x^2 - 10x + 28) = 0$; $x^4 - 10x^3 + 29x^2 - 10x + 28 = 0$.

24. A quadratic eq. with roots $1 \pm i\sqrt{7}$ is $x^2 - 2x + 8 = 0$; a quadratic eq. with roots $2 \pm i$ is $x^2 - 4x + 5 = 0$; $(x^2 - 2x + 8)(x^2 - 4x + 5) = 0$;
$x^4 - 6x^3 + 21x^2 - 42x + 40 = 0$.

25. If $1 + \sqrt{3}$, $1 - \sqrt{3}$, and r are the roots, $1 + \sqrt{3} + 1 - \sqrt{3} + r = 0$; $r = -2$;

$(1 + \sqrt{3})(1 - \sqrt{3})(-2) = 4 = -\dfrac{d}{1}$; $d = -4$; $(-2)^3 + c(-2) - 4 = 0$; $-2c = 12$;

$c = -6$

26. If $2i$, $-2i$, and r are the roots, $2i(-2i)r = 4r = -\dfrac{8}{1}$; $r = -2$; $2i + (-2i) + (-2) =$

$-2 = -\dfrac{b}{1}$; $b = 2$; $(-2)^3 + 2(-2)^2 + c(-2) + 8 = 0$; $8 = 2c$; $4 = c$

27. Let d be the double root and let r be the third root. $d + d + r = 0$, so $r = -2d$.

$d \cdot d \cdot r = \dfrac{27}{4}$, so $d^2(-2d) = \dfrac{27}{4}$; $d^3 = -\dfrac{27}{8}$; $d = -\dfrac{3}{2}$; $r = -2\left(-\dfrac{3}{2}\right) = 3$;

$4(3^3) + 3c - 27 = 0$; $3c = -81$; $c = -27$

28. Let x, $-x$, and y be the roots. $x + (-x) + y = y = -\dfrac{4}{1}$; $y = -4$;

$(-4)^3 + 4(-4)^2 - 9(-4) + d = 0$; $d = -36$

29. a. If r_1, r_2, and r_3 are the roots of $P(x) = x^3 + \dfrac{b}{a}x^2 + \dfrac{c}{a}x + \dfrac{d}{a} = 0$, then $x - r_1$,

$x - r_2$, and $x - r_3$ are factors of $P(x)$; thus, $x^3 + \dfrac{b}{a}x^2 + \dfrac{c}{a}x + \dfrac{d}{a} =$

$k(x - r_1)(x - r_2)(x - r_3)$; the coefficient of x^3 is 1, so $k = 1$.

b. $(x - r_1)(x - r_2)(x - r_3) = (x^2 - (r_1 + r_2)x + r_1 r_2) \cdot (x - r_3) =$
$x^3 - (r_1 + r_2 + r_3)x^2 + (r_1 r_2 + r_1 r_3 + r_2 r_3)x - r_1 r_2 r_3$, which equals

$x^3 + \dfrac{b}{a}x^2 + \dfrac{c}{a}x + \dfrac{d}{a} = x^3 - \left(-\dfrac{b}{a}\right)x^2 + \dfrac{c}{a}x - \left(-\dfrac{d}{a}\right)$; equating the coefficients of

the quadratic and constant terms gives $r_1 + r_2 + r_3 = -\dfrac{b}{a}$ and $r_1 r_2 r_3 = -\dfrac{d}{a}$.

c. The average of the roots is $\dfrac{r_1 + r_2 + r_3}{3} = \dfrac{1}{3}\left(-\dfrac{b}{a}\right) = -\dfrac{b}{3a}$.

30. If r_1, r_2, r_3, and r_4 are the roots of $ax^4 + bx^3 + cx^2 + dx + e = 0$, then they are the

roots of $x^4 + \dfrac{b}{a}x^3 + \dfrac{c}{a}x^2 + \dfrac{d}{a}x + \dfrac{e}{a} = 0$; hence $x - r_1$, $x - r_2$, $x - r_3$, and $x - r_4$

are factors of $x^4 + \dfrac{b}{a}x^3 + \dfrac{c}{a}x^2 + \dfrac{d}{a}x + \dfrac{e}{a}$; thus $k(x - r_1)(x - r_2)(x - r_3)(x - r_4) =$

$x^4 + \dfrac{b}{a}x^3 + \dfrac{c}{a}x^2 + \dfrac{d}{a}x + \dfrac{e}{a}$, and k must $= 1$, since the coefficient of x^4 is 1;

$(x - r_1)(x - r_2)(x - r_3)(x - r_4) = (x^2 - (r_1 + r_2)x + r_1 r_2)(x^2 - (r_3 + r_4)x + r_3 r_4)$;

compute the x^3 and constant terms: $-(r_1 + r_2)x \cdot x^2 - (r_3 + r_4)x \cdot x^2 =$

$-(r_1 + r_2 + r_3 + r_4)x^3$ and $r_1 r_2 \cdot r_3 r_4 = r_1 r_2 r_3 r_4$. Equating the coefficients of the

cubic and constant terms gives $-\left(-\dfrac{b}{a}\right) = -(r_1 + r_2 + r_3 + r_4)$ or

$-\dfrac{b}{a} = r_1 + r_2 + r_3 + r_4$ and $\dfrac{e}{a} = r_1 r_2 r_3 r_4$.

31. $\overline{x + z} = \overline{(r + si) + (u + vi)} = \overline{(r + u) + (s + v)i} = (r + u) - (s + v)i =$
$(r - si) + (u - vi) = \overline{x} + \overline{z}$

32. $\overline{xz} = \overline{(r + si)(u + vi)} = \overline{(ru - sv) + (rv + su)i} = (ru - sv) - (rv + su)i$;
$\overline{x} \cdot \overline{z} = \overline{r + si} \cdot \overline{u + vi} = (r - si)(u - vi) = ru - sui - rvi + svi^2 =$
$(ru - sv) - (rv + su)i = \overline{xz}$

33. $x^2 = \overline{(r + si)^2} = \overline{(r^2 - s^2) + 2rsi} = (r^2 - s^2) - 2rsi;$

$(\overline{x})^2 = (r - si)^2 = (r^2 - s^2) - 2rsi = \overline{x^2}$ [Or: $\overline{x^2} = \overline{x \cdot x} = \overline{x} \cdot \overline{x}$ (by Ex. 32) $= (\overline{x})^2$]

34. $\overline{x^3} = \overline{x^2 \cdot x} = \overline{x^2} \cdot \overline{x}$ (from Ex. 32) $= (\overline{x})^2 \cdot \overline{x}$ (from Ex. 33) $= (\overline{x})^3$

35. a. $\overline{a} = \overline{a + 0i} = a - 0i = a$

 b. $\overline{ax^3} = \overline{(a + 0i) \cdot x^3} = \overline{(a + 0i)} \cdot \overline{x^3}$ (from Ex. 32) $= a(\overline{x})^3$ from Ex. 34 and Ex. 35(a).

36. $\overline{bx^2} = \overline{b} \cdot \overline{x^2}$ (from Ex. 32) $= b(\overline{x})^2$ from Exs. 33 and 35(a): $\overline{cx} = \overline{c} \cdot \overline{x} = c(\overline{x})$ from Exs. 32 and 35(a).

37. From Ex. 31, $\overline{ax^3 + bx^2 + cx + d} = \overline{ax^3} + \overline{bx^2} + \overline{cx} + \overline{d} = a(\overline{x})^3 + b(\overline{x})^2 + c(\overline{x}) + d$ from Exs. 35 and 36.

38. If $P(x) = ax^3 + bx^2 + cx + d = 0$, then $P(\overline{x}) = a(\overline{x})^3 + b(\overline{x})^2 + c(\overline{x}) + d = \overline{ax^3 + bx^2 + cx + d}$ (from Ex. 37) $= \overline{0} = 0$.

39. $P(1 + i) = (1 + i)^3 - (1 + i) = (1 + 2i - 1)(1 + i) - (1 + i) = 2i + 2i^2 - 1 - i = -3 + i; \overline{P(1 + i)} = \overline{-3 + i} = -3 - i; P(1 - i) = (1 - i)^3 - (1 - i) = (1 - 2i - 1)(1 - i) - (1 - i) = -2i + 2i^2 - 1 + i = -3 - i; \overline{P(1 + i)} = P(1 - i)$

40. Suppose $P(x) = 0$ has complex coefficients and positive degree n. If $P(x) = 0$ has at least one complex root, r_1, then $P(x) = (x - r_1)P_1(x)$, where $P_1(x)$ is a polynomial of degree $n - 1$. If $n - 1 > 0$, then $P_1(x)$ has at least one complex root, r_2, and $P_1(x) = (x - r_2)P_2(x)$, where $P_2(x)$ is a polynomial of degree $n - 2$; $P(x) = (x - r_1)(x - r_2)P_2(x)$. We continue in this way as long as the polynomial $P_k(x)$ has degree greater than 0. Thus, $P(x) = (x - r_1)(x - r_2)(x - r_3) \cdots (x - r_n)P_n(x)$, where $P_n(x)$ is a polynomial of degree 0 (that is, a constant). Thus, $P(x) = 0$ has exactly n roots $(r_1, r_2, \ldots, r_n)$, which is what the fundamental theorem of algebra tells us.

Chapter Test · page 93

1. a.

$$2 \,\underline{\big|}\begin{array}{cccc} 4 & 0 & -5 & -2 \\ & 8 & 16 & 22 \\ \hline 4 & 8 & 11 & \boxed{20} = P(2) \end{array}$$

b.

$$-\tfrac{1}{2} \,\underline{\big|}\begin{array}{cccc} 4 & 0 & -5 & -2 \\ & -2 & 1 & 2 \\ \hline 4 & -2 & -4 & \boxed{0} = P\left(-\tfrac{1}{2}\right) \end{array}$$

c.

$$i \,\underline{\big|}\begin{array}{cccc} 4 & 0 & -5 & -2 \\ & 4i & -4 & -9i \\ \hline 4 & 4i & -9 & \boxed{-2 - 9i} = P(i) \end{array}$$

2. $3(2)^2 + k(2) - 8 = 0; 12 + 2k - 8 = 0; 2k = -4; k = -2$

3. $P(x) = (x + 3)(x^2 - x + 5) + 2 = x^3 + 2x^2 + 2x + 17$

4.

$$2 \,\underline{\big|}\begin{array}{ccccc} 1 & -3 & -14 & 12 & 40 \\ & 2 & -2 & -32 & -40 \\ \hline 1 & -1 & -16 & -20 & \boxed{0} \end{array}$$

$$5 \,\underline{\big|}\begin{array}{cccc} 1 & -1 & -16 & -20 \\ & 5 & 20 & 20 \\ \hline 1 & 4 & 4 & \boxed{0} \end{array}$$

$x^2 + 4x + 4 = 0; (x + 2)^2 = 0;$ -2 is a double root.

5.

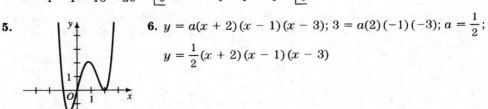

6. $y = a(x + 2)(x - 1)(x - 3); 3 = a(2)(-1)(-3); a = \dfrac{1}{2};$

$y = \dfrac{1}{2}(x + 2)(x - 1)(x - 3)$

7. If the enclosure has width x, then its length is $80 - 2x$. $A(x) = x(80 - 2x)$. The graph of $A(x)$ is a parabola with zeros 0 and 40. The maximum occurs when $x = \dfrac{0 + 40}{2} = 20$; $A(20) = 20(80 - 2 \cdot 20) = 800$ (m^2).

8. a. Use the location principle: $P(1) = 2(1)^4 - 3(1)^3 - 1 = -2 < 0$ and $P(2) = 2(2)^4 - 3(2)^3 - 1 = 7 > 0$.

   b. Use the method of Example 1 on page 75 or Example 2 on page 76; $x \approx -0.6$ or $x \approx 1.6$.

9. $2x^3 - 3x^2 - 6x + 9 = 0$; $x^2(2x - 3) - 3(2x - 3) = 0$; $(x^2 - 3)(2x - 3) = 0$; $x = \pm\sqrt{3}, \dfrac{3}{2}$

10.

$$
\begin{array}{r|rrrrr}
1 & 3 & -1 & 4 & -2 & -4 \\
  &   & 3 & 2 & 6 & 4 \\
\hline
  & 3 & 2 & 6 & 4 & \;\boxed{0}
\end{array}
$$

$3x^3 + 2x^2 + 6x + 4 = 0$ has no positive roots; the possible negative rational roots are $-\dfrac{1}{3}$, $-\dfrac{2}{3}$, -1, $-\dfrac{4}{3}$, -2, and -4. $P(0) = 4$ and $P(-1) = -3$, so there is a root between 0 and -1.

$$
\begin{array}{r|rrrr}
-\dfrac{2}{3} & 3 & 2 & 6 & 4 \\
              &   & -2 & 0 & -4 \\
\hline
              & 3 & 0 & 6 & \;\boxed{0}
\end{array}
$$

$3x^4 - x^3 + 4x^2 - 2x - 4 = (x - 1)(3x + 2)(x^2 + 2) = 0$; the roots are 1, $-\dfrac{2}{3}$, and $\pm i\sqrt{2}$.

11. Answers will vary. Students should note that the method of using technology gives *approximations* of *all real* roots, but the multiplicity of roots can remain in doubt and imaginary roots remain unknown. The method of factoring, on the other hand, may give *some* real roots and even *some* imaginary roots *exactly* (including multiplicities of roots), but there is no guarantee of finding all (or even any) roots using this method.

12. sum $= -\dfrac{-5}{2} = \dfrac{5}{2}$;

   product $= -\dfrac{8}{2} = -4$

13. The other root is $-3i$. A quadratic eq. having roots $\pm 3i$ is $x^2 + 9 = 0$, so a cubic eq. is $(x + 5)(x^2 + 9) = 0$; $x^3 + 5x^2 + 9x + 45 = 0$. (Multiples of this eq. are also correct.)

14. If a certain imaginary number is a double root, then so is the conjugate of the number. This gives the cubic eq. at least four roots, when it can have only three. For example, $x^3 - 5x^2 + 9x - 45 = 0$.

Class Exercises 3-1 · page 98

1. $x > -4$ **2.** $x < 4$ **3.** $-2x < 10; x > -5$ **4.** $-2 < x < 2$

5. $-2 < x - 5 < 2; 3 < x < 7$ **6.** $x + 3 < -1$ or $x + 3 > 1; x < -4$ or $x > -2$

7. f   **8.** a   **9.** e   **10.** c   **11.** b   **12.** d

Written Exercises 3-1 · pages 98–99

1. $7x < 21; x < 3$

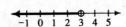

2. $8x > 24; x > 3$

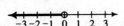

3. $15 - 6x > 15; -6x > 0; x < 0$

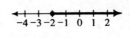

4. $8 - 11x \le 52; -11x \le 44; x \ge -4$

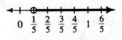

5. $3x - 3 \le 2x + 8; x \le 11$

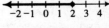

6. $15\left(\dfrac{2-x}{3}\right) < 15\left(\dfrac{3-2x}{5}\right);$

$10 - 5x < 9 - 6x; x < -1$

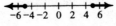

7. $8x + 3x + 3 > 5x - 9x + 6; 15x > 3; x > \dfrac{1}{5}$

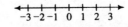

8. $12x^2 - 2x \ge 12x^2 + x - 6; -3x \ge -6; x \le 2$

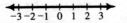

9. $12\left(\dfrac{x+2}{4} - \dfrac{2-x}{3} + \dfrac{4x-5}{6}\right) < 12 \cdot 4;$

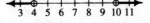

$3x + 6 - 8 + 4x + 8x - 10 < 48; 15x < 60; x < 4$

10. $6\left[\dfrac{4}{3}\left(x - \dfrac{1}{2}\right) + \dfrac{1}{2}x\right] \ge 6 \cdot \dfrac{2}{3}\left(2x - \dfrac{5}{2}\right);$

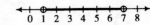

$8x - 4 + 3x \ge 8x - 10; 3x \ge -6; x \ge -2$

11. $-3 < x < 3$

12. $x \le -5$ or $x \ge 5$

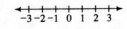

13. $|x|$ is always nonnegative; no solution

14. $|x|$ is always nonnegative; all real nos.

15. $-3 < x - 4$ and $x - 4 < 3; 1 < x$ and $x < 7; 1 < x < 7$

16. $x - 7 < -3$ or $x - 7 > 3; x < 4$ or $x > 10$

17. $x + 7 \le -3$ or $x + 7 \ge 3; x \le -10$ or $x \ge -4$

18. $-(-1) < x + 2 < -1; 1 < x + 2 < -1;$ no number is greater than 1 and less than -1, so there is no solution.

19. $x - 8 = \pm 4; x = 8 \pm 4; x = 4$ or $x = 12$

20. $x + 3 = \pm 7$; $x = -3 \pm 7$; $x = -10$ or $x = 4$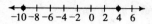

21. $-5 \leq 2x - 4$ and $2x - 4 \leq 5$; $-1 \leq 2x$ and $2x \leq 9$;

$-\dfrac{1}{2} \leq x \leq \dfrac{9}{2}$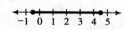

22. $3x - 9 \leq -9$ or $3x - 9 \geq 9$; $3x \leq 0$ or $3x \geq 18$;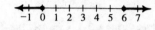
$x \leq 0$ or $x \geq 6$

23. $-9 \leq 4x + 8$ and $4x + 8 \leq 9$; $-17 \leq 4x$ and $4x \leq 1$;

$-\dfrac{17}{4} \leq x \leq \dfrac{1}{4}$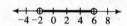

24. $-12 < 6 - 3x$ and $6 - 3x < 12$; $-18 < -3x$ and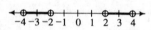
$-3x < 6$; $6 > x > -2$, or $-2 < x < 6$

25. $C = \dfrac{5}{9}(F - 32)$; $\dfrac{9}{5}C = F - 32$; $F = \dfrac{9}{5}C + 32$; $60 \leq \dfrac{9}{5}C + 32 \leq 65$; $28 \leq \dfrac{9}{5}C \leq 33$;

$28\left(\dfrac{5}{9}\right) \leq C \leq 33\left(\dfrac{5}{9}\right)$; $15\dfrac{5}{9}^{\circ} \leq C \leq 18\dfrac{1}{3}^{\circ}$

26. a. The distance from x to 0 is greater than 2 units, and the distance from x to 0 is less than 4 units.

  b. $2 < |x|$ and $|x| < 4$; $x < -2$ or $x > 2$, and
  $-4 < x < 4$; the solution set is the intersection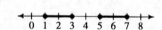
  of the graphs: $-4 < x < -2$ or $2 < x < 4$.

  c. $1 \leq |x|$ and $|x| < 5$; $x \leq -1$ or $x \geq 1$, and
  $-5 < x < 5$; the solution set is the intersection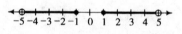
  of the graphs: $-5 < x \leq -1$ or $1 \leq x < 5$.

27. $1 \leq |x - 4|$ and $|x - 4| \leq 3$; $x - 4 \leq -1$ *or*
$x - 4 \geq 1$, and $-3 \leq x - 4$ *and* $x - 4 \leq 3$;
$x \leq 3$ or $x \geq 5$, and $1 \leq x \leq 7$; the solution set is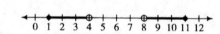
the intersection of the graphs: $1 \leq x \leq 3$ or
$5 \leq x \leq 7$.

28. $2 < |x - 6|$ and $|x - 6| \leq 5$; $x - 6 < -2$ *or*
$x - 6 > 2$, and $-5 \leq x - 6$ *and* $x - 6 \leq 5$;
$x < 4$ or $x > 8$, and $1 \leq x \leq 11$; the solution set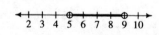
is the intersection of the graphs: $1 \leq x < 4$ or
$8 < x \leq 11$.

29. $0 \leq |x - 7|$ and $|x - 7| < 2$; $x - 7 \leq 0$ or
$x - 7 \geq 0$, and $-2 < x - 7$ and $x - 7 < 2$; $x \leq 7$
or $x \geq 7$, and $5 < x < 9$; the intersection is, thus,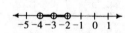
$5 < x \leq 7$ or $7 \leq x < 9$, that is, $5 < x < 9$.

30. $0 < |x + 3|$ and $|x + 3| < 1$; $x + 3 < 0$ or
$x + 3 > 0$, and $-1 < x + 3$ and $x + 3 < 1$;
$x < -3$ or $x > -3$, and $-4 < x < -2$; the
intersection is, thus, $-4 < x < -3$ or
$-3 < x < -2$.

31. If $x > 0$, $x\left(\dfrac{1}{x}\right) < x \cdot 2$; $1 < 2x$; $x > \dfrac{1}{2}$; the intersection of $x > 0$ and $x > \dfrac{1}{2}$ is $x > \dfrac{1}{2}$.

If $x < 0$, $x\left(\dfrac{1}{x}\right) > x \cdot 2$; $1 > 2x$; $x < \dfrac{1}{2}$; the intersection of $x < 0$ and $x < \dfrac{1}{2}$ is $x < 0$.

Thus, $x < 0$ or $x > \dfrac{1}{2}$.

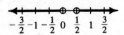   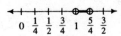

Ex. 31   **Ex. 32**

32. If $x - 1 > 0$, then $(x - 1)\left(\dfrac{1}{x - 1}\right) > (x - 1)4$; $\dfrac{1}{4} > x - 1$; $\dfrac{5}{4} > x$; the intersection of

$x > 1$ and $\dfrac{5}{4} > x$ is $1 < x < \dfrac{5}{4}$. If $x - 1 < 0$, then $(x - 1)\left(\dfrac{1}{x - 1}\right) < (x - 1)4$;

$\dfrac{1}{4} < x - 1$; $\dfrac{5}{4} < x$; the intersection of $x < 1$ and $x > \dfrac{5}{4}$ is the empty set. Thus,

$1 < x < \dfrac{5}{4}$.

33. $|x| + |x - 2| = 2$.
Case 1: If $x < 0$, $|x| = -x$ and $|x - 2| = -(x - 2) = 2 - x$; $-x + (2 - x) = 2$; $-2x = 0$; $x = 0$; this contradicts the assumption that $x < 0$, so there is no solution.
Case 2: If $x \geq 2$, then $|x| = x$ and $|x - 2| = x - 2$; $x + (x - 2) = 2$; $2x = 4$; $x = 2$; the intersection of $x \geq 2$ and $x = 2$ is $x = 2$.
Case 3: If $0 \leq x < 2$, then $|x| = x$ and $|x - 2| = -(x - 2) = 2 - x$; $x + (2 - x) = 2$; $2 = 2$, which is true for all x; thus for all x such that $0 \leq x < 2$, the equation is true.
Cases 2 and 3 yield the solution $0 \leq x \leq 2$.

34. $|x| + |x - 2| > 5$.
Case 1: If $x < 0$, $|x| = -x$ and $|x - 2| = 2 - x$; $-x + (2 - x) > 5$; $-3 > 2x$; $-\dfrac{3}{2} > x$; the intersection of $x < 0$ and $x < -\dfrac{3}{2}$ is $x < -\dfrac{3}{2}$.

Case 2: If $x \geq 2$, then $|x| = x$ and $|x - 2| = x - 2$; $x + x - 2 > 5$; $2x > 7$; $x > \dfrac{7}{2}$; the intersection of $x \geq 2$ and $x > \dfrac{7}{2}$ is $x > \dfrac{7}{2}$.

Case 3: If $0 \leq x < 2$, then $|x| = x$ and $|x - 2| = 2 - x$; $x + 2 - x > 5$; $2 > 5$, which is a contradiction, so this case has no solution.
Cases 1 and 2 yield the solution $x < -\dfrac{3}{2}$ or $x > \dfrac{7}{2}$.

35. (1) If $a > 0$ and $b > 0$, then $ab > 0$ so $|ab| = ab$. $|a| = a$ and $|b| = b$; hence $|ab| = ab = |a| \cdot |b|$. (2) If $a < 0$ and $b > 0$, then $ab < 0$ so $|ab| = -ab$. Also $|a| = -a$ and $|b| = b$; hence $|ab| = -ab = (-a) \cdot b = |a| \cdot |b|$. (3) If $a > 0$ and $b < 0$, then $ab < 0$ so $|ab| = -ab$. Also $|a| = a$ and $|b| = -b$; hence $|ab| = -ab = a(-b) = |a| \cdot |b|$. (4) If $a < 0$ and $b < 0$, then $ab > 0$, so $|ab| = ab$. Also $|a| = -a$ and $|b| = -b$; hence $|ab| = ab = (-a)(-b) = |a| \cdot |b|$.

36. a. For example, $|4 + 7| \leq |4| + |7|; |-3 + 4| \leq |-3| + |4|$, or $1 \leq 7$;
$|-5 - 12| \leq |-5| + |-12|$, or $17 \leq 17$

b. Yes, the triangle inequality holds for complex numbers.
Examples: $|(2 - 3i) + (-1 - 4i)| = |1 - 7i| = \sqrt{1 + (-7)^2} = 5\sqrt{2}; |2 - 3i| +$
$|-1 - 4i| = \sqrt{2^2 + (-3)^2} + \sqrt{(-1)^2 + (-4)^2} = \sqrt{13} + \sqrt{17}; 5\sqrt{2} < \sqrt{13} + \sqrt{17}$.
Also, $|(1 + i) + (1 - i)| = |2| = \sqrt{2^2 + 0^2} = 2; |1 + i| + |1 - i| = \sqrt{1^2 + 1^2} +$
$\sqrt{1^2 + (-1)^2} = 2\sqrt{2}; 2 < 2\sqrt{2}$.

37. a. $|a - b| = |a + (-b)| \leq |a| + |-b|$ by the triangle inequality. But $|-b| = |b|$, so
$|a - b| \leq |a| + |b|$.

b. $|a| = |(a - b) + b|$ since $(a - b) + b = a$. By the triangle inequality,
$|(a - b) + b| \leq |a - b| + |b|$. Hence $|a| \leq |a - b| + |b|$ and $|a| - |b| \leq |a - b|$.

Class Exercises 3-2 · page 103

1. $-2 < x < 2$ or $x > 4$ **2.** $x \leq 1$

3. A sign graph yields the solution $1 < x < 2$ or $x > 4$.

4. A sign graph yields the solution $x < 2$ or $4 < x < 7$.

5. The only zero is $x = 5$, so the solution is $x \leq 5$.

6. The absolute value of every number is nonnegative, so there is no solution.

Written Exercises 3-2 · pages 103–104

1. $f(x) = (x - 3)(x + 4); f(-5) = (-8)(-1) > 0; f(0) = -12 < 0; f(4) = (1)(8) > 0;$
$f(x) > 0$ when $x < -4$ or $x > 3$

2. $g(x) = (x + 7)(x + 9); g(-10) = (-3)(-1) > 0; g(-8) = (-1)(1) < 0;$
$g(-6) = (1)(3) > 0; g(x) > 0$ when $x < -9$ or $x > -7$

3. $h(x) = (x - 1)(x - 2)(x - 4); h(0) = -8 < 0; h\left(\dfrac{3}{2}\right) = \left(\dfrac{1}{2}\right)\left(-\dfrac{1}{2}\right)\left(-\dfrac{5}{2}\right) > 0; h(3) =$
$(2)(1)(-1) < 0; h(5) = (4)(3)(1) > 0; h(x) > 0$ when $1 < x < 2$ or $x > 4$

4. $F(x) = (1 - x)(x - 3)(x - 5); F(0) = 15 > 0; F(2) = (-1)(-1)(-3) < 0; F(4) =$
$(-3)(1)(-1) > 0; F(6) = (-5)(3)(1) < 0; F(x) > 0$ when $x < 1$ or $3 < x < 5$

5. $G(x) = (x - 4)(x - 2)^2(x - 3)^2; G(0) = (-4)(4)(9) < 0; 2$ is a double root, so
$G(x) < 0$ when $x < 2$ and when $2 < x < 3; 3$ is a double root, so $G(x) < 0$ when
$3 < x < 4; G(5) = 1(9)(4) > 0; G(x) < 0$ when $x < 2$ or $2 < x < 3$ or $3 < x < 4$.

6. $H(x) = (2x - 5)^2(x + 3)(x + 2); H(-4) = (-13)^2(-1)(-2) > 0; H\left(-\dfrac{5}{2}\right) =$
$(-10)^2\left(\dfrac{1}{2}\right)\left(-\dfrac{1}{2}\right) < 0; H(0) = 25(3)(2) > 0$ and $\dfrac{5}{2}$ is a double root; $H(x) < 0$ when
$-3 < x < -2$

7. $f(x) = (x + 3)(x - 5); f(-4) = (-1)(-9) > 0; f(0) = -15 < 0; f(6) = (9)(1) > 0;$
$f(x) < 0$ when $-3 < x < 5$

8. $g(x) = (x + 6)(x - 3); g(-7) = (-1)(-10) > 0; g(0) = -18 < 0; g(4) = (10)(1) > 0;$
$g(x) > 0$ when $x < -6$ or $x > 3$

9. $h(x) = (2x - 3)(x + 1); h(-2) = (-7)(-1) > 0; h(0) = -3 < 0; h(2) = (1)(3) > 0;$
$h(x) \geq 0$ when $x \leq -1$ or $x \geq \dfrac{3}{2}$

10. $F(x) = (1 - 3x)(1 + x)$; $F(-2) = (7)(-1) < 0$; $F(0) = 1 > 0$; $F(1) = (-2)(2) < 0$;

$F(x) < 0$ when $x < -1$ or $x > \dfrac{1}{3}$

11. $G(x) = (2x + 7)(x - 1)$; $G(-4) = (-1)(-5) > 0$; $G(0) = -7 < 0$;

$G(2) = (11)(1) > 0$; $G(x) \le 0$ when $-\dfrac{7}{2} \le x \le 1$

12. $H(x) = (x - 4)^2$; when $x \neq 4$, $H(x) > 0$; thus, $H(x) \le 0$ when $x = 4$

13. $f(x) = (x^2 - 5)(x^2 + 2)$; zeros are $\pm\sqrt{5} \approx \pm 2.2$; $f(-3) = f(3) = (4)(11) > 0$; $f(0) = -10 < 0$; $f(x) > 0$ when $x < -\sqrt{5}$ or $x > \sqrt{5}$

14. $g(x) = x(3x - 2)(x + 3)$; $g(-4) = (-4)(-14)(-1) < 0$; $g(-1) = (-1)(-5)(2) > 0$;

$g\left(\dfrac{1}{2}\right) = \dfrac{1}{2}\left(-\dfrac{1}{2}\right)\left(\dfrac{7}{2}\right) < 0$; $g(1) = 1(1)(4) > 0$; $g(x) \le 0$ when $x \le -3$ or $0 \le x \le \dfrac{2}{3}$

15. $h(a) = a^2(a + 2) - 4(a + 2) = (a + 2)^2(a - 2)$; $h(-3) = (-1)^2(-5) < 0$; -2 is a double root so $h(a) < 0$ when $a < -2$ and when $-2 < a < 2$; $h(3) = 5^2(1) > 0$; $h(a) > 0$ when $a > 2$

16. $F(b) = (b^2 + 4)(b^2 - 4)$; zeros are ± 2; $F(-3) = F(3) = (13)(5) > 0$;
$F(0) = 4(-4) < 0$; $F(b) < 0$ when $-2 < b < 2$

17. $G(n) = (n - 1)(n - 2)(n + 3)$; $G(-4) = (-5)(-6)(-1) < 0$; $G(0) = 6 > 0$; $G\left(\dfrac{3}{2}\right) = $

$\left(\dfrac{1}{2}\right)\left(-\dfrac{1}{2}\right)\left(\dfrac{9}{2}\right) < 0$; $G(3) = (2)(1)(6) > 0$; $G(n) < 0$ when $n < -3$ or $1 < n < 2$

18. $H(y) = (2y - 1)(y + 1)^2$; $H(-2) = (-5)(-1)^2 < 0$; -1 is a double root so $H(y) < 0$

when $y < -1$ and when $-1 < y < \dfrac{1}{2}$; $H(1) = (1)(2^2) > 0$; $H(y) \ge 0$ when $y = -1$

or $y \ge \dfrac{1}{2}$

19. $f(x) = 2x^3 + x^2 - 5x + 2 = (x + 2)(2x - 1)(x - 1)$; $f(-3) = (-1)(-7)(-4) < 0$;
$f(0) = 2 > 0$; $f(0.75) = (2.75)(0.5)(-0.25) < 0$; $f(2) = (4)(3)(1) > 0$; $f(x) < 0$ when
$2x^3 + x^2 - 5x + 2 < 0$, or $2x^3 + x^2 - 5x < -2$; this is true when $x < -2$ or

$\dfrac{1}{2} < x < 1$.

20. $g(r) = r^3 - 8r^2 - 9r = r(r + 1)(r - 9) > 0$; $g(-2) = -2(-1)(-11) < 0$;
$g(-0.5) = -0.5(0.5)(-9.5) > 0$; $g(1) = 1(2)(-8) < 0$; $g(10) = 10(11)(1) > 0$;
$r^3 - 9r > 8r^2$ when $-1 < r < 0$ or $r > 9$

21. $h(y) = y^4 + y^3 - 4y - 16 = (y + 2)(y - 2)(y^2 + y + 4)$; roots are ± 2;
$h(-3) = 50 > 0$; $h(0) = -16 < 0$; $h(3) = 80 > 0$; $y^4 + y^3 < 4(y + 4)$ when
$-2 < y < 2$

22. $F(x) = x^4 - 5x^2 + 6 = (x^2 - 3)(x^2 - 2)$; roots are $\pm\sqrt{3} \approx \pm 1.732$ and

$\pm\sqrt{2} \approx \pm 1.414$; $F(-2) = F(2) = (1)(2) > 0$; $F\left(-\dfrac{3}{2}\right) = F\left(\dfrac{3}{2}\right) = \left(-\dfrac{3}{4}\right)\left(\dfrac{1}{4}\right) < 0$;

$F(0) = 6 > 0$; $F(x) < 0$ when $x^2 + 6 < 5x^2$; this is true when $-\sqrt{3} < x < -\sqrt{2}$ or
$\sqrt{2} < x < \sqrt{3}$.

23. $G(x) = (x + 1)(x - 1)(2x - 1)^2$; $G(-2) = (-1)(-3)(-5)^2 > 0$; $G(0) = -1 < 0$; $\dfrac{1}{2}$

is a double root so $G(x) < 0$ when $-1 < x < \dfrac{1}{2}$ and when $\dfrac{1}{2} < x < 1$;

$G(2) = (3)(1)(9) > 0$; $G(x) > 0$ when $x < -1$ or $x > 1$

24. $h(y) = (y - 1)(y + 3)(2y - 3)$; $h(-4) = (-5)(-1)(-11) < 0$; $h(0) = 9 > 0$;

$h(1.1) = (0.1)(4.1)(-0.8) < 0$; $h(2) = 1(5)(1) > 0$; $h(y) < 0$ when $y < -3$ or

$1 < y < 1.5$

25. The zeros of the linear factors are 3, 4, 5, and 6 and the sign changes at each zero

except 6; $f(0) = \dfrac{(-3)(-4)}{(-5)(36)} < 0$; thus, $f(x) < 0$ when $x < 3$ or $4 < x < 5$.

26. The zeros of the linear factors are -1, 3, and 5; the sign changes at $x = -1$ only;

$f(0) = \dfrac{1 \cdot 9}{25} > 0$; thus, $f(x) > 0$ when $-1 < x < 3$, $3 < x < 5$, or $x > 5$.

27. $f(x) = \dfrac{(2x - 5)^3}{(x - 7)(x + 4)}$; the sign of $f(x)$ changes at the zero of each linear factor: $-4, \dfrac{5}{2}$,

and 7; $f(0) = \dfrac{-125}{-28} > 0$; thus, $f(x) \geq 0$ when $-4 < x \leq \dfrac{5}{2}$ or $x > 7$.

28. $g(n) = \dfrac{(3n - 12)^2}{3n^2 - 12} = \dfrac{9(n - 4)^2}{3(n^2 - 4)} = \dfrac{3(n - 4)^2}{(n + 2)(n - 2)}$; the sign of $g(n)$ changes at 2 and -2,

but not at 4; $g(0) = \dfrac{3 \cdot 16}{-4} < 0$; thus, $g(n) \leq 0$ when $-2 < n < 2$ or $n = 4$.

29. $2x^2 + 7x + 8$ and $x^2 + 1$ both have imaginary roots; thus, the numerator and the

denominator are positive for all real x; hence, $\dfrac{2x^2 + 7x + 8}{x^2 + 1} > 0$ for all real x.

30. $f(n) = \dfrac{(n + 2)^2}{n(n + 4)}$; $f(n)$ changes sign at 0 and -4, but not at -2; $f(1) = \dfrac{9}{5} > 0$; thus,

$f(n) > 0$ when $n < -4$ or $n > 0$.

31. $x^2(x + 2) - 3(x + 2) > 0$; $(x + \sqrt{3})(x - \sqrt{3})(x + 2) > 0$; from the graph, the

solution is $-2 < x < -\sqrt{3}$ or $x > \sqrt{3}$ (or $-2 < x < -1.73$ or $x > 1.73$)

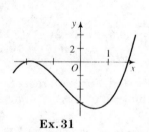

Ex. 31

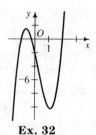

Ex. 32

32. The graph suggests roots at $x = -1$ and at $x = 2$. Synthetic substitution gives roots

-1, 2, and -0.25; $(4x + 1)(x - 2)(x + 1) \leq 0$ when $x \leq -1$ or $-0.25 \leq x \leq 2$.

33. Testing the possible rational roots $\left(\pm 1, \pm \dfrac{1}{2}\right)$ shows that $\dfrac{1}{2}$ is a root.

$(2x - 1)(x^2 - 2x - 1) = 0$ when $x = \dfrac{1}{2}$ and when $x = 1 \pm \sqrt{2}$. Thus,

$(2x - 1)(x^2 - 2x - 1) \geq 0$ when $1 - \sqrt{2} \leq x \leq \dfrac{1}{2}$ or $x \geq 1 + \sqrt{2}$ (or $-0.4 \leq x \leq 0.5$

or $x \geq 2.4$).

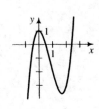

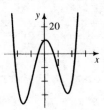

Ex. 33   **Ex. 34**

34. The possible rational roots are ± 1, ± 2, ± 5, ± 10, $\pm \dfrac{1}{2}$, $\pm \dfrac{5}{2}$, $\pm \dfrac{1}{3}$, $\pm \dfrac{2}{3}$, $\pm \dfrac{5}{3}$, $\pm \dfrac{10}{3}$, $\pm \dfrac{1}{6}$,

and $\pm \dfrac{5}{6}$. The graph shows that there is one root between -1 and 0, and another

between 0 and 1. Using synthetic substitution we have $(2x + 1)(3x - 2)(x^2 - 5) < 0$;

using the graph we have $-\sqrt{5} < x < -\dfrac{1}{2}$ or $\dfrac{2}{3} < x < \sqrt{5}$ (or $-2.2 < x < -0.5$ or

$0.7 < x < 2.2$).

35. $-x^2 - 2x + 6 = x + 2$; $x^2 + 3x - 4 = 0$; $(x + 4)(x - 1) = 0$; $P(x)$ and $Q(x)$ intersect
at $x = -4$ and $x = 1$; $P(x) > Q(x)$ when $x < -4$ or $x > 1$

36. $-x^3 + 11x^2 - 20x + 4 = x^3 - 13x^2 + 42x + 4$; $0 = 2x^3 - 24x^2 + 62x = $

$2x(x^2 - 12x + 31)$; $P(x)$ and $Q(x)$ intersect when $x = 0$ or $x = \dfrac{12 \pm \sqrt{144 - 4(1)(31)}}{2(1)} = $

$6 \pm \sqrt{5}$; $P(x) > Q(x)$ when $x < 0$ or $6 - \sqrt{5} < x < 6 + \sqrt{5}$ (or $3.8 < x < 8.2$)

37.

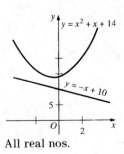

All real nos.

38.

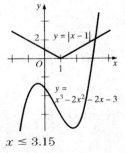

$x \leq 3.15$

39.

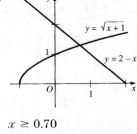

$x \geq 0.70$

40.

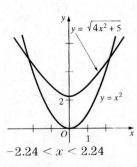

$-2.24 < x < 2.24$

41.

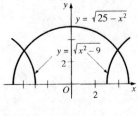

$-4.12 \leq x \leq -3$
or $3 \leq x \leq 4.12$

42.

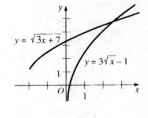

$0 \leq x < 2.62$

43. $|2x - 5| \geq 0$ for all real x; also, $x^2 + 4 \geq 0$ for all real x; thus, there is no value for which the expression is negative; no solution.

44. $|5x^3 - 10| \geq 0$ for all real x; $6x^2 + 13x - 5 = (3x - 1)(2x + 5)$;

$(3x - 1)(2x + 5) < 0$ when $-\dfrac{5}{2} < x < \dfrac{1}{3}$; $5x^3 - 10 = 0$ when $x = \sqrt[3]{2} > 1$, which is

outside the interval; thus, the solution is $-\dfrac{5}{2} < x < \dfrac{1}{3}$.

Class Exercises 3-3 · page 106

1. above   **2.** below   **3.** on   **4.** on   **5.** above   **6.** below

7. a. $x > 0$ and $y > 0$ **b.** $x < 0$ and $y > 0$; $x < 0$ and $y < 0$; $x > 0$ and $y < 0$

8. a. $x \geq 0$, $y \geq 0$, $y \leq -x + 2$ **9.** $x \geq 0$, $0 \leq y \leq 3$, $y \leq 4 - 2x$

10. $y < x + 2$, $y \geq x^2$

Written Exercises 3-3 · pages 106–108

1.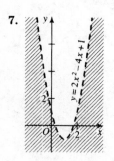

2.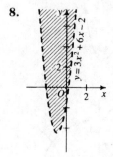

3.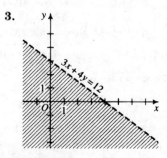

4.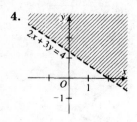

5.

6.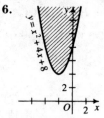

7.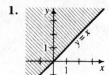

8.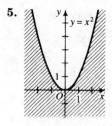

9.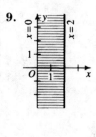

10.

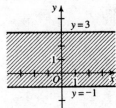

11.

12.

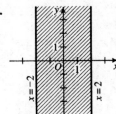

13.

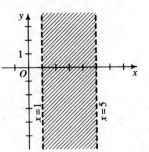

14.

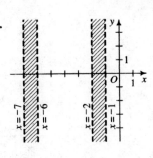

15.

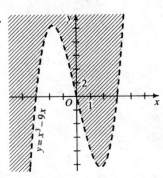

16.

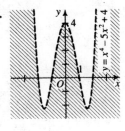

17.

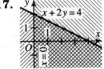

18.

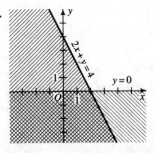

19.

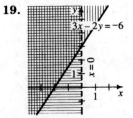

20.

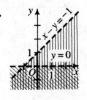

21.

22.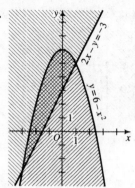

23.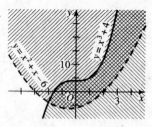

24.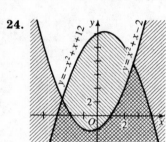

25.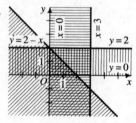

26.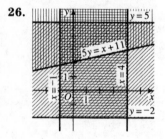

27.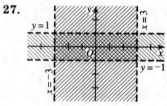

28.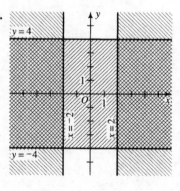

29.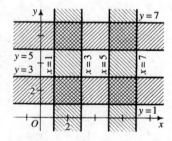

30.

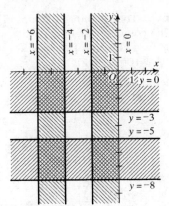

31. $1.5 \le x \le 5.5$; slanted boundary lines have eqs. $y = 10x + 30$ and $y = 10x + 50$, so $10x + 30 \le y \le 10x + 50$.

32. a. $1.9 \le x \le 2.1$
$2.03 \le y \le 2.13$

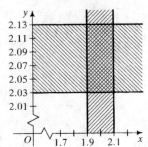

b. Also, $y > x$

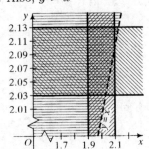

33.

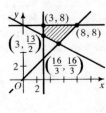

34.

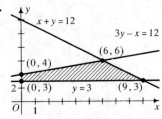

35.

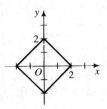

36.

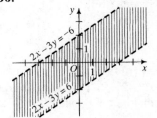

Written Exercises 3-4 · pages 112–114

1. a. Profit $= 80x + 55y$; evaluate $P(x, y)$ at each corner pt.; (1) at $(0,0)$: $80 \cdot 0 + 55 \cdot 0 = 0$; (2) at $(0,8)$: $80 \cdot 0 + 55 \cdot 8 = 440$; (3) at $(2,7)$: $80 \cdot 2 + 55 \cdot 7 = 545$; (4) at $(5,4)$: $80 \cdot 5 + 55 \cdot 4 = 620$; (5) at $(6,0)$: $80 \cdot 6 + 55 \cdot 0 = 480$; the maximum profit, \$620, occurs with 5 consoles and 4 portables.

b. If $P = 90x + 20y$, at $(0,0)$, $P = 0$; at $(0,8)$, $P = 160$; at $(2,7)$, $P = 320$; at $(5,4)$, $P = 530$; and at $(6,0)$, $P = 540$; the maximum profit, \$540, occurs with 6 consoles and 0 portables.

2. a. The number of pills of either brand cannot be negative. The combined amount of vitamin A must be greater than or equal to 4 mg. The combined amount of vitamin B must be greater than or equal to 11 mg. The combined amount of vitamin C must be greater than or equal to 100 mg (from $25x + 50y \geq 100$, $x + 2y \geq 4$).

b. Cost = $C = 3x + 9y$ cents; evaluate C at each corner pt.; (1) at $(0,4)$: $3 \cdot 0 + 9 \cdot 4 = 36$; (2) at $(1,2)$: $3 \cdot 1 + 9 \cdot 2 = 21$; (3) at $(3,0.5)$: $3 \cdot 3 + 9(0.5) = 13.5$; (4) at $(4,0)$: $3 \cdot 4 + 9 \cdot 0 = 12$; the minimum cost, 12¢, occurs with 4 Brand X and 0 Brand Y pills.

c. If $C = 10x + 4y$: at $(0,4)$, $C = 16$; at $(1,2)$, $C = 18$; at $(3,0.5)$, $C = 32$; at $(4,0)$, $C = 40$; the minimum cost, 16¢, occurs with 0 Brand X pills and 4 Brand Y pills.

3. Let x = number of consoles and y = number of portables. We have $x \geq 0$, $y \geq 0$, $x + 3y \leq 18$, $x + y \leq 8$, and $3x + y \leq 18$, as shown. $C = 70x + 40y$: at $(0,0)$, $C = 0$; at $(0,6)$, $C = 240$; at $(3,5)$, $C = 410$; at $(5,3)$, $C = 470$; at $(6,0)$, $C = 420$. The maximum profit, \$470, occurs with 5 consoles and 3 portables produced per day.

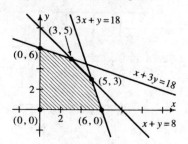

4. Let x = number of Brand X pills and y = number of Brand Y pills. We have $x \geq 0$, $y \geq 0$, $4x + 2y \geq 10$ or $2x + y \geq 5$, $6x + 6y \geq 24$ or $x + y \geq 4$, and $25x + 50y \geq 125$ or $x + 2y \geq 5$, as shown. $C = 12x + 15y$ cents: at $(0,5)$, $C = 75$; at $(1,3)$, $C = 57$; at $(3,1)$, $C = 51$; at $(5,0)$, $C = 60$. The minimum cost, 51¢, occurs with 3 Brand X pills and 1 Brand Y pill.

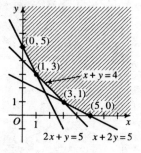

5. At $(0,2)$, $P = x + 3y = 0 + 3(2) = 6$; at $(2,1)$, $P = 2 + 3(1) = 5$; at $(3,0)$, $P = 3 + 3(0) = 3$. The maximum, 6, occurs at $(0,2)$.

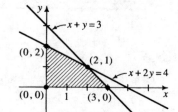

6. At $(0,3)$, $C = 3x + 4y = 3(0) + 4(3) = 12$; at $(4,1)$, $C = 3(4) + 4(1) = 16$; at $(7,0)$, $C = 3(7) + 4(0) = 21$. The minimum, 12, occurs at $(0,3)$.

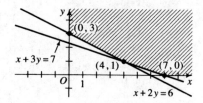

7.

Corner pt.	$5x + 6y$	$5x + 5y$
$(0, 12)$	72	60
$(2, 8)$	58	50
$(7, 3)$	53	50
$(16, 0)$	80	80

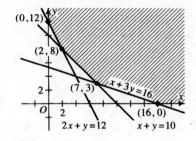

a. The minimum, 53, occurs at $(7, 3)$.

b. The minimum, 50, occurs at $(2, 8)$ and at $(7, 3)$.

8.

Corner pt.	$3x + 3y$	$3x + 4y$
$(0, 0)$	0	0
$(0, 9)$	27	36
$(4, 6)$	30	36
$(6, 3)$	27	30
$(6, 0)$	18	18

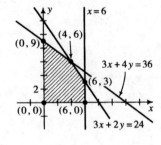

a. The maximum, 30, occurs at $(4, 6)$.

b. The maximum, 36, occurs at $(0, 9)$ and at $(4, 6)$.

9.

	Alfalfa	Corn	Hours Available
Planting time (h/ton)	1	1	500
Harvest time (h/ton)	2	3	1200
Profit (dollars/ton)	250	350	

Let x represent the number of tons of alfalfa and y the number of tons of corn to be grown.

$x \geq 0$

$y \geq 0$

$x + y \leq 500$

$2x + 3y \leq 1200$ Profit $= P = 250x + 350y$

At $(0, 400)$, $P = 140,000$.

At $(300, 200)$, $P = 145,000$.

At $(500, 0)$, $P = 125,000$.

The maximum profit, \$145,000, occurs if 300 tons of alfalfa and 200 tons of corn are grown.

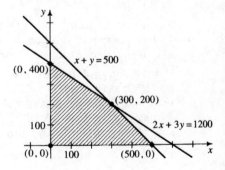

10.

	Economy	Deluxe	No. Available
No. of woofers	0	1	20
No. of tweeters	1	2	45
No. of mid-ranges	1	1	35
Income per model	$50	$200	

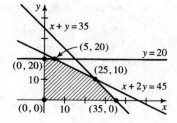

Let x represent the number of economy models and y the number of deluxe models produced.

$x \geq 0$

$y \geq 0$

$y \leq 20$

$x + 2y \leq 45$

$x + y \leq 35$

Income $= I = 50x + 200y$

At $(0, 20)$, $I = 4000$.

At $(5, 20)$, $I = 4250$.

At $(25, 10)$, $I = 3250$.

At $(35, 0)$, $I = 1750$.

The maximum income, $4250, occurs when 5 economy models and 20 deluxe models are manufactured.

11. a. Let $x =$ number of pairs of racing skis and $y =$ number of pairs of free-style skis.

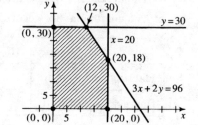

$0 \leq x \leq 20$

$0 \leq y \leq 30$

$3x + 2y \leq 96$

Profit $= P = 30x + 40y$

At $(0, 30)$, $P = 1200$.

At $(12, 30)$, $P = 1560$.

At $(20, 18)$, $P = 1320$.

At $(20, 0)$, $P = 600$.

The maximum profit, $1560, occurs when 12 pairs of racing skis and 30 pairs of free-style skis are produced each day.

b. Use the digram for part (a) with $P = 30x + 20y$; at $(0, 30)$, $P = 600$; at $(12, 30)$, $P = 960$; at $(20, 18)$, $P = 960$; at $(20, 0)$, $P = 600$. The maximum profit, $960, occurs at $(12, 30)$ and at $(20, 18)$. It will also occur at any other point on the line joining $(12, 30)$ and $(20, 18)$. Only integer solutions are possible so the additional solutions occur at $(14, 27)$, $(16, 24)$, and $(18, 21)$.

12.

	Board	Mechanical	Hours Available
Hours to manufacture	0.5	1	40
Hours to assemble	0.5	0.5	32
Hours to inspect & package	0.25	0.5	18
Profit per model	$10	$15	

Let x and y respectively represent the number of board games and the number of mechanical games produced each week.

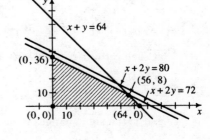

$x \geq 0$
$y \geq 0$
$0.5x + y \leq 40$, or $x + 2y \leq 80$
$0.5x + 0.5y \leq 32$, or $x + y \leq 64$
$0.25x + 0.5y \leq 18$, or $x + 2y \leq 72$
Profit $= P = 10x + 15y$

At $(64, 0)$, $P = 640$.
At $(56, 8)$, $P = 680$.
At $(0, 36)$, $P = 540$.

The maximum profit, $680, occurs when 56 board games and 8 mechanical games are produced each week.

13. The corner pts. are $(0, 13)$, $(5.5, 7.5)$, and $(8, 0)$. Since x and y must be integers, check the nearby solution points with integral coordinates, which are $(5, 8)$ and $(6, 6)$.

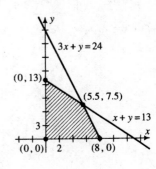

points	$x + y$	$4x + y$	$2x + y$
$(0, 13)$	13	13	13
$(5, 8)$	13	28	18
$(6, 6)$	12	30	18
$(8, 0)$	8	32	16

maximum value: **a.** 13   **b.** 32   **c.** 18

Chapter Test · page 115

1. a. $24 - 12x < 9 - 7x$; $-5x < -15$; $x > 3$

b. $4x - 12 \leq 3x - 6$; $x \leq 6$

2. a. $-5 < x - 3 < 5$; $-2 < x < 8$

b. $x + 6 < -4$ or $x + 6 > 4$; $x < -10$ or $x > -2$

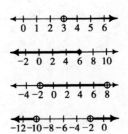

(Continued)

78　　　　　　　　　　　　　　　　　　

c. $3x + 4 = 4$ or $3x + 4 = -4$; $3x = 0$

or $3x = -8$; $x = 0$ or $x = -\dfrac{8}{3}$

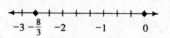

d. $-2 \le 7 - 5x$ and $7 - 5x \le 2$; $-9 \le -5x$ and

$-5x \le -5$; $\dfrac{9}{5} \ge x$ and $x \ge 1$; $1 \le x \le \dfrac{9}{5}$

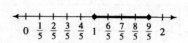

3. a. The zeros of the linear factors are -4, $\dfrac{1}{2}$, and 3; the sign changes at -4 and $\dfrac{1}{2}$, but

not at 3; $f(0) = (-1)(4)(9) < 0$; thus $f(x) \ge 0$ when $x \le -4$ or $x \ge \dfrac{1}{2}$.

b. $(4x - 3)(x + 2) < 0$; the zeros of the linear factors are -2 and $\dfrac{3}{4}$; the sign changes

at each zero; $f(0) = -6 < 0$; thus, $f(x) < 0$ when $-2 < x < \dfrac{3}{4}$.

c. $2x^3 - x^2 - x > 0$; $x(2x + 1)(x - 1) > 0$; the zeros of the linear factors are $-\dfrac{1}{2}$, 0,

and 1; the sign changes at each zero; $f(2) = 10 > 0$; thus, $f(x) > 0$ when

$-\dfrac{1}{2} < x < 0$ or $x > 1$.

d. The zeros of the linear factors are -1, 3, and 6; the sign changes at -1 and 6, but

not at 3; $f(0) = \dfrac{-6(1)}{9} < 0$; thus, $f(x) \le 0$ when $-1 \le x < 3$ or $3 < x \le 6$.

4. The points below the line $y = 2x - 5$

5. a.

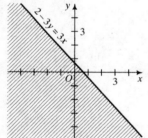

b.

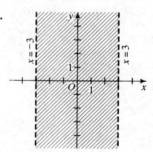

6. a.

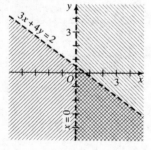

b.

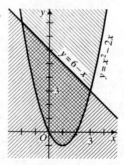

7. For example, $1 \le x \le 3$, $y \ge 2$, $y \le -\dfrac{1}{2}x + \dfrac{9}{2}$

8.

	Trucks	Autos	No. of person-days avail. per week
Shop 1	6	3	150
Shop 2	4	4	120
Profit	$500	$350	

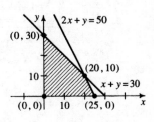

Let x = number of trucks and y = number of automobiles. We have $x \geq 0$, $y \geq 0$, $6x + 3y \leq 150$ or $2x + y \leq 50$, and $4x + 4y \leq 120$ or $x + y \leq 30$. Evaluate $P = 500x + 350y$ at each vertex: at $(0, 30)$, $P = 10{,}500$; at $(20, 10)$, $P = 13{,}500$; at $(25, 0)$, $P = 12{,}500$; at $(0, 0)$, $P = 0$; the maximum profit, \$13,500, occurs with 20 trucks and 10 automobiles per week.

Cumulative Review · page 116

1. $AB = \sqrt{(-7 + 4)^2 + (-2 - 5)^2} = \sqrt{58}$; midpt. $= \left(\dfrac{-4 - 7}{2}, \dfrac{5 - 2}{2}\right) = \left(-\dfrac{11}{2}, \dfrac{3}{2}\right)$

2. a. Slope is $\dfrac{-4 + 2}{2 - 4} = 1$; $y + 2 = 1(x - 4)$; $y = x - 6$ or $x - y = 6$

   b. $2x - 3y = k$; $2(0) - 3(5) = k$; $k = -15$; $2x - 3y = -15$

3. a. $P(d) = 12.5d - 1027$

   b. $500 = 12.5d - 1027$; $12.5d = 1527$; $d = 122.16$; 123 donations

4. a. $i + 5i - 8i = -2i$ **b.** $(2\sqrt{3}\,i)(6i) = -12\sqrt{3}$

   c. $\dfrac{(3 - i)}{2(3 + i)} \cdot \dfrac{(3 - i)}{(3 - i)} = \dfrac{9 - 6i + i^2}{2(9 - i^2)} = \dfrac{8 - 6i}{20} = \dfrac{2}{5} - \dfrac{3}{10}i$

5. a. $x^2 - 6x + 9 = -18 + 9$; $(x - 3)^2 = -9$; $x - 3 = \pm 3i$; $x = 3 \pm 3i$

   b. $x = \dfrac{12 \pm \sqrt{144 - 4(4)(7)}}{2(4)} = \dfrac{12 \pm \sqrt{32}}{8} = \dfrac{3 \pm \sqrt{2}}{2}$

   c. $2x + 1 = (x - 3) + 4\sqrt{x - 3} + 4$; $x = 4\sqrt{x - 3}$; $x^2 = 16(x - 3)$; $x^2 - 16x + 48 = 0$; $(x - 12)(x - 4) = 0$; $x = 12$ or $x = 4$

6. $y = 2x + 4$ and $y = x^2 - 4$; $2x + 4 = x^2 - 4$; $x^2 - 2x - 8 = 0$; $(x - 4)(x + 2) = 0$; $x = 4$ and $y = 2(4) + 4 = 12$; or $x = -2$ and $y = 2(-2) + 4 = 0$; $(4, 12)$ and $(-2, 0)$

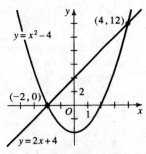

7. The x-intercepts are 3 and 7, so the x-coordinate of the vertex is 5. $f(5) = 8$, so the vertex is $(5, 8)$; $f(x) = a(x - 5)^2 + 8$; $0 = a(7 - 5)^2 + 8 = 4a + 8$; $a = -2$; $f(x) = -2(x - 5)^2 + 8$ or $f(x) = -2x^2 + 20x - 42$

8. $f(0) = a(0) + b(0) + c = c = -6$; $f(2) = a(4) + b(2) - 6 = -3$, so $4a + 2b = 3$; $f(4) = a(16) + b(4) - 6 = 2$, so $4a + b = 2$. Subtracting gives $b = 1$, $4a + 1 = 2$, so $a = \dfrac{1}{4}$; $f(x) = \dfrac{1}{4}x^2 + x - 6$.

9. a.
$$\underline{2}\;\begin{array}{cccc} 2 & -1 & 0 & 3 \\ & 4 & 6 & 12 \\ \hline 2 & 3 & 6 & \boxed{15} = P(2) \end{array}$$

b.
$$\underline{-1}\;\begin{array}{cccc} 2 & -1 & 0 & 3 \\ & -2 & 3 & -3 \\ \hline 2 & -3 & 3 & \boxed{0} = P(-1) \end{array}$$

c.
$$-\dfrac{3}{2}\;\begin{array}{cccc} 2 & -1 & 0 & 3 \\ & -3 & 6 & -9 \\ \hline 2 & -4 & 6 & \boxed{-6} = P\!\left(-\dfrac{3}{2}\right) \end{array}$$

d.
$$\underline{i}\;\begin{array}{cccc} 2 & -1 & 0 & 3 \\ & 2i & -2-i & -2i+1 \\ \hline 2 & 2i-1 & -2-i & \boxed{4-2i} = P(i) \end{array}$$

10. a.
$$\underline{-1}\;\begin{array}{cccc} 1 & 2 & -5 & -6 \\ & -1 & -1 & 6 \\ \hline 1 & 1 & -6 & \boxed{0} \end{array}$$

$P(x) = (x+1)(x^2+x-6) = (x+1)(x+3)(x-2)$;
the other roots are $x = -3$ and $x = 2$.

b.
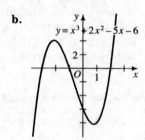

11. $y = a(x+2)^2(x-3);\; -2 = a(0+2)^2(0-3) = -12a;\; a = \dfrac{1}{6};\; y = \dfrac{1}{6}(x+2)^2(x-3)$

12. The length of the rectangle is x and the width is $\dfrac{100-2x}{2} = 50 - x$; area $= x(50-x)$;

the graph is a parabola with x-intercepts 0 and 50; by symmetry, the vertex and
maximum value occur at $x = 25$; maximum area is $25(50-25) = 625\text{ cm}^2$.

13. Use the method of Example 1 or Example 2 in Section 2-5. $P(-2) = -3, P(-1) = 1,$
$P(0) = -1, P(1) = -3,$ and $P(2) = 1$. There are real roots between -2 and -1, -1 and
0, and 1 and 2; $x \approx -1.5, -0.3, 1.9$.

14. $x^3 - 3x^2 + 4x - 12 = x^2(x-3) + 4(x-3) = (x^2+4)(x-3);\; x = \pm 2i$ or $x = 3$

15. Because the equation has integral coefficients, $2 - 2i$ is also a root;
$(2+2i) + (2-2i) = 4$ and $(2+2i)(2-2i) = 4+4 = 8$, so $x^2 - 4x + 8 = 0$ is a
quadratic equation with roots $2 \pm 2i$. Thus, $(3x+1)(x^2-4x+8) = 0$;
$3x^3 - 11x^2 + 20x + 8 = 0$.

16. a. $x - 5 < -12$ or $x - 5 > 12;\; x < -7$ or
$x > 17$

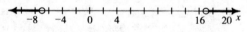

b. The zeros of the linear factors are -3,
1, and 2; the sign changes at -3 and 1,
but not at 2; $f(0) = 4(3)(-1) < 0$; thus,
$f(x) \geq 0$ when $x \leq -3$ or $x \geq 1$.

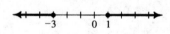

17.

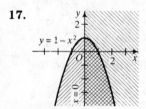

CHAPTER 4 · Functions

Class Exercises 4-1 · page 122

1. a. domain: $\{g \mid g \geq 0\}$; range: $\{C \mid C \geq 0\}$ **b.**

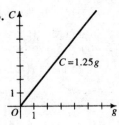

2. a. $x - 2 \neq 0$; $\{x \mid x \neq 2\}$

   b. $\{t \mid t \geq 0\}$

   c. $s - 4 \geq 0$; $\{s \mid s \geq 4\}$

3. a. $f(2) = 3(2)^2 - 12 = 12 - 12 = 0$

   b. $3(x^2 - 4) = 3(x + 2)(x - 2) = 0$; -2 is another zero.

   c. Because $x^2 \geq 0$, $3x^2 \geq 0$ and $3x^2 - 12 \geq -12$; $f(x) \geq -12$

4. domain: {all real numbers}; range: $\{y \mid y \geq -2\}$; zeros: $-3, 0, 4$

5. $x^2y = 4x - y$ passes the vert.-line test; $y^2 = 4x^2$ fails the vert.-line test.

6. Solving $y^2 - x = 1$ for y gives $y = \pm\sqrt{x + 1}$, so there are two values of y for each $x > -1$.

7. Yes

8. Yes

9. No; for every positive value of x, there are two values of y.

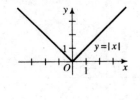

10. No; the graph of $y^2 = x^2$ is a pair of lines, $y = x$ and $y = -x$; for each nonzero value of x, there are two values of y.

11. No; fails vert.-line test.   **12.** Yes; passes vert.-line test.   **13.** No; fails vert.-line test.

Written Exercises 4-1 · pages 122–124

1. Yes; domain: $\{x \mid -2 \leq x \leq 5\}$; range: $\{y \mid -1 \leq y \leq 2\}$ **2.** No; fails vert.-line test.

3. Yes; domain: $\{0, 2, 4, 6\}$; range: $\{0, 1, 2, 3\}$

4. Yes; domain: $\{u \mid -2 \leq u \leq 6\}$; range: $\{v \mid -1 \leq v < 3\}$ **5.** No; fails vert.-line test.

6. Yes; domain: {all real nos.}; range: $\{s \mid -2 \leq s \leq 2\}$

7. Solving for y gives $y = \pm\sqrt{1 - x^2}$, so there are two values of y for each x between -1 and 1.

8. Solving for y gives $y = \sqrt[3]{1 - x^3}$, so there is only one value of y for each value of x.

9. a. $\{x \mid x \neq 0\}$ **b.** $x - 9 \neq 0$; $\{x \mid x \neq 9\}$ **c.** $x^2 - 4 \neq 0$; $x^2 \neq 4$; $\{x \mid x \neq \pm 2\}$

10. a. $t + 3 \neq 0$; $\{t \mid t \neq -3\}$ **b.** $(t + 3)(t + 2) \neq 0$; $\{t \mid t \neq -3, -2\}$

   c. $t(t + 3)(t - 3) \neq 0$; $\{t \mid t \neq 0, \pm 3\}$

11. a. domain: {all real nos.}; range: $\{f(x) \mid f(x) \geq 0\}$; zero: 0

   b. domain: {all real nos.}; range: $\{g(x) \mid g(x) \geq 0\}$; zero: 2

   c. domain: {all real nos.}; range: $\{h(x) \mid h(x) \geq -2\}$, since $|x| \geq 0$ for all x; $h(x) = 0$ when $|x| = 2$; zeros: ± 2

12. a. domain: $\{t \,|\, t \geq 0\}$; range: $\{f(t) \,|\, f(t) \geq 0\}$; zero: 0

b. $9 - t \geq 0$; domain: $\{t \,|\, t \leq 9\}$; range: $\{g(t) \,|\, g(t) \geq 0\}$; zero: 9

c. $(3 + t)(3 - t) \geq 0$; domain: $\{t \,|\, -3 \leq t \leq 3\}$; $0 \leq 9 - t^2 \leq 9$, so $0 \leq \sqrt{9 - t^2} \leq 3$; range: $\{h(t) \,|\, 0 \leq h(t) \leq 3\}$; zeros: ± 3

13. $f(x) = (x - 2)(x - 4)$; range: $\{f(x) \,|\, f(x) \geq -1\}$; zeros: 2, 4

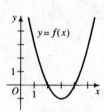

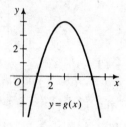

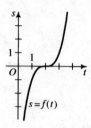

Ex. 13   **Ex. 14**   **Ex. 15**

14. $g(x) = 4 - (x - 3)^2$; range: $\{g(x) \,|\, g(x) \leq 4\}$; zeros: 1, 5

15. $f(t) = (t - 2)^3$; range: {all real nos.}; zero: 2

16. $g(t) = (t + 4)(t + 1)(t - 1)$; range: {all real nos.}; zeros: -4, ± 1

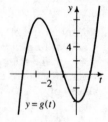

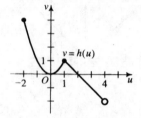

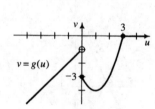

Ex. 16   **Ex. 17**   **Ex. 18**

17. $h(u) = u^2$ if $-2 \leq u < 1$, $h(u) = 2 - u$ if $1 \leq u < 4$; range: $\{h(u) \,|\, -2 < h(u) \leq 4\}$; zeros: 0, 2

18. $g(u) = u - 1$ if $u < 0$, $g(u) = u^2 - 2u - 3$ if $0 \leq u \leq 3$, $g(u) = 0$ if $u > 3$; range: $\{g(u) \,|\, g(u) \leq 0\}$; zeros: all $u \geq 3$

19. a. The volume of a cylinder is given by $V = \pi r^2 h$, so $V(C) = \pi(3)^2(4) = 36\pi$.

b. domain: {all solids}; range: {all nonnegative real numbers}

20. a. S is the sum of the measures of the interior angles of a polygon with n sides.

b. Yes; domain: {all integers greater than 2}; range: {positive multiples of $180°$}

21. a.

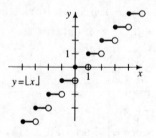

b. domain: {all real nos.}
range: {all integers}

22. a.

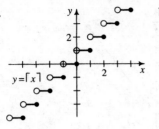

$y = \lceil x \rceil$

b.

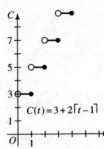

$C(t) = 3 + 2\lceil t - 1 \rceil$

23. Two functions are equal if and only if they have the same domain and the function values for any value in the domain are equal. Thus, $f(x) = g(x)$ because f and g produce the same functional values for all real x. However, $f(x) \neq h(x)$ because the domain of $f(x)$ is {all real nos.} and the domain of $h(x)$ is {nonnegative real nos.}.

24. An equation of the form $f(x, y) = 0$ is an *implicit function*. Such a function implicitly defines y as a function of x.

a. $y = \pm\sqrt{x}$ **b.** $y = \pm\sqrt{1 - x^2}$ **c.** $y = \pm x$

25. a. $f(a + b) = (a + b)^2 = a^2 + 2ab + b^2; f(a) + f(b) = a^2 + b^2; f(a + b) \neq f(a) + f(b)$

b. $f(a + b) = \dfrac{1}{a + b}; f(a) + f(b) = \dfrac{1}{a} + \dfrac{1}{b} = \dfrac{b + a}{ab}; f(a + b) \neq f(a) + f(b)$

c. $f(a + b) = 4(a + b) + 1 = 4a + 4b + 1;$
$f(a) + f(b) = (4a + 1) + (4b + 1) = 4a + 4b + 2; f(a + b) \neq f(a) + f(b)$

d. $f(a + b) = 4(a + b) = 4a + 4b = f(a) + f(b)$

26. a. $f(ab) = (ab)^2 = a^2 b^2 = f(a) \cdot f(b)$

b. $f(ab) = \dfrac{1}{ab} = \dfrac{1}{a} \cdot \dfrac{1}{b} = f(a) \cdot f(b)$

c. $f(ab) = 4ab + 1; f(a) \cdot f(b) = (4a + 1)(4b + 1) = 16ab + 4a + 4b + 1;$
$f(ab) \neq f(a) \cdot f(b)$

d. $f(ab) = 4ab; f(a) \cdot f(b) = 4a \cdot 4b = 16ab; f(ab) \neq f(a) \cdot f(b)$

27. Let $a = 0$. Then $f(0 + b) = f(0) + f(b); f(b) = f(0) + f(b); 0 = f(0)$.

28. Let $a = 1$ and $f(b) \neq 0$. Then $f(1 \cdot b) = f(1) \cdot f(b); f(b) = f(1) \cdot f(b);$
$f(1) = f(b) \div f(b) = 1$.

Class Exercises 4-2 · pages 127–128

1. a. $f(0) + g(0) = 5 + 0 = 5$
 b. $f(3) + g(3) = 4 + 4 = 8$

2. a. $f(x) - g(x) = 0$ when $f(x) = g(x); x = \pm 3$
 b. $f(x) - g(x) > 0$ when $f(x) > g(x); -3 < x < 3$

3. a.

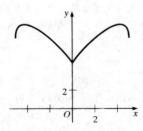

b.

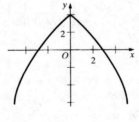

4. a. 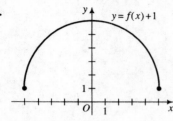 **b.**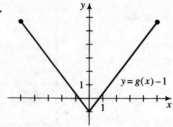

5. $f(x) + g(x) = x^2 + x + x + 1 = x^2 + 2x + 1$

6. $f(x) - g(x) = x^2 + x - (x + 1) = x^2 - 1$

7. $f(x) \cdot g(x) = (x^2 + x)(x + 1) = x^3 + 2x^2 + x$

8. $\dfrac{f(x)}{g(x)} = \dfrac{x^2 + x}{x + 1} = \dfrac{x(x + 1)}{x + 1} = x$ for $x \neq -1$

9. a. $f(g(2)) = f(3) = 12$ **b.** $f(g(x)) = f(x + 1) = (x + 1)^2 + (x + 1) = x^2 + 3x + 2$

10. a. $g(f(2)) = g(6) = 7$ **b.** $g(f(x)) = g(x^2 + x) = x^2 + x + 1$

11. a. 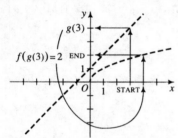 **b.**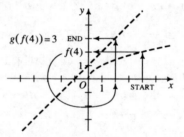

12. a. $F(M(X)); M(F(X))$ **13.** $B \circ F$ represents the birthday of the father of someone.

Written Exercises 4-2 · pages 128-131

1. 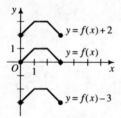 **2.**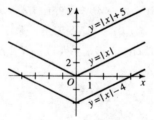

3. a. $f(1) - g(1) = 2 - 1 = 1$

  b. $f(x) - g(x) > 0$ when $f(x) > g(x)$; $0 < x < 4$; $f(x) - g(x) < 0$ when $f(x) < g(x)$; $-1 \leq x < 0$ or $4 < x \leq 6$; $f(x) - g(x) = 0$ when $f(x) = g(x)$; $x = 0$ or $x = 4$

  c. Maximum value occurs at $x = 3$; $f(3) - g(3) = 2 - 0 = 2$.

4. a. $f(1) - g(1) = 0 - 2 = -2$

  b. $f(x) - g(x) > 0$ when $f(x) > g(x)$; $-1 \leq x < 0$ or $2 < x < 5$; $f(x) - g(x) < 0$ when $f(x) < g(x)$; $0 < x < 2$ or $5 < x \leq 6$; $f(x) - g(x) = 0$ when $f(x) = g(x)$; $x = 0, 2, 5$

  c. Maximum value occurs when $x = -1$ or $x = 3$; $f(-1) - g(-1) = f(3) - g(3) = 2 - 0 = 2$.

5. $f(x) + g(x) = x^3 - 1 + x - 1 = x^3 + x - 2$

6. $f(x) - g(x) = x^3 - 1 - (x - 1) = x^3 - x$

7. $f(x) \cdot g(x) = (x^3 - 1)(x - 1) = x^4 - x^3 - x + 1$

8. $\dfrac{f(x)}{g(x)} = \dfrac{x^3 - 1}{x - 1} = \dfrac{(x - 1)(x^2 + x + 1)}{x - 1} = x^2 + x + 1$ for $x \ne 1$

9. a. $f(g(2)) = f(1) = 0$ **b.** $f(g(x)) = f(x - 1) = (x - 1)^3 - 1 = x^3 - 3x^2 + 3x - 2$

10. a. $g(f(2)) = g(7) = 6$ **b.** $g(f(x)) = g(x^3 - 1) = x^3 - 1 - 1 = x^3 - 2$

11.
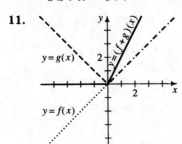
$y = g(x)$
$y = f(x)$

12.
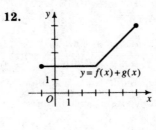
$y = f(x) + g(x)$

13. Graph $y = (f - g)(x)$ and find the x-intercepts.

14. $x^3 = x + 1$ when $x \approx 1.32$
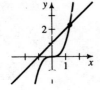

15. $\sqrt{x + 1} = 2x$ when $x \approx 0.64$

16. $\sqrt{1 - x^2} = |x|$ when $x \approx \pm 0.71$
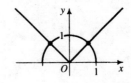

17. a. $f(g(x)) = f\left(\dfrac{x + 3}{2}\right) = 2\left(\dfrac{x + 3}{2}\right) - 3 = x + 3 - 3 = x;$

$g(f(x)) = g(2x - 3) = \dfrac{2x - 3 + 3}{2} = \dfrac{2x}{2} = x; f(g(x)) = g(f(x)) = x$ for all x

b. $f(h(x)) = f(3x + 2) = 2(3x + 2) - 3 = 6x + 1;$
$h(f(x)) = h(2x - 3) = 3(2x - 3) + 2 = 6x - 7;$ for all x, $6x + 1 \ne 6x - 7$, so
$f(h(x)) \ne h(f(x))$ for any x.

18. a. $f(g(x)) = f(\sqrt[3]{x}) = (\sqrt[3]{x})^3 = x; g(f(x)) = g(x^3) = \sqrt[3]{x^3} = x; f(g(x)) = g(f(x)) = x$
for all x

b. $f(h(x)) = f(3x) = (3x)^3 = 27x^3; h(f(x)) = h(x^3) = 3x^3; f(h(x)) = h(f(x))$ when
$27x^3 = 3x^3;$ that is, when $24x^3 = 0$, or $x = 0$

19. a. $f(g(h(6))) = f(g(2)) = f(9) = 3$

b. $f(g(h(x))) = f\left(g\left(\dfrac{x}{3}\right)\right) = f\left(6\left(\dfrac{x}{3}\right) - 3\right) = f(2x - 3) = \sqrt{2x - 3}$

20. a. $h(g(f(4))) = h(g(2)) = h(9) = 3$

b. $h(g(f(x))) = h(g(\sqrt{x})) = h(6\sqrt{x} - 3) = \dfrac{6\sqrt{x} - 3}{3} = 2\sqrt{x} - 1$

21. a. $h\left(f\left(g\left(\dfrac{1}{2}\right)\right)\right) = h(f(0)) = h(0) = 0$

    b. $h(f(g(x))) = h(f(6x - 3)) = h(\sqrt{6x - 3}) = \dfrac{\sqrt{6x - 3}}{3}$

22. a. $g(h(f(9))) = g(h(3)) = g(1) = 3$

    b. $g(h(f(x))) = g(h(\sqrt{x})) = g\left(\dfrac{\sqrt{x}}{3}\right) = 6\left(\dfrac{\sqrt{x}}{3}\right) - 3 = 2\sqrt{x} - 3$

23. $2(x - 4)^3 = 2(h(x))^3 = 2f(h(x)) = j(f(h(x)))$

24. $\sqrt{(x - 4)^3} = \sqrt{(h(x))^3} = \sqrt{f(h(x))} = g(f(h(x)))$

25. $(2x - 8)^3 = (2(x - 4))^3 = (2(h(x)))^3 = (j(h(x)))^3 = f(j(h(x)))$

26. $\sqrt{x^3 - 4} = \sqrt{f(x)} - 4 = \sqrt{h(f(x))} = g(h(f(x)))$

27. a. Speed is 1.5 m/s; consumption at 1.5 m/s is about 9 L/min.

    b. Consumption of 15 L/min occurs at speed 2 m/s; the elapsed time is about 30 s.

28. a. Speed of 55 mi/h gives fuel economy of 30 mi/gal, so the fuel cost is 4 cents/mi.

    b. Fuel cost of 4 cents/mi or less requires fuel economy of 30 mi/gal or higher; this occurs at speeds between 25 mi/h and 55 mi/h.

29. a. $C = 2\pi r$, so $r = \dfrac{C}{2\pi}$.   **b.** $A = \pi r^2$; $A = \pi \left(\dfrac{C}{2\pi}\right)^2 = \dfrac{C^2}{4\pi}$

30. a. $A = \dfrac{1}{2}\pi r^2$; $P = \dfrac{1}{2}(2\pi r) + 2r = \pi r + 2r = (\pi + 2)r$

    b. $r = \dfrac{P}{\pi + 2}$; $A = \dfrac{1}{2}\pi\left(\dfrac{P}{\pi + 2}\right)^2$

31. $s(F) = 331 + 0.6\left(\dfrac{5}{9}(F - 32)\right) = 331 + \dfrac{1}{3}(F - 32)$

32. a. $r^2 = \dfrac{A}{4\pi}$; $r = \dfrac{\sqrt{A}}{2\sqrt{\pi}}$; $r(A) = \dfrac{\sqrt{A}}{2\sqrt{\pi}}$

    b. $V(A) = \dfrac{4}{3}\pi\left(\dfrac{\sqrt{A}}{2\sqrt{\pi}}\right)^3 = \dfrac{4}{3}\pi\dfrac{A\sqrt{A}}{8\pi\sqrt{\pi}} = \dfrac{A\sqrt{A}}{6\sqrt{\pi}}$

33. $f(g(x)) = f(\sqrt{16 - x^2}) = 2\sqrt{16 - x^2}$ with domain $\{x \mid -4 \le x \le 4\}$;

    $g(f(x)) = g(2x) = \sqrt{16 - (2x)^2} = 2\sqrt{4 - x^2}$ with domain $\{x \mid -2 \le x \le 2\}$

34. $f(g(x)) = f\left(\dfrac{1}{x - 4}\right) = \sqrt{\dfrac{1}{x - 4}} = \dfrac{1}{\sqrt{x - 4}} = \dfrac{\sqrt{x - 4}}{x - 4}$ with domain $\{x \mid x > 4\}$;

    $g(f(x)) = g(\sqrt{x}) = \dfrac{1}{\sqrt{x} - 4} \cdot \dfrac{\sqrt{x} + 4}{\sqrt{x} + 4} = \dfrac{\sqrt{x} + 4}{x - 16}$ with domain $\{x \mid x \ge 0 \text{ and } x \ne 16\}$

35. $f(g(x)) = f(\sqrt{1 - x}) = (\sqrt{1 - x})^2 = 1 - x$ with domain $\{x \mid x \le 1\}$;

    $g(f(x)) = g(x^2) = \sqrt{1 - x^2}$ with domain $\{x \mid -1 \le x \le 1\}$

36. $f(g(x)) = f(\sqrt{16 - x^2}) = (\sqrt{16 - x^2})^2 = 16 - x^2$ with domain $\{x \mid -4 \le x \le 4\}$;

    $g(f(x)) = g(x^2) = \sqrt{16 - (x^2)^2} = \sqrt{16 - x^4}$ with domain $\{x \mid -2 \le x \le 2\}$

37. $g(g(1)) = g(2) = 2.5$; $g(g(g(1))) = g(2.5) = 2.75$; $g(g(g(g(1)))) = g(2.75) = 2.875$

38. $f(f(x)) = f(2x - 1) = 2(2x - 1) - 1 = 4x - 2 - 1 = 4x - 3$;

    $f(f(f(x))) = f(4x - 3) = 2(4x - 3) - 1 = 8x - 6 - 1 = 8x - 7$

39. a. $r = 1$ m/s and $t =$ number of seconds; $d = 1 + rt = 1 + 1 \cdot t = t + 1$;

$$E = \frac{I}{d^2} = \frac{130}{(t + 1)^2}$$

b. When $t = 0$, $E = 130$; $1.3 = \dfrac{130}{(t + 1)^2}$; $(t + 1)^2 = 100$; $t \geq 0$, so $t + 1 = 10$; $t = 9$;

after 9 s

Activities 4-3 · page 131

1. a.

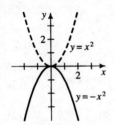

b.

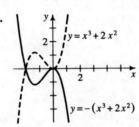

c. The graph of $y = -f(x)$ is the reflection of the graph of $y = f(x)$ in the x-axis.

2. a.

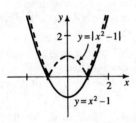

b.

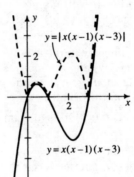

c. The graph of $y = |f(x)|$ is the same as the graph of $y = f(x)$ when $f(x) \geq 0$ and is the same as the graph of $y = -f(x)$ when $f(x) < 0$.

3. a.

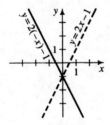

b.

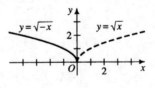

c. The graph of $y = f(-x)$ is the reflection of the graph of $y = f(x)$ in the y-axis.

4. a.

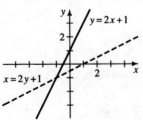

b.

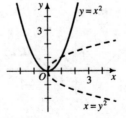

c. When the variables of an equation are interchanged, the graph of the resulting equation is the reflection of the graph of the original equation in the line $y = x$.

Class Exercises 4-3 · pages 135–136

1. a.   **b.**   **c.**

2. a. $x = 3$ **b.** $y = x$ and $y = -x$; $(0,0)$ **c.** $x = 0$ and $y = 0$; $(0,0)$

3. a. (i) yes; $x^4 + (-y)^4 = 1$ is equivalent to $x^4 + y^4 = 1$. (ii) yes; $(-x)^4 + y^4 = 1$ is
equivalent to $x^4 + y^4 = 1$. (iii) yes; $y^4 + x^4 = 1$ is equivalent to $x^4 + y^4 = 1$.
(iv) yes; $(-x)^4 + (-y)^4 = 1$ is equivalent to $x^4 + y^4 = 1$.

b. (i) no; $x(-y)^3 = 1$ is not equivalent to $xy^3 = 1$. (ii) no; $(-x)y^3 = 1$ is not equivalent
to $xy^3 = 1$. (iii) no; $yx^3 = 1$ is not equivalent to $xy^3 = 1$. (iv) yes; $(-x)(-y)^3 = 1$
is equivalent to $xy^3 = 1$.

c. (i) no; $x(x - y) = 1$ is not equivalent to $x(x + y) = 1$. (ii) no; $-x(-x + y) = 1$ is
not equivalent to $x(x + y) = 1$. (iii) no; $y(y + x) = 1$ is not equivalent to
$x(x + y) = 1$. (iv) yes; $-x(-x + (-y)) = (-1)(-1)(x)(x + y) = x(x + y) = 1$ is
equivalent to $x(x + y) = 1$.

4. The graph must also have symmetry in the origin. If (x,y) is on the graph, then $(-x,y)$
is on the graph (y-axis symmetry) and if $(-x,y)$ is on the graph, then $(-x, -y)$ is on the
graph (x-axis symmetry). Thus, if (x,y) is on the graph, then $(-x, -y)$ is on the graph
and the graph has symmetry in the origin.

5. No; if (x,y) is on the graph, then $(x, -y)$ is also on the graph. Therefore, the graph fails
the vertical-line test.

6. The graph of $y = \sqrt{|x|}$ is identical to the graph of $y = \sqrt{x}$ for $x \geq 0$ and is the
reflection of the graph of $y = \sqrt{x}$ in the y-axis for $x < 0$.

7. a. $x = -\dfrac{b}{2a} = -\dfrac{-8}{2(1)}$; $x = 4$ **b.** $x = -\dfrac{8}{2(-4)}$; $x = 1$ **c.** $x = -\dfrac{0}{2(1)}$; $x = 0$

8. a. $x = -\dfrac{b}{3a} = -\dfrac{-6}{3(1)} = 2$; $y = f(2) = 1$; $(2,1)$

b. $x = -\dfrac{6}{3(2)} = -1$; $y = g(-1) = -5$; $(-1, -5)$

c. $x = -\dfrac{0}{3(3)} = 0$; $y = h(0) = 7$; $(0,7)$

Written Exercises 4-3 · pages 136–138

1. a. 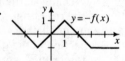   **b.** 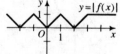   **c.**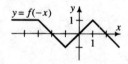

2. a. 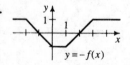   **b.** 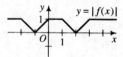   **c.**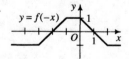

3. a.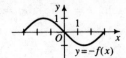

b.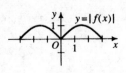

c.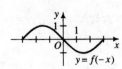

4. a.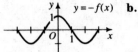

b.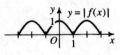

c.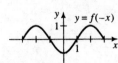

5.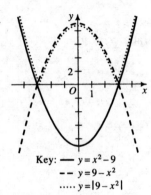

Key: —— $y = x^2 - 9$
 - - - $y = 9 - x^2$
 ..... $y = |9 - x^2|$

6.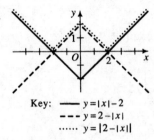

Key: —— $y = |x| - 2$
 - - - $y = 2 - |x|$
 ..... $y = |2 - |x||$

7. reflected graph:
$x = 3y - 4$, or $y = \dfrac{x + 4}{3}$

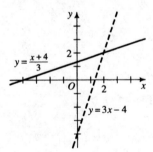

8. reflected graph: $x = \dfrac{1}{2}y + 1$,
 or $y = 2x - 2$

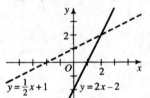

9. reflected graph: $x = y^2 - 2y$,

 or $y = \pm\sqrt{x + 1} + 1$

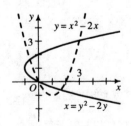

10. reflected graph: $x = y^2 + 3y$,

 or $y = -\dfrac{3}{2} \pm \sqrt{x + \dfrac{9}{4}}$

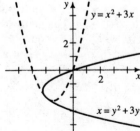

11. reflected graph: $x = y^3$, or $y = \sqrt[3]{x}$

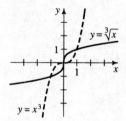
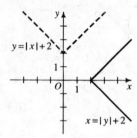

12. reflected graph: $x = \sqrt{y}$, or $y = x^2$ for $x \geq 0$

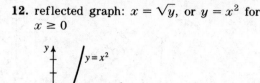
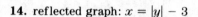

13. reflected graph: $x = |y| + 2$

14. reflected graph: $x = |y| - 3$

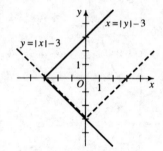

15. **a.** (i) no; $(-y)^2 - x(-y) = 2$ is not equiv. to $y^2 - xy = 2$.
(ii) no; $y^2 - (-x)y = 2$ is not equiv. to $y^2 - xy = 2$.
(iii) no; $x^2 - yx = 2$ is not equiv. to $y^2 - xy = 2$.
(iv) yes; $(-y)^2 - (-x)(-y) = 2$ is equiv. to $y^2 - xy = 2$.

b. (i) yes; $x^2 + (-y)^2 = 1$ is equiv. to $x^2 + y^2 = 1$.
(ii) yes; $(-x)^2 + y^2 = 1$ is equiv. to $x^2 + y^2 = 1$.
(iii) yes; $y^2 + x^2 = 1$ is equiv. to $x^2 + y^2 = 1$.
(iv) yes; $(-x)^2 + (-y)^2 = 1$ is equiv. to $x^2 + y^2 = 1$.

c. (i) no; $-y = x|x|$ is not equiv. to $y = x|x|$.
(ii) no; $y = -x|-x|$ is not equiv. to $y = x|x|$.
(iii) no; $x = y|y|$ is not equiv. to $y = x|x|$.
(iv) yes; $-y = -x|-x|$ is equiv. to $y = x|x|$.

16. **a.** (i) no; $x^2 + x(-y) = 4$ is not equiv. to $x^2 + xy = 4$.
(ii) no; $(-x)^2 + (-x)y = 4$ is not equiv. to $x^2 + xy = 4$.
(iii) no; $y^2 + yx = 4$ is not equiv. to $x^2 + xy = 4$.
(iv) yes; $(-x)^2 + (-x)(-y) = 4$ is equiv. to $x^2 + xy = 4$.

b. (i) yes; $|x| + |-y| = 1$ is equiv. to $|x| + |y| = 1$.
(ii) yes; $|-x| + |y| = 1$ is equiv. to $|x| + |y| = 1$.
(iii) yes; $|y| + |x| = 1$ is equiv. to $|x| + |y| = 1$.
(iv) yes; $|-x| + |-y| = 1$ is equiv. to $|x| + |y| = 1$.

c. (i) no; $-y = \dfrac{x}{|x|}$ is not equiv. to $y = \dfrac{x}{|x|}$. (ii) no; $y = \dfrac{-x}{|-x|}$ is not equiv. to $y = \dfrac{x}{|x|}$.
(iii) no; $x = \dfrac{y}{|y|}$ is not equiv. to $y = \dfrac{x}{|x|}$. (iv) yes; $-y = \dfrac{-x}{|-x|}$ is equiv. to $y = \dfrac{x}{|x|}$.

17. symmetries: x-axis, y-axis, origin, and line $y = x$

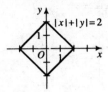

18. symmetries: x-axis, y-axis, origin, and line $y = x$

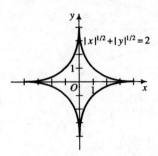

19. symmetries: y-axis only

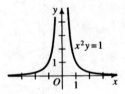

20. symmetries: x-axis, y-axis, origin, and line $y = x$

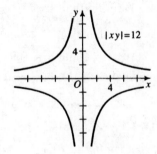

21.

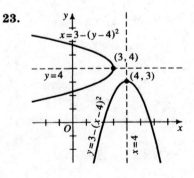

22.

23.

24.

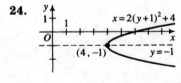

25.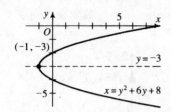

26.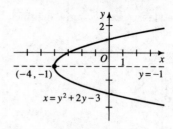

27. Since $(0,4)$ is a symmetry point, it is the midpoint of $(5,-3)$ and (a,b); $\dfrac{5+a}{2} = 0$ and $\dfrac{-3+b}{2} = 4$; $5 + a = 0$ and $-3 + b = 8$; $a = -5$ and $b = 11$; $(-5,11)$ is a local maximum.

28. a. The given cubic equation has a symmetry point at $x = \dfrac{-b}{3a} = \dfrac{-15}{3(-1)} = 5$; the point of symmetry is $(5, f(5))$, or $(5, 55)$.

   b. $(5, 55)$ is the midpoint of $(2, 1)$ and (a, b); $\dfrac{2+a}{2} = 5$ and $\dfrac{1+b}{2} = 55$; $2 + a = 10$ and $1 + b = 110$; $a = 8$ and $b = 109$; $(8, 109)$ is a local maximum.

29. a. See graph at right; $(0,0)$ is a local minimum.

   b. $y = 3x^2 - x^3$ has a point of symmetry at $x = \dfrac{-3}{3(-1)} = 1$; $y = 3(1) - 1 = 2$; $(1,2)$ is a pt. of symmetry and the midpoint of the segment joining $(0,0)$ and (a,b); $\dfrac{0+a}{2} = 1$ and $\dfrac{0+b}{2} = 2$; $a = 2$ and $b = 4$; $(2,4)$ is a local maximum.

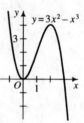

30. a. See graph at right; $(-3,0)$ is a local minimum.

   b. There is a point of symmetry at $x = \dfrac{-b}{3a} = \dfrac{6}{3(-1)} = -2$; $y = -(-2)^3 - 6(-2)^2 - 9(-2) = 2$; $(-2,2)$ is a point of symmetry and the midpoint of $(-3,0)$ and (a,b); $\dfrac{-3+a}{2} = -2$ and $\dfrac{0+b}{2} = 2$; $-3 + a = -4$ and $b = 4$; $(-1,4)$ is a local maximum.

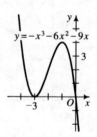

31. a. $f(-x) = (-x)^2 = x^2 = f(x)$; even   **b.** $f(-x) = (-x)^3 = -x^3 = -f(x)$; odd

   c. $f(-x) = (-x)^2 - (-x) = x^2 + x$; neither

   d. $f(-x) = (-x)^4 + 2(-x)^2 = x^4 + 2x^2 = f(x)$; even

   e. $f(-x) = (-x)^3 + 3(-x)^2 = -x^3 + 3x^2$; neither

   f. $f(-x) = (-x)^5 - 4(-x)^3 = -x^5 + 4x^3 = -(x^5 - 4x^3) = -f(x)$; odd

32. A polynomial function is even when only even powers of x occur and is odd when only odd powers of x occur.

33. a. If $f(-x) = f(x)$, then $(-x, y)$ is on the graph when (x, y) is; such a graph has symmetry in the y-axis.

   b. If $f(-x) = -f(x)$, then $(-x, -y)$ is on the graph when (x, y) is; such a graph has symmetry in the origin.

34. The graph in Ex. 3 has symmetry in the origin, so the function is odd. The graph in Ex. 4 has symmetry in the y-axis, so the function is even.

35. If $f(x)$ and $g(x)$ are odd functions and $h(x) = f(x) \cdot g(x)$, then
$h(-x) = f(-x) \cdot g(-x) = -f(x) \cdot [-g(x)] = f(x) \cdot g(x) = h(x)$; since $h(-x) = h(x)$,
$h(x)$ is an even function.

36. If $f(x)$ is even, $g(x)$ is odd, and $h(x) = f(x) \cdot g(x)$, then $h(-x) = f(-x) \cdot g(-x) =$
$f(x)[-g(x)] = -[f(x) \cdot g(x)] = -h(x)$; since $h(-x) = -h(x)$, $h(x)$ is an odd function.

37. a. increasing when $m > 0$; decreasing when $m < 0$; neither when $m = 0$
b. The graph of an increasing function rises from left to right. The graph of a decreasing function falls from left to right.

38. a.
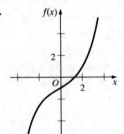

b.
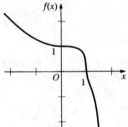

c.
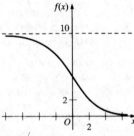

39. Consider, for example, $y = x^2$ for $x > 0$, which is an increasing function.
a. The reflection in the x-axis is decreasing.
b. The reflection in the y-axis is decreasing.
c. The reflection in the line $y = x$ is increasing.

40.
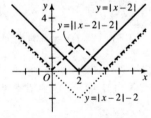

41. The graph has symmetry in the x-axis, in the y-axis, in the line $y = x$, and in the origin.

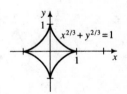

Class Exercises 4-4 · page 142

1. a. $f(x + 4) = f(x)$ for all x; the period is 4. **b.** amplitude $= \dfrac{M - m}{2} = \dfrac{1 - 0}{2} = \dfrac{1}{2}$

c. $f(25) = f(4 \cdot 6 + 1) = f(1) = 1$; $f(-25) = f(-1 + 4(-6)) = f(-1) = 0$

2.
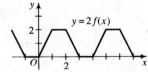

$y = 2f(x)$ has period 4 and amplitude $\dfrac{2 - 0}{2} = 1$; $y = \dfrac{1}{2}f(x)$ has period 4 and amplitude $\dfrac{\frac{1}{2} - 0}{2} = \dfrac{1}{4}$.

3. 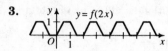 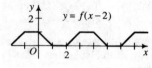 $y = f(2x)$ has period 2 and amplitude $\frac{1}{2}$; $y = f\left(\frac{1}{2}x\right)$ has period 8 and amplitude $\frac{1}{2}$.

4. 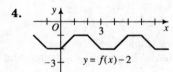 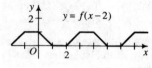   Both functions have period 4 and amplitude $\frac{1}{2}$.

5. Let $(a, 0)$ be a point on the graph of $y = f(x)$. If the graph is vertically stretched or shrunk, the equation of the new graph is $y = cf(x)$. At $x = a$, $y = c \cdot f(a) = c \cdot 0 = 0$. Thus, $(a, 0)$ is also on the new graph.

6. Let $(0, b)$ be a point on the graph of $y = f(x)$. If the graph is horizontally stretched or shrunk, the equation of the new graph is $y = f(cx)$. At $x = 0$, $y = f(c \cdot 0) = f(0) = b$. Thus, $(0, b)$ is also on the new graph.

Written Exercises 4-4 · pages 143–146

1. yes; $p = 6$; $A = \dfrac{1 - (-1)}{2} = 1$; $f(1000) = f(166 \cdot 6 + 4) = f(4) = -1$;
$f(-1000) = f(2 - 167 \cdot 6) = f(2) = 1$

2. yes; $f(x + 0.8) = f(x)$, so the period is 0.8; amplitude $= \dfrac{1 - 0}{2} = 0.5$;
$f(1000) = f(0 + 0.8(1250)) = f(0) = 0$; $f(-1000) = f(0 + 0.8(-1250)) = f(0) = 0$

3. yes; $f(x + 3) = f(x)$, so the period is 3; amplitude $= \dfrac{3 - 2}{2} = \dfrac{1}{2}$;
$f(1000) = f(1 + 3 \cdot 333) = f(1) = 2$; $f(-1000) = (-1 - 3 \cdot 333) = f(-1) = f(2) = 3$

4. no

5. a. 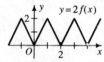   **b.** 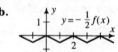   **c.**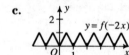

d. 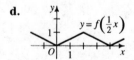   **e.** 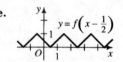   **f.**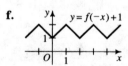

6. a.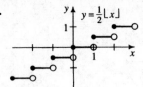

b.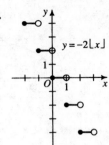

c.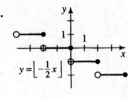

d.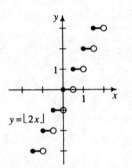

e.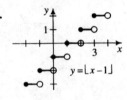

f.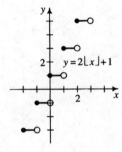

7. a.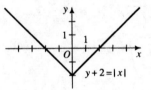

b.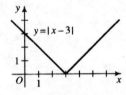

c.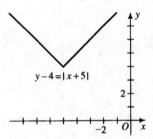

d.

e.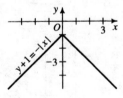

f.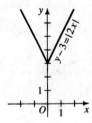

8. a.

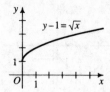

$y - 1 = \sqrt{x}$

b.

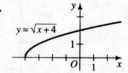

$y = \sqrt{x + 4}$

c.

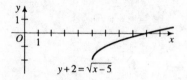

$y + 2 = \sqrt{x - 5}$

d.

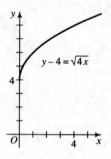

$y = 2\sqrt{x - 3}$

e.

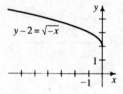

$y - 2 = \sqrt{-x}$

f.

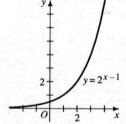

$y - 4 = \sqrt{4x}$

9. a.

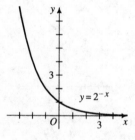

$y = 2^{-x}$

b.

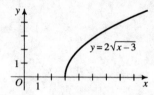

$y = 2^{x-1}$

c.

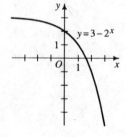

$y = 3 - 2^{x}$

d.

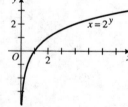

$x = 2^{y}$

10. a.

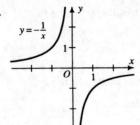

$y = -\dfrac{1}{x}$

b.

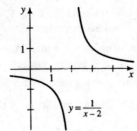

$y = \dfrac{1}{x-2}$

c.

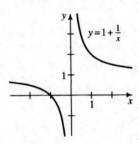

$y = 1 + \dfrac{1}{x}$

d.

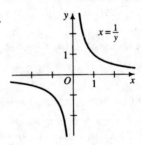

$x = \dfrac{1}{y}$

11. $(x - 8)^2 + (y - 4)^2 = 9$

12. a. See graph at right.

b. The graph is twice as wide as it was originally. The area of the region enclosed by the graph in part (a) is twice the area of the circle with eq. $x^2 + y^2 = 9$; $2 \cdot 9\pi = 18\pi$.

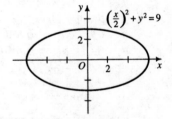

$\left(\dfrac{x}{2}\right)^2 + y^2 = 9$

Ex. 12(a)

13. a. The graph has shrunk horizontally by a factor of 2, so the x-intercepts are 0 and 3 and the local maximum is $(2, 32)$.

b. The graph has been stretched vertically by a factor of 2, so the x-intercepts are 0 and 6 and the local maximum is $(4, 64)$.

c. The graph has been translated 2 units right, so the x-intercepts are 2 and 8 and the local maximum is $(6, 32)$.

d. The graph has been translated 2 units left, so the x-intercepts are -2 and 4 and the local maximum is $(2, 32)$.

e. $f(x) = ax^2(x - 6)$, because f is tangent to the x-axis at $x = 0$ and crosses the x-axis at $x = 6$. Since $f(4) = 32$, $32 = a(16)(-2)$; $a = -1$; $f(x) = -x^2(x - 6)$.

14. a.

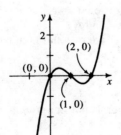

b.

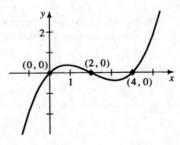

15. a. The graph shown is the translation of the given graph 4 units to the right and 2 units up. So, $y = |x - 4| + 2$ is an equation of the graph.

b. The graph shown is a translation 2 units up of the reflection of the given graph in the x-axis. So, $y = -|x| + 4$ is an equation of the graph.

c. Horizontally stretch the given graph by a factor of 2, then translate the result down 2 units. So, $y = \frac{1}{2}|x| - 2$ is an equation of the graph shown.

d. Vertically stretch the given graph by a factor of 2. Then reflect the graph that results in the x-axis. So, $y = -2|x|$ is an equation of the graph.

e. Horizontally translate the given graph 4 units to the left. Then vertically stretch the resulting graph by a factor of 2. So, $y = 2|x + 4|$ is an equation of the graph.

16. a. $f(-x) = a(-x)^3 + p(-x) = -ax^3 - px = -(ax^3 + px) = -f(x)$

b. Part (a) shows that for each pt. $(x, f(x))$ on the graph of f, the pt. $(-x, -f(x))$ is also on the graph. Therefore, the origin is a symmetry pt. of the graph.

c. The graph of $y = ax^3 + px + q$ is obtained by translating the graph of $f(x) = y = ax^3 + px$ q units vertically; in particular, the symmetry point of $y = ax^3 + px$, $(0,0)$, is translated to $(0, q)$.

d. The graph of $y = a(x - h)^3 + p(x - h) + q$ is obtained by translating the graph of $y = ax^3 + px + q$ h units horizontally; in particular, the symmetry point of $y = ax^3 + px + q$, $(0, q)$, is translated to (h, q).

e. Since $a(x - h)^3 + p(x - h) + q$ and $ax^3 + bx^2 + cx + d$ are equivalent, $a(x - h)^3 + p(x - h) + q = ax^3 + bx^2 + cx + d$. Expanding the left-hand side, $ax^3 - 3ahx^2 + 3ah^2x - ah^3 + px - ph + q = ax^3 + bx^2 + cx + d$; the coefficients of the quadratic terms are equal, so $-3ah = b$; $h = -\dfrac{b}{3a}$; from part (d), (h, q) is a point of symmetry of $y = a(x - h)^3 + p(x - h) + q$, so the symmetry point occurs when $x = h = -\dfrac{b}{3a}$.

17. If f and g both have fundamental period p, then for all x in the domain of h, $f(x + p) = f(x)$ and $g(x + p) = g(x)$. Also, $h(x + p) = (f + g)(x + p) = f(x + p) + g(x + p) = f(x) + g(x) = (f + g)(x) = h(x)$.

18. If $h = f + g$ has fundamental period p, then $h(x + p) = f(x + p) + g(x + p) = f(x) + g(x) = h(x)$. Since f has fundamental period 2, p must be a multiple of 2. Since g has fundamental period 3, p must be a multiple of 3. The smallest p which is a multiple of both 2 and 3 is 6.

19. Let $f(x) = c$ where c is a constant. Since for any positive real number p, $f(x + p) = c = f(x)$, f is periodic. However, since there is no smallest positive real number p, f does not have a fundamental period.

20. Let p be any rational number. If x is rational, then $x + p$ is rational and $f(x + p) = 1 = f(x)$. If x is irrational, then $x + p$ is irrational and $f(x + p) = 0 = f(x)$. Thus, $f(x + p) = f(x)$ for all x. Since there is no smallest positive rational number p, f does not have a fundamental period.

Class Exercises 4-5 · pages 148–149

1. No; more than one person is born each day. Since the function is not one-to-one, it does not have an inverse.

2. a. 2   **b.** 3   **c.** 2

3. a. $g^{-1}(x) = \dfrac{1}{4}x$ **b.** $x = 3y + 2; g^{-1}(x) = \dfrac{x-2}{3}$ **c.** $x = 2y - 1; g^{-1}(x) = \dfrac{x+1}{2}$

  d. $x = 4 - 5y; y = \dfrac{x-4}{-5}; g^{-1}(x) = \dfrac{4-x}{5}$

4. a. For example, $(8,2), (1,1), (0,0), (-1,-1),$ and $(-8,-2);$ see graph at right.

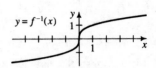

$y = f^{-1}(x)$

Ex. 4(a)

  b. $x = y^3; y = \sqrt[3]{x}; f^{-1}(x) = \sqrt[3]{x}$

5. a. From Ex. 4, $f(x) = x^3$ has an inverse and is one-to-one. Since the graph of $y = x^3 - 1$ is a vertical translation of the graph of $y = x^3$, $y = f(x) = x^3 - 1$ is also a one-to-one function.

  b. No; for example, $g(0) = g(1) = g(-1) = 0$.

6. Solve for the desired variable. For example, solve $C = \dfrac{5}{9}(F - 32)$ for F to get

  $F = \dfrac{9}{5}C + 32.$

7. $g(x)$ is one-to-one and has an inverse.

8. The functions in (a), and (c) are not one-to-one, so they have no inverses; the functions in (b) and (d) are one-to-one and have inverses.

9. Since 3 letters of the alphabet are paired with each of the digits 2–9 (for example, $T(A) = T(B) = T(C) = 2$), T is not a one-to-one function, and therefore does not have an inverse.

10. To obtain the graph of the inverse of a function f, we reflect the graph of f in the line $y = x$. This reflection transforms horizontal lines to vertical lines. Thus, if the graph of a function f passes the horizontal-line test, then the inverse graph passes the vertical-line test and is the graph of a function.

11. The point (a, b) is on the graph of a function f (so that a is an element of the domain of f and b is an element of the range of f) if and only if (b, a) is on the graph of f^{-1} (so that b is an element of the domain of f^{-1} and a is an element of the range of f^{-1}). Thus, the domain of f is the range of f^{-1} and the domain of f^{-1} is the range of f.

Written Exercises 4-5 · pages 149–150

1. a. $f^{-1}(6) = f^{-1}(f(2)) = 2$ **b.** 3 **c.** 7 **2. a.** $f^{-1}(-1) = f^{-1}(f(0)) = 0$ **b.** 0 **c.** 2

3. Since there are two x-values, 3 and -1, that correspond to the y-value 5, g is not one-to-one.

4. f is not one-to-one; for example, $f(-1) = f(0) = 0$.

5. a.

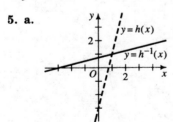

6. a.

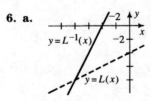

  b. $x = \dfrac{1}{2}y - 4; \dfrac{1}{2}y = x + 4;$
  $L^{-1}(x) = 2x + 8$

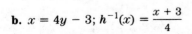

  b. $x = 4y - 3; h^{-1}(x) = \dfrac{x+3}{4}$

7. Yes   **8.** No   **9.** No   **10.** Yes

11. Yes; let $x = 3y - 5$; $f^{-1}(x) = \dfrac{x + 5}{3}$; $f(f^{-1}(x)) = f\left(\dfrac{x + 5}{3}\right) = 3\left(\dfrac{x + 5}{3}\right) - 5 =$

$(x + 5) - 5 = x$; $f^{-1}(f(x)) = f^{-1}(3x - 5) = \dfrac{(3x - 5) + 5}{3} = \dfrac{3x}{3} = x$.

12. No (for example, $f(2) = f(-2) = 0$)

13. Yes; $y = \sqrt[4]{x}$ has domain $x \geq 0$ and range $y \geq 0$; let $x = \sqrt[4]{y}$; then $f^{-1}(x) = x^4$ for $x \geq 0$; $f(f^{-1}(x)) = f(x^4) = \sqrt[4]{x^4} = x$ (because $x \geq 0$); $f^{-1}(f(x)) = f^{-1}(\sqrt[4]{x}) = (\sqrt[4]{x})^4 = x$

14. Yes; let $x = \dfrac{1}{y}$; $f^{-1}(x) = \dfrac{1}{x}$; $f(f^{-1}(x)) = f\left(\dfrac{1}{x}\right) = \dfrac{1}{\frac{1}{x}} = x$; $f^{-1}(f(x)) = f^{-1}\left(\dfrac{1}{x}\right) = \dfrac{1}{\frac{1}{x}} = x$

15. No (for example, $f(1) = f(-1) = 1$)

16. Yes; $f(x) = \sqrt{5 - x}$ has domain $x \leq 5$ and range $y \geq 0$; let $x = \sqrt{5 - y}$; $x^2 = 5 - y$; $y = 5 - x^2$; $f^{-1}(x) = 5 - x^2$ for $x \geq 0$; $f(f^{-1}(x)) = f(5 - x^2) = \sqrt{5 - (5 - x^2)} = \sqrt{x^2} = x$ (because $x \geq 0$); $f^{-1}(f(x)) = f^{-1}(\sqrt{5 - x}) = 5 - (\sqrt{5 - x})^2 = 5 - (5 - x) = x$

17. No (for example, $f(2) = f(-2) = 0$)   **18.** No (for example, $f(1) = f(-1) = 2$)

19. Yes; let $x = \sqrt[3]{1 + y^3}$; $x^3 = 1 + y^3$; $y^3 = x^3 - 1$; $y = \sqrt[3]{x^3 - 1}$; $f^{-1}(x) = \sqrt[3]{x^3 - 1}$; $f(f^{-1}(x)) = f(\sqrt[3]{x^3 - 1}) = \sqrt[3]{1 + (\sqrt[3]{x^3 - 1})^3} = \sqrt[3]{1 + x^3 - 1} = \sqrt[3]{x^3} = x$; $f^{-1}(f(x)) = f^{-1}(\sqrt[3]{1 + x^3}) = \sqrt[3]{(\sqrt[3]{1 + x^3})^3 - 1} = \sqrt[3]{1 + x^3 - 1} = \sqrt[3]{x^3} = x$

20. $y = x^2 + 2$, $x \geq 0$ has range $y \geq 2$; let $x = y^2 + 2$; $y = g^{-1}(x) = \sqrt{x - 2}$ for $x \geq 2$

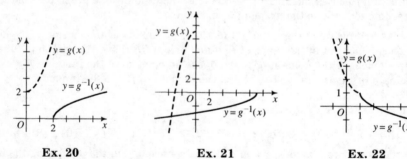

Ex. 20   Ex. 21   Ex. 22

21. $g(x) = 9 - x^2$, $x \leq 0$ has range $y \leq 9$; let $x = 9 - y^2$; $y^2 = 9 - x$; $y = g^{-1}(x) = -\sqrt{9 - x}$, $x \leq 9$

22. $g(x)$ has domain $x \leq 1$ and range $y \geq 1$; let $x = (y - 1)^2 + 1$; $y - 1 = \pm\sqrt{x - 1}$; $y = 1 \pm \sqrt{x - 1}$; $y \leq 1$, so $g^{-1}(x) = 1 - \sqrt{x - 1}$, $x \geq 1$

23. $g(x)$ has domain $x \geq 4$ and range $y \geq -1$; let $x = (y - 4)^2 - 1$; $x + 1 = (y - 4)^2$; $\pm\sqrt{x + 1} = y - 4$; $y = 4 \pm \sqrt{x + 1}$; $y \geq 4$, so $g^{-1}(x) = 4 + \sqrt{x + 1}$, $x \geq -1$

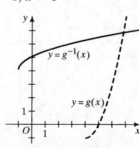

24. Let $x = \sqrt[3]{1 - y^3}$; $x^3 = 1 - y^3$;
$y^3 = 1 - x^3$; $y = \sqrt[3]{1 - x^3}$;
thus $h^{-1}(x) = h(x)$.

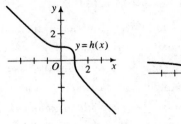

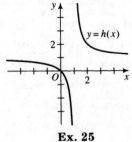

25. Let $x = \dfrac{y}{y - 1}$; $xy - x = y$;

$(x - 1)y = x$; $y = \dfrac{x}{x - 1}$;

thus, $h^{-1}(x) = h(x)$.

Ex. 24   **Ex. 25**

26. Let $x = \sqrt{1 - y^2}$, $y \geq 0$; $x^2 = 1 - y^2$; $y^2 = 1 - x^2$; $y \geq 0$, so
$y = \sqrt{1 - x^2}$; the range of $h(x)$ is $0 \leq y \leq 1$, so $h^{-1}(x) = \sqrt{1 - x^2}$,
$0 \leq x \leq 1$; thus, $h^{-1}(x) = h(x)$.

27. a. When $h^{-1}(x) = h(x)$, the graph of $h(x)$ is symmetric in the line $y = x$.

  b. For example, $h(x) = -x$

Ex. 26

28. a. According to the definition on page 138, every increasing function must be
one-to-one, and hence have an inverse that is a function. That is, if x_1 and x_2 are
distinct domain values and the function f is increasing, then $f(x_1) \neq f(x_2)$. This
implies that f is one-to-one.

  b. The reflection of the graph of an increasing function in the line $y = x$ is also the
graph of an increasing function (see Exercise 39(c) on page 138). For example,
$y = x^3$ and its inverse $y = \sqrt[3]{x}$ are both increasing functions, and $y = 3x$ and its

  inverse $y = \dfrac{1}{3}x$ are both increasing functions.

29. (ii) Proofs may vary. $[(f \circ g) \circ (g^{-1} \circ f^{-1})](x) = (f \circ g)[(g^{-1} \circ f^{-1})(x)] =$
$(f \circ g)[g^{-1}(f^{-1}(x))] = f(g(g^{-1}(f^{-1}(x)))) = f(f^{-1}(x)) = x$; $[(g^{-1} \circ f^{-1}) \circ (f \circ g)](x) =$
$(g^{-1} \circ f^{-1})[(f \circ g)(x)] = (g^{-1} \circ f^{-1})[f(g(x))] = g^{-1}(f^{-1}(f(g(x)))) = g^{-1}(g(x)) = x$.
Since $[(f \circ g) \circ (g^{-1} \circ f^{-1})](x) = x = [(g^{-1} \circ f^{-1}) \circ (f \circ g)](x)$, the inverse of $f \circ g$ is
$g^{-1} \circ f^{-1}$.

30. If $f(x) = mx + k$, $m \neq 0$, then $f^{-1}(x) = \dfrac{x - k}{m}$. $f(x + 2) - f(x) = 6$;

  $m(x + 2) + k - (mx + k) = 6$; $2m = 6$; $m = 3$.

  $f^{-1}(x + 2) - f^{-1}(x) = \dfrac{(x + 2) - k}{m} - \dfrac{x - k}{m} = \dfrac{2}{m} = \dfrac{2}{3}$.

31. If $P(x) = ax^2 + bx + c$, $a \neq 0$, and $x \leq -\dfrac{b}{2a}$, then let $x = ay^2 + by + c$;

  $ay^2 + by + (c - x) = 0$; $y = \dfrac{-b \pm \sqrt{b^2 - 4a(c - x)}}{2a}$; $y \leq -\dfrac{b}{2a}$, so

  $P^{-1}(x) = \dfrac{-b - \sqrt{b^2 - 4a(c - x)}}{2a}$, $b^2 - 4a(c - x) \geq 0$.

Calculator Exercises • page 150

  1. If you enter a number x, press e^x, then press $\ln x$, the result will be x. If you enter a
number x, press $\ln x$, then press e^x, the result will be x. Thus, f and g are inverse
functions.

  2. Yes; the domain of $f(x) = e^x$ is the set of all real numbers.

  3. The domain of $g(x) = \ln x$ is the set of positive real numbers. (Note that the domain of
$g(x)$ is the same as the range of $f(x) = e^x$.)

Class Exercises 4-6 • pages 153–154

1. The area A of a rectangle is a function of the length l and width w.

2. Distance d is a function of rate r and time t.

3. The volume V of a cone is a function of the base radius r and the height h.

4. Force F is a function of mass m and acceleration a.

5. Density D is a function of mass m and volume V.

6. The surface area A of a cylinder is a function of the base radius r and the height h.

7. First-quadrant rays with endpoints at the origin.

8. **a.** Examples: Tampa, San Francisco

   b. Any cities in which the pressure is 75.90 or less. Examples: Chihuahua, Ciudad Juarez, Phoenix

   c. Any cities in which the pressure is 77.10 or higher. Examples: Quebec, Montreal, Boston

9. For example, the driver's age, driving record, and place of residence.

10. **a.** $f(3,4) = \sqrt{3^2 + 4^2} = 5; f(-4,3) = \sqrt{(-4)^2 + 3^2} = 5; f(0,5) = \sqrt{0^2 + 5^2} = 5$

10. b.

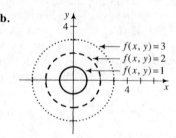

11. $a - b \neq 0$, so $(-1,-1)$, $(0,0)$, and $(2,2)$ are examples of pairs not in the domain of f.

Written Exercises 4-6 • pages 154–157

1. **a.** r and t **b.** $A = 100(1.12)^t$; $A = 100(1.16)^t$

   c. Find the value of t on each graph for which $A = 200$; about 9 yr, about 6 yr, and about 5 yr

2. **a.** speed and weight

   b. 50 mi/h gives a fuel efficiency of about 30 mi/gal, as does a speed of 25 mi/h.

   c. A 2400 lb car has a fuel efficiency of about 28 mi/gal when traveling 55 mi/h; a 3600 lb car has a fuel efficiency of about 28 mi/gal when traveling about 32 mi/h or 47 mi/h.   **d.** about 40 mi/h

3. **a.** The wind-chill equivalent temperature is a function of the recorded temperature and the wind speed.   **b.** about 16 mi/h

   c. The wind-chill equivalent temperature of the former is below zero, while the wind-chill equivalent temperature of the latter is above zero. Most people would prefer the latter since it feels warmer.   **d.** between 5 mi/h and 10 mi/h

4. **a.** The number of calories burned from walking is a function of the walker's walking speed and weight.   **b.** about 4.6 mi/h

   c. A 180 lb person walking at 2 mi/h burns approximately 250 Cal/h. In order to burn about 250 Cal/h, a 100 lb person must walk at 4 mi/h.

   d. The graph is steeper when the walking speed increases from 4 mi/h to 4.5 mi/h, so there is a greater change in that interval.

5. a. $A(8,3) = \frac{1}{2} \cdot 8 \cdot 3 = 12; A(16,6) = \frac{1}{2} \cdot 16 \cdot 6 = 48$

   b. The area A of a triangle is a function of the base b and the height h.

   c. Examples: $(2,5)$, $(1,10)$, $(20,0.5)$

6. a. $V(2,6) = \frac{1}{3} \pi (2^2) 6 = 8\pi; V(4,12) = \frac{1}{3} \pi (4^2) 12 = 64\pi$

   b. The volume V of a cone is a function of the radius r of the base and the height h.

   c. Examples: $(5,9)$, $(3,25)$, $\left(10, \frac{9}{4}\right)$

7. a. $A(3b,3h) = \frac{1}{2} \cdot 3b \cdot 3h = 9 \cdot \frac{1}{2} bh = 9 \cdot A(b,h)$

   b. If the base and height of a triangle are each tripled, the area is multiplied by 9.

8. a. $V(2r,2h) = \pi(2r)^2 \cdot 2h = \pi(4r^2)2h = 8\pi r^2 h = 8 \cdot V(r,h)$

   b. If the base radius and the height of a cylinder are each doubled, then the volume is multiplied by 8.

9. a. $d(r,t) = rt$ **10. a.** $V(s,h) = s^2 h$ **11.**

   b.

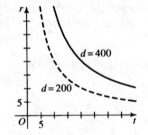

   b. Graphs may vary. For example:

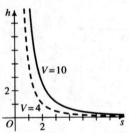

12. Answers will vary.   **13.** $x^2 - y^2 \neq 0$; $x^2 \neq y^2$; $x \neq \pm y$; $\{(x,y) \mid x \neq \pm y\}$

14. $4 - x^2 \neq 0$; $x^2 \neq 4$; $x \neq \pm 2$; $\{(x,y) \mid x \neq \pm 2\}$

15.

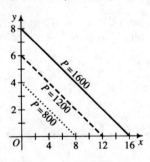

16. Graphs may vary. For example:

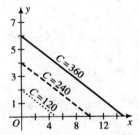

17. a. $A(4,3,5) = 2(4)(3) + 2(3)(5) + 2(4)(5) = 24 + 30 + 40 = 94$;
       $A(6,4,7) = 2(6)(4) + 2(4)(7) + 2(6)(7) = 48 + 56 + 84 = 188$

   b. The surface area A of a rectangular prism is a function of the length l, the width w, and the height h.

18. a. $A(3,4,2) = \dfrac{1}{2}(3 + 4)(2) = 7$; $A(5,7,3) = \dfrac{1}{2}(5 + 7)(3) = 18$

b. The area A of a trapezoid is a function of the bases b_1 and b_2 and the height h.

19. Answers will vary. Examples are given. Physics: Newton's Law of gravitation, $F = \dfrac{Gm_1 m_2}{d^2}$, which states that the gravitational force between two particles is directly proportional to the product of their masses, m_1 and m_2, and inversely proportional to the square of the distance, d, between them; G is the gravitational constant, 6.67×10^{-8} dyn cm^2/g^2. Biology: Fick equation for oxygen consumption, $\dot{V}o_2 = \dot{Q}t \times [Cao_2 - C\bar{v}o_2]$, which states that the volume of oxygen consumed per minute, $\dot{V}o_2$, is the product of $\dot{Q}t$, the volume of oxygenated blood pumped by the heart, and $Cao_2 - C\bar{v}o_2$, the arteriovenous oxygen content difference, which is a measure of the muscle cells' use of oxygen absorbed from the arterial blood.

20. Answers will vary. Examples are given. Psychology: $BI = w_1 A_B + w_2 SN$, an equation used to predict an individual's behavior; the equation states that a person's behavioral intention, BI, is a function of A_B, a measure of the person's attitude toward performing the behavior, and SN, a measure of the "subjective norm", a rating of expected behaviors as determined by society in general. These factors are multiplied by weights, w_1 and w_2, determined by regression. Economics: $R = n(p - c)$, an equation for calculating the revenues received from the sale of a product; in the equation, $R =$ total revenues, $n =$ number of items sold, $p =$ price charged per item, and $c =$ cost to produce each item.

21. $R = 0.9\left(\dfrac{10\sqrt{37}}{12\sqrt[3]{8.5}} + \dfrac{10 + \sqrt{37}}{4}\right) \approx 5.85$; $R > 5.5$, so the yacht does not qualify.

Class Exercises 4-7 · pages 160–161

1. a. $d^2 = 2s^2$, so $d(s) = s\sqrt{2}$ **b.** $s^2 = \dfrac{d^2}{2}$, so $s(d) = \dfrac{d\sqrt{2}}{2}$ **c.** $A = s^2 = \dfrac{d^2}{2}$; $A(d) = \dfrac{d^2}{2}$

2. a. $V(e) = e \cdot e \cdot e = e^3$ **b.** $d = e\sqrt{3}$; $e = \dfrac{d}{\sqrt{3}}$; $V = e^3 = \left(\dfrac{d}{\sqrt{3}}\right)^3 = \dfrac{d^3}{3\sqrt{3}}$; $V(d) = \dfrac{d^3\sqrt{3}}{9}$

3. a. $AP + PB = \sqrt{x^2 + 16} + 8 - x$ **b.** $\{x \mid 0 \le x \le 8\}$

4. a. $d(x,y) = \sqrt{(x - 2)^2 + y^2}$ **b.** $y = x^2$, so $y^2 = x^4$; $d(x) = \sqrt{(x - 2)^2 + x^4}$

c. Use a computer or graphing calculator to graph $y = \sqrt{(x - 2)^2 + x^4}$. Then find the approximate coordinates of the lowest point on the graph. The y-coordinate of this point is the minimum value of the function.

5. $d = \sqrt{a^2 + b^2} = \sqrt{(6t)^2 + (8t)^2} = \sqrt{100t^2} = 10t$

Written Exercises 4-7 · pages 161–165

1. If the hypotenuse has length h, then the legs have lengths $\dfrac{h}{2}$ and $\dfrac{h\sqrt{3}}{2}$;

$A(h) = \dfrac{1}{2} \cdot \dfrac{h}{2} \cdot \dfrac{h\sqrt{3}}{2} = \dfrac{h^2\sqrt{3}}{8}$.

2. If each side has length s, then $P = 3s$ and $A = \dfrac{s^2}{4}\sqrt{3}$; $s = \dfrac{P}{3}$, so

$A(P) = \dfrac{\sqrt{3}}{4}\left(\dfrac{P}{3}\right)^2 = \dfrac{P^2\sqrt{3}}{36}$.

3. $t = \dfrac{d}{r}$; $t(n) = \dfrac{n}{4} + \dfrac{2n}{36} = \dfrac{11n}{36}$

4. Each second, 2 m of string is released and the kite's altitude increases by 1 m. (The leg opposite the 30° angle is $\dfrac{1}{2}$ as long as the hyp.) Thus, $h(t) = t + 1$.

5. Total profit $= P(n) =$ total sales $-$ total cost $= 0.75(12n) - 3n = 6n$.

6. The total cost per hour in cents is $3000 + x^3$; the boat's speed is x km/h, so the cost in cents per kilometer is $C(x) = \dfrac{3000 + x^3}{x}$.

7. $h = 2d = 2(2r) = 4r$, so $r = \dfrac{1}{4}h$;

$A = 2\pi r^2 + 2\pi rh = 2\pi\left(\dfrac{1}{4}h\right)^2 + 2\pi\left(\dfrac{1}{4}h\right)h = \dfrac{1}{8}\pi h^2 + \dfrac{1}{2}\pi h^2 = \dfrac{5}{8}\pi h^2$; $A(h) = \dfrac{5}{8}\pi h^2$

8. $d = 2r = 2h$; $r = h$; $V = \dfrac{1}{3}\pi r^2 h = \dfrac{1}{3}\pi(h^2)h = \dfrac{1}{3}\pi h^3$; $V(h) = \dfrac{1}{3}\pi h^3$

9. Using similar triangles, $\dfrac{1.8}{3} = \dfrac{s}{s+d}$; $3s = 1.8s + 1.8d$; $s = \dfrac{1.8d}{1.2} = 1.5d$; $s(d) = 1.5d$.

10. Using similar triangles, $\dfrac{1.75}{h} = \dfrac{d}{15}$; $h = \dfrac{26.25}{d} = \dfrac{105}{4d}$; $h(d) = \dfrac{105}{4d}$

11. $A = 3 = 2w^2 + 4wh$; $h = \dfrac{3 - 2w^2}{4w}$; $V = w^2 h = w^2\left(\dfrac{3 - 2w^2}{4w}\right)$; $V(w) = \dfrac{3w - 2w^3}{4}$

12. $V = 6 = w^2 h$; $h = \dfrac{6}{w^2}$; $A = w^2 + 4wh = w^2 + 4w\left(\dfrac{6}{w^2}\right) = w^2 + \dfrac{24}{w}$; $A(w) = \dfrac{w^3 + 24}{w}$

13. a. $C = \pi d = \pi t$; $C(t) = \pi t$

　　b. $d = 2r = t$; $r = \dfrac{t}{2}$; $A = \pi r^2 = \pi\left(\dfrac{t}{2}\right)^2 = \dfrac{1}{4}\pi t^2$; $A(t) = \dfrac{\pi}{4}t^2$

14. a. $V(t) = 100 + 20t$

　　b. $V = \dfrac{4}{3}\pi r^3$, so $\dfrac{4}{3}\pi r^3 = 100 + 20t$; $r^3 = \dfrac{3}{4\pi}(100 + 20t) = \dfrac{75 + 15t}{\pi}$;

　　$r(t) = \sqrt[3]{\dfrac{75 + 15t}{\pi}}$

15. a. $V = 8 = w^2 h$; $h = \dfrac{8}{w^2}$. The base costs $8w^2$ dollars and the sides cost

　　$6(4wh) = 24w\left(\dfrac{8}{w^2}\right) = \dfrac{192}{w}$ dollars. Thus, $C(w) = 8w^2 + \dfrac{192}{w} = \dfrac{8w^3 + 192}{w}$.

　　b. about \$126

16. a. $V = 400\pi = \pi r^2 h$; $h = \dfrac{400}{r^2}$. The top and bottom cost $2(2\pi r^2) = 4\pi r^2$ cents and the

　　vertical surface costs $1(2\pi rh) = 2\pi r\left(\dfrac{400}{r^2}\right) = \dfrac{800\pi}{r}$ cents. Thus,

　　$C(r) = 4\pi r^2 + \dfrac{800\pi}{r} = \dfrac{4\pi(r^3 + 200)}{r}$.　　**b.** about \$8.12

17. a. t hours after 2 P.M., bike A is $(4 - 16t)$ km north of C and bike B is $(2 + 12t)$ km east of C; $\triangle ABC$ is a right triangle, so $d(t) = AB = \sqrt{(AC)^2 + (BC)^2} = $
$\sqrt{(4 - 16t)^2 + (2 + 12t)^2} = \sqrt{16 - 128t + 256t^2 + 4 + 48t + 144t^2} = $
$\sqrt{400t^2 - 80t + 20}$.

b. $d(t)$ is at a minimum when $400t^2 - 80t + 20$ is minimum; this occurs when
$t = \dfrac{-(-80)}{2(400)} = 0.1$ hours, or at 2:06 P.M.

c. $d(0.1) = \sqrt{400(0.01) - 80(0.1) + 20} = \sqrt{16} = 4$; 4 km

18. a. t hours after 11:33 A.M., the first car is $70t$ km south of Oak Corners and the second car is $(65 - 90t)$ km west of Oak Corners. By the Pythagorean Theorem, $d(t) = $
$\sqrt{(70t)^2 + (65 - 90t)^2} = \sqrt{4900t^2 + 8100t^2 - 11{,}700t + 4225} = $
$\sqrt{13{,}000t^2 - 11{,}700t + 4225} = 5\sqrt{520t^2 - 468t + 169}$.

b. $d(t)$ is at a minimum when $520t^2 - 468t + 169$ is minimum; this occurs when
$t = -\dfrac{-468}{2(520)} = 0.45$; 0.45 h = 27 min, so the cars are closest 27 min after 11:33, or at noon.

19. a. Let r be the radius of the water surface when the height of the water is h. By using similar triangles, $\dfrac{r}{h} = \dfrac{30}{120} = \dfrac{1}{4}$; $r = \dfrac{1}{4}h$. $V = \dfrac{1}{3}\pi r^2 h = \dfrac{1}{3}\pi\left(\dfrac{1}{4}h\right)^2 h$; $V(h) = \dfrac{\pi}{48}h^3$.

b. The volume in the tank after t seconds have elapsed is $V = 5t = \dfrac{\pi}{48}h^3$; $h^3 = \dfrac{240t}{\pi}$;
$h(t) = \sqrt[3]{\dfrac{240t}{\pi}}$.

20. a. Let x be the base of the triangle with height h. Each triangular end has height
$\sqrt{1^2 - (0.6)^2} = 0.8$ m. By using similar triangles, $\dfrac{h}{0.8} = \dfrac{x}{1.2}$; $x = 1.5h$. The
triangular prism has volume $V = Bh = \left(\dfrac{1}{2}xh\right)2 = 1.5h(h)$; $V(h) = 1.5h^2$.

b. 6 L/min = 0.006 m³/min; $V = 0.006t = 1.5h^2$; $h^2 = 0.004t$; $h(t) = \sqrt{0.004t}$

21. a. $d(x,y) = \sqrt{x^2 + y^2}$; $y = 10 - 2x$, so $y^2 = 4x^2 - 40x + 100$;
$d(x) = \sqrt{x^2 + 4x^2 - 40x + 100} = \sqrt{5x^2 - 40x + 100}$

b. Since $5x^2 - 40x + 100 > 0$ for all x, the domain is {all real nos.}; the minimum
value occurs when $x = -\dfrac{-40}{2(5)} = 4$; $d(4) = \sqrt{20} = 2\sqrt{5}$; therefore, the range is
$\{d(x) \mid d(x) \geq 2\sqrt{5}\}$.

22. a. $d(x,y) = \sqrt{(x - 0)^2 + (y - 1)^2} = \sqrt{x^2 + (y - 1)^2} = \sqrt{y + (y - 1)^2}$;
$d(y) = \sqrt{y^2 - y + 1}$

b. $d(y)$ is minimum when $y^2 - y + 1$ is minimum; this occurs when $y = \dfrac{-(-1)}{2(1)} = \dfrac{1}{2}$;
$d\left(\dfrac{1}{2}\right) = \sqrt{\dfrac{3}{4}} = \dfrac{\sqrt{3}}{2}$

23. a. $P(x,y) = 4x + 2y$; $y = 9 - x^2$; $P(x) = 4x + 2(9 - x^2) = -2x^2 + 4x + 18$

b. Since $A(x,y)$ is on the parabola and in Quadrant I, the domain is $\{x \mid 0 < x < 3\}$.

c. The maximum value of $P(x)$ occurs when $x = -\dfrac{4}{2(-2)} = 1$.

24. a. The trapezoidal base has legs of length x, bases of lengths $(60 - 2x)$ and

$60 - 2x + \dfrac{x}{2} + \dfrac{x}{2} = 60 - x$, and a height of length $\dfrac{x}{2}\sqrt{3}$; area of

$\text{base} = \dfrac{1}{2}h(b_1 + b_2) = \dfrac{1}{2}\left(\dfrac{x}{2}\sqrt{3}\right)(60 - 2x + 60 - x) = \dfrac{x\sqrt{3}}{4}(120 - 3x);$

$V(x) = Bh = \dfrac{x\sqrt{3}}{4}(120 - 3x) \cdot 1000 = 750x\sqrt{3}(40 - x).$

b. The quadratic function $V(x)$ has zeros at 0 and 40, so the maximum value occurs

when $x = \dfrac{1}{2}(0 + 40) = 20$.

25. a. The lifeguard swims a distance of $\sqrt{x^2 + 50^2}$ m and runs $(100 - x)$ m; $t = \dfrac{d}{r}$, so

the total time in seconds is $t(x) = \dfrac{\sqrt{x^2 + 50^2}}{1} + \dfrac{100 - x}{3} =$

$\sqrt{2500 + x^2} + \dfrac{1}{3}(100 - x).$ **b.** about 80.5 s

26. a. The underwater cable has length $\sqrt{x^2 + 60^2}$ m and the cable on land has length

$200 - x$. Thus $C(x) = 25\sqrt{x^2 + 3600} + 20(200 - x)$.

b. The minimum cost, \$4900, occurs when $x = 80$.

27. a. Since the area is 60 m², the side opposite the wall has length $\dfrac{60}{x}$. Then

$C(x) = 6x + 6x + 8\left(\dfrac{60}{x}\right) = \dfrac{12x^2 + 480}{x}$.   **b.** about \$152

28. a. If the base has radius r, then the height of the cylinder is $h = 2\sqrt{1 - r^2}$ (see the

diagram below); $V(r) = \pi r^2 h = \pi r^2(2\sqrt{1 - r^2}) = 2\pi r^2\sqrt{1 - r^2}.$

b. $r \approx 0.82; h \approx 1.14$

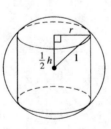

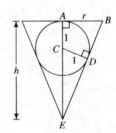

Ex. 28   Ex. 29

29. a. From the cross-sectional diagram above, $\triangle EDC$ is similar to $\triangle EAB$; hence

$\dfrac{DE}{AE} = \dfrac{CD}{AB}$; since $DE = \sqrt{(h - 1)^2 - 1^2} = \sqrt{h^2 - 2h}$, $AE = h$, $CD = 1$, and $AB = r$,

$\dfrac{\sqrt{h^2 - 2h}}{h} = \dfrac{1}{r}; \dfrac{h^2 - 2h}{h^2} = \dfrac{1}{r^2}; 1 - \dfrac{2}{h} = \dfrac{1}{r^2}; \dfrac{2}{h} = 1 - \dfrac{1}{r^2}; \dfrac{2}{h} = \dfrac{r^2 - 1}{r^2}; h = \dfrac{2r^2}{r^2 - 1}.$

Since $V = \dfrac{1}{3}\pi r^2 h$, $V(r) = \dfrac{2\pi r^4}{3(r^2 - 1)}$.   **b.** $r \approx 1.41; h \approx 4.02$

30. a. $d(t) = \sqrt{(90)^2 + (9 + 27t)^2} = 9\sqrt{9t^2 + 6t + 101}$

    b. When $t = 3$, the runner reaches second base; domain: $\{t \mid 0 \le t \le 3\}$; $d(t)$ is steadily increasing from $t = 0$ to $t = 3$, so the range is $\{d(t) \mid 9\sqrt{101} \le d(t) \le 90\sqrt{2}\}$.

31. a. When $0 \le t \le 3$, $d(t) = 30t$; when $3 \le t \le 6$,
$d(t) = \sqrt{90^2 + [30(t - 3)]^2}$; when
$6 \le t \le 9$, $d(t) = \sqrt{90^2 + [90 - 30(t - 6)]^2} = $
$\sqrt{90^2 + (270 - 30t)^2}$; when $9 \le t \le 12$,
$d(t) = 90 - 30(t - 9) = 360 - 30t$

    b.

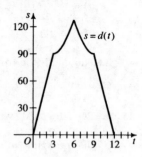

Chapter Test · pages 166–167

1. $9 - x^2 \ge 0$; $x^2 \le 9$; domain: $\{x \mid -3 \le x \le 3\}$; when $-3 \le x \le 3$, $0 \le f(x) \le 3$; range: $\{f(x) \mid 0 \le f(x) \le 3\}$; $f(x) = 0$ when $9 - x^2 = 0$; zeros: ± 3

2. a.

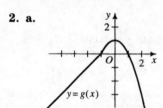

    b. Range: $\{g(x) \mid g(x) \le 1\}$; zeros: $x = \pm 1$

3. a. $f(x) + g(x) = x^2 + 2x + x + 2 = x^2 + 3x + 2$

    b. $f(x) - g(x) = x^2 + 2x - (x + 2) = x^2 + x - 2$

    c. $f(x) \cdot g(x) = (x^2 + 2x)(x + 2) = x^3 + 4x^2 + 4x$

    d. $\dfrac{f(x)}{g(x)} = \dfrac{x^2 + 2x}{x + 2} = \dfrac{x(x + 2)}{(x + 2)} = x$ for $x \ne -2$

4. a. $f(g(x)) = f(x + 2) = (x + 2)^2 + 2(x + 2) = x^2 + 6x + 8$

    b. $g(f(x)) = g(x^2 + 2x) = x^2 + 2x + 2$

5. (i) No; $x^2 - x(-y) = 4$ is not equivalent to $x^2 - xy = 4$. (ii) No; $(-x)^2 - (-x)y = 4$ is not equivalent to $x^2 - xy = 4$. (iii) No; $y^2 - yx = 4$ is not equivalent to $x^2 - xy = 4$. (iv) Yes; $(-x)^2 - (-x)(-y) = 4$ is equivalent to $x^2 - xy = 4$.

6. a.

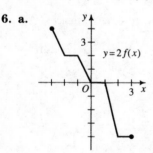

    b.

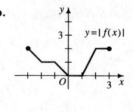

    c.

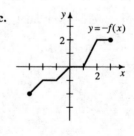

d.

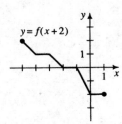

$y = f(x+2)$

7. a. 4 **b.** 4; 1 **c.** $\dfrac{4-1}{2} = \dfrac{3}{2}$

8. If the graph of a periodic function with period p is horizontally translated p units, the translated graph coincides with the original graph. For example, if you shift the graph of g in Ex. 7 four units to the right or left, the new graph is identical to the graph of g. The definition of a periodic function says that $f(x + p) = f(x)$ for all x and some positive number p. Thus, the graphs of $y = f(x + p)$ and $y = f(x)$ are identical (that is, the graph of f translated p units to the left coincides with the graph of f). Since replacing x with $x - p$ in the definition gives $f(x) = f(x - p)$ for all x, the graphs of $y = f(x)$ and $y = f(x - p)$ are also identical (that is, the graph of f translated p units to the right coincides with the graph of f).

9. **a.** g has an inverse; let $x = 3 + y$; $g^{-1}(x) = x - 3$

   b. f has no inverse because it is not one-to-one; for example, $f(1) = 2 = f(-1)$

10. **a.** $A(b, h) = \dfrac{1}{2}bh$

    b. $A(3, 4) = \dfrac{1}{2} \cdot 3 \cdot 4 = 6$; $A(6, 5) = \dfrac{1}{2} \cdot 6 \cdot 5 = 15$

    c. See diagram at right.

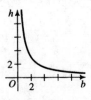

Ex. 10(c)

11. The volume of water in the tank at time t is given by $V = 10t$;

    $V = \pi r^2 h = \pi \cdot 2^2 \cdot h = 4\pi h$; thus, $10t = 4\pi h$; $h = \dfrac{5t}{2\pi}$;

    $h(t) = \dfrac{5t}{2\pi}$.

12. $A(x, y)$ is in Quadrant I and $B(-x, y)$ is in Quadrant II.

    a. The area of $\triangle AOB = \dfrac{1}{2}bh = \dfrac{1}{2}(2x)y = xy$; $y = 9 - x^2$, so $A(x) = x(9 - x^2)$.

    b. $\{x \mid 0 < x < 3\}$

    c. Use a method from Lesson 2-5 to show that the maximum area occurs when $x \approx 1.732$; the maximum area is approximately 10.4 sq. units.

Class Exercises 5-1 · page 172

1. a. $1.105 - 1 = 0.105$, or 10.5%   **b.** 1.03; 1.15; 1.046; $1 + 1.2 = 2.2$

2. a. $1 - 0.68 = 0.32$, or 32%

b. $1 - 0.12 = 0.88$; $1 - 0.075 = 0.925$; $1 - 0.8 = 0.2$; $1 - 1 = 0$

3. a. $200(1.05)^t$ **b.** $20(1.08)^t$ **c.** $1.06 - 1 = 0.06$, or 6%

4. a. $9800(0.8)^t$ **b.** $2200(0.85)^t$ **c.** $1 - 0.75 = 0.25$, or 25%

5. a. $\dfrac{1}{8}$ **b.** $\dfrac{1}{8^2} = \dfrac{1}{64}$ **6. a.** $\dfrac{3}{2}$ **b.** $\left(\dfrac{3}{2}\right)^2 = \dfrac{9}{4}$ **7. a.** $4 \cdot \dfrac{1}{3^2} = \dfrac{4}{9}$ **b.** $\dfrac{1}{(4 \cdot 3)^2} = \dfrac{1}{144}$

8. a. $2 \cdot 4 = 8$ **b.** $\left(\dfrac{1}{2} + \dfrac{1}{4}\right)^{-1} = \left(\dfrac{3}{4}\right)^{-1} = \dfrac{4}{3}$ **9.** $2^3 = 8$ **10.** $2^n \cdot 3^n = 6^n$

11. $\dfrac{4n^2}{-2n^2} = -2$ **12.** $x^2 + x^0 = x^2 + 1$ **13.** $\dfrac{18a^9}{\dfrac{1}{a}} = 18a^{10}$

14. $(3a^3 - 6a^6)a = 3a^4 - 6a^7$ **15.** $(5b^3 + 10b^6)b^2 = 5b^5 + 10b^8$

16. $(50a^3b^6)b^2 = 50a^3b^8$

Written Exercises 5-1 · pages 173–175

1. a. $\dfrac{1}{(-4)^2} = \dfrac{1}{16}$ **b.** $-\dfrac{1}{4^2} = -\dfrac{1}{16}$ **2. a.** $\left(\dfrac{1}{-3}\right)^4 = \dfrac{1}{81}$ **b.** $-\dfrac{1}{3^4} = -\dfrac{1}{81}$

3. a. $5 \cdot \dfrac{1}{2^3} = \dfrac{5}{8}$ **b.** $\dfrac{1}{(5 \cdot 2)^3} = \dfrac{1}{1000}$ **4. a.** $2 \div \dfrac{1}{4^3} = 2 \cdot 4^3 = 128$ **b.** $\left(\dfrac{1}{2}\right)^{-3} = 2^3 = 8$

5. a. x^2 **b.** $x^0 = 1$ **6. a.** $2^{-4}x^{12} = \dfrac{1}{2^4} \cdot x^{12} = \dfrac{x^{12}}{16}$ **b.** $2 \cdot \dfrac{1}{x^3} \cdot \dfrac{1}{x^4} = \dfrac{2}{x^7}$

7. a. $3^{-1} \cdot a = \dfrac{a}{3}$ **b.** $\left(3 + \dfrac{1}{a}\right)^{-1} = \left(\dfrac{3a + 1}{a}\right)^{-1} = \dfrac{a}{3a + 1}$

8. a. $(5 \cdot x^0)^2 = (5 \cdot 1)^2 = 25$

b. $25x^4 + 2 \cdot 5x^2 \cdot x^{-2} + x^{-4} = 25x^4 + 10 + \dfrac{1}{x^4}$, or $[(5x^4 + 1)(x^{-2})]^2 = \dfrac{(5x^4 + 1)^2}{x^4}$

9. a. $\left(\dfrac{1}{2^2} + \dfrac{1}{2^3}\right)^{-1} = \left(\dfrac{1}{4} + \dfrac{1}{8}\right)^{-1} = \left(\dfrac{3}{8}\right)^{-1} = \dfrac{8}{3}$ **b.** $(2^{-5})^{-1} = 2^5 = 32$

10. a. $\left(\dfrac{1}{4} - \dfrac{1}{2}\right)^2 = \left(-\dfrac{1}{4}\right)^2 = \dfrac{1}{16}$ **b.** $\left(\dfrac{1}{4} \div \dfrac{1}{2}\right)^2 = \left(\dfrac{1}{4} \cdot 2\right)^2 = \left(\dfrac{1}{2}\right)^2 = \dfrac{1}{4}$

11. a. $\left(\dfrac{1}{a} - \dfrac{1}{b}\right)^{-1} = \left(\dfrac{b - a}{ab}\right)^{-1} = \dfrac{ab}{b - a}$ **b.** $\left(\dfrac{1}{a} \cdot \dfrac{1}{b}\right)^{-1} = \left(\dfrac{1}{ab}\right)^{-1} = ab$

12. a. $\left(2 + \dfrac{1}{x^2}\right)^{-2} = \left(\dfrac{2x^2 + 1}{x^2}\right)^{-2} = \left(\dfrac{x^2}{2x^2 + 1}\right)^2 = \dfrac{x^4}{(2x^2 + 1)^2}$ **b.** $2^{-2} \cdot x^4 = \dfrac{x^4}{4}$

13. $300(1.15)^{10} \approx 300(4.046) \approx \$1214$; $300(1.15)^{20} \approx 300(16.367) \approx \$4910$

14. $35(1.08)^{10} \approx 35(2.159) \approx \$76$; $35(1.08)^{20} \approx 35(4.661) \approx \$163$

15. $1(1.07)^{10} \approx 1(1.967) \approx \$1.97$; $1(1.07)^{20} \approx 1(3.870) \approx \$3.87$

16. $12,000(1.1)^{10} \approx 12,000(2.59374) \approx \$31,125$;
$12,000(1.1)^{20} \approx 12,000(6.7275) \approx \$80,730$

17. $65,000(0.75)^3 \approx \$27,422$; $65,000(0.75)^6 \approx \$11,569$

18. $200,000(0.9)^3 \approx \$145,800$; $200,000(0.9)^6 \approx \$106,288$

19. $1(0.94)^3 \approx \$.83$; $1(0.94)^6 \approx \$.69$

20. $1(0.92)^3 \approx \$.78$; $1(0.92)^6 \approx \$.61$

21. $27a^{-6} \cdot 3a^5 = 81a^{-1} = \dfrac{81}{a}$ **22.** $16x^6 \cdot 3x^{-2} = 48x^4$ **23.** $3^{-1}n^{-2} \cdot 3^7 \cdot n^{14} = 3^6 n^{12}$

24. $2^4 r^{-4} \cdot 4^{-2} \cdot r^{-4} = 16r^{-8} \cdot \dfrac{1}{16} = \dfrac{1}{r^8}$ **25.** $(2a^{-1})^4 = 2^4 \cdot a^{-4} = \dfrac{16}{a^4}$ **26.** $\dfrac{9n^{-6}}{-9n^{-4}} = -\dfrac{1}{n^2}$

27. $\left(\dfrac{a}{b^2}\right)^{-5} = \left(\dfrac{b^2}{a}\right)^5 = \dfrac{b^{10}}{a^5}$ **28.** $(-2r)^6 = 64r^6$ **29.** $2x^2 - 4x^0 = 2x^2 - 4$

30. $x^2 y^0 - 3xy^1 = x^2 - 3xy$ **31.** $\dfrac{6a^{-2}}{3a^{-2}} + \dfrac{9a^2}{3a^{-2}} = 2 + 3a^4$ **32.** $\dfrac{8n^4}{2n^{-2}} - \dfrac{4n^{-2}}{2n^{-2}} = 4n^6 - 2$

33. $t = 1$: $V = 10,000(0.80)^1 = 8000$, $\$8000$
$t = 2$: $V = 10,000(0.80)^2 = 6400$, $\$6400$
$t = 3$: $V = 10,000(0.80)^3 = 5120$, $\$5100$
$t = 4$: $V = 10,000(0.80)^4 = 4096$, $\$4100$
$t = 5$: $V = 10,000(0.80)^5 = 3276.80$; $\$3300$

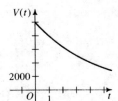

34. a. $1 - 0.85 = 0.15$; 15%   **b.** $6250 = 12,500(0.85)^t$; $\dfrac{1}{2} = (0.85)^t$; $t \approx 4$ years

35. $100(1.01)^{12} \approx \$112.68$ **36.** $(1.015)^{12} \approx 1.1956$; $1.1956 - 1 = 0.1956$; about 19.6%

37. a. $\dfrac{3^5 \cdot (3^2)^4}{(3^3)^4} = \dfrac{3^{13}}{3^{12}} = 3$ **b.** $\dfrac{(5^3)^{-3} \cdot 5^2}{5^{-8}} = \dfrac{5^{-7}}{5^{-8}} = 5$

   c. $\sqrt{\dfrac{(2^3)^n \cdot 2^7}{(2^2)^{-n}}} = \sqrt{2^{3n+7} \cdot 2^{2n}} = \sqrt{2^{5n+7}}$

38. a. $\dfrac{(2^2)^9 \cdot (2^3)^{-4}}{(2^4)^3} = \dfrac{2^{18-12}}{2^{12}} = \dfrac{1}{2^6} = \dfrac{1}{64}$ **b.** $\dfrac{3^7 \cdot (3^2)^5}{\sqrt{(3^3)^{12}}} = \dfrac{3^{17}}{\sqrt{3^{36}}} = \dfrac{3^{17}}{3^{18}} = \dfrac{1}{3}$

   c. $\sqrt[3]{\dfrac{(5^3)^n \cdot 5^{4n}}{(5^2)^{-n}}} = \sqrt[3]{5^{3n+4n+2n}} = \sqrt[3]{5^{9n}} = 5^{3n}$

39. a. $\dfrac{b^{3n}}{b^{2n}} = b^n$ **b.** $\dfrac{b^{2n}}{b^{2n+2}} = \dfrac{1}{b^2} = b^{-2}$ **c.** $\dfrac{b^{n+1}}{b^{3n}} = b^{1-2n}$

40. a. $\sqrt{b^{2n} \cdot b^{2n}} = b^{2n}$ **b.** $\dfrac{(b^{n+1})^2}{b^{2n}} = b^{2n+2-2n} = b^2$ **c.** $\sqrt{\dfrac{b^1 \cdot b^{-n}}{b^n \cdot b^{-1}}} = \sqrt{\dfrac{b^2}{b^{2n}}} = \dfrac{b}{b^n} = b^{1-n}$

41. a. $\dfrac{2^{-1}}{2^{-2} + 2^{-3}} \cdot \dfrac{2^3}{2^3} = \dfrac{2^2}{2^1 + 2^0} = \dfrac{4}{2+1} = \dfrac{4}{3}$

   b. $\dfrac{4^{-5}}{4^{-2} + 4^{-3}} \cdot \dfrac{4^5}{4^5} = \dfrac{4^0}{4^3 + 4^2} = \dfrac{1}{64 + 16} = \dfrac{1}{80}$

42. a. $\dfrac{3^{-2}}{3^{-3} + 3^{-2}} \cdot \dfrac{3^3}{3^3} = \dfrac{3}{1 + 3} = \dfrac{3}{4}$

   b. $\dfrac{2^{-1} - 2^{-2}}{2^{-1} + 2^{-2}} \cdot \dfrac{2^2}{2^2} = \dfrac{2 - 1}{2 + 1} = \dfrac{1}{3}$

43. a. $\dfrac{\dfrac{1}{x^2} - \dfrac{1}{y^2}}{\dfrac{1}{x} - \dfrac{1}{y}} = \dfrac{y^2 - x^2}{x^2 y^2} \div \dfrac{y - x}{xy} = \dfrac{(y + x)(y - x)}{x^2 y^2} \cdot \dfrac{xy}{y - x} = \dfrac{y + x}{xy}$

$\left(\text{or } \dfrac{x^{-2} - y^{-2}}{x^{-1} - y^{-1}} \cdot \dfrac{x^2 y^2}{x^2 y^2} = \dfrac{y^2 - x^2}{xy^2 - x^2 y} = \dfrac{(y + x)(y - x)}{xy(y - x)} = \dfrac{y + x}{xy} \right)$

b. $\dfrac{1 - \dfrac{1}{y}}{y - \dfrac{1}{y}} \cdot \dfrac{y}{y} = \dfrac{y - 1}{y^2 - 1} = \dfrac{1}{y + 1}$

44. a. $\dfrac{x^{-1}}{x - x^{-1}} \cdot \dfrac{x}{x} = \dfrac{1}{x^2 - 1}$ **b.** $\dfrac{4 - x^{-4}}{2 - x^{-2}} \cdot \dfrac{x^4}{x^4} = \dfrac{4x^4 - 1}{2x^4 - x^2} = \dfrac{(2x^2 - 1)(2x^2 + 1)}{x^2(2x^2 - 1)} = \dfrac{2x^2 + 1}{x^2}$

45. $24(1.06)^{370} \approx 5.5384 \times 10^{10} \approx \$55{,}000{,}000{,}000$

46. $2^x + \dfrac{8}{2^x} = 9$; let $t = 2^x$; $t + \dfrac{8}{t} = 9$; $t^2 - 9t + 8 = 0$; $(t - 8)(t - 1) = 0$; $t = 8$ or $t = 1$;
$2^x = 8 = 2^3$ or $2^x = 1 = 2^0$; $x = 3$ or $x = 0$

47. $2^x + \dfrac{1}{2^x} = \dfrac{5}{2}$; let $t = 2^x$; $t + \dfrac{1}{t} = \dfrac{5}{2}$; $2t^2 - 5t + 2 = 0$; $(2t - 1)(t - 2) = 0$; $t = \dfrac{1}{2}$ or
$t = 2$; $2^x = \dfrac{1}{2} = 2^{-1}$ or $2^x = 2 = 2^1$; $x = -1$ or $x = 1$

48. Let $y = 2^x$; $y^2 - 3 \cdot y(2) + 8 = 0$; $y^2 - 6y + 8 = 0$; $(y - 4)(y - 2) = 0$; $y = 4$ or
$y = 2$; $2^x = 4 = 2^2$ or $2^x = 2 = 2^1$; $x = 2$ or $x = 1$

49. Let $z = 3^x$; $3(3^x)^2 - 10 \cdot 3^x + 3 = 0$; $3z^2 - 10z + 3 = 0$; $(3z - 1)(z - 3) = 0$; $z = \dfrac{1}{3}$
or $z = 3$; $3^x = \dfrac{1}{3} = 3^{-1}$ or $3^x = 3 = 3^1$; $x = -1$ or $x = 1$

Class Exercises 5-2 · page 177

1. a. $\sqrt{4} = 2$ **b.** $\dfrac{1}{\sqrt{4}} = \dfrac{1}{2}$ **2. a.** $(\sqrt{4})^3 = 2^3 = 8$ **b.** $\dfrac{1}{(\sqrt{4})^3} = \dfrac{1}{2^3} = \dfrac{1}{8}$

3. a. $-\sqrt{9} = -3$ **b.** $-\dfrac{1}{\sqrt{9}} = -\dfrac{1}{3}$

4. a. $3 \cdot 5 = 15$ **b.** $(\sqrt{3} + \sqrt{5})^2 = 3 + 2\sqrt{15} + 5 = 8 + 2\sqrt{15}$

5. $\left[\left(\dfrac{49}{25} \right)^{1/2} \right]^{-1} = \left(\dfrac{7}{5} \right)^{-1} = \dfrac{5}{7}$ **6.** $\left(\sqrt{\dfrac{4}{9}} \right)^3 = \left(\dfrac{2}{3} \right)^3 = \dfrac{8}{27}$ **7.** $8^{1/3} = \sqrt[3]{8} = 2$

8. $16^{3/2} = (\sqrt{16})^3 = (4)^3 = 64$ **9.** $8x^{-1} = 8 \cdot \dfrac{1}{x} = \dfrac{8}{x}$ **10.** $\sqrt[3]{\dfrac{5^3}{x^6}} = \dfrac{5}{x^2}$

11. $\dfrac{x^{1/3 + 2/3}}{2} = \dfrac{x}{2}$ **12.** $8x^{3/2 - 1/2} = 8x$

13. a. $62(1.05)^0 = 62$; $\$62$ **b.** 2.5 **c.** $-\dfrac{9}{12} = -0.75$ **14.** $2x = 12$; $x = 6$

15. $(3^2)^x = 3^5$; $3^{2x} = 3^5$; $2x = 5$; $x = \dfrac{5}{2}$ **16.** $(x^{2/3})^{3/2} = 9^{3/2}$; $x = (\sqrt{9})^3 = 3^3 = 27$

17. $(x^{-1/2})^{-2} = 4^{-2};\ x = \dfrac{1}{16}$

18. By definition, $b^{p/q} = (\sqrt[q]{b})^p$; since $\sqrt[q]{b}$ is not a real number when q is even and b is negative, the base b is restricted to positive numbers.

Written Exercises 5-2 · pages 178–180

1. a. $\sqrt[3]{x^2}$ or $(\sqrt[3]{x})^2$ **b.** $\sqrt{x^3}$ or $(\sqrt{x})^3$ **c.** $\dfrac{\sqrt{5}}{\sqrt{x}} = \sqrt{\dfrac{5}{x}}$ **d.** $\sqrt[3]{6} \cdot \sqrt[3]{x^2} = \sqrt[3]{6x^2}$

2. a. $3\sqrt[5]{y^2}$ **b.** $\sqrt[5]{(3y)^2} = \sqrt[5]{9y^2}$ **c.** $\sqrt[7]{\dfrac{a^4}{b^4}}$ **d.** $a^{1/10} \cdot b^{-2/10} = \sqrt[10]{\dfrac{a}{b^2}}$

3. a. $x^{5/2}$ **b.** $y^{2/3}$ **c.** $(2a)^{5/6}$ **d.** $x^{1/2} \cdot x^{1/3} \cdot x^{1/6} = x^{3/6+2/6+1/6} = x$

4. a. $8^{1/3}x^{7/3} = 2x^{7/3}$ **b.** $(2x^{1/4})^3 = 8x^{3/4}$ **c.** $3x^{-2}y^{2/3} = \dfrac{3y^{2/3}}{x^2}$

   d. $\dfrac{x^{1/4} \cdot x^{1/3}}{x^{1/6}} = \dfrac{x^{7/12}}{x^{2/12}} = x^{5/12}$

5. a. $\sqrt{\dfrac{9}{25}} = \dfrac{3}{5}$ **b.** $\left[\left(\dfrac{9}{25}\right)^{1/2}\right]^{-1} = \left(\dfrac{3}{5}\right)^{-1} = \dfrac{5}{3}$ **c.** $\left[\left(\dfrac{9}{25}\right)^{1/2}\right]^5 = \left(\dfrac{3}{5}\right)^5 = \dfrac{243}{3125}$

   d. $\left[\left(\dfrac{9}{25}\right)^{1/2}\right]^{-3} = \left(\dfrac{3}{5}\right)^{-3} = \left[\left(\dfrac{3}{5}\right)^3\right]^{-1} = \left(\dfrac{27}{125}\right)^{-1} = \dfrac{125}{27}$

6. a. $\sqrt[3]{\dfrac{27}{8}} = \dfrac{3}{2}$ **b.** $\left(\sqrt[3]{\dfrac{27}{8}}\right)^2 = \left(\dfrac{3}{2}\right)^2 = \dfrac{9}{4}$

   c. $\left(\dfrac{27}{8}\right)^{-2/3} = \dfrac{1}{\left(\frac{27}{8}\right)^{2/3}} = \dfrac{1}{\frac{9}{4}}$ (from part (b)) $= \dfrac{4}{9}$ **d.** 1

7. $16^{-3/4} = (16^{1/4})^{-3} = 2^{-3} = \dfrac{1}{8}$ **8.** $25^{1/2} = 5$ **9.** $(9 - 3)^2 = 36$

10. $\left[\left(\dfrac{1}{9} + \dfrac{1}{16}\right)^{1/2}\right]^{-1} = \left[\left(\dfrac{25}{9 \cdot 16}\right)^{1/2}\right]^{-1} = \left(\dfrac{5}{12}\right)^{-1} = \dfrac{12}{5}$

11. $8^{-2/3}a^4 = (8^{2/3})^{-1} \cdot a^4 = 4^{-1}a^4 = \dfrac{a^4}{4}$ **12.** $9^{-3/2}n^{15/2} = \dfrac{n^{15/2}}{(\sqrt{9})^3} = \dfrac{n^{15/2}}{27}$

13. $4^{-1/2}x^{3/2} \cdot 4x^{1/2} = \left(\dfrac{1}{2} \cdot 4\right)x^2 = 2x^2$ **14.** $\dfrac{4^{1/3}a}{4^{-2/3}a^{-2}} = 4a^3$

15. a. $150(1.08)^{2.5} \approx \$182$ **b.** $150(1.08)^{-4.25} \approx \$108$

16. a. $2400(0.75)^{3.5} \approx \$877$ **b.** $2400(0.75)^{-5/3} \approx \$3877$

17. $a^2 - 2a$ **18.** $2n^{3/3} + 2n^0 = 2n + 2$ **19.** $x^2 - 2x$

20. $2n^{6/3} - 6n^{3/3} = 2n^2 - 6n$ **21.** $\dfrac{x^{1/2}}{x^{-1/2}} - \dfrac{2x^{-1/2}}{x^{-1/2}} = x^{1/2+1/2} - 2x^0 = x - 2$

22. $\dfrac{y^{-1/3}}{y^{-4/3}} - \dfrac{3y^{2/3}}{y^{-4/3}} = y^{-1/3+4/3} - 3y^{2/3+4/3} = y - 3y^2$ **23.** $\dfrac{2n^{1/3}}{2n^{-2/3}} - \dfrac{4n^{-2/3}}{2n^{-2/3}} = n - 2$

24. $x^{1/2}(2x^{1/2} - x^{-1/2}) = 2x - 1$

25. $\dfrac{2n^{1/3}}{2n^{-1/3}}(3n^{1/3} - 4n^{4/3}) = n^{2/3}(3n^{1/3} - 4n^{4/3}) = 3n - 4n^2$

26. $\dfrac{4ab^{-1/2}}{a^{-1}b^{-1/2}} - \dfrac{2ab^{1/2}}{a^{-1}b^{-1/2}} = 4a^2 - 2a^2b$ **27.** $\dfrac{(4a)^{2/3}}{(4a)^{1/6}} = (4a)^{1/2} = 2\sqrt{a}$

28. $\dfrac{1}{(\sqrt{2x})^4} = \dfrac{1}{(2x)^2} = \dfrac{1}{4x^2}$ **29.** $(2^3)^x = 2^6$; $2^{3x} = 2^6$; $3x = 6$; $x = 2$

30. $9^{4x} = 9^2$; $4x = 2$; $x = \dfrac{1}{2}$

31. $(2^3)^{x-1} = 2^{x+1}$; $2^{3x-3} = 2^{x+1}$; $3x - 3 = x + 1$; $2x = 4$; $x = 2$

32. $(3^2)^x = 3^{10}$; $3^{2x} = 3^{10}$; $2x = 10$; $x = 5$

33. $(2^3)^x = 2^7 \cdot (2^2)^9$; $2^{3x} = 2^{7+18}$; $3x = 25$; $x = \dfrac{25}{3}$

34. $(3^3)^{1-x} = (3^{-2})^{2-x}$; $3^{3-3x} = 3^{2x-4}$; $3 - 3x = 2x - 4$; $7 = 5x$; $x = \dfrac{7}{5}$

35. a. $[(8x)^{-3}]^{-1/3} = 64^{-1/3}$; $8x = \dfrac{1}{4}$; $x = \dfrac{1}{32}$ **b.** $x^{-3} = 8$; $(x^{-3})^{-1/3} = 8^{-1/3}$; $x = 2^{-1} = \dfrac{1}{2}$

   c. $[(8 + x)^{-3}]^{-1/3} = 64^{-1/3}$; $8 + x = 4^{-1} = \dfrac{1}{4}$; $x = -7\dfrac{3}{4} = -\dfrac{31}{4}$

36. a. $\dfrac{1}{4x^2} = 16$; $\dfrac{1}{64} = x^2$; $x = \pm\dfrac{1}{8}$ **b.** $\dfrac{1}{x^2} = 8$; $x^2 = \dfrac{1}{8}$; $x = \pm\dfrac{\sqrt{2}}{4}$

   c. $(x - 2)^{-2} = 4$; $\dfrac{1}{(x - 2)^2} = 4$; $(x - 2)^2 = \dfrac{1}{4}$; $x - 2 = \pm\dfrac{1}{2}$; $x = \dfrac{5}{2}$ or $x = \dfrac{3}{2}$

37. $150{,}000 = 50{,}000(1 + r)^{10}$; $3 = (1 + r)^{10}$; by trial-and-error, $1 + r \approx 1.12$; $r \approx 0.12$, or 12%

38. $182 = 140(1 + r)^4$; $1.3 = (1 + r)^4$; by trial-and-error, $1 + r \approx 1.07$; $r \approx 0.07$, or 7%

39. $118.3 = 100(1 + r)^4$; $1.183 = (1 + r)^4$; by trial-and-error, $1 + r \approx 1.043$; $r \approx 0.043$, or 4.3%

40. Answers may vary.

41. $18{,}500 = 14{,}000(1 + r)^4$; $(1 + r)^4 \approx 1.3214$; by trial-and-error, $1 + r \approx 1.072$; $r \approx 0.072$, or about 7%; 4 years from now the expenses will be about $18{,}500(1 + r)^4 \approx 18{,}500(1.3214) \approx \$24{,}400$.

42. Answers will vary.

43. a. $4^3 x^3 = 9^6$; $x^3 = \dfrac{(9^2)^3}{4^3}$; $x = \sqrt[3]{\dfrac{(9^2)^3}{4^3}} = \dfrac{9^2}{4} = \dfrac{81}{4}$

   b. $3^{4x} = (3^2)^6$; $3^{4x} = 3^{12}$; $4x = 12$; $x = 3$

44. a. $[(4 - x)^{1/2}]^2 = 8^2$; $4 - x = 64$; $x = -60$

   b. $(2^{-1})^{4-x} = 2^3$; $2^{x-4} = 2^3$; $x - 4 = 3$; $x = 7$

45. $2^{x^2-x} = 2^6$; $x^2 - x = 6$; $x^2 - x - 6 = 0$; $(x - 3)(x + 2) = 0$; $x = 3$ or $x = -2$

46. $\dfrac{5^{(x^2)}}{(5^x)^2} = 125$; $5^{x^2-2x} = 5^3$; $x^2 - 2x = 3$; $x^2 - 2x - 3 = 0$; $(x - 3)(x + 1) = 0$; $x = 3$ or $x = -1$

47. $\left(\dfrac{9^{x+3}}{27^x}\right)^{1/2} = 81$; $\left(\dfrac{(3^2)^{x+3}}{(3^3)^x}\right)^{1/2} = 81$; $(3^{2x+6-3x})^{1/2} = 81$; $3^{(-1/2)x+3} = 3^4$; $-\dfrac{1}{2}x + 3 = 4$; $-\dfrac{1}{2}x = 1$; $x = -2$

48. $\left(\sqrt[3]{\dfrac{8^{x+1}}{16^x}}\right)^3 = 32^3;\ \dfrac{8^{x+1}}{16^x} = 32^3;\ \dfrac{(2^3)^{x+1}}{(2^4)^x} = (2^5)^3;\ 2^{3x+3-4x} = 2^{15};\ -x + 3 = 15;\ x = -12$

49. a. $a^{3/2}b^{1/2} - a^{1/2}b^{3/2} = a^{1/2}b^{1/2}(a - b)$ **b.** $a^{1/2}b^{-1/2} - a^{3/2}b^{1/2} = a^{1/2}b^{-1/2}(1 - ab)$

50. a. $(x - 1)^{1/2} - x(x - 1)^{-1/2} = (x - 1)^{-1/2} \cdot [(x - 1) - x] = -(x - 1)^{-1/2}$

  b. $(x + 1)^{3/2} - 4(x + 1)^{1/2} = (x + 1)^{1/2} \cdot [(x + 1) - 4] = (x - 3)(x + 1)^{1/2}$

51. a. $(x^2 + 1)^{3/2} - x^2(x^2 + 1)^{1/2} = (x^2 + 1)^{1/2} \cdot [(x^2 + 1) - x^2] = (x^2 + 1)^{1/2}$

  b. $(x^2 + 2)^{1/2} - x^2(x^2 + 2)^{-1/2} = (x^2 + 2)^{-1/2} \cdot [(x^2 + 2) - x^2] = 2(x^2 + 2)^{-1/2}$

52. a. $(2x + 1)^{2/3} - 4(2x + 1)^{-1/3} = (2x + 1)^{-1/3} \cdot [(2x + 1) - 4] =$
  $(2x - 3)(2x + 1)^{-1/3}$

  b. $(1 + x^2)^{-3/2} - (1 + x^2)^{-1/2} = (1 + x^2)^{-3/2} \cdot [1 - (1 + x^2)] = -x^2(1 + x^2)^{-3/2}$

53. D: $220 \cdot 2^{5/12} \approx 293.7$ Hz; D$^\#$: $220 \cdot 2^{6/12} \approx 311.1$ Hz; E: $220 \cdot 2^{7/12} \approx 329.6$ Hz;
  F: $220 \cdot 2^{8/12} \approx 349.2$ Hz; F$^\#$: $220 \cdot 2^{9/12} \approx 370.0$ Hz; G: $220 \cdot 2^{10/12} \approx 392.0$ Hz;
  G$^\#$: $220 \cdot 2^{11/12} \approx 415.3$ Hz; A: $220 \cdot 2^{12/12} = 440$ Hz

54. a. $\dfrac{220 \cdot 2^{11/12}}{220 \cdot 2^{-1/12}} = 2^{12/12} = 2;\ 2{:}1$ **b.** $\dfrac{220 \cdot 2^{10/12}}{220 \cdot 2^{3/12}} = 2^{7/12} \approx 1.498 \approx 1.5 = \dfrac{3}{2}$

55. Let $t = x^{1/3}$; $t^2 - 7t + 12 = 0$; $(t - 4)(t - 3) = 0$; $t = 4$ or $t = 3$; $x^{1/3} = 4$ or $x^{1/3} = 3$;
  $x = 4^3 = 64$ or $x = 3^3 = 27$

56. Let $z = x^{2/3}$; $z^2 - 6z + 8 = 0$; $(z - 4)(z - 2) = 0$; $z = 4$ or $z \ne 2$; $x^{2/3} = 4$ or
  $x^{2/3} = 2$; $x = 4^{3/2} = 8$ or $x = 2^{3/2} = 2\sqrt{2}$

57. Let $z = 9^x$; $z^2 - 2z - 3 = 0$; $(z - 3)(z + 1) = 0$; $z = 3$ or $z = -1$; $9^x = 3$ or $9^x = -1$

  (no solution); $(3^2)^x = 3^1$; $2x = 1$; $x = \dfrac{1}{2}$

58. Let $z = 4^x$; $z^2 - 10z + 16 = 0$; $(z - 8)(z - 2) = 0$; $z = 8$ or $z = 2$; $4^x = 8$ or $4^x = 2$;

  $(2^2)^x = 2^3$ or $(2^2)^x = 2^1$; $2x = 3$ or $2x = 1$; $x = \dfrac{3}{2}$ or $x = \dfrac{1}{2}$

Activity 5-3 · page 180

4.6555367, 4.706965, 4.727695, 4.7287339, 4.7287859; the sequence $3^{1.4}$, $3^{1.41}$, $3^{1.414}$,...
seems to approach a fixed number that is approximately equal to 4.729.

Class Exercises 5-3 · pages 182–183

1. $(0, 7)$ is on the graph, so $7 = ab^0$; $7 = a$

2. $h(0) = 5$; $ab^0 = 5$; $a = 5$; $h(x) = 5b^x$; $h(1) = 5b^1 = 15$; $b = 3$ **3.** $b > 1$; $0 < b < 1$

4. domain, the set of all real numbers; range, the set of positive real numbers

5. $5^\pi \approx 5^3 = 125$; 156.99255; the difference is about 32.

6. $\dfrac{72}{9} = 8$ years; $\dfrac{72}{6} = 12$ mo $= 1$ yr

7. a. \$1000   **b.** $2000 = 1000(2)^{t/10}$; $2 = 2^{t/10}$; $\dfrac{t}{10} = 1$; $t = 10$ yr

8. a. 90   **b.** $9000 = 90(100)^{t/8}$; $100 = 100^{t/8}$; $\dfrac{t}{8} = 1$; $t = 8$ h

9. a. 1000   **b.** $A(4) = 1000\left(\dfrac{1}{2}\right)^{4/4} = 500$ **c.** 4 days

10. Yes; $y \approx 4(1.1)^x$; for example, \$4 invested at 10%

Written Exercises 5-3 · pages 183–186

1. $6^{\pi} \approx 278.4$; $\pi^6 \approx 961.4$ **2.** $3.6^{\sqrt{2}} \approx 6.12$; $(\sqrt{2})^{3.6} \approx 3.48$

3. $f(0) = 3$; $ab^0 = 3$; $a = 3$. $f(x) = 3b^x$; $f(1) = 3b^1 = 15$; $b = 5$; $f(x) = 3 \cdot 5^x$

4. $f(0) = 5$; $ab^0 = 5$; $a = 5$. $f(x) = 5b^x$; $f(3) = 5b^3 = 40$; $b^3 = 8$; $b = 2$; $f(x) = 5 \cdot 2^x$

5. $f(0) = 64$; $ab^0 = 64$; $a = 64$. $f(x) = 64b^x$; $f(2) = 64b^2 = 4$; $b^2 = \dfrac{1}{16}$; $b > 0$, so $b = \dfrac{1}{4}$;

$f(x) = 64\left(\dfrac{1}{4}\right)^x$

6. $f(0) = 80$; $ab^0 = 80$; $a = 80$. $f(x) = 80b^x$; $f(4) = 80b^4 = 5$; $b^4 = \dfrac{1}{16}$; $b > 0$, so

$b = \left(\dfrac{1}{16}\right)^{1/4} = \dfrac{1}{2}$; $f(x) = 80\left(\dfrac{1}{2}\right)^x$

7. $A(t) = A_0\left(\dfrac{1}{2}\right)^{t/4} = 3.2\left(\dfrac{1}{2}\right)^{t/4}$ **a.** $A(4) = 3.2\left(\dfrac{1}{2}\right)^{4/4} = 1.6$ kg

b. $A(8) = 3.2\left(\dfrac{1}{2}\right)^{8/4} = 3.2\left(\dfrac{1}{4}\right) = 0.8$ kg   **c.** $A(20) = 3.2\left(\dfrac{1}{2}\right)^{20/4} = 3.2\left(\dfrac{1}{32}\right) = 0.1$ kg

d. $A(t) = 3.2\left(\dfrac{1}{2}\right)^{t/4}$ kg

8. $A(t) = A_0\left(\dfrac{1}{2}\right)^{t/k} = 1000\left(\dfrac{1}{2}\right)^{t/1600}$

a. $A(3200) = 1000\left(\dfrac{1}{2}\right)^{3200/1600} = 1000\left(\dfrac{1}{4}\right) = 250$ g

b. $A(16{,}000) = 1000\left(\dfrac{1}{2}\right)^{16{,}000/1600} = 1000\left(\dfrac{1}{2}\right)^{10} = \dfrac{1000}{1024} \approx 0.98$ g

c. $A(800) = 1000\left(\dfrac{1}{2}\right)^{800/1600} = 1000\left(\dfrac{1}{2}\right)^{1/2} = 1000\sqrt{\dfrac{1}{2}} = 500\sqrt{2} \approx 707$ g

d. $A(t) = 1000\left(\dfrac{1}{2}\right)^{t/1600}$ g or $A(t) = \left(\dfrac{1}{2}\right)^{t/1600}$ kg

9. a. 4 days
 b. Since the half-life is 4 days, we have: 8 days, 80 g; 12 days, 40 g; 16 days, 20 g
 c. $A(t) = A_0\left(\dfrac{1}{2}\right)^{t/h}$; $A(t) = 320\left(\dfrac{1}{2}\right)^{t/4}$

10. a. Yes   **b.** The population approximately doubles from 1825 to 1900; about 75 years
 c. $P(t) = P_0 \cdot 2^{(t-1825)/k}$; $P(t) = 200 \cdot 2^{(t-1825)/75}$

11. a. $1 - 0.85 = 0.15$; 15%
 b. $2000 = 4000(0.85)^t$; $0.5 = (0.85)^t$; by trial-and-error, $t \approx 4$; in about 4 yr

12. a. $30{,}000{,}000(1.03)^{10} \approx 40{,}317{,}000$; about 40 million people   **b.** $\dfrac{72}{3} = 24$ yr

13. a. $1000(1.1)^{20} \approx \$6727$ **b.** $\dfrac{72}{10} = 7.2$; in about 7 yr

14. $P(t) = P_0(3)^{t/4}$

15. By the rule of 72, approximately $\dfrac{72}{6} = 12$ yr

16. $A(t) = A_0 \left(\dfrac{1}{2}\right)^{t/3}$; $A(30) = A_0 \left(\dfrac{1}{2}\right)^{30/3} = A_0 \left(\dfrac{1}{2}\right)^{10}$, or $\dfrac{A_0}{1024}$

17. $A(t) = A_0 \left(\dfrac{1}{2}\right)^{t/8.1}$; $A(5) = A_0 \left(\dfrac{1}{2}\right)^{5/8.1} \approx A_0 \left(\dfrac{1}{2}\right)^{0.617}$, or $0.652 A_0$

18. a. $f(-2) = 2^{-2} = \dfrac{1}{4}$; $f(-1) = 2^{-1} = \dfrac{1}{2}$; $f(0) = 2^0 = 1$; $f(1) = 2^1 = 2$; $f(2) = 2^2 = 4$

   b. See below.   **c.** f is a one-to-one function.   **d.** See below.

   e. $f(x)$ has as domain the set of all real numbers and range, the set of positive real numbers; $f^{-1}(x)$ has as domain the set of positive real numbers and range, the set of all real numbers.

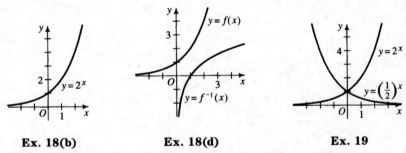

     Ex. 18(b)   **Ex. 18(d)**   **Ex. 19**

19. a. See above.   **b.** The graphs are reflections of each other in the y-axis.

20. a. See below.

   b. The graph of $y = 4^{x/2}$ is obtained by stretching the graph of $y = 4^x$ horizontally by a factor of 2.

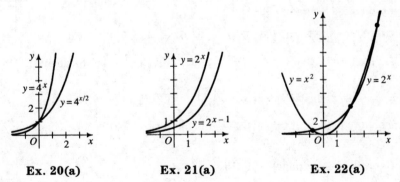

     Ex. 20(a)   **Ex. 21(a)**   **Ex. 22(a)**

21. a. See above.

   b. The graph of $y = 2^{x-1}$ is obtained by translating the graph of $y = 2^x$ one unit to the right.

22. a. See above.

   b. There are three solutions: $x \approx -0.76666$, $x = 2$, and $x = 4$.

23. Answers may vary.

a. $y = ab^x$; $76.0 = ab^{100}$ and $226.5 = ab^{180}$; $a = \dfrac{76.0}{b^{100}} = \dfrac{226.5}{b^{180}}$; $\dfrac{b^{180}}{b^{100}} = \dfrac{226.5}{76.0}$;

$b^{80} = \dfrac{2265}{760}$; $b = \left(\dfrac{2265}{760}\right)^{1/80} \approx 1.014$; $a \approx \dfrac{76.0}{b^{100}} \approx 18.924$

b. In 2000, $18.924(1.014)^{200} \approx 305.2$ million people; in 2050, $18.924(1.014)^{250} \approx 611.6$ million people.

24. a. $f(x + 1) = 2^{x+1} = 2^x \cdot 2^1 = 2 \cdot 2^x = 2f(x)$ **b.** left; vertically; 2

25. Answers will vary.

26. a. decreasing **b.** increasing **c.** Answers will vary. For example, the graph could describe a population of animals which decreases due to predators. When the number of predators decreases (at time r), the population increases again.

27. $f(x + 2) - f(x) = 10^{2(x+2)+1} - 10^{2x+1} = 10^{2x+5} - 10^{2x+1} = 10^{2x} \cdot 10^5 - 10^{2x} \cdot 10 = 10^{2x}(10^5 - 10) = 10^{2x}(99{,}990)$, which is divisible by 99.

28. a. $P(t) = 200{,}000{,}000 \cdot 2^{t/60}$

b. $300{,}000{,}000 = 200{,}000{,}000 \cdot 2^{t/60}$; $1.5 = 2^{t/60}$; by trial-and-error, $t \approx 35$; about 35 yr from now

29. To prove $f(0) = 1$, let $x = y = 0$; $f(0 + 0) = f(0) \cdot f(0)$; $f(0) = f(0) \cdot f(0)$; since $f(x) > 0$ for all x, $f(0) \neq 0$; hence, $f(0) = \dfrac{f(0)}{f(0)} = 1$. Also,

$f(2x) = f(x + x) = f(x) \cdot f(x) = [f(x)]^2$, and
$f(3x) = f(2x + x) = f(2x) \cdot f(x) = [f(x)]^2 \cdot f(x) = [f(x)]^3$. Equations may vary. For example, $f(x) = 2^x$ or $g(x) = a^x$ where $a > 0$ and $a \neq 1$.

Activity 5-4 · page 187

1. 2.59374246, 2.70481383, 2.71692393, 2.71814593, 2.71826824

2. When x becomes large, y gets closer and closer to the horizontal line $y \approx 2.7$.

Class Exercises 5-4 · pages 188–189

1. 7.3891 **2.** 24.5325 **3.** 0.0183 **4.** 4.1133 **5.** 2.7183

6. $e^\pi \approx 23.141$; $\pi^e \approx 22.459$; $e^\pi > \pi^e$ **7. a.** $\displaystyle\lim_{n \to \infty}\left(1 + \dfrac{1}{n}\right)^n$ **b.** about 2.718

8. a. $\left(1 + \dfrac{0.08}{4}\right)^4 = (1.02)^4$ **b.** $\left(1 + \dfrac{0.08}{12}\right)^{12} \approx (1.0067)^{12}$ **c.** $e^{0.08}$

9. Compounding daily at 5% yields the same interest as 5.13% compounded annually.

Written Exercises 5-4 · pages 189–190

1. a. $\left(1 + \dfrac{1}{5000}\right)^{5000} \approx 2.71801$; $\left(1 + \dfrac{1}{5{,}000{,}000}\right)^{5{,}000{,}000} \approx 2.718282$

b. They are approximately equal.

2. a. $\left(1 - \dfrac{1}{100}\right)^{100} \approx 0.36603;$ $\left(1 - \dfrac{1}{10,000}\right)^{10,000} \approx 0.36786;$

$\left(1 - \dfrac{1}{1,000,000}\right)^{1,000,000} \approx 0.36788$

b. $e^{-1} \approx 0.36788;$ the answers are approximately equal to e^{-1}. **c.** e^{-1}, or $\dfrac{1}{e}$

3. $e^{\sqrt{2}} \approx 4.1132504;$ $(\sqrt{2})^e \approx 2.5653238;$ $e^{\sqrt{2}}$ is larger.

4. a. 1.0833 **b.** 0.9231 **c.** 3.7937

5. a. $\left(1 + \dfrac{0.06}{4}\right)^4 = (1.015)^4 \approx \$1.0614$ **b.** $\left(1 + \dfrac{0.06}{12}\right)^{12} = (1.005)^{12} \approx \$1.0617$

c. $e^{0.06} \approx \$1.0618$

6. a. $\left(1 + \dfrac{0.08}{4}\right)^4 = (1.02)^4 \approx \$1.0824$ **b.** $\left(1 + \dfrac{0.08}{12}\right)^{12} = (1.00\overline{6})^{12} \approx \$1.0830$

c. $e^{0.08} \approx \$1.0833$

7. $\dfrac{107.50 - 100}{100} = \dfrac{7.5}{100} = 0.075;$ 7.5% **8.** $\dfrac{851 - 800}{800} = 0.06375;$ 6.375%

9. Plan A: $(1.06)^{10} \approx 1.7908;$ Plan B: $\left(1 + \dfrac{0.055}{4}\right)^{40} \approx 1.7268;$ an investor earns more with Plan A.

10. Plan A: $\left(1 + \dfrac{0.08}{4}\right)^{20} = (1.02)^{20} \approx 1.4859;$ Plan B: from p. 188, we use $k = 360$ for daily compounding; $\left(1 + \dfrac{0.075}{360}\right)^{1800} \approx 1.4549;$ an investor earns more with Plan A.

11. $1000 \cdot e^{(0.07)(5)} \approx \$1419.07$ **12. a.** 3000 **b.** $3000 \cdot e^{(0.01)(7)} \approx 3218$

13. a. See below. **b.** Because $f(x)$ is a one-to-one function. **c.** See below.

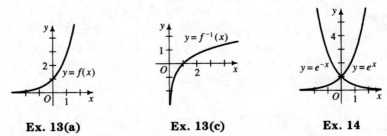

Ex. 13(a)   Ex. 13(c)   Ex. 14

14. See above. The graphs are reflections of each other in the y-axis.

15. **16.**

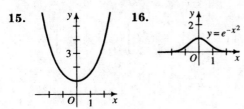

17. a. Refer to the table on p. 188. Since \$1 grows to $\left(1 + \dfrac{r}{k}\right)^k$ dollars, P dollars grows to

$P\left(1 + \dfrac{r}{k}\right)^k$ dollars.

b. $P\left(1 + \dfrac{r}{k}\right)^k = P\left(1 + \dfrac{1}{\frac{k}{r}}\right)^{(k/r)r} = P\left[\left(1 + \dfrac{1}{\frac{k}{r}}\right)^{k/r}\right]^r$; let $n = \dfrac{k}{r}$; if k is large, then $n = \dfrac{k}{r}$

is very large and we get $P\left(\left(1 + \dfrac{1}{n}\right)^n\right)^r \approx P(e)^r$ as n approaches infinity.

18. a. The number of seconds in 1 year is given by $(365)(24)(3600) = 31{,}536{,}000$;

$$10{,}000\left(1 + \dfrac{0.09}{31{,}536{,}000}\right)^{31{,}536{,}000} \approx \$10{,}941.74.$$

b. $10{,}000e^{0.09} \approx \$10{,}941.74$; the answers are approximately equal.

19. $e \approx 1 + \dfrac{1}{1!} + \dfrac{1}{2!} + \dfrac{1}{3!} + \dfrac{1}{4!} = 1 + \dfrac{1}{1} + \dfrac{1}{2} + \dfrac{1}{6} + \dfrac{1}{24} = \dfrac{24 + 24 + 12 + 4 + 1}{24} =$

$\dfrac{65}{24} \approx 2.7083$

20. $A = (0, f(0)) = (0, e^0) = (0, 1)$

a. $B = (1, f(1)) = (1, e^1)$; slope of $\overline{AB} = \dfrac{e - 1}{1 - 0} = e - 1 \approx 1.7183$

b. $B = (0.1, f(0.1)) = (0.1, e^{0.1}) \approx (0.1, 1.10517)$; slope of $\overline{AB} \approx \dfrac{1.10517 - 1}{0.1 - 0} =$

$\dfrac{0.10517}{0.1} \approx 1.0517$

c. $B = (0.01, f(0.01)) = (0.01, e^{0.01}) \approx (0.01, 1.010050)$; slope of $\overline{AB} \approx \dfrac{1.010050 - 1}{0.01 - 0} =$

$\dfrac{0.010050}{0.01} = 1.0050$

21. Let $x = \dfrac{1}{n}$. Then $\dfrac{1}{x} = n$, and if x is small, n is large. Thus, $(1 + x)^{1/x} =$

$\left(1 + \dfrac{1}{n}\right)^n \approx e$, since $\underset{n \to \infty}{\text{limit}}\left(1 + \dfrac{1}{n}\right)^n = e.$

22. $\underset{n \to \infty}{\text{limit}}\left(1 + \dfrac{1}{2n}\right)^n = \underset{n \to \infty}{\text{limit}}\left[\left(1 + \dfrac{1}{2n}\right)^{2n}\right]^{1/2}$. Let $k = 2n$. Then as $n \longrightarrow \infty$, $k \longrightarrow \infty$.

Therefore, $\underset{n \to \infty}{\text{limit}}\left[\left(1 + \dfrac{1}{2n}\right)^{2n}\right]^{1/2} = \underset{k \to \infty}{\text{limit}}\left[\left(1 + \dfrac{1}{k}\right)^k\right]^{1/2} = e^{1/2} = \sqrt{e}.$

23. Let $k = \dfrac{n}{2}$; as $n \longrightarrow \infty$, $k \longrightarrow \infty$. Therefore, $\underset{n \to \infty}{\text{limit}}\left(1 + \dfrac{2}{n}\right)^n = \underset{n \to \infty}{\text{limit}}\left[\left(1 + \dfrac{1}{\frac{n}{2}}\right)^{n/2}\right]^2 =$

$\underset{k \to \infty}{\text{limit}}\left[\left(1 + \dfrac{1}{k}\right)^k\right]^2 = e^2.$

Computer Exercise · page 190

```
100 PRINT " YIELD ON $1000 IF THE GIVEN INTEREST IS COMPOUNDED
    QUARTERLY "
110 PRINT
120 PRINT "RATE        4.5%        5.0%        5.5%        6.0%        6.5%"
130 PRINT "YEARS "
140 FOR T = 1 TO 10
150 PRINT T;
160 FOR R = .045 TO .065 STEP .005
170 LET A = 1000*(1 + R/4)^(4 * T)
175 LET A = INT(A*100 + .5)/100
180 PRINT " ";A;
190 NEXT R
200 PRINT
210 NEXT T
220 END
```

YIELD ON $1000 IF THE GIVEN INTEREST IS COMPOUNDED QUARTERLY

RATE YEARS	4.5%	5.0%	5.5%	6.0%	6.5%
1	1045.77	1050.95	1056.14	1061.36	1066.60
2	1093.62	1104.49	1115.44	1126.49	1137.64
3	1143.67	1160.75	1178.07	1195.62	1213.41
4	1196.01	1219.89	1244.21	1268.99	1294.22
5	1250.75	1282.04	1314.07	1346.85	1380.42
6	1307.99	1347.35	1387.84	1429.50	1472.36
7	1367.85	1415.99	1465.76	1517.22	1570.42
8	1430.45	1488.13	1548.06	1610.32	1675.01
9	1495.92	1563.94	1634.98	1709.14	1786.57
10	1564.38	1643.62	1726.77	1814.02	1905.56

Class Exercises 5-5 · page 194

1. a. 1.4771 **b.** 2.7686 **c.** 3.8325 **d.** -0.5229 **e.** -3.5229

2. You get an error message. Trying to find log (-7) is the same as solving $10^x = -7$. There is no solution. The domain of $f(x) = \log x$ is the set of positive real numbers.

3. a. 1.9031 **b.** 2.9031 **c.** 4.9031 **d.** $0.9030 - 1 = -0.0969$ **e.** $0.9031 - 2 = -1.0969$

4. a. 2^4 **b.** $10^{1.49}$ **c.** $10^{1.79}$

5. a. $7^x = 49; x = 2$ **b.** $2^x = 16; x = 4$ **c.** $2^x = \dfrac{1}{8}; x = -3$

  d. $5^x = \dfrac{1}{5}; x = -1$ **e.** $5^x = \sqrt{5}; x = \dfrac{1}{2}$

6. a. 0.6931 **b.** 1.0986 **c.** 0.9933 **d.** 1.0296 **e.** 1

7. a. $x = \log 50 \approx 1.70$ **b.** $x = \ln 50 \approx 3.91$

8. a. $5^2 = x$; $x = 25$ **b.** $6^2 = x$; $x = 36$ **c.** $10^2 = x$; $x = 100$ **d.** $x = e^2$

9. a. $x^2 = 121$; $x = 11$ since $x > 0$ **b.** $x^3 = 64$; $x = 4$ **c.** $x^{-1} = \dfrac{1}{2}$; $x = 2$

    d. $x^{1/2} = \sqrt{6}$; $x = 6$

Written Exercises 5-5 · pages 194–197

1. $4^2 = 16$ **2.** $4^3 = 64$ **3.** $6^{-2} = \dfrac{1}{36}$ **4.** $4^{1.5} = 8$

5. $10^3 = 1000$ **6.** $10^{1.6} \approx 40$ **7.** $e^{2.1} \approx 8$ **8.** $e^{-1.6} \approx 0.2$

9. a. $10^x = N$ **b.** (1) $x = \log 7 \approx 0.85$ (2) $x = \log 0.562 \approx -0.25$

    c. $x = \ln N$; $e^x = N$ **d.** (1) $x = \ln 12 \approx 2.48$ (2) $x = \ln 0.06 \approx -2.81$

10. a. $x = \log 170 \approx 2.23$ **b.** $x = \ln 500 \approx 6.21$

11. a. $10^2 = 100$; $\log 100 = 2$ **b.** $10^4 = 10{,}000$; $\log 10{,}000 = 4$

    c. $10^{-2} = 0.01$; $\log 0.01 = -2$ **d.** $10^{-4} = 0.0001$; $\log 0.0001 = -4$

12. a. $2^2 = 4$; $\log_2 4 = 2$ **b.** $2^5 = 32$; $\log_2 32 = 5$ **c.** $2^6 = 64$; $\log_2 64 = 6$

    d. $\log_2 (2^{10}) = 10$

13. a. $3^2 = 9$; $\log_3 9 = 2$ **b.** $3^3 = 27$; $\log_3 27 = 3$ **c.** $3^5 = 243$; $\log_3 243 = 5$

    d. $\log_3 (3^8) = 8$

14. a. $0.2 = \dfrac{1}{5} = 5^{-1}$; $\log_5 0.2 = -1$ **b.** $\dfrac{1}{125} = 5^{-3}$; $\log_5 \dfrac{1}{125} = -3$

    c. $\sqrt[3]{5} = 5^{1/3}$; $\log_5 \sqrt[3]{5} = \dfrac{1}{3}$ **d.** $\log_5 1 = \log_5 5^0 = 0$

15. a. $\log_4 64 = \log_4 4^3 = 3$ **b.** $\log_4 \dfrac{1}{64} = \log_4 4^{-3} = -3$ **c.** $\log_4 \sqrt[4]{4} = \log_4 4^{1/4} = \dfrac{1}{4}$

    d. $\log_4 1 = \log_4 4^0 = 0$

16. a. $\log_6 36 = \log_6 6^2 = 2$ **b.** $6 = \sqrt{36} = 36^{1/2}$; $\log_{36} 6 = \dfrac{1}{2}$

    c. $\log_6 6\sqrt{6} = \log_6 6^{1+(1/2)} = \dfrac{3}{2}$ **d.** $\sqrt[3]{\dfrac{1}{6}} = (6^{-1})^{1/3} = 6^{-1/3}$; $\log_6 \sqrt[3]{\dfrac{1}{6}} = -\dfrac{1}{3}$

17. a. 1 **b.** 2 **c.** $\ln e^{-1} = -1$ **d.** $\ln e^{1/2} = \dfrac{1}{2}$ **18. a.** 8 **b.** 8 **c.** 8 **d.** 8

19. If $\log 4.17 \approx 0.6201$, then $10^{0.6201} \approx 4.17$.

    a. $\log 417 = \log (4.17 \cdot 10^2) \approx \log (10^{0.6201} \cdot 10^2) = \log 10^{2.6201} = 2.6201$

    b. $\log 0.417 = \log (4.17(10^{-1})) \approx \log (10^{0.6201} \cdot 10^{-1}) = \log 10^{-0.3799} = -0.3799$

    c. $\log 0.0417 = \log (4.17(10^{-2})) \approx \log (10^{0.6201} \cdot 10^{-2}) = \log 10^{-1.3799} = -1.3799$

20. If $\log 6.92 \approx 0.8401$, then $10^{0.8401} \approx 6.92$.

    a. $\log 692 = \log (6.92(10^2)) \approx \log (10^{0.8401} \cdot 10^2) = \log 10^{2.8401} = 2.8401$

    b. $\log 0.692 = \log (6.92(10^{-1})) \approx \log (10^{0.8401} \cdot 10^{-1}) = \log 10^{-0.1599} = -0.1599$

    c. $\log 0.00692 = \log (6.92(10^{-3})) \approx \log (10^{0.8401} \cdot 10^{-3}) = \log 10^{-2.1599} = -2.1599$

21. If $\ln 10 \approx 2.3026$, then $e^{2.3026} \approx 10$.

    a. $\ln 0.1 = \ln 10^{-1} \approx \ln (e^{2.3026})^{-1} = \ln e^{-2.3026} = -2.3026$

    b. $\ln 0.01 = \ln 10^{-2} \approx \ln (e^{2.3026})^{-2} = \ln e^{-4.6052} = -4.6052$

    c. $\ln 100 = \ln 10^{2} \approx \ln (e^{2.3026})^{2} = \ln e^{4.6052} = 4.6052$

22. If $\ln 5 \approx 1.6094$, then $e^{1.6094} \approx 5$.

    a. $\ln 0.2 = \ln 5^{-1} \approx \ln (e^{1.6091})^{-1} = \ln e^{-1.6094} = -1.6094$

    b. $\ln 25 = \ln 5^{2} \approx \ln (e^{1.6094})^{2} = \ln e^{3.2188} = 3.2188$

    c. $\ln 0.04 = \ln \left(\dfrac{1}{25}\right) = \ln 5^{-2} \approx \ln (e^{1.6094})^{-2} = \ln e^{-3.2188} = -3.2188$

23. a. $10 \log_{10} 10^{6.8} = 10(6.8) = 68$ dB   **b.** $10 \log_{10} 10^{1.5} = 10(1.5) = 15$ dB

24. a. $10 \log_{10} 10^{4.1} = 10(4.1) = 41$ dB   **b.** $10 \log_{10} 10^{7.5} = 10(7.5) = 75$ dB

25. a. $I = 2 \times (10^{6.2} I_0) \approx 10^{0.3} \times (10^{6.2} I_0) = 10^{6.5} I_0$; decibel level $= 10 \log_{10} 10^{6.5} = 10(6.5) = 65$ dB

    b. $I = 3 \times (10^{6.2} I_0) \approx 10^{0.5} \times (10^{6.2} I_0) = 10^{6.7} I_0$; $10 \log_{10} 10^{6.7} = 10(6.7) = 67$ dB

26. One car has intensity $10^{8} I_0$; four will have intensity four times that amount; $I = 4 \times (10^{8} I_0) \approx 10^{0.6} \times (10^{8} I_0)$ because $\log_{10} 4 \approx 0.6$. $I = 10^{8.6} I_0$; the corresponding decibel level is $10 \log_{10} 10^{8.6} = 10(8.6) = 86$ dB.

27. a.

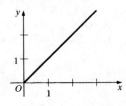

    b. The domain of f = the range of f^{-1} = {all real nos.}; the range of f = the domain of f^{-1} = {positive real nos.}.

28. a. $2^{\log_2 2^3} = 2^3 = 8$; $5^{\log_5 5^2} = 5^2 = 25$; if $x = 3^y$, then $3^{\log_3 x} = 3^{\log_3 3^y} = 3^y = x$.

    b. $f^{-1}(x) = \log_3 x$; $f(f^{-1}(x)) = f(\log_3 x) = 3^{\log_3 x} = x$

29. a. $(f \circ g)(x) = e^{\ln x} = x \ (x > 0)$; $(g \circ f)(x) = \ln e^x = x$; x is any real number

    b. $f \circ g$ is shown at the left below; $g \circ f$ is shown at the right below.

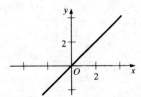

Ex. 29b and Ex. 30b

30. a. $(f \circ g)(x) = e^{-(-\ln x)} = x \ (x > 0)$; $(g \circ f)(x) = -\ln e^{-x} = -(-x) \ln e = x \ (x$ any real number)

    b. $f \circ g$ is shown at the left above; $g \circ f$ is shown at the right above.

31. a.

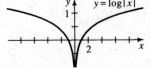

    b. For $y = \log |x|$, the domain is $\{x \,|\, x \neq 0\}$. For $y = |\log x|$, the domain is $\{x \,|\, x > 0\}$.

32. a.

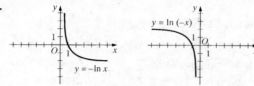

b. For $y = -\ln x$, the domain is $\{x \mid x > 0\}$ and the range is {all real numbers}.

For $y = \ln(-x)$, the domain is $\{x \mid x < 0\}$ and the range is {all real numbers}.

33. For $y = \log x + 3$, the domain is $\{x \mid x > 0\}$ and the range is {all real numbers}, and the zero is 0.001. For $y = \log(x + 3)$, the domain is $\{x \mid x > -3\}$ and the range is {all real numbers}, and the zero is -2.

34. For $y = \log_2(x - 2)$, the domain is $\{x \mid x > 2\}$ and the range is {all real nos.}. The zero is 3. For $y = \log_2 x - 2$, the domain is $\{x \mid x > 0\}$ and the range is {all real nos.}. The zero is 4.

35. a. $x = 10^3 = 1000$　**b.** $|x| = 10^3 = 1000; x = \pm 1000$

　c. $|x - 1| = 10^3 = 1000; x = 1001$ or $x = -999$

36. a. $x = 6^2 = 36$　**b.** $x = e^2$　**c.** $|x| = e^2; x = \pm e^2$

37. a. $x = 4^{3/2} = (\sqrt{4})^3 = 2^3 = 8$　**b.** $x = e^{1.5} = e^{3/2}$, or $e\sqrt{e}$　**c.** $x = e^0 = 1$

38. a. $x^2 - 1 = 10^2; x^2 = 101; x = \pm\sqrt{101}$

　b. $x^2 - 1 = e^2; x^2 = e^2 + 1; x = \pm\sqrt{e^2 + 1}$

　c. $|x| = e^1; x = \pm e$

39. a. $5^0 = \log_3 x; 1 = \log_3 x; x = 3^1 = 3$　**b.** $10^1 = \log x; x = 10^{10}$　**c.** $e^1 = \ln x; x = e^e$

40. a. $6^1 = \log_2 x; \log_2 x = 6; x = 2^6 = 64$　**b.** $e^1 = x - 2; x = e + 2$

　c. $\log x = \pm 2; x = 10^2 = 100$ or $x = 10^{-2} = 0.01$

41. a. $x = 10^{0.7} \approx 5.01$　**b.** $x = 10^{3.7} \approx 5011.87$　**c.** $x = 10^{-0.3} \approx 0.50$

42. a. $x = 10^{1.4} \approx 25.12$　**b.** $x = 10^{0.4} \approx 2.51$　**c.** $x = 10^{-0.6} \approx 0.25$

43. a. $x = e^{4.2} \approx 66.69$　**b.** $x = e^{-1.5} \approx 0.22$　**c.** $x = \ln 5 \approx 1.61$

44. a. $x = e^{1.73} \approx 5.64$　**b.** $x = e^{-0.52} \approx 0.59$　**c.** $x = \ln 16 \approx 2.77$

45. Human gastric juices: $-\log_{10} 10^{-2} = -(-2) = 2$, acidic; acid rain:
$-\log_{10}(3 \times 10^{-5}) \approx -\log_{10}(10^{0.48} \times 10^{-5}) = -\log_{10} 10^{-4.52} = -(-4.52) = 4.52$,
acidic; pure water: $-\log_{10} 10^{-7} = -(-7) = 7$, neutral; soil: $-\log_{10}(5 \times 10^{-7}) \approx$
$-\log_{10}(10^{0.70} \times 10^{-7}) = -\log_{10} 10^{-6.30} = -(-6.30) = 6.30$, acidic; sea water:
$-\log_{10} 10^{-8} = -(-8) = 8$, alkaline

46. $d = 2.5 \log_{10} r; 5 = 2.5 \log_{10} r; \log_{10} r = \dfrac{5}{2.5} = 2; r = 10^2 = 100$; one star is

100 times as bright as the other.

47. a. Since $4^2 = 16$, $\log_4 16 = 2$; since $16^{1/2} = 4$, $\log_{16} 4 = \dfrac{1}{2}$.

　b. Since $9^{3/2} = 27$, $\log_9 27 = \dfrac{3}{2}$; since $27^{2/3} = 9$, $\log_{27} 9 = \dfrac{2}{3}$.

　c. $\log_a b = \dfrac{1}{\log_b a}$. Let $\log_b a = x$ where $x \neq 0$; then $b^x = a$; also, $(b^x)^{1/x} = a^{1/x}$;

　　$b = a^{1/x}$; thus, $\log_a b = \dfrac{1}{x} = \dfrac{1}{\log_b a}$.

48. a. $\log_2 4 = 2$ since $2^2 = 4$; $\log_2 8 = 3$ since $2^3 = 8$; $\log_2 32 = 5$ since $2^5 = 32$. Hence $\log_2 4 + \log_2 8 = 2 + 3 = 5 = \log_2 32$.

   b. $\log_9 3 = \dfrac{1}{2}$ since $9^{1/2} = 3$; $\log_9 27 = \dfrac{3}{2}$ since $9^{3/2} = 27$; $\log_9 81 = 2$ since $9^2 = 81$.

   Hence $\log_9 3 + \log_9 27 = \dfrac{1}{2} + \dfrac{3}{2} = \dfrac{4}{2} = 2 = \log_9 81$.

   c. $\log_b x + \log_b y = \log_b xy$. Let $\log_b x = m$ and $\log_b y = n$. Then $b^m = x$ and $b^n = y$. Hence $xy = b^m \cdot b^n = b^{m+n}$. If $xy = b^{m+n}$, then $\log_b xy = m + n$; $\log_b xy = \log_b x + \log_b y$.

49. a. If $\log y = 1.5x - 2$, then $y = 10^{1.5x-2} = (10^{1.5x})(10^{-2}) = (10^{1.5})^x(0.01) \approx (0.01)(31.6)^x$.

   b. If $\log y = 0.5x + 1$, then $y = 10^{0.5x+1} = (10^{0.5x})(10^1) = (10^{0.5})^x(10) \approx (10)(3.16)^x$.

50. a. If $\ln y = 4x + 2$, then $y = e^{4x+2} = e^{4x} \cdot e^2 = e^2(e^4)^x \approx 7.4(54.6)^x$.

   b. If $\ln y = 1 - 0.1x$, then $y = e^{1-0.1x} = e^1 \cdot e^{-0.1x} = e(e^{-0.1})^x \approx 2.72(0.905)^x$.

51. a. (1) $\dfrac{1000}{\ln 1000} \approx 145$ (2) $\dfrac{1{,}000{,}000}{\ln 1{,}000{,}000} \approx 72{,}382$

   b. (1) $\dfrac{145}{1000} = 0.145$ (2) $\dfrac{72{,}382}{1{,}000{,}000} = 0.072382 \approx 0.072$

   c. density $= \dfrac{\dfrac{n}{\ln n}}{n} = \dfrac{n}{\ln n} \cdot \dfrac{1}{n} = \dfrac{1}{\ln n}$; as $n \longrightarrow \infty$, $\dfrac{1}{\ln n} \longrightarrow 0$.

52. Assume that $\log 2$ is rational that is, $\log 2 = \dfrac{p}{q}$ where p and q are nonzero integers and

   $\dfrac{p}{q}$ is in lowest terms. Then $10^{p/q} = 2$; $(10^{p/q})^q = 2^q$; $10^p = 2^q$; $(2 \cdot 5)^p = 2^q$;
   $2^p \cdot 5^p = 2^q$; $5^p = 2^q \div 2^p = 2^{q-p}$; hence, 5^p and 2^{q-p} must have the same prime factors; but 5^p has only 5 as a prime factor and 2^{q-p} has only 2 as a prime factor. The assumption that $\log 2$ is rational is false. It follows that $\log 2$ is irrational.

Class Exercises 5-6 · page 199

1. $2 \log M + \log N$

2. $2 \log M - \log N$

3. $\dfrac{1}{2}(\log M - \log N)$

4. $\dfrac{1}{3}(\log M + \log N)$

5. $\log M + \dfrac{1}{2} \log N$

6. $2 \log M - 3 \log N$

7. $\log_5 (2 \cdot 3) = \log_5 6$

8. $\log_3 (5 \cdot 4) = \log_3 20$

9. $\log \left(\dfrac{12}{3}\right) = \log 4$

10. $\log \left(\dfrac{3 \cdot 6}{2}\right) = \log 9$

11. $\ln 4 + \ln 3^2 = \ln (4 \cdot 3^2) = \ln 36$

12. $\ln 25^{1/2} - \ln 2 = \ln 5 - \ln 2 = \ln \dfrac{5}{2}$

13. $\log MN^2$

14. $\log \dfrac{P^2}{Q}$

15. $\log_b MNP$

16. $\log_b MN - \log_b P^3 = \log_b \dfrac{MN}{P^3}$

17. $\frac{1}{2}(\ln a - \ln b) = \frac{1}{2} \ln \frac{a}{b} = \ln \left(\frac{a}{b}\right)^{1/2}$ **18.** $\ln c + \ln d^{1/3} = \ln (cd^{1/3})$

19. Answers will vary. Example: $\log 10 + \log 1 = 1 + 0 = 1 \neq \log 11 = \log (10 + 1)$.

20. Answers will vary. Examples: $\dfrac{\log 1}{\log 10} = \dfrac{0}{1} = 0 \neq \log \left(\dfrac{1}{10}\right) = -1;\ \dfrac{\log 100}{\log 10} = \dfrac{2}{1} = 2 \neq$

$\log \left(\dfrac{100}{10}\right) = 1$.

21. It is true that $3 \log \left(\dfrac{1}{2}\right) < 2 \log \left(\dfrac{1}{2}\right)$, but $\log \left(\dfrac{1}{2}\right) < 0$. Dividing by a negative number reverses the order of the inequality.

Written Exercises 5-6 · pages 200–203

1. $\log (MN)^2 = 2 \log MN = 2(\log M + \log N)$

2. $\log \dfrac{M}{N^2} = \log M - \log N^2 = \log M - 2 \log N$

3. $\log \sqrt[3]{\dfrac{M}{N}} = \dfrac{1}{3} \log \dfrac{M}{N} = \dfrac{1}{3}(\log M - \log N)$

4. $\log M \sqrt[4]{N} = \log M + \log N^{1/4} = \log M + \dfrac{1}{4} \log N$

5. $\log M^2 \sqrt{N} = \log M^2 + \log N^{1/2} = 2 \log M + \dfrac{1}{2} \log N$

6. $\log \dfrac{1}{M} = \log M^{-1} = -\log M$ **7.** $\log (2 \cdot 3 \cdot 4) = \log 24$ **8.** $\log \dfrac{8 \cdot 5}{4} = \log 10 = 1$

9. $\log_6 9^{1/2} + \log_6 5 = \log_6 (3 \cdot 5) = \log_6 15$

10. $\log_2 48 - \log_2 27^{1/3} = \log_2 \left(\dfrac{48}{3}\right) = \log_2 16 = 4$

11. $\ln 6^2 - \ln 3 = \ln \dfrac{36}{3} = \ln 12$ **12.** $\ln 5^{1/2} + \ln 2^3 = \ln 8\sqrt{5}$

13. $\log M - \log N^3 = \log \dfrac{M}{N^3}$ **14.** $\log M^4 + \log N^{1/2} = \log M^4 (\sqrt{N})$

15. $\log A + \log B^2 - \log C^3 = \log \dfrac{AB^2}{C^3}$ **16.** $\dfrac{1}{2} \log_b \dfrac{MN}{P} = \log_b \left(\dfrac{MN}{P}\right)^{1/2} = \log_b \sqrt{\dfrac{MN}{P}}$

17. $\dfrac{1}{3}(\log_b M^2 - (\log_b N + \log_b P)) = \dfrac{1}{3}(\log_b M^2 - \log_b NP) = \dfrac{1}{3} \log_b \dfrac{M^2}{NP} =$

$\log_b \left(\dfrac{M^2}{NP}\right)^{1/3} = \log_b \sqrt[3]{\dfrac{M^2}{NP}}$

18. $5 \log_b AB - 2 \log_b C = \log_b (AB)^5 - \log_b C^2 = \log_b \dfrac{(AB)^5}{C^2}$

19. $\log \pi + \log r^2 = \log \pi r^2$ **20.** $\log \dfrac{4}{3} + \log \pi + \log r^3 = \log \dfrac{4}{3} \pi r^3$

21. $\ln (2 \cdot 6) - \ln 9^{1/2} = \ln \dfrac{12}{3} = \ln 4$

22. $\ln 10 - (\ln 5 + \ln 8^{1/3}) = \ln 10 - \ln (5 \cdot 2) = 0$

In Exercises 23–28, use the fact that e^x and $\ln x$ are inverse functions, as are 10^x and $\log x$.

23. a. 2   **b.** 3   **c.** $\ln e^{-1} = -1$ **d.** $\ln e^{1/2} = \dfrac{1}{2}$

24. a. 4   **b.** $\ln e^{-3} = -3$ **c.** $\ln e^{1/3} = \dfrac{1}{3}$ **d.** $\ln e^0 = 0$

25. a. x **b.** x **c.** $e^{2\ln x} = (e^{\ln x})^2 = x^2$ **d.** $e^{-\ln x} = \dfrac{1}{e^{\ln x}} = \dfrac{1}{x}$

26. a. $3x$ **b.** $e^{\ln x^3} = x^3$ **c.** $\sqrt{x}$ **d.** $e^{\ln x^{-1/2}} = x^{-1/2}$

27. a. 6   **b.** $10^{\log 6^2} = 6^2 = 36$ **c.** $10^3 \cdot 10^{\log 4} = 10^3 \cdot 4 = 4000$

    d. $e^3 \cdot e^{\ln 4} = e^3 \cdot 4 = 4e^3$

28. a. $10^{\log 5^3} = 5^3 = 125$ **b.** $e^{\ln 5^3} = 5^3 = 125$ **c.** $10^1 \cdot 10^{2\log x} = 10 \cdot 10^{\log x^2} = 10x^2$

    d. $e^1 \cdot e^{2\ln x} = e \cdot e^{\ln x^2} = ex^2$

29. a. $\log y = 2\log x = \log x^2$; $y = x^2$ **b.** $\log y = \log x^3 + \log 5 = \log 5x^3$; $y = 5x^3$

30. a. $\ln y = 2\ln 7 + \ln x = \ln 7^2 + \ln x = \ln 49x$; $y = 49x$

    b. $\ln y = \ln x^2 - \ln 4 = \ln \dfrac{x^2}{4}$; $y = \dfrac{x^2}{4} = 0.25x^2$

31. a. $\log y = -\log x$; $\log y = \log x^{-1} = \log \dfrac{1}{x}$; $y = \dfrac{1}{x}$

    b. $\log y = \log x^2 + \log 2 = \log 2x^2$; $y = 2x^2$

32. a. $\log y = \log 3 - \log x^{1/2} = \log \dfrac{3}{\sqrt{x}}$; $y = \dfrac{3}{\sqrt{x}}$

    b. $\ln y = \dfrac{1}{3}\ln 4x = \ln (4x)^{1/3} = \ln \sqrt[3]{4x}$; $y = \sqrt[3]{4x}$

33. a. $y = 10^{1.2x-1} = (10^{1.2})^x \cdot 10^{-1} = 0.1(10^{1.2})^x$

    b. $y = e^{1.2x-1} = (e^{1.2})^x \cdot e^{-1} = \dfrac{1}{e}(e^{1.2})^x$

34. a. $y = 10^{3-0.5x} = 10^3 \cdot (10^{-0.5})^x = 1000(10^{-0.5})^x$ **b.** $y = e^{3-0.5x} = e^3 \cdot (e^{-0.5})^x$

35. a. $\log_b \left(\dfrac{1}{x}\right) = \log_b x^{-1} = -\log_b x$,
so the graphs of $y = \log_b x$ and
$y = \log_b \left(\dfrac{1}{x}\right)$ are reflections
of each other in the x-axis.

36. a. $\log_b x^2 = 2(\log_b x)$ for $x > 0$, so the
graph of $y = \log_b x^2$ for $x > 0$ is
obtained by vertically stretching the
graph of $y = \log_b x$ by a factor of 2.
Since $\log_b x^2 = \log_b (-x)^2$, the domain
of $\log_b x^2$ is $\{x \mid x \neq 0\}$.

b.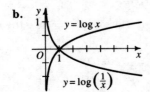

b.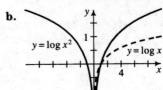

37. a. $f(2x) = \log_2 (2x) = \log_2 2 + \log_2 x = \log_2 x + 1 = f(x) + 1$

    b. Shifting the graph up one unit

38. a. $f\left(\dfrac{x}{3}\right) = \log_3\left(\dfrac{x}{3}\right) = \log_3 x - \log_3 3 = \log_3 x - 1 = f(x) - 1$

b. Part (a) shows that horizontally stretching the graph of f by a factor of 3 is equivalent to shifting it down one unit.

39. $y = 10^{ax+b} = 10^{ax}10^b = 10^b(10^a)^x$; linear; exponential

40. a. $T = \dfrac{1}{r}\ln\left(\dfrac{rR}{C} + 1\right) \approx \dfrac{1}{0.07}\ln\left(\dfrac{0.07(1691 \times 10^9)}{17 \times 10^9} + 1\right) = \dfrac{1}{0.07}\ln\left(\dfrac{118.37}{17} + 1\right) \approx 29.6$

b. $T = \dfrac{1}{r}\ln\left(\dfrac{rR}{C} + 1\right) \approx \dfrac{1}{0.07}\ln\left(\dfrac{0.07(1881 \times 10^9)}{17 \times 10^9} + 1\right) = \dfrac{1}{0.07}\ln\left(\dfrac{131.67}{17} + 1\right) \approx 31.0$

41. a. $\log_8 75 = \log_8 (5^2 \cdot 3) = \log_8 5^2 + \log_8 3 = 2\log_8 5 + \log_8 3 = 2s + r$

b. $\log_8 225 = \log_8 (3^2 \cdot 5^2) = \log_8 3^2 + \log_8 5^2 = 2\log_8 3 + 2\log_8 5 = 2r + 2s$

c. $\log_8 0.12 = \log_8\left(\dfrac{3}{25}\right) = \log_8 3 - \log_8 5^2 = \log_8 3 - 2\log_8 5 = r - 2s$

d. $\log_8 \dfrac{3}{64} = \log_8 3 - \log_8 8^2 = \log_8 3 - 2 = r - 2$

42. a. $\log_9 100 = \log_9 (5^2 \cdot 4) = \log_9 5^2 + \log_9 4 = 2\log_9 5 + \log_9 4 = 2x + y$

b. $\log_9 36 = \log_9 (9 \cdot 4) = \log_9 9 + \log_9 4 = 1 + y$

c. $\log_9 \dfrac{25}{4} = \log_9 5^2 - \log_9 4 = 2\log_9 5 - \log_9 4 = 2x - y$

d. $\log_9 3.2 = \log_9 \dfrac{16}{5} = \log_9 4^2 - \log_9 5 = 2\log_9 4 - \log_9 5 = 2y - x$

43. a. $\log_2 (x + 2) + \log_2 5 = 4$; $\log_2 5(x + 2) = 4$; $2^4 = 5x + 10$; $5x = 16 - 10$; $x = \dfrac{6}{5}$

b. $\log_4 (2x + 1) - \log_4 (x - 2) = 1$; $\log_4 \dfrac{2x + 1}{x - 2} = 1$; $4^1 = \dfrac{2x + 1}{x - 2}$; $2x = 9$; $x = \dfrac{9}{2}$

44. a. $\log_6 (x + 1) + \log_6 x = 1$; $\log_6 x(x + 1) = 1$; $6^1 = x(x + 1)$; $x^2 + x - 6 = 0$; $(x + 3)(x - 2) = 0$; $x = -3$ (reject since $x > 0$) or $x = 2$; $x = 2$

b. $\log_3 x + \log_3 (x - 2) = 1$; $\log_3 x(x - 2) = 1$; $3^1 = x(x - 2)$; $x^2 - 2x - 3 = 0$; $(x - 3)(x + 1) = 0$; $x = -1$ (reject since $x > 2$) or $x = 3$; $x = 3$

45. a. $\log_4 (x - 4) + \log_4 x = \log_4 5$; $\log_4 x(x - 4) = \log_4 5$; $x^2 - 4x = 5$; $x^2 - 4x - 5 = 0$; $(x - 5)(x + 1) = 0$; $x = -1$ (reject since $x > 4$) or $x = 5$; $x = 5$

b. $\log_2 (x^2 + 8) = \log_2 x + \log_2 6 = \log_2 6x$; $x^2 + 8 = 6x$; $x^2 - 6x + 8 = 0$; $(x - 4)(x - 2) = 0$; $x = 4$ or $x = 2$

46. a. $e^{16} = x^2$; $x = \pm\sqrt{e^{16}} = \pm e^8$ **b.** $e^{-5} = \dfrac{1}{x}$; $x = \dfrac{1}{e^{-5}} = e^5$

47. a. Since $\log_2 1 = 0$, $\log_2 M < 0$ implies that $\log_2 M < \log_2 1$. Thus, $M < 1$. Since the domain of $y = \log_2 M$ is the set of positive real numbers, $0 < M < 1$.

b. $\log_2\left(\dfrac{x - 1}{2}\right) < 0$, $0 < \dfrac{x - 1}{2} < 1$; $0 < x - 1 < 2$; $1 < x < 3$

c. $\log_3 M > 2$; $\log_3 M > \log_3 3^2$; $M > 3^2 = 9$

d. $\log_3 (x^2 + 5) > 2$; $x^2 + 5 > 9$; $x^2 > 4$; $x > 2$ or $x < -2$

e. $\log_6 (5x) > 2\log_6 x$; $\log_6 (5x) > \log_6 x^2$; $5x > x^2$; $x^2 - 5x < 0$; $x(x - 5) < 0$; $x > 0$ or $x - 5 < 0$ or $x < 0$ and $x - 5 > 0$; $x > 0$, so $0 < x < 5$

48. a. $\ln (x - 4) + \ln 3 \le 0$; $\ln (3(x - 4)) \le 0$; $0 < 3(x - 4) \le 1$; $4 < x \le \dfrac{13}{3}$

b. $\log (5 - x) - \log 7 > 0$; $\log \left(\dfrac{5 - x}{7}\right) > 0$; $\dfrac{5 - x}{7} > 1$; $5 - x > 7$; $x < -2$

49. a. $2 \log x < \log (2x - 1)$; $\log x^2 < \log (2x - 1)$; $x^2 < 2x - 1$; $x^2 - 2x + 1 < 0$; $(x - 1)^2 < 0$; no solution

b. $\ln (x + 1) - \ln 2 > 3$; $\ln \left(\dfrac{x + 1}{2}\right) > \ln e^3$; $\dfrac{x + 1}{2} > e^3$; $x + 1 > 2e^3$; $x > 2e^3 - 1$

50. a. $\log_2 (x + 5) + \log_2 (x - 2) \ge 3$; $\log_2 (x^2 + 3x - 10) \ge \log_2 8$; $x^2 + 3x - 18 > 0$; $(x + 6)(x - 3) > 0$; $x + 6 > 0$ and $x - 3 > 0$ or $x + 6 < 0$ and $x - 3 < 0$; rule out the second possibility since $x > 2$; thus, $x > 6$ and $x > 3$, that is, $x > 3$.

b. $\log_4 (x - 1) + \log_4 (x + 1) < \log_4 6$; $\log_4 (x^2 - 1) < \log_4 6$; $x^2 - 1 < 6$; $x^2 < 7$; $-\sqrt{7} < x < \sqrt{7}$; since x must be greater than 1, $1 < x < \sqrt{7}$

51. a. $\dfrac{10^{8.3}}{10^{7.1}} = 10^{1.2} \approx 16$ **b.** (1) $\dfrac{10^{7.4}}{10^6} = 10^{1.4} \approx 25$ (2) $\dfrac{10^{7.7}}{10^6} = 10^{1.7} \approx 50$

52. a. $0.67 \log (0.37E) = R - 1.46$; $\log (0.37E) = \dfrac{R - 1.46}{0.67}$; $0.37E = 10^{(R-1.46)/0.67}$;

$E = \dfrac{1}{0.37} \cdot 10^{(R-1.46)/0.67} \approx 2.7 \cdot 10^{(R-1.46)/0.67}$

b. Let $E(R) = 2.7 \cdot 10^{(R-1.46)/0.67}$. Then $E(R + 1) = 2.7 \cdot 10^{(R+1-1.46)/0.67} = 2.7 \cdot 10^{(R-1.46)/0.67} \cdot 10^{1/0.67} \approx 31 \cdot E(R)$.

53. The number of digits in 9^{9^9} is $\log 9^{9^9} = 9^9 \log 9 \approx 3.6969 \times 10^8$. At 3 digits per centimeter, the number would be $\dfrac{3.6969 \times 10^8}{3} = 1.2323 \times 10^8$ cm long, 1230 km long.

54. a. $f(1) = f(1 \cdot 1)$, and $f(1 \cdot 1) = f(1) + f(1)$. Thus $f(1) = f(1) + f(1)$, so $f(1) = 0$.

b. $f(a^2) = f(a \cdot a) = f(a) + f(a) = 2 \cdot f(a)$; $f(a^3) = f(a^2 \cdot a) = f(a^2) + f(a) = 2 \cdot f(a) + f(a) = 3 \cdot f(a)$; $f(a^n) = n \cdot f(a)$, for n a positive integer

c. $f(a) = f(\sqrt{a} \cdot \sqrt{a}) = f(\sqrt{a}) + f(\sqrt{a})$, so $2 \cdot f(\sqrt{a}) = f(a)$ and $f(\sqrt{a}) = \dfrac{1}{2} \cdot f(a)$;

$f(a) = f(\sqrt[3]{a} \cdot \sqrt[3]{a} \cdot \sqrt[3]{a}) = f(\sqrt[3]{a} \cdot \sqrt[3]{a}) + f(\sqrt[3]{a}) = f(\sqrt[3]{a}) + f(\sqrt[3]{a}) + f(\sqrt[3]{a}) = 3 \cdot f(\sqrt[3]{a})$; $f(\sqrt[3]{a}) = \dfrac{1}{3} \cdot f(a)$; $f(a^{1/n}) = \dfrac{1}{n} \cdot f(a)$, for n a positive integer

d. $f(1) = f\left(b \cdot \dfrac{1}{b}\right)$, so $f(1) = f(b) + f\left(\dfrac{1}{b}\right)$; from part (a), $0 = f(b) + f\left(\dfrac{1}{b}\right)$; $f\left(\dfrac{1}{b}\right) = -f(b)$

e. From part (d), $f\left(\dfrac{1}{b}\right) = -f(b)$. $f\left(\dfrac{a}{b}\right) = f\left(a \cdot \dfrac{1}{b}\right) = f(a) + f\left(\dfrac{1}{b}\right) = f(a) + (-f(b)) = f(a) - f(b)$

f. $f(x) = \log_b x$, $b > 0$ and $b \ne 1$

g. If $f(x) = 2$, then $f(x) = 1 + 1 = f(10) + f(10) = f(10 \cdot 10) = f(100)$, so $x = 100$; if $f(x) = 3$, then $f(x) = 1 + 2 = f(10) + f(100) = f(10 \cdot 100) = f(1000)$, so $x = 1000$.

Class Exercises 5-7 · page 205

1. $x^3 = 81$; $(x^3)^{1/3} = 81^{1/3}$; $x = \sqrt[3]{81}$, or $3\sqrt[3]{3}$ **2.** $3^x = 81$; $3^x = 3^4$; $x = 4$

3. $\log 4^x = \log 81$; $x \log 4 = \log 81$; $x = \dfrac{\log 81}{\log 4} \approx 3.17$ **4.** $x^4 = 8$; $x = \pm\sqrt[4]{8}$

5. $10^x = 3$; $\log 10^x = \log 3$; $x \log 10 = \log 3$; $x = \log 3 \approx 0.48$

6. $10^x = 8.1$; $\log 10^x = \log 8.1$; $x \log 10 = \log 8.1$; $x = \log 8.1 \approx 0.91$

7. $10^x = 256$; $\log 10^x = \log 256$; $x \log 10 = \log 256$; $x = \log 256 \approx 2.41$

8. $100^x = 302$; $\log 100^x = \log 302$; $x \log 100 = \log 302$; $x = \dfrac{1}{2} \log 302 \approx 1.24$

9. $y = \log_2 x = \dfrac{\log x}{\log 2} = \dfrac{1}{\log 2} \cdot \log x$, so graph $y = \dfrac{1}{\log 2} \log x$.

Written Exercises 5-7 · pages 205–207

1. $3^x = 12$; $\log 3^x = \log 12$; $x \log 3 = \log 12$; $x = \dfrac{\log 12}{\log 3} \approx 2.26$

2. $2^x = 100$; $\log 2^x = \log 100$; $x \log 2 = 2$; $x = \dfrac{2}{\log 2} \approx 6.64$

3. $(1.06)^x = 3$; $\log (1.06)^x = \log 3$; $x \log 1.06 = \log 3$; $x = \dfrac{\log 3}{\log 1.06} \approx 18.85$

4. $(0.98)^x = 0.5$; $\log (0.98)^x = \log 0.5$; $x \log 0.98 = \log 0.5$; $x = \dfrac{\log 0.5}{\log 0.98} \approx 34.31$

5. $e^x = 18$; $\ln e^x = \ln 18$; $x = \ln 18 \approx 2.89$

6. $e^{-x} = 0.01$; $\ln e^{-x} = \ln 0.01$; $-x = \ln 0.01$; $x = -\ln 0.01 \approx 4.61$

7. $\sqrt{e^x} = 50$; $e^x = 2500$; $\ln e^x = \ln 2500$; $x = \ln 2500 \approx 7.82$

8. $e^{3x} = 200$; $\ln e^{3x} = \ln 200$; $3x = \ln 200$; $x = \dfrac{\ln 200}{3} \approx 1.77$

9. a. $4^x = 16\sqrt{2}$; $(2^2)^x = 2^4 \cdot 2^{1/2}$; $2^{2x} = 2^{9/2}$; $2x = \dfrac{9}{2}$; $x = \dfrac{9}{4}$

b. $4^x = 20$; $\log 4^x = \log 20$; $x \log 4 = \log 20$; $x = \dfrac{\log 20}{\log 4} \approx 2.16$

10. a. $9^x = \dfrac{3}{3^x}$; $(3^2)^x \cdot 3^x = 3$; $3^{3x} = 3^1$; $3x = 1$; $x = \dfrac{1}{3}$

b. $9^x = 4$; $\log 9^x = \log 4$; $x \log 9 = \log 4$; $x = \dfrac{\log 4}{\log 9} \approx 0.63$

11. a. $25^x = (5^x)^{1/5}$; $(5^2)^x = 5^{(1/5)x}$; $2x = \dfrac{1}{5}x$; $\dfrac{9}{5}x = 0$; $x = 0$

b. $25^x = 2$; $\log 25^x = \log 2$; $x \log 25 = \log 2$; $x = \dfrac{\log 2}{\log 25} \approx 0.22$

12. a. $8^x = \left(\dfrac{2}{4^x}\right)^{1/3}$; $(2^3)^x = \dfrac{2^{1/3}}{[(2^2)^x]^{1/3}}$; $2^{3x} = 2^{1/3 - (2/3)x}$; $3x = \dfrac{1}{3} - \dfrac{2}{3}x$; $\dfrac{11}{3}x = \dfrac{1}{3}$; $x = \dfrac{1}{11}$

b. $8^x = \sqrt[3]{5}$; $\log 8^x = \log 5^{1/3}$; $x \log 8 = \dfrac{1}{3} \log 5$; $x = \dfrac{\log 5}{3 \log 8} \approx 0.26$

13. $50{,}000{,}000 = 25{,}000{,}000(1.041)^t$; $2 = (1.041)^t$; $t = \dfrac{\log 2}{\log 1.041} = 17.25$ years; about 2007

14. Note, on p. 188, that we use 360, not 365, for daily compounding. $3 = 1\left(1 + \dfrac{0.07}{360}\right)^{360t}$;

$\log 3 = \log \left(\dfrac{360.07}{360}\right)^{360t}$; $\log 3 = (360t) \log \dfrac{360.07}{360}$; $t = \log 3 \div \left(360 \log \dfrac{360.07}{360}\right) \approx 15.7$

15. a. $20,000 = 10,000r^9$; $2 = r^9$; $\log 2 = 9 \log r$; $\log r = \dfrac{\log 2}{9} \approx 0.03345$; $r \approx 1.08$;

$A(t) = 10,000(1.08)^t$

b. $30,000 = 10,000(1.08)^t$; $3 = (1.08)^t$; $t = \dfrac{\log 3}{\log 1.08} \approx 14.3$ yr

16. $P(t) = Pe^{0.01rt}$, where r is the interest rate expressed as a percent. The amount invested is doubled when $Pe^{0.01rt} = 2P$, or $e^{0.01rt} = 2$. Then $\ln e^{0.01rt} = \ln 2$; $0.01rt = \ln 2$;

$t = \dfrac{\ln 2}{0.01r} \approx \dfrac{0.693}{0.01r} = \dfrac{69.3}{r}$.

17. The amount invested is tripled when $Pe^{0.01rt} = 3P$, or $e^{0.01rt} = 3$. Then $\ln e^{0.01rt} = \ln 3$;

$0.01rt = \ln 3$; $t = \dfrac{\ln 3}{0.01r} \approx \dfrac{1.10}{0.01r} = \dfrac{110}{r}$.

18. a. $1000 = 100(1.08)^t$; $10 = (1.08)^t$; $\log 10 = t \log 1.08$; $t = \dfrac{\log 10}{\log 1.08} \approx 29.9$ years

b. $1000 = 100\left(1 + \dfrac{0.08}{4}\right)^{4t}$; $10 = (1.02)^{4t}$; $\log 10 = 4t \log (1.02)$; $t = \dfrac{\log 10}{4 \log 1.02} \approx 29.1$

c. Note on p. 188 that we use 360 for daily compounding. $1000 = 100\left(1 + \dfrac{0.08}{360}\right)^{360t}$;

$10 = \left(\dfrac{360.08}{360}\right)^{360t}$; $\log 10 = 360t \log \left(\dfrac{360.08}{360}\right)$; $t = 1 \div 360 \log \left(\dfrac{360.08}{360}\right) \approx 28.8$

19. $A(t) = 1\left(\dfrac{1}{2}\right)^{t/9.6}$ **a.** $1\left(\dfrac{1}{2}\right)^{24/9.6} \approx 0.177$ kg $= 177$ g

b. $1 = 1000\left(\dfrac{1}{2}\right)^{t/9.6}$; $\dfrac{1}{1000} = \left(\dfrac{1}{2}\right)^{t/9.6}$; $\log \dfrac{1}{1000} = \dfrac{t}{9.6} \log \dfrac{1}{2}$; $t = \dfrac{(9.6)\,(-3)}{\log \dfrac{1}{2}} \approx 95.7$ h

20. a. $\log_6 88 = \dfrac{\log 88}{\log 6} \approx 2.5$ **b.** $\log 6^x = \log 88$; $x \log 6 = \log 88$; $x = \dfrac{\log 88}{\log 6} \approx 2.5$

21. Let $\log_b c = x$; then $b^x = c$; $\log_a b^x = \log_a c$; $x \log_a b = \log_a c$; $x = \dfrac{\log_a c}{\log_a b}$;

thus, $\log_b c = \dfrac{\log_a c}{\log_a b}$.

22. a. $x^5 = 98$; $x = 98^{1/5} \approx 2.5$ **b.** $5^x = 98$; $\log 5^x = \log 98$; $x = \dfrac{\log 98}{\log 5} \approx 2.8$

23. a. You get $\log x = x \log 2$, which is no easier to solve.

b. Graph $f(x) = x$ and $g(x) = 2^x$ on the same set of axes and find the points of intersection; there are no solutions.

c. Graph $y = x + 1$ and $y = 2^x$ on the same set of axes and look for the intersection points. Graph $y = x + 2$ and $y = 2^x$ on the same set of axes and look for the intersection points. There are two solutions in each case.

d. There are 0 solutions as in part (b), or 2 solutions as in part (c).

24. Graph $y = x^x - \pi$ on a graphing calculator and find the x-intercept; $x \approx 1.85$.

25. $2^{2x} - 2^x - 6 = 0$; $(2^x - 3)(2^x + 2) = 0$; $2^x = 3$ or $2^x = -2$ (no solution);

$\log 2^x = \log 3$; $x \log 2 = \log 3$; $x = \dfrac{\log 3}{\log 2} = \log_2 3$

26. $3^{2x} - 5 \cdot 3^x + 4 = 0$; $(3^x - 4)(3^x - 1) = 0$; $3^x = 4$ or $3^x = 1$; $\log 3^x = \log 4$ or
$3^x = 3^0$; $x \log 3 = \log 4$ or $x = 0$; $x = \dfrac{\log 4}{\log 3} = \log_3 4$ or $x = 0$

27. $e^{2x} - 5e^x + 6 = 0$; $(e^x - 2)(e^x - 3) = 0$; $e^x = 2$ or $e^x = 3$; $\ln e^x = \ln 2$ or $\ln e^x =$
$\ln 3$; $x = \ln 2$ or $x = \ln 3$

28. $e^{2x} - e^x - 6 = 0$; $(e^x - 3)(e^x + 2) = 0$; $e^x = 3$ or $e^x = -2$ (no solution); $\ln e^x = \ln 3$;
$x = \ln 3$

29. $3^{2x+1} - 7 \cdot 3^x + 2 = 0$; $(3 \cdot 3^x - 1)(3^x - 2) = 0$; $3^x = \dfrac{1}{3}$ or $3^x = 2$; $3^x = 3^{-1}$ or

$\log 3^x = \log 2$; $x = -1$ or $x \log 3 = \log 2$; $x = -1$ or $x = \dfrac{\log 2}{\log 3} = \log_3 2$

30. Multiply each side by e^x: $e^{2x} + 1 = 4e^x$; $e^{2x} - 4e^x + 1 = 0$; $e^x = \dfrac{4 \pm \sqrt{16 - 4}}{2 \cdot 1} =$
$2 \pm \sqrt{3}$; $x = \ln (2 + \sqrt{3})$ or $x = \ln (2 - \sqrt{3})$

31. a. If $b^m > b^n$ and $b > 1$, then $b^{m-n} > 1 = b^0$; so, $m - n > 0$; $m > n$

   b. If $b^m < b^n$ and $0 < b < 1$, then $b^{m-n} < 1 = b^0$; so, $m - n < 0$; $m < n$

32. a. $8^x > 8^{7-x}$; $x > 7 - x$; $2x > 7$; $x > 3.5$

   b. $(0.6)^{5x} > (0.6)^{x/2}$; $5x < \dfrac{x}{2}$; $\dfrac{9x}{2} < 0$; $9x < 0$; $x < 0$

   c. $e^{3x} < e^x$; $x > 3x$; $0 > 2x$; $x < 0$

   d. $\left(\dfrac{1}{2}\right)^x > 2^{6-x}$; $2^{-x} > 2^{6-x}$; $-x > 6 - x$; No solution

33. a. $N(t) = N_0 \left(\dfrac{1}{2}\right)^{t/5700}$ **b.** $\dfrac{1}{10}N_0 = N_0 \left(\dfrac{1}{2}\right)^{t/5700}$; $0.1 = \left(\dfrac{1}{2}\right)^{t/5700}$; $\log 0.1 = \dfrac{t}{5700} \log \dfrac{1}{2}$;
$-1 = \dfrac{t}{5700}(-\log 2)$; $t = \dfrac{5700}{\log 2} \approx 18{,}900$ yr

34. $0.8N_0 = N_0 \left(\dfrac{1}{2}\right)^{t/5700}$; $0.8 = \left(\dfrac{1}{2}\right)^{t/5700}$; $\log 0.8 = \dfrac{t}{5700} \log 0.5$; $t = \dfrac{5700(\log 0.8)}{\log 0.5} \approx 1830$ yr;
since the wood is only about 1800 yr old, its date is approximately A.D. 200; it could not
be from the Hanging Garden.

35. In the formula $\log_b c = \dfrac{\log_a c}{\log_a b}$, substitute c for a: $\log_b c = \dfrac{\log_c c}{\log_c b}$; but $c^1 = c$, so $\log_c c = $
1; thus, $\log_b c = \dfrac{1}{\log_c b}$ $\left(\text{or let } x = \log_c b; c^x = b; c = b^{1/x}; \log_b c = \dfrac{1}{x} = \dfrac{1}{\log_c b}\right)$.

36. $\log_b c = \dfrac{\log_a c}{\log_a b}$; $\log_a b \log_b c = \log_a b \cdot \dfrac{\log_a c}{\log_a b} = \log_a c$

37. By Ex. 34, $\log_3 2 \cdot \log_2 27 = \log_3 27 = 3$.

38. By Ex. 34, $\log_{25} 8 \cdot \log_8 5 = \log_{25} 5 = \dfrac{1}{2}$.

39. By Ex. 33, $\dfrac{1}{\log_2 6} + \dfrac{1}{\log_3 6} = \log_6 2 + \log_6 3 = \log_6 (2 \cdot 3) = \log_6 6 = 1$.

40. By Ex. 33, $\dfrac{1}{\log_4 6} + \dfrac{1}{\log_9 6} = \log_6 4 + \log_6 9 = \log_6 (4 \cdot 9) = \log_6 36 = 2$.

41.

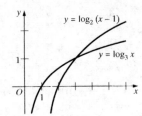

$x > 3$

42. a.

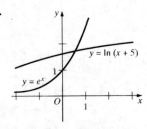

$-3.98 < x < 0.54$

b.

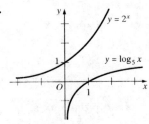

No solution

c.

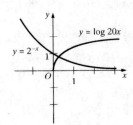

$x > 0.32$

d.

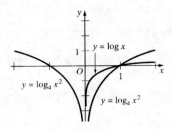

$0 < x \le 1$

43. Let $I(t) = 0.5I_0$. Then $0.5I_0 = (4^{-0.0101t})I_0$; $0.5 = 4^{-0.0101t}$; $2^{-1} = (2^2)^{-0.0101t}$;

$-1 = -0.0202t$; $t = \dfrac{1}{0.0202} \approx 49.5$ cm.   Let $I(t) = 0.02I_0$; $0.02I_0 = (4^{-0.0101t})I_0$;

$0.02 = 4^{-0.0101t}$; $\log 0.02 = \log 4^{-0.0101t}$; $\log 0.02 = -0.0101t \cdot \log 4$;

$t = \dfrac{\log 0.02}{(-0.0101) \log 4} \approx 279.4$ cm.

44. Let $x = a^{\log b}$. Then $\log x = \log (a^{\log b}) = \log b \log a$. Let $y = b^{\log a}$. Then $\log y = \log (b^{\log a}) = \log a \log b$. Since $\log x = \log y$, $x = y$. Thus, $a^{\log b} = b^{\log a}$.

45. By Ex. 35, $\dfrac{1}{\log_a ab} + \dfrac{1}{\log_b ab} = \log_{ab} a + \log_{ab} b = \log_{ab} ab$ (by Law 1, p. 197) $= 1$, since $(ab)^1 = ab$.

Chapter Test · page 209

1. $V(t) = 150{,}000(1.09)^t$

2. a. $\dfrac{2^1}{2^{-2}} = 2^3 = 8$ **b.** $\left(\dfrac{1}{25} + 1\right)^{-1} = \left(\dfrac{26}{25}\right)^{-1} = \dfrac{25}{26}$ **c.** $\dfrac{4^3}{2^{-2}} = 4^3 \cdot 2^2 = 256$

d. $\dfrac{\dfrac{1}{27} + \dfrac{1}{81}}{\dfrac{1}{27}} \cdot \dfrac{81}{81} = \dfrac{3 + 1}{3} = \dfrac{4}{3}$

3. a. $2^{6-x} = (2^2)^{2+x}$; $2^{6-x} = 2^{4+2x}$; $6 - x = 4 + 2x$; $2 = 3x$; $x = \dfrac{2}{3}$

   b. $3 \cdot \sqrt{3^3} = (3^2)^{2x}$; $3 \cdot 3^{3/2} = 3^{4x}$; $3^{5/2} = 3^{4x}$; $\dfrac{5}{2} = 4x$; $x = \dfrac{5}{8}$

4. $2.19 = 1.99(1+r)^2$; $(1+r)^2 = \dfrac{2.19}{1.99} \approx 1.1005$; $1 + r \approx 1.049$; $r \approx 0.049$; about 5%

5.

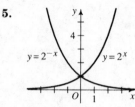

$y = 2^{-x}$ $y = 2^x$

They are reflections of each other in the y-axis.

6. Answers will vary. If the initial amount P_0 is growing at a rate of r%, then the amount present at time t is $P(t) = P_0\left(1 + \dfrac{r}{100}\right)^t$. If P_0 is decaying at a rate of r%, then $P(t) = P_0\left(1 - \dfrac{r}{100}\right)^t$. For example, $1000 invested at an annual rate of 5% will give $1000(1 + 0.05)^2$ dollars at the end of two years. A car worth $10,000 depreciating at an annual rate of 7% will be worth $10,000(1 - 0.07)^2$ dollars at the end of two years.

7. a. $\$1500\left(1 + \dfrac{0.085}{4}\right)^{4(3/2)} = \$1500(1.02125)^6 \approx \$1701.70$

   b. $\$1500e^{0.085(3/2)} \approx \$1703.98$

8. a. $\log 2.5 = \log \dfrac{25}{10} = \log 25 - \log 10 \approx 1.3979 - 1 = 0.3979$

   b. $\log (25 \cdot 100) = \log 25 + \log 100 \approx 1.3979 + 2 = 3.3979$

   c. $\log 0.04 = \log \dfrac{1}{25} = \log 1 - \log 25 = 0 - 1.3979 = -1.3979$

9. a. $2^3 = 8$; $\log_2 8 = 3$ **b.** $8^{1/3} = 2$; $\log_8 2 = \dfrac{1}{3}$

   c. $4^3 = 64$, so $\log_4 64 = 3$; $2^{\log_4 64} = 2^3 = 8$

10. a. $\log_2 y = \log_2 (2x)^2$; $y = (2x)^2 = 4x^2$

   b. $y = 2^{2 + \log_2 x} = 2^2 \cdot 2^{\log_2 x} = 4 \cdot x$; $y = 4x$ (or $\log_2 y = \log_2 4 + \log_2 x = \log_2 4x$; $y = 4x$)

11. a. $\log_b \dfrac{M^{2/3}}{N^{1/3}} = \log_b M^{2/3} - \log_b N^{1/3} = \dfrac{2}{3}\log_b M - \dfrac{1}{3}\log_b N$ **b.** $\log_b M^2 + \log_b N^3 = 2\log_b M + 3\log_b N$

12. $\log_5 x(x-4) = 1$; $x(x-4) = 5^1$; $x^2 - 4x - 5 = 0$; $(x-5)(x+1) = 0$; $x = 5$ or $x = -1$ (reject since $x > 4$); $x = 5$

13. a. $5^1 = 5$ and $5^2 = 25$; between 1 and 2   **b.** $\log_5 21 = \dfrac{\log 21}{\log 5} \approx 1.892$

14. $\log 5^x = \log 8$; $x \log 5 = \log 8$; $x = \dfrac{\log 8}{\log 5} \approx 1.29$

Class Exercises 6-1 · pages 217–218

1. a, c   **2.** a, b, and c

Written Exercises 6-1 · pages 218–219

1. Use figure (c) of Cl. Ex. 1 with $A(0,0)$, $B(2a,0)$ and $C(a,b)$. Given: Isosceles $\triangle ABC$; M and N are midpoints of $\overline{AC}$ and $\overline{BC}$, respectively. Prove: $AN = BM$.

Proof: $M = \left(\dfrac{0+a}{2}, \dfrac{0+b}{2}\right) = \left(\dfrac{a}{2}, \dfrac{b}{2}\right)$; $N = \left(\dfrac{a+2a}{2}, \dfrac{0+b}{2}\right) = \left(\dfrac{3a}{2}, \dfrac{b}{2}\right)$;

$AN = \sqrt{\left(\dfrac{3a}{2} - 0\right)^2 + \left(\dfrac{b}{2} - 0\right)^2} = \dfrac{\sqrt{9a^2 + b^2}}{2}$; $BM = \sqrt{\left(2a - \dfrac{a}{2}\right)^2 + \left(0 - \dfrac{b}{2}\right)^2} =$

$\sqrt{\left(\dfrac{3a}{2}\right)^2 + \left(-\dfrac{b}{2}\right)^2} = \dfrac{\sqrt{9a^2 + b^2}}{2}$; $AN = BM$

2. Given: In $\triangle ABC$, M and N are midpoints; $AN = BM$.
Prove: $AC = BC$. Proof: $M = (b,c)$ and $N = (a+b,c)$;
$AN = BM$, so

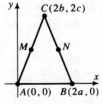

$\sqrt{(a+b-0)^2 + (c-0)^2} = \sqrt{(2a-b)^2 + (0-c)^2}$;
$(a+b)^2 + c^2 = (2a-b)^2 + c^2$; $a + b = 2a - b$
or $a + b = -(2a - b)$; $a = 2b$ or $a = 0$ (reject); thus, $B = (4b, 0)$
and $BC = \sqrt{(2b-4b)^2 + (2c-0)^2} = \sqrt{4b^2 + 4c^2} = AC$.

3. (Use the figure for Ex. 2.) Given: In $\triangle ABC$, M and N are midpoints.

Prove: $MN = \dfrac{1}{2}AB$; $\overline{MN} \parallel \overline{AB}$. Proof: $M = (b,c)$ and $N = (a+b,c)$; $\overline{MN}$ has

slope $\dfrac{c-c}{(a+b)-b} = 0$ and $\overline{AB}$ has slope $\dfrac{0-0}{2a-0} = 0$, so $\overline{MN} \parallel \overline{AB}$.

$MN = (a+b) - b = a$ and $AB = 2a - 0 = 2a$; thus, $MN = \dfrac{1}{2}AB$.

4. (Use figure (a) of Cl. Ex. 2 with $A(0,0)$, $B(a,0)$, $C(a+b,c)$, and $D(b,c)$.)
Given: $\square ABCD$. Prove $\overline{AC}$ and $\overline{BD}$ bisect each other.

Proof: $\overline{AC}$ has midpoint $\left(\dfrac{a+b}{2}, \dfrac{c}{2}\right) = $ midpoint of $\overline{BD}$. $\overline{AC}$ and $\overline{BD}$ have the same

midpoint, so they bisect each other.

5. Given: In quad. $ABDC$, $\overline{AD}$ and $\overline{BC}$ bisect each other.
Prove: $ABDC$ is a $\square$. Proof: M is the midpoint of

$\overline{AD}$ and $\overline{BC}$, so $\left(\dfrac{e}{2}, \dfrac{f}{2}\right) = \left(\dfrac{a+b}{2}, \dfrac{c}{2}\right)$; $e = a + b$

and $f = c$. $\overline{AB}$ has slope $\dfrac{0-0}{a-0} = 0$ and $\overline{CD}$ has

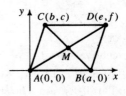

slope $\dfrac{f-c}{e-b} = \dfrac{c-c}{(a+b)-b} = 0$; $\overline{AB} \parallel \overline{CD}$. $\overline{AC}$ has slope

$\dfrac{c-0}{b-0} = \dfrac{c}{b}$ and $\overline{BD}$ has slope $\dfrac{f-0}{e-a} = \dfrac{c-0}{(a+b)-a} = \dfrac{c}{b}$;

$\overline{AC} \parallel \overline{BD}$. Thus, $ABDC$ is a $\square$. The diagonals in Cl. Ex. 2 figure (c) bisect each other (their common midpoint is $(0,0)$) so the figure is a parallelogram.

6. Given: Rectangle $ABCD$. Prove: $AC = BD$.

Proof: $AC = \sqrt{(a - 0)^2 + (b - 0)^2} = \sqrt{a^2 + b^2}$ and

$BD = \sqrt{(0 - a)^2 + (b - 0)^2} = \sqrt{a^2 + b^2}$. Thus $AC = BD$.

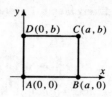

7. Given: Trap. $OPQR$ with $OR = PQ$. Prove: $OQ = PR$.

Proof: $OR = PQ$; $\sqrt{b^2 + c^2} = \sqrt{(a - d)^2 + c^2}$; $b^2 + c^2 = (a - d)^2 + c^2$; $b^2 = (a - d)^2$; $b = a - d$ or $-b = a - d$; $d < a$ and $b > 0$, so $b = a - d$; thus, $d = a - b$.

$OQ = \sqrt{d^2 + c^2} = \sqrt{(a - b)^2 + c^2} = PR$.

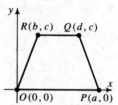

Exs. 7, 8

8. (Use the figure for Ex. 7.) Given: Trap. $OPQR$ with equal diagonals. If the diagonals are equal, $RP = OQ$; $\sqrt{(a - b)^2 + (0 - c)^2} = \sqrt{(d - 0)^2 + (c - 0)^2}$; $(a - b)^2 + c^2 = d^2 + c^2$; $(a - b)^2 = d^2$; $a - b = d$ or $a - b = -d$; $a > b$ and $d > 0$, so $a - b = d$; thus, $b = a - d$. $RO = \sqrt{b^2 + c^2} = \sqrt{(a - d)^2 + c^2} = PQ$, and so trap. $OPQR$ is isosceles.

9. Given: Quad. $JKLM$ with midpoints $W, X, Y,$ and Z.

Prove: $WXYZ$ is a $\square$.

Proof: $W = (a, 0)$, $X = (a + b, c)$,

$Y = (b + d, c + e)$, $Z = (d, e)$;

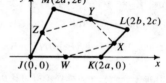

slope of $\overline{WX} = \dfrac{c}{(a + b) - a} = \dfrac{c}{b}$, slope of

$\overline{XY} = \dfrac{(c + e) - c}{(b + d) - (a + b)} = \dfrac{e}{d - a}$, slope of $\overline{YZ} = \dfrac{(c + e) - e}{(b + d) - d} = \dfrac{c}{b}$,

and slope of $\overline{WZ} = \dfrac{e}{d - a}$; opposite sides have equal slopes,

so $\overline{WX} \parallel \overline{YZ}$ and $\overline{XY} \parallel \overline{WZ}$; $WXYZ$ is a $\square$.

10. Given: Rectangle $JKLM$ with midpoints $W, X,$

$Y,$ and Z. Prove: $WXYZ$ is a rhombus.

Proof: From Ex. 9, $WXYZ$ is a $\square$. $W = (a, 0)$,

$X = (2a, b)$, and $Z = (0, b)$;

$WX = \sqrt{(2a - a)^2 + (b - 0)^2} = \sqrt{a^2 + b^2}$;

$WZ = \sqrt{(a - 0)^2 + (0 - b)^2} = \sqrt{a^2 + b^2}$;

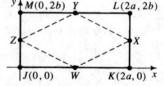

2 consecutive sides of $\square WXYZ$ are $=$, so $WXYZ$ is a rhombus.

11. (Use figure (a) of Cl. Ex. 2 with $A(0, 0)$, $B(a, 0)$, $C(a + b, c)$, and $D(b, c)$.)

Given: Parallelogram $ABCD$ with $\overline{AC} \perp \overline{BD}$. Prove: $ABCD$ is a rhombus.

Proof: To show $ABCD$ is a rhombus, we must show two adjacent sides are congruent, or

$AD = AB$. $\overline{AC} \perp \overline{BD}$, so the product of their slopes is -1; $\left(\dfrac{c}{a + b}\right)\left(\dfrac{c}{b - a}\right) = -1$;

$c^2 = (a + b)(-1)(b - a)$; $c^2 = a^2 - b^2$; $a^2 = b^2 + c^2$; $\sqrt{a^2} = \sqrt{b^2 + c^2}$.

$AB = \sqrt{(a - 0)^2 + (0 - 0)^2} = \sqrt{a^2}$ and $AD = \sqrt{(b - 0)^2 + (c - 0)^2} = \sqrt{b^2 + c^2}$, so

$AB = AD$. Hence $ABCD$ is a rhombus.

12. (Use figure (a) of Cl. Ex. 2 with $A(0,0)$, $B(a,0)$, $C(a+b,c)$, and $D(b,c)$.
  Given: Rhombus $ABCD$. Prove: $\overline{AC} \perp \overline{BD}$.
  Proof: To prove $\overline{AC} \perp \overline{BD}$, we must show that the product of their
  slopes is -1. $ABCD$ is a rhombus, so $AB = AD$;
  $\sqrt{(a-0)^2 + (0-0)^2} = \sqrt{(b-0)^2 + (c-0)^2}$; $\sqrt{a^2} = \sqrt{b^2 + c^2}$;
  thus, $a^2 = b^2 + c^2$ and $c^2 = a^2 - b^2$; slope of $\overline{AC} \cdot$ slope of $\overline{BD} =$
$$\frac{c}{a+b} \cdot \frac{c}{b-a} = \frac{c^2}{b^2-a^2} = \frac{a^2-b^2}{b^2-a^2} = -1; \text{ thus, } \overline{AC} \perp \overline{BD}.$$

13. Given: $Q(r,s)$ is in the plane of rectangle $RSTU$.
  Prove: $(QR)^2 + (QT)^2 = (QS)^2 + (QU)^2$. Proof: $(QR)^2 + (QT)^2 =$
  $(\sqrt{(r-0)^2 + (s-0)^2})^2 + (\sqrt{(r-a)^2 + (s-b)^2})^2 =$
  $r^2 + s^2 + (r-a)^2 + (s-b)^2$; $(QS)^2 + (QU)^2 =$
  $(\sqrt{(r-a)^2 + (s-0)^2})^2 + (\sqrt{(r-0)^2 + (s-b)^2})^2 =$
  $(r-a)^2 + s^2 + r^2 + (s-b)^2 = r^2 + s^2 + (r-a)^2 + (s-b)^2$;
  thus, $(QR)^2 + (QT)^2 = (QS)^2 + (QU)^2$.

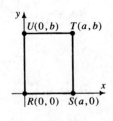

Ex. 13

14. Given: $PA = PB$. Prove: P is on the $\perp$ bis. of $\overline{AB}$. Proof: $\overline{AB}$ has slope $\dfrac{b+b}{a+a} = \dfrac{2b}{2a} = \dfrac{b}{a}$,

  so its $\perp$ bis. has slope $-\dfrac{a}{b}$ and has eq. $\dfrac{y-0}{x-0} = -\dfrac{a}{b}$, or $ax + by = 0$. $PA = PB$; let
  $P = (r,s)$; $\sqrt{(r-a)^2 + (s-b)^2} = \sqrt{(r+a)^2 + (s+b)^2}$; $(r-a)^2 + (s-b)^2 =$
  $(r+a)^2 + (s+b)^2$; $-2ar - 2bs = 2ar + 2bs$; $4ar + 4bs = 0$; $ar + bs = 0$; the
  coordinates of P satisfy the eq. of the $\perp$ bis. of $\overline{AB}$, so P lies on the $\perp$ bis. of $\overline{AB}$.

15. a. Given: $\triangle PQR$ with medians $\overline{PD}$, $\overline{QE}$, and $\overline{RF}$.
  Prove: $\overline{PD}$, $\overline{QE}$, and $\overline{RF}$ have a common point.

  Proof: D has coordinates $\left(\dfrac{c}{2}, \dfrac{b}{2}\right)$, so $\overleftrightarrow{PD}$ has equation

$$\frac{y-0}{x-a} = \frac{\frac{b}{2}}{\frac{c}{2}-a}, \text{ or } y = \frac{b(x-a)}{c-2a}. \text{ E has coordinates}$$

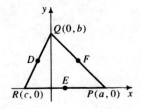

$\left(\dfrac{a+c}{2}, 0\right)$, so $\overleftrightarrow{QE}$ has equation $\dfrac{y-b}{x-0} = \dfrac{-b}{\frac{a+c}{2}}$, or $y = b - \dfrac{2bx}{a+c}$.

  Solving simultaneously, $\dfrac{b(x-a)}{c-2a} = b - \dfrac{2bx}{a+c}$; $(x-a)(a+c) =$
  $(c-2a)(a+c) - 2x(c-2a)$; $-3ax + 3cx = -a^2 + c^2$; $-3x(a-c) =$
  $-(a+c)(a-c)$; $x = \dfrac{a+c}{3}$; $y = b - \dfrac{2b}{a+c} \cdot \dfrac{a+c}{3} = \dfrac{b}{3}$;

  $\overleftrightarrow{PD}$ and $\overleftrightarrow{QE}$ intersect at $G\left(\dfrac{a+c}{3}, \dfrac{b}{3}\right)$. F has coordinates $\left(\dfrac{a}{2}, \dfrac{b}{2}\right)$, so $\overleftrightarrow{RF}$ has equation

$$\frac{y-0}{x-c} = \frac{\frac{b}{2}}{\frac{a}{2}-c}, \text{ or } y = \frac{b(x-c)}{a-2c}. \text{ The coordinates of G satisfy the equation of $\overleftrightarrow{RF}$, so}$$

  the medians of $\triangle PQR$ meet at G.

(Continued)

b. The average of the x-coordinates of P, Q, and R is $\dfrac{a + 0 + c}{3} = \dfrac{a + c}{3}$, which is the

x-coordinate of G, and the average of the y-coordinates of P, Q, and R is

$\dfrac{0 + b + 0}{3} = \dfrac{b}{3}$, which is the y-coordinate of G.

16. $PG = \sqrt{\left(a - \dfrac{a + c}{3}\right)^2 + \left(0 - \dfrac{b}{3}\right)^2} = \dfrac{1}{3}\sqrt{(2a - c)^2 + b^2};$

$GD = \sqrt{\left(\dfrac{c}{2} - \dfrac{a + c}{3}\right)^2 + \left(\dfrac{b}{2} - \dfrac{b}{3}\right)^2} = \dfrac{1}{6}\sqrt{(c - 2a)^2 + b^2} = \dfrac{1}{6}\sqrt{(2a - c)^2 + b^2};$

$PG{:}GD = 2{:}1. \quad QG = \sqrt{\left(0 - \dfrac{a + c}{3}\right)^2 + \left(b - \dfrac{b}{3}\right)^2} = \dfrac{1}{3}\sqrt{(a + c)^2 + 4b^2};$

$GE = \sqrt{\left(\dfrac{a + c}{2} - \dfrac{a + c}{3}\right)^2 + \left(0 - \dfrac{b}{3}\right)^2} = \dfrac{1}{6}\sqrt{(a + c)^2 + 4b^2}; \quad QG{:}GE = 2{:}1.$

$RG = \sqrt{\left(c - \dfrac{a + c}{3}\right)^2 + \left(0 - \dfrac{b}{3}\right)^2} = \dfrac{1}{3}\sqrt{(2c - a)^2 + b^2};$

$GF = \sqrt{\left(\dfrac{a}{2} - \dfrac{a + c}{3}\right)^2 + \left(\dfrac{b}{2} - \dfrac{b}{3}\right)^2} = \dfrac{1}{6}\sqrt{(a - 2c)^2 + b^2} = \dfrac{1}{6}\sqrt{(2c - a)^2 + b^2};$

$RG{:}GF = 2{:}1.$

17. a. Given: $\triangle PQR$ with $P(a, 0)$, $Q(0, b)$, $R(c, 0)$, and midpoints $D\left(\dfrac{c}{2}, \dfrac{b}{2}\right)$, $E\left(\dfrac{a + c}{2}, 0\right)$, and

$F\left(\dfrac{a}{2}, \dfrac{b}{2}\right)$ as in Ex. 15. Prove: The $\perp$ bisectors of $\overline{PQ}$, $\overline{QR}$, and $\overline{PR}$ meet at a point C.

Proof: The $\perp$ bisector of $\overline{PR}$ contains E and has no slope; its equation is $x = \dfrac{a + c}{2}$.

$\overline{PQ}$ has slope $-\dfrac{b}{a}$, so its $\perp$ bisector has slope $\dfrac{a}{b}$ and contains F; its equation is

$\dfrac{y - \dfrac{b}{2}}{x - \dfrac{a}{2}} = \dfrac{a}{b}$, or $y = \dfrac{a}{b}\left(x - \dfrac{a}{2}\right) + \dfrac{b}{2}$. Solving simultaneously,

$y = \dfrac{a}{b}\left(\dfrac{a + c}{2} - \dfrac{a}{2}\right) + \dfrac{b}{2}; \; y = \dfrac{ac}{2b} + \dfrac{b}{2}; \; y = \dfrac{ac + b^2}{2b}; \; C\left(\dfrac{a + c}{2}, \dfrac{ac + b^2}{2b}\right).$

$\overline{QR}$ has slope $-\dfrac{b}{c}$, so its $\perp$ bisector has slope $\dfrac{c}{b}$ and contains D; its equation

is $\dfrac{y - \dfrac{b}{2}}{x - \dfrac{c}{2}} = \dfrac{c}{b}$, or $y = \dfrac{c}{b}\left(x - \dfrac{c}{2}\right) + \dfrac{b}{2}$; the coordinates of C satisfy this equation so

the $\perp$ bisectors meet at $\left(\dfrac{a + c}{2}, \dfrac{ac + b^2}{2b}\right).$

(Continued)

b. $PC = \sqrt{\left(\dfrac{a+c}{2} - a\right)^2 + \left(\dfrac{ac+b^2}{2b} - 0\right)^2} = \sqrt{\left(\dfrac{c-a}{2}\right)^2 + \left(\dfrac{ac+b^2}{2b}\right)^2} =$

$\dfrac{1}{4}\sqrt{c^2 - 2ac + a^2 + \dfrac{a^2c^2}{b^2} + \dfrac{2acb^2}{b^2} + \dfrac{b^4}{b^2}} = \dfrac{1}{4}\sqrt{a^2 + b^2 + c^2 + \dfrac{a^2c^2}{b^2}}\,;$

$QC = \sqrt{\left(\dfrac{a+c}{2} - 0\right)^2 + \left(\dfrac{ac+b^2}{2b} - b\right)^2} = \sqrt{\left(\dfrac{a+c}{2}\right)^2 + \left(\dfrac{ac-b^2}{2b}\right)^2} =$

$\dfrac{1}{4}\sqrt{a^2 + 2ac + c^2 + \dfrac{a^2c^2}{b^2} - \dfrac{2acb^2}{b^2} + \dfrac{b^4}{b^2}} = \dfrac{1}{4}\sqrt{a^2 + b^2 + c^2 + \dfrac{a^2c^2}{b^2}}\,;$

$RC = \sqrt{\left(\dfrac{a+c}{2} - c\right)^2 + \left(\dfrac{ac+b^2}{2b} - 0\right)^2} = \sqrt{\left(\dfrac{a-c}{2}\right)^2 + \left(\dfrac{ac+b^2}{2b}\right)^2} =$

$\dfrac{1}{4}\sqrt{a^2 - 2ac + c^2 + \dfrac{a^2c^2}{b^2} + \dfrac{2acb^2}{b^2} + \dfrac{b^4}{b^2}} = \dfrac{1}{4}\sqrt{a^2 + b^2 + c^2 + \dfrac{a^2c^2}{b^2}}\,;$

$PC = QC = RC$ and thus C is equidistant from P, Q, and R.

18. From Example 3, Ex. 15, and Ex. 17, $H = \left(0, -\dfrac{ac}{b}\right)$, $G = \left(\dfrac{a+c}{3}, \dfrac{b}{3}\right)$, and

$C = \left(\dfrac{a+c}{2}, \dfrac{ac+b^2}{2b}\right).$ $\overleftrightarrow{GH}$ has slope $\dfrac{\dfrac{b}{3} + \dfrac{ac}{b}}{\dfrac{a+c}{3} - 0} = \dfrac{b^2 + 3ac}{b(a+c)}\,;$ $\overleftrightarrow{HC}$ has slope

$\dfrac{\dfrac{ac+b^2}{2b} + \dfrac{ac}{b}}{\dfrac{a+c}{2} - 0} = \dfrac{ac+b^2+2ac}{b(a+c)} = \dfrac{b^2+3ac}{b(a+c)}.$ Since $\overleftrightarrow{HC}$ and $\overleftrightarrow{GH}$ have equal

slopes, G, H, and C are collinear. $GH = \sqrt{\left(\dfrac{a+c}{3} - 0\right)^2 + \left(\dfrac{b}{3} + \dfrac{ac}{b}\right)^2} =$

$\sqrt{\left(\dfrac{a+c}{3}\right)^2 + \left(\dfrac{b^2+3ac}{3b}\right)^2} = \dfrac{1}{3}\sqrt{(a+c)^2 + \left(\dfrac{b^2+3ac}{b}\right)^2} =$

$\dfrac{1}{3}\sqrt{a^2 + 2ac + c^2 + \dfrac{b^4}{b^2} + \dfrac{6acb^2}{b^2} + \dfrac{9a^2c^2}{b^2}} = \dfrac{1}{3}\sqrt{a^2 + b^2 + c^2 + 8ac + \dfrac{9a^2c^2}{b^2}}\,;$

$GC = \sqrt{\left(\dfrac{a+c}{2} - \dfrac{a+c}{3}\right)^2 + \left(\dfrac{ac+b^2}{2b} - \dfrac{b}{3}\right)^2} = \sqrt{\left(\dfrac{a+c}{6}\right)^2 + \left(\dfrac{3ac+b^2}{6b}\right)^2} =$

$\dfrac{1}{6}\sqrt{a^2 + 2ac + c^2 + \dfrac{9a^2c^2}{b^2} + \dfrac{6acb^2}{b^2} + \dfrac{b^4}{b^2}} = \dfrac{1}{6}\sqrt{a^2 + b^2 + c^2 + 8ac + \dfrac{9a^2c^2}{b^2}}\,;$

thus, $GH : GC = \dfrac{1}{3} : \dfrac{1}{6} = 2 : 1$; $GH = 2GC$.

Class Exercises 6-2 · page 222

1. $C(0,0)$; $r = 4$ **2.** $C(2,7)$; $r = 6$ **3.** $C(4,-7)$; $r = \sqrt{7}$

4. $x^2 + y^2 + 12y + 36 = 36$; $x^2 + (y+6)^2 = 36$; $C(0,-6)$; $r = 6$

5. $x^2 - 2x + 1 + y^2 - 6y + 9 = 19$; $(x-1)^2 + (y-3)^2 = 19$; $C(1,3)$; $r = \sqrt{19}$

6. $x^2 + y^2 = 9$; $C(0,0)$; $r = 3$ **7.** $(x-7)^2 + (y-3)^2 = 36$

8. $(x+5)^2 + (y-4)^2 = 2$

9. $x^2 + y^2 = r^2$; $(-5)^2 + (12)^2 = r^2$; $r^2 = 169$; $x^2 + y^2 = 169$

10. a. All points in the plane that are less than 5 units from the origin, that is, inside the circle $x^2 + y^2 = 25$

b. All points in the plane that are more than 5 units from the origin, that is, outside the circle $x^2 + y^2 = 25$

11. (1) Solve for x in terms of y in $3y + x = 6$; (2) substitute this expression for x in $x^2 + y^2 = 10$; (3) solve for y in the resulting quadratic equation; (4) substitute these y-values in the linear equation to find the corresponding x-coordinates of the points of intersection.

12. The line and the circle do not intersect.

13. Answers may vary. Alter the scale, condensing the x-axis (or stretching the y-axis) to compensate for the distortion.

Written Exercises 6-2 · pages 222–225

1. $(x - 4)^2 + (y - 3)^2 = 4$ **2.** $(x - 5)^2 + (y + 6)^2 = 49$

3. $(x + 4)^2 + (y + 9)^2 = 9$ **4.** $(x - a)^2 + (y - b)^2 = f^2$

5. $(x - 6)^2 + y^2 = 15$ **6.** $(x + 4)^2 + (y - 2)^2 = 7$

7. $r = \sqrt{(5 - 2)^2 + (6 - 3)^2} = \sqrt{9 + 9} = 3\sqrt{2}$; $(x - 2)^2 + (y - 3)^2 = 18$

8. The center is the midpoint of the diameter: $\left(\dfrac{8 + 0}{2}, \dfrac{0 + 6}{2}\right) = (4, 3)$; $(8, 0)$ and $(4, 3)$ are endpoints of a radius: $r = \sqrt{(8 - 4)^2 + (0 - 3)^2} = \sqrt{16 + 9} = 5$; $(x - 4)^2 + (y - 3)^2 = 25$.

9. $r = $ the distance from the center to the x-axis $= |0 - (-4)| = 4$; $(x - 5)^2 + (y + 4)^2 = 16$

10. $r = $ the distance from the center to line $x = 4$, so $r = |4 - (-3)| = 7$; $(x + 3)^2 + (y - 1)^2 = 49$

11. Let $C = (h, k)$; the circle is tangent to the x-axis at $(4, 0)$, so $h = 4$. The circle contains $(0, -2)$ and $(0, -8)$, so $k = \dfrac{-2 + (-8)}{2} = -5$. $C = (4, -5)$. $r = $ the distance from C to the x-axis $= |0 - (-5)| = 5$; $(x - 4)^2 + (y + 5)^2 = 25$

12. Let $C = (h, k)$; the circle contains $(-2, 0)$ and $(-32, 0)$, so $h = \dfrac{-2 + (-32)}{2} = -17$; the circle contains $(-2, 0)$ and $(-2, 16)$, so $k = \dfrac{0 + 16}{2} = 8$; $r = \sqrt{(-17 + 2)^2 + (8 - 0)^2} = \sqrt{289} = 17$; $(x + 17)^2 + (y - 8)^2 = 289$

13. $x^2 + y^2 - 2x - 8y + 16 = 0$; $x^2 - 2x + 1 + y^2 - 8y + 16 = -16 + 1 + 16$; $(x - 1)^2 + (y - 4)^2 = 1$; $C(1, 4)$; $r = 1$

14. $x^2 + y^2 - 4x + 6y + 4 = 0$; $x^2 - 4x + 4 + y^2 + 6y + 9 = -4 + 4 + 9$; $(x - 2)^2 + (y + 3)^2 = 9$; $C(2, -3)$; $r = 3$

15. $x^2 + y^2 - 12y + 25 = 0$; $x^2 + y^2 - 12y + 36 = -25 + 36$; $x^2 + (y - 6)^2 = 11$; $C(0, 6)$; $r = \sqrt{11}$

16. $x^2 + y^2 + 14x = 0$; $x^2 + 14x + 49 + y^2 = 49$; $(x + 7)^2 + y^2 = 49$; $C(-7, 0)$; $r = 7$

17. $2x^2 + 2y^2 - 10x - 18y = 1; \; x^2 - 5x + y^2 - 9y = \dfrac{1}{2};$

$x^2 - 5x + \dfrac{25}{4} + y^2 - 9y + \dfrac{81}{4} = \dfrac{2}{4} + \dfrac{25}{4} + \dfrac{81}{4}; \; \left(x - \dfrac{5}{2}\right)^2 + \left(y - \dfrac{9}{2}\right)^2 = 27; \; C\left(\dfrac{5}{2}, \dfrac{9}{2}\right);$

$r = 3\sqrt{3}$

18. $2x^2 + 2y^2 - 5x + y = 0; \; x^2 - \dfrac{5}{2}x + y^2 + \dfrac{1}{2}y = 0;$

$x^2 - \dfrac{5}{2}x + \dfrac{25}{16} + y^2 + \dfrac{1}{2}y + \dfrac{1}{16} = \dfrac{25}{16} + \dfrac{1}{16}; \; \left(x - \dfrac{5}{4}\right)^2 + \left(y + \dfrac{1}{4}\right)^2 = \dfrac{26}{16}; \; C\left(\dfrac{5}{4}, -\dfrac{1}{4}\right);$

$r = \dfrac{\sqrt{26}}{4}$

19. $x^2 + y^2 + 6x - 4y + 8 = 0; \; (x^2 + 6x + 9) + (y^2 - 4y + 4) = 9 + 4 - 8;$
$(x + 3)^2 + (y - 2)^2 = 5; \; C(-3, 2); \; 2 = 2(-3) + 8$, so the line contains the center of the circle.

20. $x^2 + y^2 + 4x - 12y + 24 = 0; \; (x^2 + 4x + 4) + (y^2 - 12y + 36) = 4 + 36 - 24;$
$(x + 2)^2 + (y - 6)^2 = 16; \; C(-2, 6); \; 3(-2) + 2(6) = 6$, so the line contains the center of the circle.

21.

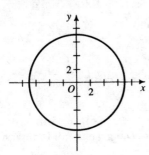

22.

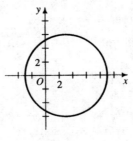

23.

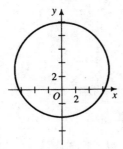

24. $y = \sqrt{9 - x^2}$

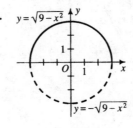

$y = -\sqrt{9 - x^2}$

25. $x = \sqrt{9 - y^2}$

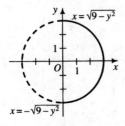

$x = -\sqrt{9 - y^2}$

26. $y = \sqrt{16 - (x - 5)^2}$

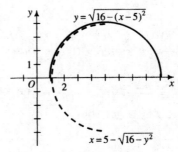

$x = 5 - \sqrt{16 - y^2}$

27. $x + y = 23$; $y = 23 - x$. $x^2 + y^2 = 289$; $x^2 + (23 - x)^2 = 289$;
　　$2x^2 - 46x + 240 = 0$; $x^2 - 23x + 120 = 0$; $(x - 15)(x - 8) = 0$; $x = 15$ and
　　$y = 23 - 15 = 8$, or $x = 8$ and $y = 23 - 8 = 15$; $(15, 8)$ and $(8, 15)$.

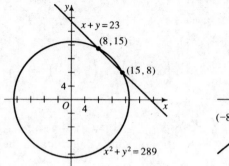

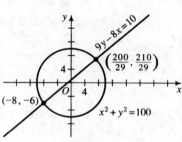

Ex. 27　　　　　　　　　　　　　　　　Ex. 28

28. $9y - 8x = 10$; $y = \dfrac{8x + 10}{9}$. $x^2 + y^2 = 100$; $x^2 + \left(\dfrac{8x + 10}{9}\right)^2 = 100$;
　　$81x^2 + 64x^2 + 160x + 100 = 8100$; $145x^2 + 160x - 8000 = 0$;
　　$29x^2 + 32x - 1600 = 0$; $(29x - 200)(x + 8) = 0$; $x = \dfrac{200}{29}$ and
　　$y = \dfrac{1}{9}\left(8 \cdot \dfrac{200}{29} + 10\right) = \dfrac{210}{29}$, or $x = -8$ and $y = \dfrac{8(-8) + 10}{9} = -6$; $\left(\dfrac{200}{29}, \dfrac{210}{29}\right)$
　　and $(-8, -6)$.

29. $2x - y = 7$; $y = 2x - 7$. $x^2 + y^2 = 7$; $x^2 + (2x - 7)^2 = 7$; $5x^2 - 28x + 42 = 0$;
　　$b^2 - 4ac = (-28)^2 - 4(5)(42) < 0$, so there is no solution; the graphs fail to intersect.

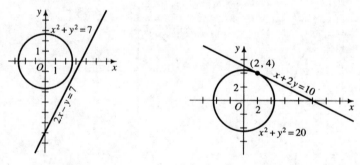

Ex. 29　　　　　　　　　　　　　　　　Ex. 30

30. $x + 2y = 10$; $x = 10 - 2y$. $x^2 + y^2 = 20$; $(10 - 2y)^2 + y^2 = 20$;
　　$5y^2 - 40y + 80 = 0$; $y^2 - 8y + 16 = 0$; $(y - 4)^2 = 0$; $y = 4$; $x = 10 - 2(4) = 2$;
　　tangent at $(2, 4)$.

31. $5x + 2y = -1$; $y = \dfrac{-5x - 1}{2}$. $x^2 + y^2 = 169$; $x^2 + \left(\dfrac{-5x - 1}{2}\right)^2 = 169$;
　　$4x^2 + 25x^2 + 10x + 1 = 676$; $29x^2 + 10x - 675 = 0$; $(29x - 135)(x + 5) = 0$;
　　$x = \dfrac{135}{29}$ and $y = \dfrac{1}{2}\left(-5 \cdot \dfrac{135}{29} - 1\right) = -\dfrac{352}{29}$, or $x = -5$ and $y = \dfrac{-5(-5) - 1}{2} = 12$;
　　$\left(\dfrac{135}{29}, -\dfrac{352}{29}\right)$ and $(-5, 12)$.

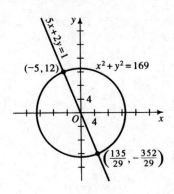

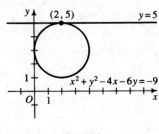

Ex. 32

Ex. 31

32. $y = 5$; $x^2 + y^2 - 4x - 6y = -9$; $x^2 + 25 - 4x - 30 = -9$; $x^2 - 4x + 4 = 0$;
$(x - 2)^2 = 0$; $x = 2$; $y = 5$; tangent at $(2, 5)$.

33. $y = \sqrt{3}x$; $x = \dfrac{y}{\sqrt{3}}$. $x^2 + (y - 4)^2 = 16$; $\dfrac{y^2}{3} + y^2 - 8y + 16 = 16$; $\dfrac{4}{3}y^2 - 8y = 0$;

$4y^2 - 24y = 0$; $4y(y - 6) = 0$; $y = 0$ and $x = 0$, or $y = 6$ and $x = \dfrac{6}{\sqrt{3}} = 2\sqrt{3}$; $(0,0)$

and $(2\sqrt{3}, 6)$.

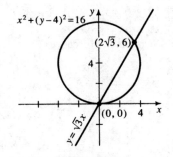

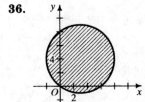

Ex. 33   **Ex. 34**

34. $x - y = 3$; $y = x - 3$. $x^2 + y^2 - 10x + 4y = -13$;
$x^2 + (x - 3)^2 - 10x + 4(x - 3) = -13$; $2x^2 - 12x + 10 = 0$; $x^2 - 6x + 5 = 0$;
$(x - 5)(x - 1) = 0$; $x = 5$ and $y = 5 - 3 = 2$, or $x = 1$ and $y = 1 - 3 = -2$; $(5, 2)$
and $(1, -2)$.

35. The graph of $x^2 + y^2 < 1$ is the set of all
points in the plane inside the circle
$x^2 + y^2 = 1$, and the graph of $x^2 + y^2 > 1$
is the set of all points in the plane outside
the circle $x^2 + y^2 = 1$.

36.

37. Center of circle is $(0, 0)$. To find A and B, solve $x - 2y = 15$ and $x^2 + y^2 = 50$ simultaneously; $x = 2y + 15$; $(2y + 15)^2 + y^2 = 50$; $5y^2 + 60y + 175 = 0$; $y^2 + 12y + 35 = 0$; $(y + 7)(y + 5) = 0$; $y = -7$ and $x = 2(-7) + 15 = 1$, or $y = -5$ and $x = 2(-5) + 15 = 5$; $A(1, -7)$ and $B(5, -5)$. $\overline{AB}$ has midpoint

$M\left(\dfrac{1 + 5}{2}, \dfrac{-7 - 5}{2}\right) = (3, -6)$ and slope $\dfrac{-5 + 7}{5 - 1} = \dfrac{1}{2}$. $\overline{OM}$ has slope $\dfrac{-6 - 0}{3 - 0} = -2$. Since

$\dfrac{1}{2}(-2) = -1$, $\overline{OM} \perp \overline{AB}$.

38. The line $y = 5$ is tangent to the circle at $P(0, 5)$, and it contains $Q(10, 5)$. $PQ = |10 - 0| = 10$.

39. a. $(x - 0)^2 + (y - 0)^2 = r^2$; $x^2 + y^2 = r^2$; $2^2 + 3^2 = r^2$; $r^2 = 13$; $x^2 + y^2 = 13$

 b. $\overline{OP}$ has slope $\dfrac{3 - 0}{2 - 0} = \dfrac{3}{2}$, so the tangent has slope $-\dfrac{2}{3}$; the tangent contains P; thus,

the tangent has eq. $\dfrac{y - 3}{x - 2} = -\dfrac{2}{3}$, or $2x + 3y = 13$.

40. $(4 - 3)^2 + (2 - 4)^2 = 1^2 + (-2)^2 = 5$, so $(4, 2)$ is on the circle; $C(3, 4)$ and $P(4, 2)$;

$\overline{CP}$ has slope $\dfrac{2 - 4}{4 - 3} = -2$, so the tangent has slope $\dfrac{1}{2}$; since the tangent contains P, its

eq. is $\dfrac{y - 2}{x - 4} = \dfrac{1}{2}$, or $y = \dfrac{x}{2}$.

41. a. The circle has eq. $(x - 2)^2 + (y - 4)^2 = 169$.
  $(14 - 2)^2 + (9 - 4)^2 = 144 + 25 = 169$, so A is on the circle.
  $(7 - 2)^2 + (16 - 4)^2 = 25 + 144 = 169$, so B is on the circle.

 b. $M = \left(\dfrac{14 + 7}{2}, \dfrac{9 + 16}{2}\right) = \left(\dfrac{21}{2}, \dfrac{25}{2}\right)$; slope of $\overline{CM} = \dfrac{4 - \frac{25}{2}}{2 - \frac{21}{2}} = 1$; slope of

$\overline{AB} = \dfrac{16 - 9}{7 - 14} = -1$; $1(-1) = -1$, so $\overline{CM} \perp \overline{AB}$.

42. a. The circle has eq. $(x + 4)^2 + y^2 = 225$. $(8 + 4)^2 + 9^2 = 144 + 81 = 225$, so A is on the circle. $(-13 + 4)^2 + 12^2 = 81 + 144 = 225$, so B is on the circle.

 b. $M = \left(\dfrac{8 - 13}{2}, \dfrac{9 + 12}{2}\right) = \left(-\dfrac{5}{2}, \dfrac{21}{2}\right)$; $\overline{AB}$ has slope $\dfrac{12 - 9}{-13 - 8} = -\dfrac{1}{7}$; the $\perp$ bis. of

$\overline{AB}$ has eq. $\dfrac{y - \frac{21}{2}}{x + \frac{5}{2}} = 7$, or $7x - y = -28$; $7(-4) - 0 = -28$, so C lies on the $\perp$ bis. of $\overline{AB}$.

 c. The perpendicular bisector of a chord of a circle contains the center of the circle.

43. a. $C = $ midpoint of $\overline{AB} = \left(\dfrac{13 - 13}{2}, \dfrac{0 + 0}{2}\right) = (0, 0)$; $r = \sqrt{(13 - 0)^2 + (0 - 0)^2} = 13$;

the circle has eq. $x^2 + y^2 = 169$. $(-5)^2 + 12^2 = 169$, so P lies on the circle.

 b. slope of $\overline{PA} = \dfrac{0 - 12}{13 + 5} = -\dfrac{2}{3}$ and slope of $\overline{PB} = \dfrac{0 - 12}{-13 + 5} = \dfrac{3}{2}$; $-\dfrac{2}{3} \cdot \dfrac{3}{2} = -1$; so

$\overline{PA} \perp \overline{PB}$.

44. a. $(x^2 - 34x + 289) + y^2 = 289$; $(x - 17)^2 + y^2 = 289$; $C = (17, 0)$, so the horizontal diameter lies on the line $y = 0$; when $y = 0$, $(x - 17)^2 + 0^2 = 17^2$; $x - 17 = \pm 17$; $x = 34$ or $x = 0$; thus, $A = (0, 0)$ and $B = (34, 0)$.

b. $(2 - 17)^2 + 8^2 = 225 + 64 = 289$, so P is on the circle. $\overline{PA}$ has slope $\dfrac{8 - 0}{2 - 0} = 4$ and

$\overline{PB}$ has slope $\dfrac{8 - 0}{2 - 34} = -\dfrac{1}{4}$; $4\left(-\dfrac{1}{4}\right) = -1$, so $\overline{PA} \perp \overline{PB}$.

45. $\overline{PO}$ has slope $\dfrac{y}{x}$ and $\overline{PN}$ has slope $\dfrac{y - 0}{x - 12} = \dfrac{y}{x - 12}$. If $\overline{PO} \perp \overline{PN}$, $\dfrac{y}{x} \cdot \dfrac{y}{x - 12} = -1$;

$y^2 = 12x - x^2$; $(x^2 - 12x + 36) + y^2 = 36$; $(x - 6)^2 + y^2 = 36$; P is on a circle with center $(6, 0)$ and radius 6.

46. $\overline{PA}$ has slope $\dfrac{y - 8}{x - 6}$ and $\overline{PB}$ has slope $\dfrac{y + 8}{x + 6}$. If $\overline{PA} \perp \overline{PB}$, $\dfrac{y - 8}{x - 6} \cdot \dfrac{y + 8}{x + 6} = -1$;

$\dfrac{y^2 - 64}{x^2 - 36} = -1$; $y^2 - 64 = 36 - x^2$; $x^2 + y^2 = 100$. If $\overline{PA} \perp \overline{PB}$, then A, P, and B are on the circle $x^2 + y^2 = 100$. Note also that right $\angle APB$ is inscribed in a semicircle.

47. a. $x^2 + y^2 = 16$; $y = \sqrt{16 - x^2}$

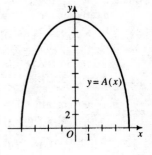

b. $A = \dfrac{1}{2}(8)(y) = 4y$, $A(x) = 4\sqrt{16 - x^2}$

c. $16 - x^2 \geq 0$; thus, the domain of $A(x) = \{x \mid -4 \leq x \leq 4\}$.

d. See diagram at right.

e. $x = 0$

$y = A(x)$

f. $A(0) = 4\sqrt{16} = 16$

48. a. $A = 4xy$, $A(x) = 4x\sqrt{16 - x^2}$

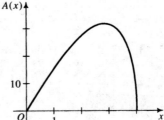

b. Maximum occurs at $x \approx 2.83$; maximum value of A is 32.

49. a. $A = \dfrac{1}{2}(b_1 + b_2)h = \dfrac{1}{2}(2x + 8)y$,

$A(x) = (x + 4)\sqrt{16 - x^2}$

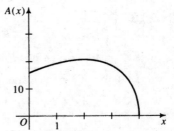

b. Maximum area occurs at $x = 2$; $A(2) = 6\sqrt{12} = 12\sqrt{3} \approx 20.8$

50. $\angle ABC$ is a right angle; thus, $\overline{AC}$ is a diameter of the circle. Midpoint of $\overline{AC}$ is $(1, 1)$ so the circle has center $C(1, 1)$ and $r = \sqrt{(1 - 0)^2 + (1 - 0)^2} = \sqrt{2}$; $(x - 1)^2 + (y - 1)^2 = 2$.

51. $\angle QPR$ is a right angle; thus $\overline{QR}$ is a diameter of the circle. Midpoint of $\overline{QR}$ is $(3, 4)$, so the circle has center $C(3, 4)$ and $r = \sqrt{(3 - 0)^2 + (4 - 0)^2} = 5$.
$(x - 3)^2 + (y - 4)^2 = 25$

52. $(x - h)^2 + (y - k)^2 = r^2$; (1) $(8 - h)^2 + (2 - k)^2 = r^2$; (2) $(1 - h)^2 + (9 - k)^2 = r^2$; (3) $(1 - h)^2 + (1 - k)^2 = r^2$; subtracting (3) from (2), $80 - 16k = 0$ and $k = 5$. Thus, $(8 - h)^2 + 9 = r^2$ and $(1 - h)^2 + 16 = r^2$; $(64 - 16h + h^2 + 9) - (1 - 2h + h^2 + 16) = 0$; $56 = 14h$; $h = 4$. $(8 - 4)^2 + 9 = r^2$; $r^2 = 25$.
$(x - 4)^2 + (y - 5)^2 = 25$.

53. $(x - h)^2 + (y - k)^2 = r^2$; (1) $(7 - h)^2 + (5 - k)^2 = r^2$; (2) $(1 - h)^2 + (-7 - k)^2 = r^2$; (3) $(9 - h)^2 + (-1 - k)^2 = r^2$. Subtracting (2) from (1): $24 - 12h - 24k = 0$; $h + 2k = 2$. Subtracting (3) from (1): $-8 + 4h - 12k = 0$; $h - 3k = 2$. Thus, $5k = 0$ and $k = 0$; $h + 2(0) = 2$ and $h = 2$.
$(x - 2)^2 + (y - 0)^2 = r^2$; $(7 - 2)^2 + 5^2 = r^2$; $r^2 = 50$; $(x - 2)^2 + y^2 = 50$.

54. $(x^2 + 2x + 1) + (y^2 + 2y + 1) = -2 + 1 + 1$; $(x + 1)^2 + (y + 1)^2 = 0$; $(-1, -1)$.

55. $(x^2 - 6x + 9) + (y^2 + 8y + 16) = -26 + 9 + 16$; $(x - 3)^2 + (y + 4)^2 = -1$; there is no point (x, y) for which $(x - 3)^2 + (y + 4)^2 < 0$.

56. $x^2 + y^2 - 1 = 0$ or $x^2 + y^2 - 4 = 0$; $x^2 + y^2 = 1$ or $x^2 + y^2 = 4$; two circles, each centered at the origin, with $r = 1$ and $r = 2$.

57. $x^3y + xy^3 - xy = 0$; $xy(x^2 + y^2 - 1) = 0$; $x = 0$, $y = 0$, or $x^2 + y^2 - 1 = 0$; the x-axis, the y-axis, and a circle centered at the origin with $r = 1$.

58. a. $x^2 + y^2 < 2$; $y > 1$

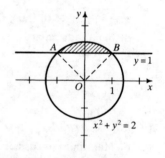

b. See diagram. $A = (-1, 1)$; $B = (1, 1)$; $AB = 2$; $OA = OB = \sqrt{2}$; since $(OA)^2 + (OB)^2 = (AB)^2$, $\triangle AOB$ is a right triangle. Area shaded region $= \dfrac{1}{4}$ area circle $-$
area $\triangle AOB = \dfrac{1}{4}\pi(\sqrt{2})^2 - \dfrac{1}{2}(\sqrt{2})(\sqrt{2}) = \dfrac{\pi}{2} - 1$

59. a.

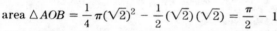

b. $\angle AOC = 60°$ ($OA = 2, OC = 1$). Area shaded region $= 4(\text{area}(\text{sector } OAB) - \text{area }(\triangle OAC)) =$
$$4\left(\frac{1}{6}\pi(2)^2 - \frac{1}{2}(\sqrt{3})(1)\right) = \frac{8\pi}{3} - 2\sqrt{3}$$

60. a. eq. of circle is $x^2 + y^2 = r^2$; since P lies on the circle, $a^2 + b^2 = r^2$; slope of $\overline{OP} = \dfrac{b}{a}$, so the slope of $l = -\dfrac{a}{b}$ and the eq. for l is $\dfrac{y - b}{x - a} = -\dfrac{a}{b}$;
$by - b^2 = -ax + a^2$; $ax + by = a^2 + b^2$; $ax + by = r^2$

b. $ax + by = r^2$; $x^2 + y^2 = r^2$; $y = \dfrac{r^2 - ax}{b}$; $x^2 + \left(\dfrac{r^2 - ax}{b}\right)^2 = r^2$;
$b^2x^2 + r^4 - 2ar^2x + a^2x^2 = b^2r^2$; $(a^2 + b^2)x^2 - 2ar^2x + r^4 - r^2b^2 = 0$;
$r^2x^2 - 2ar^2x + r^2(r^2 - b^2) = 0$; $r^2x^2 - 2ar^2x + r^2a^2 = 0$;
$r^2(x^2 - 2ax + a^2) = 0$; $r^2(x - a)^2 = 0$; since $r \neq 0$, $x = a$ is the only solution; thus, the line is tangent to the circle.

Class Exercises 6-3 · page 228

1. a. horizontal **b.** vertices: $(\pm 10, 0)$; foci: $(\pm 6, 0)$ **c.** $16 + 4 = 20$ **d.** $\dfrac{x^2}{100} + \dfrac{y^2}{64} = 1$

2. a. vertical **b.** vertices: $(0, \pm 13)$; $125 = 169 - c^2$, $c^2 = 144$, $c = 12$, foci: $(0, \pm 12)$

3. The graph of $\dfrac{x^2}{9} + y^2 < 1$ is the region inside the ellipse $\dfrac{x^2}{9} + y^2 = 1$. The graph of

$\dfrac{x^2}{9} + y^2 > 1$ is the region outside the same ellipse.

4. a. The surface of an elliptical solid **b.** the interior of an elliptical solid

Written Exercises 6-3 · pages 228–230

1. Vertices are $(\pm 6, 0)$; $c = \sqrt{a^2 - b^2} = \sqrt{36 - 16} = 2\sqrt{5}$, so foci are $(\pm 2\sqrt{5}, 0)$.

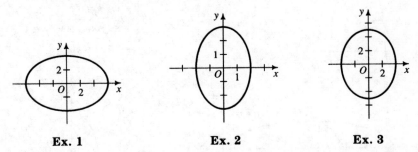

Ex. 1   **Ex. 2**   **Ex. 3**

2. Vertices are $(0, \pm 3)$; $c = \sqrt{a^2 - b^2} = \sqrt{9 - 4} = \sqrt{5}$, so foci are $(0, \pm\sqrt{5})$.

3. Vertices are $(0, \pm 5)$; $c = \sqrt{a^2 - b^2} = \sqrt{25 - 16} = \sqrt{9} = 3$, so foci are $(0, \pm 3)$.

4. $\dfrac{x^2}{25} + \dfrac{y^2}{4} = 1$; vertices are $(\pm 5, 0)$; $c = \sqrt{a^2 - b^2} = \sqrt{25 - 4} = \sqrt{21}$, so foci

are $(\pm\sqrt{21}, 0)$.

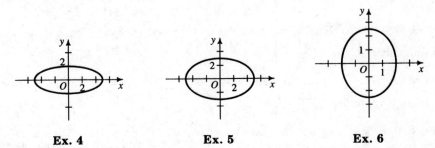

Ex. 4   **Ex. 5**   **Ex. 6**

5. $\dfrac{x^2}{25} + \dfrac{y^2}{9} = 1$; vertices are $(\pm 5, 0)$; $c = \sqrt{a^2 - b^2} = \sqrt{25 - 9} = \sqrt{16} = 4$, so foci

are $(\pm 4, 0)$.

6. $\dfrac{x^2}{2^2} + \dfrac{y^2}{(2.5)^2} = 1$; vertices are $(0, \pm 2.5)$; $c = \sqrt{a^2 - b^2} = \sqrt{6.25 - 4} = \sqrt{2.25} = 1.5$,

so foci are $(0, \pm 1.5)$.

7. a. 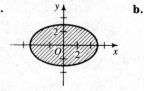 **b.**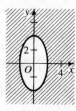

8. domain: $\{x \mid -1 \le x \le 1\}$;
range: $\{y \mid 0 \le y \le 3\}$

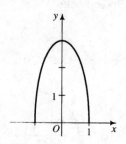

9. domain: $\{x \mid -6 \le x \le 6\}$;
range: $\{y \mid -2 \le y \le 0\}$

10. a.

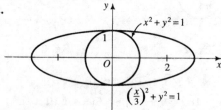

b. Area of circle $= \pi(1)^2 = \pi$;
area of ellipse $\approx 3\pi$

11. a.

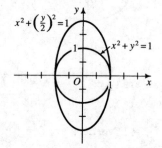

b. Area of circle $= \pi(1)^2 = \pi$;
area of ellipse $\approx 2\pi$

12. Since $(7, 0)$ is a vertex, the major axis is on the x-axis and $a = 7$; the minor axis is 2 units long, so $b = 1$; $\dfrac{x^2}{49} + y^2 = 1$.

13. Since $(0, -9)$ is a vertex, the major axis is on the y-axis and $a = 9$; the minor axis is 6 units long, so $b = 3$; $\dfrac{x^2}{9} + \dfrac{y^2}{81} = 1$.

14. Since $(0, -13)$ is a vertex, the major axis is on the y-axis and $a = 13$; since $(0, -5)$ is a focus, $c = 5$ and $c^2 = 25$; $c^2 = a^2 - b^2$, so $25 = 169 - b^2$ and $b^2 = 144$; $\dfrac{x^2}{144} + \dfrac{y^2}{169} = 1$.

15. Since $(17,0)$ is a vertex, the major axis is on the x-axis and $a = 17$; since $(8,0)$ is a focus, $c = 8$ and $c^2 = 64$; $c^2 = a^2 - b^2$, so $64 = 289 - b^2$ and $b^2 = 225$; $\dfrac{x^2}{289} + \dfrac{y^2}{225} = 1$.

16. a. Angles are congruent.

b. Light or sound waves emitted from one focus of an ellipse are reflected by the ellipse to the other focus.

17. The ellipse becomes more circular. When $F_1 = F_2$, the figure is a circle.

18. $3x + y = -3$; $y = -3x - 3$; $2y^2 = 18x^2 + 36x + 18$; $9x^2 + 18x^2 + 36x + 18 = 18$;

$27x^2 + 36x = 0$; $9x(3x + 4) = 0$; $9x = 0$ or $3x + 4 = 0$; $x = 0$ or $x = -\dfrac{4}{3}$;

$y = -3x - 3$; $y = -3$ or $y = 1$; points of intersection: $(0, -3)$, $\left(-\dfrac{4}{3}, 1\right)$

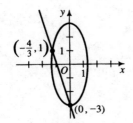

Ex. 18

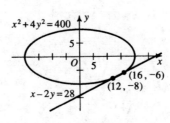

Ex. 19

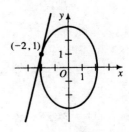

Ex. 20

19. $x - 2y = 28$; $x = 2y + 28$; $x^2 = 4y^2 + 112y + 784$; $4y^2 + 112y + 784 + 4y^2 = 400$;
$8y^2 + 112y + 384 = 0$; $y^2 + 14y + 48 = 0$; $(y + 6)(y + 8) = 0$; $y = -6$ or $y = -8$;
$x = 2y + 28$; $x = 16$ or $x = 12$; points of intersection: $(16, -6)$, $(12, -8)$

20. $y = 4x + 9$; $y^2 = 16x^2 + 72x + 81$; $2x^2 + 16x^2 + 72x + 81 = 9$;
$18x^2 + 72x + 72 = 0$; $x^2 + 4x + 4 = 0$; $(x + 2)^2 = 0$; $x = -2$; $y = 4x + 9$; $y = 1$;
point of intersection: $(-2, 1)$

21.

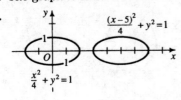

$x = \pm 2$; $x^2 = 4$; $4 + 4y^2 = 16$; $4y^2 = 12$; $y^2 = 3$;
$y = \pm\sqrt{3}$; points of intersection: $(2, \sqrt{3})$, $(2, -\sqrt{3})$,
$(-2, \sqrt{3})$, $(-2, -\sqrt{3})$

22. a. The graph is shifted 5 units to the right.

b.

23. a. The graph is shifted 6 units down.

b.

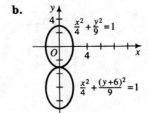

24.

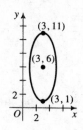

25.

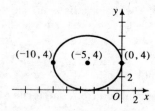

26.

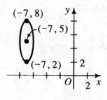

27.

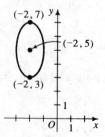

28. a. $\dfrac{x^2}{16} + \dfrac{y^2}{4} = 1$; $y^2 = 4 - \dfrac{x^2}{4}$; $y = \sqrt{4 - \dfrac{x^2}{4}}$;   **b.** Maximum area $= 16.0$

$$A(x) = 2x \cdot 2y = 4xy = 4x\sqrt{4 - \dfrac{x^2}{4}}$$

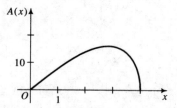

29. a. If the original ellipse is a circle, then $a = b$ and $V = \dfrac{4}{3}\pi a^3$, which is the volume of a sphere.

b. $V = \dfrac{4}{3}\pi a^2 b$

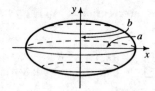

30. vertices: $(0, -3)$, $(10, -3)$; $c = \sqrt{a^2 - b^2} = 4$; foci are 4 units to the left and right of the center, $(5, -3)$; foci: $(1, -3)$, $(9, -3)$

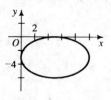

Ex. 30

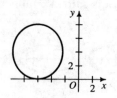

Ex. 31

31. vertices: $(-6, 0)$; $(-6, 8)$; $c = \sqrt{a^2 - b^2} = 2$; foci are 2 units above and below the center, $(-6, 4)$; foci: $(-6, 2)$, $(-6, 6)$

32. $\dfrac{(x-3)^2}{4} + \dfrac{(y+5)^2}{9} = 1$; vertices: $(3,-2)$, $(3,-8)$; $c = \sqrt{a^2 - b^2} = \sqrt{5}$; foci are $\sqrt{5}$ units above and below the center, $(3,-5)$; foci: $(3,-5+\sqrt{5})$, $(3,-5-\sqrt{5})$

Ex. 32   **Ex. 33**

33. $\dfrac{(x+1)^2}{9} + \dfrac{(y+3)^2}{\frac{9}{4}} = 1$; vertices: $(-4,-3)$, $(2,-3)$; $c = \sqrt{a^2 - b^2} = \dfrac{3\sqrt{3}}{2}$; foci are

$\dfrac{3\sqrt{3}}{2}$ units to the right and left of the center, $(-1,-3)$; foci: $\left(-1 + \dfrac{3\sqrt{3}}{2}, -3\right)$,

$\left(-1 - \dfrac{3\sqrt{3}}{2}, -3\right)$

34. $x^2 + 25y^2 - 6x - 100y + 84 = 0$ is equivalent to $(x-3)^2 + 25(y-2)^2 = 25$

(found by completing the squares), or $\dfrac{(x-3)^2}{25} + (y-2)^2 = 1$. Vertices: $(-2,2)$, $(8,2)$;

$c = \sqrt{a^2 - b^2} = \sqrt{24} = 2\sqrt{6}$; foci are $2\sqrt{6}$ units to the left and right of the center $(3,2)$; foci: $(3 - 2\sqrt{6}, 2)$, $(3 + 2\sqrt{6}, 2)$

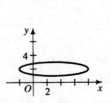   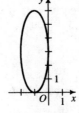

Ex. 34   **Ex. 35**

35. $9x^2 + y^2 + 18x - 6y + 9 = 0$ is equivalent to $9(x+1)^2 + (y-3)^2 = 9$ (found by

completing the squares), or $(x+1)^2 + \dfrac{(y-3)^2}{9} = 1$. Vertices: $(-1,6)$, $(-1,0)$;

$c = \sqrt{a^2 - b^2} = 2\sqrt{2}$; foci are $2\sqrt{2}$ units above and below the center $(-1,3)$;

foci: $(-1, 3 + 2\sqrt{2})$, $(-1, 3 - 2\sqrt{2})$

36. $9x^2 + 16y^2 - 18x - 64y - 71 = 0$ is equivalent to $9(x - 1)^2 + 16(y - 2)^2 = 144$
(found by completing the square), or $\dfrac{(x - 1)^2}{16} + \dfrac{(y - 2)^2}{9} = 1$. Vertices: $(-3, 2)$, $(5, 2)$;
$c = \sqrt{a^2 - b^2} = \sqrt{7}$; foci are $\sqrt{7}$ units to the right and left of the center, $(1, 2)$;
foci: $(1 - \sqrt{7}, 2)$, $(1 + \sqrt{7}, 2)$

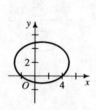

$$\text{Ex. 36} \qquad\qquad\qquad \text{Ex. 37}$$

37. a. $4x^2 + 9y^2 = 144$; $9y^2 = 144 - 4x^2$; $y^2 = 16 - \dfrac{4x^2}{9} = 4\left(4 - \dfrac{x^2}{9}\right)$; $y = \pm 2\sqrt{4 - \dfrac{x^2}{9}}$,

or $y = \pm\dfrac{2}{3}\sqrt{36 - x^2}$

b. Yes; see diagram.

38. The major axis is parallel to the x-axis. The given focus is 3 units from the center,
so $c = 3$. The given vertex is 5 units from the center, so $a = 5$. $b^2 = a^2 - c^2 = 25 - 9 = 16$. The center of the ellipse is $(3, 7)$. An equation of the ellipse is
$\dfrac{(x - 3)^2}{25} + \dfrac{(y - 7)^2}{16} = 1$.

39. The major axis is parallel to the y-axis. The given focus is 2.5 units from the center,
so $c = 2.5$. The given vertex is 4 units from the center, so $a = 4$. $b^2 = a^2 - c^2 = 16 - 6.25 = 9.75 = \dfrac{39}{4}$. The center of the ellipse is $(4, -1)$. An equation of the ellipse
is $\dfrac{4(x - 4)^2}{39} + \dfrac{(y + 1)^2}{16} = 1$.

40. The major axis is parallel to the y-axis. $a = $ half the length of the major axis $= \dfrac{9 - 1}{2} = 4$. The center is the midpoint of the major axis, $(5, 5)$. The given focus is
2 units from the center, so $c = 2$. $b^2 = a^2 - c^2 = 16 - 4 = 12$. An equation of the
ellipse is $\dfrac{(x - 5)^2}{12} + \dfrac{(y - 5)^2}{16} = 1$.

41. The major axis is parallel to the y-axis. $b = 5$ and $a = 6$. An equation of the ellipse
is $\dfrac{(x - 5)^2}{25} + \dfrac{(y - 6)^2}{36} = 1$.

42. a. $PF_1 = \sqrt{(x - 3)^2 + (y - 0)^2}$; $PF_2 = \sqrt{(x + 3)^2 + (y - 0)^2}$;
$PF_1 + PF_2 = \sqrt{(x - 3)^2 + y^2} + \sqrt{(x + 3)^2 + y^2} = 10$

b. $\sqrt{(x + 3)^2 + y^2} = 10 - \sqrt{(x - 3)^2 + y^2}$;
squaring, $(x + 3)^2 + y^2 = 100 - 20\sqrt{(x - 3)^2 + y^2} + (x - 3)^2 + y^2$;
$x^2 + 6x + 9 + y^2 = 100 - 20\sqrt{(x - 3)^2 + y^2} + x^2 - 6x + 9 + y^2$;
$12x - 100 = -20\sqrt{(x - 3)^2 + y^2}$; $3x - 25 = -5\sqrt{(x - 3)^2 + y^2}$; squaring again,
$9x^2 - 150x + 625 = 25(x^2 - 6x + 9 + y^2)$; $16x^2 + 25y^2 = 400$; $\dfrac{x^2}{25} + \dfrac{y^2}{16} = 1$

Computer Exercises 6-3 · page 231

1. a. Solve $\dfrac{x^2}{25} + \dfrac{y^2}{9} = 1$ for y: $y = \pm 0.6\sqrt{25 - x^2}$. To find y_2, evaluate the function

$y = 0.6\sqrt{25 - x^2}$ for $x = x_2 = 1 \cdot \dfrac{5}{10} = 0.5$. To find y_3, evaluate

$y = 0.6\sqrt{25 - x^2}$ for $x = x_3 = 2 \cdot \dfrac{5}{10} = 1$.

b.
```
10 LET S = 0
20 FOR I = 0 TO 9
30    LET R = .5 * .6 * SQR(25 - (I * .5)^2)
40    LET S = S + R
50 NEXT I
60 PRINT "THE APPROX. SUM OF THE RECTANGLES IS";S
70 PRINT "THE APPROX. AREA OF THE ELLIPSE IS";4 * S
80 END
```

```
THE APPROX. SUM OF THE RECTANGLES IS 12.391943
THE APPROX. AREA OF THE ELLIPSE IS 49.567772
```

2.
```
10 LET S = 0
20 FOR I = 0 TO 999
30    LET R = .005 * .6 * SQR(25 - (I * .005)^2)
40    LET S = S + R
50 NEXT I
60 PRINT "THE APPROX. SUM OF THE RECTANGLES IS";S
70 PRINT "THE APPROX. AREA OF THE ELLIPSE IS";4 * S
80 END
```

```
THE APPROX. SUM OF THE RECTANGLES IS 11.788332
THE APPROX. AREA OF THE ELLIPSE IS 47.153328
```

Class Exercises 6-4 · page 234

1. a. $V(\pm 2, 0);\ F(\pm\sqrt{13}, 0)$ **b.** Horizontal **c.** $\dfrac{x^2}{4} - \dfrac{y^2}{9} = 1$; asymptotes: $y = \pm\dfrac{3}{2}x$

2. a. $y = -\dfrac{2}{x}$ **b.** $y = \dfrac{2}{x}$

3. a. Vertical; in $\dfrac{y^2}{25} - \dfrac{x^2}{1} = 1$, there can be no points on the x-axis.

b. $V(0, \pm 5);\ c = \sqrt{a^2 + b^2} = \sqrt{26}$ so $F(0, \pm\sqrt{26})$

c. $y = \pm 5x$

d. $\dfrac{(y + 5)^2}{25} - \dfrac{(x - 6)^2}{1} = 1$

4. $x^2 - y^2 < 4$ represents the interior of the two branches of $\dfrac{x^2}{4} - \dfrac{y^2}{4} = 1$, and $x^2 - y^2 > 4$ represents the exterior.

Written Exercises 6-4 · pages 235–237

1.
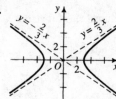

2.
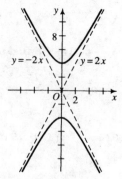

3.

4.
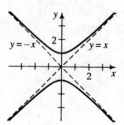

5.
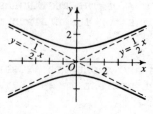

6.
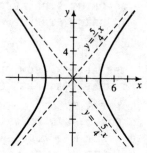

7.
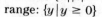

8.
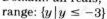

9. Domain: $\{x \mid |x| \geq 2\}$;
range: $\{y \mid y \geq 0\}$

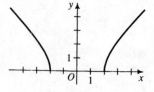

10. Domain: all reals;
range: $\{y \mid y \leq -3\}$

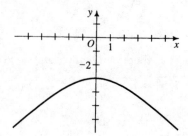

11. a.

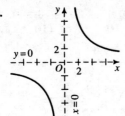

b.

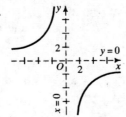

c.

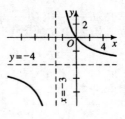

12. a.

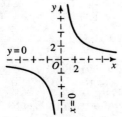

b.

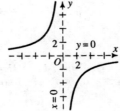

c.

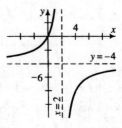

13.

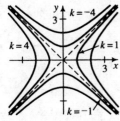

14.

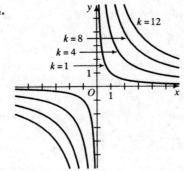

15. $y = 5 - x$; $x^2 - y^2 = 16$; $x^2 - (5 - x)^2 = 16$; $x^2 - (25 - 10x + x^2) = 16$; $10x = 41$;
$x = 4.1$; $y = 5 - 4.1 = 0.9$; $(4.1, 0.9)$

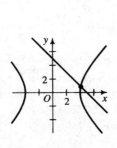

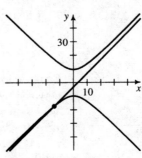

Ex. 15   **Ex. 16**

16. $y = x - 3$; $y^2 - x^2 = 94$; $(x - 3)^2 - x^2 = 94$; $x^2 - 6x + 9 - x^2 = 94$; $6x = -85$;
$x = -\dfrac{85}{6}$; $y = -\dfrac{103}{6}$; $\left(-\dfrac{85}{6}, -\dfrac{103}{6}\right)$

17. $x = 6 - 2y$; $xy = -20$; $(6 - 2y)y = -20$; $6y - 2y^2 = -20$; $y^2 - 3y - 10 = 0$;
$(y - 5)(y + 2) = 0$; $y = 5$ and $x = -4$, or $y = -2$ and $x = 10$; $(-4, 5)$, $(10, -2)$

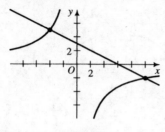

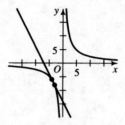

Ex. 17   **Ex. 18**

18. $y = -2x - 14$; $xy = 24$; $x(-2x - 14) = 24$; $2x^2 + 14x + 24 = 0$; $x^2 + 7x + 12 = 0$;
$(x + 4)(x + 3) = 0$; $x = -4$ and $y = -6$, or $x = -3$ and $y = -8$; $(-4, -6)$, $(-3, -8)$

19. A vertex is 6 units from the center, so $a = 6$. A focus is 10 units from the center,
so $c = 10$. $b^2 = c^2 - a^2 = 100 - 36 = 64$; $\dfrac{x^2}{36} - \dfrac{y^2}{64} = 1$.

20. A vertex is 12 units from the center, so $a = 12$. A focus is 13 units from the center, so
$c = 13$. $b^2 = c^2 - a^2 = 169 - 144 = 25$; $\dfrac{y^2}{144} - \dfrac{x^2}{25} = 1$.

21. A vertex is 2 units from the center, so $a = 2$. An asymptote has equation
$y = -x = -\dfrac{a}{b}x$, so $\dfrac{a}{b} = 1$ and $b = 2$; $\dfrac{y^2}{4} - \dfrac{x^2}{4} = 1$.

22. A vertex is 8 units from the center, so $a = 8$. An asymptote has equation
$y = \dfrac{1}{2}x = \dfrac{b}{a}x$, so $\dfrac{b}{a} = \dfrac{1}{2}$ and $b = 4$; $\dfrac{x^2}{64} - \dfrac{y^2}{16} = 1$

23. a. $\dfrac{x^2}{4} - y^2 = 1$; $y^2 = \dfrac{x^2}{4} - 1$; $y = \pm\sqrt{\dfrac{x^2 - 4}{4}} = \pm\dfrac{1}{2}\sqrt{x^2 - 4}$

    b. Vertical distance is $\dfrac{1}{2}x - \dfrac{1}{2}\sqrt{x^2 - 4} = \dfrac{1}{2}(x - \sqrt{x^2 - 4})$; when $x = 5$, the distance

    is $\dfrac{1}{2}(5 - \sqrt{21}) \approx 0.209$; when $x = 10$, the distance is $\dfrac{1}{2}(10 - \sqrt{96}) \approx 0.101$; when

    $x = 100$, the distance is $\dfrac{1}{2}(100 - \sqrt{9996}) \approx 0.010$.

24.

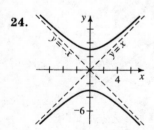

Equations of asymptotes: $y = \pm x$; if $x > 0$ and $y > 0$, distance
between hyperbola and asymptote is $\sqrt{x^2 + 9} - x$;
$\sqrt{x^2 + 9} - x < 0.001$; $\sqrt{x^2 + 9} < x + 0.001$;
$x^2 + 9 < x^2 + 0.002x + 0.000001$; $0.002x > 8.999999$;
$x > 4499.9995$, or $x > 4500$; using symmetry of the hyperbola
and its asymptotes, $|x| > 4500$.

25.

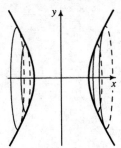

26. a.

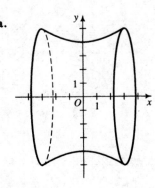

b. When $x = 3$; $y = 5$;
$V = \pi r^2 h = \pi (5)^2 (6)$
$= 150\pi$.

27. c. The angles are equal.

28. a. $\left| \sqrt{(x-5)^2 + y^2} - \sqrt{(x+5)^2 + y^2} \right| = 8$; $\sqrt{(x-5)^2 + y^2} - \sqrt{(x+5)^2 + y^2} = \pm 8$

  b. $\sqrt{(x-5)^2 + y^2} = \sqrt{(x+5)^2 + y^2} \pm 8$; $(x-5)^2 + y^2 = (x+5)^2 + y^2 \pm$
  $16\sqrt{(x+5)^2 + y^2} + 64$; $x^2 - 10x + 25 + y^2 = x^2 + 10x + 25 + y^2 \pm$
  $16\sqrt{(x+5)^2 + y^2} + 64$; $-20x - 64 = \pm 16\sqrt{(x+5)^2 + y^2}$; $-5x - 16 =$
  $\pm 4\sqrt{(x+5)^2 + y^2}$; squaring again, $25x^2 + 160x + 256 = 16[(x+5)^2 + y^2]$;
  $25x^2 + 160x + 256 = 16x^2 + 160x + 400 + 16y^2$; $9x^2 - 16y^2 = 144$; $\dfrac{x^2}{16} - \dfrac{y^2}{9} = 1$

29. a. $\left| \sqrt{x^2 + (y-10)^2} - \sqrt{x^2 + (y+10)^2} \right| = 12$;
  $\sqrt{x^2 + (y-10)^2} - \sqrt{x^2 + (y+10)^2} = \pm 12$

  b. $\sqrt{x^2 + (y-10)^2} = \sqrt{x^2 + (y+10)^2} \pm 12$; $x^2 + (y-10)^2 = x^2 + (y+10)^2 \pm$
  $24\sqrt{x^2 + (y+10)^2} + 144$; $x^2 + y^2 - 20y + 100 = x^2 + y^2 + 20y +$
  $100 \pm 24\sqrt{x^2 + (y+10)^2} + 144$; $-40y - 144 = \pm 24\sqrt{x^2 + (y+10)^2}$;
  $-5y - 18 = \pm 3\sqrt{x^2 + (y+10)^2}$; squaring again,
  $25y^2 + 180y + 324 = 9x^2 + 9y^2 + 180y + 900$; $16y^2 - 9x^2 = 576$; $\dfrac{y^2}{36} - \dfrac{x^2}{64} = 1$.

30.

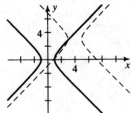

31.

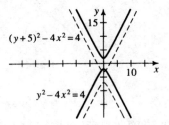

$(y+5)^2 - 4x^2 = 4$

$y^2 - 4x^2 = 4$

32. $\dfrac{(x-8)^2}{36} - \dfrac{(y-6)^2}{9} = 1$; asymptotes are $y - 6 = \pm \dfrac{1}{2}(x-8)$; $-x + 2y = 4$ and
  $x + 2y = 20$.

33. $\dfrac{y^2}{36} - \dfrac{x^2}{9} = 1$; asymptotes are $y = \pm \dfrac{a}{b}x = \pm \dfrac{6}{3}x = \pm 2x$.

34. a.

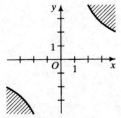

b.

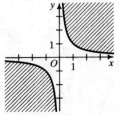

c.

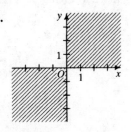

35. a. 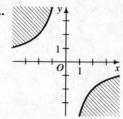   **b.**   **c.**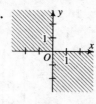

36. 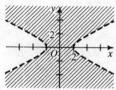   **37.** 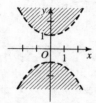   **38.**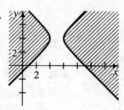

39. vertices: $(0,8), (12,8)$; $c = \sqrt{a^2 + b^2} = 10$; foci: $(6 - 10, 8) = (-4, 8)$,

$(6 + 10, 8) = (16, 8)$; asymptotes have slope $\pm \dfrac{b}{a} = \pm \dfrac{4}{3}$ and contain $(6,8)$, so equations

are $\dfrac{y - 8}{x - 6} = \pm \dfrac{4}{3}$; $-4x + 3y = 0$ and $4x + 3y = 48$.

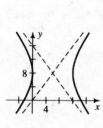   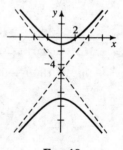   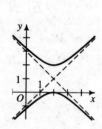

     Ex. 39   **Ex. 40**   **Ex. 41**

40. vertices: $(0, -1)$, $(0, -9)$; $c = \sqrt{a^2 + b^2} = 5$; foci: $(0, -5 - 5) = (0, -10)$,

$(0, -5 + 5) = (0, 0)$; asymptotes have slope $\pm \dfrac{a}{b} = \pm \dfrac{4}{3}$ and contain $(0, -5)$, so

equations are $\dfrac{y + 5}{x} = \pm \dfrac{4}{3}$; $y = \dfrac{4}{3}x - 5$ and $y = -\dfrac{4}{3}x - 5$.

41. $y^2 - x^2 - 2y + 4x - 4 = 0$ is equivalent to $(y - 1)^2 - (x - 2)^2 = 1$ (found by completing the squares); center: $(2, 1)$; vertices: $(2, 0)$, $(2, 2)$; $c = \sqrt{a^2 + b^2} = \sqrt{2}$; foci: $(2, 1 \pm \sqrt{2})$; asymptotes have slope ± 1 and contain $(2, 1)$, so equations are $\dfrac{y - 1}{x - 2} = \pm 1$; $y = x - 1$ and $y = -x + 3$.

42. $x^2 - 4y^2 - 2x + 16y - 19 = 0$ is equivalent to $(x - 1)^2 - 4(y - 2)^2 = 4$ (found by completing the square), or $\dfrac{(x - 1)^2}{4} - (y - 2)^2 = 1$; center: $(1, 2)$; vertices: $(-1, 2)$, $(3, 2)$; $c = \sqrt{a^2 + b^2} = \sqrt{5}$; foci: $(1 \pm \sqrt{5}, 2)$; asymptotes have slope $\pm\dfrac{1}{2}$ and pass through $(1, 2)$, so equations are $\dfrac{y - 2}{x - 1} = \pm\dfrac{1}{2}$; $x - 2y = -3$ and $x + 2y = 5$.

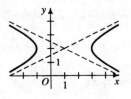

43. The major axis is horizontal. One vertex is 4 units from the center, so $a = 4$. One focus is 5 units from the center, so $c = 5$. $b^2 = c^2 - a^2 = 25 - 16 = 9$; $\dfrac{(x - 5)^2}{16} - \dfrac{y^2}{9} = 1$.

44. The major axis is vertical; $a = \dfrac{8 - 0}{2} = 4$; slope of asymptotes $= \pm\dfrac{a}{b} = \pm\dfrac{4}{b} = \pm 1$, $b = \pm 4$, $b^2 = 16$; center = midpoint of major axis = $(4, 4)$; $\dfrac{(y - 4)^2}{16} - \dfrac{(x - 4)^2}{16} = 1$.

45. The hyperbola $xy = 1$ has major axis on the line $y = x$ and its vertices are $(1, 1)$ and $(-1, -1)$. The length of the major axis is $\sqrt{(1 + 1)^2 + (1 + 1)^2} = \sqrt{8} = 2\sqrt{2}$, so $a = \sqrt{2}$. The asymptotes of $xy = 1$ are the x- and y-axes, which are perpendicular, so $\left(\dfrac{b}{a}\right)\left(-\dfrac{b}{a}\right) = -1$; $-\dfrac{b^2}{a^2} = -1$; $b^2 = a^2$; $b = a = \sqrt{2}$.

Class Exercises 6-5 · pages 239–240

1. a. 1, 1 **b.** 2, 2 **c.** 5, 5 **d.** $\sqrt{x^2 + y^2}$; $y + 2$

2. a. (1) $PF = \sqrt{(x + 4)^2 + y^2}$; (2) $PN = |x - 0| = |x|$
   b. Since P is equidistant from F and the y-axis, set the expressions found in part (a) equal to each other and simplify.

3. If the equation of the parabola has form $y = ax^2 + bx + c$, then the parabola opens up if $a > 0$ and down if $a < 0$. If the equation of the parabola has form $x = ay^2 + by + c$, then the parabola opens right if $a > 0$ and left if $a < 0$.

4. The parabola opens up. $\dfrac{1}{4} = \dfrac{1}{4p}$, so $p = 1$ and $F(0, 1)$; d: $y = -1$

5. The parabola opens down. $\dfrac{1}{4} = \dfrac{1}{4p}$, so $p = 1$; $F(0, -1)$; d: $y = 1$

6. The parabola opens right. $1 = \dfrac{1}{4p}$, so $p = \dfrac{1}{4}$; $F\left(\dfrac{1}{4}, 0\right)$; d: $x = -\dfrac{1}{4}$

7. The parabola opens left. $1 = \dfrac{1}{4p}$, so $p = \dfrac{1}{4}$; $F\left(-\dfrac{1}{4}, 0\right)$; d: $x = \dfrac{1}{4}$

8. Yes, rotate the parabola so that the axis of symmetry is a line other than a vertical or horizontal one.

9. Shift $y = x^2$ to the right h units and up k units.

Written Exercises 6-5 · pages 240–241

1. **a.** Opens up; $V(0,0)$; $\dfrac{1}{8} = \dfrac{1}{4p}$, $p = 2$; $F(0,2)$; $y = -2$

   b. Opens right; $V(0,0)$; $\dfrac{1}{8} = \dfrac{1}{4p}$, $p = 2$; $F(2,0)$; $x = -2$

2. **a.** Opens down; $V(0,0)$; $\dfrac{1}{12} = \dfrac{1}{4p}$, $p = 3$; $F(0,-3)$; $y = 3$

   b. Opens left; $V(0,0)$; $\dfrac{1}{12} = \dfrac{1}{4p}$, $p = 3$; $F(-3,0)$; $x = 3$

3. **a.** Opens down; $V(0,0)$; $\dfrac{1}{4p} = 2$, $p = \dfrac{1}{8}$; $F\left(0, -\dfrac{1}{8}\right)$; $y = \dfrac{1}{8}$

   b. Opens left; $V(0,0)$; $\dfrac{1}{4p} = 2$, $p = \dfrac{1}{8}$; $F\left(-\dfrac{1}{8},0\right)$; $x = \dfrac{1}{8}$

4. **a.** $y - 1 = x^2$; opens up; $V(0,1)$; $\dfrac{1}{4p} = 1$, $p = \dfrac{1}{4}$; $F\left(0, 1 + \dfrac{1}{4}\right) = \left(0, \dfrac{5}{4}\right)$;

   directrix: $y = k - p = 1 - \dfrac{1}{4}$, $y = \dfrac{3}{4}$

   b. $x - 1 = y^2$; opens right; $V(1,0)$; $\dfrac{1}{4p} = 1$, $p = \dfrac{1}{4}$; $F\left(1 + \dfrac{1}{4},0\right) = \left(\dfrac{5}{4},0\right)$;

   directrix: $x = h - p = 1 - \dfrac{1}{4}$, $x = \dfrac{3}{4}$

5. Opens up; $V(2,1)$; $\dfrac{1}{4p} = \dfrac{1}{4}$, $p = 1$; $F(2, 1 + 1) = (2,2)$; directrix: $y = k - p = 1 - 1$, $y = 0$

6. Opens up; $V(5,-3)$; $\dfrac{1}{4p} = \dfrac{1}{8}$, $p = 2$; $F(5, -3 + 2) = (5,-1)$;

   directrix: $y = k - p = -3 - 2$, $y = -5$

7. Opens right; $V(4,7)$; $\dfrac{1}{4p} = 1$, $p = \dfrac{1}{4}$; $F\left(4 + \dfrac{1}{4},7\right) = \left(\dfrac{17}{4},7\right)$;

   directrix: $x = h - p = 4 - \dfrac{1}{4}$, $x = \dfrac{15}{4}$

8. Opens left; $V(-2,3)$; $\dfrac{1}{4p} = 2$, $p = \dfrac{1}{8}$; $F\left(-2 - \dfrac{1}{8},3\right) = \left(-\dfrac{17}{8},3\right)$;

   directrix: $x = h + p = -2 + \dfrac{1}{8}$, $x = -\dfrac{15}{8}$

9. Opens right; $p = 4$; $x = \dfrac{1}{4p}y^2$, $x = \dfrac{1}{16}y^2$

10. The angles are equal.

11. Opens left; vertex is $(0,0)$; $p = 1$; $x = -\dfrac{1}{4}y^2$

12. Opens up; vertex is $(0,0)$; $p = \dfrac{1}{4}$; $y = x^2$

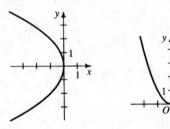

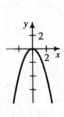

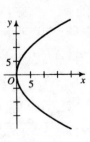

Ex. 11   **Ex. 12**   **Ex. 13**   **Ex. 14**

13. Opens down; $p = \dfrac{1}{4}$; $y = -x^2$ **14.** Opens right; $p = 5$; $x = \dfrac{1}{20}y^2$

15. Opens left; $p = 4$; $x = -\dfrac{1}{16}y^2$ **16.** Opens up; $p = 2$; $y = \dfrac{1}{8}x^2$

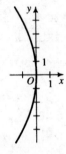

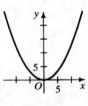

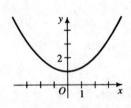

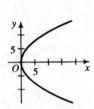

Ex. 15   **Ex. 16**   **Ex. 17**   **Ex. 18**

17. Opens up; vertex $(0, 1)$; $p = 1$; $y - 1 = \dfrac{1}{4}x^2$; $y = \dfrac{1}{4}x^2 + 1$

18. Opens right; $p = 3$; $x = \dfrac{1}{12}y^2$

19. a.   **b.**   **c.**   **d.**

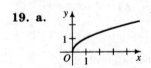 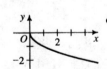 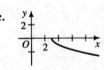 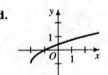

20. a.   **b.**   **c.**   **d.**

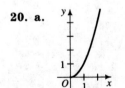   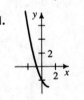

21. a. $\sqrt{x^2 + (y - 6)^2} = y$

    b. $x^2 + (y - 6)^2 = y^2$; $x^2 + y^2 - 12y + 36 = y^2$; $12y - 36 = x^2$; $12(y - 3) = x^2$;

    $y - 3 = \dfrac{1}{12}(x - 0)^2$

22. $\sqrt{(x - 0)^2 + (y - 2)^2} = |y + 2|$; $x^2 + y^2 - 4y + 4 = y^2 + 4y + 4$; $-8y = -x^2$;

    $y = \dfrac{1}{8}x^2$

23. $\sqrt{(x-0)^2+(y+3)^2} = |-2-x|$; $x^2+y^2+6y+9 = x^2+4x+4$;

$-4x = -y^2-6y-5$; $x = \dfrac{1}{4}y^2+\dfrac{3}{2}y+\dfrac{5}{4}$

24. $\sqrt{(x+1)^2+(y-3)^2} = |x|$; $x^2+2x+1+y^2-6y+9 = x^2$; $2x = -y^2+6y-10$;

$x = -\dfrac{1}{2}y^2+3y-5$

25. $\sqrt{(x+7)^2+(y+5)^2} = |-7-y|$; $x^2+14x+49+y^2+10y+25 =$

$y^2+14y+49$; $-4y = -x^2-14x-25$; $y = \dfrac{1}{4}x^2+\dfrac{7}{2}x+\dfrac{25}{4}$

26. $y-3 = 2(x^2-4x)$; $y-3+2(4) = 2(x^2-4x+4)$; $y+5 = 2(x-2)^2$; $V(2,-5)$;

$\dfrac{1}{4p} = 2$, $p = \dfrac{1}{8}$; $F\left(2,-5+\dfrac{1}{8}\right) = \left(2,-\dfrac{39}{8}\right)$; directrix: $y = k-p = -5-\dfrac{1}{8}$, $y = -\dfrac{41}{8}$

Ex. 26   **Ex. 27**   **Ex. 28**

27. $y-5 = -3(x^2+2x)$; $y-5+(-3)(1) = -3(x^2+2x+1)$; $y-8 = -3(x+1)^2$;

$V(-1,8)$; $\dfrac{1}{4p} = 3$, $p = \dfrac{1}{12}$; $F\left(-1,8-\dfrac{1}{12}\right) = \left(-1,\dfrac{95}{12}\right)$; directrix: $y = k+p = 8+\dfrac{1}{12}$,

$y = \dfrac{97}{12}$

28. $8y+1 = -(x^2-6x)$; $8y+1-9 = -(x^2-6x+9)$; $8y-8 = -(x-3)^2$;

$y-1 = -\dfrac{1}{8}(x-3)^2$; $V(3,1)$; $\dfrac{1}{4p} = \dfrac{1}{8}$, $p = 2$; $F(3,1-2) = (3,-1)$;

directrix: $y = k+p = 1+2$, $y = 3$

29. $4y-12 = x^2-8x$; $4y-12+16 = x^2-8x+16$; $4y+4(x-4)^2$;

$y+1 = \dfrac{1}{4}(x-4)^2$; $V(4,-1)$; $\dfrac{1}{4p} = \dfrac{1}{4}$, $p = 1$; $F(4,-1+1) = (4,0)$; directrix:

$y = k-p = -1-1$, $y = -2$

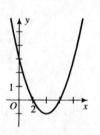

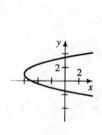

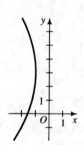

Ex. 29   **Ex. 30**   **Ex. 31**

30. $x + 5 = y^2 - 2y$; $x + 5 + 1 = y^2 - 2y + 1$; $x + 6 = (y - 1)^2$; $V(-6, 1)$; $\dfrac{1}{4p} = 1$,

$p = \dfrac{1}{4}$; $F\left(-6 + \dfrac{1}{4}, 1\right) = \left(-\dfrac{23}{4}, 1\right)$; directrix: $x = h - p = -6 - \dfrac{1}{4}$, $x = -\dfrac{25}{4}$

31. $16x + 25 = -(y^2 - 6y)$; $16x + 25 - 9 = -(y^2 - 6y + 9)$; $16x + 16 = -(y - 3)^2$;

$x + 1 = -\dfrac{1}{16}(y - 3)^2$; $V(-1, 3)$; $\dfrac{1}{4p} = \dfrac{1}{16}$, $p = 4$; $F(-1 - 4, 3) = (-5, 3)$;

directrix: $x = h + p = -1 + 4$, $x = 3$

Class Exercises 6-6 · page 244

1. Examples are given.

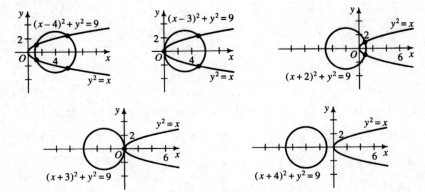

2. The graph of $x^2 + y^2 = 9$ is a circle with center $(0,0)$ and $r = 3$. $(x + y)^2 = 9$; $x + y = \pm3$, so the graph of $(x + y)^2 = 9$ is two lines: $x + y = 3$, $x + y = -3$.

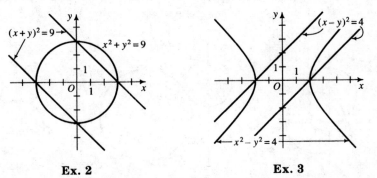

Ex. 2   Ex. 3

3. The graph of $x^2 - y^2 = 4$ is a hyperbola. $(x - y)^2 = 4$; $x - y = \pm2$, so the graph of $(x - y)^2 = 4$ is two lines: $x - y = 2$, $x - y = -2$.

Written Exercises 6-6 · pages 244–246

1. $y^2 = 16 - x^2$; $(x - 4)^2 + 16 - x^2 = 16$; $x^2 - 8x + 16 + 16 - x^2 = 16$; $8x = 16$;
   $x = 2$; $y^2 = 16 - 2^2$; $y^2 = 12$; $y = \pm 2\sqrt{3}$; $(2, 2\sqrt{3})$, $(2, -2\sqrt{3})$

2. $x^2 = 4 - y^2$; $4 - y^2 + (y - 6)^2 = 25$; $4 - y^2 + y^2 - 12y + 36 = 25$; $12y = 15$;
   $y = \dfrac{5}{4}$; $x^2 = 4 - \left(\dfrac{5}{4}\right)^2$; $x^2 = \dfrac{39}{16}$; $x = \pm\dfrac{\sqrt{39}}{4}$; $\left(\dfrac{\sqrt{39}}{4}, \dfrac{5}{4}\right)$, $\left(-\dfrac{\sqrt{39}}{4}, \dfrac{5}{4}\right)$

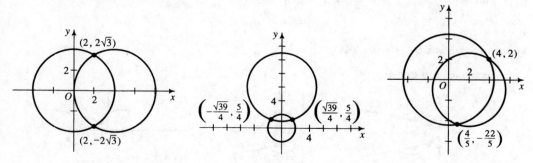

| Ex. 1 | Ex. 2 | Ex. 3 |

3. $x^2 + y^2 = 20$; $x^2 - 4x + 4 + y^2 + 2y + 1 = 13$; subtracting, $4x - 4 - 2y - 1 = 7$;
   $y = 2x - 6$; $x^2 + (2x - 6)^2 = 20$; $5x^2 - 24x + 16 = 0$; $(5x - 4)(x - 4) = 0$; $x = \dfrac{4}{5}$,

   so $y = 2\left(\dfrac{4}{5}\right) - 6 = -\dfrac{22}{5}$ or $x = 4$, so $y = 2 \cdot 4 - 6 = 2$; $\left(\dfrac{4}{5}, -\dfrac{22}{5}\right)$, $(4, 2)$

4. $x^2 + y^2 - 4y = 0$; $x^2 + y^2 - 2x = 4$; subtracting, $-4y + 2x = -4$; $x = 2y - 2$;
   $(2y - 2)^2 + y^2 - 4y = 0$; $5y^2 - 12y + 4 = 0$; $(5y - 2)(y - 2) = 0$; $y = \dfrac{2}{5}$, so

   $x = 2 \cdot \dfrac{2}{5} - 2 = -\dfrac{6}{5}$ or $y = 2$, so $x = 2 \cdot 2 - 2 = 2$; $(2, 2)$, $\left(-\dfrac{6}{5}, \dfrac{2}{5}\right)$

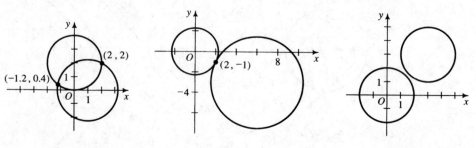

| Ex. 4 | Ex. 5 | Ex. 6 |

5. $x^2 + y^2 = 5$; $x^2 + y^2 - 12x + 6y = -25$; subtracting, $12x - 6y = 30$; $y = 2x - 5$;
   $x^2 + (2x - 5)^2 = 5$; $5x^2 - 20x + 20 = 0$; $x^2 - 4x + 4 = 0$; $(x - 2)^2 = 0$; $x = 2$, so
   $y = 2 \cdot 2 - 5 = -1$; $(2, -1)$; the circles are tangent at $(2, -1)$.

6. There is no point of intersection.

7. a. 4; the ellipses are images of each other under 90° rotations about the origin.

b. $9x^2 + 9y^2 = 162$; a circle

c. The points of intersection satisfy both equations, so they must satisfy the sum of those equations.

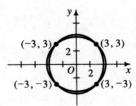

8. $5x^2 + 3y^2 = 64$; $3x^2 + 5y^2 = 64$; adding, $8x^2 + 8y^2 = 128$, so $x^2 + y^2 = 16$.

9. $x^2 + 4y^2 = 16$; $x^2 + y^2 = 4$; subtracting, $3y^2 = 12$; $y^2 = 4$; $y = \pm 2$; $x^2 + 4 = 4$; $x^2 = 0$; $x = 0$; $(0, \pm 2)$

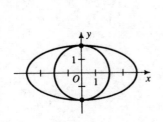

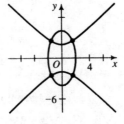

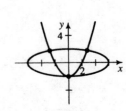

Ex. 9   **Ex. 10**   **Ex. 11**

10. $4x^2 + y^2 = 16$; $x^2 - y^2 = -4$; adding, $5x^2 = 12$; $x^2 = \dfrac{12}{5}$; $x = \pm\sqrt{\dfrac{12}{5}} = \pm\dfrac{2\sqrt{15}}{5}$;

$y^2 = \dfrac{12}{5} + 4 = \dfrac{32}{5}$; $y = \pm\sqrt{\dfrac{32}{5}} = \pm\dfrac{4\sqrt{10}}{5}$; $\left(\dfrac{2\sqrt{15}}{5}, \dfrac{4\sqrt{10}}{5}\right)$, $\left(\dfrac{2\sqrt{15}}{5}, -\dfrac{4\sqrt{10}}{5}\right)$,

$\left(-\dfrac{2\sqrt{15}}{5}, \dfrac{4\sqrt{10}}{5}\right)$, $\left(-\dfrac{2\sqrt{15}}{5}, -\dfrac{4\sqrt{10}}{5}\right)$

11. $x^2 + 9y^2 = 36$; $x^2 - 2y = 4$; subtracting, $9y^2 + 2y = 32$; $9y^2 + 2y - 32 = 0$;

$(9y - 16)(y + 2) = 0$; $y = \dfrac{16}{9}$ or $y = -2$; if $y = \dfrac{16}{9}$, $x^2 = 2 \cdot \dfrac{16}{9} + 4 = \dfrac{68}{9}$ and

$x = \pm\dfrac{2\sqrt{17}}{3}$; if $y = -2$, $x = 2(-2) + 4 = 0$; $\left(\dfrac{2\sqrt{17}}{3}, \dfrac{16}{9}\right)$, $\left(-\dfrac{2\sqrt{17}}{3}, \dfrac{16}{9}\right)$, $(0, -2)$

12. $2x + y^2 = 1$, so $y^2 = 1 - 2x$; $4x^2 + 4y^2 = 25$; $4x^2 + 4(1 - 2x) = 25$;

$4x^2 - 8x - 21 = 0$; $(2x - 7)(2x + 3) = 0$; $x = \dfrac{7}{2}$ or $x = -\dfrac{3}{2}$; if $x = \dfrac{7}{2}$,

$y^2 = 1 - 2 \cdot \dfrac{7}{2} = -6$ (no solutions); if $x = -\dfrac{3}{2}$, $y^2 = 1 - 2\left(-\dfrac{3}{2}\right) = 4$ and $y = \pm 2$;

$\left(-\dfrac{3}{2}, \pm 2\right)$

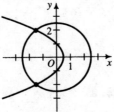

Ex. 12

13. $xy = -12$, so $y = -\dfrac{12}{x}$; $x^2 + y^2 = 25$; $x^2 + \left(-\dfrac{12}{x}\right)^2 = 25$; $x^4 - 25x^2 + 144 = 0$;

$(x^2 - 16)(x^2 - 9) = 0$; $x^2 = 16$ and $x = \pm4$, or $x^2 = 9$ and $x = \pm3$, since $y = -\dfrac{12}{x}$,

we have $(4, -3), (-4, 3), (3, -4), (-3, 4)$.

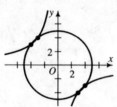

Ex. 13
            Ex. 14

14. $xy = 24$, so $y = \dfrac{24}{x}$; $y^2 - x^2 = 64$; $\left(\dfrac{24}{x}\right)^2 - x^2 = 64$; $x^4 + 64x^2 - 576 = 0$;

$(x^2 + 72)(x^2 - 8) = 0$; $x^2 = -72$ (no solutions) or $x^2 = 8$; $x = \pm2\sqrt{2}$;

$y = \dfrac{24}{\pm2\sqrt{2}} = \pm6\sqrt{2}$; $(2\sqrt{2}, 6\sqrt{2}), (-2\sqrt{2}, -6\sqrt{2})$

15. $y = -1 - x^2$ or $-x^2 - y = 1$; $x^2 - y^2 = 1$; adding, $-2y^2 = 2$, $y^2 = -1$ (no

solutions); thus, there is no point of intersection.

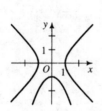
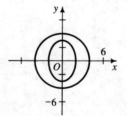

Ex. 15
            Ex. 16

16. $x^2 + y^2 = 16$, so $4x^2 + 4y^2 = 64$; $9x^2 + 4y^2 = 36$; subtracting, $5x^2 = -28$; thus,

there is no point of intersection.

17. $y = -4x^2$; $xy = 4$; $x(-4x^2) = 4$; $-4x^3 = 4$; $x^3 = -1$; $x = -1$; $y = -4(-1)^2 = -4$;

$(-1, -4)$

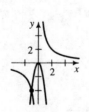
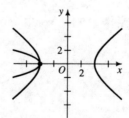

Ex. 17
            Ex. 18

18. $x + y^2 = -4$, so $y^2 = -x - 4$; $9x^2 - 16y^2 = 144$; $9x^2 - 16(-x - 4) = 144$;

$9x^2 + 16x - 80 = 0$; $(9x - 20)(x + 4) = 0$; $x = \dfrac{20}{9}$ or $x = -4$; if $x = \dfrac{20}{9}$,

$y^2 = -\dfrac{20}{9} - 4$ (no solution); if $x = -4$, $y^2 = -(-4) - 4 = 0$; $(-4, 0)$

19. $x = -y^2$, so $y^2 = -x$; $(x + 3)^2 + y^2 = 1$; $(x^2 + 6x + 9) + (-x) = 1$;
  $x^2 + 5x + 8 = 0$; discriminant is $5^2 - 4(1)(8) < 0$; no point of intersection

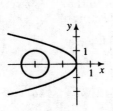

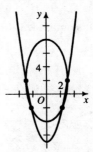

Ex. 19   **Ex. 20**

20. $x^2 + 6y^2 = 9$; so $x^2 = 9 - 6y^2$; $5x^2 + y^2 = 16$; $5(9 - 6y^2) + y^2 = 16$;
  $-29y^2 = -29$; $y^2 = 1$; $y = \pm 1$; $x^2 = 9 - 6(1) = 3$; $x = \pm\sqrt{3}$; $(\sqrt{3}, 1)$, $(-\sqrt{3}, 1)$,
  $(\sqrt{3}, -1)$, $(-\sqrt{3}, -1)$

21. $x^2 + y^2 = 9$, so $x^2 = 9 - y^2$; $x^2 + y^2 + 8x + 7 = 0$; $(9 - y^2) + y^2 + 8x + 7 = 0$;
  $8x = -16$; $x = -2$; $(-2)^2 + y^2 = 9$; $y^2 = 5$; $y = \pm\sqrt{5}$; $(-2, \pm\sqrt{5})$

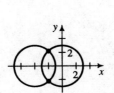

Ex. 21   **Ex. 22**

22. $x^2 - y - 7 = 0$, so $x^2 = y + 7$; $4x^2 + y^2 - 4y - 32 = 0$;
  $4(y + 7) + y^2 - 4y - 32 = 0$; $y^2 - 4 = 0$; $y^2 = 4$; $y = \pm 2$; if $y = 2$, $x^2 = 9$, and
  $x = \pm 3$; if $y = -2$, $x^2 = 5$, and $x = \pm\sqrt{5}$; $(\pm 3, 2)$, $(\pm\sqrt{5}, -2)$

23. a. $x^2 - 12x + 36 + y^2 - 6y + 9 = 5$; $x^2 + y^2 = 20$; subtracting,
  $12x - 36 + 6y - 9 = 15$; $y = -2x + 10$; $x^2 + (-2x + 10)^2 = 20$;
  $5x^2 - 40x + 80 = 0$; $x^2 - 8x + 16 = 0$; $(x - 4)^2 = 0$; $x = 4$, so $y = -2 \cdot 4 + 10$;
  $y = 2$; $(4, 2)$ is the point of tangency. The line segment containing the centers of the

  circles has slope $\dfrac{3}{6} = \dfrac{1}{2}$. The tangent line is perpendicular to the line containing the

  centers and thus has slope -2. An equation of the tangent line is $\dfrac{y - 2}{x - 4} = -2$;

  $y = -2x + 10$.

  b. $y = -2x + 10$, which is the equation of the common internal tangent.

24. a. $x^2 + y^2 = 225$; $C(0,0)$; $r = 15$; $(x - 6)^2 + (y - 8)^2 = 25$; $C(6,8)$; $r = 5$; distance
between the centers is $\sqrt{(6 - 0)^2 + (8 - 0)^2} = 10$.

b. $15 - 5 = 10$, so the circles must be internally tangent.

c. The line through the centers is perpendicular to the
common tangent. The slope of the line connecting

the centers is $\dfrac{8 - 0}{6 - 0} = \dfrac{4}{3}$, so the slope of the tangent

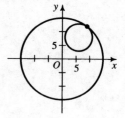

line must be $-\dfrac{3}{4}$. $x^2 + y^2 = 225$;

$(x - 6)^2 + (y - 8)^2 = 25$; $x^2 - 12x + y^2 - 16y = -75$;

subtracting, $12x + 16y = 300$; $3x + 4y = 75$; $x = -\dfrac{4}{3}y + 25$;

$\left(-\dfrac{4}{3}y + 25\right)^2 + y^2 = 225$; $25y^2 - 600y + 3600 = 0$; $y^2 - 24y + 144 = 0$;

$(y - 12)^2 = 0$; $y = 12$, so $x = -\dfrac{4}{3} \cdot 12 + 25 = 9$. The point of tangency is $(9, 12)$,

so the equation of the common tangent is $\dfrac{y - 12}{x - 9} = -\dfrac{3}{4}$, or $3x + 4y = 75$.

25. $x^2 - 9y^2 = 0$ is equivalent to $(x + 3y)(x - 3y) = 0$; $y = \dfrac{1}{3}x$ or $y = -\dfrac{1}{3}x$

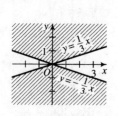

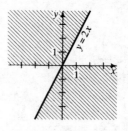

Ex. 25 **Ex. 26** **Ex. 27**

26. $y^2 - 2xy = 0$ is equivalent to $y(y - 2x) = 0$; $y = 0$ or $y = 2x$

27. $x^2 + 4y^2 - 10x + 24y + 61 = 0$ is equivalent to
$(x^2 - 10x + 25) + 4(y^2 + 6y + 9) = -61 + 25 + 36$; $(x - 5)^2 + 4(y + 3)^2 = 0$; thus,
$(x - 5)^2 + 4(y + 3)^2 \le 0$ has solution $(5, -3)$.

28. $x^2 - y^2 + 2x - 2y = 0$ is equivalent to $(x + 1)^2 - (y + 1)^2 = 0$; $(x + 1)^2 = (y + 1)^2$;
$x + 1 = \pm(y + 1)$; $x = y$ or $x + 1 = -y - 1$; $y = x$ or $y = -x - 2$

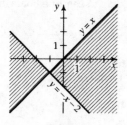

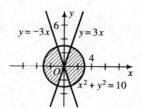

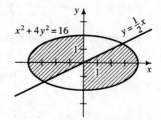

Ex. 28 **Ex. 29** **Ex. 30**

29. $9x^2 - y^2 = 0$ is equivalent to $(3x + y)(3x - y) = 0$; $y = \pm3x$; the graph of
$x^2 + y^2 \le 10$ is the interior of and the points on the related circle.

30. $x^2 - 2xy = 0$ is equivalent to $x(x - 2y) = 0$; $x = 0$ or $y = \dfrac{1}{2}x$; the graph of

$\dfrac{x^2}{16} + \dfrac{y^2}{4} \leq 1$ is the interior of and the points on the related ellipse.

31. Width of board $= w$. In $\triangle AEH$, $(80 - y)^2 + (60 - x)^2 =$
90^2 (Pythagorean Th.). $\triangle AEH \cong \triangle CGF$ by AAS Theorem;
$\therefore AE = GC = 80 - y$ and thus $EB = DG = y$.
$\triangle HDG \sim \triangle EAH$ by AA Postulate;

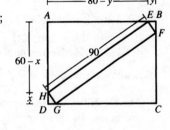
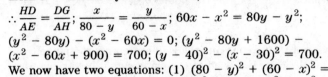

$\therefore \dfrac{HD}{AE} = \dfrac{DG}{AH}$; $\dfrac{x}{80 - y} = \dfrac{y}{60 - x}$; $60x - x^2 = 80y - y^2$;
$(y^2 - 80y) - (x^2 - 60x) = 0$; $(y^2 - 80y + 1600) -$
$(x^2 - 60x + 900) = 700$; $(y - 40)^2 - (x - 30)^2 = 700$.
We now have two equations: (1) $(80 - y)^2 + (60 - x)^2 =$
90^2, which can be written $(y - 80)^2 + (x - 60)^2 = 90^2$.
This is a circle. (2) $(y - 40)^2 - (x - 30)^2 = 700$. This is a hyperbola. By graphing
these two curves, the 4 points of intersection can be approximated. Only 1 pt.
$(x \approx 8, y \approx 6)$ is a possible solution for this problem, since x and y must both be greater
than 0. By using a calculator, a closer approximation can be determined; $x \approx 8.7, y \approx 6$.
The width of the board is $\sqrt{x^2 + y^2} \approx 10.6$ cm.

32. a. The equations of the two circles are $x^2 + y^2 = r_1^2$, and $(x - h)^2 + (y - k)^2 = r_2^2$.
Multiply out the second equation and subtract the first equation to get the difference,
$-2hx + h^2 - 2ky + k^2 = r_2^2 - r_1^2$, or $-2hx - 2ky = r_2^2 - r_1^2 - h^2 - k^2$.

If $T(x,y)$ is on this line, then $TP = \sqrt{(\sqrt{x^2 + y^2})^2 - r_1^2}$ and
$TQ = \sqrt{(\sqrt{(x - h)^2 + (y - k)^2})^2 - r_2^2}$; $TP = \sqrt{x^2 + y^2 - r_1^2}$;
$TQ = \sqrt{x^2 - 2hx + h^2 + y^2 - 2ky + k^2 - r_2^2}$;
$TQ = \sqrt{x^2 + y^2 + h^2 + k^2 - r_2^2 - 2hx - 2ky}$, and since (x,y) is on the line we
can substitute for $-2hx - 2ky$ to get
$TQ = \sqrt{x^2 + y^2 + h^2 + k^2 - r_2^2 + r_2^2 - r_1^2 - h^2 - k^2} = \sqrt{x^2 + y^2 - r_1^2} = TP$.

b. Yes

Class Exercises 6-7 · page 249

1. a. dist. from $A = \sqrt{(7 - 3)^2 + (2 - 2)^2} = 4$; dist. from
$B = \sqrt{(-3 - 3)^2 + (4 - 4)^2} = 6$; dist. from $C = \sqrt{(-5 - 3)^2 + (-1 + 1)^2} = 8$;
dist. from $P = \sqrt{(x - 3)^2 + (y - y)^2} = |x - 3|$

b. dist. from $A = \sqrt{(7 - 7)^2 + (2 + 6)^2} = 8$; dist. from
$B = \sqrt{(-3 + 3)^2 + (4 + 6)^2} = 10$; dist. from $C = \sqrt{(-5 + 5)^2 + (-1 + 6)^2} = 5$;
dist. from $P = \sqrt{(x - x)^2 + (y + 6)^2} = |y + 6|$

2. a. $\sqrt{x^2 + y^2} = 2\sqrt{(x - 7)^2 + (y - 2)^2}$

b. $\sqrt{x^2 + y^2} = \dfrac{1}{2}\sqrt{(x + 3)^2 + (y - 4)^2}$

c. $\sqrt{(x + 5)^2 + (y + 1)^2} = |y + 6|$

d. $\sqrt{(x + 3)^2 + (y - 4)^2} = 2|y + 6|$

e. $\sqrt{(x - 7)^2 + (y - 2)^2} = \dfrac{1}{2}|y + 6|$

3. A branch of a hyperbola

4. a. $B^2 - 4AC = 1$; hyperbola   **b.** $B^2 - 4AC = -8$; ellipse   **c.** $B^2 - 4AC = 0$; parabola

   d. $B^2 - 4AC < 0$, $B = 0$, and $A = C$; circle

5. The point $(0, 0)$

6. $(x - 3y)(x + 3y) = 0$; $x - 3y = 0$ or $x + 3y = 0$; two lines

7. $x^2 + 9y^2 = -1$; no graph

8. If $A = 0$, the graph is the line $y = 0$. If $C = 0$, the graph is the line $x = 0$. If A and C are both positive or both negative, the graph is the point $(0, 0)$. If $A \cdot C < 0$, then the graph is two lines: $y = \pm \left| \dfrac{A}{C} \right| x$

9. The flashlight beam represents a single cone and the wall represents the plane that cuts the cone. In this experiment, the cone is moved so the plane cuts it in various ways.

   a. A circle results when the flashlight is held $\perp$ to the wall.

   b. The parabola is difficult to find since the light must be adjusted until one edge of the cone is $\parallel$ to the wall.

   c. A hyperbola is formed by placing the flashlight parallel to the wall. Since we have a single cone, only one branch of this hyperbola is seen.

Written Exercises 6-7 · pages 250–251

1. a. $\dfrac{x^2}{4} + \dfrac{y^2}{3} = 1$ **b.** $c^2 = a^2 - b^2 = 1$; $F = (1, 0)$

   c. $\sqrt{(x - 1)^2 + y^2} = \dfrac{1}{2} |x - 4|$; $x^2 - 2x + 1 + y^2 = \dfrac{1}{4} x^2 - 2x + 4$; $\dfrac{3}{4} x^2 + y^2 = 3$;

   $\dfrac{x^2}{4} + \dfrac{y^2}{3} = 1$

2. $\sqrt{x^2 + (y - 8)^2} = \dfrac{4}{5} \left| y - \dfrac{25}{2} \right|$; $x^2 + y^2 - 16y + 64 = \dfrac{16}{25} \left(y^2 - 25y + \dfrac{625}{4} \right)$;

   $x^2 + y^2 - 16y + 64 = \dfrac{16}{25} y^2 - 16y + 100$; $x^2 + \dfrac{9}{25} y^2 = 36$; $\dfrac{x^2}{36} + \dfrac{y^2}{100} = 1$

3. $\sqrt{(x + 6)^2 + y^2} = 2 \left| x + \dfrac{3}{2} \right|$; $x^2 + 12x + 36 + y^2 = 4 \left(x^2 + 3x + \dfrac{9}{4} \right)$;

   $x^2 + 12x + 36 + y^2 = 4x^2 + 12x + 9$; $-3x^2 + y^2 = -27$; $\dfrac{x^2}{9} - \dfrac{y^2}{27} = 1$

4. $B^2 - 4AC = 8 > 0$; hyperbola

5. $B^2 - 4AC = -7 < 0$ and $B \neq 0$; ellipse

6. $y = x - \dfrac{1}{x}$; $xy = x^2 - 1$; $x^2 - xy = 1$; $B^2 - 4AC = 1 > 0$; hyperbola

7. $B^2 - 4AC = 5 > 0$; hyperbola

8. $B^2 - 4AC = -3 < 0$ and $B \neq 0$; ellipse

9. $B^2 - 4AC = -3 < 0$ and $B \neq 0$; ellipse

10. $4x^2 - y^2 = 0$; $(2x - y)(2x + y) = 0$; two lines: $2x - y = 0$, $2x + y = 0$

11. No graph

12. $y(y - 3x) = 0$; two lines: $y = 0$, $y = 3x$

13. $2a = 9.14 \times 10^7 + 9.44 \times 10^7 = 18.58 \times 10^7$; $a = 9.29 \times 10^7$;
$c = (9.44 - 9.29) \times 10^7$; $c = 0.15 \times 10^7$; $e = (0.15 \times 10^7)/(9.29 \times 10^7) \approx 0.016$

14. $2a = 8.8 \times 10^7 + 5.282 \times 10^9 \approx 537 \times 10^7$;
$a = 268.5 \times 10^7$; $c = (268.5 - 8.8) \times 10^7$;
$c = 259.7 \times 10^7$;
$e = (259.7 \times 10^7)/(268.5 \times 10^7) \approx 0.967$

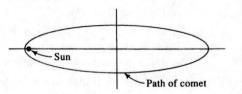

15. $x^2 + 2xy + y^2 = 9$; $x^2 + y^2 = 9$; subtracting, $2xy = 0$; either $x = 0$ or $y = 0$; if $x = 0$, then $0^2 + y^2 = 9$, so $y = \pm 3$; if $y = 0$, then $x^2 + 0^2 = 9$, so $x = \pm 3$; $(0, \pm 3)$, $(\pm 3, 0)$

16. $2x^2 + 3xy - 2y^2 = 0$; $(2x - y)(x + 2y) = 0$; $2x = y$ or $x = -2y$; if $2x = y$, $x^2 + 2(2x)^2 = 6$; $9x^2 = 6$; $x = \pm \dfrac{\sqrt{6}}{3}$; if $x = \dfrac{\sqrt{6}}{3}$, $y = \dfrac{2\sqrt{6}}{3}$; if $x = -\dfrac{\sqrt{6}}{3}$, $y = -\dfrac{2\sqrt{6}}{3}$; if $x = -2y$, $(-2y)^2 + 2y^2 = 6$; $6y^2 = 6$; $y^2 = 1$; $y = \pm 1$; if $y = 1$, $x = -2$; if $y = -1$, $x = 2$; $\left(\dfrac{\sqrt{6}}{3}, \dfrac{2\sqrt{6}}{3}\right)$, $\left(-\dfrac{\sqrt{6}}{3}, -\dfrac{2\sqrt{6}}{3}\right)$, $(-2, 1)$, $(2, -1)$

17. $x^2 + 2xy - 3y^2 = 0$; $(x + 3y)(x - y) = 0$; $x = -3y$ or $x = y$; if $x = -3y$, then $9y^2 + y^2 = 40$; $y^2 = 4$; $y = \pm 2$; if $y = 2$, $x = -6$; if $y = -2$, $x = 6$; if $x = y$, then $y^2 + y^2 = 40$; $y^2 = 20$, $y = \pm 2\sqrt{5}$; if $y = 2\sqrt{5}$, $x = 2\sqrt{5}$; if $y = -2\sqrt{5}$, $x = -2\sqrt{5}$; $(6, -2)$, $(-6, 2)$, $(2\sqrt{5}, 2\sqrt{5})$, $(-2\sqrt{5}, -2\sqrt{5})$

18. $3x^2 - 4xy - 4y^2 = 0$; $(3x + 2y)(x - 2y) = 0$; $x = -\dfrac{2}{3}y$ or $x = 2y$;

$x - 2y^2 + 4 = 0$; if $x = -\dfrac{2}{3}y$, $-\dfrac{2}{3}y - 2y^2 + 4 = 0$; $3y^2 + y - 6 = 0$; using the

quadratic formula, $y = \dfrac{-1 \pm \sqrt{73}}{6}$, so $x = \dfrac{1 \mp \sqrt{73}}{9}$; if $x = 2y$, $2y - 2y^2 + 4 = 0$;

$y^2 - y - 2 = 0$; $(y + 1)(y - 2) = 0$; $y = -1$ or $y = 2$; if $y = -1$, $x = -2$; if $y = 2$, $x = 4$; $\left(\dfrac{1 - \sqrt{73}}{9}, \dfrac{-1 + \sqrt{73}}{6}\right)$, $\left(\dfrac{1 + \sqrt{73}}{9}, \dfrac{-1 - \sqrt{73}}{6}\right)$, $(-2, -1)$, $(4, 2)$

19. $x^2 + xy + y^2 = 3$ and $x^2 - y^2 = 3$; subtracting, $xy + 2y^2 = 0$; $y(x + 2y) = 0$; $y = 0$ or $y = -\dfrac{1}{2}x$; if $y = 0$, $x^2 - 0 = 3$ and $x = \pm\sqrt{3}$; if $y = -\dfrac{1}{2}x$, $x^2 - \dfrac{1}{4}x^2 = 3$;

$\dfrac{3}{4}x^2 = 3$; $x^2 = 4$; $x = \pm 2$; $y = \mp 1$; $(\pm\sqrt{3}, 0)$, $(2, -1)$, $(-2, 1)$

20. $x^2 + 3y^2 = 3$ and $3x^2 - xy = 6$; $2x^2 + 6y^2 = 6$ and $3x^2 - xy = 6$; subtracting, $x^2 - xy - 6y^2 = 0$; $(x - 3y)(x + 2y) = 0$; $x = 3y$ or $x = -2y$; if $x = 3y$,

$9y^2 + 3y^2 = 3$, $y^2 = \dfrac{3}{12} = \dfrac{1}{4}$, $y = \pm\dfrac{1}{2}$, and $x = \pm\dfrac{3}{2}$; if $x = -2y$, $4y^2 + 3y^2 = 3$,

$y^2 = \dfrac{3}{7}$, $y = \pm\sqrt{\dfrac{3}{7}} = \pm\dfrac{\sqrt{21}}{7}$, and $x = \mp\dfrac{2\sqrt{21}}{7}$; $\left(\dfrac{3}{2}, \dfrac{1}{2}\right)$, $\left(-\dfrac{3}{2}, -\dfrac{1}{2}\right)$, $\left(\dfrac{2\sqrt{21}}{7}, -\dfrac{\sqrt{21}}{7}\right)$,
$\left(-\dfrac{2\sqrt{21}}{7}, \dfrac{\sqrt{21}}{7}\right)$

21. a. $\overline{PA}$ and $\overline{PM}$ are tangents from an exterior point to the sphere containing circle C_1 so $PA = PM$. $\overline{PB}$ and $\overline{PN}$ are tangents from an exterior point to the sphere containing circle C_2 so $PB = PN$.

b. For any point P on the ellipse, $PA = PM$ and $PB = PN$, so $PA + PB = PM + PN$. But P is a point on $\overleftrightarrow{MN}$ between M and N, so $PM + PN = MN$, a constant. Thus, the sum $PA + PB$ is constant, and A and B are the foci of the ellipse.

22. $\overline{PA}$ and $\overline{PM}$ are tangents from an exterior point to the sphere containing circle C_1 so $PA = PM$. $\overline{PB}$ and $\overline{PN}$ are tangents from an exterior point to the sphere containing circle C_2 so $PB = PN$. Then $PA - PB = PM - PN$. But N is a point on $\overleftrightarrow{PM}$ between P and M so $PN + NM = PM$ or $PM - PN = NM$, a constant. Therefore $PA - PB = NM$, a constant, and A and B are the foci of the hyperbola.

Chapter Test · page 253

1. Given: Square $ABCD$. Prove: $\overline{AC} \perp \overline{BD}$; $AC = BD$.

Proof: $\overline{AC}$ has slope $\dfrac{a - 0}{a - 0} = 1$, and $\overline{BD}$ has slope

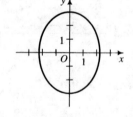

$\dfrac{a - 0}{0 - a} = -1$; $1(-1) = -1$, so $\overline{AC} \perp \overline{BD}$.

$AC = \sqrt{(a - 0)^2 + (a - 0)^2} = a\sqrt{2}$;
$BD = \sqrt{(a - 0)^2 + (0 - a)^2} = \sqrt{a^2 + a^2} = a\sqrt{2}$; $AC = BD$.

2. $y = x - 2$; $x^2 + y^2 = 4$; $x^2 + (x - 2)^2 = 4$; $2x^2 - 4x = 0$; $2x(x - 2) = 0$; either $x = 0$ and $y = 0 - 2 = -2$ or $x = 2$ and $y = 2 - 2 = 0$; $(0, -2)$, $(2, 0)$

3. Center is the midpoint of the segment with endpoints $(2, 5)$ and $(-2, -1)$; center is $\left(\dfrac{2 - 2}{2}, \dfrac{5 - 1}{2} \right) = (0, 2)$; radius is $\sqrt{(2 - 0)^2 + (5 - 2)^2} = \sqrt{2^2 + 3^2} = \sqrt{13}$; $x^2 + (y - 2)^2 = 13$

4. $\dfrac{x^2}{5} + \dfrac{y^2}{9} = 1$; center $(0, 0)$;
vertices $(0, \pm 3)$; $c^2 = a^2 - b^2 = 9 - 5 = 4$;
$c = 2$; foci $(0, \pm 2)$

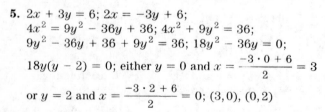

5. $2x + 3y = 6$; $2x = -3y + 6$;
$4x^2 = 9y^2 - 36y + 36$; $4x^2 + 9y^2 = 36$;
$9y^2 - 36y + 36 + 9y^2 = 36$; $18y^2 - 36y = 0$;
$18y(y - 2) = 0$; either $y = 0$ and $x = \dfrac{-3 \cdot 0 + 6}{2} = 3$

Ex. 4

or $y = 2$ and $x = \dfrac{-3 \cdot 2 + 6}{2} = 0$; $(3, 0)$, $(0, 2)$

6. $a = 4$, $b = 3$, major axis is vertical; $\dfrac{x^2}{9} + \dfrac{y^2}{16} = 1$

7. a. Major axis is vertical; $a = 3$, $c = \sqrt{13}$, $b^2 = c^2 - a^2 = 13 - 9 = 4$; $\dfrac{y^2}{9} - \dfrac{x^2}{4} = 1$

b. Major axis is horizontal; $a = 2$, $b = 4$; $\dfrac{x^2}{4} - \dfrac{y^2}{16} = 1$

8.

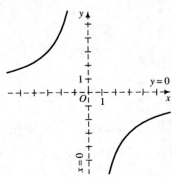

9. $y = \frac{1}{6}x^2$; $y = \frac{1}{4p}x^2$; $4p = 6$; $p = \frac{3}{2}$; parabola

opens up; $V(0,0)$, $F\left(0, \frac{3}{2}\right)$, $d: y = -\frac{3}{2}$

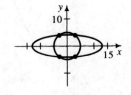

Ex. 11

10. Parabola opens right; vertex is $(0,0)$; $p = 3$; $x = \frac{1}{12}y^2$

11. $x^2 = 25 - y^2$; $x^2 + 10y^2 = 169$; $25 - y^2 + 10y^2 = 169$;
$9y^2 = 144$; $y^2 = 16$; $y = \pm 4$, $x = \pm 3$;
$(3, 4)$, $(-3, 4)$, $(3, -4)$, $(-3, -4)$

12. a. $B^2 - 4AC = -64 < 0$, $A = C$, and $B = 0$, so the graph is a circle.

b. $B^2 - 4AC = -8 < 0$, and $B \neq 0$, so the graph is an ellipse.

c. $B^2 - 4AC = 0$, so the graph is a parabola.

13. $\sqrt{(x - 1)^2 + (y - 3)^2} = |y + 4|$; squaring, $(x - 1)^2 + y^2 - 6y + 9 = y^2 + 8y + 16$;
$14y + 7 = (x - 1)^2$; $14\left(y + \frac{1}{2}\right) = (x - 1)^2$; $y + \frac{1}{2} = \frac{1}{14}(x - 1)^2$

14. If $A = B = C = 0$, the graph is the entire xy-plane. If $A = B = 0$ and $C \neq 0$, the graph
is the line $y = 0$. If $B = C = 0$ and $A \neq 0$, the graph is the line $x = 0$. If $A = C = 0$
and $B \neq 0$, the graph is two lines: $y = 0$, $x = 0$. If only $A = 0$, the graph is two lines:
$y = 0$, $Bx + Cy = 0$. If only $C = 0$, the graph is two lines: $x = 0$, $Ax + By = 0$.

Cumulative Review • pages 254–255

1.

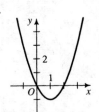

Range: $\{y \,|\, y \geq -1\}$;
zeros: $0, 2$

2. a. $x - 1 + 1 - x = 0$

b. $x - 1 - (1 - x) = 2x - 2$

c. $(x - 1)(1 - x) = -x^2 + 2x - 1$

d. $\frac{x - 1}{1 - x} = \frac{-(1 - x)}{1 - x} = -1$, $x \neq 1$

e. $f(g(x)) = f(1 - x) = (1 - x) - 1 = -x$

f. $g(f(x)) = g(x - 1) = 1 - (x - 1) = 2 - x$

3. a. (i) $x(-y)^2 + x^2(-y) = 4$; $xy^2 - x^2y = 4$; no
  (ii) $-xy^2 + (-x)^2y = 4$; $-xy^2 + x^2y = 4$; no
  (iii) $yx^2 + y^2x = 4$; $xy^2 + x^2y = 4$; yes
  (iv) $(-x)(-y)^2 + (-x)^2(-y) = 4$; $-xy^2 - x^2y = 4$; no

  b. (i) $-y = x^4 + 2x^2 + 1$; $y = -x^4 - 2x^2 - 1$; no
  (ii) $y = (-x)^4 + 2(-x)^2 + 1$; $y = x^4 + 2x^2 + 1$; yes
  (iii) $x = y^4 + 2y^2 + 1$; no
  (iv) $-y = (-x)^4 + 2(-x)^2 + 1$; $y = -x^4 - 2x^2 - 1$; no

(Continued)

c. (i) $(-y)^2 = |x| + 1$; $y^2 = |x| + 1$; yes

(ii) $y^2 = |-x| + 1$; $y^2 = |x| + 1$; yes

(iii) $x^2 = |y| + 1$; no

(iv) $(-y)^2 = |-x| + 1$; $y^2 = |x| + 1$; yes

4. a. $6; \dfrac{2 - (-2)}{2} = 2$ **b.** $4; \dfrac{3 - (-1)}{2} = 2$

5. a.

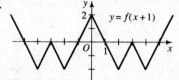

b.

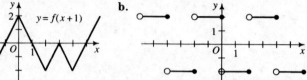

6. a. f is one-to-one, so it has an inverse; $x = 8 - 2y$; $2y = 8 - x$; $y = \dfrac{8 - x}{2}$;

$f^{-1}(x) = \dfrac{8 - x}{2}$; $f(f^{-1}(x)) = f\left(\dfrac{8 - x}{2}\right) = 8 - 2\left(\dfrac{8 - x}{2}\right) = 8 - (8 - x) = x$;

$f^{-1}(f(x)) = f^{-1}(8 - 2x) = \dfrac{8 - (8 - 2x)}{2} = \dfrac{2x}{2} = x$

b. f is not one-to-one, so it has no inverse.

c. f is one-to-one, so it has an inverse; $x = \dfrac{1}{y - 2}$; $xy - 2x = 1$; $xy = 2x + 1$;

$y = \dfrac{2x + 1}{x}$; $f^{-1}(x) = \dfrac{2x + 1}{x}$; $f(f^{-1}(x)) = f\left(\dfrac{2x + 1}{x}\right) =$

$\dfrac{1}{\dfrac{2x + 1}{x} - 2} = \dfrac{1}{\dfrac{2x + 1 - 2x}{x}} = \dfrac{1}{\dfrac{1}{x}} = x$; $f^{-1}(f(x)) = f^{-1}\left(\dfrac{1}{x - 2}\right) =$

$\dfrac{2\left(\dfrac{1}{x - 2}\right) + 1}{\dfrac{1}{x - 2}} = \dfrac{\dfrac{2}{x - 2} + \dfrac{x - 2}{x - 2}}{\dfrac{1}{x - 2}} = \dfrac{\dfrac{x}{x - 2}}{\dfrac{1}{x - 2}} = x$

7. a. $S(x, y) = 6x + 4y$

b. $S(40, 25) = 6(40) + 4(25) = 340$;
$S(32, 48) = 6(32) + 4(48) = 384$

c.

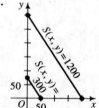

8. a. $\left(\dfrac{1}{4} + \dfrac{1}{2}\right)^2 = \left(\dfrac{3}{4}\right)^2 = \dfrac{9}{16}$ **b.** $(8y^4)(2^{-2}y^6) = (8y^4)\left(\dfrac{1}{4}y^6\right) = 2y^{10}$

c. $\dfrac{3^{-1}x^{-2}}{6x^{-3}} = \dfrac{3^{-1}x^{-2}}{6x^{-3}} \cdot \dfrac{3x^3}{3x^3} = \dfrac{x}{18}$ **d.** $\left(\dfrac{64}{81}\right)^{1/2} = \dfrac{8}{9}$ **e.** $27^{2/3} = (\sqrt[3]{27})^2 = 3^2 = 9$

f. $a^0 - a^1 = 1 - a$

9. a. $3^{4x+1} = 3^4$; $4x + 1 = 4$; $4x = 3$; $x = \dfrac{3}{4}$

    b. $x^{2/3} = 100$; $x = 100^{3/2} = (\sqrt{100})^3 = 10^3 = 1000$

    c. $x - 1 = \pm 25^{-1/2}$; $x - 1 = \pm\dfrac{1}{5}$; $x = \dfrac{6}{5}$ or $x = \dfrac{4}{5}$

10. $f(x) = a \cdot b^x$; $a \cdot b^0 = \dfrac{1}{2}$; $a = \dfrac{1}{2}$; $ab^4 = \dfrac{9}{2}$; $\dfrac{1}{2}b^4 = \dfrac{9}{2}$; $b^4 = 9$; $b^4 = 3^2$;

    $b = 3^{2/4} = 3^{1/2} = \sqrt{3}$; $f(x) = \dfrac{1}{2}(\sqrt{3})^x$

11. a. $N(0) = 150$, $N(3) = 300$, so the doubling time is 3 hours.

    b. $N(t) = 150 \cdot 2^{t/3}$

    c. $N(24) = 150 \cdot 2^{24/3} = 150 \cdot 2^8 = 150 \cdot 256 = 38{,}400$

12. a. $100\left(1 + \dfrac{0.07}{4}\right)^4 = 100(1.0175)^4 \approx \$107.19$

    b. $100\left(1 + \dfrac{0.07}{12}\right)^{12} = 100(1.00583\overline{3})^{12} \approx \$107.23$

    c. $100e^{0.07} \approx 100(1.0725082) \approx \$107.25$

13. Plan A: $\left(1 + \dfrac{0.081}{4}\right)^4 \approx 1.0834938$

    Plan B: $e^{0.08} \approx 1.0832871$

    Plan A yields the greater return.

14. a. $\log 10^{-2} = -2$ **b.** $\log_2 2^4 = 4$ **c.** $125^x = 5$; $5^{3x} = 5$; $3x = 1$; $x = \dfrac{1}{3}$

    d. $\ln e^{-5} = -5$ **e.** $e^{\ln 3^2} = 3^2 = 9$ **f.** $10^1 \cdot 10^{\log 5} = 10 \cdot 5 = 50$

15. a. $\log M^2 + \log N^3 = 2 \log M + 3 \log N$

    b. $\log\left(\dfrac{M^3}{N}\right)^{1/2} = \dfrac{1}{2}\log\dfrac{M^3}{N} = \dfrac{1}{2}[\log M^3 - \log N] = \dfrac{1}{2}(3 \log M - \log N) =$

    $\dfrac{3}{2}\log M - \dfrac{1}{2}\log N$

    c. $\log 100 + \log M + \log N^{1/2} = \log 10^2 + \log M + \dfrac{1}{2}\log N = 2 + \log M + \dfrac{1}{2}\log N$

16. a. $y = 10^{\log x + 2} = 10^{\log x} \cdot 10^2 = x \cdot 100 = 100x$

    b. $\ln y = \ln 2 - \ln x^3$; $\ln y = \ln\dfrac{2}{x^3}$; $y = \dfrac{2}{x^3}$

    c. $y = 10^{2.5x+1} = 10^{2.5x} \cdot 10^1 = 10 \cdot 10^{2.5x}$

17. a. $e^x = 8$; $x = \ln 8 \approx 2.08$

    b. $2 \cdot 2^x = 50$; $2^x = 25$; $\log 2^x = \log 25$; $x \log 2 = \log 25$; $x = \dfrac{\log 25}{\log 2} \approx 4.64$

    c. $\log 1.04^x = \log 2$; $x \log 1.04 = \log 2$; $x = \dfrac{\log 2}{\log 1.04} \approx 17.67$

18. Given: $PQRS$ is a rhombus; A, B, C, D are midpoints of its
sides. Prove: $ABCD$ is a rectangle. Proof: $A = (a, b)$,
$B = (-a, b)$, $C = (-a, -b)$, $D = (a, -b)$; slope of
$\overline{AB} = \dfrac{b - b}{a + a} = 0$; slope of $\overline{BC} = \dfrac{b + b}{-a + a}$, which is
undefined; thus $\overline{AB} \perp \overline{BC}$; $AB = DC = 2a$ and
$BC = AD = 2b$, so $ABCD$ is a parallelogram with
one right angle ($\angle B$), so $ABCD$ is a rectangle.

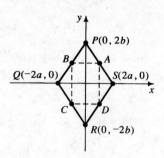

19. $x^2 - 4x + y^2 + 14y = -28$; $x^2 - 4x + 4 + y^2 + 14y + 49 = -28 + 4 + 49$;
$(x - 2)^2 + (y + 7)^2 = 25$; $C(2, -7)$; $r = 5$

20. $C(0, 0)$; $a^2 = 16$, $b^2 = 9$, so $c^2 = 7$; $c = \sqrt{7}$; $V(0, \pm 4)$ (major axis is vertical); $F(0, \pm\sqrt{7})$

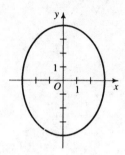

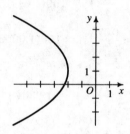

Ex. 20   **Ex. 22**

21. $\dfrac{x^2}{a^2} - \dfrac{y^2}{b^2} = 1$; $a = 2$, $c = \sqrt{5}$, $b^2 = c^2 - a^2 = 5 - 4 = 1$, so $b = 1$ and the equation is
$\dfrac{x^2}{4} - y^2 = 1$.

22. $V(-2, 1)$; $\dfrac{1}{4} = \dfrac{1}{4p}$, so $p = 1$; opens left; $F(-3, 1)$; $d: x = -1$

23. $-2x^2 - 2y^2 = -50$; $2x^2 - 3y = 6$; adding, $-2y^2 - 3y = -44$; $2y^2 + 3y - 44 = 0$;
$(2y + 11)(y - 4) = 0$; $y = -\dfrac{11}{2}$ (gives no solution) or $y = 4$, so $x = \pm 3$; $(3, 4)$
or $(-3, 4)$

24. $B^2 - 4AC = 4 - 4(1)(4) = -12 < 0$ and $B = -2$, so the curve is an ellipse.

Class Exercises 7-1 · page 260

1. a. $180° \cdot \dfrac{\pi}{180°} = \pi$ **b.** $90° \cdot \dfrac{\pi}{180°} = \dfrac{\pi}{2}$ **c.** $315° \cdot \dfrac{\pi}{180°} = \dfrac{7\pi}{4}$ **d.** $60° \cdot \dfrac{\pi}{180°} = \dfrac{\pi}{3}$

   e. $120° \cdot \dfrac{\pi}{180°} = \dfrac{2\pi}{3}$ **f.** $240° \cdot \dfrac{\pi}{180°} = \dfrac{4\pi}{3}$ **g.** $30° \cdot \dfrac{\pi}{180°} = \dfrac{\pi}{6}$ **h.** $1° \cdot \dfrac{\pi}{180°} = \dfrac{\pi}{180}$

2. a. $2\pi \cdot \dfrac{180°}{\pi} = 360°$ **b.** $\pi \cdot \dfrac{180°}{\pi} = 180°$ **c.** $\dfrac{\pi}{2} \cdot \dfrac{180°}{\pi} = 90°$ **d.** $\dfrac{\pi}{4} \cdot \dfrac{180°}{\pi} = 45°$

   e. $\dfrac{3\pi}{4} \cdot \dfrac{180°}{\pi} = 135°$ **f.** $\dfrac{5\pi}{3} \cdot \dfrac{180°}{\pi} = 300°$ **g.** $\dfrac{11\pi}{6} \cdot \dfrac{180°}{\pi} = 330°$ **h.** $\dfrac{5\pi}{6} \cdot \dfrac{180°}{\pi} = 150°$

3. Answers may vary. **a.** $10° + 360° = 370°$; $10° - 360° = -350°$

   b. $100° + 360° = 460°$; $100° - 360° = -260°$

   c. $-5° + 360° = 355°$; $-5° - 360° = -365°$

   d. $400° - 360° = 40°$; $40° - 360° = -320°$ **e.** $\pi + 2\pi = 3\pi$; $\pi - 2\pi = -\pi$

   f. $\dfrac{\pi}{2} + 2\pi = \dfrac{5\pi}{2}$; $\dfrac{\pi}{2} - 2\pi = -\dfrac{3\pi}{2}$ **g.** $-\dfrac{\pi}{3} + 2\pi = \dfrac{5\pi}{3}$; $-\dfrac{\pi}{3} - 2\pi = -\dfrac{7\pi}{3}$

   h. $4\pi - 2\pi = 2\pi$; $2\pi - 2 \cdot 2\pi = -2\pi$

4. a. $60°$ **b.** $\theta = \dfrac{\pi}{3} + 2n\pi$, n an integer

5. a. $\dfrac{s}{r} = \dfrac{2}{2} = 1$ **b.** $\dfrac{s}{r} = \dfrac{4}{2} = 2$ **c.** $\dfrac{s}{r} = \dfrac{1.5}{2} = 0.75$

6. a. $1\dfrac{2}{3} \cdot 360° = 600°$ **b.** $2\dfrac{3}{4} \cdot (-360°) = -990°$

Written Exercises 7-1 · pages 261–262

1. a. $315° \cdot \dfrac{\pi}{180°} = \dfrac{7\pi}{4}$ **b.** $225° \cdot \dfrac{\pi}{180°} = \dfrac{5\pi}{4}$ **c.** $15° \cdot \dfrac{\pi}{180°} = \dfrac{\pi}{12}$ **d.** $-45° \cdot \dfrac{\pi}{180°} = -\dfrac{\pi}{4}$

2. a. $-90° \cdot \dfrac{\pi}{180°} = -\dfrac{\pi}{2}$ **b.** $135° \cdot \dfrac{\pi}{180°} = \dfrac{3\pi}{4}$ **c.** $-180° \cdot \dfrac{\pi}{180°} = -\pi$

   d. $-225° \cdot \dfrac{\pi}{180°} = -\dfrac{5\pi}{4}$

3. a. $-120° \cdot \dfrac{\pi}{180°} = -\dfrac{2\pi}{3}$ **b.** $-240° \cdot \dfrac{\pi}{180°} = -\dfrac{4\pi}{3}$ **c.** $300° \cdot \dfrac{\pi}{180°} = \dfrac{5\pi}{3}$

   d. $360° \cdot \dfrac{\pi}{180°} = 2\pi$

4. a. $210° \cdot \dfrac{\pi}{180°} = \dfrac{7\pi}{6}$ **b.** $-135° \cdot \dfrac{\pi}{180°} = -\dfrac{3\pi}{4}$ **c.** $-210° \cdot \dfrac{\pi}{180°} = -\dfrac{7\pi}{6}$

   d. $-315° \cdot \dfrac{\pi}{180°} = -\dfrac{7\pi}{4}$

5. a. $-\dfrac{\pi}{2} \cdot \dfrac{180°}{\pi} = -90°$ **b.** $\dfrac{4\pi}{3} \cdot \dfrac{180°}{\pi} = 240°$ **c.** $-\dfrac{3\pi}{4} \cdot \dfrac{180°}{\pi} = -135°$

   d. $-\dfrac{\pi}{6} \cdot \dfrac{180°}{\pi} = -30°$

6. a. $-\dfrac{5\pi}{6} \cdot \dfrac{180°}{\pi} = -150°$ **b.** $-2\pi \cdot \dfrac{180°}{\pi} = -360°$ **c.** $\dfrac{5\pi}{4} \cdot \dfrac{180°}{\pi} = 225°$

   d. $-\dfrac{\pi}{3} \cdot \dfrac{180°}{\pi} = -60°$

7. a. $\pi \cdot \dfrac{180°}{\pi} = 180°$ **b.** $-\dfrac{3\pi}{2} \cdot \dfrac{180°}{\pi} = -270°$ **c.** $\dfrac{2\pi}{3} \cdot \dfrac{180°}{\pi} = 120°$

   d. $\dfrac{7\pi}{6} \cdot \dfrac{180°}{\pi} = 210°$

8. a. $-\dfrac{\pi}{4} \cdot \dfrac{180°}{\pi} = -45°$ **b.** $\dfrac{7\pi}{4} \cdot \dfrac{180°}{\pi} = 315°$ **c.** $4\pi \cdot \dfrac{180°}{\pi} = 720°$

   d. $\dfrac{11\pi}{6} \cdot \dfrac{180°}{\pi} = 330°$

9. a. $\theta = \dfrac{s}{r} = \dfrac{6}{5} = 1.2$ **b.** $\theta = \dfrac{s}{r} = \dfrac{6}{8} = 0.75$

10. a. $\theta = \dfrac{s}{r} = \dfrac{5}{4} = 1.25$ **b.** $\theta = \dfrac{s}{r} = \dfrac{15}{6} = 2.5$

11. a. $95° = 95 \times \dfrac{\pi}{180} \approx 1.66$ **b.** $110° = 110 \times \dfrac{\pi}{180} \approx 1.92$

   c. $95°10' \approx 95.167 \times \dfrac{\pi}{180} \approx 1.66$

   d. $119.2° = 119.2 \times \dfrac{\pi}{180} \approx 2.08$

12. a. $212° = 212 \times \dfrac{\pi}{180} \approx 3.70$

   b. $365° = 365 \times \dfrac{\pi}{180} \approx 6.37$

   c. $200°40' \approx 200.667 \times \dfrac{\pi}{180} \approx 3.50$

   d. $240.8° = 240.8 \times \dfrac{\pi}{180} \approx 4.20$

13. a. $1.6 = 1.6 \times \dfrac{180}{\pi} \approx 91.7°;\ 91°40'$

   b. $1.7 = 1.7 \times \dfrac{180}{\pi} \approx 97.4°;\ 97°20'$

   c. $1.21 = 1.21 \times \dfrac{180}{\pi} \approx 69.3°;\ 69°20'$

   d. $1.32 = 1.32 \times \dfrac{180}{\pi} \approx 75.6°;\ 75°40'$

14. a. $2.2 = 2.2 \times \dfrac{180}{\pi} \approx 126.1°;\ 126°$

   b. $3.7 = 3.7 \times \dfrac{180}{\pi} \approx 212.0°;\ 212°$

   c. $2.82 = 2.82 \times \dfrac{180}{\pi} \approx 161.6°;\ 161°30'$

   d. $3.41 = 3.41 \times \dfrac{180}{\pi} \approx 195.4°;\ 195°20'$

15. Estimate: about 0.5 radians; $30° = \dfrac{\pi}{6}$

16. Estimate: about 1.75 radians; using tables, $100° = 90° + 10° \approx 1.5708 +$
   $0.1745 = 1.7453$

17-22. Answers may vary. Examples are given.

17. a. $500° - 360° = 140°;\ 140° - 360° = -220°$

   b. $-60° + 360° = 300°;\ -60° - 360° = -420°$

   c. $\dfrac{\pi}{4} + 2\pi = \dfrac{9\pi}{4};\ \dfrac{\pi}{4} - 2\pi = -\dfrac{7\pi}{4}$ **d.** $-\dfrac{2\pi}{3} + 2\pi = \dfrac{4\pi}{3};\ -\dfrac{2\pi}{3} - 2\pi = -\dfrac{8\pi}{3}$

18. a. $1000° - 2(360°) = 280°; 280° - 360° = -80°$

b. $-100° + 360° = 260°; -100° - 360° = -460°$

c. $\dfrac{4\pi}{3} + 2\pi = \dfrac{10\pi}{3}; \dfrac{4\pi}{3} - 2\pi = -\dfrac{2\pi}{3}$ **d.** $-\dfrac{\pi}{6} + 2\pi = \dfrac{11\pi}{6}; -\dfrac{\pi}{6} - 2\pi = -\dfrac{13\pi}{6}$

19. a. $28.5° + 360° = 388.5°; 28.5° - 360° = -331.5°$

b. $116.3° + 360° = 476.3°; 116.3° - 360° = -243.7°$

c. $-60.4° + 360° = 299.6°; -60.4° - 360° = -420.4°$

d. $-315.3° + 360° = 44.7°; -315.3° - 360° = -675.3°$

20. a. $38.4° + 360° = 398.4°; 38.4° - 360° = -321.6°$

b. $127.6° + 360° = 487.6°; 127.6° - 360° = -232.4°$

c. $-50.8° + 360° = 309.2°; -50.8° - 360° = -410.8°$

d. $-320.7° + 360° = 39.3°; -320.7° - 360° = -680.7°$

21. a. $360°30' - 360° = 0°30'; 0°30' - 360° = -359°30'$

b. $-90°40' + 360° = 269°20'; -90°40' - 360° = -450°40'$

c. $3°21' + 360° = 363°21'; 3°21' - 360° = -356°39'$

d. $115°15' + 360° = 475°15'; 115°15' - 360° = -244°45'$

22. a. $180°20' + 360° = 540°20'; 180°20' - 360° = -179°40'$

b. $-270°30' + 360° = 89°30'; -270°30' - 360° = -630°30'$

c. $11°44' + 360° = 371°44'; 11°44' - 360° = -348°16'$

d. $172°11' + 360° = 532°11'; 172°11' - 360° = -187°49'$

23. $(29.7 + n \cdot 360)°$, n an integer   **24.** $-116°10' + n \cdot 360°$, n an integer

25. a. $35 \cdot 360° = 12{,}600°$ per minute   **b.** $35 \cdot 2\pi = 70\pi \approx 219.91$ radians per minute

26. a. $27 \cdot 360° = 9720°$ per minute   **b.** $27 \cdot 2\pi = 54\pi \approx 169.65$ radians per minute

27. a. $2.5(360°) = 900°$ per minute   **b.** $2.5(2\pi) = 5\pi \approx 15.71$ radians per minute

28. a. $6.5(360°) = 2340°$ per minute   **b.** $6.5(2\pi) = 13\pi \approx 40.84$ radians per minute

29. a. $14.6(360°) = 5256°$ per minute   **b.** $14.6(2\pi) = 29.2\pi \approx 91.73$ radians per minute

30. a. $19.8(360°) = 7128°$ per minute   **b.** $19.8(2\pi) = 39.6\pi \approx 124.41$ radians per minute

31. $\dfrac{s}{r} = \theta$; $35 \cdot \dfrac{5280 \cdot 12}{60} \cdot \dfrac{1}{14} \cdot \dfrac{1}{2\pi}$ rev. per min ≈ 420 rpm

32. $\dfrac{s}{r} = \theta$; $\dfrac{5}{15} \dfrac{\text{mi}}{\text{min}} = \dfrac{1}{3} \cdot 5280 \cdot 12\dfrac{\text{in.}}{\text{min}} = 4 \cdot 5280 \cdot \dfrac{1}{13.5} \cdot \dfrac{1}{2\pi}\dfrac{\text{rev}}{\text{min}} \approx 249$ rpm

33. Radius of Earth ≈ 3963 miles

a. $s = r\theta$; $s = 3963 \cdot \dfrac{\pi}{180} \approx 69$ miles

b. Each latitude has a different radius as you move away from the equator, whereas every longitude is a great circle.

34. On a star map, the right ascension of a star is measured east from a meridian that passes through the celestial pole (an extension into space of Earth's north-south pole) and contains the spring equinox, a point now in the constellation of Pisces on the celestial equator (a projection into space of Earth's equator). Because Earth rotates once every 24 hours, angles of right ascension are measured in hours, minutes, and seconds (where 24 hours = 360 degrees).

The declination of a star is measured north and south from the celestial equator. Angles of declination are measured in degrees, minutes, and seconds.

Examples are given below.

Star	Right ascension	Declination
Alpha Centauri	14 h 36.2 min	−60°38′
Sirius	6 h 42.9 min	−16°39′
Polaris	1 h 48.8 min	+89°02′
Vega	18 h 35.2 min	+38°44′

Class Exercises 7-2 · page 264

1. $s = \dfrac{\theta}{360} \cdot 2\pi r = \dfrac{45}{360} \cdot 2\pi \cdot 4 = \pi$; $K = \dfrac{\theta}{360} \cdot \pi r^2 = \dfrac{45}{360} \cdot \pi \cdot 4^2 = 2\pi$

2. $s = r\theta = 6 \cdot \dfrac{2\pi}{3} = 4\pi$; $K = \dfrac{1}{2} r^2 \theta = \dfrac{1}{2} \cdot 6^2 \cdot \dfrac{2\pi}{3} = 12\pi$

3. $s = r\theta = 2 \cdot 4 = 8$; $K = \dfrac{1}{2} r^2 \theta = \dfrac{1}{2} \cdot 2^2 \cdot 4 = 8$

4. a. height $\approx s = r\theta = 2(0.05) = 0.1$ km or 100 m

   b. Because in a large circle, a small arc is very nearly a straight line.

Written Exercises 7-2 · pages 264–267

1. $s = r\theta = 6(0.5) = 3$ cm; $K = \dfrac{1}{2} rs = \dfrac{1}{2} \cdot 6 \cdot 3 = 9$ cm^2

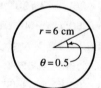

2. $s = r\theta = 5 \cdot 3 = 15$ cm; $K = \dfrac{1}{2} rs = \dfrac{1}{2} \cdot 5 \cdot 15 = 37.5$ cm^2

3. $r = \dfrac{s}{\theta} = \dfrac{11}{2.2} = 5$ cm; $K = \dfrac{1}{2} rs = \dfrac{1}{2} \cdot 5 \cdot 11 = 27.5$ cm^2

4. $r = \dfrac{s}{\theta} = \dfrac{2.0}{0.4} = 5$ cm; $K = \dfrac{1}{2} rs = \dfrac{1}{2} \cdot 5 \cdot 2 = 5$ cm^2

5. $K = \dfrac{1}{2} r^2 \theta$; $25 = \dfrac{1}{2} (r^2)(0.5)$; $r^2 = 100$; $r = 10$ cm; $s = r\theta = 10(0.5) = 5$ cm

6. $K = \dfrac{1}{2} r^2 \theta$; $90 = \dfrac{1}{2} (r^2)(0.2)$; $r^2 = 900$; $r = 30$ cm; $s = r\theta = 30(0.2) = 6$ cm

7. $s = \dfrac{\theta}{360} \cdot 2\pi r;\ 3.5 = \dfrac{30}{360} \cdot 2\pi r;\ r = \dfrac{21}{\pi};\ K = \dfrac{1}{2}rs = \dfrac{1}{2} \cdot \dfrac{21}{\pi}(3.5) \approx 12 \text{ cm}^2$

8. $s = \dfrac{\theta}{360} \cdot 2\pi r;\ 8.4 = \dfrac{24}{360} \cdot 2\pi r;\ r = \dfrac{63}{\pi};\ K = \dfrac{1}{2}rs = \dfrac{1}{2}\left(\dfrac{63}{\pi}\right)(8.4) \approx 84 \text{ cm}^2$

9. $P = 2r + s$, so $7 = 2r + s$, or $s = 7 - 2r;\ K = \dfrac{1}{2}rs$, so $3 = \dfrac{1}{2}rs$ and $rs = 6$;

substituting, $r(7 - 2r) = 6;\ 7r - 2r^2 = 6;\ 2r^2 - 7r + 6 = 0;\ (2r - 3)(r - 2) = 0;$
$r = 1.5$ cm or $r = 2$ cm

10. $P = 2r + s$, so $12 = 2r + s$ and $s = 12 - 2r;\ K = \dfrac{1}{2}rs$, so $8 = \dfrac{1}{2}rs$ and $rs = 16$;

substituting, $r(12 - 2r) = 16;\ 12r - 2r^2 = 16;\ r^2 - 6r + 8 = 0;\ (r - 4)(r - 2) = 0;$
$r = 4$ cm or $r = 2$ cm

11. $r = \dfrac{s}{\theta} = \dfrac{3500}{0.0087} \approx 402{,}000$ km   **12.** $s = r\theta = 56{,}000{,}000(0.00012) \approx 6720$ km

13. a. 200 rpm; $200(360°) = 72{,}000°;\ 200(2\pi) = 400\pi$ radians ≈ 1257 radians

b. $s = r\theta;\ s = \dfrac{11.9}{2} \cdot 400\pi \approx 7477$ cm   **c.** $7477\dfrac{\text{cm}}{\text{min}} = 7477 \cdot \dfrac{\text{cm}}{60 \text{ s}} \approx 125$ cm/s

14. $r = 6.5$ in., 120 rpm $= 240\pi$ rad/min; $s = r\theta = (6.5)(240\pi) \approx 4901$ in.

15. $3500 = (4 \times 10^5)\theta$ and $d = (1.5 \times 10^8)\theta$, so $\dfrac{3500}{4 \times 10^5} = \theta = \dfrac{d}{1.5 \times 10^8}.$

$\dfrac{3500}{4 \times 10^5} = \dfrac{d}{1.5 \times 10^8};\ d = \dfrac{3500 \times 1.5 \times 10^8}{4 \times 10^5} = 1312.5 \times 10^3 \approx 1.3 \times 10^6$ km

16. a. $\overline{OS}$ is parallel to the sun's rays, so $\theta = m\angle AOS$ because when parallel lines are cut by a transversal, alternate interior angles are congruent.

b. $s = r\theta;\ 5000 = r\left(7\dfrac{1}{5} \cdot \dfrac{\pi}{180}\right);\ r = 5000 \cdot \dfrac{180}{\pi} \cdot \dfrac{5}{36} = \dfrac{125{,}000}{\pi}.$

$C = 2\pi r = 2\pi \cdot \dfrac{125{,}000}{\pi} = 250{,}000$ stadia.

c. $250{,}000$ stadia $\approx 250{,}000(0.168) = 42{,}000$ km; $\dfrac{42{,}000 - 40{,}067}{40{,}067} \times 100 \approx 4.8\%$

17. The cow has the following grazing areas: one semicircle with radius 30 m; one quarter-circle with radius 20 m; two quarter-circles with radius 10 m. Total
area $= \dfrac{1}{2} \cdot 30^2 \cdot \pi + \dfrac{1}{2} \cdot 20^2 \cdot \dfrac{\pi}{2} + 2\left(\dfrac{1}{2} \cdot 10^2 \cdot \dfrac{\pi}{2}\right) = 450\pi + 100\pi + 50\pi =$
$600\pi \text{ m}^2 \approx 1885 \text{ m}^2$

18. $s = r\theta;\ 1 = 80\theta;\ \theta = \dfrac{1}{80}$ radian; about $0.716°$ or 0.0125 radians

19. Let d be the distance in kilometers to the mountain at the first sighting. Then, since the
car travels $80 \cdot \dfrac{1}{4} = 20$ km in 15 minutes, the distance at the second sighting is $d - 20$.

Let s be the height of the mountain. $s = d\left(\dfrac{1}{2} \cdot \dfrac{\pi}{180}\right)$ and

$s = (d - 20)\left(1 \cdot \dfrac{\pi}{180}\right)$, so $d \cdot \dfrac{\pi}{360} = (d - 20) \cdot \dfrac{\pi}{180};\ d = 2(d - 20);\ d = 2d - 40;$

$d = 40$; hence $s = d \cdot \dfrac{\pi}{360} = \dfrac{40\pi}{360} = \dfrac{\pi}{9}$ km or about 0.35 km (350 m).

20. Let d_1 be the distance to the lighthouse at the first sighting and d_2 be the distance at the second sighting. $s = r\theta$, so $20 = d_1(0.005)$ and $20 = d_2(0.010)$. Hence, $d_1 = 4000$ and $d_2 = 2000$. In 10 minutes, the boat travels $4000 - 2000 = 2000$ m, or 2 km. Since $d = rt$, $2 = r \cdot \dfrac{1}{6}$; $r = 12$ km/h.

21. $s = $ diameter of circular path $= 2(1.5 \times 10^8)$ km $= 3 \times 10^8$ km; $\theta = 1.5'' = \dfrac{1.5°}{3600} \approx$

7.2722×10^{-6} radians; $r = \dfrac{s}{\theta} \approx \dfrac{3 \times 10^8}{7.2722 \times 10^{-6}} \approx 4.125 \times 10^{13}$ km $\approx 4.1 \times 10^{13}$ km

22. Light travels at 3.00×10^8 m/s $= 60^2 \cdot 24 \cdot 365 \cdot (3 \times 10^8)$ m/yr $\approx 9.46 \times 10^{15}$ m/yr $= 9.46 \times 10^{12}$ km/yr; distance from Earth to the star is about $\dfrac{4.125 \times 10^{13}}{9.46 \times 10^{12}} \approx 4.4$; 4.4 light yr

23. The angular speed is the same for any two points on a rotating line. The linear speed varies, however: the greater the distance from the center, the greater the linear speed. Thus, the farther a skater is from the center, the faster he or she must skate in order to keep the line straight.

24. a. $P = 2r + s$, so $20 = 2r + s$. Since $s = r\theta$, we have $20 = 2r + r\theta$, or $\dfrac{20}{r} = 2 + \theta$.

Thus $\theta = \dfrac{20}{r} - 2$. $K = \dfrac{1}{2}r^2\theta = \dfrac{1}{2}r^2\left(\dfrac{20}{r} - 2\right) = 10r - r^2$.

b. The graph of $K = 10r - r^2$ is a parabola opening downward with vertex at $(5, 25)$. Hence the radius giving maximum area is $r = 5$ cm.

c. If $r = 5$, $\theta = \dfrac{20}{5} - 2 = 2$ (radians).

25. a. The area of each triangle is $\dfrac{1}{2}bh$. If there are n triangles, the total area is

$n\left(\dfrac{1}{2}bh\right) = \dfrac{1}{2}nbh.$

b. r; s

c. $A = \dfrac{1}{2}nbh$. As n increases, the value of nb approaches s and the value of h

approaches r. Hence $\dfrac{1}{2}nbh$ approaches $\dfrac{1}{2}sr$, so $K = \dfrac{1}{2}rs$.

d. $K = \dfrac{1}{2}rs$ and $s = r\theta$, so $K = \dfrac{1}{2}r \cdot r\theta = \dfrac{1}{2}r^2\theta$. In a circle, $\theta = 2\pi$, so

$K = \dfrac{1}{2}r^2 \cdot 2\pi = \pi r^2.$

Activity 7-3 · page 269

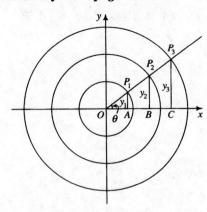

d. Ratios are equal.

e. $\triangle OP_1A \sim \triangle OP_2B \sim \triangle OP_3C$ by AA. Corres. sides are proportional; $\dfrac{y_1}{r_1} = \dfrac{y_2}{r_2} = \dfrac{y_3}{r_3} = \sin\theta$

Class Exercises 7-3 · page 271

1. $\sin\theta = \dfrac{\sqrt{10}}{10}$; $\cos\theta = -\dfrac{3\sqrt{10}}{10}$ **2.** $\sin\theta = -\dfrac{2\sqrt{5}}{5}$; $\cos\theta = \dfrac{\sqrt{5}}{5}$

3. $\sin\theta = -\dfrac{8}{2\sqrt{41}} = -\dfrac{4\sqrt{41}}{41}$; $\cos\theta = -\dfrac{10}{2\sqrt{41}} = -\dfrac{5\sqrt{41}}{41}$

4. a. Pos.   **b.** Neg.   **c.** Neg.   **d.** Pos.   **e.** Pos.   **f.** Neg.   **g.** Neg.   **h.** Pos.   **i.** Pos.
   j. Neg.   **k.** Neg.   **l.** Neg.

5. a. Dec.   **b.** Dec.   **c.** Inc.   **d.** Inc.

6. a. Inc.   **b.** Dec.   **c.** Dec.   **d.** Inc.

7. $x^2 + y^2 = 1$; $\sin\theta = \dfrac{y}{1} = y$, $\cos\theta = \dfrac{x}{1} = x$;
   $(\cos\theta)^2 + (\sin\theta)^2 = x^2 + y^2$; $(\cos\theta)^2 + (\sin\theta)^2 = 1$

8. $\theta = 90° + n\cdot 180°$, n an integer; $\pm 90°$, $\pm 270°$

9. a. θ is coterminal with an angle of $45°$.

   b. $\theta = \dfrac{1}{4}\pi + 2n\pi$, n an integer

Written Exercises 7-3 · pages 272–274

1. a. 0   **b.** -1 **c.** -1 **d.** 0   **2. a.** -1 **b.** 0   **c.** 0   **d.** 1

3. a. 0   **b.** -1 **c.** -1 **d.** 0

4. a. 1   **b.** -1 **c.** $\sin 3\pi = \sin(3\pi - 2\pi) = \sin\pi = 0$ **d.** 0

5. a. II   **b.** III   **6. a.** IV   **b.** I

7. a. $\theta = \dfrac{\pi}{2} + 2n\pi$ **b.** $\theta = \pi + 2n\pi$ **c.** $\theta = n\pi$ **d.** no solution

8. a. $\theta = 2n\pi$ **b.** $\theta = \dfrac{3\pi}{2} + 2n\pi$ **c.** $\theta = \dfrac{\pi}{2} + n\pi$ **d.** no solution

9. a. 0 **b.** N **c.** N **d.** N   **10. a.** N **b.** P **c.** N **d.** 0   **11. a.** P **b.** N **c.** P **d.** P

12. a. P **b.** P **c.** N **d.** P   **13. a.** N **b.** N **c.** 0 **d.** P   **14. a.** P **b.** P **c.** N **d.** P

15. a. P **b.** N **c.** 0 **d.** P   **16. a.** P **b.** N **c.** 0 **d.** N

17. $\sin \theta = \dfrac{4}{5}$; $\cos \theta = \dfrac{3}{5}$ **18.** $\sin \theta = -\dfrac{1}{\sqrt{10}} = -\dfrac{\sqrt{10}}{10}$; $\cos \theta = -\dfrac{3}{\sqrt{10}} = -\dfrac{3\sqrt{10}}{10}$

19. $x = -\sqrt{13^2 - 12^2} = -5$; $\sin \theta = \dfrac{12}{13}$; $\cos \theta = -\dfrac{5}{13}$

20. $y = -\sqrt{(\sqrt{5})^2 - 2^2} = -1$; $\sin \theta = -\dfrac{1}{\sqrt{5}} = -\dfrac{\sqrt{5}}{5}$; $\cos \theta = \dfrac{2}{\sqrt{5}} = \dfrac{2\sqrt{5}}{5}$

21. $\cos \theta = \dfrac{4}{5}$ **22.** $\cos \theta = -\dfrac{12}{13}$ **23.** $\sin \theta = -\dfrac{7}{25}$ **24.** $\sin \theta = -\dfrac{8}{17}$

25. $\cos \theta = -\dfrac{2\sqrt{6}}{5}$ **26.** $\cos \theta = -\dfrac{2\sqrt{10}}{7}$ **27.** $\sin \theta = -\dfrac{\sqrt{7}}{4}$ **28.** $\cos \theta = -\dfrac{4\sqrt{5}}{9}$

29. a. $x^2 + y^2 = 1$ and $y = \dfrac{1}{2}$; $x^2 + \dfrac{1}{4} = 1$; $x^2 = \dfrac{3}{4}$; $x = \pm\dfrac{\sqrt{3}}{2}$; $P\left(\dfrac{\sqrt{3}}{2}, \dfrac{1}{2}\right)$, $Q\left(-\dfrac{\sqrt{3}}{2}, \dfrac{1}{2}\right)$

b. On the unit circle, $y = \sin \theta$ and $x = \cos \theta$. Thus, "If $y = \dfrac{1}{2}$, then $x = \pm\dfrac{\sqrt{3}}{2}$" is equivalent to "If $\sin \theta = \dfrac{1}{2}$, then $\cos \theta = \pm\dfrac{\sqrt{3}}{2}$."

30. a. $x^2 + y^2 = 1$ and $x = -\dfrac{1}{2}$; $\dfrac{1}{4} + y^2 = 1$; $y^2 = \dfrac{3}{4}$; $y = \pm\dfrac{1}{2}\sqrt{3}$; $P\left(-\dfrac{1}{2}, \dfrac{\sqrt{3}}{2}\right)$, $Q\left(-\dfrac{1}{2}, -\dfrac{\sqrt{3}}{2}\right)$

b. On the unit circle, $y = \sin \theta$ and $x = \cos \theta$. Thus, "If $x = -\dfrac{1}{2}$, then $y = \pm\dfrac{\sqrt{3}}{2}$" is equivalent to "If $\cos \theta = -\dfrac{1}{2}$, then $\sin \theta = \pm\dfrac{\sqrt{3}}{2}$."

31. a. Length of arc $PQ = r\theta = 1 \cdot z = z$

    b. Since $r = 1$, $\sin \theta = y$ for every θ; $PA = y$-coord. of point $P = \sin z$

    c. $\sin z \approx z$ for small z; for example, $\sin 0.25 \approx 0.2474$, $\sin 0.1 \approx 0.0998$.

32. a. Since $r = 1$, $\cos \theta = x$ for every θ; $OA = x$-coord. of point $P = \cos z$.

    b. $1 - (PA)^2 = (OA)^2$; $OA = \sqrt{1 - (PA)^2}$; $\cos z = \sqrt{1 - \sin^2 z} \approx \sqrt{1 - z^2}$

    c. For example, $\cos 0.1 \approx 0.9950$ and $\sqrt{1 - (0.1)^2} \approx 0.9950$

33. $\sin 40° > \sin 30°$ **34.** $\cos 40° < \cos 30°$ **35.** $\sin 172° = \sin 8°$

36. $\sin 310° = \sin 230°$ **37.** $\sin 130° = \sin 50°$ **38.** $\cos 50° = \cos(-50°)$

39. $\cos 214° > \cos 213°$ **40.** $\sin 169° < \sin 168°$

41. 1 radian $\approx 57.3°$; 2 radians $\approx 114.6°$; 3 radians $\approx 171.9°$; 4 radians $\approx 229.2°$. Thus $\sin 4 < \sin 3 < \sin 1 < \sin 2$.

42. See Ex. 41; $\cos 3 < \cos 4 < \cos 2 < \cos 1$

43. a. $W(2) = (0, 1)$; $W(3) = (0, 0)$; $W(4) = (1, 0)$; $W(5) = (1, 1)$

    b. W is a periodic function with fundamental period 4 because $W(t + 4) = W(t)$ for all t.

    c. $c(t + 4) = x$-coordinate of $W(t + 4) = x$-coordinate of $W(t) = c(t)$; thus, c is periodic. $s(t + 4) = y$-coordinate of $W(t + 4) = y$-coordinate of $W(t) = s(t)$; thus, s is periodic.

    d.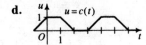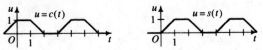

44. Consider the function W that wraps a vertical number line whose origin is at $R(1, 0)$ around the unit circle. With each real number t on the vertical number line, W associates a point $P(x, y)$ on the circle. For example, $W\left(\dfrac{\pi}{2}\right) = (0, 1)$ and $W\left(-\dfrac{\pi}{2}\right) = (0, -1)$. From W we can define the sine and cosine functions as follows: $\cos t = x$-coordinate of P and $\sin t = y$-coordinate of P.

Computer Exercises · page 274

1.
```
10 PRINT "WHAT IS THE VALUE OF X"
15 INPUT X
20 LET Y=X-X^3/(3*2)+X^5/(5*4*3*2)-X^7/(7*6*5*4*3*2)+
   X^9/(9*8*7*6*5*4*3*2)
25 LET Y=INT (100000 * Y + .5)/100000
26 LET S=INT (100000 * SIN(X) + .5)/100000
30 PRINT "WHEN X = ";X;" THE SUM IS ";Y;" AND SIN(X) = ";S
40 END

WHAT IS THE VALUE OF X
?1
WHEN X = 1 THE SUM IS .84147 AND SIN(X) = .84147
WHAT IS THE VALUE OF X
?2
WHEN X = 2 THE SUM IS .90935 AND SIN(X) = .9093
WHAT IS THE VALUE OF X
?1.5708
WHEN X = 1.5708 THE SUM IS 1 AND SIN(X) = 1.
```

2. Make these changes in the program for Ex. 1.
```
20 LET Y=1-X^2/2+X^4/(4*3*2)-X^6/(6*5*4*3*2)+X^8/(8*7*6*5*4*3*2)
26 LET C=INT (100000 * COS(X) + .5)/100000
30 PRINT "WHEN X = ";X;" THE SUM IS ";Y;" AND COS(X) = ";C
WHAT IS THE VALUE OF X
?1
WHEN X = 1 THE SUM IS .5403 AND COS(X) = .5403
WHAT IS THE VALUE OF X
?2
WHEN X = 2 THE SUM IS -.41587 AND COS(X) = -.41615
WHAT IS THE VALUE OF X
?3.1416
WHEN X = 3.1416 THE SUM IS -.97602 AND COS(X) = -1
```

Class Exercises 7-4 • pages 278–279

1. **a.** $180° - 170° = 10°$ **b.** $360° - 310° = 50°$ **c.** $205.1° - 180° = 25.1°$
   d. $\pi - 3 \approx 0.14$

2. $110°$ 3. $320°$

4. **a.** $\sin 170° = \sin(180° - 170°) = \sin 10°$

   b. $\sin 330° = -\sin(360° - 330°) = -\sin 30°$

   c. $\sin(-15°) = -\sin 15°$

   d. $\sin 400° = \sin(400° - 360°) = \sin 40°$

5. **a.** $\cos 160° = -\cos(180° - 160°) = -\cos 20°$

   b. $\cos 182° = -\cos(182° - 180°) = -\cos 2°$

   c. $\cos(-100°) = \cos(-100° + 360°) = \cos 260° = -\cos(260° - 180°) = -\cos 80°$

   d. $\cos 365° = \cos(365° - 360°) = \cos 5°$

6. **a.** -0.1392 **b.** -0.9397 **c.** 0.9848 **d.** -0.6428

7. **a.** 0.9842 **b.** 0.4787 **c.** -0.8161 **d.** 0.8471

8. **a.** 0.1411 **b.** -0.6536 **c.** -0.8415 **d.** -0.4161

9. **a.** -0.8011 **b.** 0.7452 **c.** 0.2555 **d.** 0.3802

10. **a.** $\dfrac{\sqrt{2}}{2}$ **b.** $\dfrac{\sqrt{2}}{2}$ **c.** $-\dfrac{\sqrt{2}}{2}$ **d.** $-\dfrac{\sqrt{2}}{2}$ 11. **a.** $\dfrac{1}{2}$ **b.** $-\dfrac{1}{2}$ **c.** $-\dfrac{1}{2}$ **d.** $\dfrac{1}{2}$

12. **a.** $\dfrac{1}{2}$ **b.** $-\dfrac{1}{2}$ **c.** $\dfrac{\sqrt{3}}{2}$ **d.** $\dfrac{\sqrt{3}}{2}$ 13. **a.** $-\dfrac{1}{2}$ **b.** $\dfrac{\sqrt{3}}{2}$ **c.** $-\dfrac{1}{2}$ **d.** $-\dfrac{\sqrt{3}}{2}$

14. **a.** $\dfrac{\sqrt{2}}{2}$ **b.** $-\dfrac{\sqrt{3}}{2}$ **c.** $-\dfrac{\sqrt{3}}{2}$ **d.** $\dfrac{\sqrt{3}}{2}$

15.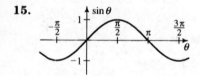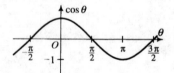

16. The graph of $y = \cos(\theta - 90°)$ is the graph of $y = \cos\theta$ translated to the right 90°. This coincides with the graph of $y = \sin\theta$.

17. **a.** Origin; each point $(x, \sin x)$ has an image point $(-x, -\sin x)$.

    b. y-axis; each point $(x, \cos x)$ has an image point $(-x, \cos x)$.

18. Since the cosine graph has symmetry in the y-axis, $\cos(-\theta) = \cos\theta$ for all θ. Thus, $\cos(90° - \theta) = \cos(-(\theta - 90°)) = \cos(\theta - 90°)$. Since $\cos(\theta - 90°) = \sin\theta$, we also have $\cos(90° - \theta) = \sin\theta$ for all θ.

Written Exercises 7-4 • pages 279–282

1. **a.** $\sin 128° = \sin 52°$ **b.** $\cos 128° = -\cos 52°$ **c.** $\sin(-37°) = -\sin 37°$

   d. $\cos 500° = -\cos 40°$

2. **a.** $\sin 310° = -\sin 50°$ **b.** $\cos 310° = \cos 50°$ **c.** $\cos(-53°) = \cos 53°$

   d. $\sin 1000° = -\sin 80°$

3. a. $\cos 224.5° = -\cos 44.5°$ **b.** $\cos 658° = \cos 62°$ **c.** $\sin 145.7° = \sin 34.3°$

     d. $\sin(-201°) = \sin 21°$

4. a. $\cos 107.9° = -\cos 72.1°$ **b.** $\sin 271.3° = -\sin 88.7°$ **c.** $\sin 834° = \sin 66°$

     d. $\cos(-132°) = -\cos 48°$

5. a. 0.4695 **b.** 0.7193 **c.** 0.4179 **d.** 0.4706

6. a. 0.9877 **b.** 1.0000 **c.** 0.3902 **d.** 0.8957

7. a. 0.3420 **b.** -0.9041 **c.** 0.5314 **d.** -0.4586

8. a. -0.5299 **b.** 0.6820 **c.** -0.9627 **d.** -0.6704

9. a. 0.9320 **b.** 0.8085 **c.** -0.9365 **d.** 0.2837

10. a. -0.3880 **b.** 0.9602 **c.** -0.9093 **d.** -0.9900

11. a. $\dfrac{\sqrt{2}}{2}$ **b.** $-\dfrac{\sqrt{2}}{2}$ **c.** $-\dfrac{1}{2}$ **d.** $-\dfrac{\sqrt{3}}{2}$ **12. a.** $-\dfrac{\sqrt{2}}{2}$ **b.** $\dfrac{\sqrt{2}}{2}$ **c.** $-\dfrac{\sqrt{2}}{2}$ **d.** $\dfrac{\sqrt{2}}{2}$

13. a. $\dfrac{1}{2}$ **b.** $-\dfrac{1}{2}$ **c.** $-\dfrac{\sqrt{2}}{2}$ **d.** $\dfrac{\sqrt{3}}{2}$ **14. a.** $-\dfrac{\sqrt{3}}{2}$ **b.** 0 **c.** $-\dfrac{\sqrt{3}}{2}$ **d.** $\dfrac{\sqrt{2}}{2}$

15. a. $\dfrac{\sqrt{3}}{2}$ **b.** $\dfrac{\sqrt{3}}{2}$ **c.** $-\dfrac{1}{2}$ **d.** $\dfrac{\sqrt{2}}{2}$ **16. a.** $\dfrac{\sqrt{2}}{2}$ **b.** $-\dfrac{\sqrt{2}}{2}$ **c.** $-\dfrac{\sqrt{3}}{2}$ **d.** $-\dfrac{\sqrt{3}}{2}$

17. a. 1 **b.** $-\dfrac{1}{2}$ **c.** $-\dfrac{\sqrt{3}}{2}$ **d.** $-\dfrac{\sqrt{2}}{2}$ **18. a.** $\dfrac{1}{2}$ **b.** 0 **c.** $-\dfrac{\sqrt{2}}{2}$ **d.** $-\dfrac{1}{2}$

19.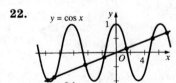

20.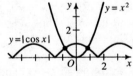

21. a.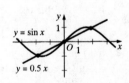

     b. 3 solutions $-1.90, 0, 1.90$

22.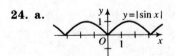

     b. 7 solutions 1.4

23. a.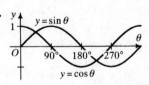

     b. 2 solutions $-0.82, 0.82$

24. a.

25. a. 2   **b.** $45°, 225°$ **c.**

$y = \sin x$: domain $= \mathcal{R}$;
     range $= \{y \mid -1 \le y \le 1\}$; $p = 2\pi$
$y = |\sin x|$: domain $= \mathcal{R}$;
     range $= \{y \mid 0 \le y \le 1\}$; $p = \pi$

26. The angle determined by the North Pole and Durham is $90° - 36° = 54°$, so

$$s = r\theta = 3963 \cdot 54° \cdot \frac{\pi}{180°} \approx 3735 \text{ mi.}$$

27. The angle determined by the South Pole and Lima is $90° - 12° = 78°$, so

$$s = r\theta = 3963 \cdot 78° \cdot \frac{\pi}{180°} \approx 5395 \text{ mi.}$$

28. The angle determined by the cities is $39°55' + 31°58' = 71°53'$. $s = r\theta =$

$$3963 \cdot (71°53') \cdot \frac{\pi}{180°} \approx 4972 \text{ mi.}$$

29. $\theta = 35°6' - 30° = 5°6'$; $s = r\theta = 3963 \, (5°6') \cdot \frac{\pi}{180°} \approx 353 \text{ mi}$

30. **a.** The circle at latitude $45°$ has radius $r = 3963 \cdot \cos 45° \approx 2802.26$ mi. Its circumference is $C = 2\pi r = 2\pi(2802.26) \approx 17607.12$ mi; rotational

speed $\approx \dfrac{17607.12}{24} \approx 734$ mi/h.

b. $r = 3963 \cdot \cos 30° = 3963 \cdot \dfrac{\sqrt{3}}{2}$; $C = 3963 \dfrac{\sqrt{3}}{2} \cdot 2\pi$; rotational speed $=$

$\dfrac{3963\sqrt{3} \cdot \pi}{24} \approx 899$ mi/h

31. **a.** $r = 3963 \cos 61°$; $C = 2\pi \cdot 3963 \cos 61°$; rotational

speed $= \dfrac{2\pi \cdot 3963 \cos 61°}{24} \approx 503$ mi/h

b. $r = 3963 \cos 23°$; $C = 2\pi \cdot 3963 \cos 23°$; rotational speed $= \dfrac{2\pi \cdot 3963 \cos 23°}{24} \approx$ 955 mi/h

32. **a.** The circle at latitude L has radius $r = 3963 \cos L$. The circumference of this circle is $C = 2\pi r = 2\pi(3963 \cos L) = 7926\pi \cos L$. Thus, the rotational speed $= \dfrac{7926\pi \cos L}{24} \approx 1038 \cos L$ mi/h.

b. At the North Pole, $L = 90°$; $1038 \cos 90° = 0$; speed $= 0$

33. The plane stays with the sun if it travels the same speed as Earth rotates along the given latitude. By Ex. 32a, the rotational speed of Earth at the 42°N latitude is $1038 \cos 42° \approx 771$ mi/h.

34. At time t, the coordinates of P are $(\cos t, \ \sin t)$ and the coordinates of Q are $(\cos t + \sqrt{16 - (\sin t)^2}, 0)$.

Class Exercises 7-5 · page 285

1. a. $90° + n \cdot 180°$, n an integer   **b.** $90° + n \cdot 360°$ **c.** $135° + n \cdot 180°$ **d.** $n \cdot 180°$

2. a. $\sec(-15°) \approx 1.035$ **b.** $\sec 165° \approx -1.035$ **c.** $\sec 195° \approx -1.035$
   d. $\sec 345° \approx 1.035$

3. a. $\tan 2 \approx -2.185$ **b.** $\cot 185° \approx 11.43$ **c.** $\csc 3 \approx 7.086$ **d.** $\sec(-22°) \approx 1.079$

4. quadrant III

5. a. $\cos \theta = \dfrac{x}{r} = -\dfrac{3}{5}$ **b.** $\tan \theta = \dfrac{y}{x} = -\dfrac{4}{3}$ **c.** $\cot \theta = \dfrac{x}{y} = -\dfrac{3}{4}$

   d. $\sec \theta = \dfrac{1}{\cos \theta} = -\dfrac{5}{3}$ **e.** $\csc \theta = \dfrac{1}{\sin \theta} = \dfrac{5}{4}$

Written Exercises 7-5 · pages 285–286

1. a. $\tan 100° = -\tan 80° \approx -5.671$ **b.** $\cot 276° = -\cot 84° \approx -0.1051$

  c. $\csc 5 = -\csc (2\pi - 5) \approx -\csc 1.28 \approx -1.044$ (or -1.043)

  d. $\sec 2.14 = -\sec (\pi - 2.14) \approx -\sec 1 \approx -1.851$ (or -1.855)

2. a. $\sec (-11°) = \sec 11° \approx 1.019$ **b.** $\csc 233° = -\csc 53° \approx -1.252$

  c. $\tan 3 = -\tan (\pi - 3) \approx -\tan 0.14 \approx -0.1409$ (or -0.1425)

  d. $\cot 7.28 = \cot (7.28 - 2\pi) \approx \cot 1 \approx 0.6421$ (or 0.6466)

3. a. $\sin 195° = -\sin 15°$ **b.** $\sec 280° = \sec 80°$ **c.** $\tan (-140°) = \tan 220° = \tan 40°$

  d. $\sec 2 = -\sec (\pi - 2) \approx -\sec 1.14$

4. a. $\cot 285° = -\cot 75°$ **b.** $\sec (-105°) = \sec 255° = -\sec 75°$

  c. $\csc 600° = \csc 240° = -\csc 60°$ **d.** $\tan 3 = -\tan (\pi - 3) \approx -\tan 0.14$

5. a. $\tan 820° = \tan 100° = -\tan 80°$ **b.** $\sec 290° = \sec 70°$

  c. $\cot 185° = \cot 5°$ **d.** $\csc 4 = -\csc (4 - \pi) \approx -\csc 0.86$

6. a. $\tan 160° = -\tan 20°$ **b.** $\csc 115° = \csc 65°$ **c.** $\sec 235° = -\sec 55°$

  d. $\cot 5 = -\cot (2\pi - 5) \approx -\cot 1.28$

7. a. $x = n\pi$ **b.** none **c.** $x = \dfrac{\pi}{2} + 2n\pi$ **d.** $x = \dfrac{3\pi}{2} + 2n\pi$

8. a. $x = n\pi$ **b.** $x = \dfrac{\pi}{2} + n\pi$ **c.** $x = \dfrac{\pi}{4} + n\pi$ **d.** $x = \dfrac{3\pi}{4} + n\pi$

9.

10.

domain: all reals except $n\pi$; range $\{y \mid y \le -1 \text{ or } y \ge 1\}$
vertical asymptotes $x = n\pi$

local minima at $\dfrac{\pi}{2} + n\pi$,

$n = \pm 1, \pm 5, \ldots$

local maxima at $\dfrac{\pi}{2} + n\pi$,

$n = \pm 3, \pm 7, \ldots$
period: 2π

11.

$\tan x = 2x$ has infinitely many solutions.

12.

$\sec x = \sin x$ has no solution.

13. $\sin x = \dfrac{5}{13}$; $-\sqrt{13^2 - 5^2} = -12$; $\cos x = -\dfrac{12}{13}$, $\tan x = -\dfrac{5}{12}$, $\cot x = -\dfrac{12}{5}$,

$\sec x = -\dfrac{13}{12}$, $\csc x = \dfrac{13}{5}$

14. $\cos x = \dfrac{24}{25}$; $-\sqrt{25^2 - 24^2} = -7$; $\sin x = -\dfrac{7}{25}$, $\tan x = -\dfrac{7}{24}$, $\cot x = -\dfrac{24}{7}$,

$\sec x = \dfrac{25}{24}$, $\csc x = -\dfrac{25}{7}$

15. $\tan x = \dfrac{3}{4}$; $\sqrt{3^2 + 4^2} = 5$; $\sin x = -\dfrac{3}{5}$, $\cos x = -\dfrac{4}{5}$, $\cot x = \dfrac{4}{3}$,

$\sec x = -\dfrac{5}{4}$, $\csc x = -\dfrac{5}{3}$

16. $\cot x = \dfrac{-12}{5}$; $\sqrt{(-12)^2 + 5^2} = 13$; $\sin x = \dfrac{5}{13}$, $\cos x = -\dfrac{12}{13}$, $\tan x = -\dfrac{5}{12}$,

$\sec x = -\dfrac{13}{12}$, $\csc x = \dfrac{13}{5}$

17. $\sec x = \dfrac{3}{-1}$; $\sqrt{3^2 - (-1)^2} = 2\sqrt{2}$; $\sin x = \dfrac{2\sqrt{2}}{3}$, $\cos x = -\dfrac{1}{3}$,

$\tan x = -2\sqrt{2}$, $\cot x = -\dfrac{1}{2\sqrt{2}} = -\dfrac{\sqrt{2}}{4}$, $\csc x = \dfrac{3}{2\sqrt{2}} = \dfrac{3\sqrt{2}}{4}$

18. $\csc x = \dfrac{5}{-1}$; $-\sqrt{5^2 - (-1)^2} = -2\sqrt{6}$; $\sin x = -\dfrac{1}{5}$, $\cos x = -\dfrac{2\sqrt{6}}{5}$,

$\tan x = \dfrac{1}{2\sqrt{6}} = \dfrac{\sqrt{6}}{12}$, $\cot x = 2\sqrt{6}$, $\sec x = -\dfrac{5}{2\sqrt{6}} = -\dfrac{5\sqrt{6}}{12}$

19. $\cot x$ is undefined for $x = n\pi$, n an integer. Moreover, as x gets close to a multiple of π, $|\cot x|$ gets large without bound. Thus, the cotangent graph has vertical asymptotes at multiples of π.

20. The functions are defined in terms of $x, y,$ and r, where x and y are the coordinates of a point on the terminal ray of θ and $r = \sqrt{x^2 + y^2}$.

Function	Def.	Domain	Range	Period
$\sin \theta$	$\dfrac{y}{r}$	all reals	$-1 \le \sin \theta \le 1$	2π
$\cos \theta$	$\dfrac{x}{r}$	all reals	$-1 \le \cos \theta \le 1$	2π
$\tan \theta$	$\dfrac{y}{x}$	$\theta \ne \dfrac{\pi}{2} + n\pi$	all reals	π
$\cot \theta$	$\dfrac{x}{y}$	$\theta \ne n\pi$	all reals	π
$\sec \theta$	$\dfrac{r}{x}$	$\theta \ne \dfrac{\pi}{2} + n\pi$	$|\sec \theta| \ge 1$	2π
$\csc \theta$	$\dfrac{r}{y}$	$\theta \ne n\pi$	$|\csc \theta| \ge 1$	2π

The graphs of these functions are shown at the top of the next page.

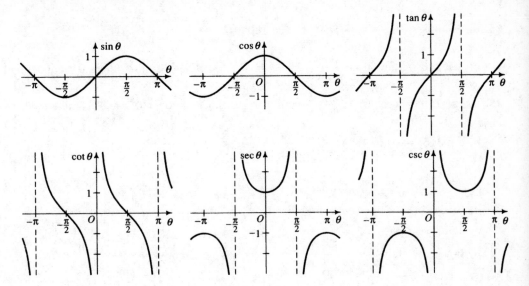

21. a. $1 + \tan^2 \dfrac{\pi}{3} = 1 + (\sqrt{3})^2 = 4$; $\sec^2 \dfrac{\pi}{3} = 2^2 = 4$. So, $1 + \tan^2 \dfrac{\pi}{3} = \sec^2 \dfrac{\pi}{3}$.

   b. True for all $x \neq \dfrac{\pi}{2} + n\pi$; n an integer

22. a.

x-value	$\dfrac{\pi}{2}$	$\dfrac{3\pi}{4}$	$\dfrac{7\pi}{6}$
$1 + \cot^2 x$	$1 + 0^2 = 1$	$1 + (-1)^2 = 2$	$1 + (\sqrt{3})^2 = 4$
$\csc^2 x$	$1^2 = 1$	$(\sqrt{2})^2 = 2$	$(-2)^2 = 4$

   b. $1 + \cot^2 x = \csc^2 x$; proof: $\sin^2 x + \cos^2 x = 1$; $\dfrac{\sin^2 x}{\sin^2 x} + \dfrac{\cos^2 x}{\sin^2 x} = \dfrac{1}{\sin^2 x}$ (provided

   that $\sin x \neq 0$); $1 + \left(\dfrac{\cos x}{\sin x}\right)^2 = \left(\dfrac{1}{\sin x}\right)^2$; $1 + \cot^2 x = \csc^2 x$

23. a. $\cos 120° = -\cos 60° = -\dfrac{1}{2}$ **b.** $\sec 120° = -\sec 60° = -\dfrac{1}{\cos 60°} = -\dfrac{1}{\frac{1}{2}} = -2$

   c. $\sin 120° = \sin 60° = \dfrac{\sqrt{3}}{2}$ **d.** $\tan 120° = -\tan 60° = -\dfrac{\sin 60°}{\cos 60°} = -\dfrac{\frac{\sqrt{3}}{2}}{\frac{1}{2}} = -\sqrt{3}$

24. a. $\sin 225° = -\sin 45° = -\dfrac{\sqrt{2}}{2}$ **b.** $\csc 225° = -\csc 45° = -\dfrac{1}{\sin 45°} = -\dfrac{1}{\frac{\sqrt{2}}{2}} = -\sqrt{2}$

   c. $\tan 225° = \tan 45° = \dfrac{\sin 45°}{\cos 45°} = \dfrac{\frac{\sqrt{2}}{2}}{\frac{\sqrt{2}}{2}} = 1$

   d. $\sec 225° = -\sec 45° = -\dfrac{1}{\cos 45°} = -\dfrac{1}{\frac{\sqrt{2}}{2}} = -\sqrt{2}$

25. a. $\csc 90° = \dfrac{1}{\sin 90°} = \dfrac{1}{1} = 1$ **b.** $\sec 180° = \dfrac{1}{\cos 180°} = \dfrac{1}{-1} = -1$

     c. $\tan 240° = \tan 60° = \dfrac{\sin 60°}{\cos 60°} = \dfrac{\frac{\sqrt{3}}{2}}{\frac{1}{2}} = \sqrt{3}$ **d.** Since $\sin 0° = 0$, $\cot 0°$ is undefined.

26. a. $\csc 150° = \csc 30° = \dfrac{1}{\sin 30°} = \dfrac{1}{\frac{1}{2}} = 2$ **b.** Since $\sin 0° = 0$, $\csc 0°$ is undefined.

     c. $\tan 315° = -\tan 45° = -\dfrac{\sin 45°}{\cos 45°} = -\dfrac{\frac{\sqrt{2}}{2}}{\frac{\sqrt{2}}{2}} = -1$

     d. $\sec 315° = \sec 45° = \dfrac{1}{\cos 45°} = \dfrac{1}{\frac{\sqrt{2}}{2}} = \sqrt{2}$

27. a. Since $\sin \pi = 0$, $\csc \pi$ is undefined.

     b. $\tan \dfrac{2\pi}{3} = -\tan \dfrac{\pi}{3} = -\dfrac{\sin \frac{\pi}{3}}{\cos \frac{\pi}{3}} = -\dfrac{\frac{\sqrt{3}}{2}}{\frac{1}{2}} = -\sqrt{3}$ **c.** $\cot \dfrac{\pi}{2} = \dfrac{\cos \frac{\pi}{2}}{\sin \frac{\pi}{2}} = \dfrac{0}{1} = 0$

     d. $\sec \dfrac{5\pi}{6} = -\sec \dfrac{\pi}{6} = -\dfrac{1}{\cos \frac{\pi}{6}} = -\dfrac{1}{\frac{\sqrt{3}}{2}} = -\dfrac{2}{\sqrt{3}} = -\dfrac{2\sqrt{3}}{3}$

28. a. Since $\cos \dfrac{\pi}{2} = 0$, $\tan \dfrac{\pi}{2}$ is undefined.

     b. $\cot \dfrac{7\pi}{4} = -\cot \dfrac{\pi}{4} = -\dfrac{\cos \frac{\pi}{4}}{\sin \frac{\pi}{4}} = -\dfrac{\frac{\sqrt{2}}{2}}{\frac{\sqrt{2}}{2}} = -1$

     c. $\sec (-3\pi) = \sec (4\pi - 3\pi) = \sec \pi = \dfrac{1}{\cos \pi} = \dfrac{1}{-1} = -1$

     d. $\csc \dfrac{7\pi}{6} = -\csc \dfrac{\pi}{6} = -\dfrac{1}{\sin \frac{\pi}{6}} = -\dfrac{1}{\frac{1}{2}} = -2$

Class Exercises 7-6 · page 289

1. $50.2°$ **2.** $-17.5°$ **3.** $115.2°$ **4.** -1.24 **5.** 0.85 **6.** 1.51

7. Error message; domain of $y = \text{Sin}^{-1} x$ is $\{x \mid -1 \le x \le 1\}$; 1.7 is not in the domain.

8. a. $\dfrac{3}{5}$ **b.** $\dfrac{4}{3}$ **c.** $\dfrac{3}{4}$

Written Exercises 7-6 · pages 289–291

1. a. $64.2°$ **b.** $-64.2°$ **c.** $41.4°$ **d.** $138.6°$

2. a. $66.8°$ **b.** $-66.8°$ **c.** $66.4°$ **d.** $113.6°$ **3. a.** 0.23 **b.** -0.23 **c.** 1.22 **d.** 1.92

4. a. 0.38 **b.** -0.38 **c.** 0.59 **d.** 2.56 **5. a.** 0 **b.** $\dfrac{\pi}{2}$ **c.** $\dfrac{\pi}{4}$ **d.** $-\dfrac{\pi}{4}$

6. a. $\dfrac{\pi}{2}$ **b.** $-\dfrac{\pi}{2}$ **c.** 0 **d.** π **7. a.** $\dfrac{\pi}{6}$ **b.** $-\dfrac{\pi}{6}$ **c.** $\dfrac{\pi}{3}$ **d.** $\dfrac{2\pi}{3}$

8. a. $\dfrac{\pi}{4}$ **b.** $-\dfrac{\pi}{4}$ **c.** $\dfrac{\pi}{4}$ **d.** $\dfrac{3\pi}{4}$

9. $\text{Cos}^{-1}\,3$ is meaningless since 3 is not in the domain of $y = \text{Cos}^{-1}\,x$; error message.

10. For any x between -1 and 1 inclusive, we can define the inverse cosine function, denoted Arccos (read "arc cosine"), as follows:

$$\text{Arccos } x = s \text{ provided } \cos s = x$$

where s is the length of the arc with endpoints $R(1,0)$ and $P(x,y)$ on the unit circle. (Note that P must be on the upper half of the circle.) Since the definition describes the range of the inverse cosine function as a set of arc lengths, the function's denotation (Arccos) is appropriate.

11. a. Draw a triangle in Quadrant I with $\cos \theta = \dfrac{12}{13}$; $\tan \theta = \dfrac{\text{opp.}}{\text{adj.}} = \dfrac{5}{12}$; 0.42

b. Draw a triangle in Quadrant II with $\cos \theta = \dfrac{-12}{13}$; $\tan \theta = \dfrac{\text{opp.}}{\text{adj.}} = \dfrac{5}{-12} = -\dfrac{5}{12}$; -0.42

12. a. Draw a triangle in Quadrant I with $\cos \theta = \dfrac{1}{5}$; $\sin \theta = \dfrac{\text{opp.}}{\text{hyp.}} = \dfrac{2\sqrt{6}}{5}$; 0.98

b. Draw a triangle in Quadrant II with $\cos \theta = -\dfrac{1}{5}$; $\sin \theta = \dfrac{\text{opp.}}{\text{hyp.}} = \dfrac{2\sqrt{6}}{5}$; 0.98

13. a. Draw a triangle in Quadrant I with $\tan \theta = 1.05 = \dfrac{21}{20}$; $\csc \theta = \dfrac{\text{hyp.}}{\text{opp.}} = \dfrac{29}{21}$; 1.38

b. Draw a triangle in Quadrant IV with $\sin \theta = -\dfrac{1}{2}$; $\sec \theta = \dfrac{\text{hyp.}}{\text{adj.}} = \dfrac{2}{\sqrt{3}} = \dfrac{2\sqrt{3}}{3}$; 1.15

14. Draw a triangle in Quadrant IV with $\tan \theta = -0.2 = \dfrac{-1}{5}$.

a. $\cos \theta = \dfrac{\text{adj.}}{\text{hyp.}} = \dfrac{5}{\sqrt{26}} = \dfrac{5\sqrt{26}}{26}$; 0.98 **b.** $\sin \theta = \dfrac{\text{opp.}}{\text{hyp.}} = \dfrac{-1}{\sqrt{26}} = -\dfrac{\sqrt{26}}{26}$; -0.20

15. a.

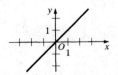

b.

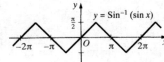

domain: $-1 \le x < 1$; range:
$-1 \le y \le 1$; $y = x$ for $-1 \le x \le 1$

c. No; The function $y = \sin x$ is not one-to-one and is not the inverse of $y = \text{Sin}^{-1}$; $y = \text{Sin}^{-1}$ has inverse $y = \text{Sin } x$, defined on page 287.

16. a.

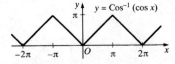

The high points have coordinates $(\pi + 2\pi n, \pi)$, where n is an integer. The low points have coordinates $(2\pi n, 0)$, where n is an integer.

b.

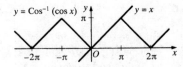

Equality occurs when $0 \le x \le \pi$.

c. The function $y = \cos x$ is not one-to-one and is not the inverse of $y = \text{Cos}^{-1}$. $\text{Cos}^{-1}(\cos x) = x$ is true for $0 \le x \le \pi$, the values of x in the domain of $y = \text{Cos } x$ (defined on page 288), which is the inverse of $y = \text{Cos}^{-1}$.

17. The graph of $y = \tan(\text{Tan}^{-1}x)$ is the line $y = x$, while the graph of $y = \text{Tan}^{-1}(\tan x)$ is a set of parallel line segments, each with slope 1 and with endpoints that lie on the lines $y = -\dfrac{\pi}{2}$ and $y = \dfrac{\pi}{2}$. Because $\tan x$ is undefined when $x = \dfrac{\pi}{2} + n\pi$, these endpoints, $\left(-\dfrac{\pi}{2} + n\pi, -\dfrac{\pi}{2}\right)$ and $\left(-\dfrac{\pi}{2} + n\pi, -\dfrac{\pi}{2}\right)$, are not on the graph of $y = \text{Tan}^{-1}(\tan x)$.

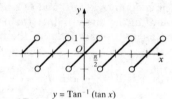
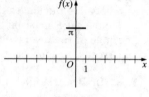

$y = \tan(\text{Tan}^{-1}x)$ $y = \text{Tan}^{-1}(\tan x)$

Ex. 17   **Ex. 18**

18. domain: $-1 < x < 1$; range $y = \pi$

19. a. True; the domain of $y = \text{Tan}^{-1}x$ is $-\dfrac{\pi}{2} < x < \dfrac{\pi}{2}$ and in this domain $\tan x$ and $\text{Tan}^{-1}x$ are inverses.

    b. Not true. Let $x = \pi$. $\text{Tan}^{-1}(\tan \pi) = \text{Tan}^{-1}0 = 0$.

20. a. True   **b.** Not true. Let $x = 1$; $\text{Cos}^{-1}(-1) = \pi$, but $-\text{Cos}^{-1}(1) = 0$.

21. a. True   **b.** Not true. Let $x = -1$; $\text{Tan}^{-1}\left(\dfrac{1}{-1}\right) = -\dfrac{\pi}{4}$, but $\dfrac{\pi}{2} - \text{Tan}^{-1}(-1) = \dfrac{3\pi}{4}$.

22. Domain is all reals, range is $\{y \mid 0 < y < \pi\}$

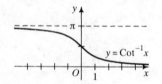
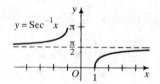
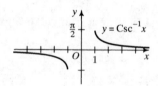

      Ex. 22   **Ex. 23**   **Ex. 24**

23. Domain is $\{x \mid x \le -1 \text{ or } x \ge 1\}$, range is $\left\{y \mid 0 \le y \le \pi, y \ne \dfrac{\pi}{2}\right\}$

24. Domain is $\{x \mid x \le -1 \text{ or } x \ge 1\}$, range is $\left\{y \mid -\dfrac{\pi}{2} \le y \le \dfrac{\pi}{2}, y \ne 0\right\}$

25. a. $\text{Sec}^{-1}2 = \dfrac{\pi}{3}$, $\text{Cos}^{-1}\dfrac{1}{2} = \dfrac{\pi}{3}$ **b.** $\text{Sec}^{-1}x = \text{Cos}^{-1}\dfrac{1}{x}$

26. $\text{Csc}^{-1}x = \text{Sin}^{-1}\dfrac{1}{x}$

27. Since we can replace x with $-x$ without altering the expression $\text{Cos}^{-1}x + \text{Cos}^{-1}(-x)$, we only need to consider $0 \le x \le 1$.

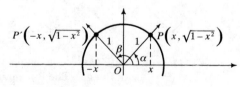

Case 1: If $x = 0$, then $\text{Cos}^{-1}0 + \text{Cos}^{-1}0 = \dfrac{\pi}{2} + \dfrac{\pi}{2} = \pi$.

Case 2: If $x = 1$, then $\text{Cos}^{-1}1 + \text{Cos}^{-1}(-1) = 0 + \pi = \pi$.

Case 3: If $0 < x < 1$, then let $\alpha = \text{Cos}^{-1}x$ and $\beta = \text{Cos}^{-1}(-x)$, as shown in the diagram. Since rays OP and OP' are reflections in the vertical axis, α is the reference angle for β; $\beta = \pi - \alpha$, or $\alpha + \beta = \pi$; $\text{Cos}^{-1}x + \text{Cos}^{-1}(-x) = \alpha + \beta = \pi$.

28. Since we can replace x with $-x$ without altering the expression $\text{Sin}^{-1} x + \text{Sin}^{-1} (-x)$, we only need to consider $0 \le x \le 1$.

Case 1: If $x = 0$, then $\text{Sin}^{-1} 0 + \text{Sin}^{-1} 0 = 0 + 0 = 0$.

Case 2: If $x = 1$, then $\text{Sin}^{-1} 1 + \text{Sin}^{-1} (-1) = \dfrac{\pi}{2} +$

$\left(-\dfrac{\pi}{2}\right) = 0$.

Case 3: If $0 < x < 1$, then let $\alpha = \text{Sin}^{-1} x$ and $\beta = \text{Sin}^{-1} (-x)$, as shown in the diagram. Since rays OP and OP' are reflections in the horizontal axis, α is the reference angle for β. That is, $\beta = -\alpha$, or $\alpha + \beta = 0$. Thus, $\text{Sin}^{-1} x + \text{Sin}^{-1} (-x) = \alpha + \beta = 0$.

29. $\text{Tan}^{-1} x = x - \dfrac{x^3}{3} + \dfrac{x^5}{5} - \dfrac{x^7}{7} + \cdots, \ |x| \le 1$; let $x = 1$;

$\text{Tan}^{-1} 1 = 1 - \dfrac{1}{3} + \dfrac{1}{5} - \dfrac{1}{7} + \cdots,$ but $\text{Tan}^{-1} 1 = \dfrac{\pi}{4}$;

$\dfrac{\pi}{4} = 1 - \dfrac{1}{3} + \dfrac{1}{5} - \dfrac{1}{7} + \cdots,$ or $\pi = 4\left(1 - \dfrac{1}{3} + \dfrac{1}{5} - \dfrac{1}{7} + \cdots\right)$

30. Let $\text{Cos}^{-1} x = A$ and $\text{Cos}^{-1} y = B$; then $\cos A = x$, $\cos B = y$ and $A + B = \dfrac{\pi}{2}$, or

$B = \dfrac{\pi}{2} - A$; from Ex. 18, p. 279, $y = \cos B = \cos\left(\dfrac{\pi}{2} - A\right) = \sin A$; from Ex. 7, p. 271, $(\cos A)^2 + (\sin A)^2 = 1$, or $x^2 + y^2 = 1$

Chapter Test · page 293

1. a. $270° \cdot \dfrac{\pi}{180°} = \dfrac{3}{2}\pi$ **b.** $192° \cdot \dfrac{\pi}{180°} = 3.35$

2. a. $\dfrac{5\pi}{3} \cdot \dfrac{180°}{\pi} = 300°$ **b.** $(2.5)\left(\dfrac{180°}{\pi}\right) = 143°10'$ or $143.2°$

3. Answers may vary. **a.** $-200° + 360° = 160°$; $-200° - 360° = -560°$

b. $313.2° + 360° = 673.2°$; $313.2° - 360° = -46.8°$

c. $\dfrac{5\pi}{6} + 2\pi = \dfrac{17\pi}{6}; \dfrac{5\pi}{6} - 2\pi = -\dfrac{7\pi}{6}$

d. $142°10' + 360° = 502°10'$; $142°10' - 360 = -217°50'$

4. $48° = 48° \cdot \dfrac{\pi}{180°} = \dfrac{4}{15}\pi$; $s = r\theta = 5 \cdot \dfrac{4}{15}\pi = \dfrac{4}{3}\pi \approx 4.2$ cm; $A = \dfrac{1}{2}(5)^2\left(\dfrac{4}{15}\pi\right) \approx$ 10.5 cm^2

5. $s = r\theta = (4 \times 10^5)(0.0087) = 3480$; diameter ≈ 3480 km

6. a. $r = \sqrt{2}$; $\sin\theta = \dfrac{1}{\sqrt{2}} = \dfrac{\sqrt{2}}{2}$; $\cos\theta = -\dfrac{1}{\sqrt{2}} = -\dfrac{\sqrt{2}}{2}$

b. $r = 2$; $\sin\theta = -\dfrac{\sqrt{3}}{2}$; $\cos\theta = -\dfrac{1}{2}$

c. $r = 1$; $\sin\theta = 1$; $\cos\theta = 0$

7. a. $\sin 60° < \sin 65°$ **b.** $\cos 60° > \cos 65°$ **c.** $\sin 20° = \sin 160°$

d. $\cos 184° < \cos 185°$

8. a. $-\dfrac{\sqrt{2}}{2}$ **b.** 0 **c.** $\dfrac{1}{2}$ **d.** $\dfrac{\sqrt{3}}{2}$ **9. a.** -1 **b.** $-\dfrac{\sqrt{3}}{3}$ **c.** 2 **d.** undefined

10. $\sin x = \dfrac{\sqrt{10}}{10}$; $\cos x = -\dfrac{3\sqrt{10}}{10}$; $\cot x = -3$; $\sec x = -\dfrac{\sqrt{10}}{3}$;

$\csc x = \sqrt{10}$

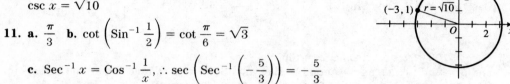

11. a. $\dfrac{\pi}{3}$ **b.** $\cot\left(\operatorname{Sin}^{-1}\dfrac{1}{2}\right) = \cot\dfrac{\pi}{6} = \sqrt{3}$

c. $\operatorname{Sec}^{-1} x = \operatorname{Cos}^{-1}\dfrac{1}{x}$, $\therefore \sec\left(\operatorname{Sec}^{-1}\left(-\dfrac{5}{3}\right)\right) = -\dfrac{5}{3}$

12. This restriction is necessary because $f(x) = \sin x$ is not a one-to-one function. In order to consider an inverse function, it is necessary to restrict the domain in such a way that the function is one-to-one and the range is preserved. The interval $-\dfrac{\pi}{2} \le x \le \dfrac{\pi}{2}$ does this as does $\dfrac{\pi}{2} \le x \le \dfrac{3\pi}{2}$; however, it is logical to include $x = 0$ in our interval and hence the selection $-\dfrac{\pi}{2} \le x \le \dfrac{\pi}{2}$. The interval $0 \le x \le \pi$ could not have been used because f is not one-to-one on this interval.

Class Exercises 8-1 · page 299

1. $\cos \theta = \dfrac{1}{2}$; ref. $\angle = 60°$; $\cos \theta > 0$ for θ in quadrants I and IV; $60°$, $360° - 60° = 300°$

2. $\sin \theta = -4$; $-1 \le \sin \theta \le 1$, no solution

3. $\csc \theta = 2$; ref. $\angle = 30°$; $\csc \theta > 0$ for θ in quadrants I and II; $30°$, $180° - 30° = 150°$

4. $\cot \theta = -1$; ref. $\angle = 45°$; $\cot \theta < 0$ for θ in quadrants II and IV; $180° - 45° = 135°$, $360° - 45° = 315°$

5. $\cos x = -\dfrac{\sqrt{3}}{2}$; ref. $\angle = \dfrac{\pi}{6}$; $\cos x < 0$ for x in quadrants II and III; $\pi - \dfrac{\pi}{6} = \dfrac{5\pi}{6}$, $\pi + \dfrac{\pi}{6} = \dfrac{7\pi}{6}$

6. $\cot x = 1$; ref. $\angle = \dfrac{\pi}{4}$; $\cot x > 0$ for x in quadrants I and III; $\dfrac{\pi}{4}$, $\pi + \dfrac{\pi}{4} = \dfrac{5\pi}{4}$

7. $\tan x = -\sqrt{3}$; ref. $\angle = \dfrac{\pi}{3}$; $\tan x < 0$ for x in quadrants II and IV; $\pi - \dfrac{\pi}{3} = \dfrac{2\pi}{3}$, $2\pi - \dfrac{\pi}{3} = \dfrac{5\pi}{3}$

8. $\sec x = \dfrac{1}{2}$; $\sec x \le -1$ or $\sec x \ge 1$; no solution

9. $\cos \theta = -1$; $\theta = (180 + 360n)°$, n an integer

10. $\sin \theta = -\dfrac{\sqrt{2}}{2}$; ref. $\angle = 45°$; $\sin \theta < 0$ in quadrants III and IV; $(180 + 45 + 360n)° = (225 + 360n)°$, $(360 - 45 + 360n)° = (315 + 360n)°$, n an integer

11. $\tan \theta = 1$; ref. $\angle = 45°$; $\tan \theta > 0$ in quadrants I and III; $(45 + 180n)°$, n an integer

12. $m = \tan \alpha$; $m = \tan 140° = -0.8391$

13. $m = \dfrac{3}{5}$; $\tan \alpha = \dfrac{3}{5}$, $\alpha = 31°$

14. $\tan 2\alpha = \dfrac{B}{A - C} = \dfrac{2}{5 - (-1)} = \dfrac{1}{3}$; $2\alpha = 18.4°$; $\alpha = 9°$

15. You can graph only a *function* on most graphing calculators. Since a conic is a relation, solving a conic's equation for y in terms of x usually results in two equations (involving a positive and a negative root).

Written Exercises 8-1 · pages 299–301

1. $\sin \theta = -0.7$; ref. $\angle = 44.4°$; $\theta = 224.4°$, $315.6°$

2. $\cos \theta = 0.42$; ref. $\angle = 65.2°$; $\theta = 65.2°$, $294.8°$

3. $\tan \theta = 1.2$; ref. $\angle = 50.2°$; $\theta = 50.2°$, $230.2°$

4. $\cot \theta = -0.3$; ref. $\angle = 73.3°$; $\theta = 106.7°$, $286.7°$

5. $\sec \theta = -5$; ref. $\angle = 78.5°$; $\theta = 101.5°$, $258.5°$

6. $\csc \theta = 14$; ref. $\angle = 4.1°$; $\theta = 4.1°$, $175.9°$

7. $3 \cos \theta = 1$; $\cos \theta = \dfrac{1}{3}$; ref. $\angle = 70.5°$; $\theta = 70.5°$, $289.5°$

8. $4 \sin \theta = 3$; $\sin \theta = \dfrac{3}{4}$; ref. $\angle = 48.6°$; $\theta = 48.6°$, $131.4°$

9. $5 \sec \theta + 6 = 0$; $\sec \theta = -\dfrac{6}{5}$; ref. $\angle = 33.6°$; $\theta = 146.4°$, $213.6°$

10. $2 \tan \theta + 1 = 0$; $\tan \theta = -\dfrac{1}{2}$; ref. $\angle = 26.6°$; $\theta = 153.4°$, $333.4°$

11. $6 \csc \theta - 9 = 0$; $\csc \theta = \dfrac{3}{2}$; ref. $\angle = 41.8°$; $\theta = 41.8°$, $138.2°$

12. $4 \cot \theta - 5 = 0$; $\cot \theta = \dfrac{5}{4}$; ref. $\angle = 38.7°$; $\theta = 38.7°$, $218.7°$

13. $\tan x = -1.5$; ref. $\angle = 0.98$; $x = 2.16$, 5.30

14. $\sec x = 2.5$; ref. $\angle = 1.16$; $x = 1.16$, 5.12

15. $\csc x = -1.4$; ref. $\angle = 0.80$; $x = 3.94$, 5.49

16. $\cos x = -0.8$; ref. $\angle = 0.64$; $x = 2.50$, 3.79

17. $\cot x = 6$; ref. $\angle = 0.17$; $x = 0.17$, 3.31

18. $3 \sin x + 2 = 4$; $\sin x = \dfrac{2}{3}$; ref. $\angle = 0.73$; $x = 0.73$, 2.41

19. $8 = 9 \cos x + 2$; $\cos x = \dfrac{2}{3}$; ref. $\angle = 0.83$; $x = 0.83$, 5.44

20. $\dfrac{5 \csc x}{3} = \dfrac{9}{4}$; $\csc x = \dfrac{27}{20}$; ref. $\angle = 0.83$; $x = 0.83$, 2.31

21. $\dfrac{3 \cot x}{4} + 1 = 0$; $\cot x = -\dfrac{4}{3}$; ref. $\angle = 0.64$; $x = 2.50$, 5.64

22. $m = \tan 45°$; $m = 1$, $b = 4$; $y = x + 4$

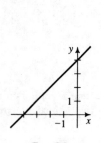
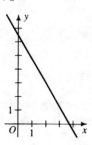
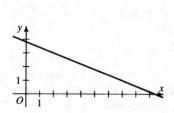

| Ex. 22 | Ex. 23 | Ex. 24 |

23. $m = \tan 120°$; $m = -\sqrt{3}$; $y - 3 = -\sqrt{3}(x - 2)$; $y = -\sqrt{3}x + 3 + 2\sqrt{3}$

24. $m = \tan 158°$; $m = -0.40$; $y - 5 = -0.40(x + 3)$; $y = -0.40x + 3.80$

25. $3x + 5y = 8$; $m = -\dfrac{3}{5}$; $\tan \alpha = -\dfrac{3}{5}$; $\alpha = 149°$

26. $x - 4y = 7$; $m = \dfrac{1}{4}$; $\tan \alpha = \dfrac{1}{4}$; $\alpha = 14°$

27. $m = \dfrac{1 - 2}{4 - (-1)} = -\dfrac{1}{5}$; $\tan \alpha = -\dfrac{1}{5}$; $\alpha = 169°$

28. $m = \dfrac{2 - 1}{4 - (-1)} = \dfrac{1}{5}$; $\tan \alpha = \dfrac{1}{5}$; $\alpha = 11°$

29. $4x + 3y = 12$, line has slope $-\dfrac{4}{3}$; perpendicular has slope $\dfrac{3}{4}$; $\tan \alpha = \dfrac{3}{4}$; $\alpha = 37°$

30. $2x + 3y = -6$; $\parallel$ lines have equal slopes; $m = -\dfrac{2}{3}$; $\tan \alpha = -\dfrac{2}{3}$; $\alpha = 146°$

31. $|\csc x| = 1$; $\csc x = 1$ or $\csc x = -1$; $x = \dfrac{\pi}{2}, \dfrac{3\pi}{2}$

32. $|\sec x| = \sqrt{2}$; $\sec x = \pm\sqrt{2}$; ref. $\angle = \dfrac{\pi}{4}$; $x = \dfrac{\pi}{4}, \dfrac{3\pi}{4}, \dfrac{5\pi}{4}, \dfrac{7\pi}{4}$

33. $\log_2(\sin x) = 0$; $\sin x = 2^0 = 1$; $x = \dfrac{\pi}{2}$

34. $\log_2(\cos x) = -1$; $\cos x = 2^{-1} = \dfrac{1}{2}$; ref. $\angle = \dfrac{\pi}{3}$; $x = \dfrac{\pi}{3}, \dfrac{5\pi}{3}$

35. $\log_3(\tan x) = \dfrac{1}{2}$; $\tan x = 3^{1/2} = \sqrt{3}$; ref. $\angle = \dfrac{\pi}{3}$; $x = \dfrac{\pi}{3}, \dfrac{4\pi}{3}$

36. $\log_{\sqrt{3}}(\cot x) = 1$; $\cot x = (\sqrt{3})^1 = \sqrt{3}$; ref. $\angle = \dfrac{\pi}{6}$; $x = \dfrac{\pi}{6}, \dfrac{7\pi}{6}$

37. 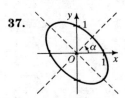 $x^2 + xy + y^2 = 1$; $A = C$, so $\alpha = \dfrac{\pi}{4}$; $B^2 - 4AC = 1 - 4 < 0$; the graph is an ellipse. For graphing, use the quadratic formula: $y^2 + (x)y + (x^2 - 1) = 0$, $a = 1$, $b = x$, $c = x^2 - 1$; $y = \dfrac{-x \pm \sqrt{x^2 - 4(1)(x^2 - 1)}}{2(1)}$; $y = \dfrac{-x \pm \sqrt{4 - 3x^2}}{2}$

38. 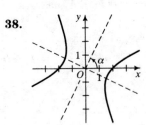 $x^2 - 2xy - y^2 = 4$; $B^2 - 4AC = 4 + 4 > 0$; the graph is a hyperbola; $\tan 2\alpha = \dfrac{-2}{2} = -1$; $2\alpha = \dfrac{3\pi}{4}$; $\alpha = \dfrac{3\pi}{8}$; from the quadratic formula, $y = -x \pm \sqrt{2x^2 - 4}$.

39. 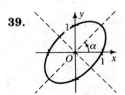 $x^2 - xy + y^2 = 1$; $B^2 - 4AC = 1 - 4 < 0$; the graph is an ellipse; $A = C$, so $\alpha = \dfrac{\pi}{4}$; from the quadratic formula, $y = \dfrac{x \pm \sqrt{4 - 3x^2}}{2}$. (See Ex. 37.)

40. 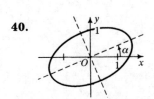 $x^2 - xy + 2y^2 = 2$; $B^2 - 4AC = 1 - 8 < 0$; the graph is an ellipse; $\tan 2\alpha = \dfrac{-1}{-1} = 1$; $2\alpha = \dfrac{\pi}{4}$; $\alpha = \dfrac{\pi}{8}$; from the quadratic formula, $y = \dfrac{x \pm \sqrt{16 - 7x^2}}{4}$.

41.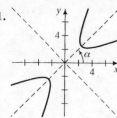

$x^2 - 3xy + y^2 = -6$; $B^2 - 4AC = 9 - 4 > 0$; the graph is a hyperbola; $A = C$, so $\alpha = \dfrac{\pi}{4}$; from the quadratic formula,

$$y = \frac{3x \pm \sqrt{5x^2 - 24}}{2}.$$

42.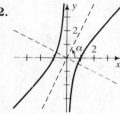

$y = x - \dfrac{1}{x}$; $xy = x^2 - 1$; $x^2 - xy + 1 = 0$;

$B^2 - 4AC = 1 > 0$; the graph is a hyperbola; $\tan 2\alpha = \dfrac{-1}{1}$;

$2\alpha = \dfrac{3\pi}{4}$; $\alpha = \dfrac{3\pi}{8}$.

43.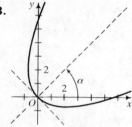

$x^2 - 2xy + y^2 - 4\sqrt{2}x - 4\sqrt{2}y = 0$; $B^2 - 4AC = 4 - 4 = 0$; the graph is a parabola; $A = C$, so $\alpha = \dfrac{\pi}{4}$; from the quadratic formula, $y = (x + 2\sqrt{2}) \pm 2\sqrt{2\sqrt{2}x + 2}$.

44.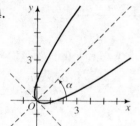

$x^2 - 2xy + y^2 - \sqrt{2}x - \sqrt{2}y = 0$; $B^2 - 4AC = 4 - 4 = 0$; the graph is a parabola; $A = C$, so $\alpha = \dfrac{\pi}{4}$; from the quadratic formula, $y = \dfrac{1}{2}((2x + \sqrt{2}) \pm \sqrt{8\sqrt{2}x + 2})$.

45. a. $\dfrac{\text{speed of light in air}}{\text{speed of light in water}} = \dfrac{3.00 \times 10^8}{2.25 \times 10^8} = \dfrac{4}{3} = \dfrac{\sin \alpha}{\sin \beta}$. Therefore

$\sin \beta = \dfrac{3}{4} \sin \beta = 0.75 \sin \alpha$. $\mathrm{Sin}^{-1}(\sin \beta) = \beta = \mathrm{Sin}^{-1}(0.75 \sin \alpha)$. The domain is

$0° \le \alpha \le 90°$ and the range is about $0° \le \beta \le 48.6°$.

b.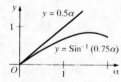

No

46. great circle arc $AN = 90° - 41.53° = 48.47°$; great circle arc $BN = 90° - 42.20° = 47.80°$; $\theta = 71.05°\text{W} - 12.30°\text{E} = 71.05° + 12.30° = 83.35°$; $\cos n = \cos 48.47° \cdot \cos 47.80° + \sin 48.47° \cdot \sin 47.80° \cdot \cos 83.35°$; $\cos n \approx 0.5096$; $n \approx 59.4°$

Class Exercises 8-2 · page 304

1. per. $= \dfrac{2\pi}{2} = \pi$; amplitude $= 4$ 2. per. $= \dfrac{2\pi}{\frac{1}{2}} = 4\pi$; amp. $= 3$

3. per. $= \dfrac{2\pi}{\frac{2\pi}{7}} = 7$; amp. $= 5$ 4. per. $= \dfrac{2\pi}{\frac{2\pi}{3}} = 3$; amp. $= 6$

5. per. $= 4$; amp. $= 3$; $B = \dfrac{2\pi}{4} = \dfrac{\pi}{2}$; $y = 3 \sin \dfrac{\pi}{2}x$

6. per. $= 4\pi$; amp. $= 2$; $B = \dfrac{2\pi}{4\pi} = \dfrac{1}{2}$; $y = -2 \cos \dfrac{1}{2}x$

7. a. 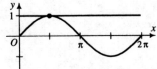   b.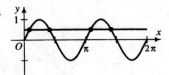

       $\sin x = 1$; one solution $\sin 2x = \dfrac{1}{2}$; 4 solutions

c.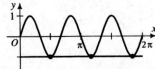

       $\sin 3x = -1$; 3 solutions

Written Exercises 8-2 · pages 305–308

1.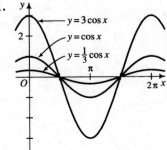

2.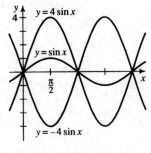

3.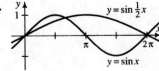

4.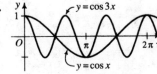

5. amp. = 2; per. = $\dfrac{2\pi}{3}$ **6**. amp. = 4; per. = π

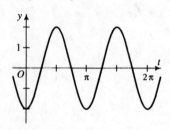
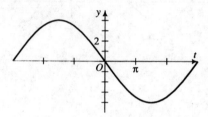

7. amp. = 2; per. = π **8.** amp. = 4; per. = 6π

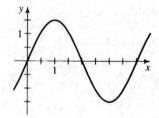

9. amp. = $\dfrac{1}{2}$; per. = 1   **10.** amp. = 1.5; per. = 4

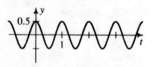
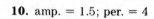

11. amp. = 4; per. = π; $B = \dfrac{2\pi}{\pi} = 2$; $y = 4 \sin 2x$

12. amp. = 2; per. = 24π; $B = \dfrac{2\pi}{24\pi} = \dfrac{1}{12}$; $y = 2 \cos \dfrac{x}{12}$

13. amp. = 3; per. = 4π; $B = \dfrac{2\pi}{4\pi} = \dfrac{1}{2}$; $y = 3 \cos \dfrac{x}{2}$

14. amp. = 2; per. = 8; $B = \dfrac{2\pi}{8} = \dfrac{\pi}{4}$; $y = -2 \sin \dfrac{\pi}{4} x$

15. amp. = 3; per. = 4; $B = \dfrac{2\pi}{4} = \dfrac{\pi}{2}$; $y = 3 \sin \dfrac{\pi}{2} x$

16. amp. = 5; per. = 2; $B = \dfrac{2\pi}{2} = \pi$; $y = -5 \cos \pi x$

17. amp. = $\dfrac{4 - (-4)}{2} = 4$; $B = \dfrac{2\pi}{12} = \dfrac{\pi}{6}$; $y = \pm 4 \sin \dfrac{\pi}{6} x$

18. amp. $= \dfrac{9 - (-9)}{2} = 9$; $B = \dfrac{2\pi}{5}$; $y = \pm 9 \cos \dfrac{2\pi}{5} x$

19. a. $\cos x = -1$; $x = \pi$ or 3.14

    b. $\cos 2x = -1$; $2x = \pi + 2\pi k$; $x = \dfrac{\pi}{2}, \dfrac{3\pi}{2}$ or 1.57, 4.71

    c. $\cos 3x = -1$; $3x = \pi + 2\pi k$; $x = \dfrac{\pi}{3}, \pi, \dfrac{5\pi}{3}$ or 1.05, 3.14, 5.24

20. a. $2 \sin x = 1$; $\sin x = \dfrac{1}{2}$; $x = \dfrac{\pi}{6}, \dfrac{5\pi}{6}$ or 0.52, 2.62

    b. $2 \sin 2x = 1$; $\sin 2x = \dfrac{1}{2}$; $2x = \dfrac{\pi}{6} + 2\pi k, \dfrac{5\pi}{6} + 2\pi k$; $x = \dfrac{\pi}{12}, \dfrac{5\pi}{12}, \dfrac{13\pi}{12}, \dfrac{17\pi}{12}$ or 0.26,

    1.31, 3.40, 4.45

    c. $2 \sin \dfrac{1}{2} x = 1$; $\sin \dfrac{1}{2} x = \dfrac{1}{2}$; $\dfrac{1}{2} x = \dfrac{\pi}{6}, \dfrac{5\pi}{6}$; $x = \dfrac{\pi}{3}, \dfrac{5\pi}{3}$ or 1.05, 5.24

21. $8 \cos 2x = 1$; $\cos 2x = \dfrac{1}{8}$; $2x = \pm 1.45 + 2\pi k$; $x = \pm \dfrac{1.45}{2} + \pi k$;

    $x = 0.72, 2.42, 3.86, 5.56$

22. $5 \sin 3x = -2$; $\sin 3x = -\dfrac{2}{5}$; $3x = 3.55 + 2\pi k, 5.87 + 2\pi k$;

    $x = 1.18 + 2.09k, 1.96 + 2.09k$; $x = 1.18, 1.96, 3.28, 4.05, 5.37, 6.15$

23. $3 \sin \dfrac{x}{2} = -1$; $\sin \dfrac{x}{2} = -\dfrac{1}{3}$; $\dfrac{x}{2} = 3.48, 5.94$; $x > 2\pi$; no solution

24. $1.5 \cos \dfrac{x}{2} = \dfrac{1}{2}$; $\cos \dfrac{x}{2} = \dfrac{1}{3}$; $\dfrac{x}{2} = 1.23, 5.05$; $x = 2.46$

25. $4 \sin \dfrac{\pi}{2} x = 1$; $\sin \dfrac{\pi}{2} x = \dfrac{1}{4}$; $\dfrac{\pi}{2} x = 0.25 + 2\pi k, 2.89 + 2\pi k$; $x = 0.16 + 4k, 1.84 + 4k$;

    $x = 0.16, 1.84, 4.16, 5.84$

26. $-3 \cos \dfrac{\pi}{4} x = 1$; $\cos \dfrac{\pi}{4} x = -\dfrac{1}{3}$; $\dfrac{\pi}{4} x = 1.91, 4.37$; $x = 2.43, 5.57$

27.

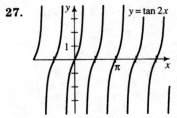
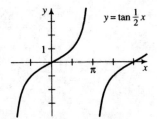

28.

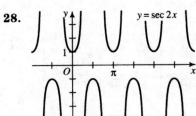
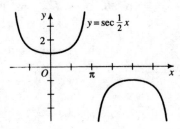

29. Maximum and minimum values of the tangent and secant functions are not defined, so amplitude is not defined.

30. a. frequency $= \dfrac{1}{\text{period}}$; per. $= \dfrac{1}{800(1000)} = 1.25 \times 10^{-6}$ s

    b. per. $= \dfrac{2\pi}{1.85\pi \times 10^6} = \dfrac{2}{1.85 \times 10^6}$; freq. $= \dfrac{1850 \times 10^3}{2} = 925$ kHz

31. a. $f(t) = 200{,}000{,}000 + 10{,}000 \sin 500\pi t$ is the varying frequency; $200{,}000{,}000 - 10{,}000 \le f(t) \le 200{,}000{,}000 + 10{,}000$; $f(t)$ remains near 200×10^6, so assigned frequency is 200 MHz.

    b. $\sin 500\pi t$ varies between 1 and -1; $f(t)$ varies between $200{,}000{,}000 + 10{,}000$ and $200{,}000{,}000 - 10{,}000$; 200.01 MHz and 199.99 MHz

32. a. 88 MHz $-$ 108 MHz

    b. VHF: 54 MHz $-$ 72 MHz, 76 MHz $-$ 88 MHz, and 174 MHz $-$ 216 MHz
UHF: 470 MHz $-$ 806 MHz

    c. Answers may vary. UV light: 8×10^{14} Hz to 3×10^{16} Hz; IR light: 3×10^{11} Hz to 4×10^{14} Hz

    d. Concert (440 Hz $=$ A) scale: 261.625 Hz; scientific (just) scale: 256 Hz

33. $d = -A \cos\left[\left(\sqrt{\dfrac{k}{m}}\right)t\right]$

    a. per. $= \dfrac{2\pi}{\sqrt{\dfrac{k}{m}}} = 2\pi\sqrt{\dfrac{m}{k}}$

    b. per. $= 1.1$, $m = 100$; $1.1 = 2\pi\sqrt{\dfrac{100}{k}}$; $\dfrac{k}{100} = \left(\dfrac{2\pi}{1.1}\right)^2$; $k \approx 3263$

    c. per. $= 1.4$, $k = 3263$; $1.4 = 2\pi\sqrt{\dfrac{m}{3263}}$; $m \approx 162$ g

34. a. $D \approx A \sin\left[\left(\sqrt{\dfrac{980}{100}}\right)t\right] = A \sin(\sqrt{9.8}\,t)$;

    period $= \dfrac{2\pi}{\sqrt{9.8}} = 2.007$; sine is a maximum $\dfrac{1}{4}$ through a cycle; earliest time for max. D is $t = 0.5018$ s.

    b. period $= \dfrac{2\pi}{\sqrt{\dfrac{980}{l}}}$; $1 = 2\pi\sqrt{\dfrac{l}{980}}$; $l = \left(\dfrac{1}{2\pi}\right)^2 \cdot 980$; $l = 24.8$ cm

35.

36. $D = B(2^{-t}) \sin 880\pi t$

    a. 2^{-t}; 2^{-t} is very small for $t > 5$.

    b. The sine is first at a maximum at $\dfrac{1}{4}$ of the period;

        per. $= \dfrac{1}{\text{freq.}} = \dfrac{1}{440}$; $t = \dfrac{1}{4} \cdot \dfrac{1}{440} = 5.68 \times 10^{-4}$ s

    c. per. $= \dfrac{1}{220} = \dfrac{2\pi}{b}$; $b = 440\pi$; $D = B(2^{-t}) \sin 440\pi t$

Class Exercises 8-3 · page 312

1. amp. $= \dfrac{7-3}{2} = 2$ 2. per. $= 4$ 3. Axis is $y = 5$.

4. a. horizontal $= 0$; vertical $= 5$ b. $y = 5 + 2 \sin \dfrac{\pi}{2} x$; note $B = \dfrac{2\pi}{\text{per.}} = \dfrac{2\pi}{4} = \dfrac{\pi}{2}$

5. a. horizontal $= 1$; vertical $= 5$ b. $y = 5 + 2 \cos \dfrac{\pi}{2}(x - 1)$; note $B = \dfrac{2\pi}{\text{per.}} = \dfrac{2\pi}{4} = \dfrac{\pi}{2}$

6. $y - 6 = 4 \sin 3(x + 5)$

   3: Period is $\dfrac{2\pi}{3}$.

   4: Amplitude is 4.
   5: Horizontal shift is 5 units left.
   6: Vertical shift is 6 units up.

Written Exercises 8-3 · pages 313–316

1. max. at $x = 0$, $h = 0$; $A = \dfrac{4 - 0}{2} = 2$; per. $= 4$, $B = \dfrac{2\pi}{\text{per.}} = \dfrac{2\pi}{4} = \dfrac{\pi}{2}$; $k = \dfrac{4 + 0}{2} = 2$;

   $y = 2 + 2 \cos \dfrac{\pi}{2} x$ or $y = 2 + 2 \sin \dfrac{\pi}{2}(x - 3)$

2. max. at $x = 2$, $h = 2$; $A = \dfrac{4 - 2}{2} = 1$; per. $= 8$, $B = \dfrac{2\pi}{8} = \dfrac{\pi}{4}$; $k = \dfrac{4 + 2}{2} = 3$;

   $y = 3 + \cos \dfrac{\pi}{4}(x - 2)$ or $y = 3 + \sin \dfrac{\pi}{4} x$

3. max. at $x = 1$, $h = 1$; $A = \dfrac{3 - 1}{2} = 1$; per. $= 6$, $B = \dfrac{2\pi}{6} = \dfrac{\pi}{3}$; $k = \dfrac{3 + 1}{2} = 2$;

   $y = 2 + \cos \dfrac{\pi}{3}(x - 1)$ or $y = 2 + \sin \dfrac{\pi}{3}\left(x + \dfrac{1}{2}\right)$

4. max. at $x = 2$, $h = 2$; $A = \dfrac{2 - (-1)}{2} = \dfrac{3}{2}$; per. $= 10$, $B = \dfrac{2\pi}{10} = \dfrac{\pi}{5}$; $k = \dfrac{2 + (-1)}{2} = \dfrac{1}{2}$;

   $y = \dfrac{1}{2} + \dfrac{3}{2} \cos \dfrac{\pi}{5}(x - 2)$ or $y = \dfrac{1}{2} + \dfrac{3}{2} \sin \dfrac{\pi}{5}\left(x + \dfrac{1}{2}\right)$

5. $A = \dfrac{3 - (-3)}{2} = 3$; per. $= \pi$, $B = \dfrac{2\pi}{\pi} = 2$; $k = \dfrac{3 + (-3)}{2} = 0$; the curve intersects its axis

   at $x = \dfrac{\pi}{6}$; $y = 3 \sin 2\left(x - \dfrac{\pi}{6}\right)$ or $y = 3 \cos 2\left(x - \dfrac{5\pi}{12}\right)$

6. max.; at $x = -\dfrac{\pi}{4}$, $h = -\dfrac{\pi}{4}$; $A = \dfrac{2 - (-2)}{2} = 2$; per. $= 2\pi$, $B = 1$; $k = \dfrac{2 + (-2)}{2} = 0$;

   $y = 2 \cos\left(x + \dfrac{\pi}{4}\right)$ or $y = 2 \sin\left(x + \dfrac{3\pi}{4}\right)$

7. max. at $x = \dfrac{\pi}{8}$, $h = \dfrac{\pi}{8}$; $A = \dfrac{5 - 1}{2} = 2$; per. $= \dfrac{\pi}{4}$, $B = \dfrac{2\pi}{\frac{\pi}{4}} = 8$; $k = \dfrac{5 + 1}{2} = 3$;

   $y = 3 + 2 \cos 8\left(x - \dfrac{\pi}{8}\right)$ or $y = 3 + 2 \sin 8\left(x - \dfrac{\pi}{16}\right)$

8. max. at $x = \dfrac{\pi}{2}$, $h = \dfrac{\pi}{2}$; $A = \dfrac{4 - (-1)}{2} = \dfrac{5}{2}$; per. $= \pi$, $B = \dfrac{2\pi}{\pi} = 2$; $k = \dfrac{4 + (-1)}{2} = \dfrac{3}{2}$;

$y = \dfrac{3}{2} + \dfrac{5}{2} \cos 2\left(x - \dfrac{\pi}{2}\right)$ or $y = \dfrac{3}{2} + \dfrac{5}{2} \sin 2\left(x - \dfrac{\pi}{4}\right)$

9.

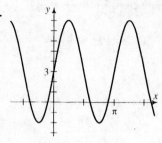

10.

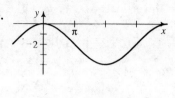

11.

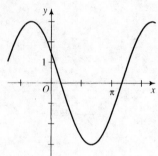

12.

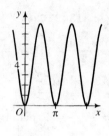

13.

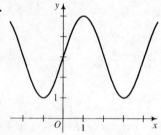

14.

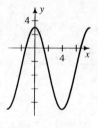

15.

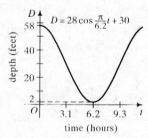

$k = \dfrac{\text{max.} + \text{min.}}{2} = \dfrac{58 + 2}{2} = 30$;

per. $= 12$ h 24 min $= 12.4$ h, $B = \dfrac{2\pi}{12.4} = \dfrac{\pi}{6.2}$;

amp. $= \dfrac{58 - 2}{2} = 28$; $D = 28 \cos \dfrac{\pi}{6.2}t + 30$

16.

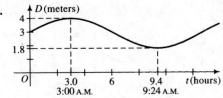

a. $A = \dfrac{4 - 1.8}{2} = 1.1$; vert. shift $= \dfrac{4 + 1.8}{2} =$

2.9; horiz. shift $= 3$; $\dfrac{p}{2} = 9.4 - 3.0$;

$p = 12.8$; $B = \dfrac{2\pi}{12.8} \approx 0.491$;

$D = 2.9 + 1.1 \cos (0.491(t - 3))$

b. Depth at noon, let $t = 12$; $2.9 + 1.1 \cos (0.491(9)) = 2.58$ m

c. $3 = 2.9 + 1.1 \cos (0.491(t - 3))$; $\cos (0.491(t - 3)) = 0.0909$;
$t = 3 \pm 3.014 + 12.8k$; $t = -0.014 + 12.8$ or $t = 6.014 + 12.8$;
$12.786 < t < 18.814$; between 12:47 P.M. and 6:49 P.M.

17. a. per. $= 365$; amp. $= \dfrac{15 - 9}{2} = 3$; $y = k + A \sin (B(x - h))$; $k = \dfrac{15 + 9}{2} = 12$;

$B = \dfrac{2\pi}{365}$; $A = 3$; $h = 80$; $y = 12 + 3 \sin \dfrac{2\pi}{365}(x - 80)$

b. Let $x = 1$; $y = 12 + 3 \sin \dfrac{2\pi}{365}(-79) \approx 9.1$ h.

Let $x = 185$; $y \approx 14.9$ h.

c. $12 + 3 \sin\left(\dfrac{2\pi}{365}(x - 80)\right) \geq 14$; $123 \leq x \leq 220$, the 123rd day (May 3) through the 220th day (Aug. 8)

d. Greater; the farther from the equator a region is, the greater the variation in the length of the day over a year.

18. Since the seasons in the southern hemisphere are opposite to the seasons in the northern hemisphere, the graph at $x = 80$ would move down instead of up. The only change is that $A = -3$;

$y = 12 - 3 \sin \dfrac{2\pi}{365}(x - 80)$.

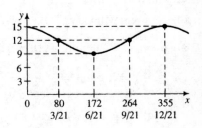

19. a. $T = k + A \cos B(x - h)$; $k = 20$, $A = \dfrac{28 - 12}{2} = 8$; $h = 197$; $p = 365$, $B = \dfrac{2\pi}{365}$;

$T = 20 + 8 \cos \dfrac{2\pi}{365}(x - 197)$

b. $20 + 8 \cos\left(\dfrac{2\pi}{365}(x - 197)\right) \leq 15$; $1 \leq x \leq 66$ or $328 \leq x \leq 365$; The first day (Jan. 1) through the 66th day (Mar. 7), the 328th day (Nov. 24) through the 365th day (Dec. 31)

20. a. amp. $= \dfrac{26 - (-14)}{2} = 20$; vert. shift $= \dfrac{26 + (-14)}{2} = 6$; $T = 6 + 20 \sin \dfrac{2\pi}{365}(x - 105)$

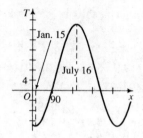

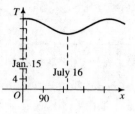

Ex. 20(a)   **Ex. 20(b)**

b. vert. shift $= \dfrac{28 + 22}{2} = \dfrac{50}{2} = 25$; amp. $= \dfrac{28 - 22}{2} = 3$; $T = 25 + 3 \cos \dfrac{2\pi}{365}(x - 15)$

21. a. $(1, 18.35)$, $(29, 17.5)$; $m = \dfrac{17.5 - 18.35}{29 - 1} = -0.03$; $y - 18.35 = -0.03(x - 1)$;

$y = -0.03x + 18.38$

b. amp. $= (19.47 - 16.18)/2 \approx 1.65$; vert. shift $= \dfrac{19.47 + 16.18}{2} \approx 17.83$;

$y = 17.83 + 1.65 \cos \dfrac{2\pi}{365}(x - 172)$

c. $17.83 + 1.65 \cos\left(\dfrac{2\pi}{365}(x - 172)\right) > 19$; $127 < x < 217$; the 127th day (May 7)

through the 217th day (Aug. 5)

22. Answers will vary.

23. Answers will vary.

24. a. max. $= 20$, min. $= 18$; vert. shift $= \dfrac{18 + 20}{2} = 19$; amp. $= \dfrac{20 - 18}{2} = 1$;

$\dfrac{1}{2}p = 15$ min $= \dfrac{1}{4}$ h; $p = \dfrac{1}{2}$ h; $B = \dfrac{2\pi}{\frac{1}{2}} = 4\pi$; max. at $x = 10.5$;

$y = 19 + \cos 4\pi(x - 10.5)$

b. Answers may vary. Examples: A furnace can heat a house faster than the cold air
outside cools it. Thus, the heating and cooling times will not be equal. Also, since
the temperature outdoors is variable, the heating function will not be regular.

25. max. $= 13$, min. $= 1$; amp. $= \dfrac{13 - 1}{2} = 6$; vert. shift $=$

$\dfrac{13 + 1}{2} = 7$; $p = \dfrac{1}{4}$ min $= 15$ s; $B = \dfrac{2\pi}{15}$; lowest point at

$t = 0$ s, $t = 15$ s; highest point at $t = 7.5$ s;

$y = 7 + 6 \cos \dfrac{2\pi}{15}(t - 7.5)$

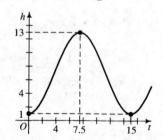

26. amp. $= \dfrac{55 - 15}{2} = 20$; vert. shift $= \dfrac{55 + 15}{2} = 35$;

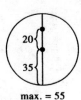
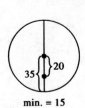

max. = 55   min. = 15

1 cycle = circumference = 70π cm;

10 km/h $= \dfrac{10(1000)(100) \text{ cm}}{3600 \text{ s}} = \dfrac{2500}{9}$ cm/s;

$70\pi \div \dfrac{2500}{9} = \dfrac{63\pi}{250}$ seconds for a cycle;

per. $= \dfrac{63\pi}{250}$; $B = \dfrac{2\pi}{\frac{63\pi}{250}} = \dfrac{500}{63}$; $h = 35 + 20 \cos \dfrac{500}{63}t$

Activity 1, 8-4 · page 318

1. a. $\sin 50° = 0.766$, $\cos 40° = 0.766$

  b. $\sin 25° = 0.423$, $\cos 65° = 0.423$

  c. $\cos 11° = 0.982$, $\sin 79° = 0.982$

  d. $\sin 83° = 0.993$, $\cos 7° = 0.993$

2. a. $\sin 18° = \cos 72°$

  b. $\cos 89° = \sin 1°$

  c. $\sin \theta = \cos (90° - \theta)$

  d. $\cos \theta = \sin (90° - \theta)$

Activity 2, 8-4 · page 319

1. a.

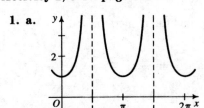

b. Yes; the graphs coincide.

2. a.

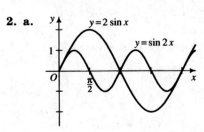

$y = 2 \sin x$

$y = \sin 2x$

b. No; the graphs are different.

3. a.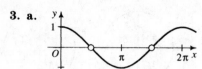

The domain of the sine function is all real numbers. The domain of both the secant and tangent functions is all real numbers except $\dfrac{n\pi}{2}$, n an odd integer.

b.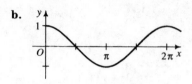

The domain of the cosine function is all real numbers.

c. The graphs coincide for remaining values of x. For $x \neq \dfrac{n\pi}{2}$, n an odd integer, $\cos x = \sec x - \sin x \tan x$.

Class Exercises 8-4 · page 320

1. a. $\cos^2 \theta$ **b.** $\sin^2 \theta$ **2. a.** $\tan^2 \theta$ **b.** 1 **3. a.** $\cot^2 \theta$ **b.** 1

4. a. $\tan \theta \cdot \cos \theta = \dfrac{\sin \theta}{\cos \theta} \cdot \cos \theta = \sin \theta$ **b.** $\tan (90° - A) = \cot A$

 c. $\cos \left(\dfrac{\pi}{2} - x \right) = \sin x$

5. a. $(1 - \sin x)(1 + \sin x) = 1 - \sin^2 x = \cos^2 x$ **b.** $\sin^2 x - 1 = -\cos^2 x$
 c. $(\sec x - 1)(\sec x + 1) = \sec^2 x - 1 = \tan^2 x$

6. a. $\tan A \cdot \cot A = 1$ **b.** $\cot y \cdot \sin y = \dfrac{\cos y}{\sin y} \cdot \sin y = \cos y$ **c.** $\cot^2 x - \csc^2 x = -1$

7. a. $\sin^2 \dfrac{5\pi}{6} + \cos^2 \dfrac{5\pi}{6} = 1$ **b.** $\sec^2 \pi - \tan^2 \pi = 1$ **c.** $\csc^2 135° - \cot^2 135° = 1$

8. a. Multiply the numerator and denominator by t; multiply the numerator and denominator by $\tan A$.

 b. Multiply the first fraction by $\dfrac{a}{a}$ and the second fraction by $\dfrac{b}{b}$.

 Multiply the first fraction by $\dfrac{\sec \theta}{\sec \theta}$ and the second fraction by $\dfrac{\tan \theta}{\tan \theta}$.

 c. Multiply the numerator and denominator by xy; multiply the numerator and denominator by $\cos \theta \sin \theta$.

 d. Multiply the numerator and denominator by x; multiply the numerator and denominator by $\cos \theta$.

Written Exercises 8-4 · pages 321–322

1. a. $\cos^2 \theta + \sin^2 \theta = 1$ **b.** $(1 - \cos \theta)(1 + \cos \theta) = 1 - \cos^2 \theta = \sin^2 \theta$
 c. $(\sin \theta - 1)(\sin \theta + 1) = \sin^2 \theta - 1 = -\cos^2 \theta$

2. a. $1 + \tan^2 \theta = \sec^2 \theta$ **b.** $(\sec x - 1)(\sec x + 1) = \sec^2 x - 1 = \tan^2 x$
 c. $\tan^2 x - \sec^2 x = -1$

3. a. $1 + \cot^2 A = \csc^2 A$ **b.** $(\csc A - 1)(\csc A + 1) = \csc^2 A - 1 = \cot^2 A$

 c. $\dfrac{1}{\sin^2 A} - \dfrac{1}{\tan^2 A} = \csc^2 A - \cot^2 A = 1$

4. a. $\dfrac{1}{\cos(90° - \theta)} = \dfrac{1}{\sin\theta} = \csc\theta$ **b.** $1 - \dfrac{\sin^2\theta}{\tan^2\theta} = 1 - \dfrac{\sin^2\theta}{\dfrac{\sin^2\theta}{\cos^2\theta}} = 1 - \cos^2\theta = \sin^2\theta$

c. $\dfrac{1}{\cos^2\theta} - \dfrac{1}{\cot^2\theta} = \sec^2\theta - \tan^2\theta = 1$

5. a. $\cos\theta\cot(90° - \theta) = \cos\theta\cdot\tan\theta = \cos\theta\cdot\dfrac{\sin\theta}{\cos\theta} = \sin\theta$

b. $\csc^2 x(1 - \cos^2 x) = \dfrac{1}{\sin^2 x}\cdot\sin^2 x = 1$

c. $\cos\theta(\sec\theta - \cos\theta) = 1 - \cos^2\theta = \sin^2\theta$

6. a. $\cot A\sec A\sin A = \dfrac{\cos A}{\sin A}\cdot\dfrac{1}{\cos A}\cdot\sin A = 1$

b. $\cos^2 A(\sec^2 A - 1) = \cos^2 A\sec^2 A - \cos^2 A = 1 - \cos^2 A = \sin^2 A$

c. $\sin\theta(\csc\theta - \sin\theta) = \sin\theta\csc\theta - \sin^2\theta = 1 - \sin^2\theta = \cos^2\theta$

7. $\sin A\tan A + \sin(90° - A) = \sin A\cdot\dfrac{\sin A}{\cos A} + \cos A = \dfrac{\sin^2 A + \cos^2 A}{\cos A} = \dfrac{1}{\cos A} = \sec A$

8. $\csc A - \cos A\cot A = \dfrac{1}{\sin A} - \cos A\cdot\dfrac{\cos A}{\sin A} = \dfrac{1 - \cos^2 A}{\sin A} = \dfrac{\sin^2 A}{\sin A} = \sin A$

9. $(\sec B - \tan B)(\sec B + \tan B) = \sec^2 B - \tan^2 B = 1$

10. $(1 - \cos B)(\csc B + \cot B) = (1 - \cos B)\left(\dfrac{1}{\sin B} + \dfrac{\cos B}{\sin B}\right) = \dfrac{(1 - \cos B)(1 + \cos B)}{\sin B} =$
$\dfrac{1 - \cos^2 B}{\sin B} = \dfrac{\sin^2 B}{\sin B} = \sin B$

11. $(\csc x - \cot x)(\sec x + 1) = \left(\dfrac{1}{\sin x} - \dfrac{\cos x}{\sin x}\right)\left(\dfrac{1}{\cos x} + 1\right) = \left(\dfrac{1 - \cos x}{\sin x}\right)\left(\dfrac{1 + \cos x}{\cos x}\right) =$
$\dfrac{1 - \cos^2 x}{\sin x\cos x} = \dfrac{\sin^2 x}{\sin x\cos x} = \dfrac{\sin x}{\cos x} = \tan x$

12. $(1 - \cos x)(1 + \sec x)\cos x = (1 - \cos x)(\cos x + 1) = 1 - \cos^2 x = \sin^2 x$

13. $\dfrac{\sin x\cos x}{1 - \cos^2 x} = \dfrac{\sin x\cos x}{\sin^2 x} = \dfrac{\cos x}{\sin x} = \cot x$

14. $\dfrac{\tan x + \cot x}{\sec^2 x} = \dfrac{\tan x}{\tan x}\left(\dfrac{\tan x + \cot x}{\sec^2 x}\right) = \dfrac{\tan^2 x + 1}{\tan x\sec^2 x} = \dfrac{\sec^2 x}{\tan x\sec^2 x} = \dfrac{1}{\tan x} = \cot x$

15. $(\sin x + \cos x)^2 + (\sin x - \cos x)^2 = (\sin^2 x + 2\sin x\cos x + \cos^2 x) +$
$(\sin^2 x - 2\sin x\cos x + \cos^2 x) = 1 + 2\sin x\cos x + 1 - 2\sin x\cos x = 2$

16. $(\sec^2\theta - 1)(\csc^2\theta - 1) = \tan^2\theta\cdot\cot^2\theta = 1$

17. $\dfrac{\cot^2\theta}{1 + \csc\theta} + \sin\theta\csc\theta = \dfrac{\csc^2\theta - 1}{1 + \csc\theta} + 1 = \dfrac{(\csc\theta + 1)(\csc\theta - 1)}{1 + \csc\theta} + 1 =$
$\csc\theta - 1 + 1 = \csc\theta$

18. $\dfrac{\tan^2\theta}{\sec\theta + 1} + 1 = \dfrac{\sec^2\theta - 1}{\sec\theta + 1} + 1 = \dfrac{(\sec\theta + 1)(\sec\theta - 1)}{\sec\theta + 1} + 1 = \sec\theta - 1 + 1 = \sec\theta$

19. $\cos^3 y + \cos y\sin^2 y = \cos y(\cos^2 y + \sin^2 y) = \cos y$

20. $\dfrac{\sec y + \csc y}{1 + \tan y} = \dfrac{\dfrac{1}{\cos y} + \dfrac{1}{\sin y}}{1 + \dfrac{\sin y}{\cos y}} = \dfrac{\dfrac{\sin y + \cos y}{\sin y \cos y}}{\dfrac{\cos y + \sin y}{\cos y}} = \dfrac{\sin y + \cos y}{\sin y \cos y} \cdot \dfrac{\cos y}{\cos y + \sin y} = \dfrac{1}{\sin y} = \csc y$

21. $\dfrac{\cos \theta}{1 + \sin \theta} + \dfrac{1 + \sin \theta}{\cos \theta} = \dfrac{\cos^2 \theta + (1 + \sin \theta)^2}{\cos \theta (1 + \sin \theta)} = \dfrac{\cos^2 \theta + 1 + 2 \sin \theta + \sin^2 \theta}{\cos \theta (1 + \sin \theta)} =$

$\dfrac{2 + 2 \sin \theta}{\cos \theta (1 + \sin \theta)} = \dfrac{2(1 + \sin \theta)}{\cos \theta (1 + \sin \theta)} = \dfrac{2}{\cos \theta} = 2 \sec \theta$

22. $\dfrac{\sin \theta \cot \theta + \cos \theta}{2 \tan (90° - \theta)} = \dfrac{\sin \theta \cdot \dfrac{\cos \theta}{\sin \theta} + \cos \theta}{2 \cot \theta} = \dfrac{\cos \theta + \cos \theta}{\dfrac{2 \cos \theta}{\sin \theta}} = \dfrac{2 \cos \theta}{\dfrac{2 \cos \theta}{\sin \theta}} = \sin \theta$

23. $\dfrac{\sin^4 \theta - \cos^4 \theta}{\sin^2 \theta - \cos^2 \theta} = \dfrac{(\sin^2 \theta + \cos^2 \theta)(\sin^2 \theta - \cos^2 \theta)}{\sin^2 \theta - \cos^2 \theta} = \sin^2 \theta + \cos^2 \theta = 1$

24. $\dfrac{\sin^2 \theta}{1 + \cos \theta} = \dfrac{1 - \cos^2 \theta}{1 + \cos \theta} = \dfrac{(1 + \cos \theta)(1 - \cos \theta)}{1 + \cos \theta} = 1 - \cos \theta$

25. $\sin^2 \theta + \cos^2 \theta = 1$; divide both sides by $\cos^2 \theta$; $\dfrac{\sin^2 \theta}{\cos^2 \theta} + \dfrac{\cos^2 \theta}{\cos^2 \theta} = \dfrac{1}{\cos^2 \theta}$;

$\tan^2 \theta + 1 = \sec^2 \theta$

26. $\sin^2 \theta + \cos^2 \theta = 1$; divide both sides by $\sin^2 \theta$; $\dfrac{\sin^2 \theta}{\sin^2 \theta} + \dfrac{\cos^2 \theta}{\sin^2 \theta} = \dfrac{1}{\sin^2 \theta}$;

$1 + \cot^2 \theta = \csc^2 \theta$

27. domain is all real numbers; $y = \sin^4 x + 2 \sin^2 x \cos^2 x + \cos^4 x = (\sin^2 x + \cos^2 x)^2 = 1$; $y = 1$

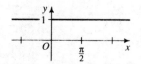

Ex. 27

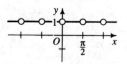

Ex. 28

28. domain is all real numbers except $\dfrac{n \pi}{2}$, where n is an integer;

$y = (\sin x \div \cos x) \div \tan x = \dfrac{\sin x}{\cos x} \div \tan x = \tan x \div \tan x = 1$; $y = 1$, not defined

when $x = \dfrac{n \pi}{2}$, n an integer.

29. $\cot^2 \theta + \cos^2 \theta + \sin^2 \theta = \cot^2 \theta + 1 = \csc^2 \theta$

30. $\dfrac{\cot \theta - \tan \theta}{\sin \theta \cos \theta} = \dfrac{\dfrac{\cos \theta}{\sin \theta}}{\sin \theta \cos \theta} - \dfrac{\dfrac{\sin \theta}{\cos \theta}}{\sin \theta \cos \theta} = \dfrac{1}{\sin^2 \theta} - \dfrac{1}{\cos^2 \theta} = \csc^2 \theta - \sec^2 \theta$

31. $\dfrac{\sin \theta}{\csc \theta} + \dfrac{\cos \theta}{\sec \theta} = \dfrac{\sin \theta}{\dfrac{1}{\sin \theta}} + \dfrac{\cos \theta}{\dfrac{1}{\cos \theta}} = \sin^2 \theta + \cos^2 \theta = 1 = \sin \theta \csc \theta$

32. $\dfrac{1 - \sin^2 \theta}{1 + \cot^2 \theta} = \dfrac{\cos^2 \theta}{\csc^2 \theta} = \dfrac{\cos^2 \theta}{\dfrac{1}{\sin^2 \theta}} = \sin^2 \theta \cos^2 \theta$

33. $\tan^2 x - \sin^2 x = \dfrac{\sin^2 x}{\cos^2 x} - \sin^2 x = \sin^2 x \left(\dfrac{1}{\cos^2 x} - 1 \right) = \sin^2 x (\sec^2 x - 1) =$
$\tan^2 x \sin^2 x$

34. $\dfrac{\tan^2 x}{1 + \tan^2 x} = \dfrac{\tan^2 x}{\sec^2 x} = \dfrac{\dfrac{\sin^2 x}{\cos^2 x}}{\dfrac{1}{\cos^2 x}} = \sin^2 x$

35. $\dfrac{\tan \theta}{1 + \tan \theta} = \dfrac{\dfrac{\sin \theta}{\cos \theta}}{1 + \dfrac{\sin \theta}{\cos \theta}} = \dfrac{\sin \theta}{\cos \theta + \sin \theta} = \dfrac{\sin \theta}{\sin \theta + \cos \theta}$

36. $\dfrac{\tan x}{1 + \sec x} + \dfrac{1 + \sec x}{\tan x} = \dfrac{\tan^2 x + (1 + \sec x)^2}{(1 + \sec x)(\tan x)} = \dfrac{(\sec^2 x - 1) + (1 + 2 \sec x + \sec^2 x)}{(1 + \sec x) \left(\dfrac{\sin x}{\cos x} \right)} =$

$\dfrac{2 \sec^2 x + 2 \sec x}{(1 + \sec x)} \cdot \dfrac{\cos x}{\sin x} = \dfrac{2 \sec x (\sec x + 1) \cdot \cos x}{(1 + \sec x) \cdot \sin x} = \dfrac{2 \sec x \cos x}{\sin x} = \dfrac{2}{\sin x} = 2 \csc x$

37. a. $\angle OQP = \angle OBA = 90°$ and $\angle AOB = \angle POQ = \theta$; $\triangle OPQ \sim \triangle OAB$ by AA similarity.

b. Since lengths of corresponding sides are proportional, $\dfrac{PQ}{AB} = \dfrac{OQ}{OB}$ and $\dfrac{OP}{OA} = \dfrac{OQ}{OB}$;

thus, $\dfrac{PQ}{OQ} = \dfrac{AB}{OB}$ and $\dfrac{OP}{OQ} = \dfrac{AO}{BO}$.

c. $OB = 1$, $PQ = \sin \theta$, and $OQ = \cos \theta$; $AB = OB \cdot \dfrac{PQ}{OQ} = 1 \cdot \dfrac{\sin \theta}{\cos \theta} = \tan \theta$. Also, since

$OP = 1$, $AO = BO \cdot \dfrac{OP}{OQ} = 1 \cdot \dfrac{1}{\cos \theta} = \sec \theta$.

d. $AB = \dfrac{\sin \theta}{\cos \theta}$ and $\overleftrightarrow{AB}$ is tangent to the circle at B. $AO = \dfrac{1}{\cos \theta}$ and $\overleftrightarrow{AO}$ is a secant of
the circle.

e. In rt. $\triangle OAB$, $(AB)^2 + (OB)^2 = (AO)^2$; from part (c), $AB = \tan \theta$ and $AO = \sec \theta$;
since $OB = 1$, $\tan^2 \theta + 1 = \sec^2 \theta$; $\sec^2 \theta = 1 + \tan^2 \theta$.

f. $AC = AO + 1 = \sec \theta + 1$ and $AP = AO - OP = \sec \theta - 1$; $(AB)^2 = AP \cdot AC$;
$\tan^2 \theta = (\sec \theta - 1)(\sec \theta + 1)$.

38. Draw $\overline{PQ} \perp \overline{OD}$ at point Q. By AA Similarity, $\triangle OQP \sim \triangle ODC$, so $\dfrac{PQ}{CD} = \dfrac{OQ}{OD}$ and

$\dfrac{PO}{CO} = \dfrac{QO}{DO}$; thus, $\dfrac{PQ}{OQ} = \dfrac{CD}{OD}$ and $\dfrac{PO}{QO} = \dfrac{CO}{DO}$; $PQ = \cos \theta$, $OQ = \sin \theta$, and $DO = PO = 1$;

$CD = OD \cdot \dfrac{PQ}{OQ} = 1 \cdot \dfrac{\cos \theta}{\sin \theta} = \cot \theta$ and $CO = DO \cdot \dfrac{PO}{QO} = 1 \cdot \dfrac{1}{\sin \theta} = \csc \theta$.

39. This happened because $\sqrt{x^2} = |x|$, which means $\sqrt{\sec^2 x} = \sec x$ only when $\sec x$ is
greater than zero. If Jon graphs $y = |\sec x|$, he will see that its graph is the same as
that of $y = \sqrt{1 + \tan^2 x}$.

40. $\tan \theta = \dfrac{\sin \theta}{\cos \theta} = \pm \dfrac{\sqrt{1 - \cos^2 \theta}}{\cos \theta}$

41. $\sec \theta = \dfrac{1}{\cos \theta} = \pm \dfrac{1}{\sqrt{1 - \sin^2 \theta}}$

42. $\sqrt{\dfrac{1 - \sin x}{1 + \sin x}} = \sqrt{\dfrac{(1 - \sin x)}{(1 + \sin x)} \cdot \dfrac{(1 - \sin x)}{(1 - \sin x)}} = \sqrt{\dfrac{(1 - \sin x)^2}{1 - \sin^2 x}} = \sqrt{\dfrac{(1 - \sin x)^2}{\cos^2 x}} =$

$\left| \dfrac{1 - \sin x}{\cos x} \right| = \left| \dfrac{1}{\cos x} - \dfrac{\sin x}{\cos x} \right| = |\sec x - \tan x|$; the identity is true for all real numbers

x except $x = \dfrac{n\pi}{2}$, n an odd integer.

Computer Exercise · page 322

```
10 PRINT "X = ";
20 INPUT X
30 PRINT "SIN X = ";SIN(X)
40 PRINT "COS X = "; (1 - (SIN(X))^2)^.5
50 PRINT "TAN X = "; SIN(X)/((1 - (SIN(X))^2)^.5)
60 PRINT "CSC X = "; 1/SIN(X)
70 PRINT "SEC X = "; 1/((1 - (SIN(X))^2)^.5)
80 PRINT "COT X = "; ((1 - (SIN(X))^2)^.5)/SIN(X)
90 END
```

Written Exercises 8-5 · pages 326–327

1. $\sec^2 \theta = 9$; $\sec \theta = \pm 3$; $70.5°$, $109.5°$, $250.5°$, $289.5°$

2. $\tan^2 \theta = 1$; $\tan \theta = \pm 1$; $45°$, $135°$, $225°$, $315°$

3. $1 - \csc^2 \theta = -3$; $\csc^2 \theta = 4$; $\csc \theta = \pm 2$; $30°$, $150°$, $210°$, $330°$

4. $8 \cos^2 \theta - 3 = 1$; $\cos^2 \theta = \dfrac{1}{2}$; $\cos \theta = \pm \dfrac{\sqrt{2}}{2}$; $45°$, $135°$, $225°$, $315°$

5. $6 \sin^2 \theta - 7 \sin \theta + 2 = 0$; $(3 \sin \theta - 2)(2 \sin \theta - 1) = 0$; $\sin \theta = \dfrac{2}{3} \approx 0.6667$ or

$\sin \theta = \dfrac{1}{2}$; $\theta = 41.8°$, $138.2°$ or $\theta = 30°$, $150°$

6. $2 \tan^2 \theta - 3 \tan \theta + 1 = 0$; $(2 \tan \theta - 1)(\tan \theta - 1) = 0$; $\tan \theta = \dfrac{1}{2}$ or $\tan \theta = 1$;

$\theta = 26.6°$, $206.6°$ or $\theta = 45°$, $225°$

7. $6 \sin^2 \theta = 7 - 5 \cos \theta$; $6(1 - \cos^2 \theta) = 7 - 5 \cos \theta$; $6 - 6 \cos^2 \theta = 7 - 5 \cos \theta$;

$6 \cos^2 \theta - 5 \cos \theta + 1 = 0$; $(3 \cos \theta - 1)(2 \cos \theta - 1) = 0$; $\cos \theta = \dfrac{1}{3}$ or $\cos \theta = \dfrac{1}{2}$;

$\theta = 70.5°$, $289.5°$ or $\theta = 60°$, $300°$

8. $\cos^2 \theta - 3 \sin \theta = 3$; $(1 - \sin^2 \theta) - 3 \sin \theta = 3$; $\sin^2 \theta + 3 \sin \theta + 2 = 0$;

$(\sin \theta + 2)(\sin \theta + 1) = 0$; $\sin \theta = -2$ (no solution) or $\sin \theta = -1$; $\theta = 270°$

9. $\cos x \tan x - \cos x = 0$; $\cos x(\tan x - 1) = 0$; $\cos x = 0$ or $\tan x = 1$; $x = \dfrac{\pi}{2}, \dfrac{3\pi}{2}$

or $x = \dfrac{\pi}{4}, \dfrac{5\pi}{4}$; checking these values in the original equation gives $x = \dfrac{\pi}{4}$ or $x = \dfrac{5\pi}{4}$.

10. $\sec x \sin x - 2 \sin x = 0$; $\sin x(\sec x - 2) = 0$; $\sin x = 0$ or $\sec x = 2$;

$x = 0$, π or $x = \dfrac{\pi}{3}, \dfrac{5\pi}{3}$

11. $\sin^2 x - \sin x = 0$; $\sin x(\sin x - 1) = 0$; $\sin x = 0$ or $\sin x = 1$; $x = 0$, π or $x = \dfrac{\pi}{2}$

12. $\tan^2 x - \tan x = 0$; $\tan x(\tan x - 1) = 0$; $\tan x = 0$ or $\tan x = 1$; $x = 0$, π or

$x = \dfrac{\pi}{4}, \dfrac{5\pi}{4}$

13. $2\cos^2 x - \cos x = 0$; $\cos x(2\cos x - 1) = 0$; $\cos x = 0$ or $\cos x = \dfrac{1}{2}$; $x = \dfrac{\pi}{2}, \dfrac{3\pi}{2}$

or $x = \dfrac{\pi}{3}, \dfrac{5\pi}{3}$

14. $3\sin x = \cos x$; $\dfrac{\sin x}{\cos x} = \dfrac{1}{3}$; $\tan x = \dfrac{1}{3} \approx 0.3333$; $x = 0.32, 3.46$

15. $\sin x + \cos x = 0$; $\sin x = -\cos x$; $\dfrac{\sin x}{\cos x} = -1$; $\tan x = -1$; $x = \dfrac{3\pi}{4}, \dfrac{7\pi}{4}$

16. $\sec x = 2\csc x$; $\dfrac{1}{\cos x} = \dfrac{2}{\sin x}$; $\sin x = 2\cos x$; $\dfrac{\sin x}{\cos x} = 2$; $\tan x = 2$; $x = 1.11, 4.25$

17. $\tan^2 x - 2\tan x \sin x = 0$; $\tan x(\tan x - 2\sin x) = 0$; $\tan x = 0$ or $\tan x = 2\sin x$.

If $\tan x = 2\sin x$, then $\dfrac{\sin x}{\cos x} = 2\sin x$; $\sin x = 2\sin x \cos x$; $\sin x(1 - 2\cos x) = 0$;

$\sin x = 0$ or $\cos x = \dfrac{1}{2}$. Thus, we have $\tan x = 0$, or $\sin x = 0$, or $\cos x = \dfrac{1}{2}$;

$x = 0$, π or $x = \dfrac{\pi}{3}, \dfrac{5\pi}{3}$.

18. $2\sin x \cos x = \tan x$; $2\sin x \cos x = \dfrac{\sin x}{\cos x}$; $2\sin x \cos^2 x = \sin x$;

$\sin x(2\cos^2 x - 1) = 0$; $\sin x = 0$ or $2\cos^2 x = 1$; $\sin x = 0$ or $\cos^2 x = \dfrac{1}{2}$; $\sin x = 0$

or $\cos x = \pm\dfrac{\sqrt{2}}{2}$; $x = 0$, π or $x = \dfrac{\pi}{4}, \dfrac{3\pi}{4}, \dfrac{5\pi}{4}, \dfrac{7\pi}{4}$

19. $2\csc^2 x = 3\cot^2 x - 1$; $2(1 + \cot^2 x) = 3\cot^2 x - 1$; $\cot^2 x = 3$; $\cot x = \pm\sqrt{3}$;

$x = \dfrac{\pi}{6}, \dfrac{5\pi}{6}, \dfrac{7\pi}{6}, \dfrac{11\pi}{6}$

20. $2\sec^2 x + \tan x = 5$; $2(1 + \tan^2 x) + \tan x = 5$; $2\tan^2 x + \tan x - 3 = 0$;

$(2\tan x + 3)(\tan x - 1) = 0$; $\tan x = -1.5$ or $\tan x = 1$; $x = 2.16, 5.30$ or $x = \dfrac{\pi}{4}, \dfrac{5\pi}{4}$

21. $\sin^2 x + \sin x - 1 = 0$; $\sin x = \dfrac{-1 \pm \sqrt{1 - 4(1)(-1)}}{2 \cdot 1} = \dfrac{-1 \pm \sqrt{5}}{2}$; $\sin x = \dfrac{-1 - \sqrt{5}}{2} < -1$

(no solution) or $\sin x = \dfrac{-1 + \sqrt{5}}{2} \approx 0.618$; $x = 0.67, 2.48$

22. $\cos^2 x - 2\cos x - 1 = 0$; $\cos x = \dfrac{-(-2) \pm \sqrt{(-2)^2 - 4(1)(-1)}}{2 \cdot 1} = \dfrac{2 \pm 2\sqrt{2}}{2} = 1 \pm \sqrt{2}$;

$\cos x = 1 + \sqrt{2} > 1$ (no solution) or $\cos x = 1 - \sqrt{2} \approx -0.414$; $x = 2.00, 4.29$

23. $3 \cos x \cot x + 7 = 5 \csc x$; $3 \cos x \cdot \dfrac{\cos x}{\sin x} + 7 = \dfrac{5}{\sin x}$; $3 \cos^2 x + 7 \sin x = 5$;

$3(1 - \sin^2 x) + 7 \sin x = 5$; $3 \sin^2 x - 7 \sin x + 2 = 0$; $(3 \sin x - 1)(\sin x - 2) = 0$;

$\sin x = \dfrac{1}{3}$ or $\sin x = 2$ (no solution); $x = 0.34,\ 2.80$

24. $2 \sin^3 x - \sin^2 x - 2 \sin x + 1 = 0$; $\sin^2 x(2 \sin x - 1) - (2 \sin x - 1) = 0$;
$(\sin^2 x - 1)(2 \sin x - 1) = 0$; $(\sin x + 1)(\sin x - 1)(2 \sin x - 1) = 0$; $\sin x = -1$,

or $\sin x = 1$, or $\sin x = \dfrac{1}{2}$; $x = \dfrac{3\pi}{2}$, or $x = \dfrac{\pi}{2}$, or $x = \dfrac{\pi}{6}, \dfrac{5\pi}{6}$

25. $2 \cos^2 x - \cos x = 2 - \sec x$; $2 \cos^2 x - \cos x = 2 - \dfrac{1}{\cos x}$;

$2 \cos^3 x - \cos^2 x - 2 \cos x + 1 = 0$; $(\cos x - 1)(2 \cos^2 x + \cos x - 1) = 0$;

$(\cos x - 1)(2 \cos x - 1)(\cos x + 1) = 0$; $\cos x = 1$, or $\cos x = \dfrac{1}{2}$, or $\cos x = -1$;

$x = 0$, or $x = \dfrac{\pi}{3}, \dfrac{5\pi}{3}$, or $x = \pi$

26. $\csc^2 x - 2 \csc x = 2 - 4 \sin x$; $\csc^2 x - 2 \csc x = 2 - \dfrac{4}{\csc x}$;

$\csc^3 x - 2 \csc^2 x - 2 \csc x + 4 = 0$; $\csc^2 x(\csc x - 2) - 2(\csc x - 2) = 0$;
$(\csc^2 x - 2)(\csc x - 2) = 0$; $\csc^2 x = 2$ or $\csc x = 2$; $\csc x = \pm\sqrt{2}$ or $\csc x = 2$;

$\sin x = \pm\dfrac{\sqrt{2}}{2}$ or $\sin x = \dfrac{1}{2}$; $x = \dfrac{\pi}{4}, \dfrac{3\pi}{4}, \dfrac{5\pi}{4}, \dfrac{7\pi}{4}$ or $x = \dfrac{\pi}{6}, \dfrac{5\pi}{6}$

27. $2 \cos x = 2 + \sin x$; $4 \cos^2 x = 4 + 4 \sin x + \sin^2 x$;
$4(1 - \sin^2 x) = 4 + 4 \sin x + \sin^2 x$; $5 \sin^2 x + 4 \sin x = 0$; $\sin x(5 \sin x + 4) = 0$;

$\sin x = 0$ or $\sin x = -\dfrac{4}{5}$; $x = 0, \pi$ or $x = 4.07,\ 5.36$; checking these values in the

original equation gives $x = 0$ or $x = 5.36$.

28. $1 - \sin x = 3 \cos x$; $1 - 2 \sin x + \sin^2 x = 9 \cos^2 x$;
$1 - 2 \sin x + \sin^2 x = 9(1 - \sin^2 x)$; $10 \sin^2 x - 2 \sin x - 8 = 0$;
$5 \sin^2 x - \sin x - 4 = 0$; $(5 \sin x + 4)(\sin x - 1) = 0$; $\sin x = -0.8$ or $\sin x = 1$;

$x = 4.07,\ 5.36$ or $x = \dfrac{\pi}{2}$; checking these values in the original equation gives $x = 5.36$

or $x = \dfrac{\pi}{2}$.

29. 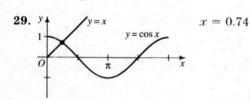 $x = 0.74$ **30.** 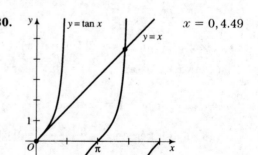 $x = 0, 4.49$

31. 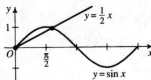 $x = 0, 1.90$ **32.** 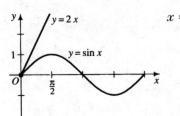 $x = 0$

33. 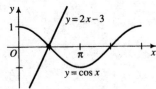 $x = 1.52$ **34.** 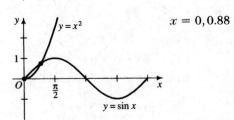 $x = 0, 0.88$

35. a. W is on $y = x$ so x-coord. of $W = y$-coord. of $W = y$-coord. of $A = \cos x$; then x-coord. of $B = \cos x$. Similarly, since the y-coord. of B is $\cos(\cos x)$, the x-coord. of C is $\cos(\cos x)$.

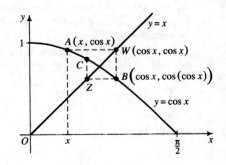

b. You approach the pt. of intersection of $y = x$ and $y = \cos x$.

c. The x-coords. approach the solution of the equation $\cos x = x$.

d. Enter a guess; continue pressing the cos key until the values agree to the desired number of places.

e. 0.739 **f.** No

Chapter Test · pages 328–329

1. $x - 3y = 9$; $m = \dfrac{1}{3}$; $\tan \alpha = \dfrac{1}{3}$; $\alpha = 18°$

2. $5 \cos \theta = -1$; $\cos \theta = -\dfrac{1}{5}$; ref. $\angle = 78.5°$; $\theta = 101.5°, 258.5°$

3. $3 - \csc x = 7$; $\csc x = -4$; ref. $\angle = 0.25$; $x = 3.39, 6.03$

4. $2 \sin 3x = \sqrt{2}$; $\sin 3x = \dfrac{\sqrt{2}}{2}$; ref. $\angle = \dfrac{\pi}{4}$; $3x = \dfrac{\pi}{4} + 2\pi k$ or $3x = \dfrac{3\pi}{4} + 2\pi k$,

k an integer; $x = \dfrac{\pi}{12} + \dfrac{2\pi}{3}k$ or $x = \dfrac{\pi}{4} + \dfrac{2\pi}{3}k$; $x = \dfrac{\pi}{12}, \dfrac{\pi}{4}, \dfrac{3\pi}{4}, \dfrac{11}{12}\pi, \dfrac{17}{12}\pi, \dfrac{19}{12}\pi$

5. 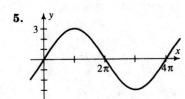 $y = 3 \sin \dfrac{x}{2}$; amp. $= 3$; per. $= \dfrac{2\pi}{\frac{1}{2}} = 4\pi$

6. amp. $= \dfrac{3 - (-1)}{2} = 2$; vert. shift $= \dfrac{3 + (-1)}{2} = 1$; horiz. shift $= 0$; per. $= 4$,

$B = \dfrac{2\pi}{4} = \dfrac{\pi}{2}$; $y = 1 + 2 \sin \dfrac{\pi}{2}x$ or $y = 1 + 2 \cos \dfrac{\pi}{2}(x - 1)$

7. a.

$D = 1.4 \cos\left(\frac{2\pi}{13}(t-2)\right) + 3.6$

b. Answers may vary. The difference in time between high and low tides is half the period. The difference between the heights of the high and low tides is twice the amplitude. The average of the high and low tides is the axis of the function, or the vertical shift. The time of the first high tide after midnight would be the horizontal shift if the function is expressed in terms of cosine.

8. a. $\cot A(\sec A - \cos A) = \dfrac{\cos A}{\sin A} \cdot \sec A - \dfrac{\cos A}{\sin A} \cdot \cos A = \dfrac{1}{\sin A} - \dfrac{\cos^2 A}{\sin A} = \dfrac{1 - \cos^2 A}{\sin A} =$

$\dfrac{\sin^2 A}{\sin A} = \sin A$

b. $\dfrac{\cot \theta}{\sin (90° - \theta)} = \dfrac{\dfrac{\cos \theta}{\sin \theta}}{\cos \theta} = \dfrac{\cos \theta}{\cos \theta \sin \theta} = \dfrac{1}{\sin \theta} = \csc \theta$

c. $(\sec x + \tan x)(1 - \sin x) = \left(\dfrac{1}{\cos x} + \dfrac{\sin x}{\cos x}\right)(1 - \sin x) = \dfrac{(1 + \sin x)}{\cos x}(1 - \sin x) =$

$\dfrac{1 - \sin^2 x}{\cos x} = \dfrac{\cos^2 x}{\cos x} = \cos x$

d. $\dfrac{\cot \alpha + \tan \alpha}{\csc^2 \alpha} = \dfrac{\dfrac{\cos \alpha}{\sin \alpha} + \dfrac{\sin \alpha}{\cos \alpha}}{\dfrac{1}{\sin^2 \alpha}} = \dfrac{\cos^2 \alpha + \sin^2 \alpha}{\sin \alpha \cos \alpha} \cdot \sin^2 \alpha = \dfrac{1}{\sin \alpha \cos \alpha} \cdot \sin^2 \alpha =$

$\dfrac{\sin \alpha}{\cos \alpha} = \tan \alpha$

9. a. $2 \cos x = \sin x$; since $\cos x = 0$ is not a solution, divide both sides by $\cos x$;

$2 = \dfrac{\sin x}{\cos x}$; $\tan x = 2$; $x = 1.1$, 4.25

b. $\sin x = \csc x$; since $\sin x = 0$ is not a solution, multiply both sides by $\sin x$;

$\sin^2 x = 1$; $\sin x = \pm 1$; $x = \dfrac{\pi}{2}, \dfrac{3\pi}{2}$

10. a. $2 \cos^2 \theta + 3 \sin \theta - 3 = 0$; $2(1 - \sin^2 \theta) + 3 \sin \theta - 3 = 0$;

$2 - 2 \sin^2 \theta + 3 \sin \theta - 3 = 0$; $2 \sin^2 \theta - 3 \sin \theta + 1 = 0$;

$(2 \sin \theta - 1)(\sin \theta - 1) = 0$; $\sin \theta = \dfrac{1}{2}$ or $\sin \theta = 1$; $\theta = 30°, 90°, 150°$

b. $\cos \theta \cot \theta = 2 \cos \theta$; $\cos \theta \cot \theta - 2 \cos \theta = 0$; $\cos \theta(\cot \theta - 2) = 0$; $\cos \theta = 0$ or

$\cot \theta = 2$; $\theta = 90°, 270°$ or $\theta = 26.6°, 206.6°$

Class Exercises 9-1 · page 333

1. $\sin = \dfrac{\text{opposite}}{\text{hypotenuse}}$; $\cos = \dfrac{\text{adjacent}}{\text{hypotenuse}}$; $\tan = \dfrac{\text{opposite}}{\text{adjacent}}$; $\cot = \dfrac{\text{adjacent}}{\text{opposite}}$;

$\sec = \dfrac{\text{hypotenuse}}{\text{adjacent}}$; $\csc = \dfrac{\text{hypotenuse}}{\text{opposite}}$

2. **a.** $\sin \alpha = \dfrac{x}{z}$, $\cos \alpha = \dfrac{y}{z}$, $\tan \alpha = \dfrac{x}{y}$ **b.** $\sin \beta = \dfrac{y}{z}$, $\cos \beta = \dfrac{x}{z}$, $\tan \beta = \dfrac{y}{x}$

c. $\sin \alpha = \dfrac{x}{z}$; $\cos \beta = \dfrac{x}{z}$; $\sin \alpha = \cos \beta$ is a true statement. $\tan \alpha = \dfrac{x}{y}$;

$\cot \beta = \dfrac{x}{y}$; $\tan \alpha = \cot \beta$ is a true statement. $\sec \alpha = \dfrac{z}{y}$; $\csc \beta = \dfrac{z}{y}$;

$\sec \alpha = \csc \beta$ is a true statement.

3. $\cot 28°$; the calculation with $\cot 28°$ will require multiplication instead of division.

4. **a.** $\dfrac{c}{40} = \csc 28°$; $c = 40 \cdot \dfrac{1}{\sin 28°} = 85.2$ **b.** Use the Pythagorean Theorem.

5. **a.** $\sin 50° = \dfrac{x}{10}$; $\csc 50° = \dfrac{10}{x}$ **b.** $\cos 32° = \dfrac{5}{x}$; $\sec 32° = \dfrac{x}{5}$

c. $\tan 55° = \dfrac{x}{8}$; $\cot 55° = \dfrac{8}{x}$ **d.** $\cos 70° = \dfrac{3}{x}$; $\sec 70° = \dfrac{x}{3}$

6. **a.** $\tan \theta = \dfrac{3}{2}$; $\cot \theta = \dfrac{2}{3}$ **b.** $\cos \theta = \dfrac{5}{8}$; $\sec \theta = \dfrac{8}{5}$

c. $\sin \theta = \dfrac{7}{12}$; $\csc \theta = \dfrac{12}{7}$ **d.** $\cos \theta = \dfrac{7}{20}$; $\sec \theta = \dfrac{20}{7}$

Written Exercises 9-1 · pages 334–338

1. $\sin 25° = \dfrac{b}{18}$; $b = 18(\sin 25°) \approx 7.61$; $\cos 25° = \dfrac{c}{18}$; $c = 18(\cos 25°) \approx 16.3$

2. $\sin 64° = \dfrac{q}{27}$; $q = 27(\sin 64°) \approx 24.3$; $\cos 64° = \dfrac{r}{27}$; $r = 27(\cos 64°) \approx 11.8$

3. $\csc 12° = \dfrac{d}{9}$; $d = 9 \csc 12° \approx 43.3$; $\cot 12° = \dfrac{f}{9}$; $f = 9 \cot 12° \approx 42.3$

4. $\sec 37° = \dfrac{x}{25}$; $x = 25 \sec 37° \approx 31.3$; $\tan 37° = \dfrac{y}{25}$; $y = 25 \tan 37° \approx 18.8$

5. $c = \sqrt{5^2 + 12^2} = 13$; **a.** $\sin A = \dfrac{5}{13}$ **b.** $\cos B = \dfrac{5}{13}$ **c.** $\tan A = \dfrac{5}{12}$

d. $\cot B = \dfrac{5}{12}$ **e.** $\sec A = \dfrac{13}{12}$ **f.** $\csc B = \dfrac{13}{12}$

6. a. $\sin A = \dfrac{a}{c}$; $\cos B = \dfrac{a}{c}$; they are equal.

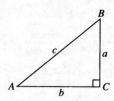

b. $\tan A = \dfrac{a}{b}$; $\cot B = \dfrac{a}{b}$; they are equal.

c. $\sec A = \dfrac{c}{b}$; $\csc B = \dfrac{c}{b}$; they are equal.

7. $\sin A = \dfrac{3}{5}$;

$\angle A \approx 36.9°$;

$\angle B \approx 90° - 36.9° = 53.1°$

8. Let $\angle A$ and $\angle B$ be the acute angles.

$\tan A = \dfrac{9}{16} = 0.5625$; $\angle A \approx 29.4°$;

$\angle B \approx 90° - 29.4° = 60.6°$

9. a. From the Pythagorean Thm., $(\text{hyp.})^2 = 1^2 + 1^2$; hyp. $= \sqrt{2}$.

b. (1) $\tan 45° = \dfrac{1}{1} = 1$; (2) $\sin 45° = \dfrac{1}{\sqrt{2}} = \dfrac{\sqrt{2}}{2}$; (3) $\cos 45° = \dfrac{1}{\sqrt{2}} = \dfrac{\sqrt{2}}{2}$

c. (1) $\tan 45° = 1$; (2) $\sin 45° \approx 0.7071$; (3) $\cos 45° \approx 0.7071$

10. a. In a 30°-60°-90° $\triangle$, the length of the shorter leg is half the length of the hyp.; the longer leg is $\sqrt{3} \cdot$ (the shorter leg). $\therefore$ the lengths are 1, $\sqrt{3}$, and 2.

b. (1) $\sin 30° = \dfrac{1}{2}$ (2) $\sin 60° = \dfrac{\sqrt{3}}{2}$ (3) $\tan 30° = \dfrac{1}{\sqrt{3}} = \dfrac{\sqrt{3}}{3}$ (4) $\tan 60° = \sqrt{3}$

c. (1) $\sin 30° = 0.5$ (2) $\sin 60° \approx 0.866$ (3) $\tan 30° \approx 0.5774$ (4) $\tan 60° \approx 1.732$

11. The length of each segment is the value of the tangent of the corresponding angle.
For example, $\tan 10° = \dfrac{CB_1}{AC} = \dfrac{CB_1}{1} = CB_1$.

12. No; no; using a calculator to find $\tan 10°$, $\tan 20°$, and $\tan 40°$, $CB_1 = 0.1763$, $CB_2 = 0.3640$, and $CB_4 = 0.8391$; $CB_2 > 2 \cdot CB_1$ and $CB_4 > 4 \cdot CB_1$.

13.

a. $\dfrac{35,000}{y} = \tan 6°$;

$y = \dfrac{35,000}{\tan 6°} \approx 333,000$ ft

b. $\dfrac{35,000}{x} = \sin 6°$; $x = \dfrac{35,000}{\sin 6°} \approx 335,000$ ft; it is assumed that the plane was flying over a level stretch of ground.

14.

$\dfrac{19}{x} = \tan 3°$; $x = \dfrac{19}{\tan 3°} \approx 363$ m

15.

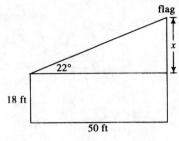

flag

$\dfrac{x}{50} = \tan 22°$; $x = 50 \tan 22° \approx 20.2$; the

flagpole's height is $18 + 20.2 = 38.2$ ft.

16. a. $s = r\theta = 250 \cdot 5 \cdot \dfrac{\pi}{180} \approx \dfrac{1250(3.14)}{180} \approx 21.8$;

21.8 m

b. $\tan 5° = \dfrac{BC}{250}$; $BC = 250 \tan 5° \approx 21.9$;

21.9 m; the two answers vary by 0.1 m.

17. Let θ be the measure of a base angle. $\cos \theta = \dfrac{4}{6}$; $\theta \approx 48.2°$; each base angle is 48.2° and the vertex angle is $180° - 2(48.2°) = 83.6°$. To find the height h, $h = \sqrt{6^2 - 4^2} = 2\sqrt{5}$; $A = \dfrac{1}{2}bh = \dfrac{1}{2} \cdot 8 \cdot 2\sqrt{5} = 8\sqrt{5} \approx 17.9$ sq. units.

18. The altitude to the base bisects the vertex angle and the base; let b be the length of the base; $\sin 26° = \dfrac{\frac{1}{2}b}{21}$; $\dfrac{1}{2}b = 21(\sin 26°)$; $b \approx 42(0.4384) \approx 18.4$ cm.

19. Draw $\overline{PO}$. Since $\overline{PA}$ and $\overline{PB}$ are tangents, $\triangle OAP$ and $\triangle OBP$ are congruent rt. $\triangle$s with hypotenuse $\overline{OP}$. $\angle APO = 21°$; $\cot 21° = \dfrac{PA}{6}$; $PA = 6 \cot 21° \approx 15.6$; $PA = PB = 15.6$

20. In the diagram, the measure of θ is half the required measure. The circle has center $C(6, 8)$ and radius 3. $OC = \sqrt{6^2 + 8^2} = 10$. Since $\overrightarrow{OT}$ is a tangent, $\triangle OCT$ is a rt. $\triangle$; $\sin \theta = \dfrac{3}{10} = 0.3$; $\theta \approx 17.46°$; $2\theta \approx 34.9°$.

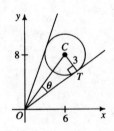

21. In the right triangle formed by the altitude, one leg has length $\dfrac{1}{2}(14x - 8x) = 3x$; thus, $\sin \alpha = \dfrac{3x}{5x} = 0.6$; $\alpha \approx 36.9°$; each obtuse angle has measure $\alpha + 90° \approx 126.9°$.

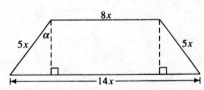

22. In the diagram, the measure of α is half the required measure. $\tan \alpha = \dfrac{7}{24} \approx 0.2917$; $\alpha \approx 16.26°$; $2\alpha \approx 32.5°$

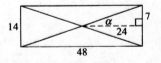

23. Let $DC = h$ and $BC = x$. $\tan 54° = \dfrac{h}{x}$ and $\tan 40° = \dfrac{h}{10 + x}$; $h = x \tan 54°$ and $h = (10 + x) \tan 40°$; $x \tan 54° = 10 \tan 40° + x \tan 40°$; $x(\tan 54° - \tan 40°) = 10 \tan 40°$; $x = \dfrac{10 \tan 40°}{\tan 54° - \tan 40°} \approx 15.6$; $h = x \tan 54° \approx 21.5$ m

24. In 20 min, the ship travels $\frac{1}{3}(4.5)$knots = 1.5 knots =

1.5(6080)ft = 9120 ft; $\tan 1.4° = \frac{h}{x}$ and

$\tan 1.1° = \frac{h}{9120 + x}$; $h = x \tan 1.4°$ and

$h = (9120 + x)\tan 1.1°$; $x \tan 1.4° = 9120 \tan 1.1° +$

$x \tan 1.1°$; $x(\tan 1.4° - \tan 1.1°) = 9120 \tan 1.1°$; $x = \dfrac{9120 \tan 1.1°}{\tan 1.4° - \tan 1.1°} \approx 33{,}428;$

$h = x \tan 1.4° \approx 33{,}428(0.0244) \approx 817$ ft

25. a. $\frac{HB}{180} = \tan 23°$; $HB = 180 \tan 23° \approx 76.4$ ft

   b. $AH = \sqrt{180^2 + 76.4^2} \approx 196$ ft; $HC = \sqrt{60^2 + 76.4^2} \approx 97.1$ ft

   c. $r = \dfrac{d}{t}$; $r = \dfrac{196}{23.5} \approx 8.3$ ft/s

   d. $r = \dfrac{d}{t}$; $r = \dfrac{97.1}{2.8} = 34.7$ ft/s; the approximation is too small.

26.

   a. $\tan 10° = \dfrac{y}{1.5}$ and $\tan 50° = \dfrac{x + y}{1.5}$; $y = 1.5 \tan 10°$ and
$y = 1.5 \tan 50° - x$; $1.5 \tan 10° = 1.5 \tan 50° - x$;
$x = 1.5(\tan 50° - \tan 10°) \approx 1.52$ mi

   b. $\dfrac{1.52}{10} = 0.152$ mi/s

27. a.

   $AC = \sqrt{13^2 - 5^2} = 12$; $\sec A = \dfrac{13}{12}$; $\csc A = \dfrac{13}{5}$

   b. If $\sin A = x$, then $\cos A = \sqrt{1 - x^2}$;

   $\sec A = \dfrac{1}{\sqrt{1 - x^2}}$.

28. a.

$PQ = \sqrt{1 + 4^2} = \sqrt{17}$; $\sin P = \dfrac{1}{\sqrt{17}} = \dfrac{\sqrt{17}}{17}$;

$\cos Q = \dfrac{1}{\sqrt{17}} = \dfrac{\sqrt{17}}{17}$

   b. If $\tan P = x$, then $PQ = \sqrt{1 + x^2}$; $\sin P = \dfrac{x}{\sqrt{1 + x^2}}$.

29. a. Since the line is tangent to the circle at P, $\overline{ST} \perp \overline{PO}$. In $\triangle OPT$, $\tan \theta = \dfrac{PT}{1}$ and

   $\sec \theta = \dfrac{OT}{1}$; thus, $PT = \tan \theta$ and $OT = \sec \theta$.

   b. In right triangle OPT, $(OT)^2 = (OP)^2 + (PT)^2$; substituting, $\sec^2\theta = 1 + \tan^2\theta$.

   c. As θ increases from 0° to 90°, PT increases from 0 to $+\infty$ and OT increases from 1 to
$+\infty$. As θ increases from 90° to 180°, PT increases from $-\infty$ to 0 and OT increases
from $-\infty$ to -1. As θ increases from 180° to 270°, PT increases from 0 to $+\infty$ and
OT decreases from -1 to $-\infty$. As θ increases from 270° to 360°, PT increases from
$-\infty$ to 0 and OT decreases from $+\infty$ to 1.

30. a. Since the line is tangent to the circle at P, $\overline{ST} \perp \overline{PO}$. In $\triangle OPS$, $\angle SOP = 90° - \theta$ and $\angle OPS = 90°$, so $\angle OSP = \theta$. In $\triangle OPS$, $\cot \theta = \dfrac{PS}{1}$, or $\cot \theta = PS$. In $\triangle OPS$,

$\csc \theta = \dfrac{OS}{1}$, or $\csc \theta = OS$.

b. Since $PS = \cot \theta$, $OS = \csc \theta$, and $\triangle OPS$ is a right triangle, by the Pythagorean Theorem, $(OS)^2 = (OP)^2 + (PS)^2$; $\csc^2\theta = 1^2 + \cot^2\theta$; $\csc^2\theta = 1 + \cot^2\theta$.

c. As θ increases from $0°$ to $90°$, PS decreases from $+\infty$ to 0 and OS decreases from $+\infty$ to 1. As θ increases from $90°$ to $180°$, PS decreases from 0 to $-\infty$ and OS increases from 1 to $+\infty$. As θ increases from $180°$ to $270°$, PS decreases from $+\infty$ to 0 and OS increases from $-\infty$ to -1. As θ increases from $270°$ to $360°$, PS decreases from 0 to $-\infty$ and OS decreases from -1 to $-\infty$.

31. If the given altitude of the quadrilateral has length t, then $\sin \beta = \dfrac{t}{a}$; $t = a \sin \beta$.

$\sin \alpha = \dfrac{t}{y} = \dfrac{a \sin \beta}{y}$; $y = \dfrac{a \sin \beta}{\sin \alpha} = a \sin \beta \csc \alpha$. Also, $\tan \alpha = \dfrac{t}{x} = \dfrac{a \sin \beta}{x}$;

$x = \dfrac{a \sin \beta}{\tan \alpha} = a \sin \beta \cot \alpha$.

32. a. Since $\sin A = \dfrac{h}{b}$, $h = b \sin A$. Also, since $\sin B = \dfrac{h}{a}$, $h = a \sin B$. Thus, $h = b \sin A = a \sin B$.

b. Area of $\triangle ABC = \dfrac{1}{2}$ (base)(height) $= \dfrac{1}{2}ch$. Since $h = b \sin A = a \sin B$,

area of $\triangle ABC = \dfrac{1}{2}ac \sin B = \dfrac{1}{2}bc \sin A$.

33. In rt. $\triangle OAC$, $\tan 28° = \dfrac{AC}{12}$; $AC = 12 \tan 28°$; area $\triangle OAC = \dfrac{1}{2} \cdot OA \cdot AC = \dfrac{1}{2} \cdot 12 \cdot$

$12 \tan 28° = 72 \tan 28°$; the area of the sector is $K = \dfrac{\theta}{360} \pi r^2 = \dfrac{28}{360} \cdot \pi \cdot 12^2 = 11.2\pi$;

the required area is $72 \tan 28° - 11.2\pi \approx 3$ cm^2.

34. a. $\triangle PQR$ is a right triangle; $\sin \alpha = \dfrac{3}{5} = 0.6$; $\alpha \approx 0.6435$;

$\beta = \dfrac{\pi}{2} - \alpha \approx 0.9273$; $\angle RPS = 2\alpha \approx 1.29$;

$\angle RQS = 2\beta \approx 1.85$.

b. $PR = \sqrt{5^2 - 3^2} = 4$;

area of $\triangle PRQ = \dfrac{1}{2} \cdot 4 \cdot 3 = 6$ sq. units;

area of sector $PUT = \dfrac{\alpha}{2\pi} \cdot \pi \cdot 2^2 = 2\alpha \approx 1.2870$; area of sector $QRT = \dfrac{\beta}{2\pi} \cdot \pi \cdot 3^2 = 4.5\beta \approx 4.1729$; area of shaded region $= 2$(area of $\triangle PRQ -$ area of sectors) $= 2(6 - (1.2870 + 4.1729)) \approx 1.08$ sq. units

35. $\sin\dfrac{\theta}{2} = \dfrac{AD}{8}$; $\sin\dfrac{\theta}{2} = \dfrac{b}{16}$; $\theta = 2\,\text{Sin}^{-1}\!\left(\dfrac{b}{16}\right)$, where b is the length in centimeters of the

triangle's base and θ is the measure of the triangle's vertex angle; Domain: $0 < b < 16$; range: $0° < \theta < 180°$.

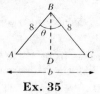

Ex. 35

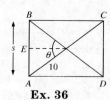

Ex. 36

36. $\sin\dfrac{\theta}{2} = \dfrac{BE}{10}$; $\sin\dfrac{\theta}{2} = \dfrac{s}{20}$; $\theta = 2\,\text{Sin}^{-1}\!\left(\dfrac{s}{20}\right)$, where s is the length in centimeters of the

rectangle's shorter side and θ is the measure of the acute angle formed by the rectangle's diagonals; Domain: $0 < s < 10\sqrt{2}$; range $0° < \theta < 90°$.

37. a. $\tan\theta = \dfrac{30{,}000}{d}$; $\theta = \text{Tan}^{-1}\!\left(\dfrac{30{,}000}{d}\right)$

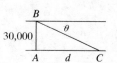

b. $\tan\theta = \dfrac{30{,}000}{60 \times 5{,}280} = 0.0947$; $\theta = 5.4°$

38. a. $a = 2 \times 4000 \times \theta = 8000\theta$; $\cos\theta = \dfrac{4000}{4000 + d}$; $\theta = \text{Cos}^{-1}\!\left(\dfrac{4000}{4000 + d}\right)$

b. $\theta = \dfrac{a}{8000}$; so $a = 8000\,\text{Cos}^{-1}\!\left(\dfrac{4000}{4000 + d}\right)$

c. Domain $d > 0$; range $0 < a < 4000\pi$

d. Yes;

$\text{Cos}^{-1}\!\left(\dfrac{4000}{4000 + 22{,}300}\right) = 11{,}344.9 > 7726$

39. If the cube has sides of length s, the diagonal of a face has length $\sqrt{s^2 + s^2} = s\sqrt{2}$. A

diagonal of the cube has length $\sqrt{2s^2 + s^2} = s\sqrt{3}$. $\cos\theta = \dfrac{\text{adjacent}}{\text{hypotenuse}} = \dfrac{s\sqrt{2}}{s\sqrt{3}} = \dfrac{\sqrt{6}}{3} \approx 0.8165$. Thus $\theta \approx 35.3°$.

40. Using the result of Ex. 39, $\alpha = 180° - 2\theta = 180° - 2\left(\text{Cos}^{-1}\dfrac{\sqrt{6}}{3}\right) \approx 109.5°$.

$\left(\text{Alternatively, } \tan\dfrac{\alpha}{2} = \dfrac{\text{opp.}}{\text{adj.}} = \dfrac{\frac{1}{2}s\sqrt{2}}{\frac{1}{2}s} = \sqrt{2} \approx 1.414, \text{ so } \dfrac{\alpha}{2} \approx 54.73° \text{ and } \alpha \approx 109.5°.\right)$

41. The diagonals divide the polygon into $n \cong \triangle$. Let h be the height and b be the base of each triangle. $\cos\theta = \dfrac{h}{r}$, so $h = r\cos\theta$. $\sin\theta = \dfrac{\frac{b}{2}}{r}$, so $\dfrac{b}{2} = r\sin\theta$ and $b = 2r\sin\theta$.

Area of $1\,\triangle = \dfrac{1}{2}bh = \dfrac{1}{2}(2r\sin\theta)(r\cos\theta) = r^2\sin\theta\cos\theta$; there are n triangles,

so the total area is $A = nr^2\sin\theta\cos\theta$.

42. When diagonals are drawn in, the polygon will be divided into $n \cong \triangle$. Let h be the height and b be the base of each triangle; $h = r$. $\tan\theta = \dfrac{\frac{b}{2}}{r}$, so $\dfrac{b}{2} = r\tan\theta$ and

$b = 2r\tan\theta$. Area of $1\triangle = \dfrac{1}{2}bh = \dfrac{1}{2}(2r\tan\theta)\cdot r = r^2\tan\theta$; there are n triangles,

so the total area is $K = nr^2\tan\theta$.

Computer Exercise · page 338

```
10 PRINT "A = AREA OF INSCRIBED POLYGON"
20 PRINT "K = AREA OF CIRCUMSCRIBED POLYGON"
30 PRINT "RADIUS = 1"
40 PRINT
50 LET PI = 3.14159
60 FOR N = 10 TO 100 STEP 10
70 PRINT N;" SIDES:"
80 LET T = PI/N
90 LET A = N * SIN(T) * COS(T)
100 LET K = N * TAN(T)
110 PRINT " A = ";A
120 PRINT " K = ";K
130 PRINT " K - A = ";K-A
140 NEXT N
150 END
```

As n gets very large, the values of A and K approach each other and therefore the area of the circle.

Class Exercises 9-2 · page 341

1. a. $K = \dfrac{1}{2}\cdot 5\cdot 8\cdot\sin 30° = 20\cdot\dfrac{1}{2} = 10 \text{ cm}^2$

  b. $K = \dfrac{1}{2}\cdot 5\cdot 8\cdot\sin 150° = 20\cdot\dfrac{1}{2} = 10 \text{ cm}^2$

2. $K = \dfrac{1}{2}\cdot 4\cdot 10\cdot\sin 30° = 20\cdot\dfrac{1}{2} = 10 \text{ sq. units}$

3. $K = \dfrac{1}{2}\cdot 4\cdot 10\cdot\sin 150° = 20\cdot\dfrac{1}{2} = 10 \text{ sq. units}$

4. $K = \dfrac{1}{2} \cdot 2 \cdot 5 \cdot \sin 45° = 5 \cdot \dfrac{\sqrt{2}}{2} = \dfrac{5\sqrt{2}}{2}$ sq. units

5. a. $5 = \dfrac{1}{2} \cdot 4 \cdot 5 \cdot \sin \theta$; $\sin \theta = \dfrac{1}{2}$ **b.** $\theta = 30°$ or $\theta = 150°$

6. The sector has area $K = \dfrac{1}{2}r^2\theta = \dfrac{1}{2} \cdot 2^2 \cdot \dfrac{\pi}{6} = \dfrac{\pi}{3}$. The triangle has area

$A = \dfrac{1}{2} \cdot r^2 \cdot \sin \theta = \dfrac{1}{2} \cdot 2^2 \cdot \sin \dfrac{\pi}{6} = \dfrac{1}{2} \cdot 2^2 \cdot \dfrac{1}{2} = 1$. The segment has area

$K - A = \dfrac{\pi}{3} - 1 \approx 0.047$.

Written Exercises 9-2 · pages 342–344

1. a. $K = \dfrac{1}{2}ab \sin C = \dfrac{1}{2} \cdot 4 \cdot 5 \cdot \sin 30° = 10 \cdot \dfrac{1}{2} = 5$ sq. units

b. $K = \dfrac{1}{2}ab \sin C = \dfrac{1}{2} \cdot 4 \cdot 5 \cdot \sin 150° = 10 \cdot \dfrac{1}{2} = 5$ sq. units

2. a. $K = \dfrac{1}{2}bc \sin A = \dfrac{1}{2} \cdot 3 \cdot 8 \cdot \sin 120° = 12 \cdot \dfrac{\sqrt{3}}{2} = 6\sqrt{3}$ sq. units

b. $K = \dfrac{1}{2}bc \sin A = \dfrac{1}{2} \cdot 3 \cdot 8 \cdot \sin 60° = 12 \cdot \dfrac{\sqrt{3}}{2} = 6\sqrt{3}$ sq. units

3. a. $K = \dfrac{1}{2}ac \sin B = \dfrac{1}{2} \cdot 6 \cdot 2 \cdot \sin 45° = 6 \cdot \dfrac{\sqrt{2}}{2} = 3\sqrt{2}$ sq. units

b. $K = \dfrac{1}{2}ac \sin B = \dfrac{1}{2} \cdot 6 \cdot 2 \cdot \sin 135° = 6 \cdot \dfrac{\sqrt{2}}{2} = 3\sqrt{2}$ sq. units

4. a. $K = \dfrac{1}{2}ab \sin C = \dfrac{1}{2} \cdot 10 \cdot 20 \cdot \sin 70° \approx 94.0$ sq. units

b. $K = \dfrac{1}{2}ab \sin C = \dfrac{1}{2} \cdot 10 \cdot 20 \cdot \sin 110° \approx 94.0$ sq. units

5. If $\angle C$ is a right angle, then the formula $K = \dfrac{1}{2}ab \sin C$ becomes

$K = \dfrac{1}{2}ab \sin 90° = \dfrac{1}{2}ab \cdot 1 = \dfrac{1}{2}ab$, or "the area of a right $\triangle$ is half the product of the legs."

6. Let point D be the vertex of the right angle in the diagram. Then $\sin \angle BCD = \dfrac{h}{a}$;

$\angle BCD = 180° - \angle BCA$, so $\sin (180° - \angle BCA) = \dfrac{h}{a}$; $\sin (180° - \angle BCA) = \sin \angle BCA$,

so $\sin \angle BCA = \dfrac{h}{a}$; $h = a \sin \angle BCA$, or $h = a \sin C$. $K = \dfrac{1}{2}bh = \dfrac{1}{2}b \cdot a \sin C = \dfrac{1}{2}ab \sin C$.

7. $K = \dfrac{1}{2}xy \sin Z = \dfrac{1}{2} \cdot 16 \cdot 25 \cdot \sin 52° \approx 158$ sq. units

8. $K = \dfrac{1}{2}rt \sin S = \dfrac{1}{2} \cdot 6 \cdot 15 \cdot \sin 125° \approx 36.9$ sq. units

9. $K = \dfrac{1}{2}ab \sin C$; $15 = \dfrac{1}{2} \cdot 12 \cdot 5 \cdot \sin C$; $\sin C = \dfrac{15}{30} = \dfrac{1}{2}$; $\angle C = 30°$ or $150°$

10. $K = \dfrac{1}{2}qr \sin P$; $9 = \dfrac{1}{2} \cdot 4 \cdot 9 \cdot \sin P$; $\sin P = \dfrac{1}{2}$; $\angle P = 30°$ or $150°$

11. In each triangle, the central angle is $\dfrac{360°}{8} = 45°$. Thus $K = \dfrac{1}{2} \cdot 40 \cdot 40 \cdot \sin 45° =$

$800 \cdot \dfrac{\sqrt{2}}{2} = 400\sqrt{2}$. The total area is $8 \cdot 400\sqrt{2} = 3200\sqrt{2}$ cm^2.

12. In each triangle, the central angle is $\dfrac{360°}{12} = 30°$. Thus $K = \dfrac{1}{2} \cdot 8 \cdot 8 \cdot \sin 30° =$

$\dfrac{1}{2} \cdot 64 \cdot \dfrac{1}{2} = 16$. The total area is $12 \cdot 16 = 192$ cm^2.

13. The area of half the parallelogram is $K = \dfrac{1}{2} \cdot 6 \cdot 7 \cdot \sin 30° = \dfrac{1}{2} \cdot 42 \cdot \dfrac{1}{2} = \dfrac{21}{2}$, so the

total area is $2 \cdot \dfrac{21}{2} = 21$ cm^2.

14. The area of half the parallelogram is $K = \dfrac{1}{2} \cdot a \cdot b \sin \theta$.

Therefore, the area of the parallelogram is $A =$

$2 \cdot \dfrac{1}{2}ab \sin \theta = ab \sin \theta$.

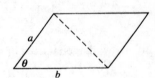

15. $K = \dfrac{1}{2}ab \sin \theta$; $K = \dfrac{1}{2} \cdot 3 \cdot 4 \cdot \sin \theta = 6 \sin \theta$;

domain $= \{\theta \mid 0 < \theta < \pi\}$;

range $= \{K \mid 0 < K \le 6\}$

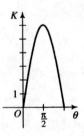

16. $K = \dfrac{1}{2}ab \sin \theta$; the maximum value of

$\sin \theta$, 1, occurs when $\theta = 90°$;

$K = \dfrac{1}{2}ab \cdot \sin 90° = \dfrac{1}{2}ab \cdot 1 = \dfrac{1}{2}ab$;

the maximum value of K is $\dfrac{1}{2}ab$. There is no minimum value.

17. a. Area of $\triangle ABC =$ area of $\triangle ABI +$ area of $\triangle ACI +$ area of $\triangle BCI = \dfrac{1}{2}(AB)r +$

$\dfrac{1}{2}(AC)r + \dfrac{1}{2}(BC)r = \dfrac{1}{2}r(AB + AC + BC)$; $\therefore$ area of $\triangle ABC = \dfrac{1}{2}r$(perimeter of

$\triangle ABC$); $r = \dfrac{2 \text{ (area of } \triangle ABC)}{\text{perimeter of } \triangle ABC}$.

b. Draw the altitude from A bisecting $\overline{BC}$ into two segments of length 8; by the

Pythagorean Thm., alt. $= \sqrt{10^2 - 8^2} = 6$; area of $\triangle ABC = \dfrac{1}{2} \cdot 16 \cdot 6 = 48$; using

the result of part (a), $r = \dfrac{2 \cdot 48}{10 + 10 + 16} = \dfrac{96}{36} = 2\dfrac{2}{3}$.

18. a. Draw the altitude from B bisecting $\overline{AC}$ into two segments of length x; by the

Pythagorean Thm., alt. $= \sqrt{100 - x^2}$; area of $\triangle ABC = \dfrac{1}{2}(2x)\sqrt{100 - x^2} =$

$x\sqrt{100 - x^2}$; using the formula of Exercise 17a, $r = \dfrac{2(\text{area of } \triangle ABC)}{\text{perimeter of } \triangle ABC} =$

$\dfrac{2x\sqrt{100 - x^2}}{10 + 10 + 2x} = \dfrac{x\sqrt{100 - x^2}}{10 + x}$.

b. Using a graphing calculator or computer, $x \approx 6.18$. The largest inscribed circle occurs when the third side is ≈ 12.36, *not* when the triangle is equilateral.

19. Draw diagonal $\overline{AC}$. In rt. $\triangle ADC$; $\tan C = \dfrac{8}{6} \approx 1.3333$;

$\angle ACD \approx 53.1°$; $\angle ACB = 115° - 53.1° = 61.9°$;

$AC = \sqrt{8^2 + 6^2} = 10$; area of $\triangle ACB = \dfrac{1}{2}ab \sin C =$

$\dfrac{1}{2} \cdot 11 \cdot 10 \cdot \sin 61.9° \approx 48.5$; area of $\triangle ADC = \dfrac{1}{2}bh =$

$\dfrac{1}{2} \cdot 8 \cdot 6 = 24$; area of quadrilateral $ABCD = 24 + 48.5 \approx 73$ sq. units

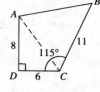

20. Draw diagonal $\overline{AC}$. In rt. $\triangle ADC$, $\tan C = \dfrac{12}{5} = 2.4$;

$\angle ACD \approx 67.4°$; $\angle ACB = 110° - 67.4° = 42.6°$;

$AC = \sqrt{12^2 + 5^2} = 13$; area of $\triangle ACB = \dfrac{1}{2}ab \sin C =$

$\dfrac{1}{2} \cdot 16 \cdot 13 \cdot \sin 42.6° \approx 70.4$; area of $\triangle ADC = \dfrac{1}{2}bh =$

$\dfrac{1}{2} \cdot 12 \cdot 5 = 30$; area of quadrilateral $ABCD = 30 + 70.4 \approx$

100 sq. units

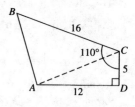

21. The sector has area $K = \dfrac{1}{2}r^2\theta = \dfrac{1}{2} \cdot 5^2 \cdot 2 = 25$; the triangle has area

$A = \dfrac{1}{2}r^2 \sin \theta = \dfrac{1}{2} \cdot 5^2 \cdot \sin 2 \approx 11.36$; the segment has area $25 - 11.36 \approx$

13.6 sq. units.

22. Let θ be the vertex angle of the isosceles triangle determined by the chord.

$\sin \dfrac{\theta}{2} = \dfrac{12}{13} \approx 0.9231$, so $\dfrac{\theta}{2} \approx 1.176$ and $\theta \approx 2.35$. The sector has area

$K = \dfrac{1}{2}r^2\theta = \dfrac{1}{2} \cdot 13^2 \cdot 2.35 \approx 198.6$; the triangle has area $A = \dfrac{1}{2}r^2 \sin \theta =$

$\dfrac{1}{2} \cdot 13^2 \cdot \sin 2.35 \approx 60.1$; the segment has area $198.6 - 60.1 \approx 139$ cm^2.

23. Area of segment $=$ area of sector $-$ area of triangle $= \dfrac{1}{2}r^2\theta - \dfrac{1}{2}r^2 \sin \theta =$

$\dfrac{1}{2}r^2(\theta - \sin \theta)$, $\therefore 95 = \dfrac{1}{2} \cdot 10^2 \cdot (\theta - \sin \theta)$; $\dfrac{95}{50} = \theta - \sin \theta$; $0 = \theta - \sin \theta - 1.9$;

using a graphing calculator or computer, $\theta \approx 2.5$ radians.

24. The area of the circle is π; $\therefore$ when the tank is $\dfrac{2}{3}$ full, the area of the

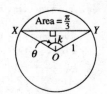

segment is $\dfrac{\pi}{3}$. Area of segment = area of sector XOY –

area of $\triangle XOY$; $\dfrac{\pi}{3} = \dfrac{1}{2} \cdot 1^2 \cdot \theta - \dfrac{1}{2} \cdot 1^2 \cdot \sin \theta = \dfrac{1}{2}(\theta - \sin \theta)$;

$\dfrac{2\pi}{3} = \theta - \sin \theta$; $0 = \theta - \sin \theta - \dfrac{2\pi}{3}$; using a graphing calculator or

computer, $\theta \approx 2.6$ radians. $k = \cos \dfrac{\theta}{2} = \cos \dfrac{2.6}{2} \approx 0.3$; the mark to show the tank is

$\dfrac{2}{3}$ full should be $1 + 0.3 = 1.3$ m from the bottom. Using the symmetry of the circle,

the mark to show the tank is $\dfrac{1}{3}$ full should be $1 - 0.3 = 0.7$ m from the bottom.

25. Measuring the amount of oil is simpler. The volume of oil is directly proportional to the depth of oil. Thus, for example, the tank is $\dfrac{1}{4}$ full when the depth of the oil is one-quarter the height of the tank.

26. Fuel tanks are generally made with rectangular cross sections which are bowed out on the ends and sides. A float is attached to an arm which is attached to a rheostat, an instrument that regulates the current sent to the fuel gauge. As the fuel is used the arm drops, and less current passes to the gauge.

27. $\cos \dfrac{\theta}{2} = \dfrac{3}{6} = 0.5$; $\dfrac{\theta}{2} = \dfrac{\pi}{3}$; $\theta = \dfrac{2\pi}{3}$; from Ex. 23, area of segment $= \dfrac{1}{2}r^2(\theta - \sin \theta) =$

$\dfrac{1}{2} \cdot 6^2 \cdot \left(\dfrac{2\pi}{3} - \sin \dfrac{2\pi}{3}\right) \approx 22.1$ sq. units

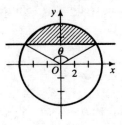

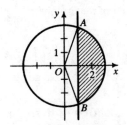

Ex. 27   **Ex. 28**

28. Let $\angle AOB = \theta$; $\cos \dfrac{\theta}{2} = \dfrac{1}{3} \approx 0.3333$; $\dfrac{\theta}{2} \approx 1.231$; $\theta \approx 2.462$; from Ex. 23, area of

segment $= \dfrac{1}{2}r^2(\theta - \sin \theta) \approx \dfrac{1}{2} \cdot 3^2 \cdot (2.462 - \sin 2.462) \approx 8.25$ sq. units

29. $x^2 + y^2 - 10x + 9 \le 0$; $x^2 - 10x + 25 +$
$y^2 \le -9 + 25$; $(x-5)^2 + y^2 \le 16$; solving the
system $x^2 + y^2 - 10x + 9 \le 0$ and $x^2 + y^2 = 9$
gives $-10x + 18 = 0$; $18 = 10x$; $x = 1.8$;
$(1.8)^2 + y^2 = 9$; $y^2 = 5.76$; $y = \pm 2.4$;
$A = (1.8, 2.4)$; $B = (1.8, -2.4)$;

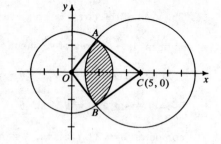

$\sin \angle AOC = \dfrac{2.4}{3} = 0.8$; $\angle AOC \approx 0.93$ and

$\angle AOB \approx 1.86$; $\sin \angle ACO = \dfrac{2.4}{4} = 0.6$; $\angle ACO \approx$

0.64 and $\angle ACB \approx 1.28$; the required region is the
sum of the segments of circles O and C; this region

has area $\dfrac{1}{2} \cdot 3^2 \cdot (1.86 - \sin 1.86) + \dfrac{1}{2} \cdot 4^2 \cdot (1.28 - \sin 1.28) \approx$

$4.5(1.86 - 0.9580) + 8(1.28 - 0.9580) \approx 6.64$ sq. units. *Note*: the formula

area of a segment $= \dfrac{1}{2} r^2 (\theta - \sin \theta)$ is derived in the solution of Ex. 23.

30. $x^2 + y^2 - 8y \le 0$; $x^2 + y^2 - 8y + 16 \le 16$; $x^2 + (y-4)^2 \le 16$;
solving the system $x^2 + y^2 - 8y = 0$ and $x^2 + y^2 = 16$ gives
$16 - 8y = 0$; $16 = 8y$; $y = 2$; $x^2 + 2^2 = 16$; $x^2 = 12$; $x = \pm 2\sqrt{3}$;

$\sin \angle AOP = \dfrac{2\sqrt{3}}{4} = \dfrac{\sqrt{3}}{2}$; $\angle AOP = \dfrac{\pi}{3}$ and $\angle AOB = \dfrac{2\pi}{3}$; similarly,

$\angle APB = \dfrac{2\pi}{3}$; since the radius of each circle is 4, the required region

is composed of two congruent segments of circles O and P; this

region has area $2 \cdot \dfrac{1}{2} \cdot 4^2 \left(\dfrac{2\pi}{3} - \sin \dfrac{2\pi}{3} \right) = 16 \left(\dfrac{2\pi}{3} - \dfrac{\sqrt{3}}{2} \right) \approx 19.7$ sq. units. *Note*: The

formula area of a segment $= \dfrac{1}{2} r^2 (\theta - \sin \theta)$ is derived in the solution of Ex. 23.

31. a. area of $\triangle BCD = \dfrac{1}{2} ax \sin \alpha = \dfrac{1}{2} ax \sin 60° = \dfrac{1}{2} ax \cdot \dfrac{\sqrt{3}}{2} = \dfrac{\sqrt{3}}{4} ax$;

area of $\triangle ACD = \dfrac{1}{2} bx \sin \beta = \dfrac{1}{2} bx \sin 60° = \dfrac{1}{2} bx \cdot \dfrac{\sqrt{3}}{2} = \dfrac{\sqrt{3}}{4} bx$

b. area of $\triangle ABC = \dfrac{1}{2} ab \sin (\alpha + \beta) = \dfrac{1}{2} ab \sin 120° = \dfrac{1}{2} ab \cdot \dfrac{\sqrt{3}}{2} = \dfrac{\sqrt{3}}{4} ab$

c. From parts (a) and (b), $\dfrac{\sqrt{3}}{4} ax + \dfrac{\sqrt{3}}{4} bx = \dfrac{\sqrt{3}}{4} ab$; $ax + bx = ab$; $x(a+b) = ab$;

$x = \dfrac{ab}{a+b}$.

d. Yes. If $a = b$, then in $\triangle ABC$, $\angle A = \angle B = \dfrac{180° - 60° - 60°}{2} = 30°$. Therefore, $\triangle BCD$

is a 30°-60°-90° triangle and $x = \dfrac{1}{2} a$. This agrees with the formula in part (c):

$x = \dfrac{ab}{a+b} = \dfrac{a \cdot a}{a+a} = \dfrac{a^2}{2a} = \dfrac{1}{2} a$.

32. a. $V = \dfrac{4}{3} \pi r^3 = \dfrac{4}{3} \pi \cdot 1^3 = \dfrac{4}{3} \pi$

b. When the tank is $\dfrac{3}{4}$ full, the volume of the cap is $\dfrac{1}{4}V = \dfrac{1}{4} \cdot \dfrac{4}{3}\pi = \dfrac{\pi}{3}$; $\dfrac{\pi}{3} =$

$\dfrac{2\pi}{3} - \pi \cos\dfrac{\alpha}{2} + \dfrac{\pi}{3}\cos^3\dfrac{\alpha}{2}$; using a graphing calculator or computer,

$\alpha \approx 2.43$ radians.

c.

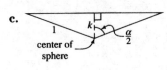

$k = \cos\dfrac{\alpha}{2} = \cos\dfrac{2.43}{2} \approx 0.35$; the mark to show the tank

is $\dfrac{3}{4}$ full should be $1 + 0.35 = 1.35$ m from the bottom.

Using the symmetry of the circle, the marks to show the

tank is $\dfrac{1}{2}$ full and $\dfrac{1}{4}$ full should be 1 m and $1 - 0.35 =$

0.65 m from the bottom, respectively.

33.

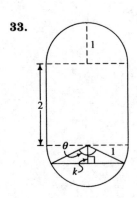

The radius of each semicircle is 1 m; $\therefore$ the area of the cross
section = area of the square + 2(area of 1 semicircle) =

$2^2 + 2\left(\dfrac{1}{2}\pi \cdot 1^2\right) = 4 + \pi$. 10% of the cross section's area is

$0.1(4 + \pi) \approx 0.7142 < \dfrac{\pi}{2}$; $\therefore$ the required cross section is a

segment of the lower semicircle. From Ex. 23, area of segment =

$\dfrac{1}{2}r^2(\theta - \sin\theta)$; $0.1(4 + \pi) = \dfrac{1}{2}\cdot 1^2(\theta - \sin\theta)$;

$0.2(4 + \pi) = \theta - \sin\theta$; $0 = \theta - \sin\theta - 0.8 - 0.2\pi$; using a

graphing calculator or computer, $\theta \approx 2.23$; $k = \cos\dfrac{\theta}{2} =$

$\cos\dfrac{2.23}{2} \approx 0.44$; the mark to show the tank is 10% full should be

$1 - 0.44 = 0.56$ m from the bottom.

34. In the diagram, circle O represents the field with
the goat tethered at point P; circle P represents
the goat's range; the region $APBM$ represents the
section of the field over which the goat can graze.

Find r so that area of $APBM = \dfrac{1}{2}\cdot\pi\cdot 1^2 = \dfrac{\pi}{2}$;

area of $APM = \dfrac{1}{2}(\text{area of } APBM) = \dfrac{1}{2}\cdot\dfrac{\pi}{2} = \dfrac{\pi}{4}$.

area of APM = area of sector APM + area of

segment $AP = \dfrac{1}{2}\cdot r^2\cdot\beta + \dfrac{1}{2}\cdot 1^2(\alpha - \sin\alpha) =$

$\dfrac{\beta r^2}{2} + \dfrac{\alpha - \sin\alpha}{2}$; $\therefore$ find r so that $\dfrac{\pi}{4} = \dfrac{\beta r^2}{2} + \dfrac{\alpha - \sin\alpha}{2}$. $\overset{\frown}{AP} = \alpha$; $\overset{\frown}{AN} = \pi - \alpha$;

$\beta = \angle APN = \dfrac{1}{2}(\pi - \alpha)$; $\therefore \dfrac{\pi}{4} = \dfrac{(\pi - \alpha)r^2}{4} + \dfrac{\alpha - \sin\alpha}{2}$; $0 = (\pi - \alpha)r^2 +$

$2(\alpha - \sin\alpha) - \pi$. In isosceles $\triangle APO$, the altitude from O to $\overline{PA}$ will bisect $\angle POA$ and

$\overline{PA}$; $\therefore \sin\dfrac{\alpha}{2} = \dfrac{r}{2}$; $r = 2\sin\dfrac{\alpha}{2}$; $r^2 = 4\sin^2\dfrac{\alpha}{2}$; $\therefore 0 = 4(\pi - \alpha)\sin^2\dfrac{\alpha}{2} +$

$2(\alpha - \sin\alpha) - \pi$; using a graphing calculator or computer, $\alpha \approx 1.24$; $r =$

$2\sin\dfrac{\alpha}{2} \approx 1.16$.

35. In order for $\triangle ORP$ to exist, $0 < \theta < \dfrac{\pi}{2}$. Find θ so that area of sector QOP = area of $\triangle ROP$ − area of sector QOP, or 2(area of sector QOP) = area of $\triangle ROP$;

$2\left(\dfrac{1}{2} \cdot 1^2 \cdot \theta\right) = \dfrac{1}{2} \cdot 1 \cdot RP$; $RP = \tan \theta$; $\therefore \theta = \dfrac{1}{2} \tan \theta$; $2\theta = \tan \theta$; $0 = \tan \theta - 2\theta$; using a graphing calculator or computer, $\theta \approx 1.17$.

36. In order for $\triangle ORP$ to exist, $0 < \theta < \dfrac{\pi}{2}$. Length of $\overarc{QP} = 1 \cdot \theta = \theta$; $\cos \theta = \dfrac{1}{OR}$;

$OR = \dfrac{1}{\cos \theta}$; $QR = OR - OQ = \dfrac{1}{\cos \theta} - 1$; $\therefore$ find θ so that $\theta = \dfrac{1}{\cos \theta} - 1$;

$\theta \cos \theta = 1 - \cos \theta$; $\theta \cos \theta + \cos \theta - 1 = 0$; using a graphing calculator or computer, $\theta \approx 1.07$.

Activity 1 · page 346

  a. No

  b. Yes; 1

  c. Yes; 2

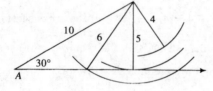

Activity 2 · page 346

  a. $\dfrac{\sin B}{b} = \dfrac{\sin A}{a}$; $\dfrac{\sin B}{10} = \dfrac{\sin 30°}{4}$; $\sin B = \dfrac{10 \cdot \frac{1}{2}}{4} = 1.25$; no solution

  b. $\dfrac{\sin B}{b} = \dfrac{\sin A}{a}$; $\dfrac{\sin B}{10} = \dfrac{\sin 30°}{5}$; $\sin B = \dfrac{10 \cdot \frac{1}{2}}{5} = 1$; $\angle B = 90°$

  c. $\dfrac{\sin B}{b} = \dfrac{\sin A}{a}$; $\dfrac{\sin B}{10} = \dfrac{\sin 30°}{6}$; $\sin B = \dfrac{10 \cdot \frac{1}{2}}{6} = \dfrac{5}{6}$; $\angle B = 56.4°, 123.6°$

Class Exercises 9-3 · pages 346–347

  1. In any triangle, the ratio of the sine of any angle to the length of the opposite side is constant.

  2. $\dfrac{\sin A}{a} = \dfrac{\frac{1}{2}}{1} = \dfrac{1}{2}$; $\dfrac{\sin B}{b} = \dfrac{\frac{\sqrt{3}}{2}}{\sqrt{3}} = \dfrac{1}{2}$; $\dfrac{\sin C}{c} = \dfrac{\sin 90°}{2} = \dfrac{1}{2}$; yes

3. $\angle B < 90°$. Since $\angle A + \angle B + \angle C = 180°$ and $\angle A \geq 90°$, $\angle B + \angle C < 90°$ and so $\angle B < 90°$.

4. $b > c$. If two angles of a triangle are unequal, then the sides opposite those angles are unequal in the same order.

5. $\angle A > \angle B$. If two sides of a triangle are unequal, then the angles opposite those sides are unequal in the same order.

6. $\dfrac{\sin 60°}{8} = \dfrac{\sin 25°}{x}$

7. $\dfrac{\sin x°}{7} = \dfrac{\sin 48°}{9}$

8. $\dfrac{\sin 112°}{7} = \dfrac{\sin 40°}{x}$

9. a.

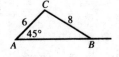

b.

c.

10. $\dfrac{\sin A}{a} = \dfrac{\sin B}{b}$; $\dfrac{\sin 45°}{8} = \dfrac{\sin B}{6}$; $\sin B = \dfrac{6 \cdot \dfrac{\sqrt{2}}{2}}{8} = \dfrac{3\sqrt{2}}{8}$; $\angle B \approx 32.0°$ or $\angle B \approx 148.0°$; however, since $\angle A = 45°$ and $\angle A + \angle B + \angle C = 180°$, $\angle B$ cannot have measure 148.0°. Thus, $\angle B \approx 32.0°$.

11. $\dfrac{\sin Y}{y} = \dfrac{\sin X}{x}$; $\dfrac{\sin Y}{8} = \dfrac{\sin 30°}{3}$; $\sin Y = \dfrac{8 \cdot \dfrac{1}{2}}{3} = \dfrac{4}{3}$; since $\sin Y$ must be ≤ 1, there is no solution for Y and $\therefore$ no $\triangle$ with the given measurements.

Written Exercises 9-3 · pages 347–349

1. a. If $a < b \sin A$, a is too short; $0\triangle$

b. If $a = b \sin A$, $\triangle ABC$ is right; $1\triangle$

c. If $b \sin A < a < b$, there is an acute $\triangle$ and an obtuse $\triangle$; $2\triangle$

d. If $a \geq b$, $\triangle ABC$ is obtuse; $1\triangle$

2. a. $b \sin A = 4 \sin 22° = 1.5 < 2 < 4$; by Ex. 1c there are $2\triangle$.

b. $c \sin B = 6 \sin 30° = 3 = b$; by Ex. 1b there is one rt. $\triangle$.

c. $7 > 5$; by Ex. 1d there is $1\triangle$.

d. $b \sin C = 4 \sin 76° = 3.9$; $3 < 3.9$; by Ex. 1a there is no $\triangle$.

3. $\dfrac{\sin 45°}{14} = \dfrac{\sin 60°}{b}$; $\dfrac{\frac{\sqrt{2}}{2}}{14} = \dfrac{\frac{\sqrt{3}}{2}}{b}$; $b = \dfrac{14\sqrt{3}}{\sqrt{2}} = 7\sqrt{6} \approx 17.1$; $\angle C = 180° - 45° - 60° = 75°$;

$\dfrac{\sin 75°}{c} = \dfrac{\sin 45°}{14}$; $c = \dfrac{14 \sin 75°}{\sin 45°} \approx 19.1$

4. $\dfrac{\sin 30°}{9} = \dfrac{\sin 45°}{c}$; $\dfrac{\frac{1}{2}}{9} = \dfrac{\frac{\sqrt{2}}{2}}{c}$; $c = 9\sqrt{2} \approx 12.7$; $\angle A = 180° - 30° - 45° = 105°$;

$\dfrac{\sin 105°}{a} = \dfrac{\sin 30°}{9}$; $a = \dfrac{9 \sin 105°}{\sin 30°} \approx 17.4$

5. $\dfrac{\sin 30°}{4} = \dfrac{\sin 135°}{a}$; $\dfrac{\frac{1}{2}}{4} = \dfrac{\frac{\sqrt{2}}{2}}{a}$; $a = 4\sqrt{2} \approx 5.66$; $\angle C = 180° - 30° - 135° = 15°$;

$\dfrac{\sin 15°}{c} = \dfrac{\sin 30°}{4}$; $c = \dfrac{4 \sin 15°}{\sin 30°} \approx 2.07$

6. $\angle C = 180° - 60° - 75° = 45°$; $\dfrac{\sin 45°}{10} = \dfrac{\sin 60°}{a}$; $\dfrac{\frac{\sqrt{2}}{2}}{10} = \dfrac{\frac{\sqrt{3}}{2}}{a}$; $a = \dfrac{10\sqrt{3}}{\sqrt{2}} = 5\sqrt{6} \approx 12.2$;

$\dfrac{\sin 45°}{10} = \dfrac{\sin 75°}{b}$; $b = \dfrac{10 \sin 75°}{\sin 45°} \approx 13.7$

7. $\dfrac{\sin 25°}{2} = \dfrac{\sin B}{3}$; $\sin B = \dfrac{3 \sin 25°}{2} \approx 0.6339$; $\angle B \approx 39.3°$ or $\angle B \approx 140.7°$.

(1) If $\angle B = 39.3°$, then $\angle A = 180° - 25° - 39.3° = 115.7°$; $\dfrac{\sin 115.7°}{a} = \dfrac{\sin 25°}{2}$;

$a = \dfrac{2 \sin 115.7°}{\sin 25°} \approx 4.26$. (2) If $\angle B = 140.7°$, then $\angle A = 180° - 25° -$

$140.7° = 14.3°$; $\dfrac{\sin 14.3°}{a} = \dfrac{\sin 25°}{2}$; $a = \dfrac{2 \sin 14.3°}{\sin 25°} \approx 1.17$.

8. $\dfrac{\sin 36°}{8} = \dfrac{\sin A}{10}$; $\sin A = \dfrac{10 \sin 36°}{8} \approx 0.7347$; $\angle A \approx 47.3°$ or $\angle A \approx 132.7°$.

(1) If $\angle A = 47.3°$, then $\angle C = 180° - 36° - 47.3° = 96.7°$; $\dfrac{\sin 96.7°}{c} = \dfrac{\sin 36°}{8}$;

$c = \dfrac{8 \sin 96.7°}{\sin 36°} \approx 13.5$. (2) If $\angle A = 132.7°$, then $\angle C = 180° - 36° - 132.7° =$

$11.3°$; $\dfrac{\sin 11.3°}{c} = \dfrac{\sin 36°}{8}$; $c = \dfrac{8 \sin 11.3°}{\sin 36°} \approx 2.67$.

9. $\dfrac{\sin 76°}{12} = \dfrac{\sin B}{4}$; $\sin B = \dfrac{4 \sin 76°}{12} \approx 0.3234$; $\angle B \approx 18.9°$. (Reject 161.1° since

$161.1° + 76° > 180°$.) $\angle C = 180° - 76° - 18.9° = 85.1°$; $\dfrac{\sin 85.1°}{c} = \dfrac{\sin 76°}{12}$;

$c = \dfrac{12 \sin 85.1°}{\sin 76°} \approx 12.3$

10. $\dfrac{\sin 130°}{15} = \dfrac{\sin C}{11}$; $\sin C = \dfrac{11 \sin 130°}{15} \approx 0.5617$; $\angle C \approx 34.2°$. (Reject 145.8° since

$130° + 145.8° > 180°$.) $\angle A = 180° - 130° - 34.2° = 15.8°$; $\dfrac{\sin 15.8°}{a} = \dfrac{\sin 130°}{15}$;

$a = \dfrac{15 \sin 15.8°}{\sin 130°} \approx 5.33$

11. $\triangle ABC$ is isosceles; $\angle B = 88°$; $\angle A = 180° - 2 \cdot 88° = 4°$; $\dfrac{\sin 4°}{a} = \dfrac{\sin 88°}{7}$;

$a = \dfrac{7 \sin 4°}{\sin 88°} \approx 0.489$

12. $\dfrac{\sin C}{10} = \dfrac{\sin 95°}{13}$; $\sin C = \dfrac{10 \sin 95°}{13} \approx 0.7663$; $\angle C \approx 50.0°$;

$\angle B = 180° - 95° - 50.0° = 35.0°$; $\dfrac{\sin 35.0°}{b} = \dfrac{\sin 95°}{13}$; $b = \dfrac{13 \sin 35°}{\sin 95°} \approx 7.48$

13. $\dfrac{\sin 40°}{6} = \dfrac{\sin A}{12}$; $\sin A = \dfrac{12 \sin 40°}{6} \approx 1.2856$; since $\sin A > 1$, there is no solution.

14. $\dfrac{\sin 112°}{5} = \dfrac{\sin A}{7}$; $\sin A = \dfrac{7 \sin 112°}{5} \approx 1.298$; since $\sin A > 1$, there is no solution.

15. $\dfrac{\sin 140°}{r} = \dfrac{\sin S}{\frac{3}{4}r}$; $\dfrac{0.6428}{1} = \dfrac{\sin S}{0.75}$; $\sin S \approx 0.4821$; $\angle S \approx 28.8°$. (Reject 151.2° since

$140° + 151.2° > 180°$.) $\angle T = 180° - 140° - 28.8° = 11.2°$.

16. $\dfrac{\sin 120°}{\frac{4}{3}e} = \dfrac{\sin E}{e}$; $\dfrac{\frac{\sqrt{3}}{2}}{\frac{4}{3}} = \sin E$; $\sin E = \dfrac{3\sqrt{3}}{8} \approx 0.6495$; $\angle E \approx 40.5°$. (Reject 139.5°.)

$\angle D = 180° - 120° - 40.5° = 19.5°$.

17. In $\triangle ABF$, $\angle F = 180° - 54° - 31° = 95°$; $\dfrac{\sin 95°}{30} = \dfrac{\sin 31°}{AF}$; $AF = \dfrac{30 \sin 31°}{\sin 95°} \approx 15.5$ km

18. Since the measure of $\angle SPQ$ is less than that of $\angle SQP$, $SQ < SP$; $\angle S = 180° -$

$44° - 66° = 70°$; $\dfrac{\sin 70°}{16} = \dfrac{\sin 44°}{SQ}$; $SQ = \dfrac{16 \sin 44°}{\sin 70°} \approx 11.8$ km

19. Draw a rt. $\triangle$ with acute $\angle A$, side opposite $\angle A = 3$, and side adjacent to $\angle A = 4$. Then

the hyp. $= 5$ and $\sin A = \dfrac{3}{5}$. Similarly, $\sin B = \dfrac{\sqrt{2}}{2}$. In $\triangle ABC$, $\dfrac{\sin A}{a} = \dfrac{\sin B}{b}$;

$\dfrac{\frac{3}{5}}{10} = \dfrac{\frac{\sqrt{2}}{2}}{b}$; $\dfrac{3}{5}b = 5\sqrt{2}$; $b = \dfrac{25\sqrt{2}}{3}$.

20. Draw a rt. $\triangle$ with acute $\angle A$, side adjacent to $\angle A = 1$, and hyp. $= 2$. The side opposite

$\angle A = \sqrt{3}$ and $\sin A = \dfrac{\sqrt{3}}{2}$. Similarly, $\sin B = \dfrac{\sqrt{15}}{4}$. In $\triangle ABC$, $\dfrac{\sin A}{a} = \dfrac{\sin B}{b}$;

$\dfrac{\frac{\sqrt{3}}{2}}{6} = \dfrac{\frac{\sqrt{15}}{4}}{b}$; $\dfrac{\sqrt{3}}{2}b = \dfrac{3\sqrt{15}}{2}$; $b = 3\sqrt{5}$.

21. a. $\angle BSP = 180° - 63° = 117°$; $\dfrac{\sin 28°}{PS} = \dfrac{\sin 117°}{3000}$; $PS = \dfrac{3000 \sin 28°}{\sin 117°} \approx 1580$ yd

b. The ship will be closest to the peninsula when $\angle SEP = 90°$; then $\sin 63° = \dfrac{PE}{1580}$;

$PE \approx 1410$ yd

c. $\angle P = 180° - 28° - 117° = 35°$; $\dfrac{\sin 35°}{BS} = \dfrac{\sin 117°}{3000}$; $BS = \dfrac{3000 \sin 35°}{\sin 117°} \approx 1930$ yd;

the ship travels 1930 yd in 10 min, or at a rate of 193 yd/min.

d. $\dfrac{193 \text{ yd/min}}{6080 \text{ ft/h}} = \dfrac{193 \cdot 3 \cdot 60 \text{ ft/h}}{6080 \text{ ft/h}} \approx 5.71$ knots

22. a. $\angle TPQ = 78° - 27° = 51°$; $\angle NPQ$ and $\angle NQP$ are supplementary; $\angle PQT = 180° -$

$78° - 43° = 59°$; $\angle T = 180° - 51° - 59° = 70°$; $\dfrac{\sin 59°}{PT} = \dfrac{\sin 70°}{180}$;

$PT = \dfrac{180 \sin 59°}{\sin 70°} \approx 164$ m

b. Point X on $\overline{PQ}$ will be closest to T when $\angle PXT = 90°$; then $\cos 51° = \dfrac{PX}{164}$;

$PX \approx 103$ m.

23. a. $\angle P$ and $\angle C$ are both inscribed angles and intercept the same arc, $\overparen{AB}$, so they have
the same measure.

b. BP is a diameter so $\angle BAP$ is a right angle. Thus in $\triangle ABP$, $\sin P = \dfrac{AB}{BP}$.

c. Since $\angle C = \angle P$, $\sin C = \sin P$, so $\sin C = \dfrac{AB}{BP}$. Thus $BP = \dfrac{AB}{\sin C} = \dfrac{c}{\sin C}$. Since

$\dfrac{\sin A}{a} = \dfrac{\sin B}{b} = \dfrac{\sin C}{c}$, $\dfrac{c}{\sin C} = \dfrac{b}{\sin B} = \dfrac{a}{\sin A}$, so each ratio $= BP$, the diameter of the
circumcircle.

24. Let $\angle A = 50°$, $\angle B = 60°$, and $\angle C = 70°$. Then, by Ex. 23, $16 = \dfrac{a}{\sin 50°} = \dfrac{b}{\sin 60°} =$

$\dfrac{c}{\sin 70°}$. $a = 16 \sin 50° \approx 12.3$ cm; $b = 16 \sin 60° \approx 13.9$ cm; $c = 16 \sin 70° \approx$
15.0 cm.

25. a. $K = \frac{1}{2}ab \sin C$ and $\frac{a}{\sin A} = \frac{b}{\sin B}$, so $b = \frac{a \sin B}{\sin A}$. Thus, $K = \frac{1}{2}a\left(\frac{a \sin B}{\sin A}\right)\sin C =$

$\frac{1}{2}\left(\frac{\sin B \sin C}{\sin A}\right)a^2$.

b. $K = \frac{1}{2}\left(\frac{\sin A \sin B}{\sin C}\right)c^2$ and $K = \frac{1}{2}\left(\frac{\sin A \sin C}{\sin B}\right)b^2$.

26. The area of $\triangle ABC$ is $K = \frac{1}{2}\left(\frac{\sin B \sin C}{\sin A}\right)a^2$ (Ex. 25) and the diameter of the

circumcircle is $d = \frac{a}{\sin A}$ (Ex. 23). Thus $r = \frac{a}{2 \sin A}$ and the area of the circle is

$A = \pi r^2 = \pi\left(\frac{a^2}{4 \sin^2 A}\right)$. $\dfrac{K}{A} = \dfrac{\frac{1}{2}\left(\frac{\sin B \sin C}{\sin A}\right)a^2}{\pi\left(\frac{a^2}{4 \sin^2 A}\right)} = \frac{1}{2\pi}\left(\frac{\sin B \sin C}{\sin A}\right)a^2 \cdot \frac{4 \sin^2 A}{a^2} =$

$\frac{2}{\pi}\sin A \sin B \sin C$.

27. a. We know $\theta = 180 - \phi$, so since $\sin \phi = \sin(180 - \phi)$ we have $\sin \phi = \sin \theta$.

b. In the triangle on the left, $\frac{\sin \theta}{a} = \frac{\sin \alpha}{x}$; thus $\frac{a}{x} = \frac{\sin \theta}{\sin \alpha}$. In the triangle on the right,

$\frac{\sin \alpha}{y} = \frac{\sin \phi}{b}$, so $\frac{b}{y} = \frac{\sin \phi}{\sin \alpha}$.

c. Since $\sin \theta = \sin \phi$, $\frac{\sin \theta}{\sin \alpha} = \frac{\sin \phi}{\sin \alpha}$. By substitution, $\frac{a}{x} = \frac{b}{y}$ and $\frac{a}{b} = \frac{x}{y}$.

28. a. Since P is the midpoint of $\overline{OQ}$, $OP = PQ$. Let $x = OP = PQ$. In $\triangle OAP$,

$\frac{\sin A}{x} = \frac{\sin O}{1}$, so $\sin O = \frac{\sin A}{x}$. In $\triangle OCQ$, $\frac{\sin C}{2x} = \frac{\sin O}{2}$, so $\frac{\sin C}{2x} = \frac{\frac{\sin A}{x}}{2}$;

$\frac{\sin C}{2x} = \frac{\sin A}{2x}$; $\sin C = \sin A$; $\angle C = \angle A$. Hence $\overline{AP} \parallel \overline{CQ}$. Since P bisects side

$\overline{OQ}$ of $\triangle OCQ$, we know A bisects side $\overline{OC}$. Therefore A is the midpoint of $\overline{OC}$.

b. B is the midpoint of $\overline{OD}$.

29. In $\triangle ADE$, $\frac{\sin \angle DAE}{DE} = \frac{\sin \angle D}{AE}$; let $\angle DAB = \alpha$; then $\angle CAE = \alpha$ and, using the

remote interior angle theorem, $\angle D = x° - \alpha$; thus, $\frac{\sin (90° + \alpha)}{DE} = \frac{\sin (x° - \alpha)}{AE}$;

$\frac{\cos \alpha}{DE} = \frac{\sin (x° - \alpha)}{AE}$; $DE = \frac{AE \cos \alpha}{\sin (x° - \alpha)}$. In $\triangle ABC$, $\frac{BC}{\sin (90° - \alpha)} = \frac{AB}{\sin y°}$, so $\frac{BC}{\cos \alpha} = \frac{AB}{\sin y°}$

and $\cos \alpha = \frac{BC \sin y°}{AB}$; substituting, $DE = \frac{AE}{\sin (x° - \alpha)} \cdot \frac{BC \sin y°}{AB}$. In rt. $\triangle ADC$,

$\sin y° = \frac{AD}{DC}$ and $\sin (x° - \alpha) = \frac{AC}{DC}$; thus, $DE = \dfrac{BC \cdot \dfrac{AE}{AB} \cdot \dfrac{AD}{DC}}{\dfrac{AC}{DC}} = BC \cdot \frac{AE}{AB} \cdot \frac{AD}{AC} =$

$BC \tan x° \tan y°$.

Class Exercises 9-4 · page 352

1. $x^2 = 5^2 + 6^2 - 2(5)(6) \cos 35°$ 2. $x^2 = 5^2 + 10^2 - 2(5)(10) \cos 115°$

3. $\cos x° = \dfrac{5^2 + 6^2 - 7^2}{2(5)(6)}$

4. $z^2 = x^2 + y^2 - 2xy \cos Z = 4^2 + 8^2 - 2(4)(8) \cos 50° \approx 38.86; z \approx 6.23$

5. $\dfrac{\sin Y}{8} = \dfrac{\sin 50°}{6.23}$; $\sin Y = \dfrac{8 \sin 50°}{6.23} \approx 0.9837$; $\angle Y \approx 79.6°$ and

   $\angle X = 180° - 50° - 79.6° = 50.4°$, or $\angle Y = 180° - 79.6° = 100.4°$ and

   $\angle X = 180° - 50° - 100.4° = 29.6°$; $z > x \therefore \angle Z > \angle X$; $\angle Y = 100.4°$, $\angle X = 29.6°$

6. $\angle X < \angle Z < \angle Y$

7. Some students may have assumed in Ex. 5 that $\angle Y$ was acute.

Written Exercises 9-4 · pages 352–354

1. $c^2 = a^2 + b^2 - 2ab \cos C = 8^2 + 5^2 - 2(8)(5) \cos 60° = 89 - 80\left(\dfrac{1}{2}\right) = 49$;

   $c = \sqrt{49} = 7$. $\dfrac{\sin 60°}{7} = \dfrac{\sin B}{5}$; $\sin B = \dfrac{5 \sin 60°}{7} \approx 0.6186$; $\angle B \approx 38.2°$.

   $\angle A = 180° - 60° - 38.2° = 81.8°$.

2. $r^2 = s^2 + t^2 - 2st \cos R = 14^2 + 16^2 - 2(14)(16) \cos 120° =$

   $452 - 448\left(-\dfrac{1}{2}\right) = 676$; $r = \sqrt{676} = 26$. $\dfrac{\sin 120°}{26} = \dfrac{\sin S}{14}$;

   $\sin S = \dfrac{14 \sin 120°}{26} \approx 0.4663$; $\angle S \approx 27.8°$. $\angle T = 180° - 120° - 27.8° = 32.2°$.

3. $\cos Z = \dfrac{x^2 + y^2 - z^2}{2xy} = \dfrac{9^2 + 40^2 - 41^2}{2(9)(40)} = 0$; $\angle Z = 90°$; $\tan X = \dfrac{\text{opp.}}{\text{adj.}} = \dfrac{9}{40} = 0.225$;

   $\angle X \approx 12.7°$. $\angle Y = 180° - 90° - 12.7° = 77.3°$.

4. $\cos B = \dfrac{a^2 + c^2 - b^2}{2ac} = \dfrac{6^2 + 7^2 - 10^2}{2(6)(7)} \approx -0.1786$; $\angle B \approx 100.3°$. $\dfrac{\sin 100.3°}{10} = \dfrac{\sin A}{6}$;

   $\sin A \approx 0.5903$; $\angle A \approx 36.2°$. $\angle C = 180° - 100.3° - 36.2° = 43.5°$.

5. $r^2 = p^2 + q^2 - 2pq \cos R = 3^2 + 8^2 - 2(3)(8) \cos 50° \approx 42.1462$;

   $r \approx \sqrt{42.1462} \approx 6.49$. $\dfrac{\sin 50°}{6.49} = \dfrac{\sin P}{3}$; $\sin P \approx 0.3541$; $\angle P \approx 20.7°$.

   $\angle Q = 180° - 50° - 20.7° = 109.3°$.

6. $f^2 = d^2 + e^2 - 2de \cos F = 5^2 + 9^2 - 2(5)(9) \cos 115° \approx 144.036$;

   $f \approx \sqrt{144.036} \approx 12.0$. $\dfrac{\sin 115°}{12.0} = \dfrac{\sin D}{5}$; $\sin D \approx 0.3776$;

   $\angle D \approx 22.2°$. $\angle E = 180° - 115° - 22.2° = 42.8°$.

7. $\cos C = \dfrac{a^2 + b^2 - c^2}{2ab} = \dfrac{8^2 + 7^2 - 13^2}{2(8)(7)} = -0.5$; $\angle C = 120°$. $\dfrac{\sin 120°}{13} = \dfrac{\sin B}{7}$;

   $\sin B \approx 0.4663$; $\angle B \approx 27.8°$. $\angle A = 180° - 120° - 27.8° = 32.2°$.

8. $\cos Z = \dfrac{x^2 + y^2 - z^2}{2xy} = \dfrac{10^2 + 11^2 - 12^2}{2(10)(11)} = 0.35$; $\angle Z \approx 69.5°$.

$\cos Y = \dfrac{10^2 + 12^2 - 11^2}{2(10)(12)} = 0.5125$; $\angle Y \approx 59.2°$. $\angle X = 180° - 69.5° - 59.2° = 51.3°$.

9. $\cos B = \dfrac{8^2 + 12^2 - 10^2}{2(8)(12)} = 0.5625$; $AD^2 = 7^2 + 8^2 - 2(7)(8) \cos B =$

$113 - 112(0.5625) = 50$; $AD = \sqrt{50} = 5\sqrt{2} \approx 7.07$

10. $\cos B = \dfrac{5^2 + 8^2 - 7^2}{2(5)(8)} = \dfrac{1}{2}$; $AD^2 = 5^2 + 5^2 - 2(5)(5) \cos B = 50 - 50\left(\dfrac{1}{2}\right) = 25$;

$AD = \sqrt{25} = 5$

11. Let D be the midpoint of $\overline{BC}$. $\cos B = \dfrac{6^2 + 8^2 - 4^2}{2(6)(8)} = 0.875$;

$AD^2 = 6^2 + 4^2 - 2(6)(4) \cos B = 52 - 48(0.875) = 10$; $AD = \sqrt{10}$

12. Let D be the midpoint of $\overline{BC}$. $\cos B = \dfrac{5^2 + 12^2 - 13^2}{2(5)(12)} = 0$;

$AD^2 = 5^2 + 6^2 - 2(5)(6) \cos B = 61 - 60(0) = 61$; $AD = \sqrt{61}$

13. Let x be the length of the shorter diagonal; $x^2 = 6^2 + 10^2 - 2(6)(10) \cos 70° \approx$ 94.96; $x = \sqrt{94.96} \approx 9.74$ cm. Let y be the length of the longer diagonal; $y^2 = 6^2 + 10^2 - 2(6)(10) \cos 110° \approx 177.04$; $y = \sqrt{177.04} \approx 13.3$ cm.

14. $BCYX$ is a rectangle; $\therefore XY = BC = 3$. $\triangle BAX \cong \triangle CDY$;

$\therefore AX = YD = \dfrac{1}{2}(7 - 3) = 2$. From the Pythagorean thm.,

$BD = \sqrt{5^2 + 4^2} = \sqrt{41} \approx 6.40$ cm.

15. Divide the quadrilateral into two triangles as shown.

$AC^2 = 10^2 + 10^2 - 2(10)(10) \cos 132° \approx 333.83$; $AC \approx 18.27$. $\triangle ADC$ is isosceles;

$\therefore \alpha = \dfrac{1}{2}(180° - 132°) = 24°$; $\beta = 108° - 24° = 84°$. Area of $\triangle ADC =$

$\dfrac{1}{2}(10)(10) \sin 132° \approx 37.16$; area of $\triangle BAC = \dfrac{1}{2}(12)(18.27) \sin 84° \approx 109.02$;

area of quadrilateral $ABCD = 37.16 + 109.02 \approx 146$ sq. units.

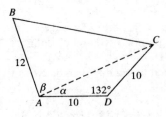

Ex. 15

16. Divide the quadrilateral into two triangles as shown.

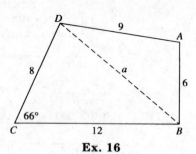

$a^2 = 8^2 + 12^2 - 2(8)(12) \cos 66° \approx 129.91$;

$a \approx 11.40$; $\cos A = \dfrac{9^2 + 6^2 - (11.40)^2}{2(9)(6)} = -0.12$;

$\angle A \approx 96.9°$; area of $\triangle ABD = \dfrac{1}{2}(9)(6) \sin 96.9° \approx$

26.8; area of $\triangle BCD = \dfrac{1}{2}(8)(12) \sin 66° \approx 43.85$;

area of quadrilateral $ABCD = 26.8 + 43.85 \approx$
$70.7 \approx 71$ sq. units

Ex. 16

17. Let D and E be the points where the altitudes from A and B, respectively, meet

the ground. $AD = \dfrac{23{,}000 \text{ ft}}{5280 \text{ ft/mi}} \approx 4.356$ mi; $BE = \dfrac{18{,}000 \text{ ft}}{5280 \text{ ft/mi}} \approx 3.409$ mi; in $\triangle TAD$,

$\sin 4° = \dfrac{4.356}{AT}$; $AT = \dfrac{4.356}{\sin 4°} \approx 62.45$ mi; in $\triangle TBE$, $\sin 2.5° = \dfrac{3.409}{BT}$;

$BT = \dfrac{3.409}{\sin 2.5°} \approx 78.15$ mi; $\angle BTA = 4° - 2.5° = 1.5°$; in $\triangle TAB$,

$x^2 = 62.45^2 + 78.15^2 - 2(62.45)(78.15) \cos 1.5° \approx 249.83$; $x \approx 15.8$ mi

18.   (not drawn to scale)

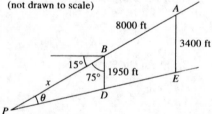

a. $\triangle PBD \sim \triangle PAE$; $\therefore$ $\dfrac{x}{1950} = \dfrac{x + 8000}{3400}$;

$3400x = 1950x + 15{,}600{,}000$;

$1450x = 15{,}600{,}000$;

$x = \dfrac{15{,}600{,}000}{1450} \approx 10{,}800$ ft

b. $PD^2 = 10{,}800^2 + 1950^2 -$
$2(10{,}800)(1950) \cos 75° \approx$
$109{,}540{,}000$; $PD \approx 10{,}466$;

in $\triangle PBD$, $\dfrac{\sin \theta}{1950} = \dfrac{\sin 75°}{10466}$;

$\sin \theta \approx 0.1800$; $\theta \approx 10.4°$

19. Given $\triangle ABC$ with sides $a = BC$, $b = AC$, and $c = AB$, if $\angle C = 0°$, then the "triangle"
collapses into a line with points A, B, and C along it. The law of cosines gives
$c^2 = a^2 + b^2 - 2ab \cos 0° = a^2 - 2ab + b^2 = (a - b)^2$. Thus, $c = \sqrt{(a - b)^2} =$
$|b - a|$, which is the distance between points A and B where the distance between the
two points is $AB = |AC - BC|$.

20. Given $\triangle ABC$ with sides $a = BC$, $b = AC$, and $c = AB$, if $\angle C = 180°$, then the "triangle"
collapses into a line with points A, B, and C along it. The law of cosines gives
$c^2 = a^2 + b^2 - 2ab \cos 180° = a^2 + b^2 - 2ab(-1) = a^2 + 2ab + b^2 = (a + b)^2$.
Thus $c = \sqrt{(a + b)^2}$, which is the distance between two points A and B where
$AB = AC + BC$.

21. a. Let $AD = x$; then $DB = 20 - x$. By Ex. 27, p. 349, $\dfrac{x}{20 - x} = \dfrac{10}{15}$; $\dfrac{x}{20 - x} = \dfrac{2}{3}$;

$3x = 40 - 2x$; $5x = 40$; $x = 8$; thus, $AD = 8$ and $DB = 20 - 8 = 12$.

b. $\cos B = \dfrac{15^2 + 20^2 - 10^2}{2(15)(20)} = 0.875$. $CD^2 = 15^2 + 12^2 - 2(15)(12) \cos B =$

$369 - 360(0.875) = 54$; $CD = \sqrt{54} = 3\sqrt{6}$

22. In $\triangle JKM$, $\cos J = \dfrac{5^2 + 4^2 - (KM)^2}{2(5)(4)} = \dfrac{25 + 16 - (KM)^2}{40} = \dfrac{41 - (KM)^2}{40}$. In $\triangle LKM$,

$\cos L = \dfrac{5^2 + 6^2 - (KM)^2}{2(5)(6)} = \dfrac{25 + 36 - (KM)^2}{60} = \dfrac{61 - (KM)^2}{60}$. Since $\cos J = -\cos L$,

$\dfrac{41 - (KM)^2}{40} = -\left(\dfrac{61 - (KM)^2}{60}\right)$; $3(41 - (KM)^2) = 2(-61 + (KM)^2)$;

$123 - 3(KM)^2 = -122 + 2(KM)^2$; $5(KM)^2 = 245$; $(KM)^2 = 49$; $KM = 7$.

In $\triangle JKL$, $\cos K = \dfrac{5^2 + 5^2 - (JL)^2}{2(5)(5)} = \dfrac{50 - (JL)^2}{50}$. In $\triangle JML$, $\cos M = \dfrac{4^2 + 6^2 - (JL)^2}{2(4)(6)} =$

$\dfrac{52 - (JL)^2}{48}$. Since $\cos K = -\cos M$, $\dfrac{50 - (JL)^2}{50} = -\left(\dfrac{52 - (JL)^2}{48}\right)$; $24(50 - (JL)^2) =$

$25(-52 + (JL)^2)$; $1200 - 24(JL)^2 = -1300 + 25(JL)^2$; $49(JL)^2 = 2500$; $(JL)^2 = \dfrac{2500}{49}$;

$JL = \dfrac{50}{7} = 7\frac{1}{7}$

23. a. In $\triangle ABC$, we have $\cos B = \dfrac{a^2 + c^2 - b^2}{2ac}$. The median divides $\overline{AB}$ into two segments,

each of length $\dfrac{c}{2}$. Thus $x^2 = a^2 + \left(\dfrac{c}{2}\right)^2 - 2a \cdot \dfrac{c}{2} \cdot \cos B =$

$a^2 + \dfrac{c^2}{4} - ac\left(\dfrac{a^2 + c^2 - b^2}{2ac}\right) = a^2 + \dfrac{c^2}{4} - \dfrac{a^2}{2} - \dfrac{c^2}{2} + \dfrac{b^2}{2} = \dfrac{a^2}{2} + \dfrac{b^2}{2} - \dfrac{c^2}{4} =$

$\dfrac{2a^2 + 2b^2 - c^2}{4}$; hence $x = \dfrac{1}{2}\sqrt{2a^2 + 2b^2 - c^2}$.

b. If $\angle C = 90°$, $a^2 + b^2 = c^2$, so $x = \dfrac{1}{2}\sqrt{2a^2 + 2b^2 - c^2} = \dfrac{1}{2}\sqrt{2c^2 - c^2} =$

$\dfrac{1}{2}\sqrt{c^2} = \dfrac{1}{2}c$.

c. In a right triangle, the length of the median drawn to the hypotenuse is half the length of the hypotenuse.

24. a. Since $\overline{CD}$ bisects $\angle ACB$, $\dfrac{b}{a} = \dfrac{p}{q}$. But $q = c - p$, so $\dfrac{b}{a} = \dfrac{p}{c - p}$; $bc - bp = ap$;

$bc = ap + bp$; $bc = p(a + b)$; $p = \dfrac{bc}{a + b}$. Since $p = c - q$, we also have $\dfrac{b}{a} = \dfrac{c - q}{q}$

and $bq = ac - aq$; $aq + bq = ac$; $(a + b)q = ac$; $q = \dfrac{ac}{a + b}$.

b. $x^2 = a^2 + q^2 - 2aq\cos B$, where $\cos B = \dfrac{a^2 + c^2 - b^2}{2ac}$. Thus, $x^2 = a^2 +$

$\left(\dfrac{ac}{a + b}\right)^2 - 2a\left(\dfrac{ac}{a + b}\right)\left(\dfrac{a^2 + c^2 - b^2}{2ac}\right) = a^2 + \left(\dfrac{ac}{a + b}\right)^2 - \dfrac{a}{a + b}(a^2 + c^2 - b^2) =$

$\dfrac{a^2(a + b)^2 + (ac)^2 - (a^2 + ab)(a^2 + c^2 - b^2)}{(a + b)^2} = \dfrac{a^3b + 2a^2b^2 - abc^2 + ab^3}{(a + b)^2} =$

$\dfrac{ab(a^2 + 2ab + b^2) - ac \cdot bc}{(a + b)^2} = ab - \dfrac{ac}{a + b} \cdot \dfrac{bc}{a + b} = ab - pq$

c. In Ex. 21, $p = \dfrac{bc}{a + b} = \dfrac{10 \cdot 20}{15 + 10} = \dfrac{200}{25} = 8$ and $q = \dfrac{ac}{a + b} = \dfrac{15 \cdot 20}{15 + 10} = \dfrac{300}{25} = 12$.

Thus, $x^2 = ab - pq = 15 \cdot 10 - 8 \cdot 12 = 54$; $x = \sqrt{54} = 3\sqrt{6}$.

25. Let θ be the measure of an angle opposite the diagonal of length y.
Then $180° - \theta$ is the measure of an angle opposite the diagonal of length x.
Thus we have $y^2 = a^2 + b^2 - 2ab \cos \theta$ and $x^2 = a^2 + b^2 - 2ab \cos (180° - \theta)$.
But $\cos (180° - \theta) = -\cos \theta$, so $x^2 = a^2 + b^2 - 2ab(-\cos \theta) = a^2 + b^2 + 2ab \cos \theta$.
Adding gives $y^2 + x^2 = a^2 + b^2 - 2ab \cos \theta + a^2 + b^2 + 2ab \cos \theta$, so
$x^2 + y^2 = 2a^2 + 2b^2$.

26. a. By the law of cosines, $\cos C = \dfrac{a^2 + b^2 - c^2}{2ab}$; $\therefore (1 + \cos C)(1 - \cos C) =$

$$\left(\frac{2ab}{2ab} + \frac{a^2 + b^2 - c^2}{2ab}\right)\left(\frac{2ab}{2ab} - \frac{a^2 + b^2 - c^2}{2ab}\right) =$$

$$\left(\frac{a^2 + 2ab + b^2 - c^2}{2ab}\right)\left(\frac{c^2 - (a^2 - 2ab + b^2)}{2ab}\right) = \frac{(a + b)^2 - c^2}{2ab} \cdot \frac{c^2 - (a - b)^2}{2ab}.$$

b. $K = \dfrac{1}{2}ab \sin C = \dfrac{1}{2}ab \sqrt{\dfrac{2s(2s - 2c)(2s - 2b)(2s - 2a)}{4a^2 b^2}} =$

$$\frac{1}{2}ab \cdot \frac{\sqrt{16s(s - c)(s - b)(s - a)}}{2ab} = \frac{1}{4} \cdot 4\sqrt{s(s - c)(s - b)(s - a)} =$$

$\sqrt{s(s - a)(s - b)(s - c)}$; thus, $K = \sqrt{s(s - a)(s - b)(s - c)}$ where $s = \dfrac{a + b + c}{2}$.

Computer Exercise · page 355

```
10 PRINT "INDICATE THE PROBLEM TYPE;"
20 PRINT "ENTER SSS, SAS, AAS, ASA, OR SSA:";
30 INPUT A$
40 PRINT
50 REM LET PI = 3.1415926353589
60 IF A$ = "SSS" THEN GOSUB 1000
70 IF A$ = "SAS" THEN GOSUB 2000
80 IF A$ = "AAS" THEN GOSUB 3000
90 IF A$ = "ASA" THEN GOSUB 4000
100 IF A$ = "SSA" THEN GOSUB 5000
110 PRINT " SIDE A =";SA
120 PRINT " SIDE B =";SB
130 PRINT " SIDE C =";SC
140 PRINT "ANGLE A =";AA
150 PRINT "ANGLE B =";AB
160 PRINT "ANGLE C =";AC
170 GOTO 6000
1000 PRINT "INPUT THE SHORTEST SIDE:";
1010 INPUT SA
1020 PRINT "INPUT THE SHORTER REMAINING SIDE:";
1030 INPUT SB
1040 PRINT "INPUT THE LONGEST SIDE:";
1050 INPUT SC
1060 PRINT
1070 LET X = (SB^2 + SC^2 - SA^2)/(2 * SB * SC)
1080 LET AA = ATN(SQR(1 - X^2)/X) * 180/PI
1090 LET X = (SA^2 + SC^2 - SB^2)/(2 * SA * SC)
```

```
1100 LET AB = ATN(SQR(1 - X^2)/X) * 180/PI
1110 LET AC =180 - AA - AB
1200 RETURN
2000 PRINT "INPUT THE ANGLE:";
2010 INPUT AC
2020 PRINT "INPUT THE SHORTER SIDE:";
2030 INPUT SA
2040 PRINT "INPUT THE LONGER SIDE:";
2050 INPUT SB
2060 PRINT
2070 LET SC = SQR(SA^2 + SB^2 -2 * SA * SB * COS(AC * PI/180))
2080 LET X = (SB^2 + SC^2 - SA^2)/(2 * SB * SC)
2090 LET AA = ATN(SQR(1 - X^2)/X) * 180/PI
2100 LET AB = 180 - AA - AC
2110 RETURN
3000 PRINT "INPUT THE SIDE:";
3010 INPUT SA
3020 PRINT "INPUT THE OPPOSITE ANGLE:";
3030 INPUT AA
3040 PRINT "INPUT THE OTHER ANGLE:";
3050 INPUT AB
3060 PRINT
3070 LET AC = 180 - AA - AB
3080 LET SB = SA * SIN(AB * PI/180)/SIN(AA * PI/180)
3090 LET SC = SA * SIN(AC * PI/180)/SIN(AA * PI/180)
3100 RETURN
4000 PRINT "INPUT THE SMALLER ANGLE:";
4010 INPUT AA
4020 PRINT "INPUT THE LARGER ANGLE:";
4030 INPUT AB
4040 PRINT "INPUT THE SIDE:";
4050 INPUT SC
4060 PRINT
4070 LET AC = 180 - AA - AB
4080 LET SA = SC * SIN(AA * PI/180)/SIN(AC * PI/180)
4090 LET SB = SC * SIN(AB * PI/180)/SIN(AC * PI/180)
4100 RETURN
5000 PRINT "INPUT THE ANGLE:";
5010 INPUT AA
5020 PRINT "INPUT THE OPPOSITE SIDE:";
5030 INPUT SA
5040 PRINT "INPUT THE OTHER SIDE:";
5050 INPUT SB
5060 PRINT
5070 IF SB * SIN(AA * PI/180)/SA <= 1.0001 THEN GOTO 5100
5080 PRINT "THERE IS NO SOLUTION."
5090 GOTO 6000
```

Continued on next page

```
5100 IF SB * SIN(AA * PI/180)/SA > 0.9999 THEN GOTO 5120
5110 IF SB * SIN(AA * PI/180)/SA < 0.9999 THEN GOTO 5160
5120 LET AB = 90
5130 LET AC = 90 - AA
5140 LET SC = SB * COS(AA * PI/180)
5150 GOTO 5500
5160 LET X = SB * SIN(AA * PI/180)/SA
5170 LET AB = ATN(X/SQR(1 - X^2)) * 180/PI
5180 LET AC = 180 - AA - AB
5190 LET SC = SA * SIN(AC * PI/180)/SIN(AA * PI/180)
5200 IF AA > 90 THEN RETURN
5210 PRINT "SOLUTION ONE"
5220 PRINT " SIDE A =";SA
5230 PRINT " SIDE B =";SB
5240 PRINT " SIDE C =";SC
5250 PRINT "ANGLE A =";AA
5260 PRINT "ANGLE B =";AB
5270 PRINT "ANGLE C =";AC
5280 PRINT
5290 PRINT "SOLUTION TWO"
5300 LET SC = SA
5310 LET SA = SB
5320 LET AC = AA
5330 LET AA = 180 - AB
5340 LET AB = 180 - AA - AC
5350 LET SB = SC * SIN(AB * PI/180)/SIN(AC * PI/180)
5500 RETURN
6000 END
```

Mixed Trigonometry Exercises · pages 355–358

1. $K = \dfrac{1}{2}qr \sin P = \dfrac{1}{2} \cdot 6 \cdot 7 \cdot \sin 50° \approx 16.1$ sq. units;

   $p^2 = q^2 + r^2 - 2qr \cos P = 6^2 + 7^2 - 2(6)(7) \cos 50° \approx 31; \; p \approx 5.57$

2. $K = \dfrac{1}{2}pq \sin R = \dfrac{1}{2} \cdot 7 \cdot 10 \cdot \sin 130° \approx 26.8$ sq. units;

   $r^2 = p^2 + q^2 - 2pq \cos R = 7^2 + 10^2 - 2(7)(10) \cos 130° \approx 238.99; \; r \approx 15.5$

3. Given $\triangle ABC$ with $BC = 3\sqrt{6}$, $AC = 6\sqrt{3}$, and $AB = 9\sqrt{2}$, $\cos C = \dfrac{a^2 + b^2 - c^2}{2ab} =$

   $\dfrac{(3\sqrt{6})^2 + (6\sqrt{3})^2 - (9\sqrt{2})^2}{2(3\sqrt{6})(6\sqrt{3})} = 0; \; \angle C = 90°$

4. Given $\triangle ABC$ with $BC = 7$, $AC = 13$, and $AB = 12$, $\cos A = \dfrac{b^2 + c^2 - a^2}{2bc} =$

   $\dfrac{13^2 + 12^2 - 7^2}{2(13)(12)} \approx 0.8462; \; \angle A \approx 32.2°$

5. $\angle T = 180° - 75° - 45° = 60°$; $\dfrac{\sin 60°}{3} = \dfrac{\sin 75°}{r}$; $r = \dfrac{3 \sin 75°}{\sin 60°} \approx 3.35$.

$\dfrac{\sin 60°}{3} = \dfrac{\sin 45°}{s}$; $\dfrac{\frac{\sqrt{3}}{2}}{3} = \dfrac{\frac{\sqrt{2}}{2}}{s}$; $s = \dfrac{3\sqrt{2}}{\sqrt{3}} = \sqrt{6} \approx 2.45$.

6. $\angle R = 180° - 100° - 30° = 50°$; $\dfrac{\sin 100°}{s} = \dfrac{\sin 50°}{8}$; $s = \dfrac{8 \sin 100°}{\sin 50°} \approx 10.3$.

$\dfrac{\sin 30°}{t} = \dfrac{\sin 50°}{8}$; $t = \dfrac{8 \sin 30°}{\sin 50°} \approx 5.22$.

7. By the law of sines, $\dfrac{\sin 60°}{4} = \dfrac{\sin B}{5}$; $\dfrac{\frac{\sqrt{3}}{2}}{4} = \dfrac{\sin B}{5}$; $\sin B = \dfrac{5\sqrt{3}}{8} \approx 1.08$;

no angle has a sine greater than 1, $\therefore$ no such triangle exists.

8. The polygon comprises 180 congruent isosceles triangles with vertex angles of

$\dfrac{360°}{180} = 2°$ and legs of length 1. Area of polygon = 180(area of each triangle) =

$180\left(\dfrac{1}{2} \cdot 1 \cdot 1 \cdot \sin 2°\right) \approx 3.141$; the value of π to three decimal places is 3.142.

9. $\sin^2 A = 1 - \cos^2 A = 1 - (-0.6)^2 = 0.64$; $\sin A = \sqrt{0.64} = 0.8$; $\tan A = \dfrac{\sin A}{\cos A} =$

$\dfrac{0.8}{-0.6} = -\dfrac{4}{3}$

10. $\cos A = \dfrac{b^2 + c^2 - a^2}{2bc} = \dfrac{10^2 + 21^2 - 17^2}{2(10)(21)} = 0.6$; $\sin^2 A = 1 - \cos^2 A = 1 - (0.6)^2 = 0.64$;

$\sin A = 0.8$

11. Let x be the horizontal length of the ramp; $\tan 8° = \dfrac{30}{x}$; $x = \dfrac{30}{\tan 8°} \approx 213.5$ in. =

$(213.5 \div 12)\text{ft} \approx 17.8$ ft

12. Let x = the length of the submarine's path; $\sin 16° = \dfrac{300}{x}$; $x = \dfrac{300}{\sin 16°} \approx 1088.4$ ft.

$r = \dfrac{d}{t} = \dfrac{1088.4}{4} = 272.1$ ft/min = $60 \cdot 272.1$ ft/h = 16,326 ft/h =

$(16,326 \div 6080)$ knots ≈ 2.69 knots

13. $\dfrac{\sin Y}{9} = \dfrac{\sin 21.1°}{6}$; $\sin Y = \dfrac{9 \sin 21.1°}{6} \approx 0.5400$; $\angle Y = 32.70°$ or $\angle Y = 180° - 32.7°$;

$\angle Y = 32.7°, 147.3°$.

14. a. $K = 2\left(\dfrac{1}{2} \cdot 8 \cdot 5 \cdot \sin 60°\right) = 40 \cdot \dfrac{\sqrt{3}}{2} = 20\sqrt{3}$ sq. units

b. $DB^2 = 5^2 + 8^2 - 2(5)(8) \cos 60° = 89 - 80\left(\dfrac{1}{2}\right) = 49$; $DB = 7$.

$AC^2 = 5^2 + 8^2 - 2(5)(8) \cos 120° = 89 - 80\left(-\dfrac{1}{2}\right) = 129$; $AC = \sqrt{129}$.

15. $K = \dfrac{1}{2}ab \sin C$; $21 = \dfrac{1}{2} \cdot 9 \cdot 14 \cdot \sin C$; $\sin C = \dfrac{21}{63} = \dfrac{1}{3}$; $\angle C \approx 19.5°$ or

$\angle C \approx 180° - 19.5° = 160.5°$

16. $\angle F = 180° - 36° - 64° = 80°$; $\dfrac{\sin 80°}{8} = \dfrac{\sin 36°}{d}$; $d = \dfrac{8 \sin 36°}{\sin 80°} \approx 4.77$.

$\dfrac{\sin 80°}{8} = \dfrac{\sin 64°}{e}$; $e \approx 7.30$.

17. a. $a^2 = 6^2 + 10^2 - 2(6)(10) \cos 120° = 136 - 120\left(-\dfrac{1}{2}\right) = 196$; $a = \sqrt{196} = 14$.

Since $\triangle ABC \sim \triangle DEF$, $\dfrac{14}{20} = \dfrac{6}{e}$; $e = \dfrac{120}{14} = \dfrac{60}{7} \approx 8.57$; $\dfrac{14}{20} = \dfrac{10}{f}$; $f = \dfrac{200}{14} = \dfrac{100}{7} \approx 14.3$.

b. The ratio of the areas of similar triangles equals the square of the ratio of

corresponding sides; $\dfrac{\text{area } \triangle ABC}{\text{area } \triangle DEF} = \left(\dfrac{BC}{EF}\right)^2 = \left(\dfrac{14}{20}\right)^2 = \left(\dfrac{7}{10}\right)^2$; $49:100$.

18. In $\square ABCD$, $AC = 8$ and $BD = 14$. Let $\overline{AC}$ and $\overline{BD}$ intersect at point X. Then, since $\overline{AC}$ and $\overline{BD}$ bisect each other, $AX = XC = 4$, $BX = XD = 7$, $\angle BXC = 60°$ (given), and so

$\angle AXB = 120°$. In $\triangle BXC$, $(BC)^2 = 4^2 + 7^2 - 2(4)(7) \cos 60° = 65 - 56\left(\dfrac{1}{2}\right) = 37$;

$BC = \sqrt{37}$. In $\triangle AXB$, $(AB)^2 = 4^2 + 7^2 - 2(4)(7) \cos 120° = 65 - 56\left(-\dfrac{1}{2}\right) = 93$;

$AB = \sqrt{93}$. Thus, perimeter $= 2\sqrt{37} + 2\sqrt{93}$.

Area $= 2\left[\dfrac{1}{2}(4)(7) \sin 60° + \dfrac{1}{2}(4)(7) \sin 120°\right] = 2\left[14 \cdot \dfrac{\sqrt{3}}{2} + 14 \cdot \dfrac{\sqrt{3}}{2}\right] = 28\sqrt{3}$ sq. units.

19. $K = \dfrac{1}{2}ab \sin C$; $12 = \dfrac{1}{2}(4)(10) \sin C$; $\sin C = \dfrac{12}{20} = 0.6$; $\angle C \approx 36.9°$ or $\angle C = 143.1°$.

(1) If $\angle C = 36.9°$, then $c^2 = 4^2 + 10^2 - 2(10)(4) \cos C = 116 - 80(0.8) = 52$;

$c = \sqrt{52} = 2\sqrt{13} \approx 7.21$. Note that the angle opposite the side of length 10 is obtuse.

(2) If $\angle C = 143.1°$, then $c^2 = 4^2 + 10^2 - 2(10)(4) \cos C = 116 - 80(-0.8) = 180$;

$c = \sqrt{180} = 6\sqrt{5} \approx 13.4$.

20. The decagon comprises 10 congruent isosceles triangles, each with base length

$\dfrac{240}{10} = 24$, vertex angle $\dfrac{360°}{10} = 36°$, base angles of $72°$, and height h. $\tan 72° = \dfrac{h}{12}$;

$h = 12 \tan 72° \approx 36.93$; area of decagon $= 10\left(\dfrac{1}{2}bh\right) = 5(24)(36.93) \approx 4430$ sq. units.

21.

$\dfrac{y}{30} = \tan 38°$; $y \approx 23.439$; $\dfrac{30}{x} = \cos 38°$; $x \approx 38.07$;

cost $= 2.50(130 + x + y) =$
$2.50(130 + 23.439 + 38.07) \approx \$478.77$

22.

$x^2 = 125^2 + 130^2 - 2(125)(130) \cos 80° \approx$

26881.4342; $x \approx 163.96$. $\dfrac{\sin \theta}{125} = \dfrac{\sin 80°}{163.96}$;

$\sin \theta \approx .7508$; $\theta \approx 48.66°$; $\alpha = 120° - 48.66° = 71.34°$.

Area $= \dfrac{1}{2}(125)(130) \sin 80° +$

$\dfrac{1}{2}(163.96)(100) \sin 71.34° \approx 15{,}769 \text{ ft}^2$;

$\dfrac{15{,}769 \text{ ft}^2}{43{,}560 \text{ ft}^2/\text{acre}} \approx 0.362 \text{ acres}$;

tax $= 115(0.362) = \$41.63$

23. Let A be the first sighting point, let B be the second sighting point, and let C be the

island. $\angle C = 28° - 15° = 13°$; in 10 min the ship travels $\dfrac{1}{6} \cdot 6 = 1$ nautical mi;

$\therefore AB = 1$; $\dfrac{\sin 15°}{BC} = \dfrac{\sin 13°}{1}$; $BC \approx 1.15$ nautical mi $\approx 1.15 \cdot \dfrac{6080 \text{ ft}}{5280 \text{ ft/mi}} \approx$

1.32 mi (1 nautical mi $= 6080$ ft).

24. a. $\overline{AH}$ is tangent to circle C; $\therefore \angle AHC$ is a rt. $\angle$; by the Pythagorean thm.,

$AH = \sqrt{4006^2 - 4000^2} \approx 219.2$ mi.

b. $\cos \angle HCS = \dfrac{4000}{4006} \approx 0.9985$; $\angle HCS \approx 0.05474$ radians; length of

$\overset{\frown}{HS} = r\theta = 4000(0.05474) \approx 219.0$ mi

25. If $\tan A = 1$, then $\angle A = 45°$ and $\sin A = \dfrac{\sqrt{2}}{2}$. In a rt. $\triangle$ with $\tan B = \dfrac{3}{4}$, opp. leg $= 3$,

adj. leg $= 4$, and hyp. $= 5$; $\sin B = \dfrac{3}{5}$. $\dfrac{\sin B}{b} = \dfrac{\sin A}{a}$; $\dfrac{\frac{3}{5}}{18} = \dfrac{\frac{\sqrt{2}}{2}}{a}$; $\dfrac{3}{5}a = 9\sqrt{2}$;

$a = 15\sqrt{2}$.

26. Since $\sec F = -\sqrt{2}$, $\cos F = -\dfrac{1}{\sqrt{2}} = -\dfrac{\sqrt{2}}{2}$; thus, $\angle F = 135°$; $\tan F = \tan 135° = -1$.

27. x is in Quadrant IV; $\sin x = \dfrac{-1}{\sqrt{26}} = -\dfrac{\sqrt{26}}{26}$ and $\cos x = \dfrac{5}{\sqrt{26}} = \dfrac{5\sqrt{26}}{26}$

28. $K = \dfrac{1}{2}qr \sin P$; $84 = \dfrac{1}{2}(14)(13) \sin P$; $\sin P = \dfrac{84}{91} = \dfrac{12}{13}$. $\sin^2 P + \cos^2 P = 1$;

$\cos P = \pm\sqrt{1 - \left(\dfrac{12}{13}\right)^2} = \pm\dfrac{5}{13}$; $p^2 = 14^2 + 13^2 - 2(14)(13)\left(\pm\dfrac{5}{13}\right)$; $p^2 = 365 \mp 140$;

$p^2 = 225$ or $p^2 = 505$; $p = \sqrt{225} = 15$, or $p = \sqrt{505}$.

29. a. $\cos C = \dfrac{5^2 + 8^2 - 7^2}{2(5)(8)} = \dfrac{1}{2}$; $\angle C = 60°$. $\cos B = \dfrac{5^2 + 7^2 - 8^2}{2(5)(7)} \approx 0.1429$; $\angle B \approx 81.8°$.

$\angle A = 180° - 60° - 81.8° = 38.2°$.

b. $K = \dfrac{1}{2}ab \sin C = \dfrac{1}{2}(5)(8)\sin 60° = 20 \cdot \dfrac{\sqrt{3}}{2} = 10\sqrt{3} \approx 17.3$ sq. units.

c. Let $\overline{BD}$ be the altitude; in $\triangle BCD$, $\sin C = \dfrac{BD}{BC}$; $\sin 60° = \dfrac{BD}{5}$;

$BD = 5 \cdot \dfrac{\sqrt{3}}{2} = \dfrac{5\sqrt{3}}{2} \approx 4.33$

d. Let x be the length of the median. By Ex. 23, p. 354, $x = \dfrac{1}{2}\sqrt{2a^2 + 2c^2 - b^2} =$

$\dfrac{1}{2}\sqrt{2(25) + 2(49) - 64} = \dfrac{1}{2}\sqrt{84} = \sqrt{21} \approx 4.58$

e. Let $y =$ the required length. By Ex. 24, p. 354, $p = \dfrac{ab}{a+c} = \dfrac{5(8)}{5+7} = \dfrac{40}{12} = \dfrac{10}{3}$ and

$q = \dfrac{bc}{a+c} = \dfrac{8(7)}{5+7} = \dfrac{56}{12} = \dfrac{14}{3}$. Then $y^2 = ac - pq = 5(7) - \dfrac{10}{3}\left(\dfrac{14}{3}\right) = 35 - \dfrac{140}{9} =$

$\dfrac{175}{9}$; $y = \sqrt{\dfrac{175}{9}} = \dfrac{5}{3}\sqrt{7} \approx 4.41$.

f. By Ex. 23, p. 349, diameter $= \dfrac{c}{\sin C} = \dfrac{7}{\sin 60°} = 7 \div \dfrac{\sqrt{3}}{2} = \dfrac{14\sqrt{3}}{3}$; $R = \dfrac{7\sqrt{3}}{3} \approx 4.04$

g. By Ex. 17, p. 342, $r = \dfrac{2(\text{area of } \triangle ABC)}{\text{perimeter of } \triangle ABC} = \dfrac{2 \cdot 10\sqrt{3}}{5+8+7}$ (from part **(b)**) $= \dfrac{20\sqrt{3}}{20} =$

$\sqrt{3} \approx 1.73$

30. a. $\cos A = \dfrac{14^2 + 15^2 - 13^2}{2(14)(15)} = 0.6$; $\angle A \approx 53.1°$; $\cos B = \dfrac{13^2 + 15^2 - 14^2}{2(13)(15)} \approx 0.5077$;

$\angle B \approx 59.5°$. $\angle C \approx 180° - 53.1° - 59.5° = 67.4°$.

b. $K = \dfrac{1}{2}bc \sin A = \dfrac{1}{2}(14)(15)(0.8) = 84$ sq. units

c. Let $\overline{BD}$ be the altitude. In $\triangle ABD$, $\sin A = \dfrac{BD}{AB}$. $0.8 = \dfrac{BD}{15}$; $BD = 12$

d. Let x be the length of the median. By Ex. 23, p. 354, $x = \dfrac{1}{2}\sqrt{2a^2 + 2c^2 - b^2} =$

$\dfrac{1}{2}\sqrt{2(13)^2 + 2(15)^2 - 14^2} = \dfrac{1}{2}\sqrt{592} = \dfrac{1}{2}(4\sqrt{37}) = 2\sqrt{37} \approx 12.2$

e. Let $y =$ the required length. By Ex. 24, p. 354, $p = \dfrac{ab}{a+c} = \dfrac{13(14)}{13+15} = \dfrac{182}{28} = 6.5$

and $q = \dfrac{bc}{a+c} = \dfrac{14(15)}{13+15} = 7.5$. Then $y^2 = ac - pq = 13(15) - (6.5)(7.5) =$

146.25; $y = \dfrac{3}{2}\sqrt{65} \approx 12.1$.

f. By Ex. 23, p. 349, diameter $= \dfrac{a}{\sin A} = \dfrac{13}{0.8} = 16.25$; $R = \dfrac{1}{2}(16.25) \approx 8.13$

g. By Ex. 17, p. 342, $r = \dfrac{2(\text{area of } \triangle ABC)}{\text{perimeter of } \triangle ABC} = \dfrac{2(84)}{13+14+15}$ (from part **(b)**) $= 4$

31. a. $c^2 = a^2 + b^2 - 2ab \cos C = 9^2 + 10^2 - 2(9)(10)\left(-\dfrac{3}{5}\right) = 181 - 180\left(-\dfrac{3}{5}\right) =$

289; $c = \sqrt{289} = 17$.

Since $\cos C = -\dfrac{3}{5} = -0.6$, $\angle C \approx 126.9°$. $\cos B = \dfrac{9^2 + 17^2 - 10^2}{2(9)(17)} \approx 0.8824$;

$\angle B \approx 28.1°$. $\angle A \approx 180° - 126.9° - 28.1° = 25.0°$.

b. $K = \dfrac{1}{2}ab \sin C = \dfrac{1}{2}(9)(10)\left(\dfrac{4}{5}\right) = 36$ sq. units

c. Let $\overline{BD}$ be the altitude. In $\triangle BCD$, $\sin C = \dfrac{BD}{BC}$; $\dfrac{4}{5} = \dfrac{BD}{9}$; $BD = 7.2$.

d. Let x be the length of the median. By Ex. 23, p. 354, $x = \dfrac{1}{2}\sqrt{2a^2 + 2c^2 - b^2} =$

$\dfrac{1}{2}\sqrt{2(9)^2 + 2(17)^2 - 10^2} = \dfrac{1}{2}\sqrt{640} = \dfrac{1}{2}(8\sqrt{10}) = 4\sqrt{10} \approx 12.6$

e. Let y = the required length. By Ex. 24, p. 354, $p = \dfrac{ab}{a + c} = \dfrac{9(10)}{9 + 17} = \dfrac{90}{26} = \dfrac{45}{13}$ and

$q = \dfrac{bc}{a + c} = \dfrac{10(17)}{9 + 17} = \dfrac{170}{26} = \dfrac{85}{13}$. Then $y^2 = ac - pq = 9(17) - \left(\dfrac{45}{13}\right)\left(\dfrac{85}{13}\right) =$

$\dfrac{22{,}032}{169} \approx 130.37$; $y \approx 11.4$.

f. By Ex. 23, p. 349, diameter $= \dfrac{c}{\sin C} = \dfrac{17}{\frac{4}{5}} = 21.25$; $R = \dfrac{1}{2}(21.25) \approx 10.6$.

g. By Ex. 17, p. 342, $\dfrac{2(\text{area of } \triangle ABC)}{\text{perimeter of } \triangle ABC} = \dfrac{2(36)}{9 + 10 + 17}$ (from part (b)) $= 2$.

32. a. Since $\sin A = \dfrac{8}{17}$, $\angle A \approx 28.1°$. Since $\sin B = \dfrac{\sqrt{2}}{2}$ and $\angle B$ is obtuse, $\angle B = 135°$.

Then $\angle C \approx 180° - 28.1° - 135° = 16.9°$. $\dfrac{\sin 16.9°}{21} = \dfrac{\frac{8}{17}}{a}$; $a \approx 34.0$.

$\dfrac{\sin 16.9°}{21} = \dfrac{\frac{\sqrt{2}}{2}}{b}$; $b \approx 51.1$.

b. $K = \dfrac{1}{2}bc \sin A = \dfrac{1}{2}(51.1)(21)\left(\dfrac{8}{17}\right) \approx 252.5$ sq. units

c. Let $\overline{BD}$ be the altitude. In $\triangle ABD$, $\sin A = \dfrac{BD}{AB}$; $\dfrac{8}{17} = \dfrac{BD}{21}$; $BD \approx 9.88$.

d. Let x be the length of the median. By Ex. 23, p. 354, $x = \dfrac{1}{2}\sqrt{2a^2 + 2c^2 - b^2} =$

$\dfrac{1}{2}\sqrt{2(34.0)^2 + 2(21)^2 - (51.1)^2} = \dfrac{1}{2}\sqrt{582.79} \approx 12.1$.

e. Let y = the required length. By Ex. 24, p. 354, $p = \dfrac{ab}{a + c} = \dfrac{(34.0)(51.1)}{34.0 + 21} \approx 31.59$ and

$q = \dfrac{bc}{a + c} = \dfrac{(51.1)(21)}{34.0 + 21} \approx 19.51$. Then $y^2 = ac - pq = (34.0)(21) -$

$(31.59)(19.51) \approx 97.68$; $y \approx 9.88$

Continued on next page

f. By Ex. 23, p. 349, diameter $= \dfrac{21}{\sin 16.9°} \approx 72.2$; $R = \dfrac{1}{2}(72.2) = 36.1$.

g. By Ex. 17, p. 342, $r = \dfrac{2(\text{area of } \triangle ABC)}{\text{perimeter of } \triangle ABC} \approx \dfrac{2(252.5)}{34.0 + 51.1 + 21}$ (from part (b)) ≈ 4.76.

33. a. In quad. $JKLM$, let $JK = 1$, $KL = 4$, $LM = 3$, and $MJ = 2$. In $\triangle JKM$,

$\cos J = \dfrac{1^2 + 2^2 - (KM)^2}{2(1)(2)} = \dfrac{5 - (KM)^2}{4}$. In $\triangle LKM$, $\cos L = \dfrac{4^2 + 3^2 - (KM)^2}{2(4)(3)} =$

$\dfrac{25 - (KM)^2}{24}$. Since $\cos J = -\cos L$, $\dfrac{5 - (KM)^2}{4} = \dfrac{(KM)^2 - 25}{24}$; $30 - 6(KM)^2 =$

$(KM)^2 - 25$; $55 = 7(KM)^2$; $(KM)^2 = \dfrac{55}{7}$; $KM = \sqrt{\dfrac{55}{7}} = \dfrac{\sqrt{385}}{7} \approx 2.80$. Using a

similar method in $\triangle JKL$ and $\triangle LMJ$, we find that $JL = \sqrt{\dfrac{77}{5}} = \dfrac{\sqrt{385}}{5} \approx 3.92$.

b. $JL \cdot KM = \dfrac{\sqrt{385}}{5} \cdot \dfrac{\sqrt{385}}{7} = \dfrac{385}{35} = 11$ and $JK \cdot LM + JM \cdot KL = 1 \cdot 3 + 2 \cdot 4 = 11$;

$11 = 11$.

34. $\angle D = 120°$ and $\angle DEF = 15°$, so $\angle DFE = 180° - 120° - 15° = 45°$. $\overset{\frown}{DE}$ and $\overset{\frown}{DF}$ are intercepted by $\angle DFE$ and $\angle DEF$, respectively, so $\overset{\frown}{DE} = 2(45°) = 90°$ and $\overset{\frown}{DF} = 2(15°) = 30°$. Thus $\overset{\frown}{EF} = 90° + 30° = 120°$ and $\angle EOF = 120°$. $\triangle OGE$ and $\triangle OGF$ are 30°-60°-90° $\triangle$s, so $EG = GF = \dfrac{3}{2}\sqrt{3}$; $EF = d = 3\sqrt{3} \approx 5.20$. $\dfrac{\sin 15°}{e} = \dfrac{\sin 120°}{3\sqrt{3}}$; $e \approx 1.55$. $\dfrac{\sin 45°}{f} = \dfrac{\sin 120°}{3\sqrt{3}}$; $f = 3\sqrt{2} \approx 4.24$.

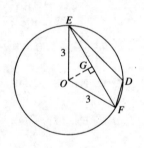

35. Let $\overline{AD}$ be the altitude with $AD = h$. If $BD = x$, then $DC = a - x$. In $\triangle ABD$,

$\cot B = \dfrac{x}{h}$; in $\triangle ACD$, $\cot C = \dfrac{a - x}{h}$. Thus, $x = h \cot B$ and $a - x = h \cot C$;

$x + (a - x) = h \cot B + h \cot C$; $a = h(\cot B + \cot C)$; $h = \dfrac{a}{\cot B + \cot C}$.

36. From geometry, it can be shown that the line joining the points O, C_1, and C_2 is a straight line that bisects $\angle O$. $\sin x = \dfrac{r_2}{h + r_1 + r_2}$;

$(h + r_1 + r_2) \sin x = r_2$;

$h \sin x + r_1 \sin x + r_2 \sin x = r_2$; $h \cdot \dfrac{r_1}{h} + r_1 \sin x = r_2 - r_2 \sin x$; $r_1(1 + \sin x) =$

$r_2(1 - \sin x)$; $\dfrac{r_1}{r_2} = \dfrac{1 - \sin x}{1 + \sin x}$

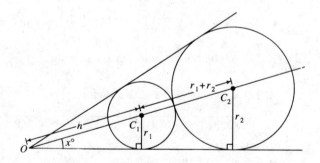

37. $a^2 = b^2 + c^2 - 2bc \cos A$, $b^2 = a^2 + c^2 - 2ac \cos B$, and $c^2 = a^2 + b^2 - 2ab \cos C$;
thus, $a^2 + b^2 + c^2 = 2a^2 + 2b^2 + 2c^2 - 2bc \cos A - 2ac \cos B - 2ab \cos C$;
$2bc \cos A + 2ac \cos B + 2ab \cos C = a^2 + b^2 + c^2$; dividing by $2ab \sin C$,

$$\frac{\cos A}{\dfrac{a}{c} \sin C} + \frac{\cos B}{\dfrac{b}{c} \sin C} + \cot C = \frac{a^2 + b^2 + c^2}{4\left(\dfrac{1}{2} ab \sin C\right)}; \text{ in any } \triangle ABC, \frac{\sin A}{a} = \frac{\sin B}{b} = \frac{\sin C}{c}; \text{ thus,}$$

$\sin B = \dfrac{b}{c} \sin C$ and $\sin A = \dfrac{a}{c} \sin C$. Substituting, $\dfrac{\cos A}{\sin A} + \dfrac{\cos B}{\sin B} + \cot C =$

$$\frac{a^2 + b^2 + c^2}{4\left(\dfrac{1}{2} ab \sin C\right)}; \cot A + \cot B + \cot C = \frac{a^2 + b^2 + c^2}{4(\text{area of } \triangle ABC)}.$$

38. If $c^2 = \dfrac{a^3 + b^3 + c^3}{a + b + c}$, then $ac^2 + bc^2 + c^3 = a^3 + b^3 + c^3$; $(a + b)c^2 = a^3 + b^3$;
$(a + b)c^2 = (a + b)(a^2 - ab + b^2)$; $c^2 = a^2 - ab + b^2$; by the law of cosines,
$c^2 = a^2 + b^2 - 2ab \cos C$; $a^2 - ab + b^2 = a^2 + b^2 - 2ab \cos C$; $-ab = -2ab \cos C$;
$ab(1 - 2 \cos C) = 0$; $ab \neq 0$, so $\cos C = \dfrac{1}{2}$; $\angle C = 60°$.

39. Let point E be the vertex of the right angle. Note that $AC = BD$. In $\triangle ACE$,
$\sin \alpha = \dfrac{x + DE}{AC}$ and $\cos \alpha = \dfrac{EA}{AC}$. In $\triangle BDE$, $\sin \beta = \dfrac{DE}{BD} = \dfrac{DE}{AC}$ and
$\cos \beta = \dfrac{EA + y}{BD} = \dfrac{EA + y}{AC}$. Thus, $\dfrac{y(\sin \alpha - \sin \beta)}{\cos \beta - \cos \alpha} =$

$$y\left(\frac{x + DE}{AC} - \frac{DE}{AC}\right) \div \left(\frac{EA + y}{AC} - \frac{EA}{AC}\right) = \frac{xy}{AC} \div \frac{y}{AC} = \frac{xy}{AC} \cdot \frac{AC}{y} = x.$$

40. $\cos \alpha = \dfrac{6400}{6400.004}$; $\alpha = 0.00112$ radians;

$\cos \beta = \dfrac{6400}{6400.03}$; $\beta = 0.00306$ radians;

$\angle BCA = \alpha + \beta = 0.00112 + 0.00306 = 0.00418$ radians; distance of ship from
shore = length of $\overset{\frown}{DE} = r\theta = 6400(0.00418) \approx 26.8$ km

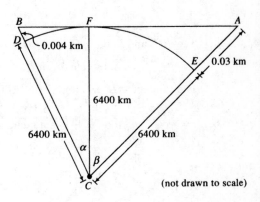

(not drawn to scale)

41. In rt. $\triangle XRY$ and rt. $\triangle XSZ$, respectively, $\cos X = \dfrac{XR}{z}$ and $\cos X = \dfrac{XS}{y}$. In $\triangle XYZ$,

$\cos X = \dfrac{z^2 + y^2 - x^2}{2yz}$. Thus, $\dfrac{XR}{z} = \dfrac{z^2 + y^2 - x^2}{2yz}$ and $\dfrac{XS}{y} = \dfrac{z^2 + y^2 - x^2}{2yz}$;

$XR = \dfrac{z^2 + y^2 - x^2}{2y}$ and $XS = \dfrac{z^2 + y^2 - x^2}{2z}$. In $\triangle XSR$,

$(RS)^2 = (XS)^2 + (XR)^2 - 2(XS)(XR)\cos X = \left(\dfrac{z^2 + y^2 - x^2}{2z}\right)^2 + \left(\dfrac{z^2 + y^2 - x^2}{2y}\right)^2 - $

$2\left(\dfrac{z^2 + y^2 - x^2}{2z}\right)\left(\dfrac{z^2 + y^2 - x^2}{2y}\right)\left(\dfrac{z^2 + y^2 - x^2}{2yz}\right) = $

$[z^2 + y^2 - x^2]^2\left[\dfrac{1}{4z^2} + \dfrac{1}{4y^2} - \dfrac{2}{4yz} \cdot \dfrac{z^2 + y^2 - x^2}{2yz}\right] = $

$[z^2 + y^2 - x^2]^2\left[\dfrac{y^2 + z^2 - z^2 - y^2 + x^2}{4y^2z^2}\right] = [z^2 + y^2 - x^2]^2 \cdot \dfrac{x^2}{4y^2z^2} = $

$x^2 \cdot \left(\dfrac{z^2 + y^2 - x^2}{2yz}\right)^2 = x^2 \cdot \cos^2 X$; since $(RS)^2 = x^2 \cdot \cos^2 X$, $RS = x \cos X$.

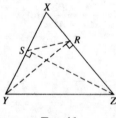

Ex. 41

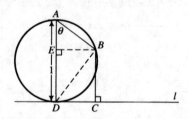

Ex. 42

42. a. Draw $\overline{BD}$, and $\overline{BE} \perp \overline{AD}$. $\angle ABD$ is inscribed in a semicircle; $\therefore \triangle ABD$ is a rt $\triangle$.

$\cos \theta = \dfrac{AB}{1}$; $\therefore AB = \cos \theta$; in rt. $\triangle ABE$, $\cos \theta = \dfrac{AE}{AB} = \dfrac{AE}{\cos \theta}$; $\therefore AE = \cos^2 \theta$. Since

l is a tangent, $\overline{AD} \perp l$, and $\therefore BEDC$ is a rectangle. Then $BC = ED = AD - AE = 1 - \cos^2 \theta$; thus, $AB + BC = \cos \theta + 1 - \cos^2 \theta = 1 + \cos \theta - \cos^2 \theta$.

b. $y = -\cos^2 \theta + \cos \theta + 1$ is a quadratic function; maximum value occurs at

$\cos \theta = -\dfrac{b}{2a} = -\dfrac{1}{2(-1)} = \dfrac{1}{2}$; $\theta = 60°$.

43. By the law of cosines, $\cos\theta =$

$\dfrac{1^2 + (2 + r)^2 - (3 - r)^2}{2(2 + r)}$ in the smaller $\triangle$

and $\cos\theta = \dfrac{3^2 + (2 + r)^2 - (1 + r)^2}{2 \cdot 3 \cdot (2 + r)}$ in

the larger $\triangle$; $\dfrac{1 + (2 + r)^2 - (3 - r)^2}{2(2 + r)} =$

$\dfrac{9 + (2 + r)^2 - (1 + r)^2}{6(2 + r)}$;

$3[1 + (2 + r)^2 - (3 - r)^2] =$
$9 + (2 + r)^2 - (1 + r)^2$;

$3[-4 + 10r] = 12 + 2r;\ r = \dfrac{6}{7}$

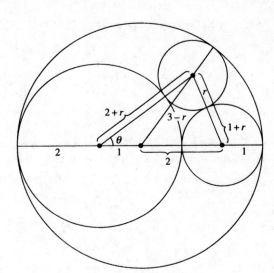

Class Exercises 9-5 · pages 361–362

1. The ship first travels on a course of 300° and then on a course of 230°. Thus, it turns through an angle of 70°. The angle X, opposite side of length x, has measure $180° - 70° = 110°$.

2. $84{,}800 \text{ ft}^2 = \dfrac{84800 \text{ ft}^2}{43560 \text{ ft}^2/\text{acre}} \approx 1.95$ acres; tax $= 75(1.95) \approx \$146$

3. 040°   **4.** 115°   **5.** 200°   **6.** 300°   **7. a.** 090°   **b.** 270°

8. 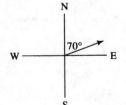   9.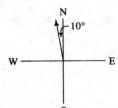

10. 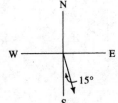   11.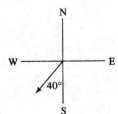

12. NE = 045°; SE = 135°; NW = 315°; SW = 225°

13. $\dfrac{315° + 360°}{2} = 337.5°$ **14.** $180° - 32° - 76° = 72°$

Written Exercises 9-5 • pages 362–364

1. N
70°

2. N
150°

3. N
340°

4. N
225°

5. See diagram below. The parallel north lines form equal alternate interior angles,
∴ $\angle ABC = 80°$. The bearing is $180° + 80° = 260°$.

6. See diagram below. $\angle AXY = 360° - 308° = 52°$; $\angle CYX$ and $\angle X$ are supplementary;
$\angle CYX = 180° - 52° = 128°$

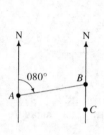

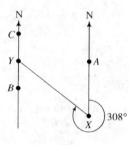

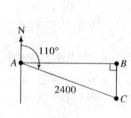

Ex. 5   **Ex. 6**   **Ex. 7**

7. See diagram above. $AC = 2 \cdot 1200 = 2400$ km; $\angle BAC = 110° - 90° = 20°$;
$\cos 20° = \dfrac{AB}{2400}$; $AB \approx 2255$ km

8. $\tan \theta = \dfrac{1}{1.5} = 0.6667$; $\theta \approx 33.7°$; course $= 180° + 33.7° = 213.7°$. We assume that
the hunter walks at a constant rate.

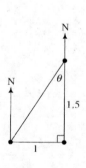

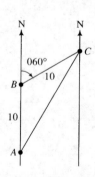

Ex. 8   **Ex. 9**

9. $\angle ABC = 180° - 60° = 120°$; $\triangle ABC$ is isosceles; ∴ $\angle BAC = \angle BCA$. The bearing of
C from A is $\angle BAC = \dfrac{180° - 120°}{2} = 030°$. $AC^2 = 10^2 + 10^2 - 2(10)(10) \cos 120°$;
$AC^2 = 200 - 200\left(-\dfrac{1}{2}\right) = 300$; $AC = \sqrt{300} \approx 17.3$ km

10. $\angle TSR = 45° + 90° = 135°$; $\triangle TSR$ is isosceles; $\therefore \angle STR = \angle R$; the bearing of R from T is $45° + \angle STR = 45° + \dfrac{180° - 135°}{2} = 45° + 22.5° = 67.5°$. The bearing is $067.5°$. $TR^2 = 4^2 + 4^2 - 2(4)(4)\cos 135°$; $TR^2 = 32 - 32\left(-\dfrac{\sqrt{2}}{2}\right) \approx 54.63$; $TR \approx 7.39$ km

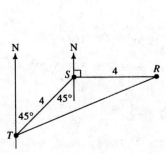

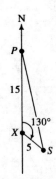

Ex. 10   **Ex. 11**

11. $PX = 10(1.5) = 15$ nautical mi; $XS = 10(0.5) = 5$ nautical mi;
$PS^2 = 15^2 + 5^2 - 2(15)(5)\cos 130° \approx 346.42$; $PS \approx 18.6$ nautical mi

12. $\angle S = \angle SPX = 205° - 180° = 25°$. $\tan 25° = \dfrac{2}{DS}$; $DS = \dfrac{2}{\tan 25°} \approx 4.29$ mi

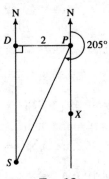

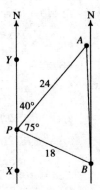

Ex. 12   **Ex. 13**

13. $\angle APB = 115° - 40° = 75°$; $AP = 2 \cdot 12 = 24$ nautical mi; $PB = 2 \cdot 9 = 18$ nautical mi;
$AB^2 = 24^2 + 18^2 - 2(24)(18)\cos 75° \approx 676.38$; $AB \approx 26$ nautical mi. $\dfrac{\sin B}{24} = \dfrac{\sin 75°}{26}$;
$\sin B = \dfrac{24 \sin 75°}{26} \approx 0.8916$; $\angle B \approx 63°$; obtuse $\angle B = \angle YPB = 115°$; the course is
$180° + 115° + 63° = 358°$.

14.

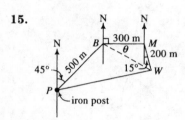

$\angle XSP = 180° - 45° - 45° = 90°$; $\tan \angle SXP = \dfrac{1}{2}$; $\angle SXP \approx 26.6°$;

obtuse $\angle X$ is the supplement of $45°$; obtuse $\angle X = 135°$;
$\angle AXP = 180° - 135° - 26.6° = 18.4°$, the course is
$180° + 18.4° = 198.4°$. $XP = \sqrt{2^2 + 1^2} = \sqrt{5} \approx 2.24$;
the return will take 2.24 h.

15.

Divide the plot into two triangles.
$\angle WMB = 90° + 15° = 105°$;
$BW^2 = 300^2 + 200^2 - 2(300)(200)\cos 105° \approx 161058$;
$BW \approx 401.3$; $\dfrac{\sin \theta}{200} = \dfrac{\sin 105°}{401.3}$;

$\sin \theta = \dfrac{200 \sin 105°}{401.3} \approx 0.4814$;

$\theta \approx 28.78°$; $\angle PBW = \angle PBM - \theta = 135° - 28.78° =$

$106.22°$; area of $\triangle PBW = \dfrac{1}{2}(500)(401.3)\sin 106.22° \approx$

$96,000$ m^2; area of $\triangle WMB = \dfrac{1}{2}(300)(200)\sin 105° = 29,000$ m^2;

area of $PBMW \approx 96,000 + 29,000 = 125,000$ m^2

16.

Divide the plot into two triangles. $RS^2 = 240^2 + 280^2 -$

$2(240)(280)\cos 40° \approx 33,044$; $RS \approx 181.8$; $\dfrac{\sin \theta}{240} = \dfrac{\sin 40°}{181.8}$;

$\sin \theta = \dfrac{240 \sin 40°}{181.8} \approx 0.8486$; $\theta \approx 58.06°$; $\angle BRS = 180° - 40° -$

$58.06° = 81.94°$; $\angle SRC = \angle BRC - \angle BRS = 135° - 81.94° =$

$53.06°$; area of $\triangle RSC = \dfrac{1}{2}(260)(181.8)\sin 53.06° \approx 18,900$ m^2;

area of $\triangle RSB = \dfrac{1}{2}(240)(280)\sin 40° \approx 21,600$ m^2;

area of $RCSB \approx 18,900 + 21,600 = 40,500$ m^2

17. Divide the plot into two triangles. $\angle XGC$ and $\angle GCZ$ are supplementary;
$\angle XGC = 180° - 78° = 102°$; $\angle CGA = 102° - 15° = 87°$.
$AC^2 = 180^2 + 250^2 - 2(180)(250)\cos 87° \approx 90190$; $AC \approx 300.3$.

$\dfrac{\sin \angle GCA}{180} = \dfrac{\sin 87°}{300.3}$; $\sin \angle GCA = \dfrac{180 \sin 87°}{300.3} \approx 0.5986$; $\angle GCA \approx 36.77°$;

$\angle ACZ = 78° - 36.77° = 41.23°$; $\angle BCZ = \angle YBC = 30°$;
$\angle ACB = 41.23° - 30° = 11.23°$. $\angle PAB$ and $\angle ABY$ are supplementary;

$\angle ABY = 180° - 78° = 102°$; $\angle ABC = 102° + 30° = 132°$; $\dfrac{\sin 132°}{300.3} = \dfrac{\sin 11.23°}{AB}$;

$AB = \dfrac{300.3 \sin 11.23°}{\sin 132°} \approx 78.7$. $\angle CAB = 180° - 132° - 11.23° = 36.77°$. Area of

$\triangle ABC = \dfrac{1}{2}(78.7)(300.3)\sin 36.77° \approx 7100$ m^2; area of $\triangle GAC =$

$\dfrac{1}{2}(180)(250)\sin 87° \approx 22,500$ m^2; area of $GCBA \approx 7100 + 22,500 = 29,600$ m^2.

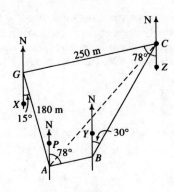

Ex. 17   Ex. 18

18. Divide the plot into two triangles. $\angle XRS = \angle RSY = 32°$; $\angle TRS = 56° + 32° = 88°$; $TS^2 = 280^2 + 320^2 - 2(280)(320) \cos 88° \approx 174{,}546$; $TS \approx 417.8$.

$\dfrac{\sin\theta}{320} = \dfrac{\sin 88°}{417.8}$; $\sin\theta = \dfrac{320 \sin 88°}{417.8} \approx 0.7655$; $\theta \approx 49.95°$.

$\angle ZTR$ and $\angle TRX$ are supplementary; $\angle ZTR = 180° - 56° = 124°$; $\angle MTS = 124° - 22° - 49.95° = 52.05°$. $\angle PMT = \angle MTZ = 22°$; $\angle TMS = 22° + 68° = 90°$; $\angle MST = 180° - 90° - 52.05° = 37.95°$. In rt. $\triangle TMS$,

$\cos 37.95° = \dfrac{MS}{417.8}$; $MS \approx 329.5$. Area of $\triangle TMS = \dfrac{1}{2}(329.5)(417.8) \sin 37.95° \approx$

$42{,}300$ m^2; area of $\triangle TRS = \dfrac{1}{2}(280)(320) \sin 88° \approx 44{,}800$ m^2; area of $TRSM \approx$

$42{,}300 + 44{,}800 = 87{,}100$ m^2.

19. "True north" is the north *geographic* pole near the center of the Arctic Ocean where all lines of longitude meet. "Magnetic north" is the north *magnetic* pole, the point toward which a north-seeking compass needle points. Although the position of true north is fixed, the north magnetic pole continually shifts; in fact, it can move many miles in just a few years.

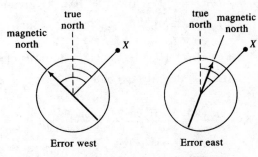

Error west   Error east

Compass variation is the angle between magnetic north and true north. When magnetic north pulls a compass needle to the left of true north, the variation is *west*; when magnetic north pulls a compass needle to the right of true north, the variation is *east*. (See illustrations.) Because the north magnetic pole shifts over time, the world's dividing lines between east and west variation also shift. Generally speaking, however, west variation occurs in the eastern parts of North America, South America, and Asia, and in most of Europe and Africa; east variation occurs in the western parts of North America, South America, and Asia, and in most of Australia. (Consult an isogonic map for more details.) In the mariner's rhyme "Error west, compass best; error east, compass least," the words "error," "best," and "least" are equivalent to "variation," "numerically greater," and "numerically smaller," respectively. As can be seen from the

Continued on next page

illustrations on page 257, when the variation is west (that is, "error west"), a compass gives a bearing for an object X that is numerically greater than its bearing from true north (that is, "compass best"); when the variation is east (that is, "error east"), a compass gives a bearing for X that is numerically smaller than its bearing from true north (that is, "compass least").

Chapter Test · page 365

1. Given $\triangle ABC$ with $AB = 10$, $BC = 10$, and $AC = 5$, let $\overline{BX}$ be the altitude from B to AC. $\overline{BX}$ bisects $\overline{AC}$; $AX = 2.5$; $\cos A = \dfrac{2.5}{10} = 0.25$; $\angle A = 75.5°$; $\angle C = 75.5°$; $\angle B = 180° - 75.5° - 75.5° = 29.0°$.

2. $\tan 28° = \dfrac{h}{100}$; $h = 100 \tan 28° \approx 53.2$ m

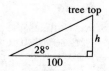

3. The pentagon is made up of 5 congruent triangles with vertex angles of $\dfrac{360°}{5} = 72°$ and legs of length 4 in. Area of pentagon = 5(area of each triangle) = $5\left(\dfrac{1}{2} \cdot 4 \cdot 4 \cdot \sin 72°\right) \approx 38.0$ in.^{2}.

4. $DB = \sqrt{7^2 + 24^2} = 25$; $\sin \alpha = \dfrac{7}{25} = 0.28$; $\alpha = 16.26°$; $\angle DBC = 120° - 16.26° = 103.74°$; area of $\triangle DBC = \dfrac{1}{2}(25)(30) \sin 103.74° \approx 364$;

area of $\triangle DAB = \dfrac{1}{2}(7)(24) = 84$; area of quadrilateral $ABCD \approx 364 + 84 = 448$

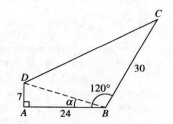

5. a. $\dfrac{\sin Q}{q} = \dfrac{\sin P}{p}$; $\dfrac{\sin 74°}{4} = \dfrac{\sin P}{5}$; $\sin P = \dfrac{5 \sin 74°}{4} \approx 1.2$; since $\sin P > 1$, no triangle can be constructed.

b. $\dfrac{\sin P}{p} = \dfrac{\sin Q}{q}$; $\dfrac{\sin 23°}{9} = \dfrac{\sin Q}{8}$; $\sin Q = \dfrac{8 \sin 23°}{9} \approx 0.3473$; $\angle Q \approx 20.3°$ or $\angle Q \approx 159.7°$; $23° + 159.7° > 180°$, so $\angle Q$ cannot be 159.7°. Thus, one triangle can be constructed.

6. $\angle C = 180° - 40° - 75° = 65°$; $\dfrac{\sin 75°}{b} = \dfrac{\sin 65°}{30}$; $b = \dfrac{30 \sin 75°}{\sin 65°} \approx 32.0$ km; $\dfrac{\sin 40°}{a} = \dfrac{\sin 65°}{30}$; $a = \dfrac{30 \sin 40°}{\sin 65°} \approx 21.3$ km

7. Given $\triangle ABC$ with $BC = 5$, $AC = 8$, and $AB = 10$, the largest angle is $\angle C$, which is opposite the longest side; $\cos C = \dfrac{5^2 + 8^2 - 10^2}{2(5)(8)}$; $\angle C = 97.9°$

8.

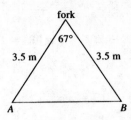

$AB^2 = 3.5^2 + 3.5^2 - 2(3.5)(3.5) \cos 67° \approx 14.93;$
$AB \approx 3.86$ mi

9. Use the law of cosines when the known parts are three sides, or two sides and the included angle. Use the law of sines when the known parts are two sides and an angle opposite one of them, or two angles and a side.

10.

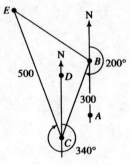

$\angle ABC = 200° - 180° = 20°$; the north lines are parallel, so
$\angle DCB = \angle ABC = 20°$; $\angle DCE = 360° - 340° = 20°$;
$\angle BCE = 20° + 20° = 40°$. $BE^2 = 500^2 + 300^2 -$
$2(500)(300) \cos 40° \approx 110{,}186.67$; $BE \approx 332$ km

Class Exercises 10-1 · page 372

1. Answers may vary. For example, let $x = 60°$, $y = 30°$; $\sin (60° + 30°) = \sin 90° = 1$; $\sin 60° + \sin 30° = \dfrac{\sqrt{3}}{2} + \dfrac{1}{2} \neq 1$.

2. Answers may vary. For example, let $a = 60°$, $b = 0°$; $\cos (60° - 0°) = \cos 60° = \dfrac{1}{2}$; $\cos 60° - \cos 0° = \dfrac{1}{2} - 1 = -\dfrac{1}{2} \neq \dfrac{1}{2}$.

3. Yes. Answers may vary. For example, let $b = 0°$; $\sin (a - b) = \sin (a - 0°) = \sin a$; $\sin a - \sin b = \sin a - \sin 0° = \sin a$.

4. Yes. Answers may vary. For example, let $b = 0°$; $\sin (a + b) = \sin (a + 0°) = \sin a$; $\sin a + \sin b = \sin a + \sin 0° = \sin a$.

5. $\sin (1° + 2°) = \sin 3°$ **6.** $\sin (20° - 15°) = \sin 5°$

7. $\cos \left(\dfrac{\pi}{4} + \dfrac{\pi}{4} \right) = \cos \dfrac{\pi}{2}$ **8.** $\cos (75° - 25°) = \cos 50°$

9. $\sin (-\alpha) = \sin (0 - \alpha) = \sin 0 \cos \alpha - \sin \alpha \cos 0 = (0) \cos \alpha - (\sin \alpha)(1) = -\sin \alpha$

Written Exercises 10-1 · pages 373–374

1. $\sin (75° + 15°) = \sin 90° = 1$ **2.** $\cos (105° - 15°) = \cos 90° = 0$

3. $\cos \left(\dfrac{5\pi}{12} + \dfrac{\pi}{12} \right) = \cos \dfrac{6\pi}{12} = \cos \dfrac{\pi}{2} = 0$ **4.** $\sin \left(\dfrac{4\pi}{3} - \dfrac{\pi}{3} \right) = \sin \dfrac{3\pi}{3} = \sin \pi = 0$

5. $\sin (3x - 2x) = \sin x$ **6.** $\cos (2x - x) = \cos x$

7. $\sin (\pi + x) = \sin \pi \cos x + \cos \pi \sin x = 0 \cdot \cos x + (-1) \sin x = -\sin x$

8. $\cos (\pi + x) = \cos \pi \cos x - \sin \pi \sin x = -1 \cdot \cos x - 0 \cdot \sin x = -\cos x$

9. $\cos \left(x + \dfrac{\pi}{2} \right) = \cos x \cos \dfrac{\pi}{2} - \sin x \sin \dfrac{\pi}{2} = \cos x \cdot 0 - \sin x \cdot 1 = -\sin x$

10. $\cos \left(\dfrac{\pi}{2} - x \right) = \cos \dfrac{\pi}{2} \cos x + \sin \dfrac{\pi}{2} \sin x = 0 \cdot \cos x + 1 \cdot \sin x = \sin x$

11. $y = \sin x$ **12.** $y = \sin x$

13. $\cos 75° = \cos (30° + 45°) = \cos 30° \cos 45° - \sin 30° \sin 45° =$
$\dfrac{\sqrt{3}}{2} \cdot \dfrac{\sqrt{2}}{2} - \dfrac{1}{2} \cdot \dfrac{\sqrt{2}}{2} = \dfrac{\sqrt{6} - \sqrt{2}}{4}$

14. $\cos 15° = \cos (45° - 30°) = \cos 45° \cos 30° + \sin 45° \sin 30° =$
$\dfrac{\sqrt{2}}{2} \cdot \dfrac{\sqrt{3}}{2} + \dfrac{\sqrt{2}}{2} \cdot \dfrac{1}{2} = \dfrac{\sqrt{6} + \sqrt{2}}{4}$

15. $\cos 105° = \cos (45° + 60°) = \cos 45° \cos 60° - \sin 45° \sin 60° =$
$\dfrac{\sqrt{2}}{2} \cdot \dfrac{1}{2} - \dfrac{\sqrt{2}}{2} \cdot \dfrac{\sqrt{3}}{2} = \dfrac{\sqrt{2} - \sqrt{6}}{4}$

16. $\sin 105° = \sin (60° + 45°) = \sin 60° \cos 45° + \cos 60° \sin 45° =$
$\dfrac{\sqrt{3}}{2} \cdot \dfrac{\sqrt{2}}{2} + \dfrac{1}{2} \cdot \dfrac{\sqrt{2}}{2} = \dfrac{\sqrt{6} + \sqrt{2}}{4}$

17. $\sin(-15°) = \sin(30° - 45°) = \sin 30° \cos 45° - \cos 30° \sin 45° =$

$\dfrac{1}{2} \cdot \dfrac{\sqrt{2}}{2} - \dfrac{\sqrt{3}}{2} \cdot \dfrac{\sqrt{2}}{2} = \dfrac{\sqrt{2} - \sqrt{6}}{4}$

18. $\cos(-165°) = \cos(60° - 225°) = \cos 60° \cos 225° + \sin 60° \sin 225° =$

$\left(\dfrac{1}{2}\right)\left(-\dfrac{\sqrt{2}}{2}\right) + \left(\dfrac{\sqrt{3}}{2}\right)\left(-\dfrac{\sqrt{2}}{2}\right) = \dfrac{-\sqrt{6} - \sqrt{2}}{4}$

19. $\sin \dfrac{7\pi}{12} = \sin\left(\dfrac{\pi}{3} + \dfrac{\pi}{4}\right) = \sin \dfrac{\pi}{3} \cos \dfrac{\pi}{4} + \cos \dfrac{\pi}{3} \sin \dfrac{\pi}{4} = \dfrac{\sqrt{3}}{2} \cdot \dfrac{\sqrt{2}}{2} + \dfrac{1}{2} \cdot \dfrac{\sqrt{2}}{2} = \dfrac{\sqrt{6} + \sqrt{2}}{4}$

20. $\sin \dfrac{11\pi}{12} = \sin\left(\dfrac{\pi}{6} + \dfrac{3\pi}{4}\right) = \sin \dfrac{\pi}{6} \cos \dfrac{3\pi}{4} + \cos \dfrac{\pi}{6} \sin \dfrac{3\pi}{4} =$

$\dfrac{1}{2}\left(-\dfrac{\sqrt{2}}{2}\right) + \dfrac{\sqrt{3}}{2} \cdot \dfrac{\sqrt{2}}{2} = \dfrac{\sqrt{6} - \sqrt{2}}{4}$

21. $\sin(30° + \theta) + \sin(30° - \theta) = \sin 30° \cos \theta + \cos 30° \sin \theta + \sin 30° \cos \theta -$

$\cos 30° \sin \theta = 2 \sin 30° \cos \theta = 2 \cdot \dfrac{1}{2} \cos \theta = \cos \theta$

22. $\cos(30° + \theta) + \cos(30° - \theta) = \cos 30° \cos \theta - \sin 30° \sin \theta + \cos 30° \cos \theta +$

$\sin 30° \sin \theta = 2 \cos 30° \cos \theta = 2 \cdot \dfrac{\sqrt{3}}{2} \cos \theta = \sqrt{3} \cos \theta$

23. $\cos\left(\dfrac{\pi}{3} + x\right) + \cos\left(\dfrac{\pi}{3} - x\right) = \cos \dfrac{\pi}{3} \cos x - \sin \dfrac{\pi}{3} \sin x + \cos \dfrac{\pi}{3} \cos x +$

$\sin \dfrac{\pi}{3} \sin x = 2 \cos \dfrac{\pi}{3} \cos x = 2 \cdot \dfrac{1}{2} \cos x = \cos x$

24. $\sin\left(\dfrac{\pi}{4} + x\right) + \sin\left(\dfrac{\pi}{4} - x\right) = \sin \dfrac{\pi}{4} \cos x + \cos \dfrac{\pi}{4} \sin x + \sin \dfrac{\pi}{4} \cos x -$

$\cos \dfrac{\pi}{4} \sin x = 2 \sin \dfrac{\pi}{4} \cos x = 2 \cdot \dfrac{\sqrt{2}}{2} \cos x = \sqrt{2} \cos x$

25. $\cos\left(\dfrac{3\pi}{2} + x\right) + \cos\left(\dfrac{3\pi}{2} - x\right) = \cos \dfrac{3\pi}{2} \cos x - \sin \dfrac{3\pi}{2} \sin x + \cos \dfrac{3\pi}{2} \cos x +$

$\sin \dfrac{3\pi}{2} \sin x = 2 \cos \dfrac{3\pi}{2} \cos x = 2 \cdot 0 \cdot \cos x = 0$

26. $\sin(\pi + x) + \sin(\pi - x) = \sin \pi \cos x + \cos \pi \sin x + \sin \pi \cos x -$
$\cos \pi \sin x = 2 \sin \pi \cos x = 2 \cdot 0 \cdot \cos x = 0$

27. Since α is acute and $\sin \alpha = \dfrac{3}{5}$, the formula $\sin^2 \alpha + \cos^2 \alpha = 1$ yields $\cos \alpha = \dfrac{4}{5}$.

$\cos \beta < 0$, so the formula yields $\cos \beta = -\dfrac{7}{25}$. Hence $\sin(\alpha + \beta) = \sin \alpha \cos \beta +$

$\cos \alpha \sin \beta = \dfrac{3}{5}\left(-\dfrac{7}{25}\right) + \dfrac{4}{5} \cdot \dfrac{24}{25} = \dfrac{3}{5}$.

28. $\cos \alpha < 0$ since α is in the second quadrant, so the formula $\sin^2 \alpha + \cos^2 \alpha = 1$ yields

$\cos \alpha = -\dfrac{3}{5}$. Likewise, $\cos \beta = -\dfrac{\sqrt{3}}{2}$. Hence $\sin(\alpha - \beta) = \sin \alpha \cos \beta -$

$\cos \alpha \sin \beta = \dfrac{4}{5}\left(-\dfrac{\sqrt{3}}{2}\right) - \left(-\dfrac{3}{5}\right)\dfrac{1}{2} = \dfrac{-4\sqrt{3}}{10} + \dfrac{3}{10} = \dfrac{3 - 4\sqrt{3}}{10}$.

29. Since α and β are in the first quadrant and $\tan^2 x + 1 = \sec^2 x$, $\sec \alpha = \dfrac{5}{3}$ and

$\sec \beta = \dfrac{13}{5}$; $\cos \alpha = \dfrac{3}{5}$ and $\cos \beta = \dfrac{5}{13}$. The formula $\sin^2 x + \cos^2 x = 1$ yields

$\sin \alpha = \dfrac{4}{5}$ and $\sin \beta = \dfrac{12}{13}$. $\cos(\alpha - \beta) = \cos \alpha \cos \beta + \sin \alpha \sin \beta =$

$\dfrac{3}{5} \cdot \dfrac{5}{13} + \dfrac{4}{5} \cdot \dfrac{12}{13} = \dfrac{63}{65}$

30. Since α is acute and $\sec \alpha = \dfrac{5}{4}$, $\cos \alpha = \dfrac{4}{5}$; the formula $\sin^2 x + \cos^2 x = 1$ yields

$\sin \alpha = \dfrac{3}{5}$. Since β is in the second quadrant and $\tan \beta = -1$, $\beta = 135°$, $\sin \beta = \dfrac{\sqrt{2}}{2}$,

and $\cos \beta = -\dfrac{\sqrt{2}}{2}$. $\cos(\alpha + \beta) = \cos \alpha \cos \beta - \sin \alpha \sin \beta = \dfrac{4}{5}\left(-\dfrac{\sqrt{2}}{2}\right) -$

$\dfrac{3}{5} \cdot \dfrac{\sqrt{2}}{2} = -\dfrac{7\sqrt{2}}{10}$.

31. $y = \sin x \cos 1 + \cos x \sin 1 = \sin(x + 1)$. The graph is the graph of $y = \sin x$ translated one unit to the left.

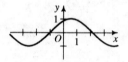   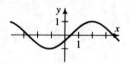

Ex. 31   **Ex. 32**

32. $y = \cos x \cos 2 + \sin x \sin 2 = \cos(x - 2)$. The graph is the graph of $y = \cos x$ translated two units to the right.

33. $\dfrac{\sin(\alpha + \beta) + \sin(\alpha - \beta)}{\cos \alpha \cos \beta} = \dfrac{\sin \alpha \cos \beta + \cos \alpha \sin \beta + \sin \alpha \cos \beta - \cos \alpha \sin \beta}{\cos \alpha \cos \beta} =$

$\dfrac{2 \sin \alpha \cos \beta}{\cos \alpha \cos \beta} = 2 \tan \alpha$

34. $\dfrac{\cos(\alpha + \beta) + \cos(\alpha - \beta)}{\sin \alpha \sin \beta} = \dfrac{\cos \alpha \cos \beta - \sin \alpha \sin \beta + \cos \alpha \cos \beta + \sin \alpha \sin \beta}{\sin \alpha \sin \beta} =$

$\dfrac{2 \cos \alpha \cos \beta}{\sin \alpha \sin \beta} = 2 \cot \alpha \cot \beta$

35. $\cos x \cos y \left(\dfrac{\sin x}{\cos x} + \dfrac{\sin y}{\cos y}\right) = \sin x \cos y + \cos x \sin y = \sin(x + y)$

36. $\sin x \sin y \left(\dfrac{\cos x}{\sin x} \cdot \dfrac{\cos y}{\sin y} - 1\right) = \cos x \cos y - \sin x \sin y = \cos(x + y)$

37. $\sin(x + y) \sec x \sec y = (\sin x \cos y + \cos x \sin y) \cdot \left(\dfrac{1}{\cos x \cos y}\right) =$

$\dfrac{\sin x}{\cos x} + \dfrac{\sin y}{\cos y} = \tan x + \tan y$

38. $\cos \left(x + \dfrac{\pi}{3}\right) + \sin \left(x - \dfrac{\pi}{6}\right) = \cos x \cos \dfrac{\pi}{3} - \sin x \sin \dfrac{\pi}{3} + \sin x \cos \dfrac{\pi}{6} -$

$\cos x \sin \dfrac{\pi}{6} = \dfrac{1}{2} \cos x - \dfrac{\sqrt{3}}{2} \sin x + \dfrac{\sqrt{3}}{2} \sin x - \dfrac{1}{2} \cos x = 0$

39. a. $\sin(\alpha + \beta) + \sin(\alpha - \beta) = \sin \alpha \cos \beta + \cos \alpha \sin \beta + \sin \alpha \cos \beta -$

$\cos \alpha \sin \beta = 2 \sin \alpha \cos \beta$; letting $\alpha = \dfrac{x + y}{2}$ and $\beta = \dfrac{x - y}{2}$ gives

$$\sin\left(\frac{x + y}{2} + \frac{x - y}{2}\right) + \sin\left(\frac{x + y}{2} - \frac{x - y}{2}\right) = 2 \sin \frac{x + y}{2} \cos \frac{x - y}{2};$$

$$\sin x + \sin y = 2 \sin \frac{x + y}{2} \cos \frac{x - y}{2}.$$

b. $\sin 40° + \sin 20° = 2 \sin \dfrac{40° + 20°}{2} \cos \dfrac{40° - 20°}{2} = 2 \sin 30° \cos 10° =$

$2 \cdot \dfrac{1}{2} \cdot \cos 10° = \cos 10°$

c. $\sin x - \sin y = \sin x + \sin(-y) = 2 \sin \dfrac{x + (-y)}{2} \cos \dfrac{x - (-y)}{2} =$

$2 \sin \dfrac{x - y}{2} \cos \dfrac{x + y}{2}$

40. a. $\cos(\alpha + \beta) + \cos(\alpha - \beta) = \cos \alpha \cos \beta - \sin \alpha \sin \beta + \cos \alpha \cos \beta +$

$\sin \alpha \sin \beta = 2 \cos \alpha \cos \beta$; letting $\alpha = \dfrac{x + y}{2}$ and $\beta = \dfrac{x - y}{2}$ gives

$$\cos\left(\frac{x + y}{2} + \frac{x - y}{2}\right) + \cos\left(\frac{x + y}{2} - \frac{x - y}{2}\right) = 2 \cos \frac{x + y}{2} \cos \frac{x - y}{2};$$

$$\cos x + \cos y = 2 \cos \frac{x + y}{2} \cos \frac{x - y}{2}$$

b. $\cos 105° + \cos 15° = 2 \cos \dfrac{105° + 15°}{2} \cos \dfrac{105° - 15°}{2} = 2 \cos 60° \cos 45° =$

$2 \cdot \dfrac{1}{2} \cdot \dfrac{\sqrt{2}}{2} = \dfrac{\sqrt{2}}{2}$

41. a. $\cos(\alpha + \beta) - \cos(\alpha - \beta) = \cos \alpha \cos \beta - \sin \alpha \sin \beta -$

$(\cos \alpha \cos \beta + \sin \alpha \sin \beta) = -2 \sin \alpha \sin \beta$; letting $\alpha = \dfrac{x + y}{2}$ and $\beta = \dfrac{x - y}{2}$ gives

$$\cos\left(\frac{x + y}{2} + \frac{x - y}{2}\right) - \cos\left(\frac{x + y}{2} - \frac{x - y}{2}\right) = -2 \sin \frac{x + y}{2} \sin \frac{x - y}{2};$$

$$\cos x - \cos y = -2 \sin \frac{x + y}{2} \sin \frac{x - y}{2}.$$

b. $\cos 75° - \cos 15° = -2 \sin \dfrac{75° + 15°}{2} \sin \dfrac{75° - 15°}{2} = -2 \sin 45° \sin 30° =$

$-2 \cdot \dfrac{\sqrt{2}}{2} \cdot \dfrac{1}{2} = -\dfrac{\sqrt{2}}{2}$

42. $\sin\left(\text{Tan}^{-1} \dfrac{1}{2} + \text{Tan}^{-1} \dfrac{1}{3}\right) = \sin\left(\text{Tan}^{-1} \dfrac{1}{2}\right) \cos\left(\text{Tan}^{-1} \dfrac{1}{3}\right) +$

$\cos\left(\text{Tan}^{-1} \dfrac{1}{2}\right) \sin\left(\text{Tan}^{-1} \dfrac{1}{3}\right) = \dfrac{1}{\sqrt{5}} \cdot \dfrac{3}{\sqrt{10}} + \dfrac{2}{\sqrt{5}} \cdot \dfrac{1}{\sqrt{10}} =$

$\dfrac{3}{\sqrt{50}} + \dfrac{2}{\sqrt{50}} = \dfrac{5}{5\sqrt{2}} = \dfrac{1}{\sqrt{2}} = \dfrac{\sqrt{2}}{2}$

43. $\cos\left(\text{Tan}^{-1}\dfrac{1}{2} + \text{Tan}^{-1} 2\right) = \cos\left(\text{Tan}^{-1}\dfrac{1}{2}\right)\cos(\text{Tan}^{-1} 2) - $

$\sin\left(\text{Tan}^{-1}\dfrac{1}{2}\right)\sin(\text{Tan}^{-1} 2) = \dfrac{2}{\sqrt{5}}\cdot\dfrac{1}{\sqrt{5}} - \dfrac{1}{\sqrt{5}}\cdot\dfrac{2}{\sqrt{5}} = \dfrac{2}{5} - \dfrac{2}{5} = 0$

44. a. An angle inscribed in a semicircle is a rt. $\angle$, so $\angle DAB = 90°$;

$\sin\alpha = \dfrac{AB}{BD} = \dfrac{AB}{1} = AB;\ \cos\alpha = \dfrac{AD}{BD} = \dfrac{AD}{1} = AD.$

b. $\angle DCB = 90°$, so $\sin\beta = \dfrac{BC}{BD} = BC$ and $\cos\beta = \dfrac{CD}{BD} = CD.$

c. Arc $AB = 2\alpha$, so arc $AD = 180° - 2\alpha$; thus $\angle DCA = \dfrac{1}{2}(180° - 2\alpha) = 90° - \alpha;$

in $\triangle ACD$, $\dfrac{\sin\angle DCA}{AD} = \dfrac{\sin(\alpha+\beta)}{AC};\ \dfrac{\sin(90°-\alpha)}{AD} = \dfrac{\sin(\alpha+\beta)}{AC};\ \dfrac{\cos\alpha}{\cos\alpha} = \dfrac{\sin(\alpha+\beta)}{AC};$

$\dfrac{\sin(\alpha+\beta)}{AC} = 1;\ \sin(\alpha+\beta) = AC$

d. $AB\cdot CD + AD\cdot BC = AC\cdot BD$; substituting, $\sin\alpha\cos\beta + \cos\alpha\sin\beta = \sin(\alpha+\beta)\cdot 1 = \sin(\alpha+\beta)$

Class Exercises 10-2 · page 377

1. a. $\tan(\alpha+\beta) = \dfrac{\tan\alpha + \tan\beta}{1 - \tan\alpha\tan\beta} = \dfrac{2+3}{1 - 2\cdot 3} = \dfrac{5}{-5} = -1$

b. $\tan(\alpha-\beta) = \dfrac{\tan\alpha - \tan\beta}{1 + \tan\alpha\tan\beta} = \dfrac{2-3}{1 + 2\cdot 3} = -\dfrac{1}{7}$

2. a. $\tan(15° + 30°) = \tan 45° = 1$ **b.** $\tan(85° - 25°) = \tan 60° = \sqrt{3}$

3. If $1 + m_1 m_2 = 0$, then $m_1 m_2 = -1$ and $l_1 \perp l_2$.

Written Exercises 10-2 · pages 377–379

1. a. $\tan(\alpha+\beta) = \dfrac{\tan\alpha + \tan\beta}{1 - \tan\alpha\tan\beta} = \dfrac{\frac{2}{3} + \frac{1}{2}}{1 - \frac{2}{3}\cdot\frac{1}{2}} = \dfrac{\frac{7}{6}}{\frac{2}{3}} = \dfrac{7}{4}$

b. $\tan(\alpha-\beta) = \dfrac{\tan\alpha - \tan\beta}{1 + \tan\alpha\tan\beta} = \dfrac{\frac{2}{3} - \frac{1}{2}}{1 + \frac{2}{3}\cdot\frac{1}{2}} = \dfrac{\frac{1}{6}}{\frac{4}{3}} = \dfrac{1}{8}$

2. a. $\tan(\alpha+\beta) = \dfrac{\tan\alpha + \tan\beta}{1 - \tan\alpha\tan\beta} = \dfrac{2 + \left(-\frac{1}{3}\right)}{1 - 2\left(-\frac{1}{3}\right)} = \dfrac{\frac{5}{3}}{\frac{5}{3}} = 1$

b. $\tan(\alpha-\beta) = \dfrac{\tan\alpha - \tan\beta}{1 + \tan\alpha\tan\beta} = \dfrac{2 - \left(-\frac{1}{3}\right)}{1 + 2\left(-\frac{1}{3}\right)} = \dfrac{\frac{7}{3}}{\frac{1}{3}} = 7$

3. $\tan(75° - 30°) = \tan 45° = 1$ **4.** $\tan(100° + 50°) = \tan 150° = -\dfrac{\sqrt{3}}{3}$

5. $\tan\left(\dfrac{2\pi}{3} + \dfrac{\pi}{12}\right) = \tan\dfrac{9\pi}{12} = \tan\dfrac{3\pi}{4} = -1$ **6.** $\tan\left(\dfrac{4\pi}{3} - \dfrac{\pi}{12}\right) = \tan\dfrac{15\pi}{12} = \tan\dfrac{5\pi}{4} = 1$

7. $\tan\left(\dfrac{\pi}{4} + \theta\right) = \dfrac{\tan\frac{\pi}{4} + \tan\theta}{1 - \tan\frac{\pi}{4}\tan\theta} = \dfrac{1 + \frac{1}{2}}{1 - 1\cdot\frac{1}{2}} = \dfrac{\frac{3}{2}}{\frac{1}{2}} = 3$

8. $\tan\left(\dfrac{3\pi}{4} - \theta\right) = \dfrac{\tan\frac{3\pi}{4} - \tan\theta}{1 + \tan\frac{3\pi}{4}\tan\theta} = \dfrac{-1 - \frac{1}{3}}{1 + (-1)\left(\frac{1}{3}\right)} = \dfrac{-\frac{4}{3}}{\frac{2}{3}} = -2$

9. $\tan(-\alpha) = \tan(0° - \alpha) = \dfrac{\tan 0° - \tan\alpha}{1 + \tan 0°\tan\alpha} = -\tan\alpha$

10. $\tan(x + \pi) = \dfrac{\tan x + \tan\pi}{1 - \tan x\tan\pi} = \dfrac{\tan x + 0}{1 - \tan x\,(0)} = \tan x$. The period of the tangent function

is π. The graphs of $y = \tan x$ and $y = \tan(x + \pi)$ coincide.

11. $\tan 75° = \tan(45° + 30°) = \dfrac{\tan 45° + \tan 30°}{1 - \tan 45°\tan 30°} = \dfrac{1 + \dfrac{1}{\sqrt{3}}}{1 - 1\cdot\dfrac{1}{\sqrt{3}}} = \dfrac{\sqrt{3} + 1}{\sqrt{3} - 1} = \dfrac{4 + 2\sqrt{3}}{2} =$

$2 + \sqrt{3}$; $\tan 165° = \tan(120° + 45°) = \dfrac{\tan 120° + \tan 45°}{1 - \tan 120°\tan 45°} = \dfrac{-\sqrt{3} + 1}{1 - (-\sqrt{3})(1)} =$

$\dfrac{1 - \sqrt{3}}{1 + \sqrt{3}} = \dfrac{4 - 2\sqrt{3}}{-2} = -2 + \sqrt{3}$

12. $\tan 15° = \tan(45° - 30°) = \dfrac{\tan 45° - \tan 30°}{1 + \tan 45°\tan 30°} = \dfrac{1 - \dfrac{1}{\sqrt{3}}}{1 + 1\cdot\dfrac{1}{\sqrt{3}}} = \dfrac{\sqrt{3} - 1}{\sqrt{3} + 1} = \dfrac{4 - 2\sqrt{3}}{2} =$

$2 - \sqrt{3}$; $\tan 105° = \tan(60° + 45°) = \dfrac{\tan 60° + \tan 45°}{1 - \tan 60°\tan 45°} = \dfrac{\sqrt{3} + 1}{1 - \sqrt{3}\cdot 1} = \dfrac{\sqrt{3} + 1}{1 - \sqrt{3}} =$

$\dfrac{4 + 2\sqrt{3}}{-2} = -2 - \sqrt{3}$

13. If $m_1 = 3$ and $m_2 = 1$, then $\tan\theta_1 = \dfrac{3 - 1}{1 + 3\cdot 1} = \dfrac{2}{4} = 0.5$; $\theta_1 \approx 26.6°$ and $\theta_2 = 180° - $

$\theta_1 \approx 153.4°$. (If $m_1 = 1$ and $m_2 = 3$, $\tan\theta_1 = -0.5$, $\theta_1 \approx 153.4°$ and $\theta_2 \approx 26.6°$.)

14. $y = -\dfrac{3}{2}x + \dfrac{5}{2}$ and $y = \dfrac{4}{3}x - \dfrac{1}{3}$. If $m_1 = -\dfrac{3}{2}$ and $m_2 = \dfrac{4}{3}$, then $\tan\theta_1 =$

$\dfrac{-\frac{3}{2} - \frac{4}{3}}{1 + \left(-\frac{3}{2}\right)\left(\frac{4}{3}\right)} = \dfrac{-\frac{17}{6}}{-1} = \dfrac{17}{6}$; $\theta_1 \approx 70.6°$ and $\theta_2 = 180° - \theta_1 \approx 109.4°$.

(If $m_1 = \dfrac{4}{3}$ and $m_2 = -\dfrac{3}{2}$, $\tan\theta_1 = -\dfrac{17}{6}$, $\theta_1 \approx 109.4°$ and $\theta_2 \approx 70.6°$.)

15. $y = \dfrac{\tan x + \tan 1}{1 - \tan x \tan 1} = \tan(x + 1)$. The graph of $y = \tan(x + 1)$ is the graph of

$y = \tan x$ translated one unit to the left.

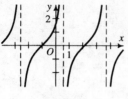

Ex. 15

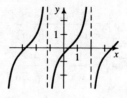

Ex. 16

16. $y = \dfrac{\tan x - \tan \dfrac{\pi}{8}}{1 + \tan x \tan \dfrac{\pi}{8}} = \tan\left(x - \dfrac{\pi}{8}\right)$. The graph of $y = \tan\left(x - \dfrac{\pi}{8}\right)$ is the graph of

$y = \tan x$ translated $\dfrac{\pi}{8}$ units to the right.

17. $\tan(\alpha + \beta) = \dfrac{\tan \alpha + \tan \beta}{1 - \tan \alpha \tan \beta} = \dfrac{\frac{1}{4} + \frac{3}{5}}{1 - \frac{1}{4} \cdot \frac{3}{5}} = \dfrac{\frac{17}{20}}{\frac{17}{20}} = 1$. Let $\alpha = \text{Tan}^{-1} \dfrac{1}{4}$ and $\beta =$

$\text{Tan}^{-1} \dfrac{3}{5}$; α and β are acute, so $0 < \alpha + \beta < \pi$; $\tan(\alpha + \beta) = 1$, so $\alpha + \beta = \dfrac{\pi}{4}$;

substituting, $\text{Tan}^{-1} \dfrac{1}{4} + \text{Tan}^{-1} \dfrac{3}{5} = \dfrac{\pi}{4}$.

18. $\tan(\alpha - \beta) = \dfrac{\tan \alpha - \tan \beta}{1 + \tan \alpha \tan \beta} = \dfrac{3 - \frac{1}{2}}{1 + 3 \cdot \frac{1}{2}} = \dfrac{\frac{5}{2}}{\frac{5}{2}} = 1$. Let $\alpha = \text{Tan}^{-1} 3$ and $\beta =$

$\text{Tan}^{-1} \dfrac{1}{2}$; α and β are acute, so $0 < \alpha - \beta < \dfrac{\pi}{2}$; $\tan(\alpha - \beta) = 1$, so $\alpha - \beta = \dfrac{\pi}{4}$;

substituting, $\text{Tan}^{-1} 3 - \text{Tan}^{-1} \dfrac{1}{2} = \dfrac{\pi}{4}$.

19. If $\alpha = \text{Tan}^{-1} 2$ and $\beta = \text{Tan}^{-1} 3$, then $\tan \alpha = 2$ and $\tan \beta = 3$; $\tan(\alpha + \beta) =$

$\dfrac{\tan \alpha + \tan \beta}{1 - \tan \alpha \tan \beta} = \dfrac{2 + 3}{1 - 2 \cdot 3} = \dfrac{5}{-5} = -1$.

20. If $\alpha = \text{Tan}^{-1} 5$ and $\beta = \text{Tan}^{-1} \dfrac{2}{3}$, then $\tan \alpha = 5$ and $\tan \beta = \dfrac{2}{3}$; $\tan(\alpha - \beta) =$

$\dfrac{\tan \alpha - \tan \beta}{1 + \tan \alpha \tan \beta} = \dfrac{5 - \frac{2}{3}}{1 + 5 \cdot \frac{2}{3}} = \dfrac{\frac{13}{3}}{\frac{13}{3}} = 1 = \tan \dfrac{\pi}{4}$; since α and β are acute with $\alpha > \beta$

(since $\tan \alpha > \tan \beta$), $0 < \alpha - \beta < \dfrac{\pi}{2}$; thus, $\alpha - \beta = \dfrac{\pi}{4}$.

21. $\tan(\alpha + \beta) = \dfrac{\tan\alpha + \tan\beta}{1 - \tan\alpha\,\tan\beta} = \tan\alpha + \tan\beta$ when $1 - \tan\alpha\,\tan\beta = 1$;

$1 - \tan\alpha\,\tan\beta = 1$ when $\tan\alpha\,\tan\beta = 0$, which occurs when α or $\beta = n\pi$ where n is an integer. Similarly, $\tan(\alpha - \beta) = \tan\alpha - \tan\beta$ when α or $\beta = n\pi$ where n is an integer.

22. Graphing shows that $\angle BAC$ is acute.

a. $m_2 = \text{slope of } \overleftrightarrow{AB} = \dfrac{-1 - 1}{14 - 3} = -\dfrac{2}{11}$; $m_1 = \text{slope of } \overleftrightarrow{AC} = \dfrac{5 - 1}{5 - 3} = 2$; $\tan \angle BAC =$

$\dfrac{2 - \left(-\frac{2}{11}\right)}{1 + 2\left(-\frac{2}{11}\right)} = \dfrac{\frac{24}{11}}{\frac{7}{11}} = \dfrac{24}{7} \approx 3.4286$; $\angle BAC \approx 73.7°$.

b. The distance formula yields $AB = \sqrt{125}$, $BC = \sqrt{117}$, and $AC = \sqrt{20}$;
$(BC)^2 = (AB)^2 + (AC)^2 - 2(AB)(AC)\cos\angle BAC$; $117 = 125 + 20 -$
$2 \cdot \sqrt{125} \cdot \sqrt{20} \cdot \cos\angle BAC$; $2 \cdot 50 \cos\angle BAC = 28$; $\cos\angle BAC = 0.28$;
$\angle BAC \approx 73.7°$.

23. $\cot(\alpha + \beta) = \dfrac{1}{\tan(\alpha + \beta)}$, where $\tan\alpha = \dfrac{1}{2}$ (since $\cot\alpha = 2$) and $\tan\beta =$

$\dfrac{3}{2}\left(\text{since } \cot\beta = \dfrac{2}{3}\right)$; $\tan(\alpha + \beta) = \dfrac{\tan\alpha + \tan\beta}{1 - \tan\alpha\,\tan\beta} = \dfrac{\frac{1}{2} + \frac{3}{2}}{1 - \frac{1}{2} \cdot \frac{3}{2}} = \dfrac{2}{\frac{1}{4}} = 8$;

$\cot(\alpha + \beta) = \dfrac{1}{8}$

24. $\cot(\alpha - \beta) = \dfrac{1}{\tan(\alpha - \beta)}$, where $\tan\alpha = \dfrac{2}{3}\left(\text{since } \cot\alpha = \dfrac{3}{2}\right)$ and $\tan\beta =$

$2\left(\text{since } \cot\beta = \dfrac{1}{2}\right)$; $\tan(\alpha - \beta) = \dfrac{\tan\alpha - \tan\beta}{1 + \tan\alpha\,\tan\beta} = \dfrac{\frac{2}{3} - 2}{1 + \frac{2}{3} \cdot 2} = \dfrac{-\frac{4}{3}}{\frac{7}{3}} = -\dfrac{4}{7}$;

$\cot(\alpha - \beta) = -\dfrac{7}{4}$

25. Since $\sin\alpha = \dfrac{3}{5}$, $\cos\alpha = \dfrac{4}{5}$ and $\tan\alpha = \dfrac{3}{4}$; $\cos\beta = \dfrac{5}{13}$ so $\sin\beta = \dfrac{12}{13}$ and $\tan\beta = \dfrac{12}{5}$.

a. $\sin(\alpha + \beta) = \sin\alpha\cos\beta + \cos\alpha\sin\beta = \dfrac{3}{5} \cdot \dfrac{5}{13} + \dfrac{4}{5} \cdot \dfrac{12}{13} = \dfrac{15}{65} + \dfrac{48}{65} = \dfrac{63}{65}$

b. $\cos(\alpha + \beta) = \cos\alpha\cos\beta - \sin\alpha\sin\beta = \dfrac{4}{5} \cdot \dfrac{5}{13} - \dfrac{3}{5} \cdot \dfrac{12}{13} = \dfrac{20}{65} - \dfrac{36}{65} = -\dfrac{16}{65}$

c. $\tan(\alpha + \beta) = \dfrac{\tan\alpha + \tan\beta}{1 - \tan\alpha\,\tan\beta} = \dfrac{\frac{3}{4} + \frac{12}{5}}{1 - \frac{3}{4} \cdot \frac{12}{5}} = \dfrac{\frac{63}{20}}{-\frac{16}{20}} = -\dfrac{63}{16}$

26. Since $\dfrac{\pi}{2} < \alpha < \beta < \pi$ and $\sin \alpha = \dfrac{4}{5}$, $\cos \alpha = -\dfrac{3}{5}$ and $\tan \alpha = -\dfrac{4}{3}$;

since $\tan \beta = -\dfrac{3}{4}$, $\sec \beta = -\dfrac{5}{4}$ so $\cos \beta = -\dfrac{4}{5}$ and $\sin \beta = \dfrac{3}{5}$.

a. $\sin (\alpha + \beta) = \sin \alpha \cos \beta + \cos \alpha \sin \beta = \dfrac{4}{5}\left(-\dfrac{4}{5}\right) + \left(-\dfrac{3}{5}\right)\dfrac{3}{5} = -\dfrac{16}{25} - \dfrac{9}{25} = -1$

b. $\cos (\alpha + \beta) = \cos \alpha \cos \beta - \sin \alpha \sin \beta = \left(-\dfrac{3}{5}\right)\left(-\dfrac{4}{5}\right) - \dfrac{4}{5}\cdot\dfrac{3}{5} = \dfrac{12}{25} - \dfrac{12}{25} = 0$

c. $\tan (\alpha + \beta) = \dfrac{\tan \alpha + \tan \beta}{1 - \tan \alpha \tan \beta} = \dfrac{-\dfrac{4}{3} + \left(-\dfrac{3}{4}\right)}{1 - \left(-\dfrac{4}{3}\right)\left(-\dfrac{3}{4}\right)} = \dfrac{-\dfrac{4}{3} - \dfrac{3}{4}}{0}$, which is not defined.

27. a. We can find the tangent of the acute angle between l and l_1: $\tan \theta_1 = \dfrac{m_1 - m}{1 + m_1 m}$.

Similarly, the tangent of the acute angle between l_2 and l: $\tan \theta_2 = \dfrac{m - m_2}{1 + m_2 m}$. Since l

bisects the angle between l_1 and l_2, $\theta_1 = \theta_2$. Hence $\tan \theta_1 = \tan \theta_2$, so $\dfrac{m_1 - m}{1 + m_1 m} =$

$\dfrac{m - m_2}{1 + m_2 m}$.

b. $m_1 = 2$ and $m_2 = 1$, so $\dfrac{2 - m}{1 + 2m} = \dfrac{m - 1}{1 + 1 \cdot m}$; $2 + m - m^2 = 2m^2 - m - 1$;

$3m^2 - 2m - 3 = 0$; $m = \dfrac{2 \pm \sqrt{4 - 4 \cdot 3(-3)}}{2 \cdot 3} = \dfrac{2 \pm \sqrt{40}}{6} = \dfrac{1 \pm \sqrt{10}}{3}$. Since l has a

positive slope, $m = \dfrac{1 + \sqrt{10}}{3}$. l contains the intersection point of l_1 and l_2, which is

$(0,0)$, so an equation for l is $\dfrac{y - 0}{x - 0} = \dfrac{1 + \sqrt{10}}{3}$, or $y = \dfrac{1 + \sqrt{10}}{3}x$.

28. $\angle A + \angle B + \angle C = 180°$; $\tan (A + B) = \dfrac{\tan A + \tan B}{1 - \tan A \tan B}$, so $\tan A + \tan B =$

$\tan (A + B) \cdot (1 - \tan A \tan B)$ if $A + B \neq 90°$; $\tan A + \tan B + \tan C =$

$\tan (A + B) \cdot (1 - \tan A \tan B) + \tan C = \tan (180° - \angle C) \cdot (1 - \tan A \tan B) +$

$\tan C = -\tan C \cdot (1 - \tan A \tan B) + \tan C = -\tan C + \tan A \tan B \tan C +$

$\tan C = \tan A \tan B \tan C$

29. Let $\alpha = \text{Tan}^{-1} \dfrac{1}{5}$ and $\beta = \text{Tan}^{-1} \dfrac{1}{239}$. $\tan 2x = \tan (x + x) = \dfrac{\tan x + \tan x}{1 - \tan x \tan x} =$

$\dfrac{2 \tan x}{1 - \tan^2 x}$, so $\tan 4x = \tan 2(2x) = \dfrac{2 \tan 2x}{1 - \tan^2 2x}$. $\tan \alpha = \dfrac{1}{5}$, so $\tan 2\alpha =$

$\dfrac{2\left(\frac{1}{5}\right)}{1 - \left(\frac{1}{5}\right)^2} = \dfrac{5}{12}$ and $\tan 4\alpha = \dfrac{2\left(\frac{5}{12}\right)}{1 - \left(\frac{5}{12}\right)^2} = \dfrac{120}{119}$. $\tan \left[4\,\text{Tan}^{-1} \dfrac{1}{5} - \text{Tan}^{-1} \dfrac{1}{239}\right] =$

$\tan (4\alpha - \beta) = \dfrac{\tan 4\alpha - \tan \beta}{1 + \tan 4\alpha \tan \beta} = \dfrac{\dfrac{120}{119} - \dfrac{1}{239}}{1 + \dfrac{120}{119}\cdot\dfrac{1}{239}} = \dfrac{\dfrac{28561}{28441}}{\dfrac{28561}{28441}} = 1$. Since $\tan \alpha < 1$,

$0 < \alpha < \dfrac{\pi}{4}$ and $0 < 4\alpha < \pi$; since $0 < \beta < \dfrac{\pi}{2}$ and $\tan \alpha > \tan \beta$, $\alpha > \beta$; thus,

$0 < 4\alpha - \beta < \pi$. Since $\dfrac{\pi}{4}$ is the only angle between 0 and π whose tangent is 1,

$\tan (4\alpha - \beta) = \tan \dfrac{\pi}{4}$; thus, $4\alpha - \beta = \dfrac{\pi}{4}$; $4 \ \text{Tan}^{-1} \dfrac{1}{5} - \text{Tan}^{-1} \dfrac{1}{239} = \dfrac{\pi}{4}$.

Communication · page 379

1. The maximum length is $n - 1$. There can be only $n - 1$ different remainders when 1 is divided by n. So, if no repetition occurs in the first $n - 1$ divisions, the nth division must produce a remainder that has occurred before.

2. Although there are an infinite number of rational and irrational numbers, there are more irrational numbers. This is because the rational numbers can be put in a one-to-one correspondence with the counting numbers and the irrational numbers cannot.

3. No, consider the graph of $y = \sin x + \sin (2x)$.

4. $b^0 = b^2 \cdot b^{-2} = b^2 \cdot \dfrac{1}{b^2} = \dfrac{b^2}{b^2} = 1$. In the example given, if $b = 0$, then we have $\dfrac{b^2}{b^2} = \dfrac{0}{0}$ which is undefined.

Activity 10-3 · page 380

1. The graphs of $y = \sin 2x$ and $y = 2 \sin x$ do not coincide; the values are different.

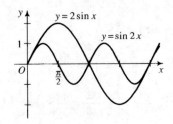

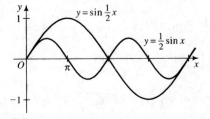

Activity 1   Activity 2

2. The graphs of $y = \sin \dfrac{1}{2} x$ and $y = \dfrac{1}{2} \sin x$ do not coincide; the values are different.

Class Exercises 10-3 · page 383

1. $\sin (2 \cdot 10°) = \sin 20°$ 2. $\cos (2 \cdot 15°) = \cos 30° = \dfrac{\sqrt{3}}{2}$ 3. $\cos (2 \cdot 35°) = \cos 70°$

4. $\cos (2 \cdot 25°) = \cos 50°$ 5. $\tan (2 \cdot 50°) = \tan 100°$ 6. $\tan (2 \cdot 40°) = \tan 80°$

7. $\cos^2 x$ 8. $\cos 2x$ 9. $2 \sin 3\alpha \cos 3\alpha = \sin 6\alpha$ 10. $\cos^2 5\theta - \sin^2 5\theta = \cos 10\theta$

11. Use the formula for $\cos \dfrac{\alpha}{2}$ with $\alpha = 70°$. $\cos 35° = \cos \dfrac{70°}{2} = \sqrt{\dfrac{1 + \cos 70°}{2}} =$

$\sqrt{\dfrac{1.342}{2}} = \sqrt{0.671}$

Written Exercises 10-3 · pages 383–385

1. $2 \cos^2 10° - 1 = \cos (2 \cdot 10°) = \cos 20°$ **2.** $2 \sin \dfrac{\alpha}{2} \cos \dfrac{\alpha}{2} = \sin \left(2 \cdot \dfrac{\alpha}{2}\right) = \sin \alpha$

3. $\dfrac{4 \tan \beta}{1 - \tan^2 \beta} = 2 \tan 2\beta$ **4.** $1 - 2 \sin^2 20° = \cos (2 \cdot 20°) = \cos 40°$

5. $2 \sin 35° \cos 35° = \sin (2 \cdot 35°) = \sin 70°$

6. $\cos^2 4A - \sin^2 4A = \cos (2 \cdot 4A) = \cos 8A$

7. $\dfrac{2 \tan 25°}{1 - \tan^2 25°} = \tan (2 \cdot 25°) = \tan 50°$ **8.** $2 \cos^2 3\alpha - 1 = \cos (2 \cdot 3\alpha) = \cos 6\alpha$

9. $1 - 2 \sin^2 \dfrac{x}{2} = \cos \left(2 \cdot \dfrac{x}{2}\right) = \cos x$

10. $\cos^2 40° - \sin^2 40° = \cos (2 \cdot 40°) = \cos 80°$

11. $\sqrt{\dfrac{1 - \cos 80°}{2}} = \sin \left(\dfrac{1}{2} \cdot 80°\right) = \sin 40°$

12. $\sqrt{\dfrac{1 + \cos 70°}{2}} = \cos \left(\dfrac{1}{2} \cdot 70°\right) = \cos 35°$

13. $\cos \left(2 \cdot \dfrac{\pi}{8}\right) = \cos \dfrac{\pi}{4} = \dfrac{\sqrt{2}}{2}$ **14.** $\tan \left(2 \cdot \dfrac{\pi}{8}\right) = \tan \left(\dfrac{\pi}{4}\right) = 1$

15. $\cos \left(2 \cdot \dfrac{\pi}{12}\right) = \cos \dfrac{\pi}{6} = \dfrac{\sqrt{3}}{2}$ **16.** $\cos \left(2 \cdot \dfrac{7\pi}{12}\right) = \cos \left(\dfrac{7\pi}{6}\right) = -\dfrac{\sqrt{3}}{2}$

17. $\dfrac{1}{2} (2 \sin 15° \cos 15°) = \dfrac{1}{2} \sin (2 \cdot 15°) = \dfrac{1}{2} \sin 30° = \dfrac{1}{2} \cdot \dfrac{1}{2} = \dfrac{1}{4}$

18. $2 \left(2 \sin \dfrac{2\pi}{3} \cos \dfrac{2\pi}{3}\right) = 2 \sin \left(2 \cdot \dfrac{2\pi}{3}\right) = 2 \sin \dfrac{4\pi}{3} = 2 \left(-\dfrac{\sqrt{3}}{2}\right) = -\sqrt{3}$

19. $\cos A = \dfrac{12}{13}$; $\tan A = \dfrac{5}{12}$; $\sin 2A = 2 \sin A \cos A = 2 \cdot \dfrac{5}{13} \cdot \dfrac{12}{13} = \dfrac{120}{169}$; $\cos 2A =$

$\cos^2 A - \sin^2 A = \dfrac{144}{169} - \dfrac{25}{169} = \dfrac{119}{169}$

20. If $\tan A = \dfrac{1}{2}$, $\sec A = \sqrt{\dfrac{5}{4}}$ and $\cos A = \dfrac{2}{\sqrt{5}}$; $\sin A = \dfrac{1}{\sqrt{5}}$; $\cos 2A = \cos^2 A -$

$\sin^2 A = \dfrac{4}{5} - \dfrac{1}{5} = \dfrac{3}{5}$; $\tan 2A = \dfrac{2 \tan A}{1 - \tan^2 A} = \dfrac{2 \cdot \frac{1}{2}}{1 - \frac{1}{4}} = \dfrac{1}{\frac{3}{4}} = \dfrac{4}{3}$

21. $\cos A = \dfrac{4}{5}$; $\sin 2A = 2 \sin A \cos A = 2 \cdot \dfrac{3}{5} \cdot \dfrac{4}{5} = \dfrac{24}{25}$; $\cos 2A = \dfrac{7}{25}$; $\sin 4A =$

$2 \sin 2A \cos 2A = 2 \cdot \dfrac{24}{25} \cdot \dfrac{7}{25} = \dfrac{336}{625}$

22. $\cos 2A = 2 \cos^2 A - 1 = 2 \left(\dfrac{1}{3}\right)^2 - 1 = -\dfrac{7}{9}$; $\cos 4A = 2 \cos^2 2A - 1 =$

$2 \left(-\dfrac{7}{9}\right)^2 - 1 = \dfrac{17}{81}$

23. $\cos 2A = 2\cos^2 A - 1 = 2\left(\dfrac{1}{5}\right)^2 - 1 = -\dfrac{23}{25}$; $\cos\dfrac{A}{2} = \sqrt{\dfrac{1 + \cos A}{2}} = \sqrt{\dfrac{1 + \dfrac{1}{5}}{2}} =$

$\sqrt{\dfrac{3}{5}} = \dfrac{\sqrt{15}}{5}$

24. $\sin A = \dfrac{\sqrt{15}}{4}$; $\sin 2A = 2\sin A\cos A = 2\cdot\dfrac{\sqrt{15}}{4}\cdot\dfrac{1}{4} = \dfrac{\sqrt{15}}{8}$; $\sin\dfrac{A}{2} = \sqrt{\dfrac{1 - \cos A}{2}} =$

$\sqrt{\dfrac{1 - \dfrac{1}{4}}{2}} = \sqrt{\dfrac{3}{8}} = \dfrac{\sqrt{6}}{4}$

25. a. $\cos(60° + 45°) = \cos 60°\cos 45° - \sin 60°\sin 45° = \dfrac{1}{2}\cdot\dfrac{\sqrt{2}}{2} - \dfrac{\sqrt{3}}{2}\cdot\dfrac{\sqrt{2}}{2} =$

$\dfrac{\sqrt{2} - \sqrt{6}}{4}$

b. $\cos\dfrac{210°}{2} = -\sqrt{\dfrac{1 + \cos 210°}{2}} = -\sqrt{\dfrac{1 - \dfrac{\sqrt{3}}{2}}{2}} = -\sqrt{\dfrac{2 - \sqrt{3}}{4}} = -\dfrac{\sqrt{2 - \sqrt{3}}}{2}$

26. a. $\sin(45° + 30°) = \sin 45°\cos 30° + \cos 45°\sin 30° = \dfrac{\sqrt{2}}{2}\cdot\dfrac{\sqrt{3}}{2} + \dfrac{\sqrt{2}}{2}\cdot\dfrac{1}{2} =$

$\dfrac{\sqrt{6} + \sqrt{2}}{4}$

b. $\sin\dfrac{150°}{2} = \sqrt{\dfrac{1 - \cos 150°}{2}} = \sqrt{\dfrac{1 + \dfrac{\sqrt{3}}{2}}{2}} = \sqrt{\dfrac{2 + \sqrt{3}}{4}} = \dfrac{\sqrt{2 + \sqrt{3}}}{2}$

27. range: $-2 \le y \le 1.13$; period $= 2\pi$ **28.** range: $-2.78 \le y \le 2.78$; period $= 2\pi$

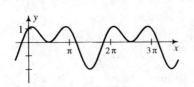

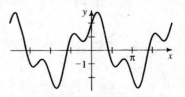

Ex. 27   Ex. 28

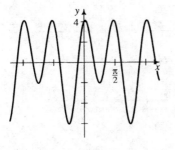

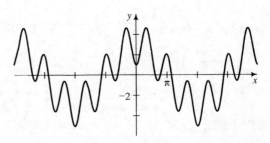

Ex. 29   Ex. 30

29. range: $-6 \le y \le 4.13$; period $= \pi$ **30.** range: $-5 \le y \le 4.62$; period $= 4\pi$

31. $\dfrac{\sin 2A}{1 - \cos 2A} = \dfrac{2 \sin A \cos A}{1 - (1 - 2 \sin^2 A)} = \dfrac{2 \sin A \cos A}{2 \sin^2 A} = \dfrac{\cos A}{\sin A} = \cot A$

32. $\dfrac{1 - \cos 2A}{1 + \cos 2A} = \dfrac{1 - (1 - 2 \sin^2 A)}{1 + 2 \cos^2 A - 1} = \dfrac{2 \sin^2 A}{2 \cos^2 A} = \left(\dfrac{\sin A}{\cos A}\right)^2 = \tan^2 A$

33. $\left(\sin \dfrac{x}{2} + \cos \dfrac{x}{2}\right)^2 = \sin^2 \dfrac{x}{2} + 2 \sin \dfrac{x}{2} \cos \dfrac{x}{2} + \cos^2 \dfrac{x}{2} = \left(\sin^2 \dfrac{x}{2} + \cos^2 \dfrac{x}{2}\right) +$

$\sin 2\left(\dfrac{x}{2}\right) = 1 + \sin x$

34. $\sin 4x = \sin 2(2x) = 2 \sin 2x \cos 2x = 2(2 \sin x \cos x) \cos 2x = 4 \sin x \cos x \cos 2x$

35. $\dfrac{1 - \tan^2 x}{1 + \tan^2 x} = \dfrac{1 - \dfrac{\sin^2 x}{\cos^2 x}}{1 + \dfrac{\sin^2 x}{\cos^2 x}} = \dfrac{\dfrac{\cos^2 x - \sin^2 x}{\cos^2 x}}{\dfrac{\cos^2 x + \sin^2 x}{\cos^2 x}} = \dfrac{\cos 2x}{\cos^2 x} \cdot \dfrac{\cos^2 x}{1} = \cos 2x$

36. $\dfrac{1 + \sin A - \cos 2A}{\cos A + \sin 2A} = \dfrac{1 + \sin A - (1 - 2 \sin^2 A)}{\cos A + 2 \sin A \cos A} = \dfrac{2 \sin^2 A + \sin A}{\cos A + 2 \sin A \cos A} =$

$\dfrac{\sin A(2 \sin A + 1)}{\cos A(1 + 2 \sin A)} = \dfrac{\sin A}{\cos A} = \tan A$

37. $\dfrac{\sin x}{1 + \cos x} = \dfrac{\sin 2\left(\dfrac{x}{2}\right)}{1 + \cos 2\left(\dfrac{x}{2}\right)} = \dfrac{2 \sin \dfrac{x}{2} \cos \dfrac{x}{2}}{1 + 2 \cos^2 \dfrac{x}{2} - 1} = \dfrac{2 \sin \dfrac{x}{2} \cos \dfrac{x}{2}}{2 \cos^2 \dfrac{x}{2}} = \dfrac{\sin \dfrac{x}{2}}{\cos \dfrac{x}{2}} = \tan \dfrac{x}{2}$

38. $\dfrac{1 - \cos x}{\sin x} = \dfrac{1 - \cos 2\left(\dfrac{x}{2}\right)}{\sin 2\left(\dfrac{x}{2}\right)} = \dfrac{1 - \left(1 - 2 \sin^2 \dfrac{x}{2}\right)}{2 \sin \dfrac{x}{2} \cos \dfrac{x}{2}} = \dfrac{2 \sin^2 \dfrac{x}{2}}{2 \sin \dfrac{x}{2} \cos \dfrac{x}{2}} = \dfrac{\sin \dfrac{x}{2}}{\cos \dfrac{x}{2}} = \tan \dfrac{x}{2}$

39. $\dfrac{1 + \cos 2x}{\cot x} = \dfrac{1 + 2 \cos^2 x - 1}{\dfrac{\cos x}{\sin x}} = 2 \cos^2 x \cdot \dfrac{\sin x}{\cos x} = 2 \cos x \sin x = \sin 2x$

40. $\dfrac{(1 + \tan^2 x)(1 - \cos 2x)}{2} = \dfrac{\sec^2 x(1 - (1 - 2 \sin^2 x))}{2} = \dfrac{2 \sin^2 x}{2 \cos^2 x} = \left(\dfrac{\sin x}{\cos x}\right)^2 = \tan^2 x$

41. $(1 - \sin^2 x)(1 - \tan^2 x) = \cos^2 x\left(1 - \dfrac{\sin^2 x}{\cos^2 x}\right) = \cos^2 x - \sin^2 x = \cos 2x$

42. $\sin x \tan x + \cos 2x \sec x = \sin x\left(\dfrac{\sin x}{\cos x}\right) + \dfrac{\cos^2 x - \sin^2 x}{\cos x} = \dfrac{\sin^2 x + \cos^2 x - \sin^2 x}{\cos x} =$

$\dfrac{\cos^2 x}{\cos x} = \cos x$

43. $\dfrac{\sin 3x}{\sin x} - \dfrac{\cos 3x}{\cos x} = \dfrac{\sin 3x \cos x - \cos 3x \sin x}{\sin x \cos x} = \dfrac{\sin (3x - x)}{\sin x \cos x} = \dfrac{\sin 2x}{\sin x \cos x} =$

$\dfrac{2 \sin x \cos x}{\sin x \cos x} = 2$

44. $\cos^2\left(\dfrac{\pi}{4} - \dfrac{x}{2}\right) - \sin^2\left(\dfrac{\pi}{4} - \dfrac{x}{2}\right) = \cos 2\left(\dfrac{\pi}{4} - \dfrac{x}{2}\right) = \cos\left(\dfrac{\pi}{2} - x\right) = \sin x$

45. a. $\angle BAO$ and $\angle BOC$ intercept the same arc. The measure of $\angle BOC$ = the measure of

the arc and $\angle BAO = \dfrac{1}{2}$ the measure of the arc $= \dfrac{1}{2}\angle BOC = \dfrac{\theta}{2}$.

b. In rt. $\triangle ABC$, $\tan \angle BAC = \dfrac{\text{opp.}}{\text{adj}} = \dfrac{\sin\theta}{1 + \cos\theta}$; $\tan \dfrac{\theta}{2} = \dfrac{\sin\theta}{1 + \cos\theta}$.

46. a. Each half of the rhombus is an isos. $\triangle$ with area $= \dfrac{1}{2}bh = \dfrac{1}{2}a \cdot a \sin 2\theta =$

$\dfrac{1}{2}a^2 \sin 2\theta$, so the area of the rhombus is $a^2 \sin 2\theta$.

b. Let $2x$ and $2y$ be the lengths of the diagonals, which are $\perp$. The rhombus consists of

$4 \cong$ rt. $\triangle$, so its area $= 4\left(\dfrac{1}{2}xy\right) = 2xy$; $x = a \sin\theta$ and $y = a \cos\theta$;

area $= 2xy = 2(a \sin\theta)(a \cos\theta) = 2a^2 \sin\theta \cos\theta$.

c. $a^2 \sin 2\theta = 2a^2 \sin\theta \cos\theta$; $\sin 2\theta = 2 \sin\theta \cos\theta$

47. $\log_2 2 + \log_2 (\sin x) + \log_2 (\cos x) = \log_2 (2 \cdot \sin x \cdot \cos x) = \log_2 (\sin 2x)$; when

$x = \dfrac{\pi}{12}$, $\log_2\left(\sin\left(2 \cdot \dfrac{\pi}{12}\right)\right) = \log_2\left(\sin\dfrac{\pi}{6}\right) = \log_2 \dfrac{1}{2} = -1$

48. $\dfrac{4^{2\cos^2\theta}}{4} = 4^{2\cos^2\theta - 1} = 4^{\cos 2\theta} = 4^{\cos(2 \cdot \pi/3)} = 4^{\cos 2\pi/3} = 4^{-1/2} = \dfrac{1}{2}$

49. $\sin 3x = \sin (2x + x) = \sin 2x \cos x + \cos 2x \sin x = 2 \sin x \cos x \cdot \cos x +$
$(1 - 2 \sin^2 x) \sin x = 2 \sin x(1 - \sin^2 x) + \sin x - 2 \sin^3 x = 2 \sin x -$
$2 \sin^3 x + \sin x - 2 \sin^3 x = 3 \sin x - 4 \sin^3 x$

50. $\cos 3x = \cos (2x + x) = \cos 2x \cos x - \sin 2x \sin x = (2 \cos^2 x - 1) \cos x -$
$2 \sin x \cos x \cdot \sin x = 2 \cos^3 x - \cos x - 2 \sin^2 x \cos x = 2 \cos^3 x - \cos x -$
$2(1 - \cos^2 x) \cos x = 2 \cos^3 x - \cos x - 2 \cos x + 2 \cos^3 x = 4 \cos^3 x - 3 \cos x$

51. a. $\angle R = \angle S = \dfrac{1}{2}(180° - 36°) = 72°$; $\angle TRS = \dfrac{1}{2}(72°) = 36°$; thus, $\angle Q = \angle TRS$ and

$\angle QRS = \angle RST$; by AA Similarity, $\triangle QRS \sim \triangle RST$

b. $\angle TRQ = \angle TQR = 36°$, so $QT = RT$, and since $\triangle RST$ is isos., $RT = x$; thus $QT = x$

and $ST = 1 - x$; $\dfrac{RS}{ST} = \dfrac{QR}{RS}$; $\dfrac{x}{1 - x} = \dfrac{1}{x}$

c. $\dfrac{x}{1 - x} = \dfrac{1}{x}$; $x^2 = 1 - x$; $x^2 + x - 1 = 0$; $x = \dfrac{-1 \pm \sqrt{1 - 4(1)(-1)}}{2 \cdot 1} = \dfrac{-1 \pm \sqrt{5}}{2}$;

$x > 0$, so $x = \dfrac{\sqrt{5} - 1}{2}$

d. The bis. of the vertex angle of an isos. $\triangle$ bisects the base and is $\perp$ to it; thus,

$\sin 18° = \cos 72° = \dfrac{\dfrac{1}{2}RS}{QR} = \dfrac{\dfrac{1}{2}x}{1} = \dfrac{1}{2}\left(\dfrac{\sqrt{5} - 1}{2}\right) = \dfrac{\sqrt{5} - 1}{4}$.

52. a. From the Law of Sines, $\dfrac{b}{\sin B} = \dfrac{c}{\sin C}$; $\dfrac{b}{\sin 2C} = \dfrac{c}{\sin C}$; $b = \dfrac{c}{\sin C} \cdot \sin 2C =$

$\dfrac{c}{\sin C} \cdot 2 \sin C \cos C = 2c \cos C$

b. $b^2 = a^2 + c^2 - 2ac \cos B = a^2 + c^2 - 2ac \cos 2C = a^2 + c^2 -$
$2ac(2 \cos^2 C - 1) = a^2 + c^2 + 2ac - 4ac \cos^2 C = a^2 + c^2 + 2ac -$
$(2c \cos C)(2a \cos C) = a^2 + c^2 + 2ac - b \cdot 2a\left(\dfrac{b^2 + a^2 - c^2}{2ab}\right)$ from part (a) and the
Law of Cosines; $b^2 = a^2 + c^2 + 2ac - b^2 - a^2 + c^2$; $2b^2 = 2c^2 + 2ac$;
$b^2 = c^2 + ac = c(a + c)$

53. a. The smallest angle, θ, is opp. the shortest side of length n and the largest angle, 2θ,
is opp. the side of length $n + 2$;

thus, $\dfrac{\sin \theta}{n} = \dfrac{\sin 2\theta}{n + 2}$; $\dfrac{\sin \theta}{n} = \dfrac{2 \sin \theta \cos \theta}{n + 2}$; $\cos \theta = \dfrac{n + 2}{2n}$.

b. $\cos \theta = \dfrac{(n + 2)^2 + (n + 1)^2 - n^2}{2(n + 2)(n + 1)} = \dfrac{n^2 + 6n + 5}{2(n + 2)(n + 1)} = \dfrac{(n + 5)(n + 1)}{2(n + 2)(n + 1)} = \dfrac{n + 5}{2(n + 2)}$

c. $\dfrac{n + 2}{2n} = \dfrac{n + 5}{2(n + 2)}$; $\dfrac{n + 2}{n} = \dfrac{n + 5}{n + 2}$; $n^2 + 4n + 4 = n^2 + 5n$; $4 = n$

Class Exercises 10-4 · page 389

1. Substitute $2 \cos^2 x - 1$ for $\cos 2x$ and solve algebraically for x.

2. Write as $\sin^2 x - \sin x = 0$ and solve algebraically for x by factoring.

3. Rewrite as $\tan x = 1$, $\cos x \neq 0$, and solve algebraically for x.

4. Rewrite as $\tan 2x = 1$, $\cos 2x \neq 0$, and solve algebraically for $2x$.

5. Rewrite as $\tan 3x = 1$, $\cos 3x \neq 0$, and solve algebraically for $3x$.

6. Solve algebraically for $x - 10°$.

7. Write the expressions in terms of $2x$ because the work is easier.

8. Write the expressions in terms of $2x$ because the work is easier.

9. a.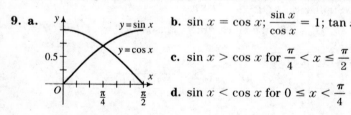

b. $\sin x = \cos x$; $\dfrac{\sin x}{\cos x} = 1$; $\tan x = 1$; $x = \dfrac{\pi}{4}$

c. $\sin x > \cos x$ for $\dfrac{\pi}{4} < x \leq \dfrac{\pi}{2}$

d. $\sin x < \cos x$ for $0 \leq x < \dfrac{\pi}{4}$

Written Exercises 10-4 · pages 389–391

1. $\cos 2x = \cos x$; $2 \cos^2 x - 1 = \cos x$; $(2 \cos x + 1)(\cos x - 1) = 0$; $\cos x = -\dfrac{1}{2}$ or 1;
$x = 0°, 120°, 240°$

2. $\sin^2 x = \sin x$; $\sin x(\sin x - 1) = 0$; $\sin x = 0$ or 1; $x = 0°, 90°, 180°$

3. $\sin x = \cos x$; $\dfrac{\sin x}{\cos x} = 1$ for $\cos x \neq 0$; $\tan x = 1$; $x = 45°, 225°$

4. $\sin 2x = \cos 2x$; $\dfrac{\sin 2x}{\cos 2x} = 1$ for $\cos 2x \neq 0$; $\tan 2x = 1$ and $0° \leq 2x < 720°$; $2x = 45°$,

225°, 405°, 585°; $x = 22.5°, 112.5°, 202.5°, 292.5°$

5. $\sin 3x = \cos 3x$; $\dfrac{\sin 3x}{\cos 3x} = 1$ for $\cos 3x \neq 0$; $\tan 3x = 1$ and $0° \leq 3x < 1080°$; $3x = 45°$,

225°, 405°, 585°, 765°, 945°; $x = 15°, 75°, 135°, 195°, 255°, 315°$

6. $\tan (x - 10°) = 1$; $x - 10° = 45°, 225°$; $x = 55°, 235°$

7. $\sin 4x = \sin 2x$; $2 \sin 2x \cos 2x = \sin 2x$; $2 \sin 2x \cos 2x - \sin 2x = 0$;

$(2 \cos 2x - 1) \sin 2x = 0$; $\cos 2x = \dfrac{1}{2}$ or $\sin 2x = 0$ and $0° \leq 2x < 720°$;

$2x = 60°, 300°, 420°, 660°$ or $0°, 180°, 360°, 540°$; $x = 0°, 30°, 90°, 150°, 180°, 210°$,
270°, 330°

8. $\cos 4x = 1 - 3 \cos 2x$; $2 \cos^2 2x - 1 = 1 - 3 \cos 2x$; $2 \cos^2 2x + 3 \cos 2x - 2 = 0$;

$(2 \cos 2x - 1)(\cos 2x + 2) = 0$; $\cos 2x = \dfrac{1}{2}$ or $\cos 2x = -2$ (no solutions);

$0° \leq 2x < 720°$; $2x = 60°, 300°, 420°, 660°$; $x = 30°, 150°, 210°, 330°$

9. a.

b. $\sin 2x = \tan x$; $2 \sin x \cos x = \dfrac{\sin x}{\cos x}$; $2 \sin x \cos^2 x - $

$\sin x = 0$; $\sin x(2 \cos^2 x - 1) = 0$; $\sin x = 0$ or $\cos x =$

$\pm\sqrt{\dfrac{1}{2}} = \pm\dfrac{\sqrt{2}}{2}$; $x = 0, \dfrac{\pi}{4}, \dfrac{3\pi}{4}, \pi, \dfrac{5\pi}{4}, \dfrac{7\pi}{4}$

c. From the graph, $\sin 2x > \tan x$ for $0 < x < \dfrac{\pi}{4}$,

$\dfrac{\pi}{2} < x < \dfrac{3\pi}{4}$, $\pi < x < \dfrac{5\pi}{4}$, $\dfrac{3\pi}{2} < x < \dfrac{7\pi}{4}$

10. a.

b. $\cos 2x = 3 \cos x$; $2\cos^2 x - 1 = 3 \cos x$;

$2 \cos^2 x - 3 \cos x - 1 = 0$; $\cos x = \dfrac{3 \pm \sqrt{9 + 8}}{4}$;

$\cos x = 1.78$ (no solutions) or $\cos x \approx -0.2808$;
$x = 1.86, 4.43$.

c. From the graph, $\cos 2x < 3 \cos x$ for
$0 \leq x < 1.86$, $4.43 < x < 2\pi$

Exercises 11–16 were solved using the method of Example 4.

11. $0 \leq x < 1.11$, $4.25 < x \leq 6.28$ **12.** $0 < x < 1.05$, $1.57 < x < 3.14$, $4.71 < x < 5.24$

13. $0.52 < x < 1.57$, $2.62 < x < 3.67$, $4.71 < x < 5.76$ **14.** $1.11 \leq x \leq 4.25$

15. $0.52 \leq x \leq 1.57$, $2.62 \leq x \leq 4.71$ **16.** No solution

17. $2 \cos (x + 45°) = 1$; $\cos (x + 45°) = \dfrac{1}{2}$; $x + 45° = 60°$ or $300°$; $x = 15°, 255°$

18. $\cot (x - 20°) = 1$; $x - 20° = 45°$ or $225°$; $x = 65°, 245°$

19. $\sin (60° - x) = 2 \sin x$; $\sin 60° \cos x - \cos 60° \sin x = 2 \sin x$; $\dfrac{\sqrt{3}}{2} \cos x - \dfrac{1}{2} \sin x =$

$2 \sin x$; $\dfrac{\sqrt{3}}{2} \cos x = \dfrac{5}{2} \sin x$; $\tan x = \dfrac{\sqrt{3}}{2} \div \dfrac{5}{2} = \dfrac{\sqrt{3}}{5}$; $x \approx 19.1°, 199.1°$

20. $2 \sin (30° + x) = 3 \cos x$; $2(\sin 30° \cos x + \cos 30° \sin x) = 3 \cos x$; $\cos x +$

$\sqrt{3} \sin x = 3 \cos x$; $\sqrt{3} \sin x = 2 \cos x$; $\tan x = \dfrac{2}{\sqrt{3}}$; $x \approx 49.1°, 229.1°$

21. $\sin x = \sin 2x$; $\sin x - \sin 2x = 0$; $\sin x - 2 \sin x \cos x = 0$; $\sin x(1 - 2 \cos x) = 0$;

$\sin x = 0$ or $\cos x = \dfrac{1}{2}$; $x = 0°, 60°, 180°, 300°$

22. $\tan^2 2x - 1 = 0$; $\tan^2 2x = 1$; $\tan 2x = \pm 1$; $2x = 45° + n \cdot 90°$, n an integer;
$x = 22.5° + n \cdot 45°$; $x = 22.5°, 67.5°, 112.5°, 157.5°, 202.5°, 247.5°, 292.5°, 337.5°$

23. $\sin x \cos x = \dfrac{1}{2}$; $2 \sin x \cos x = 1$; $\sin 2x = 1$; $2x = \dfrac{\pi}{2}, \dfrac{5\pi}{2}$; $x = \dfrac{\pi}{4}, \dfrac{5\pi}{4}$

24. $\cos (3x + \pi) = \dfrac{\sqrt{2}}{2}$; $\pi \le 3x + \pi < 7\pi$; $3x + \pi = \dfrac{7\pi}{4}, \dfrac{9\pi}{4}, \dfrac{15\pi}{4}, \dfrac{17\pi}{4}, \dfrac{23\pi}{4}, \dfrac{25\pi}{4}$; $x = \dfrac{\pi}{4}$,

$\dfrac{5\pi}{12}, \dfrac{11\pi}{12}, \dfrac{13\pi}{12}, \dfrac{19\pi}{12}, \dfrac{7\pi}{4}$

25. $\tan 2x = 3 \tan x$; $\dfrac{2 \tan x}{1 - \tan^2 x} = 3 \tan x$; $2 \tan x = 3 \tan x - 3 \tan^3 x$; $3 \tan^3 x -$

$\tan x = 0$; $\tan x (3 \tan^2 x - 1) = 0$; $\tan x = 0$ or $\pm \dfrac{\sqrt{3}}{3}$; $x = 0, \dfrac{\pi}{6}, \dfrac{5\pi}{6}, \pi, \dfrac{7\pi}{6}, \dfrac{11\pi}{6}$

26. $\tan 2x + \tan x = 0$; $\dfrac{2 \tan x}{1 - \tan^2 x} + \tan x = 0$; $2 \tan x + \tan x (1 - \tan^2 x) = 0$;

$3 \tan x - \tan^3 x = 0$; $\tan x (3 - \tan^2 x) = 0$; $\tan x = 0$ or $\pm \sqrt{3}$; $x = 0, \dfrac{\pi}{3}, \dfrac{2\pi}{3}, \pi,$

$\dfrac{4\pi}{3}, \dfrac{5\pi}{3}$

27. $\cos 2x = 5 \sin^2 x - \cos^2 x$; $\cos^2 x - \sin^2 x = 5 \sin^2 x - \cos^2 x$; $2 \cos^2 x = 6 \sin^2 x$;

$\dfrac{1}{3} = \dfrac{\sin^2 x}{\cos^2 x} = \tan^2 x$; $\tan x = \pm \dfrac{\sqrt{3}}{3}$; $x = \dfrac{\pi}{6}, \dfrac{5\pi}{6}, \dfrac{7\pi}{6}, \dfrac{11\pi}{6}$

28. $\sin 2x \sec x + 2 \cos x = 0$; $2 \sin x \cos x \cdot \dfrac{1}{\cos x} + 2 \cos x = 0$; $2(\sin x + \cos x) = 0$;

$\sin x = -\cos x$; $\tan x = -1$; $x = \dfrac{3\pi}{4}, \dfrac{7\pi}{4}$

29. $3 \sin x = 1 + \cos 2x$; $3 \sin x = 1 + 1 - 2 \sin^2 x$; $2 \sin^2 x + 3 \sin x - 2 = 0$;

$(2 \sin x - 1)(\sin x + 2) = 0$; $2 \sin x - 1 = 0$ or $\sin x + 2 = 0$; $\sin x = \dfrac{1}{2}$ or

$\sin x = -2$ (reject); $\sin x = \dfrac{1}{2}$ when $x = \dfrac{\pi}{6}, \dfrac{5\pi}{6}$.

30. $\sin 2x = 5 \cos^2 x$; $2 \sin x \cos x = 5 \cos^2 x$; $2 \sin x \cos x - 5 \cos^2 x = 0$;
$\cos x (2 \sin x - 5 \cos x) = 0$; $\cos x = 0$ or $2 \sin x - 5 \cos x = 0$; $\cos x = 0$ when

$x = \dfrac{\pi}{2}, \dfrac{3\pi}{2}$; $2 \sin x - 5 \cos x = 0$; $\dfrac{\sin x}{\cos x} = \dfrac{5}{2}$; $\tan x = \dfrac{5}{2}$; $x = 1.19, 4.33$

31. $\cos 2x = \sec x$; $2 \cos^2 x - 1 = \dfrac{1}{\cos x}$; $2 \cos^3 x - \cos x = 1$; $2 \cos^3 x - \cos x - 1 = 0$;

$(\cos x - 1)(2 \cos^2 x + 2 \cos x + 1) = 0$; $\cos x = 1$ or $\cos x = \dfrac{-2 \pm \sqrt{4 - 8}}{4}$ (reject);

$\cos x = 1$ when $x = 0$.

32. $\sin x \cos 2x = 1$; $\sin x \, (1 - 2 \sin^2 x) - 1 = 0$; $\sin x - 2 \sin^3 x - 1 = 0$; $2 \sin^3 x - \sin x + 1 = 0$; $(\sin x + 1)(2 \sin^2 x - 2 \sin x + 1) = 0$; $\sin x = -1$ or $\sin x = \dfrac{2 \pm \sqrt{4 - 8}}{2}$ (reject); $\sin x = -1$ when $x = \dfrac{3\pi}{2}$.

33. a. No   **b.**

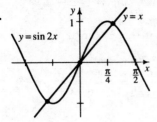

c. 3 solutions

d. 0.9

34. three solutions

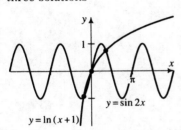

35. five solutions

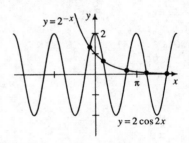

Exercises 36–39 were solved using the method of Example 4.

36. -4.49, -1.79, 1.05, 5.24 **37.** ± 5.41, ± 3.80, ± 2.14, 0

38. See the graph in Ex. 34. $x \approx -0.61$, 0, 1.14

39. See the graph in Ex. 35. $x \approx -0.42$, 0.62, 2.40, 3.91, 5.50

40. $\sin 3x = \sin 5x + \sin x$; $\sin 3x = 2 \sin \dfrac{5x + x}{2} \cos \dfrac{5x - x}{2}$; $\sin 3x = 2 \sin 3x \cos 2x$;

$\sin 3x - 2 \sin 3x \cos 2x = 0$; $\sin 3x \, (1 - 2 \cos 2x) = 0$; $\sin 3x = 0$ or $\cos 2x = \dfrac{1}{2}$;

$3x = 0,\ \pi,\ 2\pi,\ 3\pi,\ 4\pi,\ 5\pi$ or $2x = \dfrac{\pi}{3},\ \dfrac{5\pi}{3},\ \dfrac{7\pi}{3},\ \dfrac{11\pi}{3}$; $x = 0,\ \dfrac{\pi}{6},\ \dfrac{\pi}{3},\ \dfrac{2\pi}{3},\ \dfrac{5\pi}{6},\ \pi,\ \dfrac{7\pi}{6},\ \dfrac{4\pi}{3},$

$\dfrac{5\pi}{3},\ \dfrac{11\pi}{6}$

41. $\cos 3x + \cos x = \cos 2x$; $2 \cos \dfrac{3x + x}{2} \cos \dfrac{3x - x}{2} = \cos 2x$; $2 \cos 2x \cos x = \cos 2x$;

$2 \cos 2x \cos x - \cos 2x = 0$; $\cos 2x \, (2 \cos x - 1) = 0$; $\cos 2x = 0$ or $\cos x = \dfrac{1}{2}$;

$2x = \dfrac{\pi}{2},\ \dfrac{3\pi}{2},\ \dfrac{5\pi}{2},\ \dfrac{7\pi}{2}$ or $x = \dfrac{\pi}{3},\ \dfrac{5\pi}{3}$; $x = \dfrac{\pi}{4},\ \dfrac{\pi}{3},\ \dfrac{3\pi}{4},\ \dfrac{5\pi}{4},\ \dfrac{5\pi}{3},\ \dfrac{7\pi}{4}$

42. $\sin 3x - \sin x = 2 \cos 2x$; $2 \cos \dfrac{3x + x}{2} \sin \dfrac{3x - x}{2} = 2 \cos 2x$; $2 \cos 2x \sin x = 2 \cos 2x$; $2 \cos 2x \sin x - 2 \cos 2x = 0$; $2 \cos 2x \, (\sin x - 1) = 0$; $\cos 2x = 0$ or

$\sin x = 1$; $2x = \dfrac{\pi}{2},\ \dfrac{3\pi}{2},\ \dfrac{5\pi}{2},\ \dfrac{7\pi}{2}$ or $x = \dfrac{\pi}{2}$, $x = \dfrac{\pi}{4},\ \dfrac{\pi}{2},\ \dfrac{3\pi}{4},\ \dfrac{5\pi}{4},\ \dfrac{7\pi}{4}$

43. $\text{Cos}^{-1} 2x = \text{Sin}^{-1} x$; $\text{Cos} (\text{Cos}^{-1} 2x) = \text{Cos} (\text{Sin}^{-1} x)$; $2x = \pm\sqrt{1 - x^2}$; $4x^2 = 1 - x^2$;

$5x^2 = 1$; $x^2 = \dfrac{1}{5}$; reject $x = -\dfrac{\sqrt{5}}{5}$, so $x = \dfrac{\sqrt{5}}{5} \approx 0.447$

44. $\text{Tan}^{-1} 2x = \text{Sin}^{-1} x$; $\text{Tan} (\text{Tan}^{-1} 2x) = \text{Tan} (\text{Sin}^{-1} x)$; $2x = \pm\dfrac{x}{\sqrt{1 - x^2}}$; $4x^2 = \dfrac{x^2}{1 - x^2}$;

$4x^2 - 4x^4 = x^2$; $4x^4 - 3x^2 = 0$; $x^2(4x^2 - 3) = 0$; $x = 0$ or $x = \pm\dfrac{\sqrt{3}}{2}$

Computer Exercise · page 391

```
10 PRINT "TABLE OF SINES AND COSINES"
20 PRINT "ANGLES OF 1 TO 90 DEGREES"
30 LET S1 = 1.74524E-02
40 LET S = S1
50 LET C1 = SQR(1 - S1*S1)
60 LET C = C1
70 PRINT "ANGLE", "SINE", "COSINE"
80 FOR A =1 TO 90
90 PRINT A, S, C
100 LET S = S*C1 + C*S1
110 LET C = SQR(1 - S*S)
120 NEXT A
130 END
```

Chapter Test · page 393

1. a. $\cos 75° \cos 15° + \sin 75° \sin 15° = \cos (75° - 15°) = \cos 60° = \dfrac{1}{2}$

b. $\sin 75° \cos 15° + \cos 75° \sin 15° = \sin (75° + 15°) = \sin 90° = 1$

c. $\cos (30° + x) + \cos (30° - x) = \cos 30° \cos x - \sin 30° \sin x + \cos 30° \cos x +$

$\sin 30° \sin x = 2 \cos 30° \cos x = 2 \cdot \dfrac{\sqrt{3}}{2} \cdot \cos x = \sqrt{3} \cos x$

d. $\sin (45° - x) - \sin (45° + x) = \sin 45° \cos x - \cos 45° \sin x - \sin 45° \cos x -$

$\cos 45° \sin x = -2 \cos 45° \sin x = -2 \cdot \dfrac{\sqrt{2}}{2} \cdot \sin x = -\sqrt{2} \sin x$

2. $\cos (45° - 30°) = \cos 45° \cos 30° + \sin 45° \sin 30° = \dfrac{\sqrt{2}}{2} \cdot \dfrac{\sqrt{3}}{2} + \dfrac{\sqrt{2}}{2} \cdot \dfrac{1}{2} = \dfrac{\sqrt{6} + \sqrt{2}}{4}$

3. $\tan \left(\dfrac{5\pi}{4} - \theta\right) = \dfrac{\tan \dfrac{5\pi}{4} - \tan \theta}{1 + \tan \dfrac{5\pi}{4} \tan \theta} = \dfrac{1 - \left(-\dfrac{1}{3}\right)}{1 + \left(-\dfrac{1}{3}\right)} = \dfrac{\dfrac{4}{3}}{\dfrac{2}{3}} = 2$

4. $\tan (\alpha + \beta) = \dfrac{\tan \alpha + \tan \beta}{1 - \tan \alpha \tan \beta} = \dfrac{\dfrac{4}{3} + \left(-\dfrac{1}{2}\right)}{1 - \dfrac{4}{3}\left(-\dfrac{1}{2}\right)} = \dfrac{\dfrac{5}{6}}{\dfrac{5}{3}} = \dfrac{1}{2}$;

$\tan (\pi - \beta) = \dfrac{\tan \pi - \tan \beta}{1 + \tan \pi \tan \beta} = \dfrac{0 - \left(-\dfrac{1}{2}\right)}{1 + 0\left(-\dfrac{1}{2}\right)} = \dfrac{1}{2} \div 1 = \dfrac{1}{2}$

5. If one angle is θ_1 and the other is θ_2, we can use the slopes of the two lines to find the angles. $\tan \theta_1 = \dfrac{m_1 - m_2}{1 + m_1 m_2}$ and $\tan \theta_2 = \dfrac{m_2 - m_1}{1 + m_2 m_1}$. Alternatively, we could find just one angle using the slopes and find the second using the fact that $2\theta_1 + 2\theta_2 = 360°$. Dividing this equation by 2 we can see that the angles are supplementary.

6. a. $\sin A = \sqrt{1 - \left(\dfrac{4}{5}\right)^2} = \dfrac{3}{5}$ **b.** $\cos 2A = 2 \cos^2 A - 1 = 2\left(\dfrac{4}{5}\right)^2 - 1 = \dfrac{7}{25}$

    c. $\sin 2A = 2 \sin A \cos A = 2 \cdot \dfrac{3}{5} \cdot \dfrac{4}{5} = \dfrac{24}{25}$

    d. $\sin 4A = 2 \sin 2A \cos 2A = 2 \cdot \dfrac{24}{25} \cdot \dfrac{7}{25} = \dfrac{336}{625}$

7. a. $\dfrac{\sin 2x}{1 - \cos 2x} = \dfrac{2 \sin x \cos x}{1 - (1 - 2 \sin^2 x)} = \dfrac{2 \sin x \cos x}{2 \sin^2 x} = \dfrac{\cos x}{\sin x} = \cot x$

    b. $(1 + \tan^2 y)(\cos 2y - 1) = \sec^2 y \,((1 - 2 \sin^2 y) - 1) = \dfrac{-2 \sin^2 y}{\cos^2 y} = -2 \tan^2 y$

    c. $\dfrac{\tan t}{\sec t + 1} = \dfrac{\dfrac{\sin t}{\cos t}}{\dfrac{1}{\cos t} + 1} = \dfrac{\sin t}{1 + \cos t} = \tan \dfrac{t}{2}$

    d. $\cos^2 \left(\dfrac{x}{2}\right) - \sin^2 \left(\dfrac{x}{2}\right) = \cos 2\left(\dfrac{x}{2}\right) = \cos x$

8. a. $\cos \left(2 \cdot \dfrac{\pi}{12}\right) = \cos \dfrac{\pi}{6} = \dfrac{\sqrt{3}}{2}$

    b. $2\left(2 \sin \dfrac{\pi}{6} \cos \dfrac{\pi}{6}\right) = 2\left(\sin \left(2 \cdot \dfrac{\pi}{6}\right)\right) = 2 \sin \dfrac{\pi}{3} = 2\left(\dfrac{\sqrt{3}}{2}\right) = \sqrt{3}$

9. a. $(1 + \cot^2 x)(1 - \cos 2x) = \csc^2 x(1 - (1 - 2 \sin^2 x)) = \dfrac{1}{\sin^2 x}(2 \sin^2 x) = 2$

    b. $\dfrac{\sin \theta \sec \theta}{\tan \theta + \cot \theta} = \dfrac{\sin \theta}{\cos \theta} \div \left(\dfrac{\sin \theta}{\cos \theta} + \dfrac{\cos \theta}{\sin \theta}\right) = \dfrac{\sin \theta}{\cos \theta} \div \left(\dfrac{\sin^2 \theta + \cos^2 \theta}{\sin \theta \cos \theta}\right) =$

    $\dfrac{\sin \theta}{\cos \theta} \cdot \dfrac{\sin \theta \cos \theta}{1} = \sin^2 \theta = 1 - \cos^2 \theta = \cos^2 \theta + \sin^2 \theta - \cos^2 \theta =$

    $\cos^2 \theta - \cos 2\theta$

10. a.

    b. $\cos 2\theta = \sin \theta$; $1 - 2 \sin^2 \theta = \sin \theta$;

    $(2 \sin \theta - 1)(\sin \theta + 1) = 0$; $\sin \theta = \dfrac{1}{2}$ or -1;

    $\theta = 30°, 150°, 270°$

11. $\cos 2x = \cos x + 2$; $2 \cos^2 x - 1 = \cos x + 2$; $2 \cos^2 x - \cos x - 3 = 0$;

    $(2 \cos x - 3)(\cos x + 1) = 0$; $\cos x = \dfrac{3}{2}$ (reject) or $\cos x = -1$; $x = \pi$

Class Exercises 11-1 · page 399

1. Answers may vary. Examples are given. **a.** $(6, 410°)$, $(-6, -130°)$

   b. $(7, 300°)$, $(7, -60°)$ **c.** $\left(3, \dfrac{9\pi}{4}\right)$, $\left(-3, \dfrac{5\pi}{4}\right)$ **d.** $\left(4, \dfrac{11\pi}{6}\right)$, $\left(4, -\dfrac{\pi}{6}\right)$

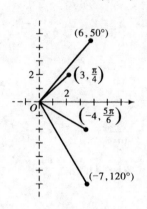

Ex. 1

Ex. 2

2. Answers may vary. Examples are given. **a.** $(5, 315°)$, $(-5, 135°)$

   b. $(2, 120°)$, $(-2, 300°)$ **c.** $\left(1, \dfrac{3\pi}{2}\right)$, $\left(-1, \dfrac{\pi}{2}\right)$ **d.** $\left(\dfrac{3}{2}, 0\right)$, $\left(-\dfrac{3}{2}, \pi\right)$

3. $r = \sqrt{1^2 + (-1)^2} = \sqrt{2}$; for example, $(\sqrt{2}, 315°)$

4. $x = -8\cos 60° = -8 \cdot \dfrac{1}{2} = -4$; $y = -8 \sin 60° = -8 \cdot \dfrac{\sqrt{3}}{2} = -4\sqrt{3}$; $(-4, -4\sqrt{3})$

5. $r = 2\cos\theta$; $r = 2 \cdot \dfrac{x}{r}$; $r^2 = 2x$; $x^2 + y^2 = 2x$; $x^2 - 2x + y^2 = 0$; $(x-1)^2 + y^2 = 1$;

   the graph is a circle with center $(1, 0)$, radius 1; the two graphs are the same.

Written Exercises 11-1 · pages 400–402

Exs. 1–6: Answers may vary. Examples are given.

1. **a.** $(2, 400°)$, $(-2, 220°)$ **b.** $(3, 260°)$, $(-3, 80°)$ **c.** $\left(5, -\dfrac{\pi}{2}\right)$, $\left(-5, \dfrac{\pi}{2}\right)$,

   d. $(4, 190°)$, $(-4, 370°)$

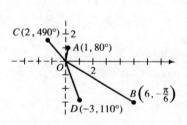

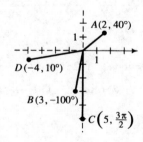

Ex. 1

Ex. 2

280

2. a. $(1, -280°), (-1, 260°)$ **b.** $\left(6, \dfrac{11\pi}{6}\right), \left(-6, \dfrac{5\pi}{6}\right)$ **c.** $(2, 130°), (-2, 310°)$

    d. $(3, 290°), (3, -70°)$

3. a. $(2\sqrt{2}, 135°)$ **b.** $(5, 0°)$ **c.** $(2, -45°)$ **d.** $(2, 150°)$

4. a. $(4\sqrt{2}, 225°)$ **b.** $(6, -60°)$ **c.** $(6, 90°)$ **d.** $(0, 0°)$

5. a. $\left(\sqrt{2}, \dfrac{5\pi}{4}\right)$ **b.** $\left(12, \dfrac{\pi}{2}\right)$ **c.** $\left(1, \dfrac{\pi}{3}\right)$ **d.** $(2, \pi)$

6. a. $\left(1, \dfrac{\pi}{4}\right)$ **b.** $\left(2\sqrt{3}, \dfrac{2\pi}{3}\right)$ **c.** $\left(4, \dfrac{3\pi}{2}\right)$ **d.** $\left(e\sqrt{2}, -\dfrac{\pi}{4}\right)$

7. a. $(-2, 2\sqrt{3})$ **b.** $(0, -3)$ **c.** $\left(-\dfrac{\sqrt{3}}{2}, \dfrac{1}{2}\right)$ **d.** $(-\sqrt{2}, \sqrt{2})$

8. a. $(-\sqrt{2}, -\sqrt{2})$ **b.** $(3\sqrt{3}, -3)$ **c.** $(0, 10)$ **d.** $(-2, -2\sqrt{3})$

9. a. $x = r \cos 20° = 1 \cos 20° = 0.940;\ y = r \sin 20° = 1 \sin 20° = 0.342;$
    $(0.940, 0.342)$

    b. $x = r \cos 20° = 2 \cos 20° = 1.879;\ y = r \sin 20° = 2 \sin 20° = 0.684;$
       $(1.879, 0.684)$

    c. $x = r \cos 2 = 1 \cos 2 = -0.416;\ y = r \sin 2 = 1 \sin 2 = 0.909;\ (-0.416, 0.909)$

    d. $x = r \cos(-2) = 1 \cos(-2) = -0.416;\ y = r \sin(-2) = 1 \sin(-2) = -0.909;$
       $(-0.416, -0.909)$

10. $x = r \cos 65° = 1 \cos 65° = 0.423;\ y = r \sin 65° = 1 \sin 65° = 0.906;\ (0.423, 0.906)$

    b. $x = r \cos 65° = 4 \cos 65° = 1.690;\ y = r \sin 65° = 4 \sin 65° = 3.625;$
       $(1.690, 3.625)$

    c. $x = r \cos 3 = 2 \cos 3 = -1.980;\ y = r \sin 3 = 2 \sin 3 = 0.282;\ (-1.980, 0.282)$

    d. $x = r \cos(-3) = 2 \cos(-3) = -1.980;\ y = r \sin(-3) = 2 \sin(-3) = -0.282;$
       $(-1.980, -0.282)$

Exs. 11, 12: Examples are given.

11. a. $r = \pm\sqrt{3^2 + 4^2} = \pm 5;\ \tan\theta = \dfrac{4}{3};\ \theta = 53.1°;\ (3, 4)$ is in Quadrant I; $(5, 53.1°)$

    b. $r = \pm\sqrt{1^2 + 2^2} = \pm\sqrt{5} \approx \pm 2.2;\ \tan\theta = \dfrac{2}{1};\ \theta = 63.4°;\ (1, 2)$ is in Quadrant I;
       $(2.2, 63.4°)$

    c. $r = \pm\sqrt{(-2)^2 + 3^2} \approx \pm 3.6;\ \tan\theta = \dfrac{3}{-2};\ \theta = -56.3°;\ (-2, 3)$ is in Quadrant II;
       $(3.6, 180° - 56.3°) = (3.6, 123.7°)$

    d. $r = \pm\sqrt{4^2 + (-6)^2} \approx \pm 7.2;\ \tan\theta = \dfrac{-6}{4};\ \theta = -56.3°;\ (4, -6)$ is in Quadrant IV;
       $(7.2, -56.3°)$

12. a. $r = \pm\sqrt{5^2 + 2^2} \approx \pm 5.4;\ \tan\theta = \dfrac{2}{5};\ \theta = 0.4;\ (5, 2)$ is in Quadrant I; $(5.4, 0.4)$

    b. $r = \pm\sqrt{8^2 + (-6)^2} = \pm 10;\ \tan\theta = \dfrac{-6}{8};\ \theta = -0.6;\ (8, -6)$ is in Quadrant IV;
       $(10, -0.6)$

(Continued)

c. $r = \pm\sqrt{(-1)^2 + (-4)^2} \approx \pm 4.1$; $\tan \theta = \dfrac{-4}{-1}$; $\theta = 1.3$; $(-1, -4)$ is in Quadrant III;

$(4.1, \pi - 1.3) = (4.1, -1.8)$

d. $r = \pm\sqrt{(-8)^2 + 6^2} = \pm 10$; $\tan \theta = \dfrac{6}{-8}$; $\theta = -0.6$; $(-8, 6)$ is in Quadrant II;

$(10, \pi - 0.6) = (10, 2.5)$

13. $r = \sin \theta$; $r = \dfrac{y}{r}$; $r^2 = y$; $x^2 + y^2 = y$; $x^2 + \left(y - \dfrac{1}{2}\right)^2 = \dfrac{1}{4}$

14. $r = \cos \theta$; $r = \dfrac{x}{r}$; $r^2 = x$; $x^2 + y^2 = x$; $\left(x - \dfrac{1}{2}\right)^2 + y^2 = \dfrac{1}{4}$

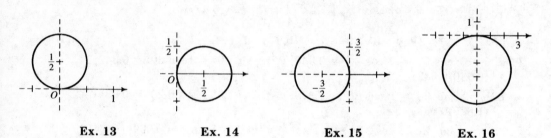

Ex. 13   **Ex. 14**   **Ex. 15**   **Ex. 16**

15. $r = -3 \cos \theta$; $r = -3 \cdot \dfrac{x}{r}$; $r^2 = -3x$; $x^2 + y^2 = -3x$; $\left(x + \dfrac{3}{2}\right)^2 + y^2 = \dfrac{9}{4}$

16. $r = -5 \sin \theta$; $r = -5 \cdot \dfrac{y}{r}$; $r^2 = -5y$; $x^2 + y^2 = -5y$; $x^2 + \left(y + \dfrac{5}{2}\right)^2 = \dfrac{25}{4}$

17. $r = 1 - \sin \theta$; $\pm\sqrt{x^2 + y^2} = 1 - \dfrac{y}{\pm\sqrt{x^2 + y^2}}$; $x^2 + y^2 = \pm\sqrt{x^2 + y^2} - y$;

$(x^2 + y^2 + y)^2 = x^2 + y^2$

18. $r = 2 + 2 \cos \theta$; $\pm\sqrt{x^2 + y^2} = 2 + \dfrac{2x}{\pm\sqrt{x^2 + y^2}}$; $x^2 + y^2 = \pm 2\sqrt{x^2 + y^2} + 2x$;

$(x^2 + y^2 - 2x)^2 = 4x^2 + 4y^2$

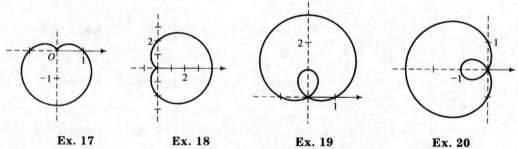

Ex. 17   **Ex. 18**   **Ex. 19**   **Ex. 20**

19. $r = 1 + 2 \sin \theta$; $\pm\sqrt{x^2 + y^2} = 1 + 2 \cdot \dfrac{y}{\pm\sqrt{x^2 + y^2}}$; $x^2 + y^2 = \pm\sqrt{x^2 + y^2} + 2y$;

$(x^2 + y^2 - 2y)^2 = x^2 + y^2$

20. $r = 1 - 2 \cos \theta$; $\pm\sqrt{x^2 + y^2} = 1 - 2 \cdot \dfrac{x}{\pm\sqrt{x^2 + y^2}}$; $x^2 + y^2 = \pm\sqrt{x^2 + y^2} - 2x$;

$(x^2 + y^2 + 2x)^2 = x^2 + y^2$

21. $r = \sec\theta$; $r = \dfrac{1}{\cos\theta}$; $\pm\sqrt{x^2 + y^2} = \dfrac{\pm\sqrt{x^2 + y^2}}{x}$; $x = 1$

Ex. 21

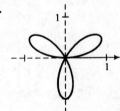

Ex. 22

22. $r = 2\csc\theta$; $r = \dfrac{2}{\sin\theta}$; $\pm\sqrt{x^2 + y^2} = 2 \cdot \dfrac{\pm\sqrt{x^2 + y^2}}{y}$; $y = 2$

23. a. By definition, $\sin\theta = \dfrac{y}{r}$ and $\cos\theta = \dfrac{x}{r}$. Multiplying both sides of each equation by

r, $y = r\sin\theta$ and $x = r\cos\theta$.

    b. By the Pythagorean Thm., $x^2 + y^2 = r^2$; thus, $r = \pm\sqrt{x^2 + y^2}$. By definition,

$$\tan\theta = \frac{\sin\theta}{\cos\theta} = \frac{\dfrac{y}{r}}{\dfrac{x}{r}} = \frac{y}{x}.$$

24. a.

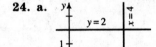

b. yes

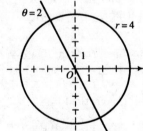

c. $x = 3$; $r\cos\theta = 3$;

    $r = \dfrac{3}{\cos\theta}$; $r = 3\sec\theta$

d. $y = 3$; $r\sin\theta = 3$;

    $r = \dfrac{3}{\sin\theta}$; $r = 3\csc\theta$

25.

26.

27.

28.

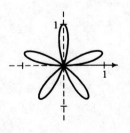

29.

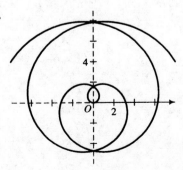

30.

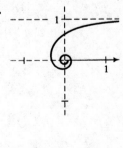

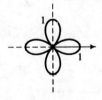

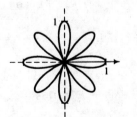

31. 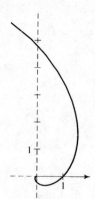   **32.** 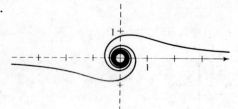   **33.**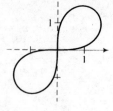

34.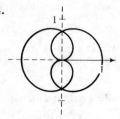

35. A *cardioid* is the locus in a plane of a fixed point on a given circle, as the circle rolls along the outside of a fixed circle of the same radius. A *limaçon* is the locus in a plane of a point on a given line at a fixed distance from the point of intersection of the line with a fixed circle, as the line rotates around a fixed point on the circle. If the fixed distance is equal to the diameter of the fixed circle, the limaçon is a cardioid. The heart-like shape of the graph of a cardioid is the source of its name. From Exs. 17–20, the equations of cardioids and limaçons appear to come in the form $r = a + b \cos \theta$ or $r = a + b \sin \theta$; if $|a| = |b|$ the curve is a cardioid.

36. Reports will vary.

37. (i) A graph has symmetry in the horizontal line $\theta = 0$ if $(r, -\theta)$ is on the graph whenever (r, θ) is on the graph; for example, $r = \cos \theta$.

(ii) A graph has symmetry in the vertical line $\theta = \dfrac{\pi}{2}$ if $(r, \pi - \theta)$ is on the graph whenever (r, θ) is on the graph; for example, $r = \sin \theta$.

(iii) A graph has symmetry in the pole if $(r, \theta + \pi)$ is on the graph whenever (r, θ) is on the graph; for example, $r^2 = 4 \sin 2\theta$.

(iv) A graph has symmetry in the pole if $(-r, \theta)$ is on the graph whenever (r, θ) is on the graph; for example, $r^2 = \sin \theta$.

38. If $P(x, y)$ is on the conic, the distance between $P(x, y)$ and the directrix D is $y + p$ and the distance between $P(x, y)$ and the focus $F(0, 0)$ is $\sqrt{x^2 + y^2}$. Thus $e = \dfrac{PF}{PD} = \dfrac{\sqrt{x^2 + y^2}}{y + p} = \dfrac{r}{r \sin \theta + p}$. Solving for r, $er \sin \theta + ep = r$; $r - er \sin \theta = ep$; $r(1 - e \sin \theta) = ep$; $r = \dfrac{ep}{1 - e \sin \theta}$.

39. a. $r = 1 + \sin\theta$ and $r = 2\sin\theta$; $1 + \sin\theta = 2\sin\theta$; $\sin\theta = 1$; $\theta = \dfrac{\pi}{2}$ and $r = 2$; $\left(2; \dfrac{\pi}{2}\right)$ is a common point.

b. See graph. The common point $(0,0)$ is generated by the cardiod $r = 1 + \sin\theta$ when $\theta = \dfrac{3\pi}{2}$ and by the circle $r = 2\sin\theta$ when $\theta = 0$ or π.

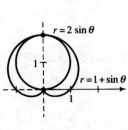

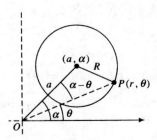

Ex. 39   **Ex. 40**

40. Let $P(r, \theta)$ be any point on the circle with center (a, α) and radius R. A triangle is formed, as shown. Using the law of cosines, we have $R^2 = a^2 + r^2 - 2ar\cos(\alpha - \theta)$; since $\cos(\theta - \alpha) = \cos[-(\alpha - \theta)] = \cos(\alpha - \theta)$, we have $R^2 = r^2 - 2ar\cos(\theta - \alpha) + a^2$.

Computer Exercises · page 402

1. a.

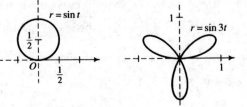

b. a rose with k leaves

c.

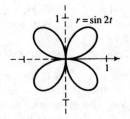

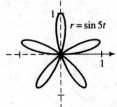

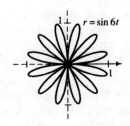

d. a rose with $2k$ leaves

e. a rose with n leaves if n is odd and $2n$ leaves if n is even

2.

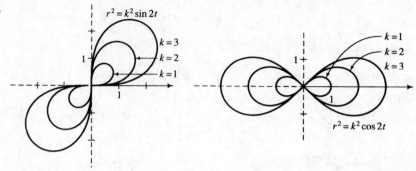

$r^2 = k^2 \sin 2t$

k = 3
k = 2
k = 1

k = 1
k = 2
k = 3

$r^2 = k^2 \cos 2t$

The lemniscates become larger as k increases; graphs of equations of the form $r^2 = k^2 \sin 2t$ are symmetric in the lines $y = \pm x$; graphs of equations of the form $r^2 = k^2 \cos 2t$ are symmetric in the lines $y = 0$ and $x = 0$.

3.

$r = \sin \frac{5}{3}t$

$r = \sin \frac{3}{5}t$

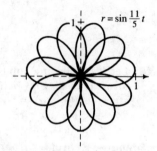

$r = \sin \frac{11}{5}t$

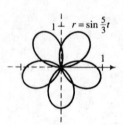

$r = \sin \frac{5}{11}t$

The graph will have a leaves and will cross the horizontal axis, $t = n\pi$, b times. If $\left|\dfrac{a}{b}\right| > 1$, the leaves have a narrow oval shape; if $\left|\dfrac{a}{b}\right| < 1$, the leaves are wider and have more of a heart shape.

4. Answers may vary. Examples are given. If $c = 0$, $|b| = 1$, and a is an integer, the graph has $2a$ non-overlapping, congruent leaves joined at the center point. If $c = 0$ and $|b| \neq 1$, the graph has $2a$ overlapping congruent leaves; as $\left|\dfrac{a}{b}\right|$ gets closer to 0 the graph becomes increasingly complex. If $b = 1$, a is an integer, and $|c| = 1$, the graph has a non-overlapping, congruent leaves joined at the center point. If $b = 1$, a is an integer, and $|c| > 1$, the graph has a separate, congruent leaves; the greater $|c|$ is, the further apart are the leaves. If $b = 1$, a is an integer, and $|c| < 1$, the graph has 2 groups of non-overlapping, congruent leaves joined at the center point; as $|c|$ gets closer to 1 the size of the smaller group of leaves decreases until they vanish when $|c| = 1$. Many other variations can be investigated.

Class Exercises 11-2 · page 405

1. $r = |1 + i| = \sqrt{1^2 + 1^2} = \sqrt{2}$; $\theta = \text{Tan}^{-1}\left(\dfrac{1}{1}\right) = 45°$; $\sqrt{2}$ cis 45°

2. $r = |i| = \sqrt{0^2 + 1^2} = 1$; $\theta = 90°$; 1 cis 90° = cis 90°

3. $r = |-3| = \sqrt{(-3)^2 + 0^2} = 3$; $\theta = 180°$; 3 cis 180°

4. $r = |\sqrt{3} - i| = \sqrt{(\sqrt{3})^2 + (1)^2} = 2$; $\text{Tan}^{-1}\left(-\dfrac{1}{\sqrt{3}}\right) = -30°$; $\theta = 360° - 30° = 330°$;

2 cis 330°

5. $2 \text{ cis } \dfrac{\pi}{4} = 2\left(\cos\dfrac{\pi}{4} + i\sin\dfrac{\pi}{4}\right)$

6. 5 cis 90° = $5(\cos 90° + i\sin 90°) = 5(0 + 1 \cdot i) = 5i$

7. 3 cis π = $3(\cos \pi + i\sin \pi) = 3(-1 + 0 \cdot i) = -3$

8. 4 cis 45° = $4(\cos 45° + i\sin 45°) = 4\left(\dfrac{\sqrt{2}}{2} + \dfrac{\sqrt{2}}{2}i\right) = 2\sqrt{2} + 2i\sqrt{2}$

9. 6 cis 30° = $6(\cos 30° + i\sin 30°) = 6\left(\dfrac{\sqrt{3}}{2} + \dfrac{1}{2}i\right) = 3\sqrt{3} + 3i$

10. $(4 \text{ cis } 25°)(6 \text{ cis } 35°) = 4 \cdot 6 \text{ cis }(25° + 35°) = 24 \text{ cis } 60°$

11. $\left(5 \text{ cis } \dfrac{\pi}{4}\right)\left(2 \text{ cis } \dfrac{3\pi}{4}\right) = 2 \cdot 5 \text{ cis }\left(\dfrac{\pi}{4} + \dfrac{3\pi}{4}\right) = 10 \text{ cis } \pi$

Written Exercises 11-2 · pages 406–407

1. $r = |-1 + i| = \sqrt{(-1)^2 + 1^2} = \sqrt{2}$; $\text{Tan}^{-1}\dfrac{1}{-1} = -45°$; $\theta = 180° - 45° = 135°$;

$\sqrt{2}$ cis 135°

2. $r = |-3i| = 3$; $\theta = 270°$; 3 cis 270°

3. $r = |1 + i\sqrt{3}| = \sqrt{1^2 + (\sqrt{3})^2} = 2$; $\theta = \text{Tan}^{-1}\dfrac{\sqrt{3}}{1} = 60°$; 2 cis 60°

4. $r = |-2 - 2i| = \sqrt{(-2)^2 + (-2)^2} = 2\sqrt{2}$; $\text{Tan}^{-1}\dfrac{-2}{-2} = 45°$; $\theta = 180° + 45°$;

$2\sqrt{2}$ cis 225°

5. $r = |-7| = 7$; $\theta = 180°$; 7 cis 180°

6. $r = |2\sqrt{3} + 2i| = \sqrt{(2\sqrt{3})^2 + 2^2} = 4$; $\theta = \text{Tan}^{-1}\left(\dfrac{2}{2\sqrt{3}}\right) = 30°$; 4 cis 30°

7. $r = |3 - 4i| = \sqrt{3^2 + (-4)^2} = 5$; $\text{Tan}^{-1}\left(\dfrac{-4}{3}\right) \approx -53°$; $\theta = 360° - 53° = 307°$;

5 cis 307°

8. $r = |5 + 12i| = \sqrt{5^2 + 12^2} = 13$; $\theta = \text{Tan}^{-1}\left(\dfrac{12}{5}\right) \approx 67°$; 13 cis 67°

9. $6(\cos 100° + i\sin 100°) \approx -1.04 + 5.91i$

10. $8(\cos 230° + i\sin 230°) \approx -5.14 - 6.13i$

11. $9\left(\cos\dfrac{4\pi}{3} + i\sin\dfrac{4\pi}{3}\right) = 9\left(-\dfrac{1}{2} - \dfrac{\sqrt{3}}{2}i\right) = -\dfrac{9}{2} - \dfrac{9\sqrt{3}}{2}i$

12. $2\left(\cos\dfrac{3\pi}{4} + i\sin\dfrac{3\pi}{4}\right) = 2\left(-\dfrac{\sqrt{2}}{2} + \dfrac{\sqrt{2}}{2}i\right) = -\sqrt{2} + i\sqrt{2}$

13. $(5 \text{ cis } 30°)(2 \text{ cis } 60°) = 5 \cdot 2 \text{ cis } (30° + 60°) = 10 \text{ cis } 90°; 10(\cos 90° + i\sin 90°) =$
$10(0 + 1 \cdot i) = 10i$

14. $(2 \text{ cis } 115°)(3 \text{ cis } 65°) = 2 \cdot 3 \text{ cis } (115° + 65°) = 6 \text{ cis } 180°;$
$6(\cos 180° + i\sin 180°) = 6(-1 + 0 \cdot i) = -6$

15. $\left(8 \text{ cis } \dfrac{\pi}{3}\right)\left(\dfrac{1}{2} \text{ cis}\left(-\dfrac{2\pi}{3}\right)\right) = 8 \cdot \dfrac{1}{2} \text{ cis } \left(\dfrac{\pi}{3} - \dfrac{2\pi}{3}\right) = 4 \text{ cis } \left(-\dfrac{\pi}{3}\right) = 4 \text{ cis } \dfrac{5\pi}{3};$

$4\left(\cos\left(\dfrac{5\pi}{3}\right) + i\sin\left(\dfrac{5\pi}{3}\right)\right) = 4\left(\dfrac{1}{2} - \dfrac{\sqrt{3}}{2}i\right) = 2 - 2i\sqrt{3}$

16. $\left(4 \text{ cis } \dfrac{\pi}{4}\right)\left(3 \text{ cis } \dfrac{\pi}{2}\right) = 4 \cdot 3 \text{ cis } \left(\dfrac{\pi}{4} + \dfrac{\pi}{2}\right) = 12 \text{ cis } \left(\dfrac{3\pi}{4}\right); 12\left(\cos\dfrac{3\pi}{4} + i\sin\dfrac{3\pi}{4}\right) =$

$12\left(-\dfrac{\sqrt{2}}{2} + \dfrac{\sqrt{2}}{2}i\right) = -6\sqrt{2} + 6i\sqrt{2}$

17. a. $z_1 z_2 = (2 + 2i\sqrt{3})(\sqrt{3} - i) = 2\sqrt{3} - 2i + 6i + 2\sqrt{3} = 4\sqrt{3} + 4i$

    b. $r_1 = |2 + 2i\sqrt{3}| = \sqrt{2^2 + (2\sqrt{3})^2} = 4; \theta = \text{Tan}^{-1}\left(\dfrac{2\sqrt{3}}{2}\right) = 60°; z_1 = 4 \text{ cis } 60°;$

    $r_2 = |\sqrt{3} - i| = \sqrt{(\sqrt{3})^2 + (-1)^2} = 2; \text{Tan}^{-1}\left(\dfrac{-1}{\sqrt{3}}\right) = -30°; \theta_2 = 360° - 30° =$

    $330°; z_2 = 2 \text{ cis } 330°; z_1 z_2 = 4 \cdot 2 \text{ cis } (60° + 330°) = 8 \text{ cis } 390° = 8 \text{ cis } 30°;$

    $8 \text{ cis } 30° = 8\left(\dfrac{\sqrt{3}}{2} + \dfrac{1}{2}i\right) = 4\sqrt{3} + 4i$ **c.** See below.

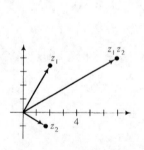

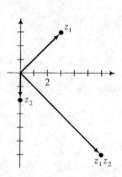

              Ex. 17(c)   **Ex. 18(c)**

18. a. $z_1 z_2 = (3 + 3i)(-2i) = -6i + 6 = 6 - 6i$

    b. $r_1 = |3 + 3i| = \sqrt{3^2 + 3^2} = 3\sqrt{2}; \theta = \text{Tan}^{-1}\left(\dfrac{3}{3}\right) = 45°; z_1 = 3\sqrt{2} \text{ cis } 45°;$

    $r_2 = |-2i| = \sqrt{0^2 + (-2)^2} = 2; \theta_2 = 270°; z_2 = 2 \text{ cis } 270°; z_1 z_2 =$

    $3\sqrt{2} \cdot 2 \text{ cis } (45° + 270°) = 6\sqrt{2} \text{ cis } 315°; 6\sqrt{2} \text{ cis } 315° = 6\sqrt{2}\left(\dfrac{\sqrt{2}}{2} - \dfrac{\sqrt{2}}{2}i\right) =$

    $6 - 6i$ **c.** See above.

19. a. $z_1 z_2 = (2 + 2i)(2 - 2i) = 4 + 4 = 8$

    b. $r_1 = |2 + 2i| = \sqrt{2^2 + 2^2} = 2\sqrt{2}; \theta_1 = \text{Tan}^{-1}\left(\dfrac{2}{2}\right) = 45°; z_1 = 2\sqrt{2} \text{ cis } 45°;$

    $r_2 = |2 - 2i| = \sqrt{2^2 + (-2)^2} = 2\sqrt{2}; \text{Tan}^{-1}\left(\dfrac{-2}{2}\right) = -45°; \theta_2 = 360° - 45° =$

    $315°; z_2 = 2\sqrt{2} \text{ cis } 315°; z_1 z_2 = 2\sqrt{2} \cdot 2\sqrt{2} \text{ cis } (45° + 315°) = 8 \text{ cis } 360° =$
    $8 \text{ cis } 0°; 8 \text{ cis } 0° = 8(1 + 0 \cdot i) = 8$ **c.** See below.

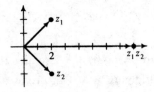
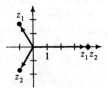

           Ex. 19(c)   **Ex. 20(c)**

20. a. $z_1 z_2 = (-1 + i\sqrt{3})(-1 - i\sqrt{3}) = 1 + 3 = 4$

    b. $r_1 = |-1 + i\sqrt{3}| = \sqrt{(-1)^2 + (\sqrt{3})^2} = 2; \text{Tan}^{-1}\left(\dfrac{\sqrt{3}}{-1}\right) = -60°;$

    $\theta_1 = 180° - 60° = 120°; z_1 = 2 \text{ cis } 120°; r_2 = |-1 - i\sqrt{3}| =$

    $\sqrt{(-1)^2 + (-\sqrt{3})^2} = 2; \text{Tan}^{-1}\left(\dfrac{-\sqrt{3}}{-1}\right) = 60°; \theta_2 = 180° + 60° = 240°;$

    $z_2 = 2 \text{ cis } 240°; z_1 z_2 = 2 \cdot 2 \text{ cis } (120° + 240°) = 4 \text{ cis } 360° = 4 \text{ cis } 0°;$
    $4 \text{ cis } 0° = 4(1 + 0 \cdot i) = 4$ **c.** See above.

21. a. $z_1 z_2 = (6 - 4i)(-5 + 2i) = -30 + 12i + 20i + 8 = -22 + 32i$

    b. $r_1 = |6 - 4i| = \sqrt{6^2 + (-4)^2} = \sqrt{52} \approx 7.211; \text{Tan}^{-1}\left(\dfrac{-4}{6}\right) \approx -33.69°; \theta_1 =$

    $360° - 33.69° = 326.31°; z_1 = 7.211 \text{ cis } 326.31°; r_2 = |-5 + 2i| = \sqrt{(-5)^2 + 2^2} =$

    $\sqrt{29} \approx 5.385; \text{Tan}^{-1}\left(\dfrac{2}{-5}\right) \approx -21.801°; \theta_2 = 180° - 21.801° = 158.199°;$

    $z_2 = 5.385 \text{ cis } 158.199°; z_1 z_2 = (7.211)(5.385) \text{ cis } (326.31° + 158.199°) \approx$
    $38.831 \text{ cis } 484.509°; 38.831 \text{ cis } 484.509° \approx 38.831(\cos 485° + i \sin 485°) \approx$
    $-22 + 32i$ **c.** See below.

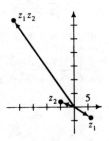

           Ex. 21(c)

22. a. $z_1 z_2 = (-3 - 5i)(4 + 7i) = -12 - 21i - 20i + 35 = 23 - 41i$

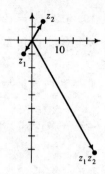

   b. $r_1 = |-3 - 5i| = \sqrt{(-3)^2 + (-5)^2} = \sqrt{34} \approx 5.831;$

     $\text{Tan}^{-1}\left(\dfrac{-5}{-3}\right) \approx 59.036°; \; \theta_1 = 180° + 59.036° = 239.036°;$

     $z_1 = 5.831 \text{ cis } 239.036°; \; r_2 = |4 + 7i| = \sqrt{4^2 + 7^2} = \sqrt{65} \approx$

     $8.062; \; \theta_2 = \text{Tan}^{-1}\left(\dfrac{7}{4}\right) \approx 60.255°; \; z_2 = 8.062 \text{ cis } 60.255°;$

     $z_1 z_2 = (5.831)(8.062) \text{ cis } (239.036° + 60.255°) \approx$
     $47.010 \text{ cis } 299.291°; \; 47.010 \text{ cis } 299.291° \approx 23 - 41i$

   c. See the figure at the right.

23. Since $1 + i = \sqrt{2}(\cos 45° + i \sin 45°)$, multiplying by $1 + i$ rotates the arrow by 45°.

24. Since $\sqrt{3} + i = 2(\cos 30° + i \sin 30°)$, multiplying by $\sqrt{3} + i$ rotates the arrow by 30°.

25. If $z_1 = z_2 = r \text{ cis } \alpha$, then $z_1 z_2 = (z_1)^2 = (r \text{ cis } \alpha)(r \text{ cis } \alpha) = r \cdot r \text{ cis } (\alpha + \alpha) = r^2 \text{ cis } 2\alpha$; thus, if $z = r \cos \theta$, $z^2 = r^2 \cos 2\theta$.

26. $(\cos \theta + i \sin \theta)^2 = \cos^2 \theta + 2i \cos \theta \sin \theta - \sin^2 \theta = (\cos^2 \theta - \sin^2 \theta) + i(2 \sin \theta \cos \theta) = \cos 2\theta + i \sin 2\theta$

27. $(\cos \theta + i \sin \theta)^3 = \cos 3\theta + i \sin 3\theta$

28. $z^2 = \cos 60° + i \sin 60°;$
    $z^3 = \cos 90° + i \sin 90°;$
    $z^4 = \cos 120° + i \sin 120°$

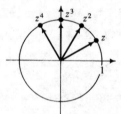

29. a.

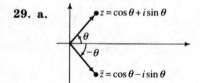

   b. For any θ, $\cos (-\theta) = \cos \theta$ and $\sin (-\theta) = -\sin \theta$. Thus, $\cos (-\theta) + i \sin (-\theta) = \cos \theta - i \sin \theta$. If $z = r \text{ cis } \theta = r \cos \theta + ri \sin \theta$, then $\bar{z} = r \cos \theta - ri \sin \theta = r(\cos \theta - i \sin \theta) = r(\cos (-\theta) + i \sin (-\theta)) = r \text{ cis } (-\theta)$.

30. $\dfrac{1}{z} = \dfrac{1}{r(\cos \theta + i \sin \theta)} = \dfrac{1}{r} \cdot \dfrac{(\cos \theta - i \sin \theta)}{(\cos \theta + i \sin \theta)(\cos \theta - i \sin \theta)} = \dfrac{1}{r} \cdot \dfrac{\cos \theta - i \sin \theta}{\cos^2 \theta + \sin^2 \theta} =$

   $\dfrac{1}{r} \cdot \dfrac{\cos (-\theta) + i \sin (-\theta)}{1} = \dfrac{1}{r} \text{ cis } (-\theta)$

31. $\dfrac{z_1}{z_2} = \dfrac{r(\cos \alpha + i \sin \alpha)}{s(\cos \beta + i \sin \beta)} = \dfrac{r}{s} \cdot \dfrac{(\cos \alpha + i \sin \alpha)(\cos \beta - i \sin \beta)}{(\cos \beta + i \sin \beta)(\cos \beta - i \sin \beta)} = \dfrac{r}{s} \cdot \dfrac{\text{cis } \alpha \text{ cis } (-\beta)}{\cos^2 \beta + \sin^2 \beta} =$

   $\dfrac{r}{s} \cdot \dfrac{\text{cis } (\alpha + (-\beta))}{1} = \dfrac{r}{s} \text{ cis } (\alpha - \beta)$

32. a. $z_1z_2 = (a + bi)(c + di) = ac + adi + bci + bdi^2 = (ac - bd) + (bc + ad)i$

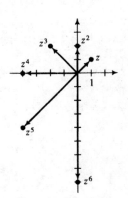

$z_1z_2 = (ac - bd) + (bc + ad)i$

b. $\dfrac{b}{a}; \dfrac{d}{c}; \dfrac{bc + ad}{ac - bd}$

c. $\tan(\alpha + \beta) = \dfrac{\tan\alpha + \tan\beta}{1 - \tan\alpha\tan\beta} = \dfrac{\dfrac{b}{a} + \dfrac{d}{c}}{1 - \dfrac{b}{a}\cdot\dfrac{d}{c}} =$

$\dfrac{\dfrac{bc + ad}{ac}}{\dfrac{ac - bd}{ac}} = \dfrac{bc + ad}{ac - bd} = \tan\theta$; hence $\alpha + \beta = \theta$.

d. $|z_2| = \sqrt{c^2 + d^2}$; $|z_1z_2| = \sqrt{(ac - bd)^2 + (bc + ad)^2} = \sqrt{a^2c^2 - 2abcd + b^2d^2 + b^2c^2 + 2abcd + a^2d^2} = \sqrt{a^2c^2 + b^2d^2 + b^2c^2 + a^2d^2}$; $|z_1|\cdot|z_2| = \sqrt{a^2 + b^2}\cdot\sqrt{c^2 + d^2} = \sqrt{(a^2 + b^2)(c^2 + d^2)} = \sqrt{a^2c^2 + b^2d^2 + a^2d^2 + b^2c^2}$; hence $|z_1z_2| = |z_1|\cdot|z_2|$.

33. Let $z_1 = r_1(\cos\theta_1 + i\sin\theta_1)$ and $z_2 = r_2(\cos\theta_2 + i\sin\theta_2)$. Then $z_1z_2 = r_1r_2(\cos(\theta_1 + \theta_2) + i\sin(\theta_1 + \theta_2))$. The angle at the origin of the lower triangle is θ_1; the angle at the origin of the upper triangle is $(\theta_1 + \theta_2) - \theta_2 = \theta_1$, so the two angles are congruent. $|z_1z_2| = r_1r_2$, $|z_2| = r_2$, and $|z_1| = r_1$. The length of the segment along the x-axis is $\sqrt{1^2 + 0^2} = 1$. Hence $\dfrac{r_1r_2}{r_1} = \dfrac{r_2}{1}$ and the lengths of corresponding sides are proportional. By SAS Similarity, the triangles are similar.

Activity 11-3 · page 409

1. $r = |1 + i| = \sqrt{1^2 + 1^2} = \sqrt{2}$; $\theta = \text{Tan}^{-1}\left(\dfrac{1}{1}\right) = 45°$;

$z = \sqrt{2}\text{ cis }45°$; $z^2 = (\sqrt{2})^2\text{ cis }(2\cdot45°) = 2\text{ cis }90°$;
$z^3 = (\sqrt{2})^3\text{ cis }(3\cdot45°) = 2\sqrt{2}\text{ cis }135°$; $z^4 = (\sqrt{2})^4\text{ cis }(4\cdot45°) = 4\text{ cis }180°$;
$z^5 = (\sqrt{2})^5\text{ cis }(5\cdot45°) = 4\sqrt{2}\text{ cis }225°$; $z^6 = (\sqrt{2})^6\text{ cis }(6\cdot45°) = 8\text{ cis }270°$. The points lie on an expanding spiral.

2. $r = \left|\dfrac{1}{2} + \dfrac{1}{2}i\right| = \sqrt{\left(\dfrac{1}{2}\right)^2 + \left(\dfrac{1}{2}\right)^2} = \dfrac{\sqrt{2}}{2}$; $\theta =$

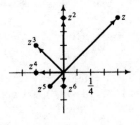

$\text{Tan}^{-1}\left(\dfrac{\dfrac{1}{2}}{\dfrac{1}{2}}\right) = 45°$; $z = \dfrac{\sqrt{2}}{2}\text{ cis }45°$; $z^2 = \dfrac{1}{2}\text{ cis }90°$;

$z^3 = \dfrac{\sqrt{2}}{4}\text{ cis }135°$; $z^4 = \dfrac{1}{4}\text{ cis }180°$; $z^5 = \dfrac{\sqrt{2}}{8}\text{ cis }225°$;

$z^6 = \dfrac{1}{8}\text{ cis }270°$. The points lie on a spiral that closes in on the pole.

3. $r = |\sqrt{3} + i| = \sqrt{(\sqrt{3})^2 + 1^2} = 2;\ \theta = \text{Tan}^{-1}\!\left(\dfrac{1}{\sqrt{3}}\right) =$

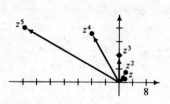

30°; $z = 2$ cis 30°; $z^2 = 4$ cis 60°; $z^3 = 8$ cis 90°; $z^4 =$
16 cis 120°; $z^5 = 32$ cis 150°; $z^6 = 64$ cis 180°. The
points lie on an expanding spiral.

Class Exercises 11-3 · page 409

1. $(2 \text{ cis } 45°)^2 = 2^2 \text{ cis } (2 \cdot 45°) = 4 \text{ cis } 90°$ **2.** $(2 \text{ cis } 45°)^3 = 2^3 \text{ cis } (3 \cdot 45°) =$
8 cis 135°

3. $(\sqrt{2} \text{ cis } (-18°))^4 = (\sqrt{2})^4 \text{ cis } (4 \cdot (-18°)) = 4 \text{ cis } (-72°)$

4. $(1 \text{ cis } 36°)^{10} = 1^{10} \text{ cis } (10 \cdot 36°) = \text{cis } 360°$

5. $\left(4 \text{ cis } \dfrac{\pi}{6}\right)^3 = 4^3 \text{ cis } \left(3 \cdot \dfrac{\pi}{6}\right) = 64 \text{ cis } \dfrac{\pi}{2}$

6. $\left(\sqrt{3} \text{ cis } \dfrac{5\pi}{6}\right)^6 = (\sqrt{3})^6 \text{ cis } \left(6 \cdot \dfrac{5\pi}{6}\right) = 27 \text{ cis } 5\pi$

7. a. $z = \sqrt{2} \text{ cis } 135°;\ z^6 = (\sqrt{2})^6 \text{ cis } (6 \cdot 135°) = 8 \text{ cis } 810° = 8 \text{ cis } 90°$

   b. $z^6 = 8 \text{ cis } 90° = 8(\cos 90° + i \sin 90°) = 8(0 + 1 \cdot i) = 8i$

8. a. $z = \sqrt{2} \text{ cis } 315°;\ z^{10} = (\sqrt{2})^{10} \text{ cis } (10 \cdot 315°) = 32 \text{ cis } 3150° = 32 \text{ cis } 270° = -32i$

   b. $z = \sqrt{2} \text{ cis } (-45°);\ z^{10} = (\sqrt{2})^{10} \text{ cis } (10 \cdot (-45°)) = 32 \text{ cis } (-450°) =$
   $32 \text{ cis } 270° = -32i$

   c. Part (b); $-450°$ is easier to simplify than $3150°$ since $3150°$ involves a larger multiple
   of 360°.

Written Exercises 11-3 · pages 410–411

1. $z = \dfrac{\sqrt{3}}{2} + \dfrac{1}{2}i = 1 \text{ cis } 30°;\ z^2 = 1^2 \text{ cis } (2 \cdot 30°) = 1 \text{ cis } 60°;\ z^3 = 1^3 \text{ cis } (3 \cdot 30°) =$
1 cis 90°

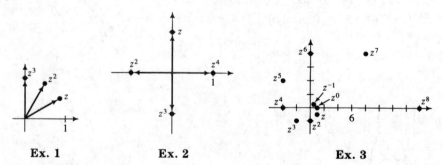

Ex. 1   Ex. 2   Ex. 3

2. $z = i = 1 \text{ cis } 90°;\ z^2 = 1^2 \text{ cis } (2 \cdot 90°) = 1 \text{ cis } 180°;\ z^3 = 1^3 \text{ cis } (3 \cdot 90°) = 1 \text{ cis } 270°;$
$z^4 = 1^4 \text{ cis } (4 \cdot 90°) = 1 \text{ cis } 360° = 1 \text{ cis } 0°$

3. a. $z = 1 - i = \sqrt{2} \text{ cis } 315°$; $z^{-1} = (\sqrt{2})^{-1} \text{ cis } (-1 \cdot 315°) = \dfrac{1}{\sqrt{2}} \text{ cis } (-315°) =$

$\dfrac{1}{\sqrt{2}} \text{ cis } 45° = \dfrac{\sqrt{2}}{2}\left(\dfrac{\sqrt{2}}{2} + i\dfrac{\sqrt{2}}{2}\right) = \dfrac{1}{2} + \dfrac{1}{2}i$; $z^0(\sqrt{2})^0 \text{ cis } (0 \cdot 315°) = 1 \text{ cis } 0° =$

$1(0 + 1 \cdot i) = 1$; $z = 1 - i$; $z^2 = (\sqrt{2})^2 \text{ cis } 630° = 2 \text{ cis } 270° = 2(0 + (-1)i) =$

$-2i$; $z^3 = (\sqrt{2})^3 \text{ cis } 945° = 2\sqrt{2} \text{ cis } 225° = 2\sqrt{2}\left(-\dfrac{\sqrt{2}}{2} + \left(-\dfrac{\sqrt{2}}{2}\right)i\right) = -2 - 2i$;

$z^4 = (\sqrt{2})^4 \text{ cis } 1260° = 4 \text{ cis } 180° = 4(-1 + 0 \cdot i) = -4$; $z^5 = (\sqrt{2})^5 \text{ cis } 1575° =$

$4\sqrt{2} \text{ cis } 135° = 4\sqrt{2}\left(-\dfrac{\sqrt{2}}{2} + \dfrac{\sqrt{2}}{2}i\right) = -4 + 4i$; $z^6 = (\sqrt{2})^6 \text{ cis } 1890° =$

$8 \text{ cis } 90° = 8(0 + 1 \cdot i) = 8i$; $z^7 = (\sqrt{2})^7 \text{ cis } 2205° = 8\sqrt{2} \text{ cis } 45° =$

$8\sqrt{2}\left(\dfrac{\sqrt{2}}{2} + \dfrac{\sqrt{2}}{2}i\right) = 8 + 8i$; $z^8 = (\sqrt{2})^8 \text{ cis } 2520° = 16 \text{ cis } 0° = 16(1 + 0 \cdot i) = 16.$

See diagram on page 292.

b. $z^{12} = (\sqrt{2})^{12} \text{ cis } (12 \cdot 315°) = 64 \text{ cis } 3780° = 64 \text{ cis } 180° = 64(-1 + 0 \cdot i) = -64$

4. a. $z = 1 + i\sqrt{3} = 2 \text{ cis } 60°$; $z^2 =$

$2^{-2} \text{ cis } (-2 \cdot 60°) = \dfrac{1}{4} \text{ cis } (-120°) = \dfrac{1}{4} \text{ cis } 240° =$

$\dfrac{1}{4}\left(-\dfrac{1}{2} - \dfrac{\sqrt{3}}{2}i\right) = -\dfrac{1}{8} - \dfrac{\sqrt{3}}{8}i$; $z^{-1} =$

$2^{-1} \text{ cis } (-1 \cdot 60°) = \dfrac{1}{2} \text{ cis } (-60°) =$

$\dfrac{1}{2}\left(\dfrac{1}{2} + \left(-\dfrac{\sqrt{3}}{2}\right)i\right) = \dfrac{1}{4} - \dfrac{\sqrt{3}}{4}i$;

$z^0 2^0 \text{ cis } (0 \cdot 60°) = 1 \text{ cis } 0^0 = 1(0 + 1 \cdot i) = 1$;

$z^1 = 1 + i\sqrt{3}$; $z^2 = 2^2 \text{ cis } (2 \cdot 60°) =$

$4 \text{ cis } 120° = 4\left(-\dfrac{1}{2} + \dfrac{\sqrt{3}}{2}i\right) = -2 + 2i\sqrt{3}$; $z^3 =$

$2^3 \text{ cis } (3 \cdot 60°) = 8 \text{ cis } 180° = 8(-1 + 0) = -8$;

$z^4 = 2^4 \text{ cis } (4 \cdot 60°) = 16 \text{ cis } 240° =$

$16\left(-\dfrac{1}{2} + \left(-\dfrac{\sqrt{3}}{2}\right)i\right) = -8 - 8i\sqrt{3}$

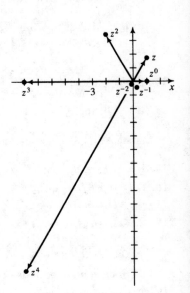

b. $z^{18} = 2^{18} \text{ cis } (18 \cdot 60°) = 2^{18} \text{ cis } 1080° =$

$2^{18} \text{ cis } 0° = 2^{18}(1 + 0 \cdot i) = 2^{18}$

5. a. $(1 - i\sqrt{3})^2(1 - i\sqrt{3}) = (1 - 2i\sqrt{3} + 3i^2)(1 - i\sqrt{3}) =$

$(-2 - 2i\sqrt{3})(1 - i\sqrt{3}) = -2 + 6i^2 = -8$

b. $(1 - i\sqrt{3})^3 = [2 \text{ cis } (-60°)]^3 = 2^3 \text{ cis } (3 \cdot (-60°)) = 8 \text{ cis } (-180°) = 8 \text{ cis } 180° =$

$8(-1 + 0 \cdot i) = -8$

6. a. $(-1 - i)^2(-1 - i) = (1 + 2i + i^2)(-1 - i) = 2i(-1 - i) = -2i^2 - 2i = 2 - 2i$

b. $(-1 - i)^3 = [\sqrt{2} \text{ cis } 225°]^3 = (\sqrt{2})^3 \text{ cis } (3 \cdot 225°) = 2\sqrt{2} \text{ cis } 675° =$

$2\sqrt{2} \text{ cis } 315° = 2\sqrt{2}\left(\dfrac{\sqrt{2}}{2} - \dfrac{\sqrt{2}}{2}i\right) = 2 - 2i$

7. $z^7 = z^6 \cdot z = 1 \cdot z = z$; $z^{14} = z^{12} \cdot z^2 = (z^6)^2 \cdot z^2 = 1^2 \cdot z^2 = z^2$; $z^{40} = z^{36} \cdot z^4 =$

$(z^6)^6 \cdot z^4 = 1^6 \cdot z^4 = z^4$; $z^{-1} = \dfrac{z^5}{z^6} = \dfrac{z^5}{1} = z^5$; $z^{-11} = \dfrac{z^1}{z^{12}} = \dfrac{z}{(z^6)^2} = \dfrac{z}{1^2} = z$

8. $z = 1 \operatorname{cis} \theta$; $z^{100} = 1^{100} \operatorname{cis} (100\theta)$; $|z^{100}| = |1^{100}| = 1$

9. a. By definition, $z^0 = 1$; by De Moivre's Theorem, $z^0 = r^0 \operatorname{cis} (0 \cdot \theta) = 1 \operatorname{cis} 0° = 1(1 + 0 \cdot i) = 1$. Thus, the theorem is true when $n = 0$.

b. $z^{-1} = \dfrac{1}{z} = \dfrac{1}{r} \operatorname{cis} (-\theta) = r^{-1} \operatorname{cis} (-1 \cdot \theta)$; thus, De Moivre's Theorem is true when $n = -1$.

10. $z^{-2} = (z^{-1})^2 = [r^{-1} \operatorname{cis} (-\theta)]^2 = (r^{-1})^2 \operatorname{cis} (2 \cdot (-\theta)) = r^{-2} \operatorname{cis} (-2\theta)$; thus De Moivre's Theorem is true when $n = -2$.

11. $-1 = 1 \operatorname{cis} 180° = 1 \operatorname{cis} 540°$. If $z = r \operatorname{cis} \theta$ and $z^2 = -1$, then $r^2 \operatorname{cis} 2\theta = 1 \operatorname{cis} 180° = 1 \operatorname{cis} 540°$; $\therefore r^2 = 1$, $2\theta = 180°$ or $2\theta = 540°$; $r = 1$, $\theta = 90°$ or $\theta = 270°$. Thus there are two possibilities for z, $z_1 = 1 \operatorname{cis} 90°$ and $z_2 = 1 \operatorname{cis} 270°$.

12. $-i = 1 \operatorname{cis} 270° = 1 \operatorname{cis} 630° = 1 \operatorname{cis} 990°$. If $z = r \operatorname{cis} \theta$ and $z^3 = -i$, then $r^3 \operatorname{cis} 3\theta = 1 \operatorname{cis} 270° = 1 \operatorname{cis} 630° = 1 \operatorname{cis} 990°$; $\therefore r^3 = 1$, $3\theta = 270°$ or $3\theta = 630°$ or $3\theta = 990°$; $r = 1$, $\theta = 90°$ or $\theta = 210°$ or $\theta = 330°$. Thus there are three possibilities for z, $z_1 = 1 \operatorname{cis} 90°$, $z_2 = 1 \operatorname{cis} 210°$, and $z_3 = 1 \operatorname{cis} 330°$.

13. Let $r = 1$ and $\theta = \pi$; $e^{i\pi} = 1 \cdot e^{i\pi} = 1 \operatorname{cis} \pi = -1 + 0 \cdot i = -1$.

14. If $r = 1$ and $\theta = \dfrac{\pi}{2}$, $1 \cdot e^{i\pi/2} = 1 \operatorname{cis} \dfrac{\pi}{2} = 0 + 1 \cdot i = i$. $i^i = (e^{i\pi/2})^i = e^{i^2\pi/2} = e^{-\pi/2}$; since e and $-\dfrac{\pi}{2}$ are real numbers, $i^i = e^{-\pi/2}$ is real and $i^i \approx 0.208$.

15. If $z = r \operatorname{cis} \theta$, then $z^n = (r \operatorname{cis} \theta)^n = (re^{i\theta})^n = r^n e^{i(n\theta)} = r^n \operatorname{cis} n\theta$.

16. $(re^{i\alpha})(se^{i\beta}) = rse^{i\alpha}e^{i\beta} = rse^{i(\alpha+\beta)} = rse^{i(\alpha+\beta)}$. Thus, $(r \operatorname{cis} \alpha)(s \operatorname{cis} \beta) = rs \operatorname{cis} (\alpha + \beta)$

17. a. $OA = r$, $OB = r^2$, $OC = r^3$, $OD = r^4$; $\angle AOB = \angle BOC = \angle COD = \theta$;
$\dfrac{OB}{OA} = \dfrac{OC}{OB} = \dfrac{OD}{OC} = r$, so by the SAS Similarity theorem $\triangle AOB \sim \triangle BOC \sim \triangle COD$.

b. Corresponding angles of similar triangles are congruent.

18. The graph is the same as the graph which appears on page 411 for Ex. 17.

Written Exercises 11-4 · pages 413–414

1. $z^3 = i = 1 \operatorname{cis} 90°$; $z = 1^{1/3} \operatorname{cis} \left(\dfrac{90°}{3} + \dfrac{k \cdot 360°}{3} \right) = 1 \operatorname{cis} (30° + k \cdot 120°)$ for $k = 0, 1, 2$;
$z_1 = 1 \operatorname{cis} 30° = \dfrac{\sqrt{3}}{2} + \dfrac{1}{2}i$; $z_2 = 1 \operatorname{cis} 150° = -\dfrac{\sqrt{3}}{2} + \dfrac{1}{2}i$; $z_3 = 1 \operatorname{cis} 270° = -i$

2. $z^3 = -i = 1 \operatorname{cis} (-90°)$; $z = 1^{1/3} \operatorname{cis} \left(\dfrac{-90°}{3} + \dfrac{k \cdot 360°}{3} \right) = 1 \operatorname{cis} (-30° + k \cdot 120°)$ for
$k = 0, 1, 2$; $z_1 = 1 \operatorname{cis} (-30°) = \dfrac{\sqrt{3}}{2} - \dfrac{1}{2}i$; $z_2 = 1 \operatorname{cis} 90° = i$; $z_3 = 1 \operatorname{cis} 210° = -\dfrac{\sqrt{3}}{2} - \dfrac{1}{2}i$

3. $z^3 = 8 = 8 \text{ cis } 0°; \; z = 8^{1/3} \text{ cis } \left(\dfrac{0°}{3} + \dfrac{k \cdot 360°}{3} \right) = 2 \text{ cis } (0° + k \cdot 120°) \text{ for } k = 0, 1, 2;$

$z_1 = 2 \text{ cis } 0° = 2; \; z_2 = 2 \text{ cis } 120° = -1 + i\sqrt{3}; \; z_3 = 2 \text{ cis } 240° = -1 - i\sqrt{3}$

4. $z^3 = -8 = 8 \text{ cis } 180°; \; z = 8^{1/3} \text{ cis } \left(\dfrac{180°}{3} + \dfrac{k \cdot 360°}{3} \right) = 2 \text{ cis } (60° + k \cdot 120°) \text{ for }$

$k = 0, 1, 2; \; z_1 = 2 \text{ cis } 60° = 1 + i\sqrt{3}; \; z_2 = 2 \text{ cis } 180° = -2; \; z_3 = 2 \text{ cis } 300° = 1 - i\sqrt{3}$

5. $z^4 = 16 = 16 \text{ cis } 0°; \; z = 16^{1/4} \text{ cis } \left(\dfrac{0°}{4} + \dfrac{k \cdot 360°}{4} \right) = 2 \text{ cis } (0° + k \cdot 90°) \text{ for }$

$k = 0, 1, 2, 3; \; z_1 = 2 \text{ cis } 0° = 2; \; z_2 = 2 \text{ cis } 90° = 2i; \; z_3 = 2 \text{ cis } 180° = -2; \; z_4 = 2 \text{ cis } 270° = -2i$

6. $z^4 = -\dfrac{1}{2} - \dfrac{\sqrt{3}}{2} i = 1 \text{ cis } 240°; \; z = 1^{1/4} \text{ cis } \left(\dfrac{240°}{4} + \dfrac{k \cdot 360°}{4} \right) = 1 \text{ cis } (60° + k \cdot 90°) \text{ for }$

$k = 0, 1, 2, 3; \; z_1 = 1 \text{ cis } 60° = \dfrac{1}{2} + \dfrac{\sqrt{3}}{2} i; \; z_2 = 1 \text{ cis } 150° = -\dfrac{\sqrt{3}}{2} + \dfrac{1}{2} i; \; z_3 =$

$1 \text{ cis } 240° = -\dfrac{1}{2} - \dfrac{\sqrt{3}}{2} i; \; z_4 = 1 \text{ cis } 330° = \dfrac{\sqrt{3}}{2} - \dfrac{1}{2} i$

7. a. $z_1 + z_2 + z_3 = (\sqrt{3} + i) + (-\sqrt{3} + i) + (-2i) = 2i + (-2i) = 0;$

$z_1 z_2 z_3 = (\sqrt{3} + i)(-\sqrt{3} + i)(-2i) = (-4)(-2i) = 8i$

b. The equation $z^3 - 8i = 0$ has no quadratic term, so the sum of its roots is 0; the

product of the roots is $-\dfrac{-8i}{1} = 8i.$

8. a. $z_1 + z_2 + z_3 + z_4 = (\sqrt{2} + i\sqrt{2}) + (-\sqrt{2} + i\sqrt{2}) + (-\sqrt{2} - i\sqrt{2}) +$

$(\sqrt{2} - i\sqrt{2}) = 2i\sqrt{2} - 2i\sqrt{2} = 0; \; z_1 z_2 z_3 z_4 =$

$(\sqrt{2} + i\sqrt{2})(-\sqrt{2} + i\sqrt{2})(-\sqrt{2} - i\sqrt{2})(\sqrt{2} - i\sqrt{2}) = -4(-4) = 16$

b. The equation $z^4 + 16 = 0$ has no cubic term, so the sum of the roots is 0; the

product of the roots is $\dfrac{16}{1} = 16.$

9. $z^3 - 8 = z^3 - 2^3 = 0; \; (z - 2)(z^2 + 2z + 4) = 0; \; z = 2 \text{ or } z = \dfrac{-2 \pm \sqrt{4 - 4(1)(4)}}{2 \cdot 1} =$

$-1 \pm i\sqrt{3}; \text{ the cube roots of 8 are } 2, -1 + i\sqrt{3}, \text{ and } -1 - i\sqrt{3}.$

10. $z^3 + 2^3 = 0; \; (z + 2)(z^2 - 2z + 4) = 0; \; z = -2 \text{ or } z = \dfrac{2 \pm \sqrt{4 - 4(1)(4)}}{2 \cdot 1} = 1 \pm i\sqrt{3};$

the cube roots of -8 are $-2, 1 + i\sqrt{3}, \text{ and } 1 - i\sqrt{3}.$

11. a. $z^4 + 4 = (z^4 + 4z^2 + 4) - 4z^2 = (z^2 + 2)^2 - (2z)^2 =$

$(z^2 + 2 + 2z)(z^2 + 2 - 2z) = (z^2 + 2z + 2)(z^2 - 2z + 2)$

b. $z = \dfrac{-2 \pm \sqrt{4 - 4(1)(2)}}{2 \cdot 1} \text{ or } z = \dfrac{2 \pm \sqrt{4 - 4(1)(2)}}{2 \cdot 1}; \text{ the fourth roots are}$

$-1 + i, -1 - i, 1 + i, \text{ and } 1 - i. \text{ See diagram at right.}$

c. $z^4 = -4 = 4 \text{ cis } 180°; \; z = 4^{1/4} \text{ cis } \left(\dfrac{180°}{4} + \dfrac{k \cdot 360°}{4} \right) =$

$\sqrt{2} \text{ cis } (45° + k \cdot 90°) \text{ for } k = 0, 1, 2, 3; \; z_1 = \sqrt{2} \text{ cis } 45° = 1 + i,$

$z_2 = \sqrt{2} \text{ cis } 135° = -1 + i, \; z_3 = \sqrt{2} \text{ cis } 225° = -1 - i, \; z_4 =$

$\sqrt{2} \text{ cis } 315° = 1 - i$

12. The n n-th roots of a complex number are the vertices of a regular n-sided polygon.

13. a. a regular pentagon inscribed in a circle with radius 10 and center $(0,0)$

    b. a regular nonagon inscribed in a circle with radius 1 and center $(0,0)$

14. $z^5 = 32 = 32 \text{ cis } 0°; z = 32^{1/5} \text{ cis } \left(\dfrac{0°}{5} + \dfrac{k \cdot 360°}{5} \right) = 2 \text{ cis } (0° + k \cdot 72°)$ for

$k = 0,1,2,3,4; z_1 = 2 \text{ cis } 0° = 2; z_2 = 2 \text{ cis } 72° = 0.6180 + 1.9021i; z_3 = 2 \text{ cis } 144° = -1.6180 + 1.1756i; z_4 = 2 \text{ cis } 216° = -1.6180 - 1.1756i; z_5 = 2 \text{ cis } 288° = 0.6180 - 1.9021i$

15. $z^3 = 3\sqrt{3} + 3i = 6 \text{ cis } 30°; z = 6^{1/3} \text{ cis } \left(\dfrac{30°}{3} + \dfrac{k \cdot 360°}{3} \right) = \sqrt[3]{6} \text{ cis } (10° + k \cdot 120°)$ for

$k = 0,1,2; z_1 = \sqrt[3]{6} \text{ cis } 10° = 1.7895 + 0.3155i; z_2 = \sqrt[3]{6} \text{ cis } 130° = -1.1680 + 1.3920i; z_3 = \sqrt[3]{6} \text{ cis } 250° = -0.6215 - 1.7075i$

16. a. $n = 0: 2^{2^0} + 1 = 2^1 + 1 = 3; n = 1: 2^{2^1} + 1 = 2^2 + 1 = 5; n = 2: 2^{2^2} + 1 = 2^4 + 1 = 17; n = 3: 2^{2^3} + 1 = 2^8 + 1 = 257$

    b. Since $32 = 2^5$, the polygon is constructible; $17 = 17 \cdot 2^0$, and since 17 is a Fermat prime, the polygon is constructible; polygons of 7 and 9 sides are not constructible; $10 = 5 \cdot 2^1$, and since 5 is a Fermat prime, the polygon is constructible; $255 = 3 \cdot 5 \cdot 17 \cdot 2^0$, and since 3, 5, and 17 are Fermat primes, the polygon is constructible.

    c. The graphs of the solutions to $z^{17} = 1$ are the vertices of a 17-sided regular polygon.

17. Let P be the perimeter of an n-sided regular polygon inscribed in a circle of radius 1. The polygon is made up of n isosceles triangles, each with vertex angle $\dfrac{360°}{n}$ and sides of lengths 1, 1, and $\dfrac{P}{n}$. $\sin \dfrac{180°}{n} = \dfrac{\frac{1}{2} \cdot \frac{P}{n}}{1}; P = 2n \sin \dfrac{180°}{n}$.

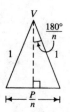

Chapter Test · page 415

1. a. $(3\sqrt{2}, 45°)$ **b.** $(0,6)$ **c.** $(2, 270°)$ **d.** $(-8, 0)$ **e.** $(-1, -\sqrt{3})$ **f.** $(2, 120°)$

2. a.

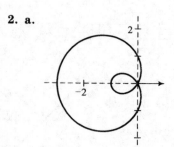

    b. $r = 1 - 2\cos\theta; \pm\sqrt{x^2 + y^2} = 1 - \dfrac{2x}{\pm\sqrt{x^2 + y^2}};$

$x^2 + y^2 = \pm\sqrt{x^2 + y^2} - 2x; (x^2 + y^2 + 2x)^2 = x^2 + y^2$

3. a–c.

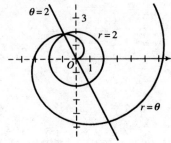

4. a. $r_1 = |\sqrt{3} - i| = \sqrt{(\sqrt{3})^2 + (-1)^2} = 2;$

$\text{Tan}^{-1}\left(\dfrac{-1}{\sqrt{3}}\right) = -30°;\ \theta_1 = 360° - 30° = 330°;$

$z_1 = 2 \text{ cis } 330°;\ r_2 = |4 + 4i| = \sqrt{4^2 + 4^2} =$

$4\sqrt{2};\ \theta_2 = \text{Tan}^{-1}\dfrac{4}{4} = 45°;\ z_2 = 4\sqrt{2} \text{ cis } 45°;$

$z_1 z_2 = 2 \cdot 4\sqrt{2} \text{ cis } (330° + 45°) =$
$8\sqrt{2} \text{ cis } 375° = 8\sqrt{2} \text{ cis } 15°$

b.

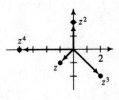

5. a. $z^2 = 2^2 \text{ cis } (2 \cdot 120°) = 4 \text{ cis } 240°;\ z = 2 \text{ cis } 120° = 2\left(-\dfrac{1}{2} + \dfrac{\sqrt{3}}{2}i\right) = -1 + i\sqrt{3};$

$z^2 = (-1 + i\sqrt{3})^2 = 1 - 2i\sqrt{3} - 3 = -2 - 2i\sqrt{3}$

b. $4 \text{ cis } 240° = 4\left(-\dfrac{1}{2} - \dfrac{\sqrt{3}}{2}i\right) = -2 - 2i\sqrt{3}$

6. a. $r = |-1 - i| = \sqrt{(-1)^2 + (-1)^2} = \sqrt{2};\ \text{Tan}^{-1}\left(\dfrac{-1}{-1}\right) = 45°;\ \theta = 180° + 45° = 225°;$

$z = \sqrt{2} \text{ cis } 225°$

b. $z = -1 - i;\ z^2 = (\sqrt{2})^2 \text{ cis } (2 \cdot 225°) = 2 \text{ cis } 450° =$
$2 \text{ cis } 90° = 2(0 + 1 \cdot i) = 2i;\ z^3 = (\sqrt{2})^3 \text{ cis } (3 \cdot 225°) =$

$2\sqrt{2} \text{ cis } 675° = 2\sqrt{2} \text{ cis } 315° = 2\sqrt{2}\left(\dfrac{\sqrt{2}}{2} - \dfrac{\sqrt{2}}{2}i\right) = 2 - 2i;$

$z^4 = (\sqrt{2})^4 \text{ cis } (4 \cdot 225°) = 4 \text{ cis } 900° = 4 \text{ cis } 180° =$
$4(-1 + 0 \cdot i) = -4$

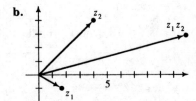

c. $z^{10} = (\sqrt{2})^{10} \text{ cis } (10 \cdot 225°) = 32 \text{ cis } 2250° = 32 \text{ cis } 90° = 32i$

7. To find the n n-th roots of a complex number, express the complex number in polar form

as $z = r \text{ cis } \theta$. Then the roots are $\sqrt[n]{r} \text{ cis }\left(\dfrac{\theta}{n} + \dfrac{k \cdot 360°}{n}\right)$ for $k = 0, 1, 2, \ldots, n - 1.$

Since $\sqrt[n]{z} = z^{1/n}$, by De Moivre's Theorem, $\sqrt[n]{z} = z^{1/n} = r^{1/n} \text{ cis }\left(\dfrac{1}{n} \cdot \theta\right) = \sqrt[n]{r} \text{ cis } \dfrac{\theta}{n}.$

8. a. $(\sqrt[6]{2} \text{ cis } 135°)^3 = (2^{1/6})^3 \text{ cis } (3 \cdot 135°) = 2^{1/2} \text{ cis } 405° = \sqrt{2} \text{ cis } 45° =$

$\sqrt{2}\left(\dfrac{\sqrt{2}}{2} + \dfrac{\sqrt{2}}{2}i\right) = 1 + i$

b. For all three cube roots, $r = \sqrt[6]{2}$. Since the number of degrees between the cube

roots is $\dfrac{360°}{3} = 120°$, the other two cube roots are $\sqrt[6]{2} \text{ cis } (135° - 120°) = \sqrt[6]{2} \text{ cis } 15°$

and $\sqrt[6]{2} \text{ cis } (135° + 120°) = \sqrt[6]{2} \text{ cis } 255°.$

Cumulative Review · pages 416–417

1. a. $-35° + 360° = 325°$; $-35° - 360° = -395°$ **b.** $\dfrac{5\pi}{6} + 2\pi = \dfrac{17\pi}{6}$, $\dfrac{5\pi}{6} - 2\pi = -\dfrac{7\pi}{6}$

2. a. $165° \cdot \dfrac{\pi}{180°} = \dfrac{11\pi}{12}$ **b.** $208° \cdot \dfrac{\pi}{180°} \approx 3.63$

3. a. $-\dfrac{7\pi}{6} \cdot \dfrac{180°}{\pi} = -210°$ **b.** $1.8 \cdot \dfrac{180°}{\pi} \approx 103°10'$ or $103.1°$

4. $A = \dfrac{1}{2}r^2\theta$; $8.1\pi = \dfrac{1}{2}r^2\left(36° \cdot \dfrac{\pi}{180°}\right)$; $r^2 = 81$; $r = 9$ cm; $s = r\theta$; $s = 9\left(36° \cdot \dfrac{\pi}{180°}\right)$;
$s = 1.8\pi$ cm

5. If θ is a third-quadrant angle and $\cos\theta = -\dfrac{2}{3} = \dfrac{x}{r}$, then $x = -2$, $r = 3$,
$y = -\sqrt{r^2 - x^2} = -\sqrt{3^2 - (-2)^2} = -\sqrt{5}$, and $\sin\theta = \dfrac{y}{r} = -\dfrac{\sqrt{5}}{3}$.

6. a. $\cos 236° = -\cos 56°$ **b.** $\sin 485° = \sin 125° = \sin 55°$ **c.** $\sin(-62°) = -\sin 62°$

7.

		sin	cos	tan	csc	sec	cot
a.	$\dfrac{\pi}{6}$	$\dfrac{1}{2}$	$\dfrac{\sqrt{3}}{2}$	$\dfrac{\sqrt{3}}{3}$	2	$\dfrac{2\sqrt{3}}{3}$	$\sqrt{3}$
b.	$225°$	$-\dfrac{\sqrt{2}}{2}$	$-\dfrac{\sqrt{2}}{2}$	1	$-\sqrt{2}$	$-\sqrt{2}$	1
c.	$-\dfrac{\pi}{2}$	-1	0	undef.	-1	undef.	0

8. In the third quadrant, $\dfrac{\text{adj.}}{\text{opp.}} = \dfrac{-5}{-12}$; $r = \sqrt{(-5)^2 + (-12)^2} = 13$; $\sin x = -\dfrac{12}{13}$;
$\cos x = -\dfrac{5}{13}$; $\tan x = \dfrac{12}{5}$; $\sec x = -\dfrac{13}{5}$; $\csc x = -\dfrac{13}{12}$.

9. a.

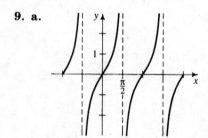

b. domain $= \left\{x \,\middle|\, x \neq \dfrac{\pi}{2} + n\pi\right\}$;
range $=$ {all real nos.};
period $= \pi$

10. a. $\dfrac{\pi}{4}$ **b.** $\dfrac{\pi}{3}$ **c.** $\dfrac{5\pi}{6}$ **d.** $\dfrac{\pi}{3}$ **e.** 0

f. Let $\theta = \text{Cos}^{-1}\dfrac{3}{5}\left(0 < \theta < \dfrac{\pi}{2}\right)$; $\cos\theta = \dfrac{3}{5} = \dfrac{x}{r}$; $y = \sqrt{5^2 - 3^2} = 4$;
$\sin\left(\text{Cos}^{-1}\dfrac{3}{5}\right) = \sin\theta = \dfrac{y}{r} = \dfrac{4}{5}$.

(Continued)

g. Let $\theta = \text{Tan}^{-1}\dfrac{5}{12}$ $\left(0 < \theta < \dfrac{\pi}{2}\right)$; $\tan\theta = \dfrac{5}{12} = \dfrac{y}{x}$; $r = \sqrt{12^2 + 5^2} = 13$;

$\sec\left(\text{Tan}^{-1}\dfrac{5}{12}\right) = \sec\theta = \dfrac{r}{x} = \dfrac{13}{12}$. **h.** $\cos(\text{Sin}^{-1} 1) = \cos\dfrac{\pi}{2} = 0$

11. $2\sec x - 5 = 0$; $\sec x = \dfrac{5}{2}$; $\cos x = \dfrac{2}{5}$; $x = \text{Cos}^{-1}\dfrac{2}{5} \approx 1.16$ or $x \approx 2\pi - 1.16 = 5.12$

12. $3x - 4y = 6$; $m = \dfrac{3}{4} = \tan\alpha$; $\alpha = $ inclination $\approx 37°$

13. a.
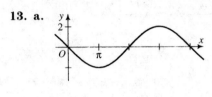
b.
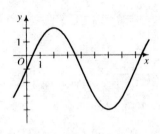

14. $\dfrac{1 - \cos\theta}{\sin^2\theta} + \tan^2\theta - \sec^2\theta = \dfrac{1 - \cos\theta}{1 - \cos^2\theta} - 1 = \dfrac{1 - \cos\theta}{(1 - \cos\theta)(1 + \cos\theta)} - 1 = \dfrac{1}{1 + \cos\theta} - $

$\dfrac{1 + \cos\theta}{1 + \cos\theta} = \dfrac{1 - 1 - \cos\theta}{1 + \cos\theta} = -\dfrac{\cos\theta}{1 + \cos\theta}$

15. $\cos x \sin x = \sin^2 x$; $\sin x(\cos x - \sin x) = 0$; $\sin x = 0$ or $\cos x = \sin x$; $\sin x = 0$ or

$\tan x = 1$ ($\cos x \neq 0$); $x = 0$, π or $x = \dfrac{\pi}{4}, \dfrac{5\pi}{4}$; $0, \dfrac{\pi}{4}, \pi, \dfrac{5\pi}{4}$

16. $\sin 32° = \dfrac{a}{14}$; $a = 14\sin 32° \approx 7.42$; $\cos 32° = \dfrac{b}{14}$; $b = 14\cos 32° \approx 11.9$

17. The shorter diagonal divides the parallelogram into 2 congruent triangles; $\therefore$ the area of

the parallelogram $= 2$(area of one $\triangle$) $= 2\left(\dfrac{1}{2} \cdot 4 \cdot 9 \cdot \sin 40°\right) \approx 23.1$ cm^2.

18. $\angle AFB = 180° - 110° - 25° = 45°$; $\dfrac{\sin 25°}{b} = \dfrac{\sin 45°}{10}$; $b \approx 6.0$ mi

19. Let x be the third side. $x^2 = 8.4^2 + 7.6^2 - 2(8.4)(7.6)\cos 82° \approx 110.55$; $x \approx 10.5$

20. Let F, S, and R be the first person, the second person, and the rock, respectively.
$\angle RFS = 90° - 25° = 65°$; $\angle RSF = 90° - 15° = 75°$; $\angle FRS = 180° - 65° - $

$75° = 40°$; $\dfrac{\sin 65°}{f} = \dfrac{\sin 40°}{100}$; $f \approx 141$ m. Let h be the length of the altitude from R to

$\overline{FS}$; $\sin 75° = \dfrac{h}{141}$; $h \approx 136$; the river is 136 m wide.

21. Since $0 < \alpha < \dfrac{\pi}{2}$ and $\sin\alpha = \dfrac{4}{5}$, $\cos\alpha = \sqrt{1 - \left(\dfrac{4}{5}\right)^2} = \dfrac{3}{5}$; since $0 < \beta < \dfrac{\pi}{2}$ and

$\sin\beta = \dfrac{5}{13}$, $\cos\beta = \sqrt{1 - \left(\dfrac{5}{13}\right)^2} = \dfrac{12}{13}$. $\cos(\alpha - \beta) = \cos\alpha\cos\beta + \sin\alpha\sin\beta = $

$\dfrac{3}{5} \cdot \dfrac{12}{13} + \dfrac{4}{5} \cdot \dfrac{5}{13} = \dfrac{56}{65}$

22. $\sin 75° = \sin(30° + 45°) = \sin 30°\cos 45° + \sin 45°\cos 30° = \dfrac{1}{2} \cdot \dfrac{\sqrt{2}}{2} + \dfrac{\sqrt{2}}{2} \cdot \dfrac{\sqrt{3}}{2} = $

$\dfrac{\sqrt{2} + \sqrt{6}}{4}$

23. $\tan\left(\dfrac{3\pi}{4}+\theta\right)=\dfrac{\tan\dfrac{3\pi}{4}+\tan\theta}{1-\tan\dfrac{3\pi}{4}\tan\theta}=\dfrac{-1-\dfrac{1}{2}}{1-(-1)\left(-\dfrac{1}{2}\right)}=-3$

24. Since $0<\alpha<\dfrac{\pi}{2}$ and $\cos\alpha=\dfrac{1}{3}$, $\sin\alpha=\sqrt{1-\left(\dfrac{1}{3}\right)^2}=\dfrac{2\sqrt2}{3}$; $\sin 2\alpha=$

$2\sin\alpha\cos\alpha=2\cdot\dfrac{2\sqrt2}{3}\cdot\dfrac{1}{3}=\dfrac{4\sqrt2}{9}$; $\cos\dfrac{1}{2}\alpha=\sqrt{\dfrac{1+\cos\alpha}{2}}=\sqrt{\dfrac{1+\dfrac{1}{3}}{2}}=\sqrt{\dfrac{2}{3}}=$

$\dfrac{\sqrt6}{3}\left(0<\alpha<\dfrac{\pi}{2},\text{ so }0<\dfrac{\alpha}{2}<\dfrac{\pi}{4}\right)$

25. $r=\sqrt{a^2+b^2}=\sqrt{(-1)^2+(\sqrt3)^2}=2$; $\text{Tan}^{-1}\left(\dfrac{\sqrt3}{-1}\right)=-60°$; $\theta=-60°+180°=$

120°; $(2,120°)$, $(2,-240°)$, $(2,480°)$

26. $a=r\cos\theta=-6\cos30°=-6\cdot\dfrac{\sqrt3}{2}=-3\sqrt3$; $b=r\sin\theta=-6\sin30°=-6\cdot\dfrac{1}{2}=$

-3; $(-3\sqrt3,-3)$

27. a.

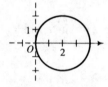

28.

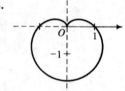

b. $r=4\cos\theta$; $r=4\cdot\dfrac{x}{r}$;

$r^2=4x$; $x^2+y^2=4x$;

$(x-2)^2+y^2=4$

cardioid

29. a. $z_1 z_2=(1-i\sqrt3)(-1+i\sqrt3)=-1+i\sqrt3+i\sqrt3+3=2+2i\sqrt3$

b. $r_1=|1-i\sqrt3|=\sqrt{1^2+(-\sqrt3)^2}=2$; $\text{Tan}^{-1}\left(\dfrac{-\sqrt3}{1}\right)=-60°$; $\theta_1=360°-60°=$

300°; $z_1=2\text{ cis }300°$; $r_2=|-1+i\sqrt3|=\sqrt{(-1)^2+(\sqrt3)^2}=2$;

$\text{Tan}^{-1}\left(\dfrac{\sqrt3}{-1}\right)=-60°$; $\theta_2=180°-60°=120°$; $z_2=2\text{ cis }120°$; $z_1 z_2=$

$2\cdot2\text{ cis }(300°+120°)=4\text{ cis }420°$

c. $4\text{ cis }420°=4\text{ cis }60°=4\left(\dfrac{1}{2}+\dfrac{\sqrt3}{2}i\right)=2+2i\sqrt3$

30. $r=|-2-2i|=\sqrt{(-2)^2+(-2)^2}=2\sqrt2$; $\text{Tan}^{-1}\left(\dfrac{-2}{-2}\right)=45°$; $\theta=180°+45°=225°$;

$z=2\sqrt2\text{ cis }225°$; $z^6=(2\sqrt2)^6\text{ cis }(6\cdot225°)=512\text{ cis }1350°=512\text{ cis }270°=$

$512(0-1\cdot i)=-512i$

31. $z^3=1-i=\sqrt2\text{ cis }315°$; $z=(\sqrt2)^{1/3}\text{ cis }\left(\dfrac{315°}{3}+\dfrac{k\cdot360°}{3}\right)=$

$\sqrt[6]{2}\text{ cis }(105°+k\cdot120°)$ for $k=0,1,2$; $z_1=\sqrt[6]{2}\text{ cis }105°$;

$z_2=\sqrt[6]{2}\text{ cis }(105°+120°)=\sqrt[6]{2}\text{ cis }225°$; $z_3=\sqrt[6]{2}\text{ cis}(105°+240°)=\sqrt[6]{2}\text{ cis }345°$

Class Exercises 12-1 · pages 422–423

1. True   **2.** False   **3.** False   **4.** True   **5.** True   **6.** True

7. True   **8.** True   **9.** Yes; no   **10.** $\overrightarrow{BC}$ **11.** 6   **12.** $\overrightarrow{BO}$ or $\overrightarrow{OD}$

13. $\overrightarrow{AC}$ **14.** $\overrightarrow{AC}$ **15.** $\overrightarrow{AB}$ **16.** $\overrightarrow{OD}$; $\overrightarrow{AD}$ **17.** $\overrightarrow{DB}$; $\overrightarrow{AB}$

18. Exercise 1 is addition of vectors. Exercise 2 is addition of real numbers.

19. The plus sign on the left side of $(k + m)\mathbf{v} = k\mathbf{v} + m\mathbf{v}$ indicates addition of real numbers. On the right, it indicates addition of vectors.

20. km is multiplication of two real numbers. That product multiplied by $\mathbf{v}$ is scalar multiplication.

21. No, because we have not defined the sum of a scalar and a vector.

Written Exercises 12-1 · pages 423–426

1. a.

b. 200 mi/h to the northwest

c. 100 mi/h to the southeast

2. a.

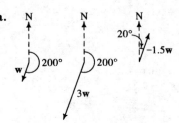

b. 600 mi on a course of 200°

c. 300 mi on a course of 20°

3. a. $\overrightarrow{PR}$ **b.** 5   **c.** 0   **4. a.** $\overrightarrow{AC}$ **b.** 14   **c.** 6   **5.**

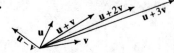

6.

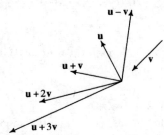

7. $2(\mathbf{u} + \mathbf{v}) = 2\mathbf{u} + 2\mathbf{v}$

8. $3(\mathbf{a} - \mathbf{b}) = 3\mathbf{a} - 3\mathbf{b}$

9.

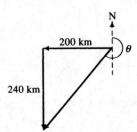

(distance)$^2 = 200^2 + 240^2$; distance $= \sqrt{200^2 + 240^2} =$ $\sqrt{97600} \approx 312.4$ km. $\text{Tan}^{-1}\left(\dfrac{240}{200}\right) \approx 50.2°$; $\theta = 270° - 50.2° = 219.8°$; bearing 219.8°

10.

The strength of the resultant sum is 10 N; $\sin \alpha = \dfrac{6}{10} = 0.6$; $\alpha \approx 36.9°$ west of north.

11. 50 knots   400 knots

about 436.8 knots; about 85.4°

12. a.

b. $(\text{speed})^2 = 2^2 + 1^2$; speed $= \sqrt{5} \approx 2.24$ km/h

c. 2 h; 2 km

13. a. **b.** **c.**

30 knots   60 knots   90 knots

$|\mathbf{v}| \approx 579$ $|\mathbf{v}| \approx 559$ $|\mathbf{v}| \approx 540$

14. a. 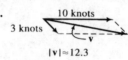 **b.** **c.**

10 knots   10 knots   10 knots

2 knots   3 knots   4 knots

$|\mathbf{v}| \approx 11.5$ $|\mathbf{v}| \approx 12.3$ $|\mathbf{v}| \approx 13.1$

15. a.

b. ≈ 10 mi

c. $3^2 + 8^2 - 2 \cdot 3 \cdot 8 \cdot \cos 120° \approx 97$; $\sqrt{97} \approx 9.85$ mi

16. a, b.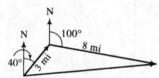

b. $|\mathbf{F}_1 + \mathbf{F}_2|^2 = 3^2 + 5^2 - 2 \cdot 3 \cdot 5 \cdot \cos 120° = 49$;

$|\mathbf{F}_1 + \mathbf{F}_2| = 7$; $\dfrac{\sin \theta}{5} = \dfrac{\sin 120°}{7}$; $\sin \theta = \dfrac{5}{7} \sin 120°$;

$\theta \approx 38°$; direction: 038°

c. $\mathbf{F}_3 = -(\mathbf{F}_1 + \mathbf{F}_2)$; $|\mathbf{F}_3| = 7$; $38° + 180° = 218°$; direction: 218°

17. a. $\overrightarrow{MN} = \mathbf{a} + \mathbf{b}$ **b.** $\overrightarrow{QR} = 2\mathbf{a} + 2\mathbf{b}$

c. The segment joining the midpoints of two sides of a triangle is parallel to the third side and is half as long as the third side.

18. a. (1) $\overrightarrow{AD}$ (2) $\overrightarrow{BC}$ (3) **0** (4) **0**

b. $\overrightarrow{AM} + \overrightarrow{BM} + \overrightarrow{MN} + \overrightarrow{MN} + \overrightarrow{ND} + \overrightarrow{NC} = \overrightarrow{AD} + \overrightarrow{BC}$; $\mathbf{0} + 2\overrightarrow{MN} = \overrightarrow{AD} + \overrightarrow{BC}$;
$2\overrightarrow{MN} = \overrightarrow{AD} + \overrightarrow{BC}$; the median of a trapezoid is parallel to the bases, and the length of the median is the mean of the lengths of the bases.

19. a. $\dfrac{2}{3}\mathbf{v}$ **b.** $\dfrac{1}{3}\mathbf{v}$ **c.** $-\dfrac{1}{3}\mathbf{v}$ **20. a.** $-\dfrac{3}{2}\mathbf{v}$ **b.** $\dfrac{5}{2}\mathbf{v}$ **c.** $-\mathbf{v}$

21. a. $\overrightarrow{AD} = \overrightarrow{AB} + \overrightarrow{BD} = \mathbf{u} + \mathbf{v}$ **b.** $\overrightarrow{AE} = \dfrac{1}{2}\overrightarrow{AD} = \dfrac{1}{2}(\mathbf{u} + \mathbf{v}) = \dfrac{1}{2}\mathbf{u} + \dfrac{1}{2}\mathbf{v}$

c. $\overrightarrow{AB} + \overrightarrow{BE} = \overrightarrow{AE}$, so $\overrightarrow{BE} = \overrightarrow{AE} - \overrightarrow{AB} = \overrightarrow{AE} + (-\overrightarrow{AB}) = \dfrac{1}{2}\mathbf{u} + \dfrac{1}{2}\mathbf{v} + (-\mathbf{u}) = \dfrac{1}{2}\mathbf{v} - \dfrac{1}{2}\mathbf{u}$

d. $\overrightarrow{BE} + \overrightarrow{EC} = \overrightarrow{BC}$, so $\overrightarrow{EC} = \overrightarrow{BC} - \overrightarrow{BE} = 4\mathbf{v} - \left(\dfrac{1}{2}\mathbf{v} - \dfrac{1}{2}\mathbf{u}\right) = \dfrac{7}{2}\mathbf{v} + \dfrac{1}{2}\mathbf{u}$

22. a. $\overrightarrow{BC} = \overrightarrow{AD} = \mathbf{y}$ **b.** $\overrightarrow{CD} = \overrightarrow{BA} = -\overrightarrow{AB} = -\mathbf{x}$ **c.** $\overrightarrow{AC} = \overrightarrow{AB} + \overrightarrow{BC} = \mathbf{x} + \mathbf{y}$

d. $\overrightarrow{AO} = \dfrac{1}{2}\overrightarrow{AC} = \dfrac{1}{2}(\mathbf{x} + \mathbf{y}) = \dfrac{1}{2}\mathbf{x} + \dfrac{1}{2}\mathbf{y}$

e. $\overrightarrow{BO} = \overrightarrow{BA} + \overrightarrow{AO} = -\mathbf{x} + \left(\dfrac{1}{2}\mathbf{x} + \dfrac{1}{2}\mathbf{y}\right) = \dfrac{1}{2}\mathbf{y} - \dfrac{1}{2}\mathbf{x}$

23. a. $\overrightarrow{DF} = \overrightarrow{DE} + \overrightarrow{EF} = -\mathbf{v} + \mathbf{w}$ **b.** $\overrightarrow{DH} = \dfrac{1}{3}\overrightarrow{DF} = \dfrac{1}{3}(-\mathbf{v} + \mathbf{w}) = -\dfrac{1}{3}\mathbf{v} + \dfrac{1}{3}\mathbf{w}$

c. $\overrightarrow{EH} = \overrightarrow{ED} + \overrightarrow{DH} = \mathbf{v} + \left(-\dfrac{1}{3}\mathbf{v} + \dfrac{1}{3}\mathbf{w}\right) = \dfrac{2}{3}\mathbf{v} + \dfrac{1}{3}\mathbf{w}$

d. $\overrightarrow{EG} = \dfrac{2}{3}(\overrightarrow{EH}) = \dfrac{2}{3}\left(\dfrac{2}{3}\mathbf{v} + \dfrac{1}{3}\mathbf{w}\right) = \dfrac{4}{9}\mathbf{v} + \dfrac{2}{9}\mathbf{w}$

e. $\overrightarrow{DG} = \overrightarrow{DE} + \overrightarrow{EG} = -\mathbf{v} + \dfrac{2}{3}\overrightarrow{EH} = -\mathbf{v} + \dfrac{2}{3}\left(\dfrac{2}{3}\mathbf{v} + \dfrac{1}{3}\mathbf{w}\right) = -\dfrac{5}{9}\mathbf{v} + \dfrac{2}{9}\mathbf{w}$

24.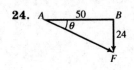

a. $|\overrightarrow{AF}| \approx 56$; $56 \div 3 \approx 18.7$ knots; about 18.7 knots at 115°

b. $\tan \theta = \dfrac{24}{50}$; $\theta \approx 25.6°$; $90° + 25.6° = 115.6°$;

$|\overrightarrow{AF}| = \sqrt{50^2 + 24^2} = \sqrt{3076} \approx 55.5$; $55.5 \div 3 = 18.5$; 18.5 knots at 115.6°

25.

$\angle CAX = 60°$; $|\overrightarrow{CX}|^2 = 450^2 + 75^2 - 2 \cdot 450 \cdot 75 \cdot \cos 60°$;

$|\overrightarrow{CX}| \approx 417.6$; $\dfrac{\sin\angle ACX}{75} = \dfrac{\sin 60°}{417.6}$; $\sin\angle ACX = \dfrac{75 \sin 60°}{417.6}$;

$\angle ACX \approx 8.9°$; $150° + 8.9° = 158.9°$

Class Exercises 12-2 · pages 428–429

1. a. $\overrightarrow{AB} = (5 - 4, 4 - 0) = (1, 4)$; $\overrightarrow{CD} = (1 - (-3), -3 - 1) = (4, -4)$

b. $|\overrightarrow{AB}| = \sqrt{1^2 + 4^2} = \sqrt{17}$; $|\overrightarrow{CD}| = \sqrt{4^2 + (-4)^2} = 4\sqrt{2}$

c. $(4, 0) + \dfrac{1}{4}(1, 4) = \left(\dfrac{17}{4}, 1\right)$ **d.** $(-3, 1) + \dfrac{3}{4}(4, -4) = (0, -2)$

2. a. $(3, 6)$ **b.** $(4, 2)$ **c.** $(-2, 2)$ **d.** $(2, 4) + (9, 0) = (11, 4)$

3. a. $\mathbf{F} = (2 \cos 150°, 2 \sin 150°) = \left(2\left(-\dfrac{\sqrt{3}}{2}\right), 2 \cdot \dfrac{1}{2}\right) = (-\sqrt{3}, 1)$ **b.** $|\mathbf{F}| = 2$

4. $2\mathbf{v} = (6 \cos 40°, 6 \sin 40°)$; $-\mathbf{v} = (-3 \cos 40°, -3 \sin 40°)$

5. a. $(3, 2) = (x, y) - (4, 0)$; $(x, y) = (3, 2) + (4, 0)$; $(x, y) = (7, 2)$

b. $(4, -1) = (8, 8) - (x, y)$; $(x, y) = (8, 8) - (4, -1)$; $(x, y) = (4, 9)$

6. Horizontal; $80 \cos 30° = 40\sqrt{3} \approx 69.3$ lb; yes, decreasing the angle increases the magnitude of the horizontal component.

Written Exercises 12-2 · pages 429–432

1. $\overrightarrow{AB} = (3 - 1, -2 - (-2)) = (2, 0); |\overrightarrow{AB}| = 2$

2. $\overrightarrow{AB} = (0 - 4, -1 - 2) = (-4, -3); |\overrightarrow{AB}| = \sqrt{(-4)^2 + (-3)^2} = 5$

3. $\overrightarrow{AB} = (-5 + 3, 1 + 5) = (-2, 6); |\overrightarrow{AB}| = \sqrt{(-2)^2 + 6^2} = \sqrt{40} = 2\sqrt{10}$

4. $\overrightarrow{AB} = (3 - 7, -2 + 2) = (-4, 0); |\overrightarrow{AB}| = 4$

5.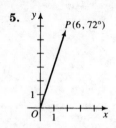

$(6 \cos 72°, 6 \sin 72°) = (1.85, 5.71)$

6.

$(8 \cos 140°, 8 \sin 140°) = (-6.13, 5.14)$

7.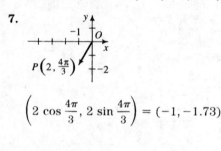

$\left(2 \cos \dfrac{4\pi}{3}, 2 \sin \dfrac{4\pi}{3}\right) = (-1, -1.73)$

8.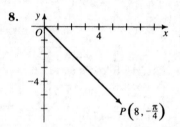

$\left(8 \cos\left(-\dfrac{\pi}{4}\right), 8 \sin\left(-\dfrac{\pi}{4}\right)\right) = (5.66, -5.66)$

9. **a.** $\mathbf{u} + \mathbf{v} = (3, 1) + (-8, 4) = (-5, 5)$ **b.** $\mathbf{u} - \mathbf{v} = (3, 1) - (-8, 4) = (11, -3)$

   c. $3\mathbf{u} + \mathbf{w} = 3(3, 1) + (-6, -2) = (3, 1)$ **d.** $|3\mathbf{u} + \mathbf{w}| = |(3, 1)| = \sqrt{3^2 + 1^2} = \sqrt{10}$

10. **a.** $\mathbf{v} + \mathbf{w} = (-8, 4) + (-6, -2) = (-14, 2)$ **b.** $2(\mathbf{v} + \mathbf{w}) = 2(-14, 2) = (-28, 4)$

    c. $2\mathbf{v} + 2\mathbf{w} = 2(-8, 4) + 2(-6, -2) = (-28, 4)$

    d. $|2\mathbf{v} + 2\mathbf{w}| = |(-28, 4)| = \sqrt{(-28)^2 + 4^2} = \sqrt{784 + 16} = \sqrt{800} = 20\sqrt{2}$

11. **a.** $\mathbf{u} + \dfrac{1}{2}\mathbf{w} = (3, 1) + \dfrac{1}{2}(-6, -2) = (0, 0)$ **b.** $\left|\mathbf{u} + \dfrac{1}{2}\mathbf{w}\right| = |(0, 0)| = 0$

    c. $|\mathbf{u}| + \left|\dfrac{1}{2}\mathbf{w}\right| = \sqrt{(3)^2 + 1^2} + \sqrt{(-3)^2 + (-1)^2} = \sqrt{10} + \sqrt{10} = 2\sqrt{10} \approx 6.325$

12. **a.** $\dfrac{7}{6}\mathbf{v} - \dfrac{2}{3}\mathbf{v} = \dfrac{7}{6}(-8, 4) - \dfrac{2}{3}(-8, 4) = \left(-\dfrac{28}{3}, \dfrac{14}{3}\right) + \left(\dfrac{16}{3}, -\dfrac{8}{3}\right) = (-4, 2)$

    b. $\left(\dfrac{7}{6} - \dfrac{2}{3}\right)\mathbf{v} = \dfrac{1}{2}\mathbf{v} = \dfrac{1}{2}(-8, 4) = (-4, 2)$

    c. $\dfrac{\mathbf{v}}{|\mathbf{v}|} = \dfrac{(-8, 4)}{\sqrt{(-8)^2 + 4^2}} = \dfrac{(-8, 4)}{\sqrt{80}} = \dfrac{(-8, 4)}{4\sqrt{5}} = \dfrac{(-2, 1)}{\sqrt{5}} = \dfrac{\sqrt{5}}{5}(-2, 1) = \left(-\dfrac{2\sqrt{5}}{5}, \dfrac{\sqrt{5}}{5}\right)$

13. $\overrightarrow{AB} = (6, 3); \dfrac{1}{2}(6, 3) = \left(3, \dfrac{3}{2}\right)$

14. $\overrightarrow{AB} = (4, -8); \overrightarrow{OA} + \dfrac{1}{4}\overrightarrow{AB} = (1, 4) + \dfrac{1}{4}(4, -8) = (2, 2); (2, 2)$

15. $\overrightarrow{AB} = (-5, 10)$; $\overrightarrow{OA} + \frac{4}{5}\overrightarrow{AB} = (7, -2) + \frac{4}{5}(-5, 10) = (3, 6)$; $(3, 6)$

16. $\overrightarrow{AB} = (-3, 5)$; $\overrightarrow{OA} + \frac{1}{6}\overrightarrow{AB} = (-3, -4) + \frac{1}{6}(-3, 5) = \left(-\frac{7}{2}, -\frac{19}{6}\right)$; $\left(-\frac{7}{2}, -\frac{19}{6}\right)$

17. $\overrightarrow{AB} = (6, 3)$; $\overrightarrow{OA} + \frac{3}{5}\overrightarrow{AB} = (-7, -4) + \frac{3}{5}(6, 3) = \left(-\frac{17}{5}, -\frac{11}{5}\right)$

18. $\overrightarrow{AB} = (2, 3)$; $\overrightarrow{OA} + \frac{2}{3}\overrightarrow{AB} = (3, -2) + \frac{2}{3}(2, 3) = \left(\frac{13}{3}, 0\right)$

19. A complex number, $z = a + bi$, can be represented as an arrow from the origin to the point (a, b). The length of the arrow is the absolute value of the complex number: $|z| = \sqrt{a^2 + b^2}$. Likewise, the absolute value of vector $\mathbf{v} = (a, b)$ is equal to the length of the arrow representing $\mathbf{v}$: $|\mathbf{v}| = \sqrt{a^2 + b^2}$.

20.

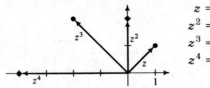

$z = (1, 1)$
$z^2 = (0, 2)$
$z^3 = (-2, 2)$
$z^4 = (-4, 0)$

21. a. $(40 \cos 60°, 40 \sin 60°) = (20, 20\sqrt{3})$

   b. Horizontal

   c. More easily; the horizontal component is greater with a 50° angle than with a 60° angle.

22. a. $\sin 20° = \frac{x}{50}$; $x = 50 \sin 20° \approx 17.1$ lb

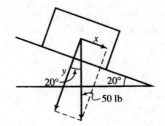

   b. $\cos 20° = \frac{y}{50}$; $y = 50 \cos 20° \approx 47.0$ lb

   c. 17.1 lb up the ramp

   d. Yes; $50 \sin 50° \approx 38.3$; 38.3 lb up the ramp

23. N–S: $y = 15 \cos 72° \approx 4.6$ knots (south);
E–W: $x = 15 \sin 72° \approx 14.3$ knots (east)

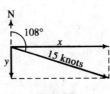

Ex. 23

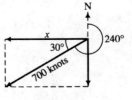

Ex. 24

24. a. $x = 700 \cos 30° = 350\sqrt{3} \approx 606.2$ knots (west)

   b. $606.2 - 50 = 556.2$ knots (west)

25.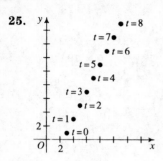

26. (1) The set of points P is the line segment $\overline{AB}$.

(2) The points P on $\overleftrightarrow{AB}$ such that B is between A and P.

(3) The points P on $\overleftrightarrow{AB}$ such that A is between B and P.

27. a.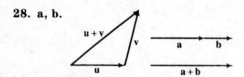

b. Let $\mathbf{v} = (a, b)$. Then $k\mathbf{v} = (ka, kb)$, and $|\mathbf{v}| = \sqrt{a^2 + b^2}$.
$|k\mathbf{v}| = \sqrt{k^2a^2 + k^2b^2} = |k|\sqrt{a^2 + b^2} = |k|\,|\mathbf{v}|$.

28. a, b.

29. Since $\overrightarrow{AB} = \overrightarrow{DC}$, $(3, 8) - (1, 2) =$
$(9, 10) - (x, y)$; $(2, 6) = (9, 10) - (x, y)$;
$(x, y) = (9, 10) - (2, 6) = (7, 4)$; $x = 7$
and $y = 4$.

30. Since $\overrightarrow{AB} = \overrightarrow{DC}$, $(-5, -3) - (-2, 1) = (3, -2) - (x, y)$;
$(-5 - (-2), -3 - 1) = (3, -2) - (x, y)$; $(-3, -4) = (3, -2) - (x, y)$;
$(x, y) = (3, -2) - (-3, -4) = (3 - (-3), -2 - (-4)) = (6, 2)$; $x = 6$ and $y = 2$

31. a. $r(1, 2) + s(3, 0) = (9, 6)$; $(r, 2r) + (3s, 0) = (9, 6)$; $(r + 3s, 2r) = (9, 6)$; $r + 3s = 9$
and $2r = 6$; $r = 3$; $3 + 3s = 9$; $3s = 6$; $s = 2$

b. The sides of the parallelogram shown in the diagram have lengths of $3|(1, 2)|$
and $2|(3, 0)|$.

32. $r(1, 2) + s(3, 0) = (-1, 4)$; $(r, 2r) + (3s, 0) = (-1, 4)$;
$(r + 3s, 2r) = (-1, 4)$; $r + 3s = -1$ and $2r = 4$; $r = 2$;
$2 + 3s = -1$; $3s = -3$; $s = -1$

33. $|r|\,|(3, 4)| = 1$; $|r| \cdot \sqrt{9 + 16} = 1$; $|r| \cdot 5 = 1$;

$|r| = \dfrac{1}{5}$; $r = \dfrac{1}{5}$ or $-\dfrac{1}{5}$

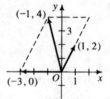

Ex. 32

34. $r|(-4, 3)| = 1$; $r\sqrt{16 + 9} = 1$; $5r = 1$; $r = \dfrac{1}{5}$; $\left(-\dfrac{4}{5}, \dfrac{3}{5}\right)$ has the same direction as

$(-4, 3)$ and has length 1.

35. Let $P = (x, y)$; $|\overrightarrow{AP}| = |(x, y) - (4, 0)| = |(x - 4, y)| = \sqrt{(x - 4)^2 + y^2}$; if $|\overrightarrow{AP}| = 2$,
$\sqrt{(x - 4)^2 + y^2} = 2$; $(x - 4)^2 + y^2 = 4$; the set of all points $P(x, y)$ is the circle with
center $A(4, 0)$ and radius 2.

36. a. Let $P = (x, y)$; $|\overrightarrow{AP}| = |(x, y) - (4, 0)| = |(x - 4, y)| = \sqrt{(x - 4)^2 + y^2}$;
$|\overrightarrow{BP}| = |(x, y) - (-4, 0)| = |(x + 4, y)| = \sqrt{(x + 4)^2 + y^2}$; if $|\overrightarrow{AP}| + |\overrightarrow{BP}| = 10$,
$\sqrt{(x - 4)^2 + y^2} + \sqrt{(x + 4)^2 + y^2} = 10$; $\sqrt{(x - 4)^2 + y^2} = -\sqrt{(x + 4)^2 + y^2} +$
10; $(x - 4)^2 + y^2 = (x + 4)^2 + y^2 - 20\sqrt{(x + 4)^2 + y^2} + 100$; $-16x - 100 =$
$-20\sqrt{(x + 4)^2 + y^2}$; $4x + 25 = 5\sqrt{(x + 4)^2 + y^2}$; $16x^2 + 200x + 625 =$
$25(x^2 + 8x + 16 + y^2)$; $9x^2 + 25y^2 = 225$; $\dfrac{x^2}{25} + \dfrac{y^2}{9} = 1$. The set of all points

$P(x, y)$ is the ellipse with foci at A and B. *(Continued)*

b. $|\overrightarrow{AP} + \overrightarrow{BP}| = |(x - 4, y) + (x + 4, y)| = |(2x, 2y)| = 2\sqrt{x^2 + y^2}$; if $|\overrightarrow{AP} + \overrightarrow{BP}| = 10$,

$2\sqrt{x^2 + y^2} = 10$; $\sqrt{x^2 + y^2} = 5$; $x^2 + y^2 = 25$; the set of all points $P(x, y)$ is the circle with radius 5 whose center is the midpoint of $\overline{AB}$, $(0, 0)$.

37. The midpoint of $\overline{AB}$ is the point halfway from A to B. $\overrightarrow{AB} = (x_2 - x_1, y_2 - y_1)$;

$\overrightarrow{OM} = \overrightarrow{OA} + \overrightarrow{AM} = (x_1, y_1) + \dfrac{1}{2}(x_2 - x_1, y_2 - y_1) = \left(\dfrac{x_1 + x_2}{2}, \dfrac{y_1 + y_2}{2}\right)$.

38. Let the coordinates be $Q(q_1, q_2)$, $R(r_1, r_2)$, $S(s_1, s_2)$, and $P(p_1, p_2)$. Using the formula

from Ex. 37, the coords. of M are $\left(\dfrac{r_1 + p_1}{2}, \dfrac{r_2 + p_2}{2}\right)$ and also $\left(\dfrac{s_1 + q_1}{2}, \dfrac{s_2 + q_2}{2}\right)$.

$\therefore r_1 + p_1 = s_1 + q_1$ and $r_2 + p_2 = s_2 + q_2$. Now $\overrightarrow{SP} = (p_1 - s_1, p_2 - s_2)$ and

$\overrightarrow{RQ} = (q_1 - r_1, q_2 - r_2)$. Since $q_1 - r_1 = p_1 - s_1$ and $q_2 - r_2 = p_2 - s_2$, $\overrightarrow{SP} = \overrightarrow{RQ}$.

Similarly, $\overrightarrow{SR} = (r_1 - s_1, r_2 - s_2)$ and $\overrightarrow{PQ} = (q_1 - p_1, q_2 - p_2)$. Since

$r_1 - s_1 = q_1 - p_1$ and $r_2 - s_2 = q_2 - p_2$, $\overrightarrow{SR} = \overrightarrow{PQ}$. If the diagonals of a quadrilateral bisect each other, then the quadrilateral is a parallelogram.

39. a.

b. $|\overrightarrow{OP} + \overrightarrow{OQ}|^2 = 10^2 + 6^2 - 2 \cdot 10 \cdot$

$6 \cos 120° = 100 + 36 - 120\left(-\dfrac{1}{2}\right) =$

$136 + 60 = 196$, so $|\overrightarrow{OP} + \overrightarrow{OQ}| = \sqrt{196} = 14$

40. a. Let the coordinates be $A(a_1, a_2)$, $B(b_1, b_2)$ and $C(c_1, c_2)$. Then the coordinates of G are $\left(\dfrac{1}{3}(a_1 + b_1 + c_1), \dfrac{1}{3}(a_2 + b_2 + c_2)\right)$.

b. $\left(\dfrac{1}{3}(-2 + 3 + 5), \dfrac{1}{3}(5 + 7 - 3)\right) = (2, 3)$

Activity · page 434

a.

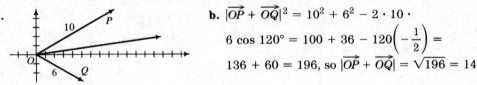

Beginning and ending at $(4, 0)$, the cursor travels around the circle of radius 4 centered at the origin, completing one counterclockwise revolution.

b. The direction is clockwise.

c. The cursor completes three revolutions (counterclockwise)

Class Exercises 12-3 · page 435

1. a. For example, $(2, -5)$, $(3, -2)$, and $(4, 1)$ **b.** For example, $(1, 3)$

c. $x = 2 + t$, $y = -5 + 3t$

2. $\overrightarrow{AB} = (5, 7) - (-3, 1) = (8, 6)$ **a.** $\mathbf{v} = \dfrac{1}{2}(8, 6) = (4, 3)$; $|\mathbf{v}| = \sqrt{4^2 + 3^2} = 5$

b. For example, $(x, y) = (-3, 1) + t(4, 3)$

Written Exercises 12-3 · pages 435–440

1–8. Answers may vary. Examples are given.

1. $(x, y) = (1, 5) + t(2, -1)$; $x = 1 + 2t$, $y = 5 - t$

2. The line contains $(4, 0)$ and $(0, -3)$; $\mathbf{v} = (4 - 0, 0 - (-3)) = (4, 3)$;
$(x, y) = (4, 0) + t(4, 3)$; $x = 4 + 4t$, $y = 3t$

3. $\mathbf{v} = (3 - 1, -4 - 0) = (2, -4);$ $(x, y) = (1, 0) + t(2, -4);$
$(x, y) = (1, 0) + (2t, -4t) = (1 + 2t, -4t);$ $x = 1 + 2t, y = -4t$

4. $\mathbf{v} = (-4 - 3, -4 - 1) = (-7, -5);$ $(x, y) = (3, 1) + t(-7, -5);$
$(x, y) = (3 - 7t, 1 - 5t);$ $x = 3 - 7t, y = 1 - 5t$

5. $\mathbf{v} = (5 - (-2), 1 - 3) = (7, -2);$ $(x, y) = (-2, 3) + t(7, -2);$
$(x, y) = (-2 + 7t, 3 - 2t);$ $x = -2 + 7t, y = 3 - 2t$

6. A line with inclination $45°$ has slope 1; thus, $\mathbf{v} = (1, 1);$ $(x, y) = (7, 5) + t(1, 1);$
$(x, y) = (7 + t, 5 + t);$ $x = 7 + t, y = 5 + t$

7. The horizontal line through (π, e) contains $(0, e)$ and $(1, e);$ $\mathbf{v} = (1 - 0, e - e) = (1, 0);$
$(x, y) = (\pi, e) + t(1, 0);$ $(x, y) = (\pi + t, e);$ $x = \pi + t, y = e$

8. The vertical line through $(\sqrt{2}, \sqrt{3})$ contains $(\sqrt{2}, 0)$ and $(\sqrt{2}, 1);$
$\mathbf{v} = (\sqrt{2} - \sqrt{2}, 1 - 0) = (0, 1);$ $(x, y) = (\sqrt{2}, \sqrt{3}) + t(0, 1);$ $(x, y) = (\sqrt{2}, \sqrt{3} + t);$
$x = \sqrt{2}, y = \sqrt{3} + t$

9. a. $(x, y) = (1, 4) + t(3, -2);$ $t = 0, (1, 4);$ $t = 1, (4, 2);$ $t = 2, (7, 0);$ $t = 3, (10, -2);$
    $t = -1, (-2, 6);$ $t = -2, (-5, 8);$ $t = -3, (-8, 10)$

   b. $\mathbf{v} = (3, -2);$ $|\mathbf{v}| = \sqrt{3^2 + (-2^2)} = \sqrt{13}$

   c. $(x, y) = (1 + 3t, 4 - 2t);$ $x = 1 + 3t, y = 4 - 2t$

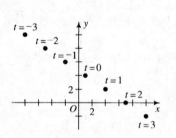

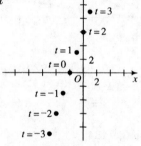

Ex. 9(a)   Ex. 10(a)

10. a. $(x, y) = (-2, 0) + t(1, 3);$ $t = 0, (-2, 0);$ $t = 1, (-1, 3);$ $t = 2, (0, 6);$ $t = 3, (1, 9);$
    $t = -1, (-3, -3);$ $t = -2, (-4, -6);$ $t = -3, (-5, -9)$

   b. $\mathbf{v} = (1, 3);$ $|\mathbf{v}| = \sqrt{1^2 + 3^2} = \sqrt{10}$ **c.** $(x, y) = (-2 + t, 0 + 3t);$ $x = -2 + t, y = 3t$

11. $(x, y) = (2, 3) + t(3, -1) = (2, 3) + (3t, -t) = (2 + 3t, 3 - t);$ $x = 2 + 3t, y = 3 - t$

12. $(x, y) = (1, -5) + t(1, -1) = (1, -5) + (t, -t) = (1 + t, -5 - t);$ $x = 1 + t, y = -5 - t$

13. For example, $(x, y) = (3 + 2t, 2 + 4t),$ so $x = 3 + 2t$ and $y = 2 + 4t;$ $t = \dfrac{x - 3}{2},$ so

   $y = 2 + 4\left(\dfrac{x - 3}{2}\right) = 2 + 2x - 6;$ $y = 2x - 4$

14. $(x, y) = (5 - t, 4 + 2t) = (5, 4) + (-t, 2t) = (5, 4) + t(-1, 2);$ $t = 5 - x,$ so
   $y = 4 + 2(5 - x);$ $y = -2x + 14$

15. a. It is a vertical line through $(2, 0).$ **b.** For example, $(0, 1)$ **c.** The slope is not defined.

16. a. It is a horizontal line through $(0, 3).$ **b.** For example, $(1, 0)$ **c.** The slope is 0.

17. a. $\dfrac{3}{2} = 1.5$ **b.** $\dfrac{6}{4} = 1.5$

   c. Each has a direction vector with a slope of 1.5 and the lines are not coincident.

   d. For example, $(x, y) = (7, 9) + t(2, 3)$

18. For example, $(x, y) = (2, 1) + t(3, 5)$

19. a. $(x, y) = (2 - 3t, -1 + 2t) = (2, -1) + t(-3, 2)$, so $\mathbf{v} = (-3, 2)$; $|\mathbf{v}| = \sqrt{13}$

    b. $x + y = 2$; $(2 - 3t) + (-1 + 2t) = 2$; $1 - t = 2$; $t = -1$;
    $x = 2 - 3t = 2 - 3(-1) = 5$ and $y = -1 + 2t = -1 + 2(-1) = -3$; it crosses
    when $t = -1$ at $(5, -3)$.

20. a. $(x, y) = (1 + 3t, 2 - 4t) = (1, 2) + t(3, -4)$, so $\mathbf{v} = (3, -4)$; $|\mathbf{v}| = 5$

    b. It crosses the x-axis when $y = 0$, or $2 - 4t = 0$; $t = 0.5$;
    $x = 1 + 3t = 1 + 3(0.5) = 2.5$; it crosses when $t = 0.5$ at $(2.5, 0)$.

21. Parametric equations are $x = 1 - t$ and $y = 1 + t$; $(x - 1)^2 + y^2 = 5$, so
$(1 - t - 1)^2 + (1 + t)^2 = 5$; $t^2 + 1 + 2t + t^2 = 5$; $2t^2 + 2t - 4 = 0$; $t^2 + t - 2 = 0$;
$(t + 2)(t - 1) = 0$; $t = -2$ or $t = 1$; when $t = -2$, $x = 1 - (-2) = 3$ and
$y = 1 + (-2) = -1$; when $t = 1$, $x = 1 - 1 = 0$ and $y = 1 + 1 = 2$. Thus the object
crosses the circle when $t = 1$ at $(0, 2)$ and when $t = -2$ at $(3, -1)$.

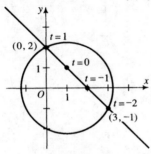
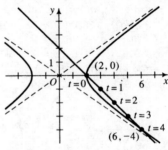

Ex. 21   **Ex. 22**

22. Parametric equations are $x = 2 + t$ and $y = -t$. $x^2 - 2y^2 = 4$; $(2 + t)^2 - 2(-t)^2 = 4$;
$4 + 4t + t^2 - 2t^2 = 4$; $t^2 - 4t = 0$; $t(t - 4) = 0$; $t = 0$ or $t = 4$; when $t = 0$,
$x = 2 + 0 = 2$ and $y = 0$; when $t = 4$, $x = 2 + 4 = 6$ and $y = -4$. Thus the object
crosses the hyperbola when $t = 0$ at $(2, 0)$ and when $t = 4$ at $(6, -4)$.

23. a. Yes, at $(-1, 3)$ **24. a.** Yes, at $(7, 0)$

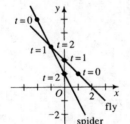
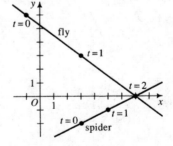

    b. They both pass through $(-1, 3)$
    but at different times. Thus,
    they do not meet.

    c. $-2 + t = 1 - t$ yields $t = 1\frac{1}{2}$;
    $5 - 2t = 1 + t$ yields $t = \frac{4}{3}$; no

    b. They meet at $(7, 0)$, that is, when $t = 2$.

    c. $3 + 2t = -1 + 4t$ yields $t = 2$;
    $-2 + t = 6 - 3t$ yields $t = 2$; yes

25. $x^2 + y^2 = r^2 \cos^2 t + r^2 \sin^2 t = r^2(\cos^2 t + \sin^2 t)$; $x^2 + y^2 = r^2$ is an equation of a
circle with center at $(0, 0)$ and radius r.

26. a. Since the object completes 2 counterclockwise revolutions, replace t by $2t$; $x = 8\cos(2t)$, $y = 8\sin(2t)$

b. Since the object completes 1 revolution clockwise, replace t by $-t$; $x = 8\cos(-t)$, $y = 8\sin(-t)$

c. Since the object begins at $(-8,0)$, replace t by $\pi + t$; $x = 8\cos(\pi + t) = -8\cos t$ and $y = 8\sin(\pi + t) = -8\sin t$

d. Since the object completes 2 clockwise revolutions beginning at $(0,-8)$, replace t by $\dfrac{3\pi}{2} - 2t$; $x = 8\cos\left(\dfrac{3\pi}{2} - 2t\right) = -8\sin 2t$ and $y = 8\sin\left(\dfrac{3\pi}{2} - 2t\right) = -8\cos 2t$

27. a. Ellipse with vertices $(0, \pm 5)$

b. $25x^2 + 4y^2 = 100$

c. If $4x^2 + y^2 = 36$, then $\dfrac{x^2}{9} + \dfrac{y^2}{36} = 1$.

Let $x = 3\cos t$ and $y = 6\sin t$.

Then $\dfrac{(3\cos t)^2}{9} + \dfrac{(6\sin t)^2}{36} =$

$\cos^2 t + \sin^2 t = 1$. So, $x = 3\cos t$ and $y = 6\sin t$ are parametric equations for $4x^2 + y^2 = 36$.

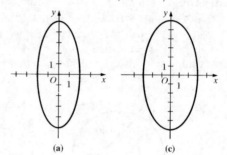

(a)   (c)

28. a.

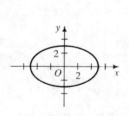

b. The graph is a circle with center $(4, -3)$ and radius 3; $(x - 4)^2 + (y + 3)^2 = 9$

29. a. $(x - 3)^2 + (y - 5)^2 = 36$

b. $x = 6\cos t$, $y = 6\sin t$; $x = 3 + 6\cos t$, $y = 5 + 6\sin t$

30. a. $\dfrac{(a\cos t)^2}{a^2} + \dfrac{(b\sin t)^2}{b^2} = \dfrac{a^2\cos^2 t}{a^2} + \dfrac{b^2\sin^2 t}{b^2} = \cos^2 t + \sin^2 t = 1$

b. See graph below.

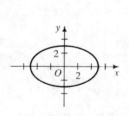

Ex. 30(b)   Ex. 31(b)

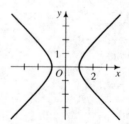

31. a. $\dfrac{(a\sec t)^2}{a^2} - \dfrac{(b\tan t)^2}{b^2} = \dfrac{a^2\sec^2 t}{a^2} - \dfrac{b^2\tan^2 t}{b^2} = \sec^2 t - \tan^2 t = 1$

b. See graph above; (1) quadrant I, (2) quadrant III, (3) quadrant II, (4) quadrant IV.

32. Since $|\sin t| \le 1$ for all values of t, the graph is the part of the line $y = x$ for which $|x| \le 1$ and $|y| \le 1$.

33. The range of the tangent function is all real numbers. Thus the equations $x = \tan t$ and $y = \tan t$ give the entire line $y = x$. However, since $|\sec t| \geq 1$ for all t, the equations $x = \sec t$ and $y = \sec t$ give the parts of the line $y = x$ for which $|x| \geq 1$ and $|y| \geq 1$.

34. 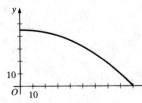 **35.** 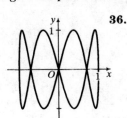 **36.** **37.**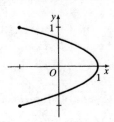

38. a. Let $y = 0$; $y = 0 = 45 - 4.9t^2$; $t \approx 3.03$; $x = 30t \approx 30(3.03) = 90.9$ m

b.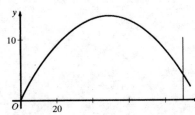

c. $x = 30t$; $t = \dfrac{x}{30}$;

$y = 45 - 4.9t^2$

$= 45 - 4.9\left(\dfrac{x}{30}\right)^2$

$y = 45 - \dfrac{49}{9000}x^2$

39. a. $t = \dfrac{x}{30\sqrt{3}} = \dfrac{x\sqrt{3}}{90}$; $y = 30t - 16\left(\dfrac{x\sqrt{3}}{90}\right) = \left(\dfrac{\sqrt{3}}{3}\right)x - \left(\dfrac{4}{675}\right)x^2$

b. No; if $x = 90$, $y \approx 3.96 < 10$

c.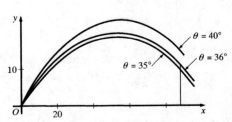

d.

Parametric equations for the goal post are $x = 90$ and $y = t$, with $0 \leq t \leq 10$.

40. a. If $\theta = 90°$, $x = 0$ and $y = vt - 5t^2$. These equations describe vertical motion.

b. If $t = \dfrac{v \sin \theta}{5}$, then $y = \dfrac{(v \sin \theta)^2}{5} - 5\left(\dfrac{v \sin \theta}{5}\right)^2 = 0$. At this moment,

$x = \dfrac{(v \cos \theta)(v \sin \theta)}{5} = \dfrac{v^2}{10}(2 \sin \theta \cos \theta) = \dfrac{v^2 \sin 2\theta}{10}$.

c. $x = \dfrac{v^2 \sin 2\theta}{10}$ is at a maximum when $\sin 2\theta = 1$. This occurs when $\theta = 45°$.

d.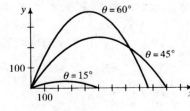

e. $t = \dfrac{x}{v \cos \theta}$; $y = (v \sin \theta)\dfrac{x}{v \cos \theta} -$

$5\left(\dfrac{x}{v \cos \theta}\right)^2 = (\tan \theta)x - \left(\dfrac{5}{v^2} \cdot \sec^2\theta\right)x^2$; yes

41. a. OA is the distance rolled. Since the circle rolls at a rate of 1 unit/s, at time t it has rolled t units. Hence $OA = t$. Arc $AP = OA$ since there is no slipping as it rolls.

$\angle ACP$, in radians, is $\dfrac{\text{arc } AP}{\text{radius}} = \dfrac{t}{1} = t$.

b. $\overrightarrow{OP} = \overrightarrow{OA} + \overrightarrow{AC} + \overrightarrow{CP}$. $\overrightarrow{OA} = (t, 0)$, $\overrightarrow{AC} = (0, 1)$, and $\overrightarrow{CP} = (-\sin t, -\cos t)$. Hence $\overrightarrow{OP} = (x, y) = (t, 0) + (0, 1) + (-\sin t, -\cos t) = (t - \sin t, 1 - \cos t)$.

c. $x = t - \sin t$ and $y = 1 - \cos t$.

Computer Exercises · page 440

1.
```
10   LET C = 0
20   LET L = (2 * 3.14159)/100
30   FOR T = L TO 100 * L STEP L
40   LET X1 = T - SIN(T)
50   LET Y1 = 1 - COS(T)
60   LET X0 = (T - L) - SIN(T - L)
70   LET Y0 = 1 - COS(T - L)
80   LET A = X1 - X0
90   LET B = Y1 - Y0
100  LET C = C + SQR(A^2 + B^2)
110  NEXT T
120  PRINT "THE LENGTH OF ONE ARCH OF A CYCLOID"
130  PRINT "WITH CIRCLE OF RADIUS 1 IS"; C
140  END

RUN

THE LENGTH OF ONE ARCH OF A CYCLOID
WITH CIRCLE OF RADIUS 1 IS 7.999672

END
```

2.
```
10   LET A = 0
20   LET L = (2*3.14159)/100
30   FOR T = L TO 100*L STEP L
40   LET X1 = T - SIN(T)
50   LET Y1 = 1 - COS(T)
60   LET X0 = (T - L) - SIN(T - L)
70   LET Y0 = 1 - COS(T - L)
80   LET A = A + .5 * (Y0 + Y1) * (X1 - X0)
90   NEXT T
100  PRINT "THE AREA UNDER ONE ARCH OF A CYCLOID"
110  PRINT "WITH CIRCLE OF RADIUS 1 IS"; A
120  END

RUN

THE AREA UNDER ONE ARCH OF A CYCLOID
WITH CIRCLE OF RADIUS 1 IS 9.42271

END
```

Class Exercises 12-4 · page 443

1. $\mathbf{u} = -2(-2,3) = -2\mathbf{v}$, so $\mathbf{u} \parallel \mathbf{v}$; $\mathbf{u} \cdot \mathbf{w} = (9 \cdot 4) + 6(-6) = 0$, so $\mathbf{u} \perp \mathbf{w}$;
   $\mathbf{v} \cdot \mathbf{w} = 9(-2) + (6 \cdot 3) = 0$, so $\mathbf{v} \perp \mathbf{w}$.

2. **a.** $\mathbf{u} \cdot \mathbf{v} = 3(-2) + 4 \cdot 2 = 2$ **b.** $2(\mathbf{u} \cdot \mathbf{v}) = 2 \cdot 2 = 4$

   c. $(2\mathbf{u}) \cdot \mathbf{v} = (6,8) \cdot (-2,2) = 6(-2) + 8 \cdot 2 = 4$

   d. $\mathbf{u} \cdot (2\mathbf{v}) = (3,4) \cdot (-4,4) = 3(-4) + 4 \cdot 4 = 4$

3. **a.** $|\mathbf{u}| = \sqrt{3^2 + 4^2} = 5$ **b.** $|\mathbf{u}|^2 = 5^2 = 25$ **c.** $\mathbf{u} \cdot \mathbf{u} = (3,4) \cdot (3,4) = 3(3) + 4(4) = 25$

4. **a.** $|\mathbf{u}| = \sqrt{7^2 + 1^2} = 5\sqrt{2}$ **b.** $|\mathbf{v}| = \sqrt{5^2 + 5^2} = 5\sqrt{2}$

   c. $\mathbf{u} \cdot \mathbf{v} = (7,1) \cdot (5,5) = 7(5) + 1(5) = 40$

   d. $\cos \theta = \dfrac{\mathbf{u} \cdot \mathbf{v}}{|\mathbf{u}||\mathbf{v}|} = \dfrac{40}{5\sqrt{2}(5\sqrt{2})} = \dfrac{4}{5} = 0.8$; $\theta \approx 37°$

5. **a.** $\cos \theta = \dfrac{\mathbf{u} \cdot \mathbf{v}}{|\mathbf{u}||\mathbf{v}|} = \dfrac{4 \cdot 1 + 3 \cdot 0}{\sqrt{4^2 + 3^2}\sqrt{1^2 + 0^2}} = \dfrac{4}{5(1)} = \dfrac{4}{5} = 0.8.$

   b. The triangle is a 3-4-5 right triangle, so $\cos \theta = \dfrac{4}{5} = 0.8$.

Written Exercises 12-4 · pages 444–446

1. **a.** $(2,3) \cdot (4,-5) = 2(4) + 3(-5) = -7$ **b.** $(3,-5) \cdot (7,4) = 3(7) + (-5)4 = 1$

2. **a.** $(-3,0) \cdot (5,7) = -3(5) + 0(7) = -15$

   b. $\left(\dfrac{3}{5},\dfrac{4}{5}\right) \cdot \left(\dfrac{1}{2},-\dfrac{3}{2}\right) = \dfrac{3}{5}\left(\dfrac{1}{2}\right) + \dfrac{4}{5}\left(-\dfrac{3}{2}\right) = -\dfrac{9}{10}$

3. **a.** $6 = 2(3)$, so $4 = 2a$; $a = 2$ **b.** $(4,6) \cdot (a,3) = 0$; $4a + 18 = 0$; $a = -\dfrac{18}{4} = -\dfrac{9}{2}$

4. **a.** $6 = \dfrac{3}{2}(4)$, so $-8 = \dfrac{3}{2}a$; $a = -\dfrac{16}{3}$ **b.** $(6,-8) \cdot (4,a) = 0$; $24 - 8a = 0$; $a = 3$

5. **a.** $(-2,3) \cdot (-2,3) = 4 + 9 = 13$ **b.** $|\mathbf{u}|^2 = (\sqrt{(-2)^2 + 3^2})^2 = (\sqrt{13})^2 = 13$

6. $\mathbf{u} \cdot \mathbf{u} = (a,b) \cdot (a,b) = a^2 + b^2$; $|\mathbf{u}|^2 = (\sqrt{a^2 + b^2})^2 = a^2 + b^2$

7. **a.** $\mathbf{u} \cdot \mathbf{v} = (5,-3) \cdot (3,7) = 15 - 21 = -6$; $\mathbf{v} \cdot \mathbf{u} = (3,7) \cdot (5,-3) = 15 - 21 = -6$

   b. $2(\mathbf{u} \cdot \mathbf{v}) = 2(-6) = -12$; $(2\mathbf{u}) \cdot \mathbf{v} = (10,-6) \cdot (3,7) = 30 - 42 = -12$

8. **a.** $\mathbf{u} \cdot \mathbf{v} = (1,3) \cdot (-4,2) = -4 + 6 = 2$; $\mathbf{v} \cdot \mathbf{u} = (-4,2) \cdot (1,3) = -4 + 6 = 2$

   b. $3(\mathbf{u} \cdot \mathbf{v}) = 3(2) = 6$; $\mathbf{u} \cdot 3\mathbf{v} = (1,3) \cdot (-12,6) = -12 + 18 = 6$

9. $\mathbf{u} \cdot (\mathbf{v} + \mathbf{w}) = (-2,5) \cdot [(1,3) + (-1,2)] = (-2,5) \cdot (0,5) = -2(0) + 5(5) = 25$;
   $\mathbf{u} \cdot \mathbf{v} + \mathbf{u} \cdot \mathbf{w} = (-2,5) \cdot (1,3) + (-2,5) \cdot (-1,2) = (-2 + 15) + (2 + 10) = 25$

10. $\mathbf{u} \cdot (\mathbf{v} + \mathbf{w}) = (1,-4) \cdot [(-2,-2) + (1,5)] = (1,-4) \cdot (-1,3) = 1(-1) - 4(3) = -13$;
    $\mathbf{u} \cdot \mathbf{v} + \mathbf{u} \cdot \mathbf{w} = (1,-4) \cdot (-2,-2) + (1,-4) \cdot (1,5) = (-2 + 8) + (1 - 20) = -13$

11. $\cos \theta = \dfrac{\mathbf{u} \cdot \mathbf{v}}{|\mathbf{u}||\mathbf{v}|} = \dfrac{(1,3) \cdot (2,1)}{\sqrt{1^2 + 3^2}\sqrt{2^2 + 1^2}} = \dfrac{2 + 3}{\sqrt{10}\sqrt{5}} = \dfrac{5}{5\sqrt{2}} = \dfrac{\sqrt{2}}{2}$; $\theta = 45°$

12. $\cos \theta = \dfrac{\mathbf{u} \cdot \mathbf{v}}{|\mathbf{u}||\mathbf{v}|} = \dfrac{(2,3) \cdot (1,-5)}{\sqrt{2^2 + 3^2}\sqrt{1^2 + (-5)^2}} = \dfrac{2 - 15}{\sqrt{13}\sqrt{26}} = \dfrac{-13}{13\sqrt{2}} = -\dfrac{\sqrt{2}}{2}$; $\theta = 135°$

13. $\cos \theta = \dfrac{(3,-4) \cdot (3,4)}{\sqrt{3^2 + (-4)^2}\sqrt{3^2 + 4^2}} = \dfrac{9 - 16}{5(5)} = -\dfrac{7}{25} = -0.28;\ \theta \approx 106.3°$

14. $\cos \theta = \dfrac{(1,3) \cdot (-8,5)}{\sqrt{1^2 + 3^2}\sqrt{(-8)^2 + 5^2}} = \dfrac{-8 + 15}{\sqrt{10}\sqrt{89}} = \dfrac{7}{\sqrt{890}} \approx 0.2346;\ \theta \approx 76.4°$

15. $\overrightarrow{AC} = (1,3)$ and $\overrightarrow{AB} = (3,1);\ \cos A = \dfrac{\overrightarrow{AB} \cdot \overrightarrow{AC}}{|\overrightarrow{AB}||\overrightarrow{AC}|} = \dfrac{(1,3) \cdot (3,1)}{\sqrt{1^2 + 3^2}\sqrt{3^2 + 1^2}} =$

$\dfrac{3 + 3}{\sqrt{10}\sqrt{10}} = \dfrac{6}{10} = 0.6;\ \angle A \approx 53.1°$

16. $\overrightarrow{AC} = (7,1)$ and $\overrightarrow{AB} = (2,2);\ \cos A = \dfrac{\overrightarrow{AC} \cdot \overrightarrow{AB}}{|\overrightarrow{AC}||\overrightarrow{AB}|} = \dfrac{(7,1) \cdot (2,2)}{\sqrt{7^2 + 1^2}\sqrt{2^2 + 2^2}} =$

$\dfrac{14 + 2}{\sqrt{50}\sqrt{8}} = \dfrac{16}{20} = 0.8;\ \angle A \approx 36.9°$

17. $\overrightarrow{PQ} = (2,1)$ and $\overrightarrow{PR} = (3,4);\ \cos P = \dfrac{\overrightarrow{PQ} \cdot \overrightarrow{PR}}{|\overrightarrow{PQ}||\overrightarrow{PR}|} = \dfrac{(2,1) \cdot (3,4)}{\sqrt{2^2 + 1^2}\sqrt{3^2 + 4^2}} =$

$\dfrac{6 + 4}{\sqrt{5}\,(5)} = \dfrac{2}{\sqrt{5}}$

18. $\overrightarrow{RS} = (2,6)$ and $\overrightarrow{RT} = (7,1);\ \cos R = \dfrac{\overrightarrow{RS} \cdot \overrightarrow{RT}}{|\overrightarrow{RS}||\overrightarrow{RT}|} = \dfrac{(2,6) \cdot (7,1)}{\sqrt{2^2 + 6^2}\sqrt{7^2 + 1^2}} =$

$\dfrac{14 + 6}{2\sqrt{10}\,(5\sqrt{2})} = \dfrac{20}{10 \cdot 2\sqrt{5}} = \dfrac{1}{\sqrt{5}}$

19. a. $\overrightarrow{CA} = (-5,-5)$ and $\overrightarrow{CB} = (-7,1);\ \cos C = \dfrac{\overrightarrow{CA} \cdot \overrightarrow{CB}}{|\overrightarrow{CA}||\overrightarrow{CB}|} = \dfrac{(-5,-5) \cdot (-7,1)}{\sqrt{(-5)^2 + (-5)^2}\sqrt{(-7)^2 + 1^2}} =$

$\dfrac{35 - 5}{5\sqrt{2}\,(5\sqrt{2})} = \dfrac{30}{50} = 0.6;\ \sin C = \sqrt{1 - (0.6)^2} = 0.8$ since $\angle C$ is acute.

b. $a = |\overrightarrow{CB}| = 5\sqrt{2}$ and $b = |\overrightarrow{CA}| = 5\sqrt{2};$ area $= \dfrac{1}{2}ab \sin C =$

$\dfrac{1}{2} \cdot 5\sqrt{2} \cdot 5\sqrt{2} \cdot 0.8 = 20$ square units

20. a. $\overrightarrow{CA} = (3,1)$ and $\overrightarrow{CB} = (1,-2);\ \cos C = \dfrac{\overrightarrow{CA} \cdot \overrightarrow{CB}}{|\overrightarrow{CA}||\overrightarrow{CB}|} = \dfrac{(3,1) \cdot (1,-2)}{\sqrt{3^2 + 1^2}\sqrt{1^2 + (-2)^2}} =$

$\dfrac{3 - 2}{\sqrt{10}\sqrt{5}} = \dfrac{1}{5\sqrt{2}} = \dfrac{\sqrt{2}}{10};\ \sin C = \sqrt{1 - \left(\dfrac{\sqrt{2}}{10}\right)^2} = \dfrac{7\sqrt{2}}{10}$ since $\angle C$ is acute.

b. $a = |\overrightarrow{CB}| = \sqrt{5}$ and $b = |\overrightarrow{CA}| = \sqrt{10};$ area $= \dfrac{1}{2}ab \sin C = \dfrac{1}{2} \cdot \sqrt{5} \cdot \sqrt{10} \cdot \dfrac{7\sqrt{2}}{10} =$

3.5 square units

21. a. $\overrightarrow{AC} = (2,4),\ \overrightarrow{AB} = (11,-2);\ \cos A = \dfrac{\overrightarrow{AC} \cdot \overrightarrow{AB}}{|\overrightarrow{AC}||\overrightarrow{AB}|} = \dfrac{(2,4) \cdot (11,-2)}{\sqrt{2^2 + 4^2}\sqrt{11^2 + (-2)^2}} =$

$\dfrac{22 - 8}{\sqrt{20} \cdot \sqrt{125}} = 0.28;\ \angle A \approx 73.7°$

b. Slope $\overrightarrow{AC} = 2;\ \tan \alpha_1 = 2,\ \alpha_1 \approx 63.4°;$ slope $\overrightarrow{AB} = -\dfrac{2}{11};\ \tan \alpha_2 = -\dfrac{2}{11},$

$\alpha_2 \approx -10.3°;\ \angle A = \alpha_1 - \alpha_2 \approx 63.4° - (-10.3°) = 73.7°$

22. a. $\overrightarrow{AB} = (2,3)$, $\overrightarrow{AC} = (1,-5)$; $\cos A = \dfrac{\overrightarrow{AB} \cdot \overrightarrow{AC}}{|\overrightarrow{AB}||\overrightarrow{AC}|} = \dfrac{(2,3) \cdot (1,-5)}{\sqrt{2^2 + 3^2}\sqrt{1^2 + (-5)^2}} =$

$\dfrac{2 - 15}{\sqrt{13} \cdot \sqrt{26}} = -\dfrac{1}{\sqrt{2}}$; $\angle A \approx 135°$

b. Slope $\overrightarrow{AB} = \dfrac{3}{2}$; $\tan \alpha_1 = \dfrac{3}{2}$, $\alpha_1 \approx 56.3°$; slope $\overrightarrow{AC} = -5$; $\tan \alpha_2 = -5$, $\alpha_2 \approx -78.7°$;

$\angle A = \alpha_1 - \alpha_2 \approx 56.3° - (-78.7°) = 135°$

23. $\mathbf{u} \cdot (\mathbf{v} + \mathbf{w}) = (x_1, y_1) \cdot [(x_2, y_2) + (x_3, y_3)] = (x_1, y_1) \cdot (x_2 + x_3, y_2 + y_3) =$
$x_1(x_2 + x_3) + y_1(y_2 + y_3)$; $\mathbf{u} \cdot \mathbf{v} + \mathbf{u} \cdot \mathbf{w} = (x_1, y_1) \cdot (x_2, y_2) + (x_1, y_1) \cdot (x_3, y_3) =$
$(x_1 x_2 + y_1 y_2) + (x_1 x_3 + y_1 y_3)$; since $x_1(x_2 + x_3) + y_1(y_2 + y_3) =$
$x_1 x_2 + x_1 x_3 + y_1 y_2 + y_1 y_3 = (x_1 x_2 + y_1 y_2) + (x_1 x_3 + y_1 y_3)$,
$\mathbf{u} \cdot (\mathbf{v} + \mathbf{w}) = \mathbf{u} \cdot \mathbf{v} + \mathbf{u} \cdot \mathbf{w}$

24. $\mathbf{u} \cdot \mathbf{v} = (x_1, y_1) \cdot (x_2, y_2) = x_1 x_2 + y_1 y_2 = x_2 x_1 + y_2 y_1 = (x_2, y_2) \cdot (x_1, y_1) = \mathbf{v} \cdot \mathbf{u}$;
also, $(k\mathbf{u}) \cdot \mathbf{v} = (k(x_1, y_1)) \cdot (x_2, y_2) = (kx_1, ky_1) \cdot (x_2, y_2) = (kx_1)x_2 + (ky_1)y_2 =$
$k(x_1 x_2) + k(y_1 y_2) = k(x_1 x_2 + y_1 y_2) = k[(x_1, y_1) \cdot (x_2, y_2)] = k(\mathbf{u} \cdot \mathbf{v})$

25. a. Work equals product of component of force in direction of motion and distance moved. Component of $\mathbf{F}$ in direction of motion $= |\mathbf{F}| \cos \theta$. Distance moved $= |\mathbf{s}|$. Therefore, Work $= |\mathbf{F}| \cos \theta \times |\mathbf{s}|$.

b. $\cos \theta = \dfrac{\mathbf{F} \cdot \mathbf{s}}{|\mathbf{F}||\mathbf{s}|}$, so $\mathbf{F} \cdot \mathbf{s} = |\mathbf{F}||\mathbf{s}| \cos \theta = |\mathbf{F}| \cos \theta \times |\mathbf{s}|$.

c. 350

26. Intuitively, work should be maximum when $\theta = 0°$ and the force is entirely in the direction of motion. Work should be minimum when $\theta = 90°$ and force is entirely perpendicular to direction of motion. From Work $= \mathbf{F} \cdot \mathbf{s} = |\mathbf{F}||\mathbf{s}| \cos \theta$, we can see this is true, because $\cos 0° = 1$ and $\cos 90° = 0$.

27. a. $(\mathbf{u} + \mathbf{v}) \cdot (\mathbf{u} + \mathbf{v}) = \mathbf{u} \cdot (\mathbf{u} + \mathbf{v}) + \mathbf{v} \cdot (\mathbf{u} + \mathbf{v}) = \mathbf{u} \cdot \mathbf{u} + \mathbf{u} \cdot \mathbf{v} + \mathbf{v} \cdot \mathbf{u} + \mathbf{v} \cdot \mathbf{v} =$
$|\mathbf{u}|^2 + 2(\mathbf{u} \cdot \mathbf{v}) + |\mathbf{v}|^2$

b. $\mathbf{u} \perp \mathbf{v}$, so $\mathbf{u} \cdot \mathbf{v} = 0$.

c. $|\overrightarrow{AC}|^2 = |\mathbf{u}|^2 + 2(\mathbf{u} \cdot \mathbf{v}) + |\mathbf{v}|^2 = |\mathbf{u}|^2 + 2(0) + |\mathbf{v}|^2 = |\mathbf{u}|^2 + |\mathbf{v}|^2$, but $|\mathbf{u}|^2 = |\overrightarrow{AB}|^2$ and
$|\mathbf{v}|^2 = |\overrightarrow{BC}|^2$, so $|\overrightarrow{AC}|^2 = |\overrightarrow{AB}|^2 + |\overrightarrow{BC}|^2$.

28. Let $\mathbf{u} = \overrightarrow{AB}$ and $\mathbf{v} = \overrightarrow{AD}$. $\overrightarrow{AC} = \overrightarrow{AB} + \overrightarrow{AD} = \mathbf{u} + \mathbf{v}$ and $\overrightarrow{BD} = \overrightarrow{BA} + \overrightarrow{AD} =$
$-\mathbf{u} + \mathbf{v} = \mathbf{v} - \mathbf{u}$. Thus $|\overrightarrow{AC}|^2 + |\overrightarrow{BD}|^2 = |\mathbf{u} + \mathbf{v}|^2 + |\mathbf{v} - \mathbf{u}|^2 =$
$(\mathbf{u} + \mathbf{v}) \cdot (\mathbf{u} + \mathbf{v}) + (\mathbf{v} - \mathbf{u}) \cdot (\mathbf{v} - \mathbf{u}) = |\mathbf{u}|^2 + 2\mathbf{u} \cdot \mathbf{v} + |\mathbf{v}|^2 + |\mathbf{v}|^2 - 2\mathbf{u} \cdot \mathbf{v} + |\mathbf{u}|^2$
(see Ex. 27) $= 2|\mathbf{u}|^2 + 2|\mathbf{v}|^2$. But $|\overrightarrow{AB}|^2 = |\mathbf{u}|^2$ and $|\overrightarrow{AD}|^2 = |\mathbf{v}|^2$, so $|\overrightarrow{AC}|^2 + |\overrightarrow{BD}|^2 =$
$2|\overrightarrow{AB}|^2 + 2|\overrightarrow{AD}|^2$.

29. a. If $|\mathbf{u}| = |\mathbf{v}|$, then $(\mathbf{u} + \mathbf{v}) \cdot (\mathbf{u} - \mathbf{v}) = \mathbf{u} \cdot (\mathbf{u} - \mathbf{v}) + \mathbf{v} \cdot (\mathbf{u} - \mathbf{v}) =$
$\mathbf{u} \cdot \mathbf{u} - \mathbf{u} \cdot \mathbf{v} + \mathbf{v} \cdot \mathbf{u} - \mathbf{v} \cdot \mathbf{v} = |\mathbf{u}|^2 - |\mathbf{v}|^2$. Since $|\mathbf{u}| = |\mathbf{v}|$, $|\mathbf{u}|^2 = |\mathbf{v}|^2$ and
$|\mathbf{u}|^2 - |\mathbf{v}|^2 = |\mathbf{u}|^2 - |\mathbf{u}|^2 = 0$. If $(\mathbf{u} + \mathbf{v}) \cdot (\mathbf{u} - \mathbf{v}) = 0$, then $|\mathbf{u}|^2 - |\mathbf{v}|^2 = 0$
and $|\mathbf{u}|^2 = |\mathbf{v}|^2$. Since $|\mathbf{u}|$ and $|\mathbf{v}|$ are nonnegative, $|\mathbf{u}| = |\mathbf{v}|$.

b. If a parallelogram is a rhombus, then adjacent sides are congruent, so $|\mathbf{u}| = |\mathbf{v}|$. But
then the dot product of the diagonals $\mathbf{u} - \mathbf{v}$ and $\mathbf{u} + \mathbf{v}$ is 0, so the diagonals are
perpendicular. On the other hand, if the diagonals $\mathbf{u} + \mathbf{v}$ and $\mathbf{u} - \mathbf{v}$ are
perpendicular, then $(\mathbf{u} + \mathbf{v}) \cdot (\mathbf{u} - \mathbf{v}) = 0$. But that means that $|\mathbf{u}| = |\mathbf{v}|$, so the
parallelogram is a rhombus.

30. (1) $\overrightarrow{OP} - \overrightarrow{OA} = \overrightarrow{AP}; \overrightarrow{OC} - \overrightarrow{OB} = \overrightarrow{BC}$. Since line $AP \perp \overline{BC}, \overrightarrow{AP} \cdot \overrightarrow{BC} = 0$ so
$(\overrightarrow{OP} - \overrightarrow{OA}) \cdot (\overrightarrow{OC} - \overrightarrow{OB}) = 0$.

(2) $(\overrightarrow{OP} - \overrightarrow{OA}) \cdot (\overrightarrow{OC} - \overrightarrow{OB}) = (\overrightarrow{OP} - \overrightarrow{OA}) \cdot \overrightarrow{OC} - (\overrightarrow{OP} - \overrightarrow{OA}) \cdot \overrightarrow{OB} =$
$\overrightarrow{OP} \cdot \overrightarrow{OC} - \overrightarrow{OA} \cdot \overrightarrow{OC} - \overrightarrow{OP} \cdot \overrightarrow{OB} + \overrightarrow{OA} \cdot \overrightarrow{OB} = 0$

(3) $\overrightarrow{OP} - \overrightarrow{OB} = \overrightarrow{BP}; \overrightarrow{OA} - \overrightarrow{OC} = \overrightarrow{CA}$. Since line $BP \perp \overline{CA}, \overrightarrow{BP} \cdot \overrightarrow{CA} = 0$ so
$(\overrightarrow{OP} - \overrightarrow{OB}) \cdot (\overrightarrow{OA} - \overrightarrow{OC}) = 0$.

(4) $(\overrightarrow{OP} - \overrightarrow{OB}) \cdot (\overrightarrow{OA} - \overrightarrow{OC}) = (\overrightarrow{OP} - \overrightarrow{OB}) \cdot \overrightarrow{OA} - (\overrightarrow{OP} - \overrightarrow{OB}) \cdot \overrightarrow{OC} =$
$\overrightarrow{OP} \cdot \overrightarrow{OA} - \overrightarrow{OB} \cdot \overrightarrow{OA} - \overrightarrow{OP} \cdot \overrightarrow{OC} + \overrightarrow{OB} \cdot \overrightarrow{OC} = 0$

(5) Adding (2) and (4) gives: $\overrightarrow{OP} \cdot \overrightarrow{OA} - \overrightarrow{OA} \cdot \overrightarrow{OC} + \overrightarrow{OB} \cdot \overrightarrow{OC} - \overrightarrow{OP} \cdot \overrightarrow{OB} = 0$;
$(\overrightarrow{OP} - \overrightarrow{OC}) \cdot \overrightarrow{OA} - (\overrightarrow{OP} - \overrightarrow{OC}) \cdot \overrightarrow{OB} = (\overrightarrow{OP} - \overrightarrow{OC}) \cdot (\overrightarrow{OA} - \overrightarrow{OB}) = 0$

(6) $\overrightarrow{OP} - \overrightarrow{OC} = \overrightarrow{CP}$ and $\overrightarrow{OA} - \overrightarrow{OB} = \overrightarrow{BA}$, so $\overrightarrow{CP} \cdot \overrightarrow{BA} = 0$ or line $CP \perp \overline{AB}$; yes, the three
lines containing the altitudes are lines AP, BP, and CP, and each line contains point P.

31. $\cos \theta = \dfrac{\mathbf{u} \cdot \mathbf{v}}{|\mathbf{u}||\mathbf{v}|}$, so $|\mathbf{u}||\mathbf{v}| \cos \theta = \mathbf{u} \cdot \mathbf{v}$. We also know $-1 \le \cos \theta \le 1$, so $-|\mathbf{u}||\mathbf{v}| \le$

$|\mathbf{u}||\mathbf{v}| \cos \theta \le |\mathbf{u}||\mathbf{v}|$. But $|\mathbf{u}||\mathbf{v}| \cos \theta = \mathbf{u} \cdot \mathbf{v}$, so $-|\mathbf{u}||\mathbf{v}| \le \mathbf{u} \cdot \mathbf{v} \le |\mathbf{u}||\mathbf{v}|$. $|\mathbf{u} + \mathbf{v}|^2 =$
$|\mathbf{u}|^2 + 2(\mathbf{u} \cdot \mathbf{v}) + |\mathbf{v}|^2$ (See Ex. 27). Since we have shown $\mathbf{u} \cdot \mathbf{v} \le |\mathbf{u}||\mathbf{v}|$,
$2(\mathbf{u} \cdot \mathbf{v}) \le 2|\mathbf{u}||\mathbf{v}|$. Hence $|\mathbf{u} + \mathbf{v}|^2 = |\mathbf{u}|^2 + 2(\mathbf{u} \cdot \mathbf{v}) + |\mathbf{v}|^2 \le |\mathbf{u}| + 2|\mathbf{u}||\mathbf{v}| + |\mathbf{v}|^2$, and
$|\mathbf{u} + \mathbf{v}|^2 \le (|\mathbf{u}| + |\mathbf{v}|)^2$. Since all quantities are nonnegative, this implies
$|\mathbf{u} + \mathbf{v}| \le |\mathbf{u}| + |\mathbf{v}|$.

Class Exercises 12-5 · page 449

1. a. $AB = \sqrt{(1 - 0)^2 + (-2 - 0)^2 + (3 - 0)^2} = \sqrt{1 + 4 + 9} = \sqrt{14}$;

midpoint: $\left(\dfrac{1 + 0}{2}, \dfrac{-2 + 0}{2}, \dfrac{3 + 0}{2}\right) = \left(\dfrac{1}{2}, -1, \dfrac{3}{2}\right)$

b. $AB = \sqrt{(5 - 3)^2 + (4 - 0)^2 + (2 - (-4))^2} = \sqrt{4 + 16 + 36} = \sqrt{56} = 2\sqrt{14}$;

midpoint: $\left(\dfrac{3 + 5}{2}, \dfrac{0 + 4}{2}, \dfrac{-4 + 2}{2}\right) = (4, 2, -1)$

2. a. $A = (4, 0, 0); B = (4, 5, 0); C = (0, 5, 0); D = (4, 0, 3); E = (0, 0, 3); F = (0, 5, 3)$

b. $|\overrightarrow{OG}| = \sqrt{4^2 + 5^2 + 3^2} = \sqrt{16 + 25 + 9} = \sqrt{50} = 5\sqrt{2}$

3. a. $x^2 + y^2 + z^2 = 25$ **b.** $(x - 1)^2 + (y - 2)^2 + (z - 3)^2 = 25$

4. a. $(3, 5, -2) + 2(1, 2, 3) = (3, 5, -2) + (2, 4, 6) = (5, 9, 4)$

b. $(3, 8, 1) \cdot (4, -1, 4) = 12 - 8 + 4 = 8$

5. If the dot product equals zero, the two vectors are perpendicular. If one vector is a scalar multiple of the other, the two vectors are parallel.

6. Answers may vary. Examples are given.

    a. $(2, 5, 1)$ and $(8, 12, 9)$

    b. $x = 2 + 6t; y = 5 + 7t; z = 1 + 8t$

    c. $(6, 7, 8)$ **d.** Both have the same direction vector.

    e. $(6, 7, 8) \cdot (4, 0, -3) = 24 + 0 + (-24) = 0.$ Since the dot product of the direction vectors is 0, the lines are perpendicular.

7. Find parametric equations for x, y, and z and substitute the parametric equations in the equation for the sphere. Solve for t and then find the corresponding points in space.

8. The vectors are parallel.

Written Exercises 12-5 · pages 450–452

1. $AB = \sqrt{(0 - 2)^2 + (3 - 5)^2 + (1 - (-3))^2} = \sqrt{4 + 4 + 16} = \sqrt{24} = 2\sqrt{6};$

    midpoint: $\left(\dfrac{2 + 0}{2}, \dfrac{5 + 3}{2}, \dfrac{-3 + 1}{2}\right) = (1, 4, -1)$

2. $AB = \sqrt{(2 - 2)^2 + (8 - 0)^2 + (-5 - (-3))^2} = \sqrt{0 + 64 + 4} = \sqrt{68} = 2\sqrt{17};$

    midpoint: $\left(\dfrac{2 + 2}{2}, \dfrac{8 + 0}{2}, \dfrac{-5 + (-3)}{2}\right) = (2, 4, -4)$

3. $AB = \sqrt{(3 - (-1))^2 + (-5 - 1)^2 + (0 - 2)^2} = \sqrt{16 + 36 + 4} = \sqrt{56} = 2\sqrt{14};$

    midpoint: $\left(\dfrac{3 - 1}{2}, \dfrac{-5 + 1}{2}, \dfrac{0 + 2}{2}\right) = (1, -2, 1)$

4. $AB = \sqrt{(4 - (-4))^2 + (1 - (-1))^2 + (2 - (-2))^2} = \sqrt{64 + 4 + 16} = \sqrt{84} = 2\sqrt{21};$

    midpoint: $\left(\dfrac{4 - 4}{2}, \dfrac{1 - 1}{2}, \dfrac{2 - 2}{2}\right) = (0, 0, 0)$

5. $A = (5, 0, 0); B = (5, 6, 0); C = (0, 6, 0); D = (5, 0, 4); E = (0, 0, 4); F = (0, 6, 4);$

    $|\overrightarrow{OG}| = \sqrt{5^2 + 6^2 + 4^2} = \sqrt{77}$

6. $A = (6, 0, 0); B = (6, 7, 0); C = (0, 7, 0); D = (6, 0, 3); E = (0, 0, 3); F = (0, 7, 3);$

    $|\overrightarrow{OG}| = \sqrt{6^2 + 7^2 + 3^2} = \sqrt{94}$

7. a. $(3, 8, -2) + 2(4, -1, 2) = (3, 8, -2) + (8, -2, 4) = (11, 6, 2)$

    b. $(1, -8, 6) \cdot (5, 2, 1) = (1)5 + (-8)2 + (6)1 = -5$

    c. $|(3, 5, 1)| = \sqrt{3^2 + 5^2 + 1^2} = \sqrt{35}$

8. a. $(8 - 2 + 1, 7 - 0 - 2, 4 - 9 + 1) = (7, 5, -4)$

    b. $1(3) + 4(1) + 2(2) = 11$ **c.** $\sqrt{6^2 + 2^2 + 3^2} = \sqrt{49} = 7$

9. $(3, -7, 1) \cdot (6, 3, 3) = 18 - 21 + 3 = 0;$ yes

10. $(2, k, -3) \cdot (4, 2, 6) = 0; 8 + 2k - 18 = 0; 2k = 10; k = 5$

11. a. $x^2 + y^2 + z^2 = 4$ **b.** $(x - 3)^2 + (y + 1)^2 + (z - 2)^2 = 4$

12. $(x - 1)^2 + (y - 5)^2 + (z - 3)^2 = 49; (7 - 1)^2 + (7 - 5)^2 + (6 - 3)^2 = 36 + 4 + 9 = 49,$ so $(7, 7, 6)$ is on the sphere.

13. $x^2 + y^2 + z^2 + 2x - 4y - 6z = 11; x^2 + 2x + 1 + y^2 - 4y + 4 + z^2 - 6z + 9 = 11 + 1 + 4 + 9; (x + 1)^2 + (y - 2)^2 + (z - 3)^2 = 25;$ center, $(-1, 2, 3); r = 5$

14. $x^2 + y^2 + z^2 - 6x + 10y + 2z = 65$; $x^2 - 6x + 9 + y^2 + 10y + 25 + z^2 + 2z + 1 =$
$65 + 9 + 25 + 1$; $(x - 3)^2 + (y + 5)^2 + (z + 1)^2 = 100$; center, $(3, -5, -1)$; $r = 10$

15. $\cos \theta = \dfrac{(8,6,0) \cdot (2,-1,2)}{\sqrt{8^2 + 6^2 + 0^2}\sqrt{2^2 + (-1)^2 + 2^2}} = \dfrac{16 + (-6) + 0}{10(3)} = \dfrac{10}{30} \approx 0.3333$; $\theta \approx 70.5°$

16. $\cos \theta = \dfrac{(2,2,1) \cdot (3,6,-2)}{\sqrt{2^2 + 2^2 + 1^2}\sqrt{3^2 + 6^2 + (-2)^2}} = \dfrac{6 + 12 - 2}{3(7)} = \dfrac{16}{21} \approx 0.7619$; $\theta \approx 40.4°$

17. a. $\overrightarrow{AB} = (3 - 1, -1 - 3, 0 - 4) = (2, -4, -4)$ and $\overrightarrow{AC} = (3 - 1, 2 - 3, 6 - 4) = $
$(2, -1, 2)$; $(2, -4, -4) \cdot (2, -1, 2) = 4 + (-4)(-1) + (-4)2 = 0$, so $\overrightarrow{AB} \perp \overrightarrow{AC}$.

b. Area $= \dfrac{1}{2}|\overrightarrow{AB}||\overrightarrow{AC}| = \dfrac{1}{2}\sqrt{2^2 + (-4)^2 + (-4)^2}\sqrt{2^2 + (-1)^2 + (2)^2} =$

$\dfrac{1}{2}\sqrt{36}\sqrt{9} = 9$ square units

18. a. $\overrightarrow{AB} = (5 - 3, 9 - 7, -4 - (-5)) = (2,2,1)$; $\overrightarrow{AC} = (7 - 3, 5 - 7, -9 - (-5)) =$
$(4, -2, -4)$; $(2,2,1) \cdot (4, -2, -4) = 2(4) + 2(-2) + 1(-4) = 0$, so $\overrightarrow{AB} \perp \overrightarrow{AC}$.

b. Area $= \dfrac{1}{2}|\overrightarrow{AB}||\overrightarrow{AC}| = \dfrac{1}{2}\sqrt{2^2 + 2^2 + 1^2}\sqrt{4^2 + (-2)^2 + (-4)^2} = \dfrac{1}{2}\sqrt{9}\sqrt{36} =$
9 square units

19. a. $\cos A = \dfrac{\overrightarrow{AB} \cdot \overrightarrow{AC}}{|\overrightarrow{AB}||\overrightarrow{AC}|} = \dfrac{(1,1,-2) \cdot (0,1,-1)}{\sqrt{1^2 + 1^2 + (-2)^2}\sqrt{0^2 + 1^2 + (-1)^2}} = \dfrac{0 + 1 + 2}{\sqrt{6}\sqrt{2}} = \dfrac{3}{2\sqrt{3}} = \dfrac{\sqrt{3}}{2}$;

$\sin A = \sqrt{1 - \left(\dfrac{\sqrt{3}}{2}\right)^2} = \dfrac{1}{2}$ since $\angle A$ is acute.

b. Area $= \dfrac{1}{2}|\overrightarrow{AB}||\overrightarrow{AC}| \sin A = \dfrac{1}{2} \cdot \sqrt{6} \cdot \sqrt{2} \cdot \dfrac{1}{2} = \dfrac{\sqrt{3}}{2}$ square units

20. a. $\overrightarrow{AB} = (0,3,4)$; $\overrightarrow{AC} = (2,1,-2)$; $\cos A = \dfrac{\overrightarrow{AB} \cdot \overrightarrow{AC}}{|\overrightarrow{AB}||\overrightarrow{AC}|} = \dfrac{(0,3,4) \cdot (2,1,-2)}{\sqrt{0^2 + 3^2 + 4^2}\sqrt{2^2 + 1^2 + (-2)^2}} =$

$\dfrac{0 + 3 - 8}{5(3)} = -\dfrac{1}{3}$; $\sin A = \sqrt{1 - \left(-\dfrac{1}{3}\right)^2} = \dfrac{2\sqrt{2}}{3}$ since $\angle A$ is obtuse.

b. Area $= \dfrac{1}{2}|\overrightarrow{AB}||\overrightarrow{AC}| \sin A = \dfrac{1}{2} \cdot 5 \cdot 3 \cdot \dfrac{2\sqrt{2}}{3} = 5\sqrt{2}$ square units

21. Answers may vary.

a. $x = -2 + 4t$, $y = -t$, $z = 1 + t$

b. $t = 0$: $(-2, 0, 1)$; $t = 1$: $(2, -1, 2)$

c. In $(-10, 2, -1)$, $-t = 2$, so $t = -2$. Thus x must equal $-2 + 4(-2) = -10$ and z
must equal $1 + 1(-2) = -1$, so $(-10, 2, -1)$ is on the line. In $(18, -2, 6)$, $-t = -2$,
so $t = 2$. Thus x must equal $-2 + 2 \cdot 4 = 6$ and z must equal $1 + 2 = 3$, so
$(18, -2, 6)$ is not on the line. In $(14, -5, 4)$, $-t = -5$ and $t = 5$, so x must be 18,
so $(14, -5, 4)$ is not on the line.

d. $(x, y, z) = (1, 2, 3) + t(4, -1, 1)$

22. Answers may vary.

   a. $x = 6 + 2t, y = -1 - t, z = t$

   b. $t = 0$: $(6, -1, 0)$; $t = 1$: $(8, -2, 1)$

   c. In $(-10, 2, -1)$, $t = -1$, so x must equal $6 + 2(-1) = 4$ and $(-10, 2, -1)$ is not on the line. In $(18, -2, 6)$, $t = 6$. Thus x must equal $6 + 2(6) = 18$ and $y = -1 - 6 = -7$, so $(18, -2, 6)$ is not on the line. In $(14, -5, 4)$, $t = 4$. Thus x must be $6 + 2 \cdot 4 = 14$ and y must be $-1 - 4 = -5$, so $(14, -5, 4)$ is on the line.

   d. $(x, y, z) = (1, 2, 3) + t(2, -1, 1)$

23. A direction vector of the line is $\overrightarrow{AB} = (6 - 4, 3 - 2, 2 - (-1)) = (2, 1, 3)$; the corresponding vector equation is $(x, y, z) = (4, 2, -1) + t(2, 1, 3)$; $x = 4 + 2t$, $y = 2 + t$, $z = -1 + 3t$

24. A direction vector of the line is $\overrightarrow{AB} = (4 - 2, -1 - 3, 3 - 1) = (2, -4, 2)$; the corresponding vector equation is $(x, y, z) = (2, 3, 1) + t(2, -4, 2)$; $x = 2 + 2t$, $y = 3 - 4t$, $z = 1 + 2t$

25. The direction vector is parallel to the xy-plane because its z-component is 0.

26. The line is parallel to the z-axis, perpendicular to the xy-plane, and goes through the point $(1, 2, 3)$; it intersects the xy-plane at the point $(1, 2, 0)$.

27. Answers may vary. Examples are given.

   a. $(x, y, z) = (1, 2, 3) + t(1, 4, -6)$; $x = 1 + t, y = 2 + 4t, z = 3 - 6t$

   b. If (a, b, c) is a direction vector of the perpendicular line, then $(a, b, c) \cdot (1, 4, -6) = 0$; $a + 4b - 6c = 0$; one such ordered triple is $(a, b, c) = (6, -3, -1)$; $(x, y, z) = (5, 3, 0) + t(6, -3, -1)$; $x = 5 + 6t, y = 3 - 3t, z = -t$

28. Answers may vary. Examples are given.

   a. $(x, y, z) = (4, 5, 6) + t(2, 1, 1)$; $x = 4 + 2t, y = 5 + t, z = 6 + t$

   b. If (a, b, c) is a direction vector of the perpendicular line, then $(a, b, c) \cdot (2, 1, 1) = 0$; $2a + b + c = 0$; one such ordered triple is $(a, b, c) = (1, -1, -1)$; $(x, y, z) = (1, 3, -2) + t(1, -1, -1)$; $x = 1 + t, y = 3 - t, z = -2 - t$

29. $|\overrightarrow{AD}|^2 + |\overrightarrow{DC}|^2 = |\overrightarrow{AC}|^2$; $|x_2 - x_1|^2 + |y_2 - y_1|^2 = |\overrightarrow{AC}|^2$; $|\overrightarrow{AC}|^2 + |\overrightarrow{CB}|^2 = |\overrightarrow{AB}|^2$; $|x_2 - x_1|^2 + |y_2 - y_1|^2 + |z_2 - z_1|^2 = |\overrightarrow{AB}|^2$; $|\overrightarrow{AB}| = \sqrt{(x_2 - x_1)^2 + (y_2 - y_1)^2 + (z_2 - z_1)^2}$

30. $\overrightarrow{OM} = \overrightarrow{OA} + \dfrac{1}{2}\overrightarrow{AB} = (x_1, y_1, z_1) + \dfrac{1}{2}(x_2 - x_1, y_2 - y_1, z_2 - z_1) =$

$\left(x_1 + \dfrac{x_2}{2} - \dfrac{x_1}{2}, y_1 + \dfrac{y_2}{2} - \dfrac{y_1}{2}, z_1 + \dfrac{z_2}{2} - \dfrac{z_1}{2}\right) = \left(\dfrac{x_1 + x_2}{2}, \dfrac{y_1 + y_2}{2}, \dfrac{z_1 + z_2}{2}\right)$;

$\therefore M = \left(\dfrac{x_1 + x_2}{2}, \dfrac{y_1 + y_2}{2}, \dfrac{z_1 + z_2}{2}\right)$

31. a. the $\perp$ bisector of $\overline{AB}$ **b.** the plane $\perp$ to $\overrightarrow{AB}$ at its midpt.

32. a. the line $\perp$ to $\overline{AB}$ at A **b.** the plane $\perp$ to $\overrightarrow{AB}$ at A

33. a. circle with diameter $\overline{AB}$ **b.** sphere with diameter $\overline{AB}$

34. a. circle with center A and radius 10   **b.** sphere with center A and radius 10

35. a. ellipse with foci at A and B, and major axis 10

   b. ellipsoid with foci at A and B, and major axis 10

36. a. the segment with endpoint B and midpoint A

b. the segment with endpoint B and midpoint A

37. Answers may vary.

a. $PA = PB$; $\sqrt{(x - 0)^2 + (y - 0)^2 + (z - 0)^2} = \sqrt{(x - 2)^2 + (y - 3)^2 + (z - 6)^2}$;
$x^2 + y^2 + z^2 = x^2 - 4x + 4 + y^2 - 6y + 9 + z^2 - 12z + 36$;
$4x + 6y + 12z = 49$

b. $M(1, 1.5, 3)$; $4(1) + 6(1.5) + 12(3) = 4 + 9 + 36 = 49$

38. a. $\overrightarrow{PA} = (3 - x, 5 - y, 2 - z)$ and $\overrightarrow{AB} = (6 - 3, 6 - 5, 4 - 2) = (3, 1, 2)$; $\overrightarrow{PA} \cdot \overrightarrow{AB} = 0$;
$(3 - x, 5 - y, 2 - z) \cdot (3, 1, 2) = 0$; $3(3 - x) + 1(5 - y) + 2(2 - z) = 0$;
$-3x - y - 2z + 18 = 0$; $3x + y + 2z = 18$

b. $A = (3, 5, 2)$; the graph of the equation is a plane $\perp$ to $\overrightarrow{AB}$ at A; A is in the plane since $3(3) + 5 + 2(2) = 18$.

39. a. $\overrightarrow{PA} = (2 - x, 2 - y, 1 - z)$ and $\overrightarrow{PB} = (-2 - x, -2 - y, -1 - z)$; $\overrightarrow{PA} \cdot \overrightarrow{PB} = 0$;
$(2 - x, 2 - y, 1 - z) \cdot (-2 - x, -2 - y, -1 - z) = 0$;
$-(4 - x^2) - (4 - y^2) - (1 - z^2) = 0$; $x^2 + y^2 + z^2 = 9$

b. Same; a sphere with $\overline{AB}$ as a diameter

40. a. The line intersects the xy-plane when $z = 0$, so $4 + 4t = 0$; $4t = -4$; $t = -1$;
thus $(x, y, z) = (6, -5, 4) + (-1)(3, -2, 4) = (6 - 3, -5 + 2, 4 - 4) = (3, -3, 0)$.

b. The equation of the yz-plane is $x = 0$; the line intersects the yz-plane when $x = 0$, so
$6 + 3t = 0$; $3t = -6$; $t = -2$; thus $(x, y, z) = (6, -5, 4) + (-2)(3, -2, 4) =$
$(6 - 6, -5 + 4, 4 - 8) = (0, -1, -4)$.

c. The equation of the xz-plane is $y = 0$; the line intersects the xz-plane when $y = 0$, so
$-5 - 2t = 0$; $2t = -5$; $t = -\dfrac{5}{2}$; thus $(x, y, z) = (6, -5, 4) + \left(-\dfrac{5}{2}\right)(3, -2, 4) =$
$\left(6 - \dfrac{15}{2}, -5 + 5, 4 - 10\right) = \left(-\dfrac{3}{2}, 0, -6\right)$.

41. Circle with center $(1, 2, 0)$ and $r = 4$; if $z = 0$, $(x - 1)^2 + (y - 2)^2 + (0 - 3)^2 = 25$, or
$(x - 1)^2 + (y - 2)^2 = 16$

42. Parametric equations for the line are $x = 3 + 2t$, $y = 4 - t$, and $z = -2 + 2t$;
substituting these values into the equation of the sphere gives $(3 + 2t - 3)^2 +$
$(4 - t - 4)^2 + (-2 + 2t + 2)^2 = 36$; $4t^2 + t^2 + 4t^2 = 36$; $9t^2 = 36$; $t^2 = 4$;
$t = 2$ or -2; $(x, y, z) = (3 + 2 \cdot 2, 4 - 2, -2 + 2 \cdot 2) = (7, 2, 2)$ or
$(x, y, z) = (3 + 2(-2), 4 - (-2), -2 + 2(-2)) = (-1, 6, -6)$

43. a. $x = 3 + t = 1 + 2s$, $y = 2t = 5 + s$, and $z = -4 + 3t = 2 + 4s$. From the first
equation, $t = 2s - 2$; substituting, $2(2s - 2) = 5 + s$; $3s = 9$; $s = 3$;
$t = 2(3) - 2 = 4$. But these values of s and t yield differing values of z, 8 and 14, so
there are no values of s and t that satisfy all three equations simultaneously. The
lines have no point in common.

b. No; there is no value of k such that $k(1, 2, 3) = (2, 1, 4)$, so the lines are skew.

44. a. If the lines have a common point (x, y, z), then the following equations must be
satisfied by the same values of s and t: $x = -1 + t = s$, $y = 5 + 2t = 1 - s$, and
$z = -t = 3 + s$. From eq. (1) and (3), $s = -1 + t$ and $s = -3 - t$; $-1 + t =$
$-3 - t$; $2t = -2$; $t = -1$; $s = -1 + (-1) = -2$. When $t = -1$, $(x, y, z) =$
$(-1, 5, 0) + (-1)(1, 2, -1) = (-2, 3, 1)$; when $s = -2$, $(x, y, z) = (0, 1, 3) +$
$(-2)(1, -1, 1) = (-2, 3, 1)$

b. $(-2, 3, 1)$

45. a. The lines intersect at $(3, -1, 5)$ and the dot product of their direction vectors equals 0, $(2, 3, 4) \cdot (2, 0, -1) = 4 + 0 - 4 = 0$.

b. Let (a, b, c) be the direction vector, so $(a, b, c) \cdot (2, 3, 4) = 0$ and $(a, b, c) \cdot (2, 0, -1) = 0$; $2a + 3b + 4c = 0$ and $2a - c = 0$; $c = 2a$; if $a = 3$, then $c = 6$; substituting in $2a + 3b + 4c = 0$, we get $6 + 3b + 24 = 0$; $b = -10$; $(x, y, z) = (3, -1, 5) + r(3, -10, 6)$

Class Exercises 12-6 • pages 454–455

1. x-intercept, 2; y-intercept, 6; z-intercept, 3

2. a. $z = 4$ **b.** $z = 0$ **c.** $x = 3$ **d.** $y = 6$

3. a. $(2, 3, 4)$ **b.** $(3, 0, -4)$ **c.** $(0, 0, 1)$

4. $x + y + z = d$; $3 + 4 + 5 = d$; $x + y + z = 12$

5. $6x + 7y + 8z = d$; $0 + 0 + 0 = d$; $6x + 7y + 8z = 0$

Written Exercises 12-6 • pages 455–458

1.
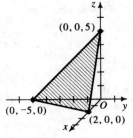

2.
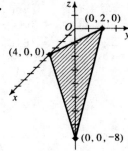

3.
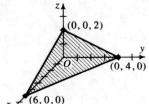

4.
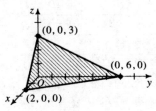

5.
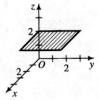

6.
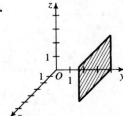

7. a. $z = 1$ **b.** $x = 4$ **c.** $y = 6$ **d.** $y = 0$

8. a. $z = 2$ **b.** $x = 5$ **c.** $y = 7$ **d.** $y = 0$

9. $(3, 4, 6)$ **10.** $(3, -5, 4)$ **11.** $(1, 1, 0)$ **12.** $(1, 0, -1)$ **13.** $(0, 0, 1)$ **14.** $(0, 1, 0)$

15. $2x + 3y + 5z = d$; $d = 2(3) + 3(1) + 5(7) = 44$; $2x + 3y + 5z = 44$

16. $x - 4y + 2z = d$; $d = 3 - 4(0) + 2(2) = 7$; $x - 4y + 2z = 7$

17. $z = d$; $d = 5$; $z = 5$ **18.** $2x + 3z = d$; $d = 2(3) + 3(-2) = 0$; $2x + 3z = 0$

19. a. $\overrightarrow{AB}$ has midpoint $M = \left(\dfrac{2+4}{2}, \dfrac{2+6}{2}, \dfrac{2+8}{2}\right) = (3, 4, 5)$;

$\overrightarrow{AB} = (4 - 2, 6 - 2, 8 - 2) = (2, 4, 6)$; $2x + 4y + 6z = d$;
$d = 2(3) + 4(4) + 6(5) = 52$; $2x + 4y + 6z = 52$; $x + 2y + 3z = 26$

b. $2 + 2(0) + 3(8) = 26$

c. $|\overrightarrow{PA}| = \sqrt{(2 - 2)^2 + (2 - 0)^2 + (2 - 8)^2} = 2\sqrt{10}$;

$|\overrightarrow{PB}| = \sqrt{(4 - 2)^2 + (6 - 0)^2 + (8 - 8)^2} = 2\sqrt{10}$

20. a. $\overrightarrow{AB}$ has midpoint $M = \left(\dfrac{-2+0}{2}, \dfrac{4+2}{2}, \dfrac{1-1}{2}\right) = (-1, 3, 0)$;

$\overrightarrow{AB} = (0 - (-2), 2 - 4, -1 - 1) = (2, -2, -2)$; $2x - 2y - 2z = d$;
$d = 2(-1) - 2(3) - 2(0) = -8$; $2x - 2y - 2z = -8$, or $x - y - z = -4$

b. Answers may vary. Example is given. $P(-4, 0, 0)$ and $Q(-2, 1, 1)$ are in the plane;
$AP = \sqrt{(-4 - (-2))^2 + (0 - 4)^2 + (0 - 1)^2} = \sqrt{21}$ and
$BP = \sqrt{(-4 - 0)^2 + (0 - 2)^2 + (0 - (-1))^2} = \sqrt{21}$;
$AQ = \sqrt{(-2 - (-2))^2 + (1 - 4)^2 + (1 - 1)^2} = \sqrt{9} = 3$ and
$BQ = \sqrt{(-2 - 0)^2 + (1 - 2)^2 + (1 - (-1))^2} = \sqrt{9} = 3$

21. a. $2^2 + 1^2 + 2^2 = 9$; $(2, 1, 2)$ is on the sphere.

b. $\overrightarrow{OA} = (2, 1, 2)$; $2x + y + 2z = d$; $d = 2(2) + 1 + 2(2) = 9$; $2x + y + 2z = 9$

22. The sphere has center $R(1, 1, 1)$ and contains $Q(7, -1, 4)$ since $(7 - 1)^2 + (-1 - 1)^2 + (4 - 1)^2 = 36 + 4 + 9 = 49$. $\overrightarrow{RQ} = (6, -2, 3)$ and $\overrightarrow{RQ} \perp$ to the tangent plane, so $6x - 2y + 3z = d$ and $d = 6(7) - 2(-1) + 3(4) = 56$; $6x - 2y + 3z = 56$.

23. When $z = 3$, $x^2 + y^2 + 3^2 = 25$; $x^2 + y^2 = 16$; the plane intersects the sphere in a circle whose radius is 4; area $= 16\pi$ sq. units

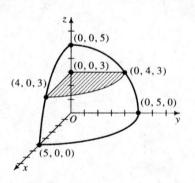

Ex. 23

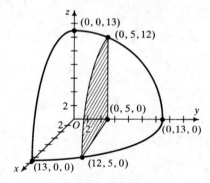

Ex. 24

24. When $y = 5$, $x^2 + 5^2 + z^2 = 169$; $x^2 + z^2 = 144$; the plane intersects the sphere in a circle whose radius is $\sqrt{144}$, or 12; area $= 144\pi$ square units

25. a. Consider the quadrilateral shown, two of whose angles are right angles. Since the angle sum for a quadrilateral is $360°$, α and the remaining angle are supplementary. Since the latter angle and θ are also supplementary, $\alpha = \theta$.

b. From Property 5 on p. 448, $\cos \alpha = \dfrac{(a_1, b_1, c_1) \cdot (a_2, b_2, c_2)}{|(a_1, b_1, c_1)||(a_2, b_2, c_2)|}$; $\alpha = \theta$, so

$$\cos \theta = \frac{(a_1, b_1, c_1) \cdot (a_2, b_2, c_2)}{|(a_1, b_1, c_1)||(a_2, b_2, c_2)|}$$

26. $(a_1, b_1, c_1) = (2, 2, -1)$ and $(a_2, b_2, c_2) = (1, 2, 1)$; $\cos \theta = \dfrac{(2, 2, -1) \cdot (1, 2, 1)}{|(2, 2, -1)||(1, 2, 1)|} = \dfrac{5}{3\sqrt{6}}$;

$\theta \approx 47.1°$

27. $(a_1, b_1, c_1) = (2, 2, -1)$ and $(a_2, b_2, c_2) = (4, -3, 2)$; $\cos \theta = \dfrac{(2, 2, -1) \cdot (4, -3, 2)}{|(2, 2, -1)||(4, -3, 2)|} =$

$\dfrac{0}{3\sqrt{29}} = 0$; $\theta = 90°$

28. a. $(3, 4, 2) \cdot (2, -1, -1) = 6 - 4 - 2 = 0$; yes

b. $(4, -5, 6) \cdot (3, 0, -2) = 12 - 12 = 0$; yes

29. a. $(2, 3, -1)$ **b.** yes   **c.** yes

d. $4x + 6y - 2z = 4$ is equiv. to
$2x + 3y - z = 2$; yes

30. Eq. (2) is equivalent to $3x + 2y - z = 4$, so planes M_1 and M_2 are parallel. Also,
$(3, 2, -1) \cdot (4, -2, 8) = 12 - 4 - 8 = 0$ and
$(6, 4, -2) \cdot (4, -2, 8) = 24 - 8 - 16 = 0$,
so plane M_3 is perpendicular to plane M_1 and plane M_2.

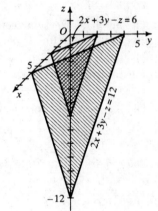

Ex. 29(c)

31–33. Answers may vary. Examples are given.

31. a. $4x + y - 3z = 0$ is a plane through the origin and parallel to M.

b. Find (a, b, c) such that $(a, b, c) \cdot (4, 1, -3) = 0$; $4a + b - 3c = 0$; for example, let $(a, b, c) = (3, 0, 4)$; then $3x + 4z = 0$ is one such plane.

32. a. $5x - 2y + z = 1$ is a plane parallel to M.

b. Find (a, b, c) such that $(a, b, c) \cdot (5, -2, 1) = 0$; $5a - 2b + c = 0$; for example, let $(a, b, c) = (2, 5, 0)$; then $2x + 5y = 1$ is one such plane.

33. a. $\mathbf{u} = (4, -5, 2)$

b. The vector $\mathbf{v} = (2, 2, 1)$ is perpendicular to the given plane; also $\mathbf{u} \cdot \mathbf{v} = (4, -5, 2) \cdot (2, 2, 1) = 8 - 10 + 2 = 0$; so $\mathbf{u} \perp \mathbf{v}$; $\mathbf{v}$ is perpendicular both to the line and to the plane; the point $(3, 1, 4)$ lies on the line but not in the plane, so the line is not contained in the plane. Thus, the line and plane are parallel.

34. a. Let θ be the angle formed by vectors (a,b,c) and (p,q,r). Then

$$\cos\theta = \frac{(a,b,c)\cdot(p,q,r)}{|(a,b,c)||(p,q,r)|}; \text{ but } (a,b,c) \text{ is perpendicular to the plane, so } \theta = 90° - \alpha$$

and $\cos\theta = \cos(90° - \alpha) = \sin\alpha$; thus, $\sin\alpha = \dfrac{(a,b,c)\cdot(p,q,r)}{|(a,b,c)||(p,q,r)|}$.

b. The line has direction vector $(p,q,r) = (0,1,-1)$ and the vector perpendicular to the

plane is $(a,b,c) = (1,2,-2)$; then $\sin\alpha = \dfrac{(1,2,-2)\cdot(0,1,-1)}{|(1,2,-2)||(0,1,-1)|} =$

$\dfrac{0+2+2}{3\sqrt{2}} = \dfrac{2\sqrt{2}}{3}$; $\alpha \approx 70.5°$

35. a. The line has direction vector $(3,4,5)$; since this vector is also perpendicular to the given plane, the line is perpendicular to the plane.

b. The line has parametric equations $x = 8 + 3t$, $y = 9 + 4t$, and $z = 10 + 5t$; substituting, $3(8 + 3t) + 4(9 + 4t) + 5(10 + 5t) = 10$; $50t = -100$; $t = -2$; $x = 8 + 3(-2) = 2$; $y = 9 + 4(-2) = 1$; $z = 10 + 5(-2) = 0$; $(2,1,0)$

36. a. The line has direction vector $(2,3,4)$; since this vector is also perpendicular to the given plane, the line is perpendicular to the plane.

b. The line has parametric equations $x = 4 + 2t$, $y = 2 + 3t$, and $z = 1 + 4t$; substituting, $2(4 + 2t) + 3(2 + 3t) + 4(1 + 4t) = 76$; $29t = 58$; $t = 2$; $x = 4 + 2(2) = 8$; $y = 2 + 3(2) = 8$; $z = 1 + 4(2) = 9$; $(8,8,9)$

37. a. The line has direction vector $(2,2,1)$; $(x,y,z) = (3,1,5) + t(2,2,1)$.

b. $x = 3 + 2t$, $y = 1 + 2t$, and $z = 5 + t$; substituting, $2(3 + 2t) + 2(1 + 2t) + 5 + t = 4$; $9t = -9$; $t = -1$; $(x,y,z) = (1,-1,4)$

c. The required distance is the distance from $(3,1,5)$ to $(1,-1,4)$:
$$d = \sqrt{(3-1)^2 + (1-(-1))^2 + (5-4)^2} = \sqrt{9} = 3.$$

38. The line through $A(x_0, y_0, z_0)$ and perpendicular to the given plane has vector equation $(x,y,z) = (x_0, y_0, z_0) + t(a,b,c)$; $x = x_0 + at$, $y = y_0 + bt$, and $z = z_0 + ct$; substituting to find the intersection point of the line and plane, $a(x_0 + at) + b(y_0 + bt) + c(z_0 + ct) + d = 0$; $(a^2 + b^2 + c^2)t =$ $-d - ax_0 - by_0 - cz_0$; $t = \dfrac{-(ax_0 + by_0 + cz_0 + d)}{a^2 + b^2 + c^2}$; thus, the point of intersection is $B(x_1, y_1, z_1) = (x_0 + at, y_0 + bt, z_0 + ct)$; the required distance is $AB =$ $\sqrt{(x_1 - x_0)^2 + (y_1 - y_0)^2 + (z_1 - z_0)^2} = \sqrt{(at)^2 + (bt)^2 + (ct)^2} = \sqrt{(a^2 + b^2 + c^2)t^2} =$ $\sqrt{(a^2 + b^2 + c^2)\left(-\dfrac{ax_0 + by_0 + cz_0 + d}{a^2 + b^2 + c^2}\right)^2} = \dfrac{|ax_0 + by_0 + cz_0 + d|}{\sqrt{a^2 + b^2 + c^2}}.$

39. $d = \dfrac{|ax_0 + by_0 + c|}{\sqrt{a^2 + b^2}}$.

Activity 12-7 · page 459

a. $-3\begin{vmatrix} 2 & 8 \\ -1 & 6 \end{vmatrix} + 4\begin{vmatrix} 5 & 8 \\ 7 & 6 \end{vmatrix} - 1\begin{vmatrix} 5 & 2 \\ 7 & -1 \end{vmatrix} = -3(20) + 4(-26) - 1(-19) =$

$-60 - 104 + 19 = -145$

b. $8\begin{vmatrix} 3 & 4 \\ 7 & -1 \end{vmatrix} - 1\begin{vmatrix} 5 & 2 \\ 7 & -1 \end{vmatrix} + 6\begin{vmatrix} 5 & 2 \\ 3 & 4 \end{vmatrix} = 8(-31) - 1(-19) + 6(14) =$

$-248 + 19 + 84 = -145$

Class Exercises 12-7 · page 460

1. $2(5) - 4(3) = -2$ **2.** $8(4) - 4(5) = 12$ **3.** $3(10) - 7(-5) = 65$

4. $(-5)(-3) - (-4)(4) = 31$ **5.** 3; 2; 4

6. $\begin{vmatrix} 5 & 7 \\ 8 & 6 \end{vmatrix}$; $\begin{vmatrix} 3 & 2 \\ 8 & 6 \end{vmatrix}$; $\begin{vmatrix} 3 & 2 \\ 5 & 7 \end{vmatrix}$

7. For example, expanding by the bottom row, $\begin{vmatrix} 1 & 2 & 3 \\ 4 & 5 & 6 \\ 0 & 0 & 1 \end{vmatrix} = 0 + 0 + \begin{vmatrix} 1 & 2 \\ 4 & 5 \end{vmatrix} = -3$

Written Exercises 12-7 · pages 460–461

1. $2(4) - 8(7) = -48$ **2.** $-3(2) - (-7)(5) = 29$

3. $\begin{vmatrix} 25 & 125 \\ 75 & 250 \end{vmatrix} = 125 \begin{vmatrix} 25 & 1 \\ 75 & 2 \end{vmatrix} = 25(125) \begin{vmatrix} 1 & 1 \\ 3 & 2 \end{vmatrix} = 25(125)(-1) = -3125$

4. $k \begin{vmatrix} a & b \\ c & d \end{vmatrix} = k(ad - bc)$; $\begin{vmatrix} ka & kb \\ c & d \end{vmatrix} = kad - kbc = k(ad - bc)$;

$\begin{vmatrix} a & b \\ kc & kd \end{vmatrix} = a(kd) - (kc)b = k(ad - bc)$

5. $\begin{vmatrix} a & b \\ c & d \end{vmatrix} = ad - bc$; $\begin{vmatrix} a - c & b - d \\ c & d \end{vmatrix} = (a - c)d - c(b - d) = ad - cd - bc + cd =$

$ad - bc = \begin{vmatrix} a & b \\ c & d \end{vmatrix}$; $\begin{vmatrix} 387 & 411 \\ 385 & 410 \end{vmatrix} = \begin{vmatrix} 387 - 385 & 411 - 410 \\ 385 & 410 \end{vmatrix} = \begin{vmatrix} 2 & 1 \\ 385 & 410 \end{vmatrix} = 435$

6. True; $\begin{vmatrix} a & b \\ c & d \end{vmatrix} = ad - bc$; $\begin{vmatrix} a - kc & b - kd \\ c & d \end{vmatrix} = (a - kc)d - c(b - kd) =$

$ad - kcd - bc + kcd = ad - bc$; thus, $\begin{vmatrix} a & b \\ c & d \end{vmatrix} = \begin{vmatrix} a - kc & b - kd \\ c & d \end{vmatrix}$.

7–10. Expansions may vary.

7. $\begin{vmatrix} 4 & -7 & 3 \\ 2 & 0 & 0 \\ 5 & 1 & 6 \end{vmatrix} = -2 \begin{vmatrix} -7 & 3 \\ 1 & 6 \end{vmatrix} = -2(-42 - 3) = 90$

8. $\begin{vmatrix} -1 & 3 & 2 \\ 4 & 0 & 1 \\ 1 & 5 & 0 \end{vmatrix} = -4 \begin{vmatrix} 3 & 2 \\ 5 & 0 \end{vmatrix} + 0 - 1 \begin{vmatrix} -1 & 3 \\ 1 & 5 \end{vmatrix} = -4(0 - 10) - 1(-5 - 3) = 40 + 8 = 48$

9. $\begin{vmatrix} 1 & -3 & 4 \\ 0 & 1 & 1 \\ 5 & -2 & 3 \end{vmatrix} = \begin{vmatrix} 1 & -3 & 3 + 4 \\ 0 & 1 & -1 + 1 \\ 5 & -2 & 2 + 3 \end{vmatrix} = \begin{vmatrix} 1 & -3 & 7 \\ 0 & 1 & 0 \\ 5 & -2 & 5 \end{vmatrix} = \begin{vmatrix} 1 & 7 \\ 5 & 5 \end{vmatrix} = 5 - 35 = -30$

10. $\begin{vmatrix} 1 & 2 & 3 \\ 2 & 4 & 6 \\ 17 & 18 & 19 \end{vmatrix}$ $\xrightarrow[\text{to row 2}]{\text{Add }(-2 \times \text{row 1})}$ $\begin{vmatrix} 1 & 2 & 3 \\ 0 & 0 & 0 \\ 17 & 18 & 19 \end{vmatrix} =$

$-0\begin{vmatrix} 2 & 3 \\ 18 & 19 \end{vmatrix} + 0\begin{vmatrix} 1 & 3 \\ 17 & 19 \end{vmatrix} - 0\begin{vmatrix} 1 & 2 \\ 17 & 18 \end{vmatrix} = 0$

11. No. Subtracting a multiple of one row (column) of a determinant from another does not change the value of the determinant. If one row is a multiple of another, i.e., if row $A = k(\text{row } B)$, then subtracting $k(\text{row } B)$ from row A produces a row of zeros. Expanding by the minors of this row shows that the value of the determinant is zero.

12. $\begin{vmatrix} 2 & -4 & 7 & 3 \\ 0 & 5 & 1 & -2 \\ -2 & 1 & 0 & -3 \\ 0 & -6 & 4 & 2 \end{vmatrix}$ $\xrightarrow[\text{to row 3}]{\text{Add row 1}}$ $\begin{vmatrix} 2 & -4 & 7 & 3 \\ 0 & 5 & 1 & -2 \\ 0 & -3 & 7 & 0 \\ 0 & -6 & 4 & 2 \end{vmatrix} =$

$2\begin{vmatrix} 5 & 1 & -2 \\ -3 & 7 & 0 \\ -6 & 4 & 2 \end{vmatrix}$ $\xrightarrow[\text{to row 1}]{\text{Add row 3}}$ $2\begin{vmatrix} -1 & 5 & 0 \\ -3 & 7 & 0 \\ -6 & 4 & 2 \end{vmatrix} = 2(2)\begin{vmatrix} -1 & 5 \\ -3 & 7 \end{vmatrix} = 4 \cdot 8 = 32$

13. $\begin{vmatrix} -2 & 1 & 5 & 0 \\ 3 & 4 & -2 & 1 \\ 0 & 0 & 1 & 2 \\ 1 & 0 & 0 & -3 \end{vmatrix}$ $\xrightarrow[\text{to col. 4}]{\text{Add }(-2) \times \text{col. 3}}$ $\begin{vmatrix} -2 & 1 & 5 & -10 \\ 3 & 4 & -2 & 5 \\ 0 & 0 & 1 & 0 \\ 1 & 0 & 0 & -3 \end{vmatrix} =$

$\begin{vmatrix} -2 & 1 & -10 \\ 3 & 4 & 5 \\ 1 & 0 & -3 \end{vmatrix}$ $\xrightarrow[\text{to col. 3}]{\text{Add 3} \times \text{col. 1}}$ $\begin{vmatrix} -2 & 1 & -16 \\ 3 & 4 & 14 \\ 1 & 0 & 0 \end{vmatrix} = \begin{vmatrix} 1 & -16 \\ 4 & 14 \end{vmatrix} = 14 + 64 = 78$

14. a. $a_1x + b_1y = c_1$ $\qquad$ $a_1a_2x + a_2b_1y = a_2c_1$
$a_2x + b_2y = c_2$ $\qquad$ $\underline{a_1a_2x + a_1b_2y = a_1c_2}$
$\qquad\qquad\qquad\qquad (a_1b_2 - a_2b_1)y = a_1c_2 - a_2c_1; \; y = \dfrac{a_1c_2 - a_2c_1}{a_1b_2 - a_2b_1};$

$a_1b_2x + b_1b_2y = b_2c_1$
$\underline{a_2b_1x + b_1b_2y = b_1c_2}$
$(a_1b_2 - a_2b_1)x = b_2c_1 - b_1c_2; \; x = \dfrac{b_2c_1 - b_1c_2}{a_1b_2 - a_2b_1}$

b. $\dfrac{\begin{vmatrix} c_1 & b_1 \\ c_2 & b_2 \end{vmatrix}}{\begin{vmatrix} a_1 & b_1 \\ a_2 & b_2 \end{vmatrix}} = \dfrac{b_2c_1 - b_1c_2}{a_1b_2 - a_2b_1} = x$ from part (a); $\dfrac{\begin{vmatrix} a_1 & c_1 \\ a_2 & c_2 \end{vmatrix}}{\begin{vmatrix} a_1 & b_1 \\ a_2 & b_2 \end{vmatrix}} = \dfrac{a_1c_2 - a_2c_1}{a_1b_2 - a_2b_1} = y$ from part (a)

Class Exercises 12-8 · page 463

1. $x = \dfrac{\begin{vmatrix} 7 & 4 \\ 8 & 6 \end{vmatrix}}{\begin{vmatrix} 3 & 4 \\ 5 & 6 \end{vmatrix}}$ and $y = \dfrac{\begin{vmatrix} 3 & 7 \\ 5 & 8 \end{vmatrix}}{\begin{vmatrix} 3 & 4 \\ 5 & 6 \end{vmatrix}}$

2. $x = \dfrac{\begin{vmatrix} 4 & 1 & -2 \\ 1 & -1 & 4 \\ 3 & 2 & 7 \end{vmatrix}}{\begin{vmatrix} 3 & 1 & -2 \\ 2 & -1 & 4 \\ 1 & 2 & 7 \end{vmatrix}}$, $y = \dfrac{\begin{vmatrix} 3 & 4 & -2 \\ 2 & 1 & 4 \\ 1 & 3 & 7 \end{vmatrix}}{\begin{vmatrix} 3 & 1 & -2 \\ 2 & -1 & 4 \\ 1 & 2 & 7 \end{vmatrix}}$, and $z = \dfrac{\begin{vmatrix} 3 & 1 & 4 \\ 2 & -1 & 1 \\ 1 & 2 & 3 \end{vmatrix}}{\begin{vmatrix} 3 & 1 & -2 \\ 2 & -1 & 4 \\ 1 & 2 & 7 \end{vmatrix}}$

3. Let $x = \dfrac{D_x}{D}$ and $y = \dfrac{D_y}{D}$; $D = \begin{vmatrix} 2 & 1 \\ 2 & 1 \end{vmatrix} = 2 - 2 = 0$; since the denominator of a fraction

cannot equal zero, the system has no solutions. The lines are parallel.

4. $\begin{vmatrix} 1 & 3 \\ 2 & 4 \end{vmatrix} = 4 - 6 = -2$; area $= |-2| = 2$ square units

Written Exercises 12-8 · pages 463–464

1. $x = \dfrac{\begin{vmatrix} 1 & -4 \\ 5 & 2 \end{vmatrix}}{\begin{vmatrix} 5 & -4 \\ 3 & 2 \end{vmatrix}} = \dfrac{2 - (-20)}{10 - (-12)} = \dfrac{22}{22} = 1$; $y = \dfrac{\begin{vmatrix} 5 & 1 \\ 3 & 5 \end{vmatrix}}{\begin{vmatrix} 5 & -4 \\ 3 & 2 \end{vmatrix}} = \dfrac{25 - 3}{10 - (-12)} = \dfrac{22}{22} = 1$; $(1, 1)$

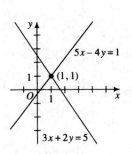

Ex. 1

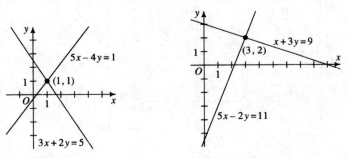

Ex. 2

2. $x = \dfrac{\begin{vmatrix} 11 & -2 \\ 9 & 3 \end{vmatrix}}{\begin{vmatrix} 5 & -2 \\ 1 & 3 \end{vmatrix}} = \dfrac{33 - (-18)}{15 - (-2)} = \dfrac{51}{17} = 3$; $y = \dfrac{\begin{vmatrix} 5 & 11 \\ 1 & 9 \end{vmatrix}}{\begin{vmatrix} 5 & -2 \\ 1 & 3 \end{vmatrix}} = \dfrac{45 - 11}{15 - (-2)} = \dfrac{34}{17} = 2$; $(3, 2)$

3. $x = \dfrac{\begin{vmatrix} -1 & 2 \\ 4 & -1 \end{vmatrix}}{\begin{vmatrix} 3 & 2 \\ 2 & -1 \end{vmatrix}} = \dfrac{1 - 8}{-3 - 4} = \dfrac{-7}{-7} = 1;\; y = \dfrac{\begin{vmatrix} 3 & -1 \\ 2 & 4 \end{vmatrix}}{\begin{vmatrix} 3 & 2 \\ 2 & -1 \end{vmatrix}} = \dfrac{12 - (-2)}{-3 - 4} = \dfrac{14}{-7} = -2;\; (1, -2)$

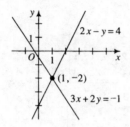

Ex. 3

Ex. 4

4. $x = \dfrac{\begin{vmatrix} 7 & 1 \\ 14 & 2 \end{vmatrix}}{\begin{vmatrix} 7 & 1 \\ -1 & 2 \end{vmatrix}} = \dfrac{14 - 14}{14 - (-1)} = \dfrac{0}{15} = 0;\; y = \dfrac{\begin{vmatrix} 7 & 7 \\ -1 & 14 \end{vmatrix}}{\begin{vmatrix} 7 & 1 \\ -1 & 2 \end{vmatrix}} = \dfrac{98 - (-7)}{14 - (-1)} = \dfrac{105}{15} = 7;\; (0, 7)$

5. $x = \dfrac{\begin{vmatrix} 1 & b \\ 1 & a \end{vmatrix}}{\begin{vmatrix} a & b \\ b & a \end{vmatrix}} = \dfrac{a - b}{a^2 - b^2} = \dfrac{1}{a + b};\; y = \dfrac{\begin{vmatrix} a & 1 \\ b & 1 \end{vmatrix}}{\begin{vmatrix} a & b \\ b & a \end{vmatrix}} = \dfrac{a - b}{a^2 - b^2} = \dfrac{1}{a + b};\; \left(\dfrac{1}{a + b}, \dfrac{1}{a + b} \right)$

6. $x = \dfrac{\begin{vmatrix} c & b \\ 4c & -2b \end{vmatrix}}{\begin{vmatrix} a & b \\ 3a & -2b \end{vmatrix}} = \dfrac{-2bc - 4bc}{-2ab - 3ab} = \dfrac{-6bc}{-5ab} = \dfrac{6c}{5a};$

$y = \dfrac{\begin{vmatrix} a & c \\ 3a & 4c \end{vmatrix}}{\begin{vmatrix} a & b \\ 3a & -2b \end{vmatrix}} = \dfrac{4ac - 3ac}{-2ab - 3ab} = \dfrac{ac}{-5ab} = -\dfrac{c}{5b};\; \left(\dfrac{6c}{5a}, -\dfrac{c}{5b} \right)$

7. a.

$x = \dfrac{D_x}{D}$ and $y = \dfrac{D_y}{D};\; D = \begin{vmatrix} 9 & -6 \\ 6 & -4 \end{vmatrix} = -36 - (-36) = 0$

so x and y are not defined and the system has no solution.

b. $\begin{vmatrix} a & b \\ d & e \end{vmatrix}$

8. a.

$$x = \frac{D_x}{D} \text{ and } y = \frac{D_y}{D}; D_x = \begin{vmatrix} -4 & 8 \\ -2 & 4 \end{vmatrix} = -16 - (-16) = 0;$$

$$D_y = \begin{vmatrix} 12 & -4 \\ 6 & -2 \end{vmatrix} = -24 - (-24) = 0;$$

$$D = \begin{vmatrix} 12 & 8 \\ 6 & 4 \end{vmatrix} = 48 - 48 = 0; \text{ since } D = 0, x \text{ and } y \text{ are}$$

not uniquely defined. Sol. is line $3x + 2y = -1$.

b. $\begin{vmatrix} a & b \\ d & e \end{vmatrix}, \begin{vmatrix} c & b \\ f & e \end{vmatrix}, \text{ and } \begin{vmatrix} a & c \\ d & f \end{vmatrix}$

9. $\overrightarrow{PQ} = (7, -1) - (4, 3) = (3, -4); \overrightarrow{PR} = (2, 3) - (4, 3) = (-2, 0);$

$\begin{vmatrix} 3 & -4 \\ -2 & 0 \end{vmatrix} = 0 - 8 = -8; \text{ area } = \frac{1}{2}|-8| = 4 \text{ sq. units}$

10. From Ex. 9, $\overrightarrow{PQ} = (3, -4); \overrightarrow{PS} = (-3, 6) - (4, 3) = (-7, 3);$

$\begin{vmatrix} 3 & -4 \\ -7 & 3 \end{vmatrix} = 9 - 28 = -19; \text{ area } = \frac{1}{2}|-19| = 9.5 \text{ sq. units}$

11. $\overrightarrow{RS} = (-3, 6) - (2, 3) = (-5, 3); \overrightarrow{RT} = (-5, 4) - (2, 3) = (-7, 1);$

$\begin{vmatrix} -5 & 3 \\ -7 & 1 \end{vmatrix} = -5 + 21 = 16; \text{ area } = \frac{1}{2}|16| = 8 \text{ sq. units}$

12. From Ex. 10, $\overrightarrow{PS} = (-7, 3); \overrightarrow{PV} = (-2, -5) - (4, 3) = (-6, -8);$

$\begin{vmatrix} -7 & 3 \\ -6 & -8 \end{vmatrix} = 56 + 18 = 74; \text{ area } = \frac{1}{2}|74| = 37 \text{ sq. units}$

13. From Exs. 9 and 10, $\overrightarrow{PR} = (-2, 0)$ and $\overrightarrow{PS} = (-7, 3); \begin{vmatrix} -2 & 0 \\ -7 & 3 \end{vmatrix} = -6 - 0 = -6;$

area $= |-6| = 6$ sq. units

14. From Ex. 10, $\overrightarrow{PS} = (-7, 3); \overrightarrow{PT} = (-5, 4) - (4, 3) = (-9, 1); \begin{vmatrix} -7 & 3 \\ -9 & 1 \end{vmatrix} = -7 + 27 = 20;$

area $= |20| = 20$ sq. units

15. The points are collinear.   **16.** The points are coplanar.

17. $D = \begin{vmatrix} 1 & -2 & 3 \\ 2 & -3 & 1 \\ 3 & -1 & 2 \end{vmatrix} = \begin{vmatrix} 1 & 0 & 0 \\ 2 & 1 & -5 \\ 3 & 5 & -7 \end{vmatrix} = \begin{vmatrix} 1 & -5 \\ 5 & -7 \end{vmatrix} = 18; D_x = \begin{vmatrix} 2 & -2 & 3 \\ 1 & -3 & 1 \\ 9 & -1 & 2 \end{vmatrix} =$

$\begin{vmatrix} 2 & 4 & 1 \\ 1 & 0 & 0 \\ 9 & 26 & -7 \end{vmatrix} = -\begin{vmatrix} 4 & 1 \\ 26 & -7 \end{vmatrix} = 54; D_y = \begin{vmatrix} 1 & 2 & 3 \\ 2 & 1 & 1 \\ 3 & 9 & 2 \end{vmatrix} = \begin{vmatrix} -3 & 2 & 1 \\ 0 & 1 & 0 \\ -15 & 9 & -7 \end{vmatrix} =$

$\begin{vmatrix} -3 & 1 \\ -15 & -7 \end{vmatrix} = 36; D_z = \begin{vmatrix} 1 & -2 & 2 \\ 2 & -3 & 1 \\ 3 & -1 & 9 \end{vmatrix} = \begin{vmatrix} 1 & 0 & 0 \\ 2 & 1 & -3 \\ 3 & 5 & 3 \end{vmatrix} = \begin{vmatrix} 1 & -3 \\ 5 & 3 \end{vmatrix} = 18;$

$x = \frac{D_x}{D} = \frac{54}{18} = 3; y = \frac{D_y}{D} = \frac{36}{18} = 2; z = \frac{D_z}{D} = \frac{18}{18} = 1; (3, 2, 1)$

18. $D = \begin{vmatrix} 4 & -2 & 3 \\ 5 & -6 & 2 \\ 3 & 4 & -5 \end{vmatrix} = 2\begin{vmatrix} 4 & -1 & 3 \\ 5 & -3 & 2 \\ 3 & 2 & -5 \end{vmatrix} = 2\begin{vmatrix} 0 & -1 & 0 \\ -7 & -3 & -7 \\ 11 & 2 & 1 \end{vmatrix} = 2\begin{vmatrix} -7 & -7 \\ 11 & 1 \end{vmatrix} = 140;$

$D_x = \begin{vmatrix} 2 & -2 & 3 \\ -1 & -6 & 2 \\ 7 & 4 & -5 \end{vmatrix} = \begin{vmatrix} 0 & -14 & 7 \\ -1 & -6 & 2 \\ 0 & -38 & 9 \end{vmatrix} = \begin{vmatrix} -14 & 7 \\ -38 & 9 \end{vmatrix} = 7\begin{vmatrix} -2 & 1 \\ -38 & 9 \end{vmatrix} = 140;$

$D_y = \begin{vmatrix} 4 & 2 & 3 \\ 5 & -1 & 2 \\ 3 & 7 & -5 \end{vmatrix} = \begin{vmatrix} 14 & 2 & 7 \\ 0 & -1 & 0 \\ 38 & 7 & 9 \end{vmatrix} = -\begin{vmatrix} 14 & 7 \\ 38 & 9 \end{vmatrix} = -7\begin{vmatrix} 2 & 1 \\ 38 & 9 \end{vmatrix} = 140;$

$D_z = \begin{vmatrix} 4 & -2 & 2 \\ 5 & -6 & -1 \\ 3 & 4 & 7 \end{vmatrix} = \begin{vmatrix} 0 & -2 & 0 \\ -7 & -6 & -7 \\ 11 & 4 & 11 \end{vmatrix} = \begin{vmatrix} 0 & -2 & 0 \\ -7 & -6 & 0 \\ 11 & 4 & 0 \end{vmatrix} = 0;$

$x = \dfrac{D_x}{D} = \dfrac{140}{140} = 1;\ y = \dfrac{D_y}{D} = \dfrac{140}{140} = 1;\ z = \dfrac{D_z}{D} = \dfrac{0}{140} = 0;\ (1,1,0)$

19. $D = \begin{vmatrix} 3 & -2 & 1 \\ 2 & 1 & -3 \\ 1 & 2 & 2 \end{vmatrix} = \begin{vmatrix} 3 & -8 & -5 \\ 2 & -3 & -7 \\ 1 & 0 & 0 \end{vmatrix} = \begin{vmatrix} -8 & -5 \\ -3 & -7 \end{vmatrix} = 41;$

$D_x = \begin{vmatrix} 7 & -2 & 1 \\ 1 & 1 & -3 \\ 4 & 2 & 2 \end{vmatrix} = \begin{vmatrix} 7 & -9 & 22 \\ 1 & 0 & 0 \\ 4 & -2 & 14 \end{vmatrix} = -\begin{vmatrix} -9 & 22 \\ -2 & 14 \end{vmatrix} = 82;$

$D_y = \begin{vmatrix} 3 & 7 & 1 \\ 2 & 1 & -3 \\ 1 & 4 & 2 \end{vmatrix} = \begin{vmatrix} 0 & -5 & -5 \\ 0 & -7 & -7 \\ 1 & 4 & 2 \end{vmatrix} = \begin{vmatrix} -5 & -5 \\ -7 & -7 \end{vmatrix} = 0;$

$D_z = \begin{vmatrix} 3 & -2 & 7 \\ 2 & 1 & 1 \\ 1 & 2 & 4 \end{vmatrix} = \begin{vmatrix} 7 & -2 & 9 \\ 0 & 1 & 0 \\ -3 & 2 & 2 \end{vmatrix} = \begin{vmatrix} 7 & 9 \\ -3 & 2 \end{vmatrix} = 41;$

$x = \dfrac{D_x}{D} = \dfrac{82}{41} = 2;\ y = \dfrac{D_y}{D} = \dfrac{0}{41} = 0;\ z = \dfrac{D_z}{D} = \dfrac{41}{41} = 1;\ (2,0,1)$

20. $\overrightarrow{DE} = (3,1,4) - (5,1,0) = (-2,0,4);\ \overrightarrow{DF} = (0,2,-1) - (5,1,0) = (-5,1,-1);$

$\overrightarrow{DG} = (5,2,0) - (5,1,0) = (0,1,0);\ \begin{vmatrix} -2 & 0 & 4 \\ -5 & 1 & -1 \\ 0 & 1 & 0 \end{vmatrix} = -\begin{vmatrix} -2 & 4 \\ -5 & -1 \end{vmatrix} = -22;$

volume $= |-22| = 22$ cubic units

21. $\overrightarrow{EF} = (0,2,-1) - (3,1,4) = (-3,1,-5);\ \overrightarrow{EG} = (5,2,0) - (3,1,4) = (2,1,-4);$

$\overrightarrow{EH} = (3,1,3) - (3,1,4) = (0,0,-1);\ \begin{vmatrix} -3 & 1 & -5 \\ 2 & 1 & -4 \\ 0 & 0 & -1 \end{vmatrix} = -1\begin{vmatrix} -3 & 1 \\ 2 & 1 \end{vmatrix} = 5;$

volume $= |5| = 5$ cubic units

22. From Ex. 20, $\overrightarrow{DF} = (-5,1,-1)$ and $\overrightarrow{DG} = (0,1,0);\ \overrightarrow{DH} = (3,1,3) - (5,1,0) = (-2,0,3);$

$\begin{vmatrix} -5 & 1 & -1 \\ 0 & 1 & 0 \\ -2 & 0 & 3 \end{vmatrix} = \begin{vmatrix} -5 & -1 \\ -2 & 3 \end{vmatrix} = -17;$ volume $= \dfrac{1}{6}|-17| = \dfrac{17}{6}$ cubic units

Class Exercises 12-9 · page 466

1. **a.** $\begin{vmatrix} \mathbf{i} & \mathbf{j} & \mathbf{k} \\ 1 & 2 & 3 \\ 4 & 5 & 6 \end{vmatrix}$ **b.** $\begin{vmatrix} \mathbf{i} & \mathbf{j} & \mathbf{k} \\ 4 & 5 & 6 \\ 7 & 8 & 9 \end{vmatrix}$

2. **a.** Vector **b.** Scalar **c.** Vector **d.** Scalar **e.** Vector **f.** Scalar

3. The right side of the equation shows the cross product of two real numbers, which makes no sense.

Written Exercises 12-9 · page 467

1. $\mathbf{v} \times \mathbf{u} = (5, -1, 0) \times (4, 0, 1) = \begin{vmatrix} \mathbf{i} & \mathbf{j} & \mathbf{k} \\ 5 & -1 & 0 \\ 4 & 0 & 1 \end{vmatrix} = (-1 - 0)\mathbf{i} - (5 - 0)\mathbf{j} + (0 + 4)\mathbf{k} =$

   $(-1, -5, 4); \mathbf{u} \times \mathbf{v} = (4, 0, 1) \times (5, -1, 0) = \begin{vmatrix} \mathbf{i} & \mathbf{j} & \mathbf{k} \\ 4 & 0 & 1 \\ 5 & -1 & 0 \end{vmatrix} = (0 + 1)\mathbf{i} - (0 - 5)\mathbf{j} +$

   $(-4 - 0)\mathbf{k} = (1, 5, -4); \mathbf{v} \times \mathbf{u} = (-1, -5, 4) = -(1, 5, -4) = -(\mathbf{u} \times \mathbf{v});$ yes

2. $\mathbf{v} \times \mathbf{w} = (5, -1, 0) \times (-3, 1, -2) = \begin{vmatrix} \mathbf{i} & \mathbf{j} & \mathbf{k} \\ 5 & -1 & 0 \\ -3 & 1 & -2 \end{vmatrix} = (2 - 0)\mathbf{i} - (-10 - 0)\mathbf{j} +$

   $(5 - 3)\mathbf{k} = (2, 10, 2); \mathbf{w} \times \mathbf{v} = (-2, -10, -2)$

3. From Ex. 1, $\mathbf{u} \times \mathbf{v} = (1, 5, -4); (\mathbf{u} \times \mathbf{v}) \cdot \mathbf{u} = (1, 5, -4) \cdot (4, 0, 1) = 4 + 0 - 4 = 0;$
   $(\mathbf{u} \times \mathbf{v}) \cdot \mathbf{v} = (1, 5, -4) \cdot (5, -1, 0) = 5 - 5 + 0 = 0;$ since the dot products are zero, the vectors are perpendicular.

4. Answers may vary. Examples are given.

   a. $\mathbf{u} \times \mathbf{w} = (4, 0, 1) \times (-3, 1, -2) = \begin{vmatrix} \mathbf{i} & \mathbf{j} & \mathbf{k} \\ 4 & 0 & 1 \\ -3 & 1 & -2 \end{vmatrix} = (-1, 5, 4);$ the vector $(-1, 5, 4)$ is

   perpendicular to the plane of $\mathbf{u}$ and $\mathbf{w}$.

   b. $\mathbf{w} \cdot \mathbf{s} = (-3, 1, -2) \cdot (x, y, z) = -3x + y - 2z = 0; (1, 3, 0)$ is perpendicular to $\mathbf{w}$.

5. By Property 3, the area of the parallelogram $= |\mathbf{u} \times \mathbf{v}| = |(1, 5, -4)|$ from
   Ex. 1 $= \sqrt{1^2 + 5^2 + (-4)^2} = \sqrt{42};$ area $= \sqrt{42}$ sq. units.

6. $\mathbf{u} \times \mathbf{w} = (4, 0, 1) \times (-3, 1, -2) = \begin{vmatrix} \mathbf{i} & \mathbf{j} & \mathbf{k} \\ 4 & 0 & 1 \\ -3 & 1 & -2 \end{vmatrix} = (0 - 1)\mathbf{i} - (-8 + 3)\mathbf{j} +$

   $(4 - 0)\mathbf{k} = (-1, 5, 4);$ area of $\triangle = \frac{1}{2}|(-1, 5, 4)| = \frac{1}{2}\sqrt{(-1)^2 + 5^2 + 4^2} = \frac{1}{2}\sqrt{42}$

7. $\mathbf{v} + \mathbf{w} = (2, 0, -2); \mathbf{u} \times (\mathbf{v} + \mathbf{w}) = (4, 0, 1) \times (2, 0, -2) = \begin{vmatrix} \mathbf{i} & \mathbf{j} & \mathbf{k} \\ 4 & 0 & 1 \\ 2 & 0 & -2 \end{vmatrix} =$

   $(0 - 0)\mathbf{i} - (-8 - 2)\mathbf{j} + (0 - 0)\mathbf{k} = (0, 10, 0);$ from Ex. 1, $\mathbf{u} \times \mathbf{v} = (1, 5, -4);$ from
   Ex. 4, $\mathbf{u} \times \mathbf{w} = (-1, 5, 4); (\mathbf{u} \times \mathbf{v}) + (\mathbf{u} \times \mathbf{w}) = (1, 5, -4) + (-1, 5, 4) =$
   $(0, 10, 0) = \mathbf{u} \times (\mathbf{v} + \mathbf{w})$

8. $\mathbf{u} \cdot (\mathbf{v} + \mathbf{w}) = (4,0,1) \cdot (2,0,-2) = 8 - 2 = 6;\ \mathbf{u} \cdot \mathbf{v} + \mathbf{u} \cdot \mathbf{w} =$
$(4,0,1) \cdot (5,-1,0) + (4,0,1) \cdot (-3,1,-2) = (20) + (-12 - 2) = 6;$
$\mathbf{u} \cdot (\mathbf{v} + \mathbf{w}) = \mathbf{u} \cdot \mathbf{v} + \mathbf{u} \cdot \mathbf{w}$

9. a. $\overrightarrow{PQ} \times \overrightarrow{PR} = (-2,-1,2) \times (1,0,1) = \begin{vmatrix} \mathbf{i} & \mathbf{j} & \mathbf{k} \\ -2 & -1 & 2 \\ 1 & 0 & 1 \end{vmatrix} = (-1 - 0)\mathbf{i} - (-2 - 2)\mathbf{j} +$

$(0 + 1)\mathbf{k} = (-1,4,1);\ (-1,4,1)$ and $(1,-4,-1)$ are perpendicular to the plane.

b. $-x + 4y + z = d;\ d = -1 + 4(1) + 0 = 3;\ -x + 4y + z = 3$

c. area of $\triangle PQR = \dfrac{1}{2}|(-1,4,1)| = \dfrac{1}{2}\sqrt{(-1)^2 + 4^2 + 1^2} = \dfrac{3}{2}\sqrt{2}$

10. a. $\overrightarrow{PQ} \times \overrightarrow{PR} = (3,-1,2) \times (4,5,-2) = \begin{vmatrix} \mathbf{i} & \mathbf{j} & \mathbf{k} \\ 3 & -1 & 2 \\ 4 & 5 & -2 \end{vmatrix} = (2 - 10)\mathbf{i} - (-6 - 8)\mathbf{j} +$

$(15 + 4)\mathbf{k} = (-8,14,19);\ (-8,14,19)$ and $(8,-14,-19)$ are perpendicular to
the plane.

b. $-8x + 14y + 19z = d;\ d = -8(0) + 14(0) + 19(0) = 0;\ -8x + 14y + 19z = 0$

c. area of $\triangle PQR = \dfrac{1}{2}|(-8,14,19)| = \dfrac{1}{2}\sqrt{(-8)^2 + 14^2 + 19^2} = \dfrac{3}{2}\sqrt{69}$

11. a. $\overrightarrow{PQ} \times \overrightarrow{PR} = (0,-2,4) \times (1,0,2) = \begin{vmatrix} \mathbf{i} & \mathbf{j} & \mathbf{k} \\ 0 & -2 & 4 \\ 1 & 0 & 2 \end{vmatrix} = (-4 - 0)\mathbf{i} - (0 - 4)\mathbf{j} +$

$(0 + 2)\mathbf{k} = (-4,4,2);\ (-4,4,2)$ and $(4,-4,-2)$ are perpendicular to the plane.

b. $4x - 4y - 2z = d;\ d = 4(2) - 4(1) - 2(0) = 4;\ 4x - 4y - 2z = 4,$ or
$2x - 2y - z = 2$

c. area of $\triangle PQR = \dfrac{1}{2}|(-4,4,2)| = \dfrac{1}{2}\sqrt{(-4)^2 + 4^2 + 2^2} = \dfrac{1}{2} \cdot 6 = 3$

12. a. $\overrightarrow{PQ} \times \overrightarrow{PR} = (1,-1,-2) \times (2,-1,0) = \begin{vmatrix} \mathbf{i} & \mathbf{j} & \mathbf{k} \\ 1 & -1 & -2 \\ 2 & -1 & 0 \end{vmatrix} = (0 - 2)\mathbf{i} - (0 + 4)\mathbf{j} +$

$(-1 + 2)\mathbf{k} = (-2,-4,1);\ (-2,-4,1)$ and $(2,4,-1)$ are perpendicular to the plane.

b. $2x + 4y - z = d;\ d = 2(0) + 4(2) - 3 = 5;\ 2x + 4y - z = 5$

c. area of $\triangle PQR = \dfrac{1}{2}|(-2,-4,1)| = \dfrac{1}{2}\sqrt{(-2)^2 + (-4)^2 + 1^2} = \dfrac{1}{2}\sqrt{21}$

13. a. $\mathbf{u} \times \mathbf{v} = (1,2,2) \times (4,3,0) = \begin{vmatrix} \mathbf{i} & \mathbf{j} & \mathbf{k} \\ 1 & 2 & 2 \\ 4 & 3 & 0 \end{vmatrix} = (0 - 6)\mathbf{i} - (0 - 8)\mathbf{j} + (3 - 8)\mathbf{k} =$

$(-6,8,-5);\ \sin\theta = \dfrac{|\mathbf{u} \times \mathbf{v}|}{|\mathbf{u}||\mathbf{v}|} = \dfrac{|(-6,8,-5)|}{|(1,2,2)||(4,3,0)|} = \dfrac{\sqrt{(-6)^2 + 8^2 + (-5)^2}}{\sqrt{1^2 + 2^2 + 2^2}\sqrt{4^2 + 3^2 + 0^2}} =$

$\dfrac{5\sqrt{5}}{3 \cdot 5} = \dfrac{1}{3}\sqrt{5}$

b. $\cos\theta = \dfrac{\mathbf{u} \cdot \mathbf{v}}{|\mathbf{u}||\mathbf{v}|} = \dfrac{(1,2,2) \cdot (4,3,0)}{3 \cdot 5}$ from part (a) $= \dfrac{4 + 6 + 0}{15} = \dfrac{2}{3}$

c. $\left(\dfrac{1}{3}\sqrt{5}\right)^2 + \left(\dfrac{2}{3}\right)^2 = \dfrac{5}{9} + \dfrac{4}{9} = 1$

14. a. $\mathbf{u} \times \mathbf{v} = (1, 2, 1) \times (1, 1, 2) = \begin{vmatrix} \mathbf{i} & \mathbf{j} & \mathbf{k} \\ 1 & 2 & 1 \\ 1 & 1 & 2 \end{vmatrix} = (4 - 1)\mathbf{i} - (2 - 1)\mathbf{j} + (1 - 2)\mathbf{k} =$

$(3, -1, -1)$; $\sin \theta = \dfrac{|\mathbf{u} \times \mathbf{v}|}{|\mathbf{u}||\mathbf{v}|} = \dfrac{|(3, -1, -1)|}{|(1, 2, 1)||(1, 1, 2)|} = \dfrac{\sqrt{3^2 + (-1)^2 + (-1)^2}}{\sqrt{1^2 + 2^2 + 1^2}\sqrt{1^2 + 1^2 + 2^2}} =$

$\dfrac{\sqrt{11}}{\sqrt{6} \cdot \sqrt{6}} = \dfrac{\sqrt{11}}{6}$

b. $\cos \theta = \dfrac{\mathbf{u} \cdot \mathbf{v}}{|\mathbf{u}||\mathbf{v}|} = \dfrac{(1, 2, 1) \cdot (1, 1, 2)}{\sqrt{6} \cdot \sqrt{6}} = \dfrac{1 + 2 + 2}{6} = \dfrac{5}{6}$

c. $\left(\dfrac{\sqrt{11}}{6}\right)^2 + \left(\dfrac{5}{6}\right)^2 = \dfrac{11 + 25}{36} = 1$

15. a. $\overrightarrow{AB} = (-19, -5, 0)$ and $\overrightarrow{AC} = (14, -4, 0)$; area $= \dfrac{1}{2}\begin{vmatrix} -19 & -5 \\ 14 & -4 \end{vmatrix} = 73$ sq. units

b. $\overrightarrow{AB} \times \overrightarrow{AC} = \begin{vmatrix} \mathbf{i} & \mathbf{j} & \mathbf{k} \\ -19 & -5 & 0 \\ 14 & -4 & 0 \end{vmatrix} = 146\mathbf{k}$; area $= \dfrac{1}{2}|\overrightarrow{AB} \times \overrightarrow{AC}| = \dfrac{1}{2}\sqrt{146^2} = 73$ sq. units

c. Area $= \sqrt{s(s - a)(s - b)(s - c)}$ where $s = \dfrac{a + b + c}{2}$ and $a = |\overrightarrow{AB}| = \sqrt{386}$,

$b = |\overrightarrow{AC}| = \sqrt{212}$ and $c = |\overrightarrow{BC}| = \sqrt{1090}$; substituting, we get Area $= 73$ sq. units.

d. Area $= \dfrac{1}{2}ab \sin \theta$. From the dot product we get $\theta = \mathrm{Cos}^{-1}\left(\dfrac{\overrightarrow{AB} \cdot \overrightarrow{AC}}{|\overrightarrow{AB}||\overrightarrow{AC}|}\right)$, and $a = |\overrightarrow{AB}|$,

and $b = |\overrightarrow{AC}|$. Substituting, we get Area $= 73$ sq. units.

16. a. $|\mathbf{u} \times \mathbf{v}| = |\mathbf{u}||\mathbf{v}| \sin \theta$; $|\mathbf{u} \times \mathbf{v}|^2 = |\mathbf{u}|^2|\mathbf{v}|^2 \sin^2\theta$;

$|\mathbf{u} \times \mathbf{v}|^2 + |\mathbf{u}|^2|\mathbf{v}|^2 \cos^2\theta = |\mathbf{u}|^2|\mathbf{v}|^2 \sin^2\theta + |\mathbf{u}|^2|\mathbf{v}|^2 \cos^2\theta$

$= |\mathbf{u}|^2|\mathbf{v}|^2(\sin^2\theta + \cos^2\theta)$

$= |\mathbf{u}|^2|\mathbf{v}|^2$;

$\therefore |\mathbf{u} \times \mathbf{v}|^2 = |\mathbf{u}|^2|\mathbf{v}|^2 - |\mathbf{u}|^2|\mathbf{v}|^2 \cos^2\theta$

$= |\mathbf{u}|^2|\mathbf{v}|^2 - (\mathbf{u} \cdot \mathbf{v})^2$

b. (1) Two three-dimensional vectors are perpendicular if and only if their dot product is zero. $(\mathbf{u} \times \mathbf{v}) \cdot \mathbf{u} = (b_1c_2 - b_2c_1, a_2c_1 - a_1c_2, a_1b_2 - a_2b_1) \cdot (a_1, b_1, c_1) =$
$(a_1b_1c_2 - a_1b_2c_1) + (a_2b_1c_1 - a_1b_1c_2) + (a_1b_2c_1 - a_2b_1c_1) = 0$;
$(\mathbf{u} \times \mathbf{v}) \cdot \mathbf{v} = (b_1c_2 - b_2c_1, a_2c_1 - a_1c_2, a_1b_2 - a_2b_1) \cdot (a_2, b_2, c_2) =$
$(a_2b_1c_2 - a_2b_2c_1) + (a_2b_2c_1 - a_1b_2c_2) + (a_1b_2c_2 - a_2b_1c_2) = 0$

(2) $\mathbf{u} \times \mathbf{v} = (b_1c_2 - b_2c_1, a_2c_1 - a_1c_2, a_1b_2 - a_2b_1)$; $\mathbf{v} \times \mathbf{u} =$

$(a_2, b_2, c_2) \times (a_1, b_1, c_1) = \begin{vmatrix} \mathbf{i} & \mathbf{j} & \mathbf{k} \\ a_2 & b_2 & c_2 \\ a_1 & b_1 & c_1 \end{vmatrix} = (b_2c_1 - b_1c_2)\mathbf{i} - (a_2c_1 - a_1c_2)\mathbf{j} +$

$(a_2b_1 - a_1b_2)\mathbf{k} = (b_2c_1 - b_1c_2, a_1c_2 - a_2c_1, a_2b_1 - a_1b_2) =$
$-(b_1c_2 - b_2c_1, a_2c_1 - a_1c_2, a_1b_2 - a_2b_1) = -(\mathbf{u} \times \mathbf{v})$

(Continued)

(3) Note that $\cos \theta = \dfrac{\mathbf{u} \cdot \mathbf{v}}{|\mathbf{u}||\mathbf{v}|}$; $|\mathbf{u} \times \mathbf{v}| = \sqrt{|\mathbf{u}|^2|\mathbf{v}|^2 - (\mathbf{u} \cdot \mathbf{v})^2} =$

$\sqrt{|\mathbf{u}|^2|\mathbf{v}|^2 - |\mathbf{u}|^2|\mathbf{v}|^2(\cos \theta)^2} = \sqrt{|\mathbf{u}|^2|\mathbf{v}|^2[1 - (\cos \theta)^2]} = \sqrt{|\mathbf{u}|^2|\mathbf{v}|^2(\sin \theta)^2} =$

$|\mathbf{u}||\mathbf{v}||\sin \theta| = |\mathbf{u}||\mathbf{v}|\sin \theta$, since $0° \le \theta \le 180°$. Let $|\mathbf{u}|$ be the length of the base of the parallelogram determined by $\mathbf{u}$ and $\mathbf{v}$. Then $|\mathbf{v}|\sin \theta$ is the height of the parallelogram. The area of the parallelogram = base $\times$ height = $|\mathbf{u}||\mathbf{v}|\sin \theta$.

(4) $\mathbf{u} \times \mathbf{v} = (b_1c_2 - b_2c_1, a_2c_1 - a_1c_2, a_1b_2 - a_2b_1)$; $\mathbf{u} \times \mathbf{w} = (a_1, b_1, c_1) \times$

$$(a_3, b_3, c_3) = \begin{vmatrix} \mathbf{i} & \mathbf{j} & \mathbf{k} \\ a_1 & b_1 & c_1 \\ a_3 & b_3 & c_3 \end{vmatrix} = (b_1c_3 - b_3c_1)\mathbf{i} - (a_1c_3 - a_3c_1)\mathbf{j} +$$

$(a_1b_3 - a_3b_1)\mathbf{k} = (b_1c_3 - b_3c_1, a_3c_1 - a_1c_3, a_1b_3 - a_3b_1)$;

$\mathbf{u} \times (\mathbf{v} + \mathbf{w}) = \mathbf{u} \times [(a_2, b_2, c_2) + (a_3, b_3, c_3)] = (a_1, b_1, c_1) \times$

$$(a_2 + a_3, b_2 + b_3, c_2 + c_3) = \begin{vmatrix} \mathbf{i} & \mathbf{j} & \mathbf{k} \\ a_1 & b_1 & c_1 \\ a_2 + a_3 & b_2 + b_3 & c_2 + c_3 \end{vmatrix} = [b_1(c_2 + c_3) -$$

$(b_2 + b_3)c_1]\mathbf{i} - [a_1(c_2 + c_3) - (a_2 + a_3)c_1]\mathbf{j} +$

$[a_1(b_2 + b_3) - (a_2 + a_3)b_1]\mathbf{k} = (b_1c_2 - b_2c_1 + b_1c_3 - b_3c_1, a_2c_1 - a_1c_2 +$

$a_3c_1 - a_1c_3, a_1b_2 - a_2b_1 + a_1b_3 - a_3b_1) = (b_1c_2 - b_2c_1, a_2c_1 - a_1c_2,$

$a_1b_2 - a_2b_1) + (b_1c_3 - b_3c_1, a_3c_1 - a_1c_3, a_1b_3 - a_3b_1) = (\mathbf{u} \times \mathbf{v}) +$

$(\mathbf{u} \times \mathbf{w})$

(5) If $\mathbf{u}$ is parallel to $\mathbf{v}$, $\mathbf{u} = k\mathbf{v}$. Then $(a_1, b_1, c_1) = k(a_2, b_2, c_2) = (ka_2, kb_2, kc_2)$, so that $a_1 = ka_2$, $b_1 = kb_2$, and $c_1 = kc_2$. $\mathbf{u} \times \mathbf{v} = (a_1, b_1, c_1) \times (a_2, b_2, c_2) =$

$$(ka_2, kb_2, kc_2) \times (a_2, b_2, c_2) = \begin{vmatrix} \mathbf{i} & \mathbf{j} & \mathbf{k} \\ ka_2 & kb_2 & kc_2 \\ a_2 & b_2 & c_2 \end{vmatrix} = (kb_2c_2 - kb_2c_2)\mathbf{i} -$$

$(ka_2c_2 - ka_2c_2)\mathbf{j} + (ka_2b_2 - ka_2b_2)\mathbf{k} = 0\mathbf{i} + 0\mathbf{j} + 0\mathbf{k} = (0, 0, 0) = \mathbf{0}$. If $\mathbf{u} \times \mathbf{v} = \mathbf{0}$, then $(b_1c_2 - b_2c_1, a_2c_1 - a_1c_2, a_1b_2 - a_2b_1) = (0, 0, 0)$. Thus,

$b_1 = \dfrac{c_1}{c_2}b_2$, $a_1 = \dfrac{c_1}{c_2}a_2$, and $\dfrac{a_1}{a_2} = \dfrac{b_1}{b_2}$, or $\dfrac{a_1}{a_2} = \dfrac{b_1}{b_2} = \dfrac{c_1}{c_2}$. That is, $a_1 = ka_2$,

$b_1 = kb_2$, and $c_1 = kc_2$, or $\mathbf{u} = k\mathbf{v}$ and $\mathbf{u}$ is parallel to $\mathbf{v}$. Thus, $\mathbf{u}$ is parallel to $\mathbf{v}$ if and only if $\mathbf{u} \times \mathbf{v} = \mathbf{0}$.

Computer Exercise · page 467

```
10 PRINT "GIVEN THREE NONCOLLINEAR POINTS,"
20 PRINT "THIS PROGRAM WILL DETERMINE THE EQUATION"
30 PRINT "OF THE PLANE CONTAINING THEM."
40 PRINT
50 PRINT "WHAT ARE THE COORDINATES OF POINT P(X1, Y1, Z1)";
60 INPUT X1, Y1, Z1
70 PRINT "COORDINATES OF POINT Q(X2, Y2, Z2)";
80 INPUT X2, Y2, Z2
90 PRINT "COORDINATES OF POINT R(X3, Y3, Z3)";
100 INPUT X3, Y3, Z3
110 LET A1 = X2 - X1
120 LET B1 = Y2 - Y1
130 LET C1 = Z2 - Z1
```

(Continued on next page)

```
140 LET A2 = X3 - X1
150 LET B2 = Y3 - Y1
160 LET C2 = Z3 - Z1
170 LET A = B1 * C2 - B2 * C1
180 LET B = A2 * C1 - A1 * C2
190 LET C = A1 * B2 - A2 * B1
200 LET D = A * X1 + B * Y1 + C * Z1
210 PRINT "THE EQUATION OF THE PLANE IS"
220 PRINT A; "X + "; B; "Y + "; C; "Z = "; D
```

Chapter Test · page 469

1. $\tan \alpha = \dfrac{12}{5}$; $\alpha \approx 67.4°$; $\theta \approx 180° - 67.4° = 112.6°$;

$|\mathbf{F}_3| = \sqrt{5^2 + 12^2} = 13$ N

2. a. $\overrightarrow{AB} = -\overrightarrow{BA} = -\mathbf{u}$ **b.** $\overrightarrow{AC} = \mathbf{v} - \mathbf{u}$ **c.** $\overrightarrow{AT} = \dfrac{1}{3}\overrightarrow{AC} = \dfrac{1}{3}(\mathbf{v} - \mathbf{u}) = \dfrac{1}{3}\mathbf{v} - \dfrac{1}{3}\mathbf{u}$

  d. $\overrightarrow{AB} + \overrightarrow{BT} = \overrightarrow{AT}$; $-\mathbf{u} + \overrightarrow{BT} = \dfrac{1}{3}\mathbf{v} - \dfrac{1}{3}\mathbf{u}$; $\overrightarrow{BT} = \dfrac{2}{3}\mathbf{u} + \dfrac{1}{3}\mathbf{v}$

3. $x = 480 \cos 50° = 308.5$ mi/h W; $y = 480 \sin 50° = 367.7$ mi/h N

4. $\overrightarrow{BC} = (5 - 0, 0 - (-3)) = (5, 3)$; $\overrightarrow{AD} = \overrightarrow{BC}$; $(x - 1, y - 3) = (5, 3)$;
  $x = 6, y = 6$; $(6, 6)$

5. a. $\mathbf{v} = (1, 4)$; $|\mathbf{v}| = \sqrt{1^2 + 4^2} = \sqrt{17}$

  b. $(x, y) = (-2 + t, -2 + 4t)$; $x = -2 + t, y = -2 + 4t$

  c. Substitute parametric eqs. into $y = 4x^2 - 6x$; $-2 + 4t = 4(-2 + t)^2 - 6(-2 + t)$;
   $-2 + 4t = 4(4 - 4t + t^2) + 12 - 6t$; $4t^2 - 26t + 30 = 0$; $2(2t - 3)(t - 5) = 0$;

   $t = \dfrac{3}{2}, 5$; at $t = \dfrac{3}{2}$, $\left(-\dfrac{1}{2}, 4\right)$ and at $t = 5$, $(3, 18)$.

6. $\mathbf{u}$ and $\mathbf{v}$ are opposite in direction. $\cos \theta = \dfrac{\mathbf{u} \cdot \mathbf{v}}{|\mathbf{u}||\mathbf{v}|}$. Thus, $\mathbf{u} \cdot \mathbf{v} = |\mathbf{u}||\mathbf{v}| \cos \theta$.

If $\mathbf{u} \cdot \mathbf{v} = -|\mathbf{u}||\mathbf{v}|$, then $\cos \theta = -1$. Therefore, $\theta = 180°$ and $\mathbf{u}$ and $\mathbf{v}$ are opposite in direction.

7. $\overrightarrow{AB} = (-3, 4)$, $\overrightarrow{AC} = (-1, -9)$; $\cos A = \dfrac{(-3, 4) \cdot (-1, -9)}{|(-3, 4)||(-1, -9)|} = \dfrac{3 - 36}{5\sqrt{82}} = \dfrac{-33}{5\sqrt{82}}$;
  $\angle A \approx 136.8°$

8. a. $AB = \sqrt{(4 + 6)^2 + 2^2 + (7 + 5)^2} = \sqrt{100 + 4 + 144} = \sqrt{248} = 2\sqrt{62}$

  b. $\left(\dfrac{4 - 6}{2}, \dfrac{0 - 2}{2}, \dfrac{-5 + 7}{2}\right) = (-1, -1, 1)$

9. a. If parallel, the direction vectors are scalar multiples of each other. The direction vector for L is $(4, -1, -3)$, so $(x, y, z) = (2, 1, 7) + t(4, -1, -3)$

  b. $(x, y, z) = (4t, -1 - t, 6 - 3t)$; $y = 0$ in xz-plane, so $-1 - t = 0$; $t = -1$; $(-4, 0, 9)$

10. $x^2 + (y + 1)^2 + (z - 3)^2 = 33$ has center $C(0, -1, 3)$. $P(2, -3, -2)$ and $\overrightarrow{PC}$ is perpendicular to the tangent plane; $\overrightarrow{PC} = (-2, 2, 5)$; $-2x + 2y + 5z = (-2)(2) + 2(-3) + 5(-2) = -20$; $2x - 2y - 5z = 20$

11. a.
$$\begin{vmatrix} 2 & 0 & 3 \\ 4 & 2 & -2 \\ -3 & 0 & 1 \end{vmatrix} = -0\begin{vmatrix} 4 & -2 \\ -3 & 1 \end{vmatrix} + 2\begin{vmatrix} 2 & 3 \\ -3 & 1 \end{vmatrix} - 0\begin{vmatrix} 2 & 3 \\ 4 & -2 \end{vmatrix} = 2\begin{vmatrix} 2 & 3 \\ -3 & 1 \end{vmatrix} =$$

$$2(2 \cdot 1 - (-3)3) = 2(2 + 9) = 22$$

b.
$$\begin{vmatrix} 0 & 0 & 0 & 5 \\ -1 & 2 & 0 & 0 \\ 3 & 1 & 2 & 0 \\ -4 & 0 & 0 & 1 \end{vmatrix} = 0 - 0 + 2\begin{vmatrix} 0 & 0 & 5 \\ -1 & 2 & 0 \\ -4 & 0 & -1 \end{vmatrix} - 0 = 2\left[0 - 0 + 5\begin{vmatrix} -1 & 2 \\ -4 & 0 \end{vmatrix}\right] =$$

$$2 \cdot 5((-1)0 - 2(-4)) = 10(0 - (-8)) = 80$$

12. $x = \dfrac{D_x}{D} = \dfrac{\begin{vmatrix} 19 & 4 \\ 14 & -10 \end{vmatrix}}{\begin{vmatrix} 7 & 4 \\ 3 & -10 \end{vmatrix}} = \dfrac{-190 - 56}{-70 - 12} = 3$; $y = \dfrac{D_y}{D} = \dfrac{\begin{vmatrix} 7 & 19 \\ 3 & 14 \end{vmatrix}}{\begin{vmatrix} 7 & 4 \\ 3 & -10 \end{vmatrix}} = \dfrac{98 - 57}{-70 - 12} = -\dfrac{1}{2}$;

$\left(3, -\dfrac{1}{2}\right)$

13. Answers may vary. Examples are given.

a. $\overrightarrow{AB} = (-6, 0, 4)$, $\overrightarrow{AC} = (4, 4, 5)$; $\overrightarrow{AB} \times \overrightarrow{AC} = \begin{vmatrix} \mathbf{i} & \mathbf{j} & \mathbf{k} \\ -6 & 0 & 4 \\ 4 & 4 & 5 \end{vmatrix} = -16\mathbf{i} + 46\mathbf{j} - 24\mathbf{k}$; $(-16, 46, -24)$

b. Area $= \dfrac{1}{2}|\overrightarrow{AB} \times \overrightarrow{AC}| = \dfrac{1}{2}|(-16, 46, -24)| = \dfrac{1}{2}\sqrt{16^2 + 46^2 + 24^2} = \dfrac{1}{2}\sqrt{2948} \approx 27.1$

Activity 13-1 · page 474

1. a.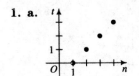

b. slope = 1

2. a.

b. slope = −4

3. a.

b. slope = 0.7

Class Exercises 13-1 · page 476

1. Arithmetic; $d = 5$ **2.** Geometric; $r = 2$ **3.** Neither

4. Arithmetic; $d = -6$ **5.** Geometric: $r = -\dfrac{2}{3}$ **6.** Neither

7. $t_1 = 7; t_2 = 12; t_3 = 17; t_4 = 22$; arithmetic

8. $t_1 = \dfrac{2}{3}; t_2 = \dfrac{3}{4}; t_3 = \dfrac{4}{5}; t_4 = \dfrac{5}{6}$; neither

9. $t_1 = 3; t_2 = 9; t_3 = 27; t_4 = 81$; geometric

10. $t_1 = 1; t_2 = 8; t_3 = 27; t_4 = 64$; neither

11. a. Yes; $d = 0$ **b.** Yes; $r = 1$

12. The graph of the sequence is a set of discrete points whose x-coordinates are positive integers; the graph of the line is continuous and so the x-coordinate of a point on the line can be any real number.

Written Exercises 13-1 · pages 476–479

1. $t_1 = 2 + 3 = 5; t_2 = 4 + 3 = 7; t_3 = 6 + 3 = 9; t_4 = 8 + 3 = 11$; arithmetic

2. $t_1 = 1 + 1 = 2; t_2 = 8 + 1 = 9; t_3 = 27 + 1 = 28; t_4 = 64 + 1 = 65$; neither

3. $t_1 = 3 \cdot 2 = 6; t_2 = 3 \cdot 4 = 12; t_3 = 3 \cdot 8 = 24; t_4 = 3 \cdot 16 = 48$; geometric

4. $t_1 = 3 - 7 = -4; t_2 = 3 - 14 = -11; t_3 = 3 - 21 = -18; t_4 = 3 - 28 = -25$; arithmetic

5. $t_1 = 1 + \dfrac{1}{1} = 2; t_2 = 2 + \dfrac{1}{2} = \dfrac{5}{2}; t_3 = 3 + \dfrac{1}{3} = \dfrac{10}{3}; t_4 = 4 + \dfrac{1}{4} = \dfrac{17}{4}$; neither

6. $t_1 = (-2)^1 = -2; t_2 = (-2)^2 = 4; t_3 = (-2)^3 = -8; t_4 = (-2)^4 = 16$; geometric

7. $t_1 = (-1)^1 \cdot 1 = -1; t_2 = (-1)^2 \cdot 2 = 2; t_3 = (-1)^3 \cdot 3 = -3; t_4 = (-1)^4 \cdot 4 = 4$; neither

8. $t_1 = 16 \cdot 2^2 = 64; t_2 = 16 \cdot 2^4 = 256; t_3 = 16 \cdot 2^6 = 1024; t_4 = 16 \cdot 2^8 = 4096$; geometric

9. $t_1 = \sin \dfrac{\pi}{2} = 1; t_2 = \sin \pi = 0; t_3 = \sin \dfrac{3\pi}{2} = -1; t_4 = \sin 2\pi = 0$; neither

10. $t_1 = \cos \pi = -1; t_2 = \cos 2\pi = 1; t_3 = \cos 3\pi = -1; t_4 = \cos 4\pi = 1$; geometric

11. $t_1 = 1; d = 3; t_n = 1 + (n - 1)3 = -2 + 3n$

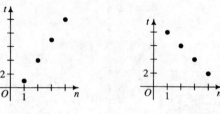

Ex. 11   Ex. 12   Ex. 13

12. $t_1 = 8; d = -2; t_n = 8 + (n - 1)(-2) = 10 - 2n$

13. $t_1 = 8; r = \dfrac{1}{2}; t_n = 8 \cdot \left(\dfrac{1}{2}\right)^{n-1}$ or $t_n = 16(0.5)^n$

14. $t_1 = 0.3; r = 3; t_n = 0.3(3^{n-1})$ or $t_n = 0.1(3^n)$

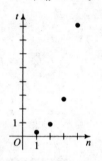 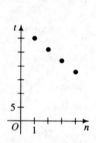 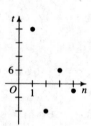

Ex. 14   Ex. 15   Ex. 16

15. $t_1 = 30; d = -4; t_n = 30 + (n - 1)4 = 34 - 4n$

16. $t_1 = 24; r = -\dfrac{1}{2}; t_n = 24\left(-\dfrac{1}{2}\right)^{n-1}$ or $t_n = -48\left(-\dfrac{1}{2}\right)^n$

17. Arithmetic; $t_n = 17 + (n - 1)4 = 4n + 13$

18. Arithmetic; $t_n = 15 + (n - 1)(-8) = 23 - 8n$

19. Geometric; $t_n = 8 \cdot \left(\dfrac{3}{2}\right)^{n-1}$ or $t_n = \dfrac{16}{3} \cdot \left(\dfrac{3}{2}\right)^n$

20. Geometric; $t_n = 100\left(-\dfrac{1}{2}\right)^{n-1}$ or $t_n = -200\left(-\dfrac{1}{2}\right)^n$

21. Neither; $t_n = n^2$ **22.** Neither; $t_n = \dfrac{n}{n + 1}$ **23.** Neither; $t_n = 10^n + 1$

24. Neither; $t_n = \dfrac{n + 1}{n^2}$

25. Arithmetic; $t_n = (2a - 2b) + (n - 1)(a + b) = n(a + b) + (a - 3b)$

26. Geometric; $t_n = \dfrac{a}{9}\left(\dfrac{a}{2}\right)^{n-1}$ or $t_n = \dfrac{2}{9} \cdot \left(\dfrac{a}{2}\right)^n$

27. Geometric; $t_n = 2^{2/3} \cdot 2^{n-1} = 2^{n-(1/3)}$ or $t_n = 2^{-1/3} \cdot 2^n$

28. Neither; $t_n = 2^{1/(n+1)}$ **29.** $d = 21 - 15 = 6$; $t_{20} = 15 + (20 - 1)6 = 129$

30. $70 = 76 + (3 - 1)d$; $2d = -6$; $d = -3$; $t_{101} = 76 + 100(-3) = -224$

31. $8 = t_1 + 2d$ and $14 = t_1 + 4d$, so $6 = 2d$; $d = 3$; $8 = t_1 + 2(3)$; $2 = t_1$;
    $t_{50} = 2 + 49(3) = 149$

32. $25 = t_1 + 7d$ and $61 = t_1 + 19d$; $36 = 12d$; $d = 3$; $25 = t_1 + 7(3)$; $t_1 = 4$

33. $r = \dfrac{2^{-3}}{2^{-4}} = 2$; $t_{12} = 2^{-4} \cdot 2^{11} = 2^7 = 128$

34. $r = \dfrac{2^{3/2}}{2^1} = 2^{1/2}$; $t_{13} = 2 \cdot (2^{1/2})^{12} = 2^7 = 128$

35. $t_2 = t_1 \cdot r^1 = 64$ and $t_5 = t_1 \cdot r^4 = -8$; $t_1 = \dfrac{64}{r} = \dfrac{-8}{r^4}$; $r^3 = -\dfrac{1}{8}$; $r = -\dfrac{1}{2}$;

    $t_1 = 64 \div \left(-\dfrac{1}{2}\right) = -128$; $t_9 = -128 \cdot \left(-\dfrac{1}{2}\right)^8 = -\dfrac{1}{2}$

36. $t_4 = t_1 \cdot r^3$; $24 = 81r^3$; $r^3 = \dfrac{8}{27}$; $r = \dfrac{2}{3}$; $t_7 = t_1 r^6 = 81\left(\dfrac{2}{3}\right)^6 = \dfrac{64}{9}$

37. $t_1 = 18$; $d = 24 - 18 = 6$; $336 = 18 + (n - 1)6$; $6n = 324$; $n = 54$

38. $t_1 = 178$; $d = 170 - 178 = -8$; $2 = 178 + (n - 1)(-8)$; $8n = 184$; $n = 23$

39. $t_1 = 35$, $t_n = 294$, and $d = 7$; $294 = 35 + (n - 1)7$; $7n = 266$; $n = 38$

40. $t_1 = 30$, $t_n = 276$, and $d = 6$; $276 = 30 + (n - 1)6$; $6n = 252$; $n = 42$

41. The LCM of 4 and 6 is 12, so find the number of 3-digit numbers divisible by 12;
    $t_1 = 108$, $t_n = 996$, and $d = 12$; $996 = 108 + (n - 1)12$; $12n = 900$; $n = 75$.

42. First find the number of 4-digit numbers divisible by 11: $t_1 = 1001$, $t_n = 9999$, and
    $d = 11$; $9999 = 1001 + (n - 1)11$; $11n = 9009$; $n = 819$. Of the 9000 4-digit
    numbers, $9000 - 819 = 8181$ are not divisible by 11.

43. False; for example, $0, \dfrac{\pi}{2}, \pi$ is arithmetic, but $\sin 0$, $\sin \dfrac{\pi}{2}$, $\sin \pi$, or $0, 1, 0$ is not.

44. True; if a, b, c is an arithmetic sequence with the common difference d, then $b = a + d$
    and $c = a + 2d$; thus 2^a, $2^b = 2^{a+d} = 2^a \cdot 2^d$, $2^c = 2^{a+2d} = (2^a \cdot 2^d)2^d$ is a geometric
    sequence with the common ratio 2^d.

45. $\log ar = \log a + \log r$; $\log ar^2 = \log a + 2 \log r$; in general, $t_n = \log ar^{n-1} = \log a +$
    $(n - 1) \log r$, and the logarithms form an arithmetic sequence with $t_1 = \log a$ and
    common difference $d = \log r$.

46. Answers will vary. Example: $F_n = F_1 + (n - 1)d$ and $P_n = P_1(r)^{n-1}$. In words, this
    idea means that population tends to increase more rapidly than food supplies. If this is
    true, either population growth must be slowed to ensure adequate food supplies, or there
    will be an inadequate supply of food for the existing population level.

47. a. $8 - 2 = 3x + 5 - 8$; $9 = 3x$; $x = 3$ **b.** $\dfrac{8}{2} = \dfrac{3x + 5}{8}$; $32 = 3x + 5$; $x = 9$

48. a. $x - 4 = \dfrac{3}{2}x - x$; $\dfrac{1}{2}x = 4$; $x = 8$ **b.** $\dfrac{x}{4} = \dfrac{\frac{3}{2}x}{x}$; $\dfrac{x}{4} = \dfrac{3}{2}$; $2x = 12$; $x = 6$

49. $2x + y - y = 7y - (2x + y) = 20 - 7y$; $2x = 6y - 2x = 20 - 7y$; $4x = 6y$, so

$x = \dfrac{3}{2}y$; $6y - 2x = 20 - 7y$, so $6y - 2\left(\dfrac{3}{2}y\right) = 20 - 7y$; $10y = 20$; $y = 2$;

$x = \dfrac{3}{2}(2) = 3$

50. $\dfrac{2xy}{2y} = \dfrac{2}{2xy} = \dfrac{\frac{1}{2}xy}{2}$; $x = \dfrac{1}{xy} = \dfrac{xy}{4}$; $x = \dfrac{1}{xy}$, so $y = \dfrac{1}{x^2}$; $x^2y^2 = 4$, so $x^2\left(\dfrac{1}{x^2}\right)^2 = 4$;

$\dfrac{1}{x^2} = 4$; $x^2 = \dfrac{1}{4}$; $x = \pm\dfrac{1}{2}$; $y = \dfrac{1}{x^2} = 4$

51. If the half-life is 2 days, $A(t) = A_0\left(\dfrac{1}{2}\right)^{t/2}$; after 2 days, $A(t) = A_0\left(\dfrac{1}{2}\right)^{2/2} = \dfrac{1}{2}A_0$, so $\dfrac{1}{2}$ is

left; after 4 days, $A(t) = A_0\left(\dfrac{1}{2}\right)^{4/2} = \dfrac{1}{4}A_0$, so $\dfrac{1}{4}$ is left; after 10 days,

$A(t) = A_0\left(\dfrac{1}{2}\right)^{10/2} = \dfrac{1}{32}A_0$, so $\dfrac{1}{32}$ is left; after d days, $A(t) = A_0\left(\dfrac{1}{2}\right)^{d/2} = \dfrac{1}{2^{d/2}}A_0$, so $\dfrac{1}{2^{d/2}}$

is left.

52. $A = P\left(1 + \dfrac{0.08}{4}\right)^{4n} = P(1.02)^{4n}$ where n is in years; after 1 quarter,

$A = P(1.02)^{4(1/4)} = 1.02P$; after 2 quarters, $A = P(1.02)^{4(2/4)} = (1.02)^2\, P$; after q
quarters, $A = P(1.02)^{4(q/4)} = P(1.02)^q$; after n years, $A = P(1.02)^{4n}$.

53. The LCM of 2, 3, and 7 is 42, so the sequence 1, 43, 85, 127, 169,... satisfies the
requirements. (Other answers are possible.)

54. The LCM of 4, 5, and 6 is 60, so the sequence 6, 66, 126, 186, 246,... satisfies the
requirements. (Other answers are possible.)

55. For A, $t_n = 3 + (n - 1)11 = 11n - 8$; for B, $t_n = 2 + (n - 1)7 = 7n - 5$; we need to

find r and s such that $11r - 8 = 7s - 5$; $\dfrac{11r - 3}{7} = s$ and r and s are positive integers;

this occurs when $r = 6, 13, 20, 27, 34,...$; substituting these values as values for n in
the formula for sequence A, we find the first 5 terms of sequence A that are also terms
of sequence B: 58, 135, 212, 289, and 366.

56. a. $t_2 = t_1 \cdot r$, so $t_1 = \dfrac{t_2}{r} = \dfrac{18}{r}$; $t_4 = t_1r^3$, so $8 = \dfrac{18}{r} \cdot r^3$; $r^2 = \dfrac{4}{9}$; $r = \pm\dfrac{2}{3}$; for $r = \dfrac{2}{3}$,

the sequence is 27, 18, 12, 8,...; for $r = -\dfrac{2}{3}$, the sequence is -27, 18, -12, 8,....

b. When both term numbers are odd or both term numbers are even

57. Let the sides have lengths a, $a + d$, and $a + 2d$; by the Pythagorean Theorem,
$(a + 2d)^2 = a^2 + (a + d)^2$; simplifying, $0 = a^2 - 2ad - 3d^2 = (a - 3d)(a + d)$;
$a = 3d$ or $a = -d$ (reject); the sides must have integral lengths, so any triangle with
sides of lengths $3d$, $4d$, and $5d$ where d is a positive integer satisfies the requirements.

58. Let the sides have lengths a, ar, and ar^2. Suppose $r > 1$. (The case for $r < 1$ is similar.) Then, by the Pythagorean Theorem, $(ar^2)^2 = a^2 + (ar)^2$; $a^2 r^4 = a^2 + a^2 r^2$; $a \neq 0$, so $r^4 - r^2 - 1 = 0$; by the quadratic formula, $r^2 = \dfrac{1 + \sqrt{5}}{2}$ (since the other root is negative) and $r = \dfrac{\sqrt{2 + 2\sqrt{5}}}{2}$. Thus, if a is an integer, ar is not an integer. It is impossible to have integral-length sides in a geometric sequence.

59. a. arithmetic mean: $\dfrac{4 + 9}{2} = \dfrac{13}{2} = 6.5$; geometric mean: $\sqrt{4 \cdot 9} = \sqrt{36} = 6$

   b. arithmetic mean: $\dfrac{5 + 10}{2} = \dfrac{15}{2} = 7.5$; geometric mean: $\sqrt{5 \cdot 10} = \sqrt{50} = 5\sqrt{2}$

60. a. $x - a = b - x$; $2x = a + b$; $x = \dfrac{a + b}{2}$ **b.** $\dfrac{x}{a} = \dfrac{b}{x}$; $x^2 = ab$; $x = \pm\sqrt{ab}$

61. a. Since $\angle A \cong \angle A$ and $\angle ADC \cong \angle ACB$, $\triangle ADC \sim \triangle ACB$ by AA Similarity. In similar triangles, lengths of corresponding sides are proportional, so $\dfrac{AD}{AC} = \dfrac{AC}{AB}$, or $\dfrac{x}{b} = \dfrac{b}{c}$. Since $b > 0$, $b = \sqrt{xc}$, so b is the geometric mean of x and c.

   b. $\triangle CDB \sim \triangle ACB$, $\dfrac{y}{a} = \dfrac{a}{c}$ and $a = \sqrt{yc}$, so a is the geometric mean of y and c.

   c. From parts (a) and (b), $b = \sqrt{xc}$ and $a = \sqrt{yc}$, so $a^2 + b^2 = yc + xc = c(y + x)$. But $y + x = c$, so $a^2 + b^2 = c \cdot c$; $a^2 + b^2 = c^2$.

62. Since $\angle ACD$ and $\angle B$ are both complements of $\angle DCB$, $\angle B \cong \angle ACD$. Also, $\angle ADC \cong \angle CDB$. Thus, $\triangle ADC \sim \triangle CDB$ by AA Similarity and so $\dfrac{AD}{CD} = \dfrac{DC}{DB}$; $\dfrac{x}{h} = \dfrac{h}{y}$; $h = \sqrt{xy}$. Therefore, h is the geometric mean of x and y.

63. We want to show that $\dfrac{a + b}{2} \geq \sqrt{ab}$. We know that $(a - b)^2 \geq 0$ since the square of any number is nonnegative. Thus $a^2 - 2ab + b^2 \geq 0$; $a^2 + 2ab + b^2 \geq 4ab$; $(a + b)^2 \geq 4ab$. Since a and b are positive, $\sqrt{(a + b)^2} \geq \sqrt{4ab}$. Hence $a + b \geq 2\sqrt{ab}$ and $\dfrac{a + b}{2} \geq \sqrt{ab}$.

64. In the inscribed triangle, the height is $\dfrac{3}{2} \cdot 6 = 9$ and each side has length $6\sqrt{3}$. Thus $A = \dfrac{1}{2} \cdot 6\sqrt{3} \cdot 9 = 27\sqrt{3}$. In the circumscribed triangle, the height is $3 \cdot 6 = 18$ and each side has length $12\sqrt{3}$, so $B = \dfrac{1}{2} \cdot 18 \cdot 12\sqrt{3} = 108\sqrt{3}$. The geometric mean of A and B is $\sqrt{AB} = \sqrt{27\sqrt{3} \cdot 108\sqrt{3}} = 54\sqrt{3}$. The area of the inscribed regular hexagon is $K = 6\left(\dfrac{1}{2} r \cdot \dfrac{r}{2}\sqrt{3}\right) = 6\left(\dfrac{1}{2} \cdot \dfrac{36}{2}\sqrt{3}\right) = 54\sqrt{3}$, so $\sqrt{AB} = K$, the area of the inscribed regular hexagon.

Activity 13-2 · page 480

1. a. $t_1 = 3; t_2 = 3 + 4 = 7; t_3 = 7 + 4 = 11; t_4 = 11 + 4 = 15; t_5 = 15 + 4 = 19;$
arithmetic

b. $t_1 = 3; d = 4; t_n = 3 + (n - 1)4 = 4n - 1$

2. a. $t_1 = 64; t_2 = \dfrac{1}{2} \cdot 64 = 32; t_3 = \dfrac{1}{2} \cdot 32 = 16; t_4 = \dfrac{1}{2} \cdot 16 = 8; t_5 = \dfrac{1}{2} \cdot 8 = 4;$
geometric

b. $t_1 = 64; r = \dfrac{1}{2}; t_n = 64 \cdot \left(\dfrac{1}{2}\right)^{n-1}$ or $t_n = 128 \cdot \left(\dfrac{1}{2}\right)^{n}$

Class Exercises 13-2 · page 481

1. $t_1 = 5; t_2 = 5 + 3 = 8; t_3 = 8 + 3 = 11; t_4 = 11 + 3 = 14$

2. $t_1 = 10; t_2 = 10 + 2 = 12; t_3 = 12 + 3 = 15; t_4 = 15 + 4 = 19$

3. $t_1 = 3; t_2 = 2 \cdot 3 = 6; t_3 = 2 \cdot 6 = 12; t_4 = 2 \cdot 12 = 24$

4. $t_1 = 4; t_2 = 2 \cdot 4 - 1 = 7; t_3 = 2 \cdot 7 - 1 = 13; t_4 = 2 \cdot 13 - 1 = 25$

5. $t_1 = 5; t_2 = 2(t_1) + 1 = 2 \cdot 5 + 1 = 11; t_3 = 2 \cdot 11 + 2 = 24; t_4 = 2 \cdot 24 + 3 = 51$

6. $t_1 = 4; t_2 = 16; t_3 = 16 + 4 = 20; t_4 = 20 + 16 = 36$

Written Exercises 13-2 · pages 481–485

1. $t_1 = 6; t_2 = 6 + 4 = 10; t_3 = 10 + 4 = 14; t_4 = 14 + 4 = 18; t_5 = 18 + 4 = 22$

2. $t_1 = 9; t_2 = \dfrac{1}{3} \cdot 9 = 3; t_3 = \dfrac{1}{3} \cdot 3 = 1; t_4 = \dfrac{1}{3} \cdot 1 = \dfrac{1}{3}; t_5 = \dfrac{1}{3} \cdot \dfrac{1}{3} = \dfrac{1}{9}$

3. $t_1 = 1; t_2 = 3(1) - 1 = 2; t_3 = 3(2) - 1 = 5; t_4 = 3(5) - 1 = 14;$
$t_5 = 3(14) - 1 = 41$

4. $t_1 = 4; t_2 = 4^2 - 10 = 6; t_3 = 6^2 - 10 = 26; t_4 = 26^2 - 10 = 666; t_5 = 666^2 - 10 =$
443,546

5. $t_1 = 1; t_2 = 1 + 2(2) - 1 = 4; t_3 = 4 + 2(3) - 1 = 9; t_4 = 9 + 2(4) - 1 = 16;$
$t_5 = 16 + 2(5) - 1 = 25$

6. $t_1 = \dfrac{1}{2}; t_2 = \dfrac{2}{2 + 1}\left(\dfrac{1}{2} + 1\right) = 1; t_3 = \dfrac{3}{3 + 1}(1 + 1) = \dfrac{3}{2}; t_4 = \dfrac{4}{4 + 1}\left(\dfrac{3}{2} + 1\right) = 2;$

$t_5 = \dfrac{5}{5 + 1}(2 + 1) = \dfrac{5}{2}$

7. $t_1 = 2; t_2 = 4; t_3 = 4 + 2 = 6; t_4 = 6 + 4 = 10; t_5 = 10 + 6 = 16$

8. $t_1 = 2; t_2 = 4; t_3 = 4 \cdot 2 = 8; t_4 = 8 \cdot 4 = 32; t_5 = 32 \cdot 8 = 256$

9. $t_1 = 5; t_2 = 8; t_3 = (8 - 5)^2 = 9; t_4 = (9 - 8)^2 = 1; t_5 = (1 - 9)^2 = 64$

10. $t_1 = 7; t_2 = 3; t_3 = 3 - 2 \cdot 7 = -11; t_4 = -11 - 2 \cdot 3 = -17; t_5 = -17 -$
$2(-11) = 5$

11. $t_1 = 6; d = 4; t_n = 6 + (n - 1)4 = 2 + 4n$

12. $t_1 = 9; r = \dfrac{1}{3}; t_n = 9 \cdot \left(\dfrac{1}{3}\right)^{n-1} = 27\left(\dfrac{1}{3}\right)^{n}$

13. $t_1 = 9; t_2 = t_1 + 4$ and $t_3 = t_2 + 4; t_n = t_{n-1} + 4$

14. $t_1 = 81;\ t_2 = \dfrac{1}{3}t_1$ and $t_3 = \dfrac{1}{3}t_2;\ t_n = \dfrac{1}{3}t_{n-1}$

15. $t_1 = 1;\ t_2 = t_1 + 2;\ t_3 = t_2 + 4;\ t_4 = t_3 + 8;\ t_n = t_{n-1} + 2^{n-1}$

16. $t_1 = 1;\ t_2 = t_1 + 2;\ t_3 = t_2 + 4;\ t_4 = t_3 + 6;\ t_5 = t_4 + 8;\ t_n = t_{n-1} + 2(n-1)$

17. $t_1 = 1;\ t_2 = t_1 + 2;\ t_3 = t_2 + 3;\ t_4 = t_3 + 4;\ t_n = t_{n-1} + n$

18. $t_1 = 1;\ t_2 = 2t_1;\ t_3 = 3t_2;\ t_4 = 4t_3;\ t_n = n \cdot t_{n-1}$

19. a. $t_1 = 3;\ t_2 = 5;\ t_3 = 5 - 3 = 2;\ t_4 = 2 - 5 = -3;\ t_5 = -3 - 2 = -5;$
$t_6 = -5 - (-3) = -2;\ t_7 = -2 - (-5) = 3;\ t_8 = 3 - (-2) = 5$

    b. $t_1 = t_7 = t_{13} = \cdots = t_{997};\ t_{1000} = t_4 = -3$

20. a. $t_1 = 4;\ t_2 = 8;\ t_3 = \dfrac{8}{4} = 2;\ t_4 = \dfrac{2}{8} = \dfrac{1}{4};\ t_5 = \dfrac{1}{4} \div 2 = \dfrac{1}{8};\ t_6 = \dfrac{1}{8} \div \dfrac{1}{4} = \dfrac{1}{2};$

$t_7 = \dfrac{1}{2} \div \dfrac{1}{8} = 4;\ t_8 = 4 \div \dfrac{1}{2} = 8$

    b. $t_1 = t_7 = t_{13} = \cdots = t_{997};\ t_{1000} = t_4 = \dfrac{1}{4}$

21. a. $P_n = 1.02P_{n-1} + 50{,}000$

    b. $P_0 = 8{,}500{,}000;\ P_1 = 1.02(8{,}500{,}000) + 50{,}000 = 8{,}720{,}000;$
$P_2 = 1.02(8{,}720{,}000) + 50{,}000 = 8{,}944{,}400;\ P_3 = 1.02(8{,}944{,}400) +$
$50{,}000 = 9{,}173{,}288;\ P_4 = 1.02(9{,}173{,}288) + 50{,}000 = 9{,}406{,}754;$
$P_5 = 1.02(9{,}406{,}754) + 50{,}000 = 9{,}644{,}889$

22. a. 8% decay yields a factor of $1 - 0.08$, or $0.92;\ Q_0 = Q$ and $Q_n = 0.92Q_{n-1}$

    b. For example, let $Q_0 = 100$ and find $Q_n = \dfrac{1}{2}Q_0 = 50;\ Q_1 = 92;\ Q_2 = 84.64;$
$Q_3 = 77.8688;\ Q_4 \approx 71.6393;\ Q_5 \approx 65.9082;\ Q_6 \approx 60.6355;\ Q_7 \approx 55.7847;$
$Q_8 \approx 51.3219;$ about 8 days.

23. If you have an n by n square array, you need to add a row
of n dots and a column of $n + 1$ dots. Thus,
$S_{n+1} = S_n + n + (n + 1) = S_n + 2n + 1.$ The case in
which $n = 3$ is shown in the figure.

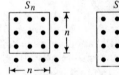

24. When an additional person enters the room, all previous handshakes are counted and
the $n - 1$ people shake hands with the nth person. Therefore, there are $n - 1$
additional handshakes; $H_n = H_{n-1} + n - 1.$

25. a. Label the vertices of the hexagon A to F, and let the new vertex be G. The additional
diagonals are $\overline{BG},\ \overline{CG},\ \overline{DG},\ \overline{EG},$ and $\overline{AF};$ 5 additional diagonals can be drawn.

    b. The nth side has $n - 3$ nonadjacent vertices, and the $(n - 1)$st side is no longer
adjacent to the first side. Thus, a total of $(n - 3) + 1 = n - 2$ additional diagonals
can be drawn; $d_4 = 2$ and $d_n = d_{n-1} + n - 2.$

26. Each new line gives an additional number of regions equal to the total number of lines
drawn; $r_1 = 2$ and $r_n = r_{n-1} + n.$

27. a. No 2 lines are parallel, so each new line intersects all previous lines;
$p_4 = 3 + 3 = 6$; $p_5 = 6 + 4 = 10$.

b. When an nth line is drawn, it intersects the $n - 1$ previous lines;
$p_1 = 0$ and $p_n = p_{n-1} + n - 1$.

28. Answers may vary. Example: You would be more likely to use a recursive definition when you need to find the term following a given list of terms in a sequence. You would be more likely to use an explicit definition when you need to find a term t_n for which n is large.

29. a. 10% yearly investment gives a factor of 1.1. $A_0 = 5000$; $A_n = 1.1A_{n-1} + 2000$.

b. Find $A_2, A_3, A_4, \ldots, A_{17}$ to find $A_{18} = \$118{,}997.93$.

30. a. $A_0 = 400{,}000$; $A_n = \left(1 + \dfrac{0.09}{12}\right)A_{n-1} - 4000 = 1.0075A_{n-1} - 4000$

b. Use a computer or programmable calculator to find $A_1, A_2, A_3, \ldots$. At the end of 185 months, there is \$2113.99 in the account, not enough for another \$4000 withdrawal.

31. a. $M_2 = 3$; $M_3 = 7$

b. To transfer n disks from A to B, first transfer the top $(n - 1)$ disks from A to C. (This takes M_{n-1} moves.) Then move the nth disk to B (one move). Finally, transfer the $(n - 1)$ disks from C to B (M_{n-1} moves). Thus, $M_n = M_{n-1} + 1 + M_{n-1} = 2M_{n-1} + 1$.

c. $M_1 = 1$; $M_2 = 2M_1 + 1 = 2(1) + 1 = 3$; $M_3 = 2M_2 + 1 = 2 \cdot 3 + 1 = 7$. Similarly, using the formula for M_n, $M_4 = 2 \cdot 7 + 1 = 15$, $M_5 = 2 \cdot 15 + 1 = 31$, $M_6 = 2 \cdot 31 + 1 = 63$, $M_7 = 2 \cdot 63 + 1 = 127$, and $M_8 = 2 \cdot 127 + 1 = 255$.

d. Notice that each term is 1 less than a power of 2; $M_n = 2^n - 1$.

e. Using the formula in part (d), $M_{64} = 2^{64} - 1$ seconds $= \dfrac{2^{64} - 1}{3600} \cdot \dfrac{1}{24} \cdot \dfrac{1}{365}$ yr $\approx$ 5.8×10^{11} yr.

32. $r_0 = 1$; $r_1 = 1$ (Newborn rabbits cannot reproduce.); $r_2 = 1 + 1 = 2$; $r_3 = 2 + 1 = 3$; $r_4 = 3 + 2 = 5$; $r_5 = 5 + 3 = 8$; $r_n = r_{n-1} + r_{n-2}$, since r_{n-1} is the number of pairs of adult rabbits n months from now and r_{n-2} is the number of pairs of newborn rabbits n months from now. The sequence is $1, 1, 2, 3, 5, 8, 13, 21, 34, 55, 89, 144, 233$; 233 pairs, or 466 rabbits.

33. a. $t_1 - t_0 = k(t_0 - R)$; $94 - 98 = k(98 - 18)$; $-4 = 80k$; $k = -0.05$. The negative sign indicates that the coffee is cooling.

b. $t_n - t_{n-1} = -0.05(t_{n-1} - 18) = -0.05t_{n-1} + 0.9$; $t_n = 0.95t_{n-1} + 0.9$

c. Use the following program:

```
10 LET T = 98
20 FOR N = 1 TO 10
30 LET T = .95 * T + .9
40 PRINT N, T
50 NEXT N
60 END
```

When the program is run, we find that after 6 min, the temperature is about 76.8°C, and after 7 min, it is about 73.9°C. The temperature of the coffee is less than 75°C after about 7 min.

34. a. $f_n - f_{n-1}$ = the number of newly exposed people on the nth day.

$P - f_{n-1}$ = the number of unexposed people on the $(n - 1)$st day.

The given equation expresses the fact that these quantities are proportional.

b. $f_1 - f_0 = k(P - f_0)$; $220 - 100 = k(2500 - 100)$; $k = \dfrac{120}{2400} = 0.05$; $f_n - f_{n-1} = 0.05(2500 - f_{n-1})$; $f_n = 0.95f_{n-1} + 125$

c. Use the following program:

```
10 LET F = 100
20 FOR N = 1 TO 200
30 LET F = .95 * F + 125
40 PRINT N, F
50 NEXT N
60 END
```

When the program is run, we find that $f_{166} = 2500$; it will take about 166 days.

35. a. Let P_n and P_{n-1} represent the population at the time of the nth and $(n - 1)$st generations, respectively. Then $P_n - P_{n-1}$ represents the population increase at the time of the nth generation. Because this increase is proportional to the population of the previous generation, $P_n - P_{n-1} = kP_{n-1}$.

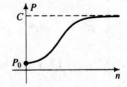

b. See right.

c. Answers may vary. For example, the change in population from one generation to the next is jointly proportional to the current population and the "room to grow" (that is, the difference between the carrying capacity and the current population).

d. Use the following program:

```
10 INPUT "INITIAL POPULATION: ";P
20 INPUT "CARRYING CAPACITY: ";C
30 INPUT "PROPORTIONALITY CONSTANT: ";K
40 PRINT "GENERATION", "POPULATION"
50 LET G = 0
60 LET P = P + K * P * (C - P)
70 LET G = G + 1
80 PRINT G,P
90 IF P < C THEN GOTO 60
100 END
```

With an initial population of 50, $C = 500$, and $k = 0.002$, a run of the program gives the following results:

GENERATION	POPULATION
1	95
2	171.95
3	284.766395
4	407.348991
5	482.831581
6	499.410491
7	499.999305
8	500

(Other initial populations give similar results.) Note that the population growth is rapid at first, then slows at it approaches the carrying capacity. These results are consistent with the graph in part (b).

Class Exercises 13-3 · page 488

1. Answers will vary. Example: $1, 3, 5, \ldots$ is an arithmetic sequence and $1 + 3 + 5 + \cdots$ is an arithmetic series.

2. The formula for the sum of a finite geometric series; $1024 = 2^{10}$, so $n = 11$;
$$S_n = \frac{1(1 - 2048)}{1 - 2} = 2047.$$

3. **a.** Neither   **b.** $S_0 = 0$; $S_n = S_{n-1} + t_n$ where $t_n = 2^n + 1$ ($n \geq 1$)

   c. $t_n = 32{,}769 = 2^n + 1$, so $2^n = 32{,}768$, and $n = 15$. Change lines 20 and 30 as follows:

$$20 \text{ FOR N } = 1 \text{ TO } 15 \qquad 30 \text{ LET T } = 2 \text{ ^ N } + 1$$

Written Exercises 13-3 · pages 489–492

1. $S_{10} = \dfrac{10(t_1 + t_{10})}{2} = 5(3 + 39) = 210$ 2. $S_{200} = \dfrac{200(18 + 472)}{2} = 49{,}000$

3. $t_{50} = 5 + (50 - 1)5 = 250$; $S_{50} = \dfrac{50(5 + 250)}{2} = 6375$

4. $t_{25} = 17 + (25 - 1)8 = 209$; $S_{25} = \dfrac{25(17 + 209)}{2} = 2825$

5. $t_1 = 5 + 3 \cdot 1 = 8$ and $t_{12} = 5 + 3 \cdot 12 = 41$; $S_{12} = \dfrac{12(8 + 41)}{2} = 294$

6. $t_3 = 11 = 5 + 2d$; $d = 3$; $t_{40} = 5 + (40 - 1)3 = 122$; $S_{40} = \dfrac{40(5 + 122)}{2} = 2540$

7. $S_{1000} = \dfrac{1000(1 + 1000)}{2} = 500{,}500$

8. $d = 7 - 3 = 4$; $99 = 3 + (n - 1)4$; $4n = 100$; $n = 25$; $S_{25} = \dfrac{25(3 + 99)}{2} = 1275$

9. **a.** $t_1 = 8$ and $r = \dfrac{1}{2}$, so $S_8 = \dfrac{8\left(1 - \left(\frac{1}{2}\right)^8\right)}{1 - \frac{1}{2}} = \dfrac{8\left(1 - \frac{1}{256}\right)}{\frac{1}{2}} = 16\left(\dfrac{255}{256}\right) = \dfrac{255}{16}$

   b. $t_1 = 8$ and $r = -\dfrac{1}{2}$, so $S_8 = \dfrac{8\left(1 - \left(-\frac{1}{2}\right)^8\right)}{1 + \frac{1}{2}} = \dfrac{8\left(1 - \frac{1}{256}\right)}{\frac{3}{2}} = \dfrac{16}{3} \cdot \dfrac{255}{256} = \dfrac{85}{16}$

10. In $1 + \dfrac{1}{3} + \cdots, r = \dfrac{1}{3}$, so $S_{10} = \dfrac{1\left(1 - \left(\frac{1}{3}\right)^{10}\right)}{1 - \frac{1}{3}} = \dfrac{1 - \left(\frac{1}{3}\right)^{10}}{\frac{2}{3}} = \dfrac{3}{2}\left(1 - \left(\dfrac{1}{3}\right)^{10}\right).$

   In $1 - \dfrac{1}{3} + \cdots, r = -\dfrac{1}{3}$, so $S_{10} = \dfrac{1\left(1 - \left(-\frac{1}{3}\right)^{10}\right)}{1 + \frac{1}{3}} = \dfrac{1 - \left(\frac{1}{3}\right)^{10}}{\frac{4}{3}} = \dfrac{3}{4}\left(1 - \left(\dfrac{1}{3}\right)^{10}\right).$

   Thus, $\dfrac{3}{2}\left(1 - \left(\dfrac{1}{3}\right)^{10}\right) = 2 \cdot \dfrac{3}{4}\left(1 - \left(\dfrac{1}{3}\right)^{10}\right).$

11. $t_1 = 1$ and $r = 2$; $S_n = \dfrac{1(1 - 2^n)}{1 - 2} = \dfrac{1 - 2^n}{-1} = 2^n - 1$

12. $t_1 = 9$ and $r = 10$; $S_n = \dfrac{9(1 - 10^n)}{1 - 10} = \dfrac{9(1 - 10^n)}{-9} = -1(1 - 10^n) = 10^n - 1$

13. $64 = \sqrt{2} \cdot (\sqrt{2})^{n-1} = (\sqrt{2})^n$; $2^6 = 2^{n/2}$; $6 = \dfrac{n}{2}$; $n = 12$; $S_{12} = \dfrac{\sqrt{2}(1 - (\sqrt{2})^{12})}{1 - \sqrt{2}} =$

$\dfrac{\sqrt{2}(1 - 64)}{1 - \sqrt{2}} = \dfrac{-63\sqrt{2}}{1 - \sqrt{2}} = \dfrac{63\sqrt{2}}{\sqrt{2} - 1}$

14. $S_{15} = \dfrac{1(1 - (3^{1/5})^{15})}{1 - 3^{1/5}} = \dfrac{1 - 3^3}{1 - 3^{1/5}} = \dfrac{-26}{1 - 3^{1/5}} = \dfrac{26}{3^{1/5} - 1}$

15. $t_1 = 1$, $t_n = n$, and $d = 1$; $S_n = \dfrac{n(1 + n)}{2} = \dfrac{n(n + 1)}{2}$

16. a. $S_1 = 1$; $S_2 = 1 + 3 = 4$; $S_3 = 1 + 3 + 5 = 9$; $S_4 = 1 + 3 + 5 + 7 = 16$

b. $S_n = n^2$; $t_1 = 1$, $t_2 = 3$, $t_3 = 5 = 2 \cdot 3 - 1$, and $t_n = 2n - 1$. Thus $S_n =$
$\dfrac{n(1 + 2n - 1)}{2} = \dfrac{n(2n)}{2} = n^2$.

17. $2 + 4 + 6 + \cdots + 2n$; $t_n = 2n$; $S_0 = 0$; $S_n = S_{n-1} + 2n$

18. $1 + 3 + 9 + \cdots + 3^{n-1}$; $t_n = 3^{n-1}$; $S_0 = 0$; $S_n = S_{n-1} + 3^{n-1}$

19. $8 + 12 + 18 + \cdots + 8\left(\dfrac{3}{2}\right)^{n-1}$; $t_n = 8\left(\dfrac{3}{2}\right)^{n-1}$; $S_0 = 0$; $S_n = S_{n-1} + 8\left(\dfrac{3}{2}\right)^{n-1}$

20. $50 + 47 + 44 + \cdots$; $t_n = 50 + (n - 1)(-3) = 53 - 3n$; $S_0 = 0$; $S_n = S_{n-1} + 53 - 3n$

21. Change lines 20 and 30 as follows:

```
20 FOR N = 1 TO 10      30 LET T = N^4
```
The sum is 25,333.

22. Change lines 20 and 30 as follows:

```
20 FOR N = 1 TO 15      30 LET T = N^N
```
The sum is $4.49317984 \times 10^{17}$.

23. Change lines 20 and 30 as follows:

```
20 FOR N = 10 TO 1000      30 LET T = N^2
```
The sum is 33,835,000.

24. Change lines 20 and 30 as follows:

```
20 FOR N = 5 TO 200 STEP 5      30 LET T = N^0.5
```
The sum is 383.744568.

25. a. A table is printed giving the sum of the first n terms for each number n from 1 to 20.

b. `20 FOR N = 1 TO 25      30 LET T = 2^(-N)`

c. $S_{25} = \dfrac{\dfrac{1}{2}(1 - 2^{-25})}{1 - \dfrac{1}{2}} = 1 - 2^{-25} \approx 0.99999997$

d. 1

26. a. Change lines 20 and 30 as follows:

     20 FOR N = 1 TO 25   30 LET T = 1/3^N

b. $S_{25} = \dfrac{\frac{1}{3}\left(1 - \left(\frac{1}{3}\right)^{25}\right)}{1 - \frac{1}{3}} = \dfrac{1 - \left(\frac{1}{3}\right)^{25}}{2} \approx \dfrac{1}{2}$

c. $\dfrac{1}{2}$

27. $S_n = n^2 + 4n$; $t_n = S_n - S_{n-1}$; $t_1 = S_1 = 5$; $t_2 = S_2 - S_1 = 12 - 5 = 7$;
$t_3 = S_3 - S_2 = 21 - 12 = 9$

28. a. $S_n = 2n^2$; $t_n = S_n - S_{n-1}$; $t_1 = S_1 = 2$; $t_2 = S_2 - S_1 = 8 - 2 = 6$;
$t_3 = S_3 - S_2 = 18 - 8 = 10$

    b. $S_n - S_{n-1}$ forms an arithmetic series with $t_1 = 2$ and $d = 4$; the nth term is
$2 + (n - 1)4 = 4n - 2$. (Or: $S_n - S_{n-1} = 2n^2 - 2(n - 1)^2 = 4n - 2$)

29. $t_1 = 3$ and $t_n = 999$; $999 = 3 + (n - 1)3$; $996 = 3(n - 1)$; $n - 1 = 332$; $n = 333$;
$S_{333} = \dfrac{333(3 + 999)}{2} = 333 \cdot 501 = 166{,}833$

30. $t_1 = 102$ and $t_n = 996$; $996 = 102 + (n - 1)6$; $894 = 6(n - 1)$; $n - 1 = 149$;
$n = 150$; $S_{150} = \dfrac{150(102 + 996)}{2} = 75 \cdot 1098 = 82{,}350$

31. $t_1 = 103$, $t_n = 993$, and $d = 10$; $993 = 103 + (n - 1)10$; $10n = 900$; $n = 90$;
$S_{90} = \dfrac{90(103 + 993)}{2} = 45 \cdot 1096 = 49{,}320$

32. $t_1 = 5$, $t_2 = 15$, $t_n = 395$, and $d = 10$; $395 = 5 + (n - 1)10$; $10n = 400$; $n = 40$;
$S_{40} = \dfrac{40(5 + 395)}{2} = 20 \cdot 400 = 8000$

33. $1 - 3 + 5 - 7 + 9 - 11 + \cdots$ is equivalent to the sum of the two series
$(1 + 5 + 9 + \cdots + 1001) + (-3 - 7 - 11 - \cdots - 999)$. For the first series,
$1001 = 1 + (n - 1)4$; $n = 251$; $S_{251} = \dfrac{251(1 + 1001)}{2} = 125{,}751$. For the second

series, $-999 = -3 + (n - 1)(-4)$; $n = 250$; $S_{250} = \dfrac{250(-3 - 999)}{2} = -125{,}250$; the

required sum is $125{,}751 + (-125{,}250) = 501$.

34. From Ex. 15, the sum of the first 299 integers is $\dfrac{299(300)}{2} = 44{,}850$. Let

$S_n = 3 + 6 + 9 + \cdots + 297$; $297 = 3 + (n - 1)3$; $n = 99$; $S_{99} = \dfrac{99(3 + 297)}{2} =$

$14{,}850$. The required sum is $44{,}850 - 14{,}850 = 30{,}000$.

35. $t_1 = 5$, $t_2 = 25$, $t_3 = 125$; $r = 5$, so $S_{12} = \dfrac{5(1 - 5^{12})}{1 - 5} = \dfrac{5(1 - 244{,}140{,}625)}{-4} =$

$305{,}175{,}780$; the process fails because (1) it requires the cooperation of millions of
people, and (2) the number of people who have not yet received letters is eventually
exhausted.

36. $t_1 = 1$, $t_2 = 2$, $t_3 = 4$; $r = 2$; $S_{14} = \dfrac{1(1 - 2^{14})}{1 - 2} = \dfrac{1 - 16{,}384}{-1} = 16{,}383$, or \$163.83

37. a. $t_1 = 2$ and $t_2 = 4$, so $r = \dfrac{4}{2} = 2$; $S_{10} = \dfrac{2(1 - 2^{10})}{1 - 2} = -2(1 - 1024) = 2046$

ancestors

b. $1{,}000{,}000 = \dfrac{2(1 - 2^n)}{1 - 2}$; $1{,}000{,}000 = -2(1 - 2^n)$; $2^n - 1 = 500{,}000$; $2^n = 500{,}001$;

$n \log 2 = \log 500{,}001$; $n = \dfrac{\log 500{,}001}{\log 2} \approx 19$ generations

38. a. $t_1 = 40{,}000$ and $r = 1.1$; $S_{10} = \dfrac{40{,}000(1 - (1.1)^{10})}{1 - 1.1} = -400{,}000(1 - (1.1)^{10}) \approx$

$\$637{,}497$

b. $1{,}000{,}000 = \dfrac{40{,}000(1 - (1.1)^n)}{1 - 1.1} = -400{,}000(1 - 1.1)^n)$; $2.5 = (1.1)^n - 1$;

$(1.1)^n = 3.5$; $n \log 1.1 = \log 3.5$; $n = \dfrac{\log 3.5}{\log 1.1} \approx 13.1$; the doctor must work 14 yr to

exceed a million dollars.

39. a. $T_1 = \dfrac{1(1 + 1)}{2} = 1$; $T_2 = \dfrac{2(2 + 1)}{2} = 3$; $T_3 = \dfrac{3(3 + 1)}{2} = 6$; $T_4 = \dfrac{4(4 + 1)}{2} = 10$;

$T_5 = \dfrac{5(5 + 1)}{2} = 15$

b. $T_1 + T_2 = 1 + 3 = 4 = 2^2$; $T_2 + T_3 = 3 + 6 = 9 = 3^2$; $T_3 + T_4 = 6 + 10 =$

$16 = 4^2$; conjecture: $T_n + T_{n+1} = (n + 1)^2$. Proof: $T_n + T_{n+1} = \dfrac{n(n + 1)}{2} +$

$\dfrac{(n + 1)[(n + 1) + 1]}{2} = \dfrac{(n + 1)[n + (n + 2)]}{2} = \dfrac{(n + 1)(2n + 2)}{2} = (n + 1)^2$.

40. $S_1 = 1$; $S_2 = 1 + 8 = 9$; $S_3 = 1 + 8 + 27 = 36$; $S_4 = 1 + 8 + 27 + 64 = 100$;

$S_n = \left(\dfrac{n(n + 1)}{2}\right)^2$

41. Since there are n^2 numbers in an n-by-n square, the total sum is $\dfrac{n^2(n^2 + 1)}{2}$, by the

formula in Ex. 15. There are n rows, so the sum in each row is $M(n) =$

$\dfrac{n^2(n^2 + 1)}{2} \div n = \dfrac{n^2(n^2 + 1)}{2n} = \dfrac{n(n^2 + 1)}{2}$. $M(3) = \dfrac{3(10)}{2} = 15$ and $M(4) = \dfrac{4(17)}{2} = 34$,

as required.

42. $t_1 = 1000$ and $r = 1.12$; $S_{10} = \dfrac{1000[1 - (1.12)^{10}]}{1 - 1.12} \approx \$17{,}548.74$

43. The first investment will yield $P(1 + r)^{n-1}$ at the time of the nth investment;

$P(1 + r)^{n-1-1} = P(1 + r)^{n-2}$ is the amount from the second investment; thus the series

is geometric with $t_1 = P(1 + r)^{n-1}$ and common ratio $= \dfrac{t_2}{t_1} = P(1 + r)^{n-2} \div$

$P(1 + r)^{n-1} = (1 + r)^{-1}$; hence $S_n = \dfrac{P(1 + r)^{n-1}(1 - [(1 + r)^{-1}]^n)}{1 - (1 + r)^{-1}} =$

$\dfrac{P(1 + r)^{n-1}[1 - (1 + r)^{-n}]}{1 - (1 + r)^{-1}} \cdot \dfrac{1 + r}{1 + r} = \dfrac{P(1 + r)^n[1 - (1 + r)^{-n}]}{(1 + r) - 1} = \dfrac{P[(1 + r)^n - 1]}{r}$ dollars.

44. a. $A_1 = (1 + r)A_0 - P = (1 + r)A - P$; $A_2 = (1 + r)A_1 - P =$

$(1 + r)[(1 + r)A - P] - P = (1 + r)^2 A - [(1 + r) + 1]P$;

$A_3 = (1 + r)A_2 - P = (1 + r)[(1 + r)^2 A - [(1 + r) + 1]P] - P =$

$(1 + r)^3 A - [(1 + r)^2 + (1 + r) + 1]P$

(Continued)

b. Rearrange the terms in the bracketed series as $1 + (1 + r) + (1 + r)^2 + \cdots + (1 + r)^{n-1}$. This is a geometric series with $t_1 = 1$ and common ratio $1 + r$;

$$S_n = \frac{1[1 - (1 + r)^n]}{1 - (1 + r)} = \frac{1 - (1 + r)^n}{-r} = \frac{(1 + r)^n - 1}{r}. \text{ Thus, substituting,}$$

$$A_n = (1 + r)^n A - \left[\frac{(1 + r)^n - 1}{r}\right] P.$$

c. After paying n installments, the loan is paid, and so $A_n = 0$. Thus, $0 = (1 + r)^n A -$

$$\left[\frac{(1 + r)^n - 1}{r}\right] P; \, P = (1 + r)^n A \div \frac{(1 + r)^n - 1}{r} = (1 + r)^n A \cdot \frac{r}{(1 + r)^n - 1} =$$

$$\frac{Ar(1 + r)^n}{(1 + r)^n - 1}.$$

45. a. $P = \dfrac{Ar(1 + r)^n}{(1 + r)^n - 1} = \dfrac{10{,}000(0.01)(1.01)^{60}}{(1.01)^{60} - 1} \approx \$222.44$

b. $A_2 = A_1 - R_1 = 9877.56 - 123.66 = 9753.90; \, I_2 = 0.01A_2 = 97.54;$
$R_2 = P - I_2 = 222.44 - 97.54 = 124.90$

c. $A_n = A_{n-1} - R_{n-1}; \, R_{n-1} = 222.44 - I_{n-1}; \, I_{n-1} = 0.01(A_{n-1});$ thus,
$A_n = A_{n-1} - (222.44 - 0.01(A_{n-1})) = 1.01A_{n-1} - 222.44.$

d. We need to prove that all values of R_n form a geometric sequence.
$A_n = A_{n-1} - R_{n-1}$, so $R_{n-1} = A_{n-1} - A_n$. Since $R_n = 222.44 - I_n$ and
$I_n = 0.01A_n$, $222.44 = R_n + 0.01A_n$. From part (c), we have $A_n = 1.01A_{n-1} -$
$222.44 = 1.01A_{n-1} - (R_n + 0.01A_n) = 1.01A_{n-1} - R_n - 0.01A_n$; thus,
$R_n = 1.01A_{n-1} - 1.01A_n = 1.01(A_{n-1} - A_n) = 1.01R_{n-1}$. Therefore the values of
R_n form a geometric sequence with ratio 1.01.

e.
```
10 PRINT "N", "A", "I", "R"
20 LET A  =  10000
30 LET R  =  0
40 FOR N  =  0 TO 60
50 LET A  =  A - R
60 LET I  =  0.01 * A
70 LET R  =  222.44 - I
80 PRINT N, INT(100 * A + 0.5)/100,
      INT(100 * I + 0.05)/100, INT(100 * R + 0.5)/10
90 NEXT N
100 END
```

Class Exercises 13-4 · page 496

1. $\dfrac{n}{n + 1} = \dfrac{1}{1 + \dfrac{1}{n}}; \, \lim\limits_{n \to \infty} \dfrac{n}{n + 1} = 1$ **2.** $\dfrac{n^2 - 1}{n^2} = \dfrac{1 - \dfrac{1}{n^2}}{1}; \, \lim\limits_{n \to \infty} \dfrac{n^2 - 1}{n^2} = 1$

3. $\dfrac{2n + 1}{3n + 1} = \dfrac{2 + \dfrac{1}{n}}{3 + \dfrac{1}{n}}; \, \lim\limits_{n \to \infty} \dfrac{2n + 1}{3n + 1} = \dfrac{2}{3}$ **4.** $\dfrac{8n^2 - 3n}{5n^2 + 7} = \dfrac{8 - \dfrac{3}{n}}{5 + \dfrac{7}{n^2}}; \, \lim\limits_{n \to \infty} \dfrac{8n^2 - 3n}{5n^2 + 7} = \dfrac{8}{5}$

5. As $n \to \infty$, $\dfrac{1}{n} \to 0; \, \lim\limits_{n \to \infty} \cos\left(\dfrac{1}{n}\right) = \cos 0 = 1.$

6. From Ex. 5, $\lim\limits_{n \to \infty} \cos\left(\dfrac{1}{n}\right) = 1$, so $\lim\limits_{n \to \infty} \log\left[\cos\left(\dfrac{1}{n}\right)\right] = \log 1 = 0$.

7. By the theorem on p. 494, $\lim\limits_{n \to \infty} (0.999)^n = 0$.

8. Each term is greater than the preceding one, so $\lim\limits_{n \to \infty} (1.001)^n = \infty$.

9. $\lim\limits_{n \to \infty} \dfrac{n^4}{2n + 1} = \lim\limits_{n \to \infty} \dfrac{n^3}{2 + \dfrac{1}{n}} \approx \lim\limits_{n \to \infty} \dfrac{1}{2}n^3 = \infty$

10. $\lim\limits_{n \to \infty} \dfrac{n^2 + 9{,}999{,}999}{n^3} = \lim\limits_{n \to \infty} \dfrac{1 + \dfrac{9{,}999{,}999}{n^2}}{n} \approx \lim\limits_{n \to \infty} \dfrac{1 + 0}{n} = 0$

11. No; the terms do not "home in" on a single value.

Written Exercises 13-4 · pages 496–498

1. $\dfrac{n + 5}{n} = \dfrac{1 + \dfrac{5}{n}}{1}$; thus, $\lim\limits_{n \to \infty} \dfrac{n + 5}{n} = 1$. **2.** $\dfrac{n^2 + 1}{n^2} = \dfrac{1 + \dfrac{1}{n^2}}{1}$; thus, $\lim\limits_{n \to \infty} \dfrac{n^2 + 1}{n^2} = 1$.

3. As $n \to \infty$, $\dfrac{(-1)^n}{n} \to 0$; thus, $\lim\limits_{n \to \infty}\left[1 + \dfrac{(-1)^n}{n}\right] = 1$.

4. $\dfrac{4n - 3}{2n + 1} = \dfrac{4 - \dfrac{3}{n}}{2 + \dfrac{1}{n}}$; thus, $\lim\limits_{n \to \infty} \dfrac{4n - 3}{2n + 1} = \dfrac{4}{2} = 2$.

5. $\dfrac{3n^2 + 5n}{8n^2} = \dfrac{3 + \dfrac{5}{n}}{8}$; thus, $\lim\limits_{n \to \infty} \dfrac{3n^2 + 5n}{8n^2} = \dfrac{3}{8}$.

6. $\dfrac{2n^4}{6n^5 + 7} = \dfrac{\dfrac{2}{n}}{6 + \dfrac{7}{n^5}}$; thus, $\lim\limits_{n \to \infty} \dfrac{2n^4}{6n^5 + 7} = \dfrac{0}{6} = 0$.

7. As $n \to \infty$, $\dfrac{1}{n} \to 0$ and $\tan \dfrac{1}{n} \to \tan 0 = 0$; thus, $\lim\limits_{n \to \infty} \tan\left(\dfrac{1}{n}\right) = 0$.

8. As $n \to \infty$, $\dfrac{1}{n} \to 0$ and $\sec \dfrac{1}{n} \to \sec 0 = 1$; thus, $\lim\limits_{n \to \infty} \sec\left(\dfrac{1}{n}\right) = 1$.

9. $\dfrac{\sqrt{n}}{n + 1} = \dfrac{\dfrac{1}{\sqrt{n}}}{1 + \dfrac{1}{n}}$; thus, $\lim\limits_{n \to \infty} \dfrac{\sqrt{n}}{n + 1} = \dfrac{0}{1} = 0$.

10. $\dfrac{5n^{2/3} - 8n}{6n - 1} = \dfrac{\dfrac{5}{\sqrt[3]{n}} - 8}{6 - \dfrac{1}{n}}$; thus, $\lim\limits_{n \to \infty} \dfrac{5n^{2/3} - 8n}{6n - 1} = \dfrac{-8}{6} = -\dfrac{4}{3}$.

11. As $n \to \infty$, $\dfrac{n + 1}{n} = \dfrac{1 + \dfrac{1}{n}}{1}$ approaches 1. Thus, $\lim\limits_{n \to \infty} \log\left(\dfrac{n + 1}{n}\right) = \log 1 = 0$.

12. $\log \sqrt[n]{10} = \log 10^{1/n} = \dfrac{1}{n} \log 10 = \dfrac{1}{n} \cdot 1 = \dfrac{1}{n}$; as $n \to \infty$, $\dfrac{1}{n} \to 0$, so $\lim\limits_{n \to \infty} \log \sqrt[n]{10} = 0$.

13. $t_n = \dfrac{1}{3}\left(-\dfrac{1}{3}\right)^{n-1} = (-1)^{n-1}\left(\dfrac{1}{3}\right)^{n}$; as $n \longrightarrow \infty$, $\left(\dfrac{1}{3}\right)^{n} \longrightarrow 0$; thus, $\lim\limits_{n \to \infty}\ (-1)^{n-1}\left(\dfrac{1}{3}\right)^{n} = 0$.

14. The limit does not exist because $|t_n|$ increases without bound.

15. $t_n = (-1)^{n+1}\cdot\dfrac{n+2}{n+1}$; the even-numbered terms have limit -1 and the odd-numbered terms have limit 1, so the limit of the sequence does not exist.

16. $t_n = (-1)^{n+1}\cdot\dfrac{n}{10^n} = (-1)^{n+1}\cdot\left(\dfrac{n^{1/n}}{10}\right)^{n}$; for all n, $\dfrac{n^{1/n}}{10} < 1$, so $\lim\limits_{n \to \infty}\ (-1)^{n+1}\cdot\dfrac{n}{10^n} = 0$.

17. $t_1 = \cos\dfrac{\pi}{2} = 0$; $t_2 = \cos\pi = -1$; $t_3 = \cos\dfrac{3\pi}{2} = 0$; $t_4 = \cos 2\pi = 1$; the sequence $0, -1, 0, 1, 0, -1, 0, 1, \ldots$ has no limit.

18. $t_1 = \sin\pi = 0$; $t_2 = \sin 2\pi = 0$; $t_3 = \sin 3\pi = 0$; the sequence $0, 0, 0, \ldots$ has the limit 0.

19. $\dfrac{5n^{5/2} - 7n}{n^2 + 10n} = \dfrac{5n^{1/2} - \dfrac{7}{n}}{1 + \dfrac{10}{n}}$; as $n \longrightarrow \infty$, $\dfrac{7}{n}$ and $\dfrac{10}{n}$ approach 0; thus, as $n \longrightarrow \infty$,

$\dfrac{5n^{1/2} - \dfrac{7}{n}}{1 + \dfrac{10}{n}} \longrightarrow 5n^{1/2}$, and so $\lim\limits_{n \to \infty}\dfrac{5n^{5/2} - 7n}{n^2 + 10n} = \lim\limits_{n \to \infty} 5n^{1/2} = \infty$.

20. $\dfrac{5n}{n^{1/2} - 3} = \dfrac{5n^{1/2}}{1 - \dfrac{3}{n^{1/2}}}$; as $n \longrightarrow \infty$, $\dfrac{3}{n^{1/2}} \longrightarrow 0$; thus, as $n \longrightarrow \infty$, $\dfrac{5n^{1/2}}{1 - \dfrac{3}{n^{1/2}}} \longrightarrow 5n^{1/2}$, and so

$\lim\limits_{n \to \infty}\dfrac{5n}{n^{1/2} - 3} = \lim\limits_{n \to \infty} 5n^{1/2} = \infty$.

21. As $n \longrightarrow \infty$, $\dfrac{1}{n} \longrightarrow 0$, and $\log\left(\dfrac{1}{n}\right) \longrightarrow -\infty$. Thus, $\lim\limits_{n \to \infty}\log\left(\dfrac{1}{n}\right) = -\infty$.

22. For all n, $-1 \le \sin n \le 1$; also, $\lim\limits_{n \to \infty}\dfrac{1}{n} = 0$; thus, $\lim\limits_{n \to \infty}\dfrac{\sin n}{n} = 0$.

23. For all n, $\cos n\pi = \pm 1$. Also, $\lim\limits_{n \to \infty}\dfrac{1}{n} = 0$. Thus, $\lim\limits_{n \to \infty}\dfrac{\cos (n\pi)}{n} = 0$.

24. $\lim\limits_{n \to \infty} e^{-n} = \lim\limits_{n \to \infty}\dfrac{1}{e^n} = \lim\limits_{n \to \infty}\left(\dfrac{1}{e}\right)^{n}$; $\dfrac{1}{e} < 1$, so $\lim\limits_{n \to \infty}\left(\dfrac{1}{e}\right)^{n} = 0$.

25. For all n, $\tan\left(\dfrac{\pi}{4} + n\pi\right) = \dfrac{\tan\dfrac{\pi}{4} + \tan n\pi}{1 - \tan\dfrac{\pi}{4}\tan n\pi} = \dfrac{1 + 0}{1 - 0} = 1$; the sequence $1, 1, 1, \ldots,$ has the limit 1.

26. The terms of the sequence $2, 4, 8, 16, 32, 64, \ldots$ increase without bound as $n \longrightarrow \infty$; thus, $\lim\limits_{n \to \infty} 2^n = \infty$.

27. By Ex. 8, $\lim\limits_{n \to \infty}\sec\left(\dfrac{1}{n}\right) = 1$; thus, $\lim\limits_{n \to \infty}\log\left[\sec\left(\dfrac{1}{n}\right)\right] = \log 1 = 0$.

28. When n is odd, $t_n = -1 - 1 = -2$ and when n is even, $t_n = 1 - 1 = 0$; thus, the limit of the sequence does not exist.

29. $\dfrac{\sqrt{n+1}}{\sqrt{n-1}} = \sqrt{\dfrac{n+1}{n-1}} = \sqrt{\dfrac{1+\dfrac{1}{n}}{1-\dfrac{1}{n}}}$; thus, $\displaystyle\lim_{n\to\infty}\dfrac{\sqrt{n+1}}{\sqrt{n-1}} = \sqrt{1} = 1.$

30. $\dfrac{\sqrt[3]{8n^2-5n+1}}{\sqrt[3]{n^2+7n-3}} = \sqrt[3]{\dfrac{8n^2-5n+1}{n^2+7n-3}} = \sqrt[3]{\dfrac{8-\dfrac{5}{n}+\dfrac{1}{n^2}}{1+\dfrac{7}{n}-\dfrac{3}{n^2}}}$; thus,

$\displaystyle\lim_{n\to\infty}\dfrac{\sqrt[3]{8n^2-5n+1}}{\sqrt[3]{n^2+7n-3}} = \sqrt[3]{8} = 2.$

31. a. $t_1 = \left(\dfrac{1}{3}\right)^{1-1} = 1$ and $t_2 = \left(\dfrac{1}{3}\right)^{2-1} = \dfrac{1}{3}$, so $r = \dfrac{1}{3}$; $S_n = \dfrac{1\left[1-\left(\dfrac{1}{3}\right)^n\right]}{1-\dfrac{1}{3}} = $

$\dfrac{1-\left(\dfrac{1}{3}\right)^n}{\dfrac{2}{3}} = \dfrac{3}{2}\left[1-\left(\dfrac{1}{3}\right)^n\right].$

b. $\displaystyle\lim_{n\to\infty}\left(\dfrac{1}{3}\right)^n = 0$, so $1-\left(\dfrac{1}{3}\right)^n$ approaches 1 as $n\longrightarrow\infty$; hence, $\displaystyle\lim_{n\to\infty} S_n = $

$\displaystyle\lim_{n\to\infty}\dfrac{3}{2}\left[1-\left(\dfrac{1}{3}\right)^n\right] = \dfrac{3}{2}\cdot 1 = \dfrac{3}{2}.$

32. $t_1 + t_1 r + \cdots + t_1 r^{n-1} = \dfrac{t_1(1-r^n)}{1-r}$; if $|r|<1$, then $\displaystyle\lim_{n\to\infty} r^n = 0$, so the sum of the

infinite series is $\displaystyle\lim_{n\to\infty}\dfrac{t_1(1-r^n)}{1-r} = \dfrac{t_1(1-0)}{1-r} = \dfrac{t_1}{1-r}.$

33. $\left(1+\dfrac{2}{n}\right)^n\left(1+\dfrac{1}{\frac{n}{2}}\right)^n = \left[\left(1+\dfrac{1}{\frac{n}{2}}\right)^{n/2}\right]^2$; as $n\longrightarrow\infty$, $\dfrac{n}{2}\longrightarrow\infty$, so, by the definition

of e, $\displaystyle\lim_{n\to\infty}\left(1+\dfrac{2}{n}\right)^n = \lim_{n\to\infty}\left[\left(1+\dfrac{1}{\frac{n}{2}}\right)^{n/2}\right]^2 = e^2.$

34. a. $\displaystyle\lim_{n\to\infty}\left(1+\dfrac{3}{n}\right)^n = \lim_{n\to\infty}\left(1+\dfrac{1}{\frac{n}{3}}\right)^n = \lim_{n\to\infty}\left[\left(1+\dfrac{1}{\frac{n}{3}}\right)^{n/3}\right]^3 = e^3$, since as $n\longrightarrow\infty$,

$\dfrac{n}{3}\longrightarrow\infty.$

b. $\left(1+\dfrac{1}{2n}\right)^n = \left[\left(1+\dfrac{1}{2n}\right)^{2n}\right]^{1/2}$; as $n\longrightarrow\infty$, $2n\longrightarrow\infty$, so $\displaystyle\lim_{n\to\infty}\left[\left(1+\dfrac{1}{2n}\right)^{2n}\right]^{1/2} = e^{1/2}$,

or $\sqrt{e}$.

35. a. $A_n = \dfrac{1}{n^4}[1^3 + 2^3 + 3^3 + \cdots + n^3] = \dfrac{1}{n^4}\left[\dfrac{n(n+1)}{2}\right]^2 = \dfrac{n^2(n+1)^2}{4n^4} = \dfrac{n^2+2n+1}{4n^2}$

b. $A_n = \dfrac{n^2+2n+1}{4n^2} = \dfrac{1+\dfrac{2}{n}+\dfrac{1}{n^2}}{4}$; as n becomes very large, $A_n\longrightarrow\dfrac{1+0+0}{4}$, or $\dfrac{1}{4}$.

c. $A = \displaystyle\lim_{n\to\infty} A_n = \lim_{n\to\infty}\dfrac{n^2+2n+1}{4n^2} = \dfrac{1}{4}.$

36. Approximate the area A under the curve $y = x^2$ between $x = 0$ and $x = 1$ by adding the areas of n rectangles.

$$A \approx A_n = \left(\frac{1}{n}\right)^2 \cdot \frac{1}{n} + \left(\frac{2}{n}\right)^2 \cdot \frac{1}{n} + \left(\frac{3}{n}\right)^2 \cdot \frac{1}{n} + \cdots + \left(\frac{n}{n}\right)^2 \cdot \frac{1}{n} =$$

$$\frac{1}{n^3}[1^2 + 2^2 + 3^2 + \cdots + n^2] = \frac{1}{n^3}\left[\frac{n(n+1)(2n+1)}{6}\right] = \frac{2n^2 + 3n + 1}{6n^2} = \frac{2 + \frac{3}{n} + \frac{1}{n^2}}{6};$$

$$A = \lim_{n \to \infty} A_n = \lim_{n \to \infty} \frac{2 + \frac{3}{n} + \frac{1}{n^2}}{6} = \frac{2 + 0 + 0}{6} = \frac{1}{3}.$$

Calculator Exercises · pages 498–499

1. a. As $n \to \infty$, $\sqrt{n+1} \approx \sqrt{n}$, so $\lim_{n \to \infty} (\sqrt{n+1} - \sqrt{n}) = 0$.

b. $(\sqrt{n+1} - \sqrt{n}) \cdot \dfrac{\sqrt{n+1} + \sqrt{n}}{\sqrt{n+1} + \sqrt{n}} = \dfrac{(n+1) - n}{\sqrt{n+1} + \sqrt{n}} = \dfrac{1}{\sqrt{n+1} + \sqrt{n}}$

c. When n becomes very large, $\sqrt{n+1} \to \infty$ and $\sqrt{n} \to \infty$, so $\dfrac{1}{\sqrt{n+1} + \sqrt{n}} \to 0$. This agrees with the answer in part (a).

2. a. $t_1 = 1$; $t_2 = \dfrac{1}{2} + \dfrac{1}{1} = 1.5$; $t_3 = \dfrac{1.5}{2} + \dfrac{1}{1.5} = 1.41\overline{6}$; $t_4 = \dfrac{1.41\overline{6}}{2} + \dfrac{1}{1.41\overline{6}} \approx 1.4142157$;

$t_5 = \dfrac{t_4}{2} + \dfrac{1}{t_4} \approx 1.4142136$

b. The terms of the sequence seem to be approaching $\sqrt{2} \approx 1.4142136$.

3. a, b. $\sqrt{1 + \sqrt{1}} \approx 1.4142136$; $\sqrt{1 + \sqrt{1 + \sqrt{1}}} \approx 1.553774$;

$\sqrt{1 + \sqrt{1 + \sqrt{1 + \sqrt{1}}}} \approx 1.5980532$; continuing, the values approach $1.618034 \approx R$, the golden ratio.

Computer Exercise · page 499

```
10 LET M = 1
20 LET P = 0
30 FOR N = 1 TO 20 STEP 2
40 LET D = .5 * (M - P)
50 LET P = P + D
60 PRINT P
70 LET P = .5 * P
80 PRINT P
90 NEXT N
100 END
```

After twenty changes of mind, the student will be approximately $\frac{1}{3}$ of the distance from home; however, after 21 changes of mind, the student will be approximately $\frac{2}{3}$ of the distance from home. The student does not approach a limiting point.

Class Exercises 13-5 · page 502

1. $S_1 = 1; S_2 = 1 + \dfrac{1}{3} = \dfrac{4}{3}; S_3 = \dfrac{4}{3} + \dfrac{1}{9} = \dfrac{13}{9}; S_4 = \dfrac{13}{9} + \dfrac{1}{27} = \dfrac{40}{27}; S = \dfrac{t_1}{1-r} = \dfrac{1}{1 - \frac{1}{3}} =$

$\dfrac{1}{\frac{2}{3}} = \dfrac{3}{2}$

2. $S_1 = \dfrac{1}{2}; S_2 = \dfrac{1}{2} - \dfrac{1}{4} = \dfrac{1}{4}; S_3 = \dfrac{1}{4} + \dfrac{1}{8} = \dfrac{3}{8}; S_4 = \dfrac{3}{8} - \dfrac{1}{16} = \dfrac{5}{16}; S = \dfrac{t_1}{1-r} =$

$\dfrac{\frac{1}{2}}{1 - \left(-\frac{1}{2}\right)} = \dfrac{\frac{1}{2}}{\frac{3}{2}} = \dfrac{1}{3}$

3. $S_1 = 1; S_2 = 1 + 3 = 4; S_3 = 4 + 9 = 13; S_4 = 13 + 27 = 40; r = 3 \geq 1$, so the series diverges.

4. $S_1 = 1; S_2 = 1.1; S_3 = 1.11; S_4 = 1.111; S = \dfrac{1}{1 - 0.1} = \dfrac{1}{0.9} = \dfrac{10}{9}$

5. a. $r = x$, so the series converges when $|r| < 1$, or $|x| < 1$; $-1 < x < 1$.

b. $r = 2x$, so the series converges when $|r| < 1$, or $|2x| < 1$; $-1 < 2x < 1$;

$-\dfrac{1}{2} < x < \dfrac{1}{2}$.

6. $0.333\ldots = 0.3 + 0.03 + 0.003 + \cdots$ **a.** $t_1 = 0.3$ **b.** $r = 0.1$

c. $S = \dfrac{0.3}{1 - 0.1} = \dfrac{0.3}{0.9} = \dfrac{1}{3}$

7. For an arithmetic series with $t_1 \neq 0$ and $d \neq 0$, $S_n = \dfrac{n(t_1 + t_n)}{2}$; as $n \longrightarrow \infty$,

$\left| \dfrac{n(t_1 + t_n)}{2} \right| \longrightarrow \infty$, so an infinite arithmetic series diverges.

8. The series diverges; each term of the series is greater than or equal to $\frac{1}{2}$, so the sequence of partial sums approaches infinity.

Written Exercises 13-5 · pages 502–505

1. $S = \dfrac{t_1}{1-r} = \dfrac{1}{1 - \frac{1}{2}} = \dfrac{1}{\frac{1}{2}} = 2$ **2.** $S = \dfrac{t_1}{1-r} = \dfrac{1}{1 - \left(-\frac{1}{3}\right)} = \dfrac{1}{\frac{4}{3}} = \dfrac{3}{4}$

3. $S = \dfrac{t_1}{1-r} = \dfrac{24}{1 - \left(-\frac{1}{2}\right)} = \dfrac{24}{\frac{3}{2}} = 16$ **4.** $S = \dfrac{t_1}{1-r} = \dfrac{\frac{1}{4}}{1 - \frac{1}{4}} = \dfrac{\frac{1}{4}}{\frac{3}{4}} = \dfrac{1}{3}$

5. $S = \dfrac{t_1}{1-r} = \dfrac{5}{1 - 5^{-2}} = \dfrac{5}{\frac{24}{25}} = \dfrac{125}{24}$

6. $r = \dfrac{\sqrt{9}}{\sqrt{27}} = \dfrac{1}{\sqrt{3}}; S = \dfrac{t_1}{1-r} = \dfrac{\sqrt{27}}{1 - \frac{1}{\sqrt{3}}} = \dfrac{\sqrt{81}}{\sqrt{3} - 1} = \dfrac{9}{\sqrt{3} - 1} \cdot \dfrac{\sqrt{3} + 1}{\sqrt{3} + 1} = \dfrac{9 + 9\sqrt{3}}{2}$

7. $t_1 = 8 \cdot 5^{-1} = \dfrac{8}{5}; \; r = 5^{-1} = \dfrac{1}{5}; \; S = \dfrac{\dfrac{8}{5}}{1 - \dfrac{1}{5}} = \dfrac{\dfrac{8}{5}}{\dfrac{4}{5}} = 2$

8. $t_1 = (-2)^{1-1} = (-2)^0 = 1; \; r = (-2)^{-1} = -\dfrac{1}{2}; \; S = \dfrac{1}{1 - \left(-\dfrac{1}{2}\right)} = \dfrac{1}{\dfrac{3}{2}} = \dfrac{2}{3}$

9. $S = \dfrac{t_1}{1-r}; \; 8 = \dfrac{4}{1-r}; \; 1 - r = \dfrac{1}{2}; \; \dfrac{1}{2} = r$

10. $S = \dfrac{t_1}{1-r}; \; 81 = \dfrac{t_1}{1 - \dfrac{1}{3}}; \; t_1 = 81\left(\dfrac{2}{3}\right) = 54; \; t_2 = 54 \cdot \dfrac{1}{3} = 18; \; t_3 = 18 \cdot \dfrac{1}{3} = 6$

11. $S = \dfrac{t_1}{1-r}; \; \dfrac{3}{5} = \dfrac{1}{1 - 2x}; \; 3 - 6x = 5; \; -2 = 6x; \; x = -\dfrac{1}{3}$

12. $S = \dfrac{t_1}{1-r}; \; \dfrac{x}{5} = \dfrac{x^2}{1 - (-x)}; \; x(1 + x) = 5x^2; \; x = 4x^2; \; x(4x - 1) = 0; \; x = 0 \text{ or } x = \dfrac{1}{4}$

13. a. $r = x^2; \; |r| < 1; \; |x^2| < 1; \; x^2 < 1 \text{ since } x^2 \geq 0 \text{ for all } x; \; -1 < x < 1$ **b.** $S = \dfrac{1}{1 - x^2}$

14. a. $r = 3x; \; |r| < 1; \; |3x| < 1; \; -1 < 3x < 1; \; -\dfrac{1}{3} < x < \dfrac{1}{3}$ **b.** $S = \dfrac{1}{1 - 3x}$

15. a. $r = x - 3; \; |r| < 1; \; |x - 3| < 1; \; -1 < x - 3 < 1; \; 2 < x < 4$

   b. $S = \dfrac{1}{1 - (x - 3)} = \dfrac{1}{4 - x}$

16. a. $r = -(x - 1) = 1 - x; \; |r| < 1; \; |1 - x| < 1; \; -1 < 1 - x < 1; \; -2 < -x < 0;$
     $2 > x > 0$

   b. $S = \dfrac{1}{1 - [-(x - 1)]} = \dfrac{1}{x}$

17. a. $r = -\dfrac{2}{x}; \; |r| < 1; \; \left|-\dfrac{2}{x}\right| < 1; \; \dfrac{2}{|x|} < 1; \; 2 < |x|; \; x < -2 \text{ or } x > 2$

   b. $S = \dfrac{1}{1 - \left(-\dfrac{2}{x}\right)} = \dfrac{1}{\dfrac{x + 2}{x}} = \dfrac{x}{x + 2}$

18. a. $r = -\dfrac{x^4}{6} \div \dfrac{x^2}{3} = -\dfrac{x^4}{6} \cdot \dfrac{3}{x^2} = -\dfrac{x^2}{2}; \; |r| < 1; \; \left|-\dfrac{x^2}{2}\right| < 1; \; x^2 \geq 0 \text{ for all } x, \text{ so } \dfrac{x^2}{2} < 1;$
     $x^2 < 2; \; -\sqrt{2} < x < \sqrt{2}$

   b. $S = \dfrac{\dfrac{x^2}{3}}{1 - \left(-\dfrac{x^2}{2}\right)} = \dfrac{2x^2}{6 + 3x^2}$

19. $t_1 = \sin^2 x \text{ and } r = \sin^2 x; \; S = \dfrac{\sin^2 x}{1 - \sin^2 x} = \dfrac{\sin^2 x}{\cos^2 x} = \tan^2 x \text{ providing } |r| < 1;$

   $|\sin^2 x| < 1; \; -1 < \sin^2 x < 1; \; x \neq \dfrac{\pi}{2} + n\pi, \; n \text{ an integer}$

20. a. $t_1 = \tan^2 x$ and $r = -\tan^2 x$; if $-\dfrac{\pi}{4} < x < \dfrac{\pi}{4}$, then $-1 < \tan x < 1$ and

$|-\tan^2 x| < 1$, so the series converges. $\quad S = \dfrac{\tan^2 x}{1 - (-\tan^2 x)} = \dfrac{\tan^2 x}{\sec^2 x} =$

$\dfrac{\sin^2 x}{\cos^2 x} \cdot \cos^2 x = \sin^2 x$

b. $|r| < 1$ when $|-\tan^2 x| < 1$, or $-1 < \tan x < 1$; this is true, for example, when

$\dfrac{3\pi}{4} < x < \dfrac{5\pi}{4}$, and in general when $-\dfrac{\pi}{4} + n\pi < x < \dfrac{\pi}{4} + n\pi$, n an integer.

21. If $t_1 = 10$ and $S = 4$, then $4 = \dfrac{10}{1 - r}$; $1 - r = \dfrac{10}{4} = \dfrac{5}{2}$; $r = -\dfrac{3}{2}$ and so $|r| > 1$; such a series diverges; the assumption that the series converges is false, there is no such infinite series.

22. If an infinite geometric series has a sum S, then $S = \dfrac{t_1}{1 - r}$ and $|r| < 1$; thus $1 - r > 0$; S is positive if and only if t_1 is positive.

23. $0.777\ldots = 0.7 + 0.07 + 0.007 + \cdots$; $t_1 = 0.7$ and $r = 0.1$, so $S = \dfrac{0.7}{1 - 0.1} = \dfrac{0.7}{0.9} = \dfrac{7}{9}$

24. $0.636363\ldots = 0.63 + 0.0063 + 0.000063 + \cdots$; $t_1 = 0.63$ and $r = 0.01$, so $S = \dfrac{0.63}{1 - 0.01} = \dfrac{0.63}{0.99} = \dfrac{63}{99} = \dfrac{7}{11}$

25. $44.444\ldots = 40 + 4 + 0.4 + \cdots$; $t_1 = 40$ and $r = 0.1$; $S = \dfrac{40}{1 - 0.1} = \dfrac{40}{0.9} = \dfrac{400}{9}$

26. $5.363636\ldots = 5 + 0.36 + 0.0036 + \cdots$; $t_1 = 0.36$ and $r = 0.01$, so $0.36 + 0.0036 + \cdots = S = \dfrac{0.36}{1 - 0.01} = \dfrac{0.36}{0.99} = \dfrac{36}{99} = \dfrac{4}{11}$; $5.3636\ldots = 5\dfrac{4}{11} = \dfrac{59}{11}$

27. $0.142857142857\ldots = 0.142857 + 0.000000142857 + \cdots$; $t_1 = 0.142857$ and $r = 0.000001$; $S = \dfrac{0.142857}{1 - 0.000001} = \dfrac{0.142857}{0.999999} = \dfrac{142857}{999999} = \dfrac{1}{7}$

28. $0.0123123123\ldots = 0.0123 + 0.0000123 + \cdots = \dfrac{0.0123}{1 - 0.001} = \dfrac{0.0123}{0.999} = \dfrac{123}{9990} = \dfrac{41}{3330}$

29. $S_1 = \dfrac{1}{1 \cdot 2} = \dfrac{1}{2}$; $S_2 = \dfrac{1}{2} + \dfrac{1}{2 \cdot 3} = \dfrac{1}{2} + \dfrac{1}{6} = \dfrac{2}{3}$; $S_3 = \dfrac{2}{3} + \dfrac{1}{3 \cdot 4} = \dfrac{9}{12} = \dfrac{3}{4}$;

$S_4 = \dfrac{3}{4} + \dfrac{1}{4 \cdot 5} = \dfrac{16}{20} = \dfrac{4}{5}$; $S_n = \dfrac{n}{n + 1}$; $\displaystyle\lim_{n \to \infty} \dfrac{n}{n + 1} = \lim_{n \to \infty} \dfrac{1}{1 + \dfrac{1}{n}} = 1$

30. $S_1 = \dfrac{1}{1 \cdot 3} = \dfrac{1}{3}$; $S_2 = \dfrac{1}{3} + \dfrac{1}{3 \cdot 5} = \dfrac{2}{5}$; $S_3 = \dfrac{2}{5} + \dfrac{1}{5 \cdot 7} = \dfrac{3}{7}$; $S_4 = \dfrac{3}{7} + \dfrac{1}{7 \cdot 9} = \dfrac{4}{9}$;

$S_n = \dfrac{n}{2n + 1}$; $\displaystyle\lim_{n \to \infty} \dfrac{n}{2n + 1} = \lim_{n \to \infty} \dfrac{1}{2 + \dfrac{1}{n}} = \dfrac{1}{2}$

31. $S_1 = \dfrac{1}{1 \cdot 4} = \dfrac{1}{4}$; $S_2 = \dfrac{1}{4} + \dfrac{1}{4 \cdot 7} = \dfrac{2}{7}$; $S_3 = \dfrac{2}{7} + \dfrac{1}{7 \cdot 10} = \dfrac{3}{10}$; $S_4 = \dfrac{3}{10} + \dfrac{1}{10 \cdot 13} = \dfrac{4}{13}$;

$S_n = \dfrac{n}{3n + 1}$; $\displaystyle\lim_{n \to \infty} \dfrac{n}{3n + 1} = \lim_{n \to \infty} \dfrac{1}{3 + \dfrac{1}{n}} = \dfrac{1}{3}$

32. $S_1 = \dfrac{3}{1 \cdot 4} = \dfrac{3}{4}$; $S_2 = \dfrac{3}{4} + \dfrac{5}{4 \cdot 9} = \dfrac{8}{9}$; $S_3 = \dfrac{8}{9} + \dfrac{7}{9 \cdot 16} = \dfrac{15}{16}$; $S_4 = \dfrac{15}{16} + \dfrac{9}{16 \cdot 25} = \dfrac{24}{25}$;

$S_n = \dfrac{(n+1)^2 - 1}{(n+1)^2} = \dfrac{n^2 + 2n}{n^2 + 2n + 1}$; $\lim\limits_{n \to \infty} \dfrac{n^2 + 2n}{n^2 + 2n + 1} = \lim\limits_{n \to \infty} \dfrac{1 + \dfrac{2}{n}}{1 + \dfrac{2}{n} + \dfrac{1}{n^2}} = \dfrac{1}{1} = 1$

33. $D = 8 + \dfrac{3}{4} \cdot 8 + \dfrac{3}{4} \cdot 8 + \dfrac{3}{4}\left(\dfrac{3}{4} \cdot 8\right) + \dfrac{3}{4}\left(\dfrac{3}{4} \cdot 8\right) + \cdots$, since after the initial drop the

ball travels up *and* down. Hence $D = 8 + S$, where $S = \dfrac{3}{4}(8 \cdot 2) + \left(\dfrac{3}{4}\right)^2 (8 \cdot 2) + \cdots$;

$t_1 = 16 \cdot \dfrac{3}{4} = 12$ and $r = \dfrac{3}{4}$, so $S = \dfrac{12}{1 - \dfrac{3}{4}} = \dfrac{12}{\dfrac{1}{4}} = 48$; $D = 8 + 48 = 56$; 56 m.

34. $D = 10 + (0.95)(10) + (0.95)(10) + 0.95[(0.95)(10)] + 0.95[(0.95)(10)] + \cdots$, since after the initial drop the ball travels up *and* down. Hence $D = 10 + S$, where

$S = 0.95(10 \cdot 2) + (0.95)^2(10 \cdot 2) + \cdots$; $t_1 = 19$ and $r = 0.95$, so $S = \dfrac{19}{1 - 0.95} =$

$\dfrac{19}{0.05} = 380$; $D = 10 + 380 = 390$; 390 m.

35. The first square has sides of length 12, the second has sides of length $6\sqrt{2}$, the third has sides of length 6, and so on.

a. The sum of the areas is $12^2 + (6\sqrt{2})^2 + 6^2 + \cdots = 144 + 72 + 36 + \cdots$; $t_1 = 144$

and $r = \dfrac{72}{144} = \dfrac{1}{2}$, so $S = \dfrac{144}{1 - \dfrac{1}{2}} = 288$; 288 square units.

b. The sum of the perimeters is $48 + 24\sqrt{2} + 24 + \cdots$; $t_1 = 48$ and $r = \dfrac{24\sqrt{2}}{48} = \dfrac{\sqrt{2}}{2}$,

so $S = \dfrac{48}{1 - \dfrac{\sqrt{2}}{2}} = \dfrac{48}{\dfrac{2 - \sqrt{2}}{2}} = \dfrac{96}{2 - \sqrt{2}} = \dfrac{96(2 + \sqrt{2})}{(2 - \sqrt{2})(2 + \sqrt{2})} = \dfrac{96(2 + \sqrt{2})}{2} =$

$96 + 48\sqrt{2}$; $96 + 48\sqrt{2}$ units.

36. The first triangle has sides of length 12, the second has sides of length 6, the third has sides of length 3, and so on.

a. The sum of the areas is $36\sqrt{3} + 9\sqrt{3} + \dfrac{9}{4}\sqrt{3} + \cdots$; $t_1 = 36\sqrt{3}$ and

$r = \dfrac{9\sqrt{3}}{36\sqrt{3}} = \dfrac{1}{4}$, so $S = \dfrac{36\sqrt{3}}{1 - \dfrac{1}{4}} = \dfrac{36\sqrt{3}}{\dfrac{3}{4}} = 48\sqrt{3}$; $48\sqrt{3}$ square units.

b. The sum of the perimeters is $36 + 18 + 9 + \cdots$; $t_1 = 36$ and $r = \dfrac{18}{36} = \dfrac{1}{2}$, so

$S = \dfrac{36}{1 - \dfrac{1}{2}} = 72$; 72 units.

37. $S = \dfrac{1}{1 - \dfrac{1}{2}} = 2$ and $S_n = \dfrac{1\left(1 - \left(\dfrac{1}{2}\right)^n\right)}{1 - \dfrac{1}{2}} = 2 - 2\left(\dfrac{1}{2}\right)^n$; $S - S_n = 2 - \left(2 - \left(\dfrac{1}{2}\right)^{n-1}\right) =$

$\dfrac{1}{2^{n-1}}$; if $S - S_n < 0.0001$, $\dfrac{1}{2^{n-1}} < 0.0001$; $\dfrac{1}{0.0001} < 2^{n-1}$ or $2^{n-1} > 10{,}000$. The smallest

n for which $2^{n-1} > 10{,}000$ is 15 since $2^{15-1} = 2^{14} = 16{,}384$ and $2^{14-1} = 2^{13} = 8192$.

38. $S = \dfrac{1}{1 - \dfrac{2}{3}} = \dfrac{1}{\dfrac{1}{3}} = 3$ and $S_n = \dfrac{1\left(1 - \left(\dfrac{2}{3}\right)^n\right)}{1 - \dfrac{2}{3}} = 3\left(1 - \left(\dfrac{2}{3}\right)^n\right); \; S - S_n =$

$3 - 3\left(1 - \left(\dfrac{2}{3}\right)^n\right) = 3\left(\dfrac{2}{3}\right)^n;$ if $S - S_n < 0.0001, \; 3\left(\dfrac{2}{3}\right)^n < 0.0001; \; \left(\dfrac{2}{3}\right)^n < \dfrac{0.0001}{3};$

$n \log \dfrac{2}{3} < \log \left(0.0001 \cdot \dfrac{1}{3}\right); \; n > \dfrac{\log \left(0.0001 \cdot \frac{1}{3}\right)}{\log \frac{2}{3}}; \; n > \dfrac{-4 - \log 3}{\log 2 - \log 3} \approx 25.4,$

so $n = 26$ is the smallest integer.

39. $\dfrac{100 \text{ m}}{10 \text{ m/s}} = 10 \text{ s} = t_1; \; \dfrac{10 \text{ m}}{10 \text{ m/s}} = 1 \text{ s} = t_2; \; \dfrac{1 \text{ m}}{10 \text{ m/s}} = 0.1 \text{ s} = t_3.$ Hence

$S = t_1 + t_2 + t_3 + \cdots = 10 + 1 + 0.1 + \cdots; \; t_1 = 10$ and $r = 0.1$, so

$S = \dfrac{10}{1 - 0.1} = \dfrac{10}{0.9} = \dfrac{100}{9}; \dfrac{100}{9}$ s.

40. Answers may vary. Example: Let $d =$ the distance between you and the door. The distances covered by walking halfway to the door are $\dfrac{1}{2}d, \dfrac{1}{4}d, \dfrac{1}{8}d$, and so on. The total distance covered is $\dfrac{1}{2}d + \dfrac{1}{4}d + \dfrac{1}{8}d + \cdots$, which is an infinite geometric series with common ratio $\dfrac{1}{2}$. Thus, $S = \dfrac{\dfrac{1}{2}d}{1 - \dfrac{1}{2}} = \dfrac{\dfrac{1}{2}d}{\dfrac{1}{2}} = d.$ The total distance covered is d, so you could eventually get to the door.

41. $d = t_2 - t_1 = (3 + 4i) - (1 + 3i) = 2 + i; \; t_3 = (3 + 4i) + (2 + i) = 5 + 5i; \; t_4 = (5 + 5i) + (2 + i) = 7 + 6i; \; t_5 = 9 + 7i; \; t_{25} = t_1 + 24d = (1 + 3i) + 24(2 + i) = 49 + 27i; \; S_{25} = \dfrac{25(t_1 + t_n)}{2} = \dfrac{25(1 + 3i + 49 + 27i)}{2} = 25(25 + 15i) = 625 + 375i$

42. $t_1 = i; \; t_2 = i \cdot 2i = 2i^2 = -2; \; t_3 = -2 \cdot 2i = -4i; \; t_4 = -4i \cdot 2i = -8i^2 = 8; \; t_5 = 8 \cdot 2i = 16i; \; S_5 = \dfrac{i(1 - (2i)^5)}{1 - 2i} =$

$\dfrac{i - 32i^6}{1 - 2i} = \dfrac{i - 32(-1)}{1 - 2i} = \dfrac{32 + i}{1 - 2i} \cdot \dfrac{1 + 2i}{1 + 2i} = \dfrac{30 + 65i}{5} = 6 + 13i$

(Or: $S_5 = i - 2 - 4i + 8 + 16i = 6 + 13i$)

43. a. $t_1 = 1$ and $r = \dfrac{i}{2}; \; S = \dfrac{1}{1 - \dfrac{i}{2}} = \dfrac{2}{2 - i} \cdot \dfrac{2 + i}{2 + i} =$

$\dfrac{4 + 2i}{5} = \dfrac{4}{5} + \dfrac{2}{5}i$

b. $t_1 = 27$ and $r = \dfrac{-9i}{27} = -\dfrac{i}{3}; \; S = \dfrac{27}{1 - \left(-\dfrac{i}{3}\right)} =$

$\dfrac{81}{3 + i} \cdot \dfrac{3 - i}{3 - i} = \dfrac{243 - 81i}{10} = \dfrac{243}{10} - \dfrac{81}{10}i$

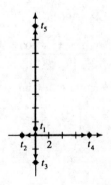

Ex. 42

44. a. $r = \dfrac{1}{2}i$

b. $t_1 = 1$, $t_2 = \dfrac{1}{2}i$, $t_3 = -\dfrac{1}{4}$, and so on; $S = \dfrac{1}{1 - \dfrac{1}{2}i} = \dfrac{2}{2 - i} \cdot \dfrac{2 + i}{2 + i} = \dfrac{4 + 2i}{5} =$

$\dfrac{4}{5} + \dfrac{2}{5}i$, or $\left(\dfrac{4}{5}, \dfrac{2}{5}\right)$.

c. $d = 1 + \dfrac{1}{2} + \dfrac{1}{4} + \dfrac{1}{8} + \cdots$; $S = \dfrac{1}{1 - \dfrac{1}{2}} = 2$; 2 units

45. $S = t_1 + t_1 r + t_1 r^2 + t_1 r^3 + \cdots$; the sum of the odd-numbered terms is $S_{\text{odd}} =$
$t_1 + t_1 r^2 + t_1 r^4 + \cdots = \dfrac{t_1}{1 - r^2}$; the sum of the even-numbered terms is $S_{\text{even}} =$
$t_1 r + t_1 r^3 + t_1 r^5 + \cdots = \dfrac{t_1 r}{1 - r^2}$; $S = \dfrac{t_1}{1 - r}$, so $\dfrac{S_{\text{odd}}}{S} = \dfrac{t_1}{1 - r^2} \div \dfrac{t_1}{1 - r} = \dfrac{1 - r}{(1 - r)(1 + r)} =$
$\dfrac{1}{1 + r}$; $\dfrac{S_{\text{even}}}{S} = \dfrac{t_1 r}{1 - r^2} \div \dfrac{t_1}{1 - r} = \dfrac{r(1 - r)}{(1 - r)(1 + r)} = \dfrac{r}{1 + r}$.

46. a. When dividing a by b, there is a remainder r. If $r = 0$, then $\dfrac{a}{b}$ is repeating with 0 as
the repeating digit. If $r \neq 0$, then $r < b$, so there are only $b - 1$ possibilities for r.
Hence the same remainder will again occur in $b - 1$ steps at most, and the same
partial quotient will also reoccur. When this happens, the quotient will repeat.

b. Every repeating decimal can be written as $a + t_1\left(\dfrac{1}{10}\right)^r + t_1\left(\dfrac{1}{10}\right)^{2r} + t_1\left(\dfrac{1}{10}\right)^{3r} + \cdots$,
where t_1 is a rational number equal to the value of the initial block of the pattern;
r is the number of digits in the pattern; and a is the value of the nonrepeating part of
the number. $\Big($For example, given the repeating decimal $3.472121\ldots$, $a = 3.47$;
$t_1 = \dfrac{21}{10000}$; and $r = 2.\Big)$ Thus $S = a + \dfrac{t_1}{1 - \left(\dfrac{1}{10}\right)^r} = a + \dfrac{t_1}{\dfrac{10^r - 1}{10^r}} = a + \dfrac{10^r t_1}{10^r - 1} =$
$\dfrac{a(10^r - 1) + 10^r t_1}{10^r - 1}$; since this expression is rational, S is rational.

47. An equilateral triangle with sides of length s has area $\dfrac{s^2\sqrt{3}}{4}$. $A_1 = \dfrac{1^2\sqrt{3}}{4} = \dfrac{\sqrt{3}}{4}$;
$A_2 = 3\left[\dfrac{\left(\frac{1}{3}\right)^2\sqrt{3}}{4}\right] = \dfrac{\sqrt{3}}{12}$; $A_3 = 12\left[\dfrac{\left(\frac{1}{9}\right)^2\sqrt{3}}{4}\right] = \dfrac{\sqrt{3}}{27}$; $A_4 = 48\left[\dfrac{\left(\frac{1}{27}\right)^2\sqrt{3}}{4}\right] = \dfrac{4\sqrt{3}}{243}$;
$S = \dfrac{\sqrt{3}}{4} + \dfrac{\sqrt{3}}{12} + \dfrac{\sqrt{3}}{27} + \dfrac{4\sqrt{3}}{243} + \cdots = \dfrac{\sqrt{3}}{4} + \dfrac{\sqrt{3}}{12}\left(1 + \dfrac{4}{9} + \left(\dfrac{4}{9}\right)^2 + \cdots\right) = \dfrac{\sqrt{3}}{4} +$
$\dfrac{\sqrt{3}}{12}\left[\dfrac{1}{1 - \frac{4}{9}}\right] = \dfrac{\sqrt{3}}{4} + \dfrac{\sqrt{3}}{12} \cdot \dfrac{9}{5} = \dfrac{8\sqrt{3}}{20} = \dfrac{2\sqrt{3}}{5}$ square units

48. Let each side of the initial equilateral triangle be 1 unit long. The sequence of
perimeters is $3, 4, \dfrac{16}{3}, \dfrac{64}{9}, \ldots$, which is an infinite geometric sequence with common
ratio $\dfrac{4}{3}$; since $\dfrac{4}{3} > 1$, the series $3 + 4 + \dfrac{16}{3} + \cdots$ diverges and the perimeter is infinite.

Calculator Exercise · page 506

a. $S_n = \dfrac{t_1(1 - r^n)}{1 - r}$; $0.33 = \dfrac{\frac{1}{5}\left[1 - \left(\frac{2}{5}\right)^n\right]}{1 - \frac{2}{5}}$; $0.99 = 1 - \left(\dfrac{2}{5}\right)^n$; $\left(\dfrac{2}{5}\right)^n = 0.01$; $n \log \dfrac{2}{5} =$

  $\log 0.01$; $n = \dfrac{-2}{\log 2 - \log 5} \approx 5.03$; 6 terms are needed.

b. $0.333 = \dfrac{\frac{1}{5}\left[1 - \left(\frac{2}{5}\right)^n\right]}{1 - \frac{2}{5}}$; $0.999 = 1 - \left(\dfrac{2}{5}\right)^n$; $\left(\dfrac{2}{5}\right)^n = 0.001$; $n \log \dfrac{2}{5} = \log 0.001$;

  $n = \dfrac{-3}{\log 2 - \log 5} \approx 7.54$; 8 terms are needed.

c. $0.3333 = \dfrac{\frac{1}{5}\left[1 - \left(\frac{2}{5}\right)^n\right]}{1 - \frac{2}{5}}$; $0.9999 = 1 - \left(\dfrac{2}{5}\right)^n$; $\left(\dfrac{2}{5}\right)^n = 0.0001$; $n \log \dfrac{2}{5} = \log 0.0001$;

  $n = \dfrac{-4}{\log 2 - \log 5} \approx 10.05$; 11 terms are needed.

Class Exercises 13-6 · page 508

1. $5 \cdot 1 + 5 \cdot 2 + 5 \cdot 3 + 5 \cdot 4$, or $5 + 10 + 15 + 20$

2. $3^2 + 4^2 + 5^2 + 6^2$, or $9 + 16 + 25 + 36$

3. $(-1)^2 + (-1)^3 + (-1)^4 + (-1)^5 + (-1)^6 + (-1)^7 + (-1)^8$, or $1 - 1 + 1 - 1 +$
$1 - 1 + 1$

4. $1 + \dfrac{1}{2} + \dfrac{1}{3} + \dfrac{1}{4} + \cdots$ **5.** $\displaystyle\sum_{n=2}^{6} n^2$ **6.** $\displaystyle\sum_{n=1}^{4} \dfrac{n}{n + 1}$ **7.** $\displaystyle\sum_{n=1}^{100} 3n$ **8.** $\displaystyle\sum_{n=1}^{\infty} \dfrac{1}{3^n}$, or $\displaystyle\sum_{n=1}^{\infty} 3^{-n}$

Written Exercises 13-6 · pages 508–510

1. $2 + 3 + 4 + 5 + 6$ **2.** $5 + 10 + 15 + 20 + \cdots + 50$ **3.** $1 + \dfrac{1}{2} + \dfrac{1}{3} + \dfrac{1}{4} + \dfrac{1}{5}$

4. $5 + 9 + 13 + 17 + 21$ **5.** $3^1 + 3^0 + 3^{-1} + 3^{-2} + \cdots = 3 + 1 + \dfrac{1}{3} + \dfrac{1}{9} + \cdots$

6. $2(1) + 3(-1) + 4(1) + 5(-1) + \cdots = 2 - 3 + 4 - 5 + \cdots$

7. $4^{-2} + 4^{-1} + 4^0 + 4^1 + 4^2 = \dfrac{1}{16} + \dfrac{1}{4} + 1 + 4 + 16$

8. $|13 - 3 \cdot 0| + |13 - 3 \cdot 1| + |13 - 3 \cdot 2| + \cdots + |13 - 3 \cdot 6| = 13 + 10 + 7 + 4 +$
$1 + 2 + 5$

9–16. Answers may vary. Examples are given.

9. $\displaystyle\sum_{k=1}^{5} 4k$ **10.** $\displaystyle\sum_{k=0}^{5} 2^k$ **11.** $\displaystyle\sum_{k=1}^{25} [5 + (k - 1)4] = \displaystyle\sum_{k=1}^{25} (4k + 1)$ **12.** $\displaystyle\sum_{k=1}^{100} 2k$ **13.** $\displaystyle\sum_{k=1}^{\infty} \dfrac{1}{k^2}$

14. $\displaystyle\sum_{k=1}^{\infty} \dfrac{1}{2k}$ **15.** $\displaystyle\sum_{k=1}^{\infty} \sin kx$ **16.** $\displaystyle\sum_{k=0}^{\infty} 48\left(\dfrac{1}{2}\right)^k$

17. $\displaystyle\sum_{t=1}^{4} \log t = \log 1 + \log 2 + \log 3 + \log 4 = \log (1 \cdot 2 \cdot 3 \cdot 4) = \log 24$

18. $\displaystyle\sum_{k=1}^{4} k \log 2 = 1 \cdot \log 2 + 2 \log 2 + 3 \log 2 + 4 \log 2 = 10 \log 2 = \log 2^{10}$

19. $\displaystyle\sum_{k=1}^{100} \cos (k\pi) = \cos \pi + \cos 2\pi + \cos 3\pi + \cos 4\pi + \cdots + \cos 99\pi + \cos 100\pi =$

$(-1 + 1) + (-1 + 1) + \cdots + (-1 + 1) = 0 + 0 + \cdots + 0 = 0$

20. $\displaystyle\sum_{k=1}^{50} \sin \left(k \cdot \frac{\pi}{2} \right) = \sin \frac{\pi}{2} + \sin \pi + \sin \frac{3\pi}{2} + \sin 2\pi + \cdots + \sin \frac{45\pi}{2} + \sin 23\pi +$

$\sin \frac{47\pi}{2} + \sin 24\pi + \sin \frac{49\pi}{2} + \sin 25\pi = (1 + 0 + (-1) + 0) + \cdots +$

$(1 + 0 + (-1) + 0) + 1 + 0 = 0 + \cdots + 0 + 1 + 0 = 1$

21. a. $\displaystyle\sum_{n=1}^{8} \left(\frac{\sqrt{2}}{2} + \frac{\sqrt{2}}{2} i \right)^n = \sum_{n=1}^{8} (\operatorname{cis} 45°)^n = \sum_{n=1}^{8} \operatorname{cis} (45° \cdot n) = \operatorname{cis} 45° + \operatorname{cis} 90° +$

$\operatorname{cis} 135° + \operatorname{cis} 180° + \cdots + \operatorname{cis} 360° = \left(\frac{\sqrt{2}}{2} + \frac{\sqrt{2}}{2} i \right) + i + \left(-\frac{\sqrt{2}}{2} + \frac{\sqrt{2}}{2} i \right) - 1 +$

$\left(-\frac{\sqrt{2}}{2} - \frac{\sqrt{2}}{2} i \right) - i + \left(\frac{\sqrt{2}}{2} - \frac{\sqrt{2}}{2} i \right) + 1 = 0$

b. By part (a), $\displaystyle\sum_{n=1}^{8} \left| \left(\frac{\sqrt{2}}{2} + \frac{\sqrt{2}}{2} i \right)^n \right| = \sum_{n=1}^{8} 1 = 8.$

22. $\displaystyle\sum_{n=1}^{\infty} \left(\frac{1}{2} + \frac{1}{2} i \right)^n = \left(\frac{1}{2} + \frac{1}{2} i \right) + \left(\frac{1}{2} + \frac{1}{2} i \right)^2 + \left(\frac{1}{2} + \frac{1}{2} i \right)^3 + \cdots = \frac{\dfrac{1}{2} + \dfrac{1}{2} i}{1 - \left(\dfrac{1}{2} + \dfrac{1}{2} i \right)} =$

$\dfrac{\dfrac{1}{2} + \dfrac{1}{2} i}{\dfrac{1}{2} - \dfrac{1}{2} i} = \dfrac{1 + i}{1 - i} \cdot \dfrac{1 + i}{1 + i} = \dfrac{2i}{2} = i$

23. $\displaystyle\sum_{k=1}^{8} (-1)^k \log k = -1 \log 1 + 1 \log 2 - 1 \log 3 + \cdots + 1 \log 8 =$

$(\log 2 + \log 4 + \log 6 + \log 8) - (\log 1 + \log 3 + \log 5 + \log 7) =$

$\log (2 \cdot 4 \cdot 6 \cdot 8) - \log (1 \cdot 3 \cdot 5 \cdot 7) = \log \left(\dfrac{2 \cdot 4 \cdot 6 \cdot 8}{1 \cdot 3 \cdot 5 \cdot 7} \right)$

24. $\displaystyle\sum_{k=1}^{100} \lfloor \sqrt{k} \rfloor = \lfloor \sqrt{1} \rfloor + \lfloor \sqrt{2} \rfloor + \lfloor \sqrt{3} \rfloor + \cdots + \lfloor \sqrt{100} \rfloor = 1 + 1 + 1 + 2 + 2 + 2 + 2 +$

$2 + \cdots + 10 = 3(1) + 5(2) + 7(3) + 9(4) + 11(5) + 13(6) + 15(7) + 17(8) +$

$19(9) + 10 = 625$

25–28. Answers may vary. Examples are given.

25. a. $\displaystyle\sum_{k=0}^{5} \left(-\frac{1}{2} \right)^k$ **b.** $\displaystyle\sum_{k=0}^{5} (-1) \left(-\frac{1}{2} \right)^k$ **26. a.** $\displaystyle\sum_{k=0}^{5} 27 \left(-\frac{1}{3} \right)^k$ **b.** $\displaystyle\sum_{k=0}^{5} (-27) \left(-\frac{1}{3} \right)^k$

27. $99 = 1 + (n - 1)2; n = 50; \displaystyle\sum_{n=1}^{50} (-1)^{n+1}[1 + (n - 1)2] = \sum_{n=1}^{50} (-1)^{n+1}(2n - 1),$ or

$\displaystyle\sum_{k=0}^{49} (-1)^k (2k + 1)$

28. $100 = 2(50),$ so $\displaystyle\sum_{n=1}^{50} (-1)^n \cdot (2n)$

29. $\displaystyle\sum_{i=1}^{n} (a_i + b_i) = (a_1 + b_1) + (a_2 + b_2) + \cdots + (a_n + b_n) = (a_1 + a_2 + \cdots + a_n) +$

$(b_1 + b_2 + \cdots + b_n) = \displaystyle\sum_{i=1}^{n} a_i + \sum_{i=1}^{n} b_i$

30. $\displaystyle\sum_{i=1}^{n} ca_i = ca_1 + ca_2 + \cdots + ca_n = c(a_1 + a_2 + \cdots + a_n) = c\sum_{i=1}^{n} a_i$

31. $\displaystyle\sum_{i=1}^{n} (a_i + b_i)^2 = \sum_{i=1}^{n} (a_i{}^2 + 2a_i b_i + b_i{}^2) = \sum_{i=1}^{n} a_i{}^2 + \sum_{i=1}^{n} 2a_i b_i + \sum_{i=1}^{n} b_i{}^2 =$

$\displaystyle\sum_{i=1}^{n} a_i{}^2 + 2\sum_{i=1}^{n} a_i b_i + \sum_{i=1}^{n} b_i{}^2$

32. $\displaystyle\sum_{x=1}^{n} (ax^2 + bx + c) = \sum_{x=1}^{n} ax^2 + \sum_{x=1}^{n} bx + \sum_{x=1}^{n} c = a\sum_{x=1}^{n} x^2 + b\sum_{x=1}^{n} x + cn$

33. $1 \cdot 2 + 2 \cdot 3 + 3 \cdot 4 + \cdots + 100 \cdot 101 = \displaystyle\sum_{k=1}^{100} k(k + 1) = \sum_{k=1}^{100} (k^2 + k) =$

$\displaystyle\sum_{k=1}^{100} k^2 + \sum_{k=1}^{100} k = \frac{100 \cdot 101 \cdot 201}{6} + \frac{100 \cdot 101}{2} = 338{,}350 + 5050 = 343{,}400$

34. $1 \cdot 3 + 3 \cdot 5 + 5 \cdot 7 + \cdots + 99 \cdot 101 = \displaystyle\sum_{k=1}^{50} (2k - 1)(2k + 1) = \sum_{k=1}^{50} (4k^2 - 1) =$

$4\displaystyle\sum_{k=1}^{50} k^2 - \sum_{k=1}^{50} 1 = 4\left[\frac{50 \cdot 51 \cdot 101}{6}\right] - 50 = 171{,}650$

35. $(1 \cdot 2 \cdot 3) + (2 \cdot 3 \cdot 4) + (3 \cdot 4 \cdot 5) + \cdots + (20 \cdot 21 \cdot 22) = \displaystyle\sum_{n=1}^{20} n(n + 1)(n + 2) =$

$\displaystyle\sum_{n=1}^{20} n^3 + 3n^2 + 2n = \sum_{n=1}^{20} n^3 + 3\sum_{n=1}^{20} n^2 + 2\sum_{n=1}^{20} n = \left[\frac{20(21)}{2}\right]^2 + 3\left[\frac{20(21)(41)}{6}\right] +$

$2\left[\dfrac{20 \cdot 21}{2}\right] = 44{,}100 + 8610 + 420 = 53{,}130$

36. $2 \cdot 4 + 6 \cdot 8 + 10 \cdot 12 + \cdots + 389 \cdot 400 = \displaystyle\sum_{k=1}^{100} 2(2k - 1) \cdot 2(2k) = \sum_{k=1}^{100} (16k^2 - 8k) =$

$16\displaystyle\sum_{k=1}^{100} k^2 - 8\sum_{k=1}^{100} k = 16\left[\frac{100 \cdot 101 \cdot 201}{6}\right] - 8\left[\frac{100 \cdot 101}{2}\right] = 5{,}413{,}600 - 40{,}400 =$

$5{,}373{,}200$

37. a. An 8-by-8 checkerboard contains 1 8-by-8 square, 4 7-by-7 squares, 9 6-by-6 squares, 16 5-by-5 squares, 25 4-by-4 squares, 36 3-by-3 squares, 49 2-by-2 squares, and

64 1-by-1 squares, or $\displaystyle\sum_{k=1}^{8} k^2 = \frac{8(9)(17)}{6} = 204$ squares in all.

b. $\displaystyle\sum_{k=1}^{n} k^2 = \frac{n(n + 1)(2n + 1)}{6}$

38. a. A 3-by-3-by-3 cube contains 1 3-by-3-by-3 cube, 8 2-by-2-by-2 cubes, and 27 1-by-1-by-1 cubes, or 36 cubes in all.

b. An n-by-n-by-n cube contains 1 n-by-n-by-n cube, 8 cubes of dimension $n - 1$, 27 cubes of dimension $n - 2$, and so on, with n^3 1-by-1-by-1 cubes;

$\displaystyle\sum_{k=1}^{n} k^3 = \left[\frac{n(n + 1)}{2}\right]^2.$

39. By looking at the diagram, we see that $t_1 = 1$, $t_2 = 3$, $t_3 = 6$, $t_4 = 10$, and $t_5 = 15$;

$$t_i = \sum_{k=1}^{i} k = \frac{i(i+1)}{2}; \quad \sum_{i=1}^{n} t_i = \sum_{i=1}^{n} \frac{i(i+1)}{2} = \frac{1}{2} \sum_{i=1}^{n} (i^2 + i) = \frac{1}{2} \sum_{i=1}^{n} i^2 + \frac{1}{2} \sum_{i=1}^{n} i =$$

$$\frac{1}{2} \left[\frac{n(n+1)(2n+1)}{6} \right] + \frac{1}{2} \left[\frac{n(n+1)}{2} \right] = \frac{2n^3 + 6n^2 + 4n}{12} = \frac{n^3 + 3n^2 + 2n}{6} =$$

$$\frac{n(n+1)(n+2)}{6}.$$

Computer Exercise · page 510

a.
```
10 LET S = 0
20 LET N = 1
30 LET S = S + 1/N
40 IF S > 3 THEN 70
50 LET N = N + 1
60 GOTO 30
70 PRINT N;" TERMS"
80 END

11 TERMS
```

b. Change line 40 as follows:
```
40 IF S > 4 THEN 70

31 TERMS
```

c. Change line 40 as follows:
```
40 IF S > 10 THEN 70

12,367 TERMS
```

Written Exercises 13-7 · pages 513–514

1. (1) For $n = 1$, $\frac{1(1+1)}{2} = \frac{2}{2} = 1$. (2) If $1 + 2 + \cdots + k = \frac{k(k+1)}{2}$, then

$(1 + 2 + \cdots + k) + (k+1) = \frac{k(k+1)}{2} + \frac{2(k+1)}{2} = \frac{k^2 + 3k + 2}{2} = \frac{(k+1)(k+2)}{2} =$

$\frac{(k+1)[(k+1)+1]}{2}$.

2. (1) For $n = 1$, $2 \cdot 1 - 1 = 1 = 1^2$. (2) If $1 + 3 + \cdots + (2k - 1) = k^2$, then
$[1 + 3 + \cdots + (2k-1)] + [2(k+1) - 1] = k^2 + 2k + 1 = (k+1)^2$.

3. (1) For $n = 1$, $2 \cdot 1 = 2 = 1^2 + 1$. (2) If $\sum_{i=1}^{k} 2i = k^2 + k$, then $\sum_{i=1}^{k+1} 2i = k^2 + k +$

$2(k+1) = (k^2 + 2k + 1) + (k+1) = (k+1)^2 + (k+1)$.

4. (1) For $n = 1$, $2^{1-1} = 2^0 = 1 = 2^1 - 1$. (2) If $\sum_{i=1}^{k} 2^{i-1} = 2^k - 1$, then $\sum_{i=1}^{k+1} 2^{i-1} =$

$(2^k - 1) + 2^{(k+1)-1} = 2^k - 1 + 2^k = 2 \cdot 2^k - 1 = 2^{k+1} - 1$.

5. (1) For $n = 1$, $1^2 = 1 = \frac{1(2)(3)}{6}$. (2) If $\sum_{i=1}^{k} i^2 = \frac{k(k+1)(2k+1)}{6}$, then $\sum_{i=1}^{k+1} i^2 =$

$\frac{k(k+1)(2k+1)}{6} + (k+1)^2 = \frac{k(k+1)(2k+1)}{6} + \frac{6(k+1)^2}{6} = \frac{(k+1)(2k^2 + 7k + 6)}{6} =$

$\frac{(k+1)(k+2)(2k+3)}{6} = \frac{(k+1)[(k+1)+1][2(k+1)+1]}{6}$.

6. (1) For $n = 1$, $1(1+1) = 2 = \frac{1(2)(3)}{3}$. (2) If $\sum_{i=1}^{k} i(i+1) = \frac{k(k+1)(k+2)}{3}$, then

$\sum_{i=1}^{k+1} i(i+1) = \frac{k(k+1)(k+2)}{3} + (k+1)[(k+1)+1] = \frac{k(k+1)(k+2)}{3} +$

$\frac{3(k+1)(k+2)}{3} = \frac{(k+1)(k+2)(k+3)}{3} = \frac{(k+1)[(k+1)+1][(k+1)+2]}{3}$.

7. (1) For $n = 1$, $(2 \cdot 1 - 1)^2 = 1^2 = 1 = \dfrac{1(1)(3)}{3}$. (2) If $\displaystyle\sum_{i=1}^{k} (2i - 1)^2 = \dfrac{k(2k - 1)(2k + 1)}{3}$,

then $\displaystyle\sum_{i=1}^{k+1} (2i - 1)^2 = \dfrac{k(2k - 1)(2k + 1)}{3} + [2(k + 1) - 1]^2 = \dfrac{k(2k - 1)(2k + 1)}{3} +$

$\dfrac{3(2k + 1)^2}{3} = \dfrac{(2k + 1)(2k^2 + 5k + 3)}{3} = \dfrac{(2k + 1)(k + 1)(2k + 3)}{3} =$

$\dfrac{(k + 1)[2(k + 1) - 1][2(k + 1) + 1]}{3}$.

8. (1) For $n = 1$, $(1 + x)^1 = 1 + x = 1 + 1 \cdot x$. (2) If $(1 + x)^k \geq 1 + kx$, then
$(1 + x)^{k+1} = (1 + x)(1 + x)^k \geq (1 + x)(1 + kx) = 1 + (k + 1)x + kx^2 \geq 1 +$
$(k + 1)x$ since $k > 0$, $x^2 \geq 0$, and $1 + x > 0$.

9. (1) For $n = 1$, $1 \cdot 2^{1-1} = 1 = 1 + (1 - 1) \cdot 2^1$. (2) If $\displaystyle\sum_{i=1}^{k} (i \cdot 2^{i-1}) = 1 + (k - 1) \cdot 2^k$,

then $\displaystyle\sum_{i=1}^{k+1} (i \cdot 2^{i-1}) = [1 + (k - 1) \cdot 2^k] + (k + 1) \cdot 2^{(k+1)-1} = [1 + (k - 1) \cdot 2^k] +$

$(k + 1) \cdot 2^k = 1 + 2k \cdot 2^k = 1 + k \cdot 2^{k+1} = 1 + [(k + 1) - 1] \cdot 2^{k+1}$.

10. (1) For $n = 1$, $18^1 - 1 = 17$, and 17 is a multiple of 17. (2) If $18^k - 1$ is a multiple of
17, then $18^{k+1} - 1 = 18 \cdot 18^k - (18 - 17) = (18 \cdot 18^k - 18) + 17 = 18(18^k - 1) +$
$17 = 18(\text{a multiple of } 17) + (\text{a multiple of } 17) = \text{a multiple of } 17$.

11. (1) For $n = 1$, $11^1 - 4^1 = 7$, and 7 is a multiple of 7. (2) If $11^k - 4^k$ is a multiple of 7,
then $11^{k+1} - 4^{k+1} = 11 \cdot 11^k - 4 \cdot 4^k = (7 + 4)11^k - 4 \cdot 4^k = 7 \cdot 11^k +$
$4(11^k - 4^k) = \text{the sum of 2 multiples of 7} = \text{a multiple of 7}$.

12. (1) For $n = 1$, $1(1^2 + 5) = 6$, and 6 is a multiple of 6. (2) If $k(k^2 + 5)$ is a multiple of
6, then $(k + 1)[(k + 1)^2 + 5] = (k^3 + 3k^2 + 3k + 1) + (5k + 5) = (k^3 + 5k) +$
$(3k^2 + 3k) + 6 = k(k^2 + 5) + 3k(k + 1) + 6$; $3k(k + 1)$ is a multiple of 6, since either
k or $k + 1$ must be even; thus, $k(k^2 + 5)$, $3k(k + 1)$, and 6 are all multiples of 6, as is
their sum.

13. For $n = 1$, $2^1 = 2 = 2 \cdot 1$; for $n = 2$, $2^2 = 4 = 2 \cdot 2$. It appears that the inequality is
true for $n \geq 3$. (1) For $n = 3$, $2^3 = 8$, $2 \cdot 3 = 6$, and $8 > 6$. (2) If $2^k > 2k$ where k is
an integer ≥ 3, then $2^{k+1} = 2 \cdot 2^k = 2^k + 2^k > 2^k + 2k > 2k + 2 = 2(k + 1)$. (Note
that $2k > 2$ because $k \geq 3$.)

14. For $n = 1$, $1! = 1 < 2^1$; for $n = 2$, $2! = 2 < 4 = 2^2$; for $n = 3$, $3! = 6 < 8 = 2^3$. It
appears that the inequality is true for $n \geq 4$. (1) For $n = 4$, $4! = 24$, $2^4 = 16$, and
$24 > 16$. (2) If $k! > 2^k$ where k is an integer ≥ 4, then $(k + 1)! = (k + 1) \cdot k! >$
$(k + 1) \cdot 2^k > 2 \cdot 2^k = 2^{k+1}$. (Note that $k + 1 > 2$ because $k \geq 4$.)

15. (1) For $n = 2$, there are two people shaking hands once with each other, and $\dfrac{2^2 - 2}{2} = 1$

handshake.

(2) Assume that a room of k people yields $\dfrac{k^2 - k}{2}$ handshakes. If one more person enters

the room, there will be k additional handshakes (one between the new person and each

of the k original people), so that a room of $k + 1$ people yields $\dfrac{k^2 - k}{2} + k = \dfrac{k^2 + k}{2} =$

$\dfrac{(k^2 + 2k + 1) - (k + 1)}{2} = \dfrac{(k + 1)^2 - (k + 1)}{2}$ handshakes.

16. Let n, $n + 1$, and $n + 2$ be the consecutive positive integers. We want to prove that $n^3 + (n + 1)^3 + (n + 2)^3$ is a multiple of 9 for every positive integer n. (1) For $n = 1$, $1^3 + 2^3 + 3^3 = 1 + 8 + 27 = 36$, and 36 is a multiple of 9. (2) If $k^3 + (k + 1)^3 + (k + 2)^3$ is a multiple of 9, then $(k + 1)^3 + [(k + 1) + 1]^3 + [(k + 1) + 2]^3 = (k + 1)^3 + (k + 2)^3 + (k + 3)^3 = (k + 1)^3 + (k + 2)^3 + (k^3 + 9k^2 + 27k + 27) = [k^3 + (k + 1)^3 + (k + 2)^3] + 9(k^2 + 3k + 3) = $ a multiple of 9 + a multiple of 9 = a multiple of 9.

17. (1) For $n = 1$, $|a_1| = |a_1|$. (2) If $|a_1 + a_2 + \cdots + a_k| \le |a_1| + |a_2| + \cdots + |a_k|$, then, by the triangle inequality, $|(a_1 + a_2 + \cdots + a_k) + a_{k+1}| \le |a_1 + a_2 + \cdots + a_k| + |a_{k+1}| \le |a_1| + |a_2| + \cdots + |a_k| + |a_{k+1}|$.

18. Let the complex number $z = r(\cos \theta + i \sin \theta)$. We want to prove that $z^n = r^n(\cos n\theta + i \sin n\theta)$ for every positive integer n. (1) For $n = 1$, $z^1 = z = r(\cos \theta + i \sin \theta) = r^1[\cos (1 \cdot \theta) + i \sin (1 \cdot \theta)]$. (2) If $z^k = r^k(\cos k\theta + i \sin k\theta)$, then $z^{k+1} = z \cdot z^k = [r(\cos \theta + i \sin \theta)][r^k(\cos k\theta + i \sin k\theta)] = r^{k+1}[(\cos \theta \cos k\theta - \sin \theta \sin k\theta) + i(\sin \theta \cos k\theta + \cos \theta \sin k\theta)] = r^{k+1}[\cos (\theta + k\theta) + i \sin (\theta + k\theta)] = r^{k+1}[\cos (k + 1)\theta + i \sin (k + 1)\theta]$.

19.

n	1	2	3	4	5	6
a_n	$\dfrac{1}{2}$	$\dfrac{1}{3}$	$\dfrac{1}{4}$	$\dfrac{1}{5}$	$\dfrac{1}{6}$	$\dfrac{1}{7}$
a_n	$\dfrac{1}{1 + 1}$	$\dfrac{1}{2 + 1}$	$\dfrac{1}{3 + 1}$	$\dfrac{1}{4 + 1}$	$\dfrac{1}{5 + 1}$	$\dfrac{1}{6 + 1}$

It appears that $a_n = \dfrac{1}{n + 1}$ for every positive integer n. (1) For $n = 1$, $a_1 = \dfrac{1}{2}$ and $\dfrac{1}{1 + 1} = \dfrac{1}{2}$. (2) If $a_k = \dfrac{1}{k + 1}$, then $a_{k+1} = a_{(k+1)-1} - \dfrac{a_{(k+1)-1}}{(k + 1) + 1} = a_k - \dfrac{a_k}{k + 2} = \dfrac{1}{k + 1} - \dfrac{1}{(k + 1)(k + 2)} = \dfrac{k + 2 - 1}{(k + 1)(k + 2)} = \dfrac{k + 1}{(k + 1)(k + 2)} = \dfrac{1}{k + 2} = \dfrac{1}{(k + 1) + 1}$.

20.

n	1	2	3	4	5	6
a_n	1	3	7	15	31	63
a_n	$2 - 1$	$4 - 1$	$8 - 1$	$16 - 1$	$32 - 1$	$64 - 1$

It appears that $a_n = 2^n - 1$ for every positive integer n. (1) For $n = 1$, $a_1 = 1$ and $2^1 - 1 = 1$. (2) If $a_k = 2^k - 1$, then $a_{k+1} = 2a_{(k+1)-1} + 1 = 2a_k + 1 = 2(2^k - 1) + 1 = 2 \cdot 2^k - 2 + 1 = 2^{k+1} - 1$.

21.

n	1	2	3	4	5	6
a_n	$\dfrac{1}{4}$	$\dfrac{2}{7}$	$\dfrac{3}{10}$	$\dfrac{4}{13}$	$\dfrac{5}{16}$	$\dfrac{6}{19}$
a_n	$\dfrac{1}{3+1}$	$\dfrac{2}{6+1}$	$\dfrac{3}{9+1}$	$\dfrac{4}{12+1}$	$\dfrac{5}{15+1}$	$\dfrac{6}{18+1}$

It appears that $a_n = \dfrac{n}{3n+1}$ for every positive integer n. (1) For $n = 1$, $a_1 = \dfrac{1}{4}$ and

$\dfrac{1}{3(1)+1} = \dfrac{1}{4}$. (2) If $a_k = \dfrac{k}{3k+1}$, then $a_{k+1} = a_{(k+1)-1} + \dfrac{1}{[3(k+1)-2][3(k+1)+1]} =$

$a_k + \dfrac{1}{(3k+1)(3k+4)} = \dfrac{k}{3k+1} + \dfrac{1}{(3k+1)(3k+4)} = \dfrac{k(3k+4)+1}{(3k+1)(3k+4)} =$

$\dfrac{(3k+1)(k+1)}{(3k+1)(3k+4)} = \dfrac{k+1}{3k+4} = \dfrac{k+1}{3(k+1)+1}$

22.

n	1	2	3	4	5
S_n	$1^3 = 1$	$1 + 2^3 = 9$	$9 + 3^3 = 36$	$36 + 4^3 = 100$	$100 + 5^3 = 225$
S_n	1^2	3^2	6^2	10^2	15^2

Note that 1, 3, 6, 10, and 15 are the first few terms in the sequence from Ex. 1. It

appears that $S_n = \displaystyle\sum_{i=1}^{n} i^3 = \left[\dfrac{n(n+1)}{2}\right]^2$ for every positive integer n. (1) For $n = 1$,

$1^3 = 1$ and $\left[\dfrac{1(1+1)}{2}\right]^2 = 1^2 = 1$. (2) If $S_k = \displaystyle\sum_{i=1}^{k} i^3 = \left[\dfrac{k(k+1)}{2}\right]^2$, then

$S_{k+1} = \displaystyle\sum_{i=1}^{k+1} i^3 = \left(\displaystyle\sum_{i=1}^{k} i^3\right) + (k+1)^3 = \left[\dfrac{k(k+1)}{2}\right]^2 + \dfrac{4(k+1)^3}{4} =$

$\dfrac{(k+1)^2[k^2 + 4(k+1)]}{4} = \dfrac{(k+1)^2(k^2 + 4k + 4)}{4} = \dfrac{(k+1)^2(k+2)^2}{4} =$

$\left[\dfrac{(k+1)((k+1)+1)}{2}\right]^2$.

23.

n	1	2	3	4	5
$n!$	1	2	6	24	120
$n \cdot n!$	1	4	18	96	600
S_n	1	$1 + 4 = 5$	$5 + 18 = 23$	$23 + 96 = 119$	$119 + 600 = 719$
S_n	$2 - 1$	$6 - 1$	$24 - 1$	$120 - 1$	$720 - 1$

(Continued)

It appears that $S_n = (n + 1)! - 1$ for every positive integer n. (1) For $n = 1$, $1 \cdot 1! = 1$ and $(1 + 1)! - 1 = 1$. (2) If $S_k = \sum_{i=1}^{k} (i \cdot i!) = (k + 1)! - 1$, then $S_{k+1} = \sum_{i=1}^{k+1} (i \cdot i!) =$

$\sum_{i=1}^{k} (i \cdot i!) + [(k + 1) \cdot (k + 1)!] = [(k + 1)! - 1] + [(k + 1) \cdot (k + 1)!] = (k + 1)! +$

$(k + 1) \cdot (k + 1)! - 1 = [1 + (k + 1)](k + 1)! - 1 = (k + 2)(k + 1)! - 1 =$
$(k + 2)! - 1 = [(k + 1) + 1]! - 1.$

24. $S_1 = (1)(2)(3) = 6;\ S_2 = S_1 + (2)(3)(4) = 6 + 24 = 30;\ S_3 = 90.$ From Ex. 1,

$\sum_{i=1}^{n} i = \frac{n(n + 1)}{2}$, and from Ex. 6, $\sum_{i=1}^{n} i(i + 1) = \frac{n(n + 1)(n + 2)}{3}$. This pattern suggests

that $S_n = \sum_{i=1}^{n} i(i + 1)(i + 2) = \frac{n(n + 1)(n + 2)(n + 3)}{4}.$

(1) For $n = 1$, $1(1 + 1)(1 + 2) = 1(2)(3) = 6$ and $\dfrac{1(1 + 1)(1 + 2)(1 + 3)}{4} =$

$\dfrac{1(2)(3)(4)}{4} = 6.$

(2) If $S_k = \sum_{i=1}^{k} i(i + 1)(i + 2) = \dfrac{k(k + 1)(k + 2)(k + 3)}{4}$, then $S_{k+1} =$

$\sum_{i=1}^{k+1} i(i + 1)(i + 2) = \left[\sum_{i=1}^{k} i(i + 1)(i + 2)\right] +$

$(k + 1)[(k + 1) + 1][(k + 1) + 2] = \dfrac{k(k + 1)(k + 2)(k + 3)}{4} +$

$\dfrac{4(k + 1)(k + 2)(k + 3)}{4} = \dfrac{(k + 1)(k + 2)(k + 3)(k + 4)}{4} =$

$\dfrac{(k + 1)[(k + 1) + 1][(k + 1) + 2][(k + 1) + 3]}{4}.$

25. (1) For $n = 3$, the polygon is a triangle and has no diagonals, and
$\dfrac{3^2 - 3(3)}{2} = \dfrac{9 - 9}{2} = 0.$

(2) Assume that a convex polygon with k sides (and therefore k vertices, $v_1, v_2, \ldots, v_k$) where $k \geq 3$ has $\dfrac{k^2 - 3k}{2}$ diagonals. If one side (and therefore one vertex, v_{k+1}) is added to the polygon, there will be 1 additional diagonal joining v_1 and v_k, and $k - 2$ additional diagonals joining v_2 and v_{k+1}, v_3 and v_{k+1}, $\ldots$, v_{k-1} and v_{k+1}.

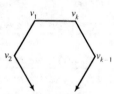

Therefore, a convex polygon with $k + 1$ sides has

$\dfrac{k^2 - 3k}{2} + 1 + (k - 2) = \dfrac{k^2 - 3k}{2} + \dfrac{2(k - 1)}{2} = \dfrac{k^2 - k - 2}{2}$

diagonals, and $\dfrac{(k + 1)^2 - 3(k + 1)}{2} = \dfrac{k^2 + 2k + 1 - 3k - 3}{2} =$

$\dfrac{k^2 - k - 2}{2}.$

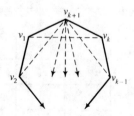

26. We want to prove that $\$n$ = a nonnegative integer $\times$ 2 + a nonnegative integer $\times$ 5 where n is an integer ≥ 4.

(1) For $n = 4$, $\$4 = 2 \cdot \$2 + 0 \cdot \$5$, and 2 and 0 are nonnegative integers.

(2) Assume that a debt of $\$k$, with $k \geq 4$, can be paid using only $2 bills and $5 bills.
(a) If $(k + 1)$ is an even number, then a debt of $\$(k + 1)$ can be paid with $2 bills.
(b) If $(k + 1)$ is odd, then k is even, and there exists some $r \geq 2$ such that $\$k = r \cdot \$2$. Then $\$(k + 1) = \$k + \$1 = (r \cdot \$2) + \$1 = (r - 2) \cdot \$2 + 2 \cdot \$2 + \$1 = (r - 2) \cdot \$2 + \$5$, where $r - 2$ is nonnegative.

Chapter Test · page 515

1. a. arithmetic; $t_n = 10 + (n - 1)(-4) = 14 - 4n$

b. neither; $t_n = n^2 + 2$ **c.** geometric; $t_n = \dfrac{2}{3}(-3)^n$ or $t_n = (-2)(-3)^{n-1}$

2. $t_2 = t_1 + d = 6$ and $t_6 = t_1 + 5d = 16$; $4d = 10$; $d = \dfrac{5}{2}$; $t_1 = 6 - \dfrac{5}{2} = \dfrac{7}{2}$;

$t_{21} = \dfrac{7}{2} + 20\left(\dfrac{5}{2}\right) = 53\dfrac{1}{2}$

3. $t_3 = t_1 r^2 = 9$ and $t_6 = t_1 r^5 = \dfrac{9}{8}$; $\dfrac{\frac{9}{8}}{r^2} = \dfrac{\frac{9}{8}}{r^5}$; $r^3 = \dfrac{1}{8}$; $r = \dfrac{1}{2}$; $t_1 = 9 \div \left(\dfrac{1}{2}\right)^2 = 36$;

$t_{12} = 36\left(\dfrac{1}{2}\right)^{11} = \dfrac{9}{512}$

4. $t_1 = 1$; $t_2 = 4$; $t_3 = 2 \cdot 4 + 1 = 9$; $t_4 = 2 \cdot 9 + 4 = 22$; $t_5 = 2 \cdot 22 + 9 = 53$; $t_6 = 2 \cdot 53 + 22 = 128$

5. $t_1 = 1$; $t_n = 4 \cdot t_{n-1} + 1$

6. a. $t_1 = 104$, $t_2 = 114$, $t_3 = 124$; this is an arithmetic sequence with $d = 10$; $t_n = 104 + (n - 1)10 = 94 + 10n$

b. $S_1 = 104$; $S_2 = 104 + 114 = 218$; $S_3 = 218 + 124 = 342$; $S_n = \dfrac{n(t_1 + t_n)}{2} = $

$\dfrac{n(104 + (94 + 10n))}{2} = n(99 + 5n) = 99n + 5n^2$

7. $r = -\dfrac{1}{3}$ and $t_1 = 27$; $S_6 = \dfrac{27\left(1 - \left(-\dfrac{1}{3}\right)^6\right)}{1 - \left(-\dfrac{1}{3}\right)} = \dfrac{3}{4}(27)\left(1 - \left(\dfrac{1}{3}\right)^6\right) = \dfrac{182}{9}$

8. Answers will vary. Example: An infinite sequence has a limit L if the nth term gets arbitrarily close to L as n becomes larger.

9. a. $\dfrac{3n^2 + 1}{4n^2 - 2n + 5} = \dfrac{3 + \dfrac{1}{n^2}}{4 - \dfrac{2}{n} + \dfrac{5}{n^2}}$; thus, $\lim\limits_{n \to \infty} \dfrac{3n^2 + 1}{4n^2 - 2n + 5} = \dfrac{3}{4}$

b. When n is even, $\cos n\pi = 1$; when n is odd, $\cos n\pi = -1$; thus, as $n \to \infty$, the even-numbered terms approach ∞ and the odd-numbered terms approach $-\infty$; the limit does not exist.

c. Since $\lim\limits_{n \to \infty} r^n = 0$ if $|r| < 1$, $\lim\limits_{n \to \infty} (0.59)^n = 0$.

10. $0.131313\ldots = 0.13 + 0.0013 + 0.000013 + \cdots ;\ t_1 = 0.13$ and $r = 0.01$;

$$S = \frac{0.13}{1 - 0.01} = \frac{13}{99}$$

11. a. The series is geometric with $r = \dfrac{3x}{2};\ |r| < 1;\ \left|\dfrac{3x}{2}\right| < 1;\ -1 < \dfrac{3x}{2} < 1;\ -\dfrac{2}{3} < x < \dfrac{2}{3}.$

b. $S = \dfrac{1}{1 - \dfrac{3x}{2}} = \dfrac{2}{2 - 3x}$

12. Answers may vary. **a.** $\displaystyle\sum_{n=0}^{4} \frac{(-1)^n}{(2n + 1)^2}$

b. $t_n = 7 + (n - 1)(-4) = 11 - 4n;\ \displaystyle\sum_{n-1}^{8} (11 - 4n)$

13. $\displaystyle\sum_{k=1}^{20} 3k(k + 2) = \sum_{k=1}^{20} (3k^2 + 6k) = 3\sum_{k=1}^{20} k^2 + 6\sum_{k=1}^{20} k = 3 \cdot \frac{20(21)(41)}{6} + 6 \cdot \frac{20(21)}{2} =$

$8610 + 1260 = 9870$

14. (1) For $n = 1,\ 2 = \dfrac{1(5 \cdot 1 - 1)}{2}$, so the statement is true when $n = 1$.

(2) Assume that $2 + 7 + 12 + \cdots + (5k - 3) = \dfrac{k(5k - 1)}{2}$. Prove that the statement is

true for $n = k + 1$: $2 + 7 + 12 + \cdots + (5k - 3) + [5(k + 1) - 3] = \dfrac{k(5k - 1)}{2} +$

$5(k + 1) - 3 = \dfrac{5k^2 - k}{2} + \dfrac{10k + 4}{2} = \dfrac{5k^2 + 9k + 4}{2} = \dfrac{(k + 1)(5k + 4)}{2} =$

$\dfrac{(k + 1)[5(k + 1) - 1]}{2}$; the statement is true for $n = k + 1$.

Class Exercises 14-1 · page 520

1. $A_{2\times1}$; $B_{1\times2}$; $C_{1\times2}$; $D_{1\times2}$ 2. A and B, A and D 3. $B = D$ 4. $A_{3\times2}$; $B_{3\times2}$; $C_{3\times3}$; $D_{2\times3}$

5. $A + B = \begin{bmatrix} 8 & 6 \\ 5 & 5 \\ -3 & 14 \end{bmatrix}$ 6. a. 2×3 b. $\begin{bmatrix} 5 & 3 & 0 \\ -1 & -4 & 6 \end{bmatrix}$

7. a. There are 10.4 g of carbohydrates in tomato juice.

   b. There are 0.002 g of sodium in a serving of orange juice.

   c. Nutritional information for two servings.

   d. N^t is a "nutrient by juice-type" matrix.

8. It may be looked at as a "student by grade" matrix.

Written Exercises 14-1 · pages 520–522

1. $\begin{bmatrix} -3 \\ 0 \end{bmatrix} + \begin{bmatrix} 1 \\ 2 \end{bmatrix} = \begin{bmatrix} -2 \\ 2 \end{bmatrix}$ 2. $[12 \quad 7] + [3 \quad -4] = [15 \quad 3]$

3. $\begin{bmatrix} 8 & 1 \\ -1 & 5 \end{bmatrix} - \begin{bmatrix} 3 & -2 \\ 4 & -1 \end{bmatrix} = \begin{bmatrix} 5 & 3 \\ -5 & 6 \end{bmatrix}$ 4. $[1 \quad -1 \quad 0] - [6 \quad -9 \quad -1] = [-5 \quad 8 \quad 1]$

5. $\begin{bmatrix} 8 & 2 & -2 \\ -3 & 1 & 14 \end{bmatrix} + \begin{bmatrix} 12 & 3 & 10 \\ 0 & 0 & -6 \end{bmatrix} = \begin{bmatrix} 20 & 5 & 8 \\ -3 & 1 & 8 \end{bmatrix}$

6. $\begin{bmatrix} 1 & -5 \\ -3 & 1 \\ 4 & 2 \end{bmatrix} - \begin{bmatrix} 0 & -2 \\ 0 & 7 \\ -8 & 3 \end{bmatrix} = \begin{bmatrix} 1 & -3 \\ -3 & -6 \\ 12 & -1 \end{bmatrix}$

7. $8\begin{bmatrix} 5 & -2 \\ 4 & 0 \end{bmatrix} = \begin{bmatrix} 40 & -16 \\ 32 & 0 \end{bmatrix}$

8. $-4\begin{bmatrix} 0 & -1 \\ -6 & 5 \end{bmatrix} = \begin{bmatrix} 0 & 4 \\ 24 & -20 \end{bmatrix}$

9. $2\begin{bmatrix} 3 & 0 \\ -4 & 1 \\ 0 & -1 \end{bmatrix} + \begin{bmatrix} -2 & -2 \\ 3 & 0 \\ 6 & 11 \end{bmatrix} = \begin{bmatrix} 6 & 0 \\ -8 & 2 \\ 0 & -2 \end{bmatrix} + \begin{bmatrix} -2 & -2 \\ 3 & 0 \\ 6 & 11 \end{bmatrix} = \begin{bmatrix} 4 & -2 \\ -5 & 2 \\ 6 & 9 \end{bmatrix}$

10. $\begin{bmatrix} 18 & 12 & 5 \\ 2 & 0 & -3 \end{bmatrix} - 5\begin{bmatrix} 4 & -1 & 3 \\ -3 & 2 & 5 \end{bmatrix} = \begin{bmatrix} 18 & 12 & 5 \\ 2 & 0 & -3 \end{bmatrix} + \begin{bmatrix} -20 & 5 & -15 \\ 15 & -10 & -25 \end{bmatrix} =$

$\begin{bmatrix} -2 & 17 & -10 \\ 17 & -10 & -28 \end{bmatrix}$

11. a. 3×4, 4×3, 4×3 b. $B + C = \begin{bmatrix} 17 & 1 & 9 \\ 7 & 3 & 8 \\ 5 & 17 & 13 \\ 12 & 4 & 0 \end{bmatrix}$

12. a. 4×3

b. $A^t = \begin{bmatrix} 3 & 0 & 3 \\ 2 & 1 & 7 \\ 4 & 8 & 9 \\ 8 & -2 & 1 \end{bmatrix}$; $A^t + B = \begin{bmatrix} 3 & 0 & 3 \\ 2 & 1 & 7 \\ 4 & 8 & 9 \\ 8 & -2 & 1 \end{bmatrix} + \begin{bmatrix} 8 & 1 & 4 \\ 3 & 0 & 7 \\ -2 & 5 & 11 \\ 6 & 4 & -3 \end{bmatrix} = \begin{bmatrix} 11 & 1 & 7 \\ 5 & 1 & 14 \\ 2 & 13 & 20 \\ 14 & 2 & -2 \end{bmatrix}$;

$C - A^t = \begin{bmatrix} 9 & 0 & 5 \\ 4 & 3 & 1 \\ 7 & 12 & 2 \\ 6 & 0 & 3 \end{bmatrix} - \begin{bmatrix} 3 & 0 & 3 \\ 2 & 1 & 7 \\ 4 & 8 & 9 \\ 8 & -2 & 1 \end{bmatrix} = \begin{bmatrix} 6 & 0 & 2 \\ 2 & 2 & -6 \\ 3 & 4 & -7 \\ -2 & 2 & 2 \end{bmatrix}$

13. $2B = \begin{bmatrix} 16 & 2 & 8 \\ 6 & 0 & 14 \\ -4 & 10 & 22 \\ 12 & 8 & -6 \end{bmatrix}$, $2B + C = \begin{bmatrix} 25 & 2 & 13 \\ 10 & 3 & 15 \\ 3 & 22 & 24 \\ 18 & 8 & -3 \end{bmatrix}$

14. $3C = \begin{bmatrix} 27 & 0 & 15 \\ 12 & 9 & 3 \\ 21 & 36 & 6 \\ 18 & 0 & 9 \end{bmatrix}$, $3C - B = \begin{bmatrix} 19 & -1 & 11 \\ 9 & 9 & -4 \\ 23 & 31 & -5 \\ 12 & -4 & 12 \end{bmatrix}$

15. a.

$$A = \begin{array}{c} \\ D \\ M \end{array} \begin{array}{ccc} l & m & h \\ \left[\begin{array}{ccc} 31 & 42 & 18 \\ 22 & 25 & 11 \end{array}\right] \end{array}$$

b. Answers may vary depending on the matrix: location by mower-type

c.

$$M = \begin{array}{c} \\ D \\ M \end{array} \begin{array}{ccc} l & m & h \\ \left[\begin{array}{ccc} 28 & 29 & 20 \\ 20 & 18 & 9 \end{array}\right] \end{array}$$

d.

$$A + M = \begin{array}{c} \\ D \\ M \end{array} \begin{array}{ccc} l & m & h \\ \left[\begin{array}{ccc} 59 & 71 & 38 \\ 42 & 43 & 20 \end{array}\right] \end{array}$$; this matrix sum represents the total sales for April and May of each type of lawnmower at each location.

e. $1.08 \cdot 18 = 19.44$, about 20; round up so that the store is not understocked; 1.08

f. 9% for April and 15% for May

16. a. Ten colonial chairs to be shipped this week; 40 colonial chairs in the warehouse

b. 4×3; 4×3

c. $T - S = \begin{bmatrix} 10 & 10 & 14 \\ 30 & 13 & 14 \\ 15 & 38 & 15 \\ 18 & 16 & 22 \end{bmatrix}$; warehouse inventory after the shipment

d. $T - S - 1.5S = T - 2.5S$

17. $\begin{bmatrix} x & 3 \\ -7 & 0 \end{bmatrix} = \begin{bmatrix} 2 & y \\ -7 & z+1 \end{bmatrix}$; $x = 2$, $y = 3$, $z + 1 = 0$; $x = 2$, $y = 3$, $z = -1$

18. $\begin{bmatrix} a-3 & 5 \\ 2-c & 4b \end{bmatrix} = \begin{bmatrix} -1 & 2x+1 \\ -8 & 12 \end{bmatrix}$; $a - 3 = -1$, $a = 2$; $5 = 2x + 1$, $x = 2$; $2 - c = -8$, $c = 10$; $4b = 12$, $b = 3$

19. $(A^t)^t = A$

20. True; for example, $A = [1 \quad 3]$; $2A = [2 \quad 6]$; $(2A)^t = \begin{bmatrix} 2 \\ 6 \end{bmatrix}$; $A^t = \begin{bmatrix} 1 \\ 3 \end{bmatrix}$; $2A^t = \begin{bmatrix} 2 \\ 6 \end{bmatrix}$, so

$(2A)^t = 2A^t$.

21. True; for example, if $A = \begin{bmatrix} -1 & 2 \\ 3 & 1 \end{bmatrix}$ and $B = \begin{bmatrix} -4 & 0 \\ 1 & -1 \end{bmatrix}$, then $(A + B)^t = \begin{bmatrix} -5 & 2 \\ 4 & 0 \end{bmatrix}^t =$

$\begin{bmatrix} -5 & 4 \\ 2 & 0 \end{bmatrix}$; $A^t + B^t = \begin{bmatrix} -1 & 3 \\ 2 & 1 \end{bmatrix} + \begin{bmatrix} -4 & 1 \\ 0 & -1 \end{bmatrix} = \begin{bmatrix} -5 & 4 \\ 2 & 0 \end{bmatrix}$; $(A + B)^t = A^t + B^t$.

22. Answers will vary. One example follows: The sum of a 2×3 matrix representing two students' grade percentages over three marking periods and a 2×3 matrix representing the students' heights in cm over the same time period would make no sense because each matrix measures different things in different units.

23. Answers may vary depending on the edition of the almanac used.

24. Answers will vary.

Class Exercises · page 525

1. Yes; no; $AB_{3\times2}$ **2.** No; yes; $BA_{2\times6}$ **3.** No; yes; $BA_{3\times4}$ **4.** Yes; no; $AB_{5\times3}$

5. No; no   **6.** Yes; yes; $AB_{3\times3}$, $BA_{4\times4}$

7. $\begin{bmatrix} 1 & 1 \\ 2 & 3 \end{bmatrix}\begin{bmatrix} 4 & 2 \\ 5 & 6 \end{bmatrix} = \begin{bmatrix} 1(4) + 1(5) & 1(2) + 1(6) \\ 2(4) + 3(5) & 2(2) + 3(6) \end{bmatrix} = \begin{bmatrix} 9 & 8 \\ 23 & 22 \end{bmatrix}$

8. $\begin{bmatrix} 1 & 3 & 1 \\ 2 & 0 & 4 \end{bmatrix}\begin{bmatrix} 2 & 1 \\ 1 & 3 \\ -1 & 0 \end{bmatrix} = \begin{bmatrix} 1(2) + 3(1) + 1(-1) & 1(1) + 3(3) + 1(0) \\ 2(2) + 0(1) + 4(-1) & 2(1) + 0(3) + 4(0) \end{bmatrix} = \begin{bmatrix} 4 & 10 \\ 0 & 2 \end{bmatrix}$

9. *CM*

10. No; yes; suppose matrix A is a "car model by profit" matrix and matrix B represents the number of car models sold for each of 4 months, then $A^t \cdot B$ represents the profit for each of 4 months.

Written Exercises 14-2 · pages 526–529

1. $\begin{bmatrix} 4 & 3 \\ -1 & -2 \end{bmatrix}\begin{bmatrix} 5 \\ 1 \end{bmatrix} = \begin{bmatrix} 4(5) + 3(1) \\ -1(5) + (-2)(1) \end{bmatrix} = \begin{bmatrix} 23 \\ -7 \end{bmatrix}$ **2.** Not defined

3. $\begin{bmatrix} 1 & -5 \\ 2 & 3 \end{bmatrix}\begin{bmatrix} 4 & -4 \\ 0 & 1 \end{bmatrix} = \begin{bmatrix} 1(4) - 5(0) & 1(-4) - 5(1) \\ 2(4) + 3(0) & 2(-4) + 3(1) \end{bmatrix} = \begin{bmatrix} 4 & -9 \\ 8 & -5 \end{bmatrix}$

4. $\begin{bmatrix} -2 & 3 \\ 4 & 2 \end{bmatrix}\begin{bmatrix} 0 & 3 \\ -6 & 5 \end{bmatrix} = \begin{bmatrix} -2(0) + 3(-6) & -2(3) + 3(5) \\ 4(0) + 2(-6) & 4(3) + 2(5) \end{bmatrix} = \begin{bmatrix} -18 & 9 \\ -12 & 22 \end{bmatrix}$

5. Not defined   **6.** $[7 \quad 1 \quad -3 \quad 4]\begin{bmatrix} 4 & 1 \\ -3 & 8 \\ 9 & 5 \\ -2 & 6 \end{bmatrix} =$

$[7(4) + 1(-3) + (-3)(9) + 4(-2) \quad 7(1) + 1(8) - 3(5) + 4(6)] = [-10 \quad 24]$

7. $\begin{bmatrix} 9 & -4 & 4 \\ 2 & -1 & -6 \end{bmatrix} \begin{bmatrix} 2 & -1 & 0 \\ 0 & 1 & -3 \\ 3 & 5 & 2 \end{bmatrix} =$

$\begin{bmatrix} 9(2) - 4(0) + 4(3) & 9(-1) - 4(1) + 4(5) & 9(0) - 4(-3) + 4(2) \\ 2(2) - 1(0) - 6(3) & 2(-1) - 1(1) - 6(5) & 2(0) - 1(-3) - 6(2) \end{bmatrix} =$

$\begin{bmatrix} 30 & 7 & 20 \\ -14 & -33 & -9 \end{bmatrix}$

8. Not defined

9. $\begin{bmatrix} 1 & 0 & 0 \\ 0 & 1 & 0 \\ 0 & 0 & 1 \end{bmatrix} \begin{bmatrix} a & b & c \\ d & e & f \\ g & h & i \end{bmatrix} = \begin{bmatrix} 1 \cdot a + 0 + 0 & 1 \cdot b + 0 + 0 & 1 \cdot c + 0 + 0 \\ 0 + 1 \cdot d + 0 & 0 + 1 \cdot e + 0 & 0 + 1 \cdot f + 0 \\ 0 + 0 + 1 \cdot g & 0 + 0 + 1 \cdot h & 0 + 0 + 1 \cdot i \end{bmatrix} =$

$\begin{bmatrix} a & b & c \\ d & e & f \\ g & h & i \end{bmatrix}$

10. $\begin{bmatrix} 9 & 4 \\ 3 & 1 \\ 2 & 8 \\ 1 & 5 \end{bmatrix} \begin{bmatrix} 4 & 2 & 1 \\ 3 & 0 & 2 \end{bmatrix} = \begin{bmatrix} 9(4) + 4(3) & 9(2) + 4(0) & 9(1) + 4(2) \\ 3(4) + 1(3) & 3(2) + 1(0) & 3(1) + 1(2) \\ 2(4) + 8(3) & 2(2) + 8(0) & 2(1) + 8(2) \\ 1(4) + 5(3) & 1(2) + 5(0) & 1(1) + 5(2) \end{bmatrix} = \begin{bmatrix} 48 & 18 & 17 \\ 15 & 6 & 5 \\ 32 & 4 & 18 \\ 19 & 2 & 11 \end{bmatrix}$

11. Answers will vary. Examples are given.

a. Let $A = \begin{bmatrix} 1 & 0 \\ 0 & 2 \end{bmatrix}$ and $B = \begin{bmatrix} 3 & -1 \\ -2 & 1 \end{bmatrix}$; $AB = \begin{bmatrix} 3 & -1 \\ -4 & 2 \end{bmatrix}$ and $BA = \begin{bmatrix} 3 & -2 \\ -2 & 2 \end{bmatrix}$.

b. Yes. The sum of two matrices that have the same dimensions is the matrix whose elements are the sums of the corresponding elements of the matrices being added. Changing the order in which matrices are added only changes the order in which elements are added. Since addition of real numbers is commutative, addition of matrices is commutative.

12. $CD = \begin{bmatrix} 5 & 2 \\ 7 & 3 \end{bmatrix} \begin{bmatrix} 3 & -2 \\ -7 & 5 \end{bmatrix} = \begin{bmatrix} 1 & 0 \\ 0 & 1 \end{bmatrix}$ and $DC = \begin{bmatrix} 3 & -2 \\ -7 & 5 \end{bmatrix} \begin{bmatrix} 5 & 2 \\ 7 & 3 \end{bmatrix} = \begin{bmatrix} 1 & 0 \\ 0 & 1 \end{bmatrix}$

13. a. Answers will vary. For example, let $A = \begin{bmatrix} 1 & 1 \\ 1 & 1 \end{bmatrix}$, $B = \begin{bmatrix} 2 & 0 \\ -3 & 1 \end{bmatrix}$, and $C = \begin{bmatrix} -1 & 1 \\ 1 & -2 \end{bmatrix}$;

$BC = \begin{bmatrix} -2 & 2 \\ 4 & -5 \end{bmatrix}$; $AB = \begin{bmatrix} -1 & 1 \\ -1 & 1 \end{bmatrix}$; $A(BC) = \begin{bmatrix} 1 & 1 \\ 1 & 1 \end{bmatrix} \begin{bmatrix} -2 & 2 \\ 4 & -5 \end{bmatrix} = \begin{bmatrix} 2 & -3 \\ 2 & -3 \end{bmatrix} =$

$\begin{bmatrix} -1 & 1 \\ -1 & 1 \end{bmatrix} \begin{bmatrix} -1 & 1 \\ 1 & -2 \end{bmatrix} = (AB)C$

b. Yes

c. Matrix addition is associative. Examples will vary.

14. a. Answers will vary. For example, using matrices A, B, and C from the solution to

Ex. 13, $AC = \begin{bmatrix} 0 & -1 \\ 0 & -1 \end{bmatrix}$; $A(B + C) = \begin{bmatrix} 1 & 1 \\ 1 & 1 \end{bmatrix} \begin{bmatrix} 1 & 1 \\ -2 & -1 \end{bmatrix} = \begin{bmatrix} -1 & 0 \\ -1 & 0 \end{bmatrix}$ and

$AB + AC = \begin{bmatrix} -1 & 1 \\ -1 & 1 \end{bmatrix} + \begin{bmatrix} 0 & -1 \\ 0 & -1 \end{bmatrix} = \begin{bmatrix} -1 & 0 \\ -1 & 0 \end{bmatrix}$; $A(B + C) = AB + AC$

b. Yes

15. $PS = [37{,}200 \quad 35{,}050]$, which represents the profit for each dealer in the month of March.

16. a. $NP = \begin{bmatrix} 272 \\ 360.5 \\ 180.75 \end{bmatrix}$ **b.** 360.5 mg

17. a. PM is defined because the number of columns in P equals the number of rows in M. PM is not meaningful because the columns of P do not represent the same things as the rows of M.

b.
$$
\begin{array}{c}
 \\ A \\ B \\ C
\end{array}
\begin{array}{ccc}
\text{Jan.} & \text{Feb.} & \text{Mar.} \\
\left[\begin{array}{ccc} 1600 & 2200 & 2400 \\ 1200 & 2300 & 2800 \\ 2400 & 3800 & 4400 \end{array}\right]
\end{array} ; 2400
$$

c. There are 3 columns in M and only 2 rows in R.

d.
$$
\begin{array}{c}
 \\ x \\ y
\end{array}
\begin{array}{ccc}
s & b & g \\
\left[\begin{array}{ccc} 17 & 21 & 37 \\ 22 & 28 & 50 \end{array}\right]
\end{array} ;
$$ number of relays of each type needed for each type of calculator

e.
$$
\begin{array}{c}
 \\ x \\ y
\end{array}
\begin{array}{ccc}
\text{Jan.} & \text{Feb.} & \text{Mar.} \\
\left[\begin{array}{ccc} 16{,}400 & 25{,}500 & 29{,}200 \\ 21{,}600 & 33{,}800 & 38{,}800 \end{array}\right]
\end{array} ;
$$ number of relays of each type needed per month

18. a. $C^t D$ gives the cost of each menu item per day; $D^t C$ gives the cost per day of each menu item. $C^t D$ and $D^t C$ give the same information, but one matrix is the transpose of the other.

b. $CA = \begin{bmatrix} 120 \\ 100 \\ 144 \\ 124 \end{bmatrix}$, which gives the cost of one serving of each of the menus.

d. $BD = [222 \quad 229 \quad 229 \quad 235 \quad 213]$, which gives the number of meals served per day.

19. a. $(0.3)8 + (0.2)7 + (0.5)9 = 8.3$

b. Let
$$
\begin{array}{c}
\text{precision} \\ \text{finish} \\ \text{design}
\end{array}
\begin{array}{c}
\text{weight} \\
\left[\begin{array}{c} 0.3 \\ 0.2 \\ 0.5 \end{array}\right]
\end{array} = W \text{ and }
\begin{array}{c}
 \\ S \\ J \\ E \\ F \\ A \\ D
\end{array}
\begin{array}{ccc}
\text{precision} & \text{finish} & \text{design} \\
\left[\begin{array}{ccc} 8 & 7 & 9 \\ 8 & 6 & 10 \\ 6 & 8 & 10 \\ 9 & 10 & 7 \\ 10 & 10 & 6 \\ 8 & 7 & 8 \end{array}\right]
\end{array} = S;
$$

the total score for each student $= SW = \begin{bmatrix} 8.3 \\ 8.6 \\ 8.4 \\ 8.2 \\ 8.0 \\ 7.8 \end{bmatrix}$;
1st: Joyce
2nd: Eduardo
3rd: Stephanie

$$\begin{array}{c} \\ \textbf{20. cost } [2 \quad 5 \end{array} \begin{array}{ccc} \text{s} & \text{a} & \text{s.c.} \\ & & 4 \ ], \end{array} \begin{array}{c} \\ \begin{array}{c} \text{s} \\ \text{a} \\ \text{s.c.} \end{array} \end{array} \begin{array}{c} \text{Fri.} \quad \text{Sat.} \\ \begin{bmatrix} 121 & 183 \\ 164 & 140 \\ 32 & 25 \end{bmatrix} \end{array}$$; product = [1190 \quad 1166]; Fri.: \$1190, Sat.: \$1166

21. a. $\begin{array}{c} \\ \text{R\&D} \\ \text{Sales} \end{array} \begin{array}{c} \text{f} \quad \text{b} \quad \text{c} \\ \begin{bmatrix} 5 & 3 & 2 \\ 0 & 4 & 8 \end{bmatrix} \end{array} = T$

b. $\begin{array}{c} \\ \text{f} \\ \text{b} \\ \text{c} \end{array} \begin{array}{cccc} \text{A} & \text{B} & \text{C} & \text{D} \\ \begin{bmatrix} 1280 & 1400 & 1320 & 1450 \\ 922 & 1024 & 905 & 1050 \\ 676 & 728 & 654 & 734 \end{bmatrix} \end{array}$; $\begin{array}{c} \\ \text{R\&D} \\ \text{Sales} \end{array} \begin{array}{cccc} \text{A} & \text{B} & \text{C} & \text{D} \\ \begin{bmatrix} 10{,}518 & 11{,}528 & 10{,}623 & 11{,}868 \\ 9096 & 9920 & 8852 & 10{,}072 \end{bmatrix} \end{array} = TP$

c. \$10,072 **d.** airline A

22. $N = \begin{bmatrix} 1.05 & 0 & 0 & 0 \\ 0 & 1.04 & 0 & 0 \\ 0 & 0 & 1.06 & 0 \\ 0 & 0 & 0 & 1.02 \end{bmatrix}$; $PN = \begin{array}{c} \\ \text{f} \\ \text{b} \\ \text{c} \end{array} \begin{array}{cccc} \text{A} & \text{B} & \text{C} & \text{D} \\ \begin{bmatrix} 1.05a_1 & 1.04b_1 & 1.06c_1 & 1.02d_1 \\ 1.05a_2 & 1.04b_2 & 1.06c_2 & 1.02d_2 \\ 1.05a_3 & 1.04b_3 & 1.06c_3 & 1.02d_3 \end{bmatrix} \end{array}$

Class Exercises · page 534

1. $-B = \begin{bmatrix} -7 & 1 \\ -6 & -2 \end{bmatrix}$, $-C = \begin{bmatrix} -3 & -2 & -5 \\ -4 & -3 & 0 \end{bmatrix}$ **2.** $A^{-1} = \begin{bmatrix} 2 & 3 \\ 5 & 8 \end{bmatrix}$, $B^{-1} = \begin{bmatrix} \dfrac{1}{10} & \dfrac{1}{20} \\ -\dfrac{3}{10} & \dfrac{7}{20} \end{bmatrix}$

3. C is not a square matrix and $|D| = 0$.

4. $A^2 = \begin{bmatrix} 79 & -30 \\ -50 & 19 \end{bmatrix}$, both calculations are correct because multiplication of matrices is associative.

5. Multiplication of matrices is not commutative.

6. $3X = \begin{bmatrix} -9 & -6 \\ -3 & -15 \end{bmatrix}$ Simplify by performing the indicated subtraction.

$X = \begin{bmatrix} -3 & -2 \\ -1 & -5 \end{bmatrix}$ Simplify by using scalar multiplication.

Written Exercises · pages 534–537

1. a. $\begin{bmatrix} -\dfrac{7}{3} & \dfrac{4}{3} \\ 2 & -1 \end{bmatrix}$ **b.** $\begin{bmatrix} 1 & 0 \\ 0 & 1 \end{bmatrix}$ **2. a.** $\begin{bmatrix} \dfrac{7}{2} & \dfrac{11}{2} \\ -\dfrac{3}{2} & -\dfrac{5}{2} \end{bmatrix}$ **b.** $\begin{bmatrix} 1 & 0 \\ 0 & 1 \end{bmatrix}$

3. a. $\begin{bmatrix} -9 & -3 \\ -12 & -4 \end{bmatrix}$ **b.** not defined **4. a.** not defined **b.** $\begin{bmatrix} -2 & 0 & 8 & -1 \\ -5 & -7 & -3 & -9 \end{bmatrix}$

5. a. $\begin{bmatrix} -3 & -1 \\ -4 & -\dfrac{4}{3} \end{bmatrix}$ **b.** not defined **6. a.** not defined **b.** $\begin{bmatrix} 26 & 66 \\ -18 & -46 \end{bmatrix}$

7. a. $DX = E$; $D^{-1}(DX) = D^{-1}E$; $(D^{-1}D)X = D^{-1}E$; $IX = D^{-1}E$; $X = D^{-1}E$

b. $XD = E$; $(XD)D^{-1} = ED^{-1}$; $X(DD^{-1}) = ED^{-1}$; $XI = ED^{-1}$; $X = ED^{-1}$

8. a. $TX - S = R$; $TX = R + S$; $X = T^{-1}(R + S)$

b. $XT - S = R$; $XT = R + S$; $X = (R + S)T^{-1}$

9. $X + 3\begin{bmatrix} 2 & 5 \\ 1 & 4 \end{bmatrix} = \begin{bmatrix} 12 & 7 \\ 0 & -8 \end{bmatrix}$; $X + \begin{bmatrix} 6 & 15 \\ 3 & 12 \end{bmatrix} = \begin{bmatrix} 12 & 7 \\ 0 & -8 \end{bmatrix}$; $X = \begin{bmatrix} 12 & 7 \\ 0 & -8 \end{bmatrix} - \begin{bmatrix} 6 & 15 \\ 3 & 12 \end{bmatrix} =$

$\begin{bmatrix} 6 & -8 \\ -3 & -20 \end{bmatrix}$

10. $2X - \begin{bmatrix} 1 & 5 \\ 9 & 2 \end{bmatrix} = \begin{bmatrix} 7 & 11 \\ -1 & 12 \end{bmatrix}$; $2X = \begin{bmatrix} 7 & 11 \\ -1 & 12 \end{bmatrix} + \begin{bmatrix} 1 & 5 \\ 9 & 2 \end{bmatrix}$; $2X = \begin{bmatrix} 8 & 16 \\ 8 & 14 \end{bmatrix}$; $X = \begin{bmatrix} 4 & 8 \\ 4 & 7 \end{bmatrix}$

11. $\begin{bmatrix} 3 & 2 \\ 7 & 5 \end{bmatrix}X = \begin{bmatrix} 9 & 3 \\ 9 & 1 \end{bmatrix}$; $\begin{bmatrix} 5 & -2 \\ -7 & 3 \end{bmatrix}\begin{bmatrix} 3 & 2 \\ 7 & 5 \end{bmatrix}X = \begin{bmatrix} 5 & -2 \\ -7 & 3 \end{bmatrix}\begin{bmatrix} 9 & 3 \\ 9 & 1 \end{bmatrix}$; $X = \begin{bmatrix} 27 & 13 \\ -36 & -18 \end{bmatrix}$

12. $\begin{bmatrix} 5 & 1 \\ 7 & 2 \end{bmatrix}X = \begin{bmatrix} 3 & 0 & 5 \\ 1 & 4 & 2 \end{bmatrix}$; $\frac{1}{3}\begin{bmatrix} 2 & -1 \\ -7 & 5 \end{bmatrix}X = \frac{1}{3}\begin{bmatrix} 2 & -1 \\ -7 & 5 \end{bmatrix}\begin{bmatrix} 3 & 0 & 5 \\ 1 & 4 & 2 \end{bmatrix}$;

$X = \begin{bmatrix} \frac{5}{3} & -\frac{4}{3} & \frac{8}{3} \\ -\frac{16}{3} & \frac{20}{3} & -\frac{25}{3} \end{bmatrix}$

13. $\begin{bmatrix} 4 & 2 \\ 1 & 1 \end{bmatrix}X + \begin{bmatrix} 2 & 7 \\ 6 & 3 \end{bmatrix} = \begin{bmatrix} 2 & 3 \\ 0 & 1 \end{bmatrix}$; $\begin{bmatrix} 4 & 2 \\ 1 & 1 \end{bmatrix}X = \begin{bmatrix} 2 & 3 \\ 0 & 1 \end{bmatrix} - \begin{bmatrix} 2 & 7 \\ 6 & 3 \end{bmatrix}$; $\begin{bmatrix} 4 & 2 \\ 1 & 1 \end{bmatrix}X = \begin{bmatrix} 0 & -4 \\ -6 & -2 \end{bmatrix}$;

$\frac{1}{2}\begin{bmatrix} 1 & -2 \\ -1 & 4 \end{bmatrix}\begin{bmatrix} 4 & 2 \\ 1 & 1 \end{bmatrix}X = \frac{1}{2}\begin{bmatrix} 1 & -2 \\ -1 & 4 \end{bmatrix}\begin{bmatrix} 0 & -4 \\ -6 & -2 \end{bmatrix}$; $X = \begin{bmatrix} 6 & 0 \\ -12 & -2 \end{bmatrix}$

14. $\begin{bmatrix} 5 & 3 \\ 3 & 2 \end{bmatrix}X - \begin{bmatrix} 1 & 3 \\ 2 & 1 \end{bmatrix} = \begin{bmatrix} 1 & 2 \\ -1 & 3 \end{bmatrix}$; $\begin{bmatrix} 5 & 3 \\ 3 & 2 \end{bmatrix}X = \begin{bmatrix} 1 & 2 \\ -1 & 3 \end{bmatrix} + \begin{bmatrix} 1 & 3 \\ 2 & 1 \end{bmatrix}$;

$\begin{bmatrix} 5 & 3 \\ 3 & 2 \end{bmatrix}X = \begin{bmatrix} 2 & 5 \\ 1 & 4 \end{bmatrix}$; $\begin{bmatrix} 2 & -3 \\ -3 & 5 \end{bmatrix}\begin{bmatrix} 5 & 3 \\ 3 & 2 \end{bmatrix}X = \begin{bmatrix} 2 & -3 \\ -3 & 5 \end{bmatrix}\begin{bmatrix} 2 & 5 \\ 1 & 4 \end{bmatrix}$; $X = \begin{bmatrix} 1 & -2 \\ -1 & 5 \end{bmatrix}$

15. a. $\begin{bmatrix} 8 & -2 \\ 12 & 3 \end{bmatrix}\begin{bmatrix} x \\ y \end{bmatrix} = \begin{bmatrix} -14 \\ 9 \end{bmatrix}$; $\frac{1}{48}\begin{bmatrix} 3 & 2 \\ -12 & 8 \end{bmatrix}\begin{bmatrix} 8 & -2 \\ 12 & 3 \end{bmatrix}\begin{bmatrix} x \\ y \end{bmatrix} = \frac{1}{48}\begin{bmatrix} 3 & 2 \\ -12 & 8 \end{bmatrix}\begin{bmatrix} -14 \\ 9 \end{bmatrix}$;

$\begin{bmatrix} x \\ y \end{bmatrix} = \frac{1}{48}\begin{bmatrix} -24 \\ 240 \end{bmatrix} = \begin{bmatrix} -\frac{1}{2} \\ 5 \end{bmatrix}$; $x = -\frac{1}{2}$, $y = 5$

b. $\begin{bmatrix} 8 & -2 \\ 12 & 3 \end{bmatrix}\begin{bmatrix} x \\ y \end{bmatrix} = \begin{bmatrix} 38 \\ 39 \end{bmatrix}$; $\frac{1}{48}\begin{bmatrix} 3 & 2 \\ -12 & 8 \end{bmatrix}\begin{bmatrix} 8 & -2 \\ 12 & 3 \end{bmatrix}\begin{bmatrix} x \\ y \end{bmatrix} = \frac{1}{48}\begin{bmatrix} 3 & 2 \\ -12 & 8 \end{bmatrix}\begin{bmatrix} 38 \\ 39 \end{bmatrix}$;

$\begin{bmatrix} x \\ y \end{bmatrix} = \frac{1}{48}\begin{bmatrix} 192 \\ -144 \end{bmatrix}$; $\begin{bmatrix} x \\ y \end{bmatrix} = \begin{bmatrix} 4 \\ -3 \end{bmatrix}$; $x = 4$, $y = -3$

16. a. $\begin{bmatrix} 12 & -4 \\ 17 & -5 \end{bmatrix}\begin{bmatrix} x \\ y \end{bmatrix} = \begin{bmatrix} 3 \\ 8 \end{bmatrix}$; $\dfrac{1}{8}\begin{bmatrix} -5 & 4 \\ -17 & 12 \end{bmatrix}\begin{bmatrix} 12 & -4 \\ 17 & -5 \end{bmatrix}\begin{bmatrix} x \\ y \end{bmatrix} = \dfrac{1}{8}\begin{bmatrix} -5 & 4 \\ -17 & 12 \end{bmatrix}\begin{bmatrix} 3 \\ 8 \end{bmatrix}$;

$\begin{bmatrix} x \\ y \end{bmatrix} = \dfrac{1}{8}\begin{bmatrix} 17 \\ 45 \end{bmatrix}$; $x = \frac{17}{8}$, $y = \frac{45}{8}$

b. $\begin{bmatrix} 12 & -4 \\ 17 & -5 \end{bmatrix}\begin{bmatrix} x \\ y \end{bmatrix} = \begin{bmatrix} 0 \\ -16 \end{bmatrix}$; $\dfrac{1}{8}\begin{bmatrix} -5 & 4 \\ -17 & 12 \end{bmatrix}\begin{bmatrix} 12 & -4 \\ 17 & -5 \end{bmatrix}\begin{bmatrix} x \\ y \end{bmatrix} = \dfrac{1}{8}\begin{bmatrix} -5 & 4 \\ -17 & 12 \end{bmatrix}\begin{bmatrix} 0 \\ -16 \end{bmatrix}$;

$\begin{bmatrix} x \\ y \end{bmatrix} = \dfrac{1}{8}\begin{bmatrix} -64 \\ -192 \end{bmatrix}$; $\begin{bmatrix} x \\ y \end{bmatrix} = \begin{bmatrix} -8 \\ -24 \end{bmatrix}$; $x = -8$, $y = -24$

17. a. It cannot be done because the coefficient matrix has no inverse.

b.   There is no solution because the lines are parallel.

18. a. It cannot be done because the coefficient matrix has no inverse.

b.   There are infinitely many solutions because the graphs are coincident.

19. If we assume that A is an $m \times n$ matrix where $m \neq n$, then the equation $A_{m \times n} \cdot I = A_{m \times n}$ implies that I must be an $n \times n$ identity matrix. If this is the case, however, then the product $I_{n \times n} \cdot A_{m \times n}$ is undefined. Thus, the identity property for matrix multiplication applies only to a square matrix A. A similar argument can be made for the inverse property for matrix multiplication.

20. a. $\begin{bmatrix} 1 & 2 \\ 3 & 4 \end{bmatrix}\begin{bmatrix} 0 & -5 \\ 7 & 9 \end{bmatrix} + \begin{bmatrix} 1 & 2 \\ 3 & 4 \end{bmatrix}\begin{bmatrix} 4 & 1 \\ -1 & -3 \end{bmatrix} = \begin{bmatrix} 1 & 2 \\ 3 & 4 \end{bmatrix}\left(\begin{bmatrix} 0 & -5 \\ 7 & 9 \end{bmatrix} + \begin{bmatrix} 4 & 1 \\ -1 & -3 \end{bmatrix}\right) =$

$\begin{bmatrix} 1 & 2 \\ 3 & 4 \end{bmatrix}\begin{bmatrix} 4 & -4 \\ 6 & 6 \end{bmatrix} = \begin{bmatrix} 16 & 8 \\ 36 & 12 \end{bmatrix}$

b. $X\begin{bmatrix} 5 & 0 \\ 2 & 3 \end{bmatrix} - X\begin{bmatrix} 4 & -7 \\ 2 & 2 \end{bmatrix} = X\left(\begin{bmatrix} 5 & 0 \\ 2 & 3 \end{bmatrix} - \begin{bmatrix} 4 & -7 \\ 2 & 2 \end{bmatrix}\right) = X\begin{bmatrix} 1 & 7 \\ 0 & 1 \end{bmatrix}$; $X\begin{bmatrix} 1 & 7 \\ 0 & 1 \end{bmatrix} =$

$\begin{bmatrix} 2 & 3 \\ -5 & 7 \end{bmatrix}$; $X\begin{bmatrix} 1 & 7 \\ 0 & 1 \end{bmatrix}\begin{bmatrix} 1 & -7 \\ 0 & 1 \end{bmatrix} = \begin{bmatrix} 2 & 3 \\ -5 & 7 \end{bmatrix}\begin{bmatrix} 1 & -7 \\ 0 & 1 \end{bmatrix}$; $X = \begin{bmatrix} 2 & -11 \\ -5 & 42 \end{bmatrix}$

21. $X - AX = D$; $IX - AX = D$; $(I - A)X = D$; $(I - A)^{-1}(I - A)X = (I - A)^{-1}D$
$X = (I - A)^{-1}D$

22. $X - XA = D$; $XI - XA = D$; $X(I - A) = D$; $X(I - A)(I - A)^{-1} = D(I - A)^{-1}$;
$X = D(I - A)^{-1}$

23. $\begin{bmatrix} 3 & 5 & -9 \\ 4 & 7 & 2 \\ 6 & -9 & -8 \end{bmatrix}\begin{bmatrix} x \\ y \\ z \end{bmatrix} = \begin{bmatrix} 26 \\ -7 \\ 3 \end{bmatrix}$; $x = -2$, $y = 1$, $z = -3$

24. $\begin{bmatrix} 3 & 4 & -5 & -11 \\ 5 & -2 & -7 & 8 \\ 2 & -4 & -8 & 16 \\ 0 & 3 & 2 & 8 \end{bmatrix} \begin{bmatrix} a \\ b \\ c \\ d \end{bmatrix} = \begin{bmatrix} 5 \\ 6 \\ -3 \\ 4 \end{bmatrix}$; $a = \dfrac{5}{2}$, $b = \dfrac{3}{5}$, $c = \dfrac{5}{6}$, $d = \dfrac{1}{15}$

25. $\begin{bmatrix} 3 & 2 & 2 & 6 \\ 2 & 3 & 5 & 0 \\ 8 & 6 & 4 & 7 \\ 5 & 5 & 8 & 6 \end{bmatrix} \begin{bmatrix} w \\ x \\ y \\ z \end{bmatrix} = \begin{bmatrix} 78 \\ 67 \\ 146 \\ 153 \end{bmatrix}$; $w \approx 4$, $x \approx 6$, $y \approx 8$, $z \approx 6$;

I: 4 oz, II: 6 oz, III: 8 oz, IV: 6 oz

26. a. $\begin{bmatrix} 120 & 105 & 125 \\ 40 & 65 & 110 \\ 80 & 90 & 125 \end{bmatrix} \begin{bmatrix} t \\ c \\ d \end{bmatrix} = \begin{bmatrix} 20{,}250 \\ 12{,}070 \\ 17{,}000 \end{bmatrix}$; 25 tables, 150 chairs, 12 desks

b. $\begin{bmatrix} 120 & 105 & 125 \\ 40 & 65 & 110 \\ 80 & 90 & 125 \end{bmatrix} \begin{bmatrix} t \\ c \\ d \end{bmatrix} = \begin{bmatrix} 14{,}960 \\ 8970 \\ 12{,}590 \end{bmatrix}$; 18 tables, 110 chairs, 10 desks

27. $\left. \begin{array}{l} 16a + 4b + c = 11 \\ 36a - 6b + c = -9 \\ 64a - 8b + c = 61 \end{array} \right\}$ $\begin{bmatrix} 16 & 4 & 1 \\ 36 & -6 & 1 \\ 64 & -8 & 1 \end{bmatrix} \begin{bmatrix} a \\ b \\ c \end{bmatrix} = \begin{bmatrix} 11 \\ -9 \\ 61 \end{bmatrix}$;

$a = \dfrac{3}{4}$; $b = \dfrac{7}{2}$, $c = -15$; $y = \dfrac{3}{4}x^2 + \dfrac{7}{2}x - 15$

28. No. For example, let $A = \begin{bmatrix} 1 & 0 \\ 0 & 0 \end{bmatrix}$ and $B = \begin{bmatrix} 0 & 0 \\ 0 & 1 \end{bmatrix}$. Then $AB = 0$, but $A \neq 0$ and $B \neq 0$.

29. $A = \begin{bmatrix} a & b \\ c & d \end{bmatrix}$, $A^{-1} = \begin{bmatrix} w & x \\ y & z \end{bmatrix}$; $AA^{-1} = \begin{bmatrix} aw + by & ax + bz \\ cw + dy & cx + dz \end{bmatrix} = \begin{bmatrix} 1 & 0 \\ 0 & 1 \end{bmatrix}$;

① $aw + by = 1$, ② $ax + bz = 0$, ③ $cw + dy = 0$, ④ $cx + dz = 1$.

Solve ① and ③ simultaneously for w and y. Multiply ① by $(-c)$ and ③ by a and add the results. $ady - bcy = -c$; $y(ad - bc) = -c$; $y = -\dfrac{c}{ad - bc}$. Multiply ① by d and ③ by $(-b)$ and add the results. $adw - bcw = d$; $w(ad - bc) = d$; $w = \dfrac{d}{ad - bc}$. In a similar fashion, solve ② and ④ simultaneously for x and z. Multiply ② by d and ④ by $(-b)$ and add. $adx - bcx = -b$; $x(ad - bc) = -b$; $x = -\dfrac{b}{ad - bc}$. Multiply ② by $(-c)$ and ④ by a and add. $-bcz + adz = a$; $z(ad - bc) = a$; $z = \dfrac{a}{ad - bc}$.

Class Exercises · page 540

1. C and D   **2.** C   **3.**

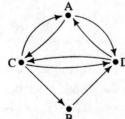

4. Number of ways a message can be sent using one relay

5. No; a station can send a message to itself using another station as a relay.

Written Exercises · pages 540–543

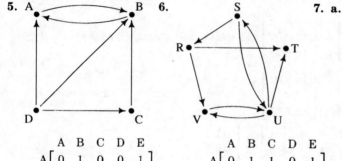

1. $\begin{array}{c}\\A\\B\\C\end{array}\begin{array}{ccc}A&B&C\\ \left[\begin{array}{ccc}0&1&0\\1&0&0\\1&1&0\end{array}\right]\end{array}$

2. $\begin{array}{c}\\D\\E\\F\\G\end{array}\begin{array}{cccc}D&E&F&G\\ \left[\begin{array}{cccc}0&1&0&0\\0&0&1&1\\1&0&0&0\\0&1&0&0\end{array}\right]\end{array}$

3. $\begin{array}{c}\\A\\B\\C\\D\\E\end{array}\begin{array}{ccccc}A&B&C&D&E\\ \left[\begin{array}{ccccc}0&0&1&0&1\\1&0&0&0&0\\1&0&0&1&0\\0&1&0&0&0\\1&1&0&0&0\end{array}\right]\end{array}$

4. $\begin{array}{c}\\P\\Q\\R\\S\\T\end{array}\begin{array}{ccccc}P&Q&R&S&T\\ \left[\begin{array}{ccccc}0&1&1&0&1\\1&0&0&0&1\\1&1&0&1&0\\1&0&1&0&1\\0&0&0&0&0\end{array}\right]\end{array}$

5. (graph A B C D) **6.** (graph S R T V U) **7. a.**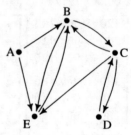

b. $M=\begin{array}{c}\\A\\B\\C\\D\\E\end{array}\begin{array}{ccccc}A&B&C&D&E\\ \left[\begin{array}{ccccc}0&1&0&0&1\\0&0&1&0&1\\0&1&0&1&1\\0&0&1&0&0\\0&1&0&0&0\end{array}\right]\end{array}$

c. $M^2=\begin{array}{c}\\A\\B\\C\\D\\E\end{array}\begin{array}{ccccc}A&B&C&D&E\\ \left[\begin{array}{ccccc}0&1&1&0&1\\0&2&0&1&1\\0&1&2&0&1\\0&1&0&1&1\\0&0&1&0&1\end{array}\right]\end{array}$

Ship E cannot send a message to Ship D using only 1 relay.

d. $M+M^2=\begin{array}{c}\\A\\B\\C\\D\\E\end{array}\begin{array}{ccccc}A&B&C&D&E\\ \left[\begin{array}{ccccc}0&2&1&0&2\\0&2&1&1&2\\0&2&2&1&2\\0&1&1&1&1\\0&1&1&0&1\end{array}\right]\end{array}$

e. The last element in row 4 of M^3 represents the number of ways Ship D can send messages to Ship E using exactly two ships as relays. From the diagram, there is exactly one such path, D-C-B-E. Thus, the last element in row 4 of M^3 is 1.

8. a. Answers may vary. **b.**

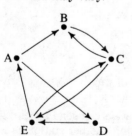

$M=\begin{array}{c}\\A\\B\\C\\D\\E\end{array}\begin{array}{ccccc}A&B&C&D&E\\ \left[\begin{array}{ccccc}0&1&0&1&0\\0&0&1&0&0\\0&1&0&0&1\\0&0&0&0&1\\1&0&1&0&0\end{array}\right]\end{array}$; $M+M^2=\left[\begin{array}{ccccc}0&1&1&1&1\\0&1&1&0&1\\1&1&2&0&1\\1&0&1&0&1\\1&2&1&1&1\end{array}\right]$;

the number of ways rangers can communicate using at most one relay

9. a. transmitters: Bob, Kim; receivers: Ron, Sue; relays: Ted, Dot, Les, Neal

b. A transmitter has at least one 1 in its row and all 0's in its column; a receiver has all 0's in its row and at least one 1 in its column.

10. Each operator can reach any other operator or itself by using 1 relay.

11. There are zeros on the main diagonal only, so there are n zeros. Since there are n^2 elements and n are zeros, there are $n^2 - n$ ones.

12. a.

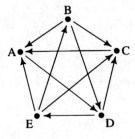

b. Bella and Enrico are tied because each has won 3 and lost 1.

c. $\begin{bmatrix} 0 & 0 & 1 & 0 & 1 \\ 1 & 0 & 1 & 1 & 1 \\ 0 & 0 & 0 & 1 & 0 \\ 2 & 1 & 1 & 0 & 0 \\ 2 & 0 & 1 & 2 & 0 \end{bmatrix}$ **d.** $\begin{bmatrix} 0 & 0 & 1 & 1 & 1 \\ 2 & 0 & 2 & 2 & 1 \\ 1 & 0 & 0 & 1 & 0 \\ 2 & 1 & 2 & 0 & 1 \\ 3 & 1 & 2 & 2 & 0 \end{bmatrix} = M^2 + M$; Enrico; Bella, Denise

13. a.
$$\begin{array}{c} \\ c \\ g \\ p \\ r \\ s \\ w \end{array} \begin{array}{c} \begin{array}{cccccc} c & g & p & r & s & w \end{array} \\ \begin{bmatrix} 0 & 0 & 1 & 0 & 0 & 1 \\ 1 & 0 & 0 & 1 & 0 & 0 \\ 0 & 0 & 0 & 0 & 0 & 0 \\ 0 & 0 & 0 & 0 & 0 & 1 \\ 0 & 0 & 1 & 0 & 0 & 0 \\ 0 & 0 & 0 & 0 & 0 & 0 \end{bmatrix} \end{array}$$
b. $\begin{bmatrix} 0 & 0 & 0 & 0 & 0 & 0 \\ 0 & 0 & 1 & 0 & 0 & 2 \\ 0 & 0 & 0 & 0 & 0 & 0 \\ 0 & 0 & 0 & 0 & 0 & 0 \\ 0 & 0 & 0 & 0 & 0 & 0 \\ 0 & 0 & 0 & 0 & 0 & 0 \end{bmatrix}$; wolves feed on 2 animals that feed on grass.

c. Put a "1" in the grass row, salmon column.

14. Answers will vary.

15. Any row of M gives the one-step paths from a particular point to all points. Any column of M gives the one-step paths from all points ending at a particular point. So when we multiply M by itself, the sum of the products of a row of M and a column of M links the one-step paths starting from a particular point with the one-step paths ending at a particular point, thus giving all the two-step paths. By a similar argument, multiplying M^2 by M links all the two-step paths with the one-step paths, giving all the three-step paths.

Class Exercises · pages 545–546

1.

$\underline{(0.4)(0.05)}$	$+$	$\underline{(0.3)(0.7)}$	$+$	$\underline{(0.3)(0.2)}$	$=$	$\underline{0.29}$
40% use Highlight and 5% of these will change to Silky Shine.		30% use Silky Shine and 70% of these will buy it next month.		30% use another brand and 20% of these will change to Silky Shine.		29% will use Silky Shine next month.

$\underline{(0.4)(0.05)}$	$+$	$\underline{(0.3)(0.2)}$	$+$	$\underline{(0.3)(0.6)}$	$=$	$\underline{0.26}$
40% use Highlight and 5% of these will change to another brand.		30% use Silky Shine and 20% of these will change to another brand.		30% use another brand and 60% of these will buy another brand next month.		26% will use another brand next month.

2. Associative property

3. a.

$$\begin{array}{cc} & \text{c} \quad\;\; \text{s} \\ \begin{array}{c}\text{c}\\\text{s}\end{array} & \left[\begin{array}{cc} 0.9 & 0.1 \\ 0.03 & 0.97 \end{array}\right] \end{array}$$

b. $\begin{array}{cc}\text{c}&\text{s}\end{array}$ [0.6 0.4]

c. [0.552 0.448]; 55.2% of the metropolitan population will live in the city next year and 44.8% will live in the suburbs.

d. the population distribution in 2 years

e. Calculate M_n for large values of n to see if its values stabilize.

Written Exercises · pages 546–550

1. a.

$$\begin{array}{cc} & \text{c} \quad\;\; \text{p.t.} \\ \begin{array}{c}\text{c}\\\text{p.t.}\end{array} & \left[\begin{array}{cc} 0.82 & 0.18 \\ 0.05 & 0.95 \end{array}\right] \end{array}$$

b. $\begin{array}{cc}\text{c}&\text{p.t.}\end{array}$ [0.4 0.6]

c. [0.358 0.642]; next year, 36% will commute by car and 64% by public transportation.

d. [0.326 0.674]; in two years, 33% will commute by car and 67% by public transportation.

2. a.

$$\begin{array}{cc} & \text{u} \quad\;\; \text{s} \\ \begin{array}{c}\text{u}\\\text{s}\end{array} & \left[\begin{array}{cc} 0.97 & 0.03 \\ 0.02 & 0.98 \end{array}\right] \end{array}$$

b. $\begin{array}{cc}\text{u}&\text{s}\end{array}$ $\left[\begin{array}{cc} \frac{2}{3} & \frac{1}{3} \end{array}\right]$

c. In 1 yr, 65% of the population will be living in the urban area and 35% will be in the suburbs. In 2 yr, 64% of the population will be urban and 36% will be suburban.

d. [0.4 0.6] $\left[\begin{array}{cc} 0.97 & 0.03 \\ 0.02 & 0.98 \end{array}\right]$ = [0.4 0.6]; 40% urban and 60% suburban.

e. [0.4 0.6]

3. a.

$$\begin{array}{cc} & \text{D} \quad\;\; \text{S} \\ \begin{array}{c}\text{D}\\\text{S}\end{array} & \left[\begin{array}{cc} 0.9 & 0.1 \\ 0.2 & 0.8 \end{array}\right] \end{array} = T;$$ $\begin{array}{cc}\text{D}&\text{S}\end{array}$ [0.6 0.4] $= M_0$; $M_0 T =$ [0.62 0.38] $= M_1$; 62% ate at Dixie's on Tuesday; $M_1 T =$ [0.634 0.366] $= M_2$; about 63% ate at Dixie's on Wednesday.

b. $\left[\begin{array}{cc} \frac{2}{3} & \frac{1}{3} \end{array}\right] T = \left[\begin{array}{cc} \frac{2}{3} & \frac{1}{3} \end{array}\right]$; $\frac{2}{3}$ at Dixie's and $\frac{1}{3}$ at Sargent's

c. $\left[\begin{array}{cc} \frac{2}{3} & \frac{1}{3} \end{array}\right]$

4. a.

$$\begin{array}{cc} & 1 \quad\;\; 2 \\ \begin{array}{c}1\\2\end{array} & \left[\begin{array}{cc} 0.7 & 0.3 \\ 0.5 & 0.5 \end{array}\right] \end{array} = T$$

b. [0.625 0.375]T = [0.625 0.375]; 62.5% will go to room 1 on the 21st trial.

c. [0.625 0.375]

5. a.
$$\begin{array}{c} \\ s \\ c \\ r \end{array}\begin{array}{ccc} s & c & r \\ \left[\begin{array}{ccc} 0.8 & 0.1 & 0.1 \\ 0 & 0.4 & 0.6 \\ 1 & 0 & 0 \end{array}\right]\end{array}$$

b. $T^2 = \begin{array}{c} \\ s \\ c \\ r \end{array}\begin{array}{ccc} s & c & r \\ \left[\begin{array}{ccc} 0.74 & 0.12 & 0.14 \\ 0.6 & 0.16 & 0.24 \\ 0.8 & 0.1 & 0.1 \end{array}\right]\end{array}$ there is a 14% chance ; that Wed. will be rainy if Mon. is sunny.

c. $T^3 = \begin{array}{c} \\ s \\ c \\ r \end{array}\begin{array}{ccc} s & c & r \\ \left[\begin{array}{ccc} 0.732 & 0.122 & 0.146 \\ 0.72 & 0.124 & 0.156 \\ 0.74 & 0.12 & 0.14 \end{array}\right]\end{array}$

there is a 12.2% chance
that Thurs. will be cloudy
if Mon. is sunny.

d. $T^{30} = \begin{array}{c} \\ s \\ c \\ r \end{array}\begin{array}{ccc} s & c & r \\ \left[\begin{array}{ccc} 0.732 & 0.122 & 0.146 \\ 0.732 & 0.122 & 0.146 \\ 0.732 & 0.122 & 0.146 \end{array}\right]\end{array}$;

There is a 73% chance of a sunny day 30 days
from now if today is sunny; 73% chance
sunny if cloudy; 73% chance sunny if rainy.

e. same answers as in part (d)

f. 73% are sunny; 12% are cloudy; 15% are rainy.

6. a.
$$\begin{array}{c} \\ A \\ B \\ C \end{array}\begin{array}{ccc} A & B & C \\ \left[\begin{array}{ccc} 0.5 & 0.25 & 0.25 \\ 0.3 & 0.6 & 0.1 \\ 0.25 & 0.3 & 0.45 \end{array}\right]\end{array} = T$$

b. (1) $[0.42 \quad 0.37 \quad 0.21] \begin{bmatrix} 0.5 & 0.25 & 0.25 \\ 0.3 & 0.6 & 0.1 \\ 0.25 & 0.3 & 0.45 \end{bmatrix} = [0.3735 \quad 0.39 \quad 0.2365];$

A: 37.35%, B: 39%, C: 23.65%

(2) $[0.42 \quad 0.37 \quad 0.21]T^{10} = [0.3602 \quad 0.4028 \quad 0.237];$ 40.28%

(3)
$$\begin{bmatrix} 0.3602 & 0.4028 & 0.237 \\ 0.3602 & 0.4028 & 0.237 \\ 0.3602 & 0.4028 & 0.237 \end{bmatrix}$$

7. a.
$$\begin{array}{c} \\ w \\ f \\ i \end{array}\begin{array}{ccc} w & f & i \\ \left[\begin{array}{ccc} 0.8 & 0.2 & 0 \\ 0 & 0.5 & 0.5 \\ 0 & 0 & 1 \end{array}\right]\end{array} = T$$

b. $\begin{array}{ccc} w & f & i \end{array}$ $[0.85 \quad 0.05 \quad 0.1] = M_0;\ M_0T = [0.68 \quad 0.195 \quad 0.125];$ 19.5% flu, 12.5% immune

c. $M_0T^2 = [0.54 \quad 0.23 \quad 0.22];$ after 10 days: 54% well, 23% flu, 22% immune
$M_0T^6 = [0.22 \quad 0.14 \quad 0.64];$ after 30 days: 22% well, 14% flu, 64% immune
$M_0T^{10} = [0.09 \quad 0.06 \quad 0.85];$ after 50 days: 9% well, 6% flu, 85% immune

8. a. $\begin{bmatrix} \dfrac{1}{3} & \dfrac{2}{3} \end{bmatrix} \begin{bmatrix} \dfrac{3}{5} & \dfrac{2}{5} \\ \dfrac{1}{5} & \dfrac{4}{5} \end{bmatrix} = \begin{bmatrix} \dfrac{1}{3}\cdot\dfrac{3}{5} + \dfrac{2}{3}\cdot\dfrac{1}{5} & \dfrac{1}{3}\cdot\dfrac{2}{5} + \dfrac{2}{3}\cdot\dfrac{4}{5} \end{bmatrix} = \begin{bmatrix} \dfrac{1}{3} & \dfrac{2}{3} \end{bmatrix}$

b. *S* is unaffected by right-multiplication by *T*. That is, the transition matrix does not
change *S*. Thus, *S* must be the steady-state matrix.

9. $[x \quad 1-x]\begin{bmatrix} 0.3 & 0.7 \\ 0.1 & 0.9 \end{bmatrix} = [0.3x + 0.1 - 0.1x \quad 0.7x + 0.9 - 0.9x] =$

$[0.2x + 0.1 \quad -0.2x + 0.9] = [x \quad 1-x];\ 0.2x + 0.1 = x;\ -0.8x = -0.1;\ x = 0.125;$
$S = [0.125 \quad 0.875]$

10. a. $[x \quad 1-x]\begin{bmatrix} 0.25 & 0.75 \\ 0.45 & 0.55 \end{bmatrix} = [0.25x + 0.45 - 0.45x \quad 0.75 + 0.55 - 0.55x] =$

$[x \quad 1-x];\ -0.20x + 0.45 = x;\ -1.20x = -0.45;\ x = 0.375;\ S = [0.375 \quad 0.625]$

b. $[x \quad 1-x]\begin{bmatrix} \dfrac{1}{3} & \dfrac{2}{3} \\[2mm] \dfrac{4}{5} & \dfrac{1}{5} \end{bmatrix} = \left[\dfrac{1}{3}x + \dfrac{4}{5} - \dfrac{4}{5}x \quad \dfrac{2}{3}x + \dfrac{1}{5} - \dfrac{1}{5}x\right] = [x \quad 1-x];$

$\dfrac{1}{3}x + \dfrac{4}{5} - \dfrac{4}{5}x = x;\ 5x + 12 - 12x = 15x;\ -22x = -12;\ x = \dfrac{12}{22} = \dfrac{6}{11};$

$S = \begin{bmatrix} \dfrac{6}{11} & \dfrac{5}{11} \end{bmatrix}$

11. $[x \quad y \quad 1-x-y]\begin{bmatrix} 0.4 & 0.3 & 0.3 \\ 0.5 & 0.5 & 0 \\ 0.3 & 0.3 & 0.4 \end{bmatrix} = [x \quad y \quad 1-x-y];\ x = \dfrac{5}{12},\ y = \dfrac{3}{8};$

$S = \begin{bmatrix} \dfrac{5}{12} & \dfrac{3}{8} & \dfrac{5}{24} \end{bmatrix}$

12. a. $[0.6 \quad 0.3 \quad 0.1]T^n = [0.59 \quad 0.22 \quad 0.19]$

b. $[0.8 \quad 0.1 \quad 0.1]T^n = [0.59 \quad 0.22 \quad 0.19]$
The matrix is independent of M_0.

13. a.

	1	2	3	4	5
1	0	$\frac{1}{3}$	0	$\frac{1}{3}$	$\frac{1}{3}$
2	$\frac{1}{3}$	0	$\frac{1}{3}$	$\frac{1}{3}$	0
3	0	$\frac{1}{2}$	0	$\frac{1}{2}$	0
4	$\frac{1}{4}$	$\frac{1}{4}$	$\frac{1}{4}$	0	$\frac{1}{4}$
5	$\frac{1}{2}$	0	0	$\frac{1}{2}$	0

b. 21.4%; 21.4%; 14.3%; 28.6%; 14.3%

c. 21.4%; 21.4%; 14.3%; 28.6%; 14.3%

d. No

14. a. 0.5 means that 50% of those in their second year survive to the third; 0.3 means that 30% of those in their third year survive to the fourth; 20 means that for each of those in their fourth year, 20 will be born into their first year.

b. [2000   320   150   60]; 2000 in their first year, 320 in their second year, 150 in their third year, and 60 in their fourth year.

c. [1200   800   160   45]

d. [1728   1152   230   65]

15. a.

$$
\begin{array}{c c c c c}
 & 1 & 2 & 3 & 4 \\
\begin{matrix} 1 \\ 2 \\ 3 \\ 4 \end{matrix} &
\left[\begin{matrix}
0 & 0.4 & 0 & 0 \\
0 & 0 & 0.5 & 0 \\
20 & 0 & 0 & 0.3 \\
20 & 0 & 0 & 0
\end{matrix}\right] & & & = T
\end{array}
$$

b. [6000　320　150　60]

c. [4200　2400　160　45]

d. [444,768　45,824　25,190　6286]

16.

$$
\begin{array}{c c c c c c}
 & A_1 & A_2 & A_3 & A_4 & A_5 \\
\begin{matrix} A_1 \\ A_2 \\ A_3 \\ A_4 \\ A_5 \end{matrix} &
\left[\begin{matrix}
B_1 & S_1 & 0 & 0 & 0 \\
B_2 & 0 & S_2 & 0 & 0 \\
B_3 & 0 & 0 & S_3 & 0 \\
B_4 & 0 & 0 & 0 & S_4 \\
B_5 & 0 & 0 & 0 & 0
\end{matrix}\right] & & & & = T
\end{array}
$$

17. a.
$$
\begin{array}{c c c c c}
 & A_1 & A_2 & A_3 & A_4 & A_5 \\
\end{array}
$$
$M_0 T^{10} = [329\quad 128\quad 70\quad 48\quad 13]$

b. $M_0 T^{18} \begin{bmatrix} 1 \\ 1 \\ 1 \\ 1 \\ 1 \end{bmatrix} = [949]; \; M_0 T^{19} \begin{bmatrix} 1 \\ 1 \\ 1 \\ 1 \\ 1 \end{bmatrix} = [1007]$; the population will be greater than or equal to 1000 after 38 yr.

Activity 14-6 · page 553

a. 7 sq. units, 14 sq. units　**b.** 2　**c.** $|T| = 2(-1) - 0(0) = -2$　**d.** They are opposites.

Class Exercises · page 556

1. $(g \circ f)(-1) = g(f(-1)) = g(5) = 3$

2. $(g \circ f)(x) = g(f(x)) = g(-x + 1) = 2(-x + 1) - 4 = -2x - 2$

3. $\begin{bmatrix} 3 & -2 \\ 4 & 1 \end{bmatrix} \begin{bmatrix} x \\ y \end{bmatrix} = \begin{bmatrix} x' \\ y' \end{bmatrix}$　**4.** $T = \begin{bmatrix} 3 & -2 \\ 4 & 1 \end{bmatrix}$; $|T| = 3(1) - 4(-2) = 11$

5. $\dfrac{\text{area } \triangle A'B'C'}{\text{area } \triangle ABC} = 11$　**6.** Same

	Transformation	Changes orientation	Changes size	Changes shape
7.	Translation	No	No	No
8.	Rotation	No	No	No
9.	Dilation	No	Yes	No

Written Exercises 14-6 · pages 557–559

1. a. See below. **b.** $T = \begin{bmatrix} 2 & 0 \\ 0 & 3 \end{bmatrix}$; $|T| = 6$ **c.** 6 **d.** same

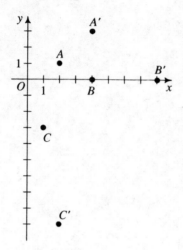

Ex. 1(a)

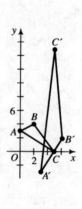

Exs. 2(a), (c)

2. a. See above. **b.** $M = \begin{bmatrix} 0 & 2 & 5 \\ 3 & 4 & 0 \end{bmatrix}$; $TM = \begin{bmatrix} 3 & 6 & 5 \\ -3 & 2 & 15 \end{bmatrix}$; $A'(3, -3), B'(6, 2), C'(5, 15)$

   c. See above.

   d. $|T| = -4$; the ratio of the area of $\triangle A'B'C'$ to the area of $\triangle ABC$ is 4 and the △ have opposite orientations.

3. a, b. See below.

   c. $T = \begin{bmatrix} 2 & 1 \\ 3 & 1 \end{bmatrix}$; $|T| = -1$; the areas are equal and the △ have opposite orientations.

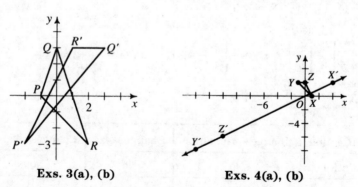

Exs. 3(a), (b)   Exs. 4(a), (b)

4. a, b. See above.

   c. $T = \begin{bmatrix} 4 & -6 \\ 2 & -3 \end{bmatrix}$; $|T| = 0$; the images of X, Y, and Z are mapped onto a line through the origin.

5. a. $T(P) = \begin{bmatrix} 4 & 1 \\ 0 & 5 \end{bmatrix}\begin{bmatrix} 1 \\ 2 \end{bmatrix} = \begin{bmatrix} 6 \\ 10 \end{bmatrix} = P'$; $S(P') = \begin{bmatrix} -1 & 3 \\ 4 & -2 \end{bmatrix}\begin{bmatrix} 6 \\ 10 \end{bmatrix} = \begin{bmatrix} 36 \\ 4 \end{bmatrix} = P''$

b. $ST = \begin{bmatrix} 1 & 3 \\ 4 & -2 \end{bmatrix}\begin{bmatrix} 4 & 1 \\ 0 & 5 \end{bmatrix} = \begin{bmatrix} 4 & 16 \\ 16 & -6 \end{bmatrix}$; $\begin{bmatrix} 4 & 16 \\ 16 & -6 \end{bmatrix}\begin{bmatrix} 1 \\ 2 \end{bmatrix} = \begin{bmatrix} 36 \\ 4 \end{bmatrix} = P''$

6. $ST = \begin{bmatrix} 3 & 7 \\ 1 & 4 \end{bmatrix}\begin{bmatrix} 1 & 2 \\ 2 & 5 \end{bmatrix} = \begin{bmatrix} 17 & 41 \\ 9 & 22 \end{bmatrix}$; $(S \circ T)(P) = \begin{bmatrix} 17 & 41 \\ 9 & 22 \end{bmatrix}\begin{bmatrix} x \\ y \end{bmatrix} = \begin{bmatrix} 17x + 41y \\ 9x + 22y \end{bmatrix}$

7. $y = \dfrac{1}{2}x$; $T = \begin{bmatrix} 4 & 2 \\ 2 & 1 \end{bmatrix}$, $|T| = 0$ **8.** $y = -\dfrac{1}{5}x$; $T = \begin{bmatrix} 5 & -15 \\ -1 & 3 \end{bmatrix}$, $|T| = 0$

9. Right 5, down 1   **10.** Right 6, down 1   **11.** Down 2   **12.** $\begin{bmatrix} x \\ y \end{bmatrix} + \begin{bmatrix} 1 \\ 5 \end{bmatrix} = \begin{bmatrix} x' \\ y' \end{bmatrix}$

13. $\begin{bmatrix} x \\ y \end{bmatrix} + \begin{bmatrix} 0 \\ -4 \end{bmatrix} = \begin{bmatrix} x' \\ y' \end{bmatrix}$ **14.** $\begin{bmatrix} x \\ y \end{bmatrix} + \begin{bmatrix} 6 \\ 0 \end{bmatrix} = \begin{bmatrix} x' \\ y' \end{bmatrix}$ **15.** $\begin{bmatrix} 1 & 0 \\ 0 & -1 \end{bmatrix}$ **16.** $\begin{bmatrix} 0 & 1 \\ 1 & 0 \end{bmatrix}$

17. $\begin{bmatrix} 0 & -1 \\ -1 & 0 \end{bmatrix}$ **18.** $\begin{bmatrix} -1 & 0 \\ 0 & -1 \end{bmatrix}$ **19.** $\begin{bmatrix} 0 & -1 \\ 1 & 0 \end{bmatrix}$ **20.** $\begin{bmatrix} 0 & 1 \\ -1 & 0 \end{bmatrix}$

21. $XY = \begin{bmatrix} 1 & 0 \\ 0 & -1 \end{bmatrix}\begin{bmatrix} -1 & 0 \\ 0 & 1 \end{bmatrix} = \begin{bmatrix} -1 & 0 \\ 1 & 0 \end{bmatrix} = R_{180}$

22. $LX = \begin{bmatrix} 0 & 1 \\ 1 & 0 \end{bmatrix}\begin{bmatrix} 1 & 0 \\ 0 & -1 \end{bmatrix} = \begin{bmatrix} 0 & -1 \\ 1 & 0 \end{bmatrix} = R_{90}$

23. $R_{90}R_{-90} = \begin{bmatrix} 0 & -1 \\ 1 & 0 \end{bmatrix}\begin{bmatrix} 0 & 1 \\ -1 & 0 \end{bmatrix} = \begin{bmatrix} 1 & 0 \\ 0 & 1 \end{bmatrix}$; $I_{2\times 2}$

24. $LM = \begin{bmatrix} 0 & 1 \\ 1 & 0 \end{bmatrix}\begin{bmatrix} 0 & -1 \\ -1 & 0 \end{bmatrix} = \begin{bmatrix} -1 & 0 \\ 0 & -1 \end{bmatrix} = R_{180}$

25. $(R_{90})^2 = R_{90} \cdot R_{90} = \begin{bmatrix} 0 & -1 \\ 1 & 0 \end{bmatrix}\begin{bmatrix} 0 & -1 \\ 1 & 0 \end{bmatrix} = \begin{bmatrix} -1 & 0 \\ 0 & -1 \end{bmatrix} = R_{180}$

26. $(R_{-90})^{51} = R_{90} = \begin{bmatrix} 0 & -1 \\ 1 & 0 \end{bmatrix}$

27. $X^2 = L^2 = M^2 = \begin{bmatrix} 1 & 0 \\ 0 & 1 \end{bmatrix}$; because $X^2 = L^2 = M^2$, X, L, and M are their own inverses.

Visually thinking, this conclusion should be intuitively clear. It means roughly that a reflection of a reflection yields the original figure. Imagine your reflection in a mirror—left appears right and right appears left. If you use another mirror to reflect the reflection, you'll see yourself unchanged—left is left and right is right.

28. a. $(-\cos(90° - \alpha), \sin(90° - \alpha)) = (-\sin\alpha, \cos\alpha)$

b. $R_\alpha: (1,0) \to (\cos\alpha, \sin\alpha)$ and
$R_\alpha: (0,1) \to (-\sin\alpha, \cos\alpha)$, so

$$T = \begin{bmatrix} \cos\alpha & -\sin\alpha \\ \sin\alpha & \cos\alpha \end{bmatrix}$$

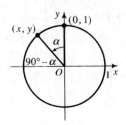

c. $|T| = \cos^2\alpha - (-\sin^2\alpha) = \cos^2\alpha + \sin^2\alpha = 1$; the areas are equal.

29. a. $R_\beta = \begin{bmatrix} \cos\beta & -\sin\beta \\ \sin\beta & \cos\beta \end{bmatrix}$

b. $(R_\alpha)^2 = \begin{bmatrix} \cos\alpha & -\sin\alpha \\ \sin\alpha & \cos\alpha \end{bmatrix} \begin{bmatrix} \cos\alpha & -\sin\alpha \\ \sin\alpha & \cos\alpha \end{bmatrix} = $

$\begin{bmatrix} \cos^2\alpha - \sin^2\alpha & -2\cos\alpha\sin\alpha \\ 2\cos\alpha\sin\alpha & -\sin^2\alpha + \cos^2\alpha \end{bmatrix} = \begin{bmatrix} \cos 2\alpha & -\sin 2\alpha \\ \sin 2\alpha & \cos 2\alpha \end{bmatrix};$

$R_\alpha R_\beta = \begin{bmatrix} \cos\alpha & -\sin\alpha \\ \sin\alpha & \cos\alpha \end{bmatrix} \begin{bmatrix} \cos\beta & -\sin\beta \\ \sin\beta & \cos\beta \end{bmatrix} = $

$\begin{bmatrix} \cos\alpha\cos\beta - \sin\alpha\sin\beta & -\cos\alpha\sin\beta - \sin\alpha\cos\beta \\ \sin\alpha\cos\beta + \cos\alpha\sin\beta & -\sin\alpha\sin\beta + \cos\alpha\cos\beta \end{bmatrix} = $

$\begin{bmatrix} \cos(\alpha+\beta) & -\sin(\alpha+\beta) \\ \sin(\alpha+\beta) & \cos(\alpha+\beta) \end{bmatrix};$ $(R_\alpha)^2$ has the same matrix as $R_{2\alpha}$ because it is a

rotation through α followed by another rotation through α; $R_\alpha R_\beta$ has the same matrix as $R_{\alpha+\beta}$ or $R_{\beta+\alpha}$ because it is a rotation through β followed by a rotation through α.

30. a. $T = \begin{bmatrix} a & b \\ c & d \end{bmatrix};$ $T \begin{bmatrix} 0 & 1 & 1 & 0 \\ 0 & 0 & 1 & 1 \end{bmatrix} = \begin{bmatrix} 0 & a & a+b & b \\ 0 & c & c+d & d \end{bmatrix}$

b. Solutions will vary. An example is given: Refer to the

figure at the right. Area $\triangle$I $=$ area $\triangle$VI $= \dfrac{1}{2}ac;$

area $\triangle$III $=$ area $\triangle$IV $= \dfrac{1}{2}bd;$ area rect. II $=$

area rect. V $= bc;$ area parallelogram $=$
area large rect. $-$ area (regions I to VI) $=$

$(a+b)(c+d) - 2\cdot\dfrac{1}{2}ac - 2\cdot\dfrac{1}{2}bd - 2\cdot bc = $

$ac + bc + ad + bd - ac - bd - 2bc = ad - bc$

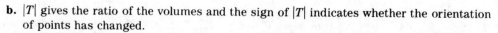

31. a. $T = \begin{bmatrix} 2 & 0 & 0 \\ 0 & 3 & 0 \\ 0 & 1 & 4 \end{bmatrix}$

b. $|T|$ gives the ratio of the volumes and the sign of $|T|$ indicates whether the orientation of points has changed.

c. They are the coordinates of the images of $(1,0,0)$, $(0,1,0)$, and $(0,0,1)$.

32. a. $M: (1,0,0) \longrightarrow (0,1,0)$, $M: (0,1,0) \longrightarrow (-1,0,0)$, and $M: (0,0,1) \longrightarrow (0,0,1)$,

so $M = \begin{bmatrix} 0 & -1 & 0 \\ 1 & 0 & 0 \\ 0 & 0 & 1 \end{bmatrix}.$

b. $N: (1,0,0) \longrightarrow (-1,0,0)$, $N: (0,1,0) \longrightarrow (0,1,0)$, and $N: (0,0,1) \longrightarrow (0,0,-1)$,

so $N = \begin{bmatrix} -1 & 0 & 0 \\ 0 & 1 & 0 \\ 0 & 0 & -1 \end{bmatrix}.$

c. $NM = \begin{bmatrix} 0 & 1 & 0 \\ 1 & 0 & 0 \\ 0 & 0 & -1 \end{bmatrix}$; a 90° counterclockwise rotation about the z-axis followed by a

180° rotation about the y-axis.

d. No. Consider $(1,0,0)$. NM: $(1,0,0) \longrightarrow (0,1,0)$ and MN: $(1,0,0) \longrightarrow (0,-1,0)$;

$MN = \begin{bmatrix} 0 & -1 & 0 \\ -1 & 0 & 0 \\ 0 & 0 & -1 \end{bmatrix} \neq NM$

33. a. $Y = \begin{bmatrix} -1 & 0 \\ 0 & 1 \end{bmatrix}$, $G = \begin{bmatrix} h \\ k \end{bmatrix}$, $P = \begin{bmatrix} x \\ y \end{bmatrix}$, $P' = \begin{bmatrix} -x+h \\ y+k \end{bmatrix}$ **b.** $P' = YP + G$

Chapter Test • pages 560–561

1. a.
$\begin{array}{c} \\ M \\ P \\ B \end{array} \begin{array}{ccc} s & m & l \\ \end{array}$
$\begin{array}{c} M \\ P \\ B \end{array} \begin{bmatrix} 7 & 14 & 10 \\ 8 & 20 & 9 \\ 15 & 20 & 12 \end{bmatrix}$

b. 21 Mama, 30 Papa, and 30 Baby Bears; $1.5S$; $1.5S = $ $\begin{array}{c} M \\ P \\ B \end{array} \begin{bmatrix} 10.5 & 21 & 15 \\ 12 & 30 & 13.5 \\ 22.5 & 30 & 18 \end{bmatrix}$ with columns s, m, l

2. $2A - B = 2\begin{bmatrix} 1 & 2 & 0 \\ -4 & 1 & 5 \end{bmatrix} - \begin{bmatrix} -3 & 4 & 6 \\ -2 & -1 & 2 \end{bmatrix} = \begin{bmatrix} 2 & 4 & 0 \\ -8 & 2 & 10 \end{bmatrix} - \begin{bmatrix} -3 & 4 & 6 \\ -2 & -1 & 2 \end{bmatrix} =$

$\begin{bmatrix} 5 & 0 & -6 \\ -6 & 3 & 8 \end{bmatrix}$

3. Let $A = \begin{bmatrix} 1 & 2 \\ 3 & -1 \end{bmatrix}$, $B = \begin{bmatrix} 0 & -1 \\ 1 & 2 \end{bmatrix}$; $AB = \begin{bmatrix} 2 & 3 \\ -1 & -5 \end{bmatrix}$, $BA = \begin{bmatrix} -3 & 1 \\ 7 & 0 \end{bmatrix}$

4. a. $NP = \begin{bmatrix} 12 & 4 \end{bmatrix}\begin{bmatrix} 0.40 & 0.25 & 0.15 & 0.20 \\ 0.25 & 0.25 & 0.25 & 0.25 \end{bmatrix} = \begin{bmatrix} 5.8 & 4 & 2.8 & 3.4 \end{bmatrix}$ **b.** 2.8 lb

5. $\begin{bmatrix} 3 & 5 \\ 2 & -7 \end{bmatrix}\begin{bmatrix} x \\ y \end{bmatrix} = \begin{bmatrix} -10 \\ 45 \end{bmatrix}$; $-\frac{1}{31}\begin{bmatrix} -7 & -5 \\ -2 & 3 \end{bmatrix}\begin{bmatrix} 3 & 5 \\ 2 & -7 \end{bmatrix}\begin{bmatrix} x \\ y \end{bmatrix} = -\frac{1}{31}\begin{bmatrix} -7 & -5 \\ -2 & 3 \end{bmatrix}\begin{bmatrix} -10 \\ 45 \end{bmatrix}$;

$\begin{bmatrix} x \\ y \end{bmatrix} = -\frac{1}{31}\begin{bmatrix} -155 \\ 155 \end{bmatrix}$; $\begin{bmatrix} x \\ y \end{bmatrix} = \begin{bmatrix} 5 \\ -5 \end{bmatrix}$; $x = 5$, $y = -5$

6. a.
$\begin{array}{c} \\ B \\ L \\ P \\ S \\ W \end{array} \begin{array}{ccccc} B & L & P & S & W \\ \end{array}$
$\begin{array}{c} B \\ L \\ P \\ S \\ W \end{array}\begin{bmatrix} 0 & 1 & 0 & 0 & 0 \\ 1 & 0 & 1 & 0 & 0 \\ 0 & 0 & 0 & 1 & 1 \\ 0 & 1 & 0 & 0 & 0 \\ 0 & 0 & 1 & 1 & 0 \end{bmatrix} = M$

b.
$\begin{array}{c} \\ B \\ L \\ P \\ S \\ W \end{array}\begin{array}{ccccc} B & L & P & S & W \\ \end{array}$
$\begin{array}{c} B \\ L \\ P \\ S \\ W \end{array}\begin{bmatrix} 1 & 0 & 1 & 0 & 0 \\ 0 & 1 & 0 & 1 & 1 \\ 0 & 1 & 1 & 1 & 0 \\ 1 & 0 & 1 & 0 & 0 \\ 0 & 1 & 0 & 1 & 1 \end{bmatrix} = M^2$

Lookout Peak, Park Entrance, Scenic Overlook, and Waterfall each have two two-step paths to other sites.

7. a. $\begin{array}{cc} & c \quad\quad s \\ \begin{matrix}c\\s\end{matrix} & \begin{bmatrix} 0.92 & 0.08 \\ 0.03 & 0.97 \end{bmatrix} \end{array} = T$ **b.** $M_0 = [0.6 \quad 0.4]$

c. $M_0T = \begin{array}{cc} c \quad\quad\quad s \\ [0.564 \quad 0.436] \end{array}$; 56.4% in the city, 43.6% in the suburbs

8. a, b.

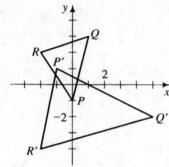

c. $T = \begin{bmatrix} 2 & 1 \\ 1 & -1 \end{bmatrix}$, $|T| = -3$; area of $\triangle P'Q'R'$ is 3 times the area of $\triangle PQR$, and the triangles have opposite orientations.

Cumulative Review · pages 562–563

1. a.

b. distance$^2 = 120^2 + 300^2 - 2 \cdot 120 \cdot 300 \cdot \cos 160°$; distance ≈ 414.8 km; $\dfrac{\sin P}{120} = \dfrac{\sin 160°}{414.8}$; $\angle P \approx 5.7°$; course $\approx 180° + 5.7° = 185.7°$

2. a. $2(-3,4) = (-6,8)$ **b.** $3(2,-1) + (-3,4) = (3,1)$

   c. $|(2,-1) - (-3,4)| = |(5,-5)| = \sqrt{25 + 25} = \sqrt{50} = 5\sqrt{2}$

3. a. $(x,y) = (-2,-4) + t(-2,5)$ **b.** $x = -2 - 2t, y = -4 + 5t$

4. a. $\mathbf{v} = (2,-4)$; speed $= |\mathbf{v}| = \sqrt{2^2 + (-4)^2} = \sqrt{20} = 2\sqrt{5}$

   b. $2x - y = 11$; $2(1 + 2t) - (-1 - 4t) = 11$; $2 + 4t + 1 + 4t = 11$; $t = 1$; $x = 1 + 2(1) = 3$; $y = -1 - 4(1) = -5$; $(3,-5)$

5. $\overrightarrow{BA} = (3,8)$; $\overrightarrow{BC} = (8,1)$; $\cos B = \dfrac{(3,8) \cdot (8,1)}{|(3,8)| |(8,1)|} = \dfrac{32}{\sqrt{73}\sqrt{65}} \approx 0.4645$; $\angle B \approx 62.3°$

6. Answers may vary; $\overrightarrow{AB} = (-5,-1,-1)$; $(x,y,z) = (2,0,5) + t(-5,-1,-1)$; $x = 2 - 5t$, $y = -t, z = 5 - t$

7. Midpoint of $\overline{AB} = M = \left(\dfrac{1-3}{2}, \dfrac{7+5}{2}, \dfrac{3+5}{2}\right) = (-1,6,4)$;

$\overrightarrow{MA} = (1 - (-1), 7 - 6, 3 - 4) = (2,1,-1)$; $2x + y - z = d$; $2(-1) + 6 - 4 = d = 0$; $2x + y - z = 0$

8. $\begin{vmatrix} 10 & 4 & 3 \\ 0 & 2 & 5 \\ -1 & -7 & 0 \end{vmatrix} = 10 \begin{vmatrix} 2 & 5 \\ -7 & 0 \end{vmatrix} - 0 \begin{vmatrix} 4 & 3 \\ -7 & 0 \end{vmatrix} + (-1) \begin{vmatrix} 4 & 3 \\ 2 & 5 \end{vmatrix} =$

$10(0 - (-35)) + 0 - 1(20 - 6) = 350 - 14 = 336$

9. $4x - 5y = 2$, $5x - 7y = 1$; $D = \begin{vmatrix} 4 & -5 \\ 5 & -7 \end{vmatrix} = -3$, $D_x = \begin{vmatrix} 2 & -5 \\ 1 & -7 \end{vmatrix} = -9$,

$D_y = \begin{vmatrix} 4 & 2 \\ 5 & 1 \end{vmatrix} = -6$; $x = \dfrac{D_x}{D} = \dfrac{-9}{-3} = 3$, $y = \dfrac{D_y}{D} = \dfrac{-6}{-3} = 2$; $(3, 2)$

10. Answers may vary. $\overrightarrow{PQ} = (-5, 3, -3)$, $\overrightarrow{PR} = (0, -1, -9)$;

$\overrightarrow{PQ} \times \overrightarrow{PR} = \begin{vmatrix} \mathbf{i} & \mathbf{j} & \mathbf{k} \\ -5 & 3 & -3 \\ 0 & -1 & -9 \end{vmatrix} = \mathbf{i}\begin{vmatrix} 3 & -3 \\ -1 & -9 \end{vmatrix} - \mathbf{j}\begin{vmatrix} -5 & -3 \\ 0 & -9 \end{vmatrix} + \mathbf{k}\begin{vmatrix} -5 & 3 \\ 0 & -1 \end{vmatrix} =$

$30\mathbf{i} - 45\mathbf{j} + 5\mathbf{k} = -5(6\mathbf{i} + 9\mathbf{j} - \mathbf{k})$; $(6, 9, -1)$; $6x + 9y - z = d$;
$d = 6(1) + 9(0) - 5 = 1$; $6x + 9y - z = 1$

11. a. Geometric; $t_n = -\dfrac{4}{9}\left(-\dfrac{3}{2}\right)^n$

    b. Neither; $t_n = \dfrac{2n}{2n + 1}$

    c. Arithmetic; $t_n = n - \dfrac{1}{3}$

12. a. $-2, 0, 4, 12, 28$ **b.** $t_n = 2^n - 4$ **c.** $t_{10} = 2^{10} - 4 = 1020$

13. $t_1 = 10$; $t_5 = 10 + 4d = 2$, $d = -2$; $t_{20} = 10 + (20 - 1)(-2) = -28$;

$S_{20} = \dfrac{20(10 + (-28))}{2} = -180$

14. $S_6 = \dfrac{t_1(1 - r^6)}{1 - r} = \dfrac{81\left(1 - \left(\dfrac{1}{3}\right)^6\right)}{1 - \dfrac{1}{3}} = \dfrac{81\left(1 - \dfrac{1}{729}\right)}{\dfrac{2}{3}} = \dfrac{364}{3}$

15. a. 1   **b.** Does not exist.

16. a. $\displaystyle\lim_{n \to \infty} \dfrac{3n^2 + 2n + 7}{n^2 - 9} = \lim_{n \to \infty} \dfrac{\dfrac{3n^2}{n^2} + \dfrac{2n}{n^2} + \dfrac{7}{n^2}}{\dfrac{n^2}{n^2} - \dfrac{9}{n^2}} = \lim_{n \to \infty} \dfrac{3 + \dfrac{2}{n} + \dfrac{7}{n^2}}{1 - \dfrac{9}{n^2}} = 3$

    b. $\displaystyle\lim_{n \to \infty} \dfrac{5n^3 - 2n}{n^2 + 1} = \lim_{n \to \infty} \dfrac{5n - \dfrac{2}{n}}{1 + \dfrac{1}{n^2}} = \lim_{n \to \infty} 5n = \infty$

    c. $1 + (-1)^n = 2, 0, 2, 0, \ldots$; the limit does not exist.

17. $r = -\dfrac{x}{4}$; $\left|-\dfrac{x}{4}\right| < 1$; $-4 < x < 4$; $S = \dfrac{t_1}{1 - r} = \dfrac{2}{1 - \left(-\dfrac{x}{4}\right)} = \dfrac{8}{4 + x}$

18. $t_1 = 0.345$, $r = 0.001$; $S = \dfrac{t_1}{1 - r} = \dfrac{0.345}{1 - 0.001} = \dfrac{345}{999} = \dfrac{115}{333}$

19. $7 + 12 + 17 + \cdots + 102 = \displaystyle\sum_{n=1}^{20}(5n + 2) = 5\sum_{n=1}^{20} n + \sum_{n=1}^{20} 2 = 5\left[\dfrac{20(1 + 20)}{2}\right] +$

$20(2) = 1090$

20. (1) For $n = 1$, $\dfrac{1}{(2 \cdot 1 - 1)(2 \cdot 1 + 1)} = \dfrac{1}{1 \cdot 3} = \dfrac{1}{3} = \dfrac{1}{2 \cdot 1 + 1}$.

(2) If $\displaystyle\sum_{i=1}^{k} \dfrac{1}{(2i - 1)(2i + 1)} = \dfrac{k}{2k + 1}$, then $\displaystyle\sum_{i=1}^{k+1} \dfrac{1}{(2i - 1)(2i + 1)} =$

$\dfrac{k}{2k + 1} + \dfrac{1}{(2(k + 1) - 1)(2(k + 1) + 1)} = \dfrac{k}{2k + 1} + \dfrac{1}{(2k + 1)(2k + 3)} = \dfrac{k(2k + 3) + 1}{(2k + 1)(2k + 3)} =$

$\dfrac{2k^2 + 3k + 1}{(2k + 1)(2k + 3)} = \dfrac{(2k + 1)(k + 1)}{(2k + 1)(2k + 3)} = \dfrac{k + 1}{2k + 3} = \dfrac{k + 1}{2(k + 1) + 1}$.

21. a. $2A^t - B = 2\begin{bmatrix} -1 & 3 \\ 6 & 4 \\ -2 & 0 \end{bmatrix} - \begin{bmatrix} 2 & -1 \\ -3 & 0 \\ 5 & 4 \end{bmatrix} = \begin{bmatrix} -2 & 6 \\ 12 & 8 \\ -4 & 0 \end{bmatrix} + \begin{bmatrix} -2 & 1 \\ 3 & 0 \\ -5 & -4 \end{bmatrix} = \begin{bmatrix} -4 & 7 \\ 15 & 8 \\ -9 & -4 \end{bmatrix}$

b. $AB = \begin{bmatrix} -1 & 6 & -2 \\ 3 & 4 & 0 \end{bmatrix}\begin{bmatrix} 2 & -1 \\ -3 & 0 \\ 5 & 4 \end{bmatrix} = \begin{bmatrix} -30 & -7 \\ -6 & -3 \end{bmatrix}$

$BA = \begin{bmatrix} 2 & -1 \\ -3 & 0 \\ 5 & 4 \end{bmatrix}\begin{bmatrix} -1 & 6 & -2 \\ 3 & 4 & 0 \end{bmatrix} = \begin{bmatrix} -5 & 8 & -4 \\ 3 & -18 & 6 \\ 7 & 46 & -10 \end{bmatrix}$; A^tB is not defined.

22. $\begin{bmatrix} 3 & 4 \\ -3 & -3 \end{bmatrix}X - \begin{bmatrix} 2 & -5 \\ 0 & 1 \end{bmatrix} = \begin{bmatrix} 0 & 4 \\ -2 & 1 \end{bmatrix}$; $\begin{bmatrix} 3 & 4 \\ -3 & -3 \end{bmatrix}X = \begin{bmatrix} 2 & -1 \\ -2 & 2 \end{bmatrix}$;

$X = \dfrac{1}{3}\begin{bmatrix} -3 & -4 \\ 3 & 3 \end{bmatrix}\begin{bmatrix} 2 & -1 \\ -2 & 2 \end{bmatrix}$; $X = \dfrac{1}{3}\begin{bmatrix} 2 & -5 \\ 0 & 3 \end{bmatrix} = \begin{bmatrix} \dfrac{2}{3} & -\dfrac{5}{3} \\ 0 & 1 \end{bmatrix}$; $X = \begin{bmatrix} \dfrac{2}{3} & -\dfrac{5}{3} \\ 0 & 1 \end{bmatrix}$

23. $\begin{bmatrix} 3 & 5 \\ 1 & 6 \end{bmatrix}\begin{bmatrix} x \\ y \end{bmatrix} = \begin{bmatrix} 4 \\ 10 \end{bmatrix}$; $\begin{bmatrix} x \\ y \end{bmatrix} = \dfrac{1}{13}\begin{bmatrix} 6 & -5 \\ -1 & 3 \end{bmatrix}\begin{bmatrix} 4 \\ 10 \end{bmatrix}$; $\begin{bmatrix} x \\ y \end{bmatrix} = \dfrac{1}{13}\begin{bmatrix} -26 \\ 26 \end{bmatrix} = \begin{bmatrix} -2 \\ 2 \end{bmatrix}$;

$x = -2, y = 2$

24. a.
$$\begin{array}{c} \\ A \\ B \\ C \\ D \\ E \end{array}\begin{array}{cccccc} A & B & C & D & E \\ \begin{bmatrix} 0 & 1 & 0 & 0 & 1 \\ 0 & 0 & 1 & 1 & 1 \\ 0 & 1 & 0 & 1 & 0 \\ 0 & 0 & 1 & 0 & 1 \\ 1 & 0 & 0 & 0 & 0 \end{bmatrix} = M \end{array}$$

b.

$M + M^2 = \begin{bmatrix} 0 & 1 & 0 & 0 & 1 \\ 0 & 0 & 1 & 1 & 1 \\ 0 & 1 & 0 & 1 & 0 \\ 0 & 0 & 1 & 0 & 1 \\ 1 & 0 & 0 & 0 & 0 \end{bmatrix} + \begin{bmatrix} 1 & 0 & 1 & 1 & 1 \\ 1 & 1 & 1 & 1 & 1 \\ 0 & 0 & 2 & 1 & 2 \\ 1 & 1 & 0 & 1 & 0 \\ 0 & 1 & 0 & 0 & 1 \end{bmatrix} = \begin{bmatrix} 1 & 1 & 1 & 1 & 2 \\ 1 & 1 & 2 & 2 & 2 \\ 0 & 1 & 2 & 2 & 2 \\ 1 & 1 & 1 & 1 & 1 \\ 1 & 1 & 0 & 0 & 1 \end{bmatrix}$

25. a.

$$\begin{array}{c}\\ A \\ B \\ C \end{array}\begin{array}{ccc} A & B & C \end{array}$$

$$\begin{array}{c} A \\ B \\ C \end{array}\left[\begin{array}{ccc} 1 & 0 & 0 \\ 0.35 & 0.65 & 0 \\ 0.45 & 0 & 0.55 \end{array}\right] = T$$

b.

$$M_0 T = [0.20 \quad 0.50 \quad 0.30]\left[\begin{array}{ccc} 1 & 0 & 0 \\ 0.35 & 0.65 & 0 \\ 0.45 & 0 & 0.55 \end{array}\right] = M_1 = [0.51 \quad 0.325 \quad 0.165];$$

51% Brand A, 32.5% Brand B, 16.5% Brand C

c. $M_1 T = [0.698 \quad 0.211 \quad 0.091]$; 69.8%, 21.1%, 9.1%, respectively.

26. a, b.

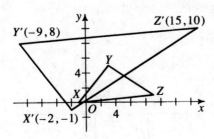

c. $T = \begin{bmatrix} 2 & -3 \\ 1 & 1 \end{bmatrix}$; $|T| = 5$; the area of $\triangle X'Y'Z'$ is 5 times the area of $\triangle XYZ$; since $|T| > 0$; the $\triangle$ have the same orientation.

Class Exercises 15-1 · page 568

1. The set of freshmen who are music majors; 15

2. The set of students who are either freshmen or music majors; $285 + 15 + 50 = 350$

3. The set of students who are not freshmen; $650 + 50 = 700$

4. The set of students who are not music majors; $285 + 650 = 935$

5. The set of music majors who are not freshmen; 50

6. The set of freshmen who are not music majors; 285

7. The set of students who are neither freshmen nor music majors; 650

8. The set of students who are either freshmen or not music majors; $285 + 15 + 650 = 950$

9. **a.** $B \cap \overline{A}$ **b.** $A \cap B$ **c.** $\overline{A \cup B}$ or $\overline{A} \cap \overline{B}$

10. **a.** A **b.** $\emptyset$ **c.** U **d.** $\emptyset$ **e.** U **f.** A

Written Exercises 15-1 · pages 568–571

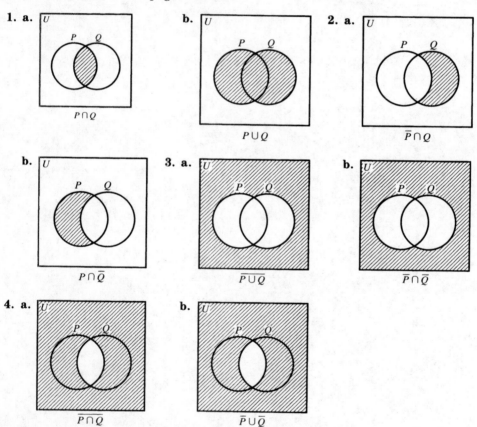

1. a. $P \cap Q$ **b.** $P \cup Q$ **2. a.** $\overline{P} \cap Q$

b. $P \cap \overline{Q}$ **3. a.** $\overline{P \cup Q}$ **b.** $\overline{P} \cap \overline{Q}$

4. a. $\overline{P \cap Q}$ **b.** $\overline{P} \cup \overline{Q}$

5–10. Names of specific teachers may vary.

5. a. Teachers who teach either math or physics

   b. Teachers who teach math but not physics

6. a. Teachers who teach both physics and chemistry

  b. Teachers who teach neither physics nor chemistry

7. a. Teachers who teach either biology or both physics and chemistry

  b. Teachers who teach both biology and chemistry

8. a. Teachers who teach math and either biology or chemistry

  b. Teachers who teach math and either biology or chemistry

9. a. Teachers who teach neither math, nor physics, nor chemistry

  b. Teachers who teach neither math, nor physics, nor chemistry

10. a. Teachers who do not teach any of the following: math, biology, physics, chemistry

  b. Teachers who do not teach any of the following: math, biology, physics, chemistry

11. See figure below; $x + 18 = 52$ and $y + 18 = 38$; $x = 34$ and $y = 20$; $z = 100 - 34 - 18 - 20 = 28$; 28 voted for neither.

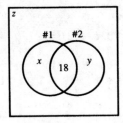

Ex. 11

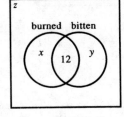

Ex. 12

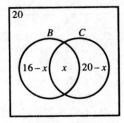

Ex. 13

12. See figure above; $x + 12 = 17$ and $y + 12 = 25$; $x = 5$ and $y = 13$; $z = 30 - 5 - 12 - 13 = 0$; 0 had neither.

13. See figure above; since 48 students were surveyed, $(16 - x) + x + (20 - x) + 20 = 48$; $56 - x = 48$; $x = 8$; $20 - 8$, or 12, liked classical music but not bluegrass.

14. See figure below; since there are 52 teachers in all, $(27 - x) + x + (25 - x) + 12 = 52$; $x = 64 - 52 = 12$; $25 - 12$, or 13, enjoy fishing but not sailing.

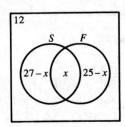

Ex. 14

15. a.

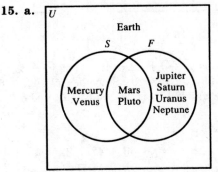

b. $\bar{S} \cap F$

16. a. Arizona, New Mexico, Texas **b.** California **c.** 42

17. a. Bolivia, Paraguay

b.

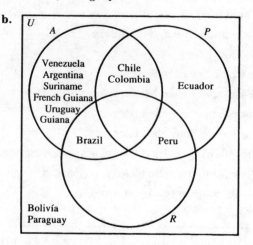

18.

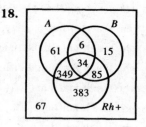

19. See figure below. $b + 21 = 26, b = 5; c + 21 = 29, c = 8; a + 21 = 27, a = 6;$
$d = 45 - (a + b + 21) = 45 - (6 + 5 + 21) = 13; e = 43 - (b + c + 21) =$
$43 - (5 + 8 + 21) = 9; f = 46 - (a + c + 21) = 46 - (6 + 8 + 21) = 11; g =$
$85 - (a + b + c + d + e + f + 21) = 85 - (6 + 5 + 8 + 13 + 9 + 11 + 21) = 12$
a. 11 **b.** 12

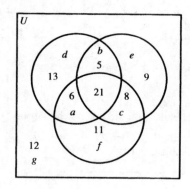

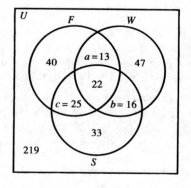

Ex. 19   Ex. 20

20. See figure above. Since $a + c + 22 = 40 = 100, a + c = 38.$ Since
$a + b + 22 + 47 = 98, a + b = 29.$ Since $b + c + 22 + 33 = 96, b + c = 41.$
Then $(a + b) - (b + c) = 29 - 41; a - c = -12$ and $a + c = 38; 2a = 26; a = 13;$
$b = 16; c = 25.$

a. $n(F \cap W \cap \overline{S}) = a = 13$

b. $n(\overline{F \cup W \cup S}) = 415 - (40 + 33 + 47 + 22 + 13 + 16 + 25) = 415 - 196 = 219$

21. The universal set is the set of all elements. $\overline{U}$ represents the set of all elements not in U.
There are no such elements, so $\overline{U} = \emptyset.$

22. $\overline{\overline{A}}$ is the set of all elements not in set $\overline{A}$; since A is the set of all elements not in set $\overline{A}$,
$\overline{\overline{A}} = A.$

23. $A \cap (B \cup C)$ $(A \cap B) \cup (A \cap C)$ **24.** $A \cup (B \cap C)$

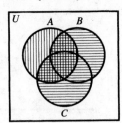   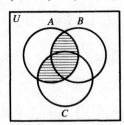   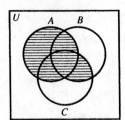

$(A \cup B) \cap (A \cup C)$ **25. a.** $\overline{A \cup B}$ $\overline{A} \cap \overline{B}$

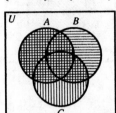   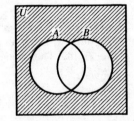   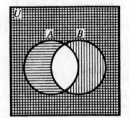

Yes, union distributes   Since the same regions are shaded, $\overline{A \cup B} = \overline{A} \cap \overline{B}$.
over intersection.

b. $\overline{A \cap B}$ $\overline{A} \cup \overline{B}$ **c.** From part (b), $\overline{A \cap (B \cup C)} =$

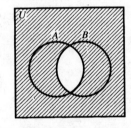   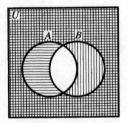

$\overline{A} \cup \overline{(B \cup C)}$; from part (a),

$\overline{A} \cup \overline{(B \cup C)} = \overline{A} \cup (\overline{B} \cap \overline{C})$;

thus, $\overline{A \cap (B \cup C)} =$

$\overline{A} \cup (\overline{B} \cap \overline{C})$.

Since the same regions are shaded,
$\overline{A \cap B} = \overline{A} \cup \overline{B}$.

26. a. From Ex. 23, $A \cap (\overline{A} \cup B) = (A \cap \overline{A}) \cup (A \cap B) = \emptyset \cup (A \cap B) = A \cap B$

   b. From Ex. 25(a), $\overline{A \cup B} = \overline{A} \cap \overline{B}$; $A \cup (\overline{A} \cup B) =$

   $A \cup (\overline{A} \cap \overline{B}) = (A \cup \overline{A}) \cap (A \cap \overline{B})$ (from Ex. 24) $= U \cap (A \cup \overline{B}) = A \cup \overline{B}$

27. $n(A \cup B \cup C) = n(A) + n(B) + n(C) - n(A \cap B) - n(A \cap C) - n(B \cap C) +$
   $n(A \cap B \cap C)$; consider the Venn diagram for sets A, B, and C. In counting the
   elements of A and the elements of B the elements in $A \cap B$ were counted twice, so they
   should be subtracted once. Repeat the process for $A \cap C$ and $B \cap C$. The elements in
   $A \cap B \cap C$ were counted 3 times and subtracted 3 times, so they haven't been counted;
   add these elements. More formally $= n(A \cup B \cup C) = n(A \cup (B \cup C)) =$
   $n(A) + n(B \cup C) - n(A \cap (B \cup C)) = n(A) + n(B) + n(C) - n(B \cap C) -$
   $n[(A \cap B) \cup (A \cap C)] = n(A) + n(B) + n(C) - n(B \cap C) - [n(A \cap B) +$
   $n(A \cap C) - n((A \cap B) \cap (A \cap C))] = n(A) + n(B) + n(C) - n(A \cap B) -$
   $n(A \cap C) - n(B \cap C) + n(A \cap B \cap C)$.

Class Exercises 15-2 • page 574

1. a. $2 \cdot 1 = 2$ **b.** $3 \cdot 2 \cdot 1 = 6$ **c.** $4 \cdot 3 \cdot 2 \cdot 1 = 24$

2. $10! = 10 \cdot 9! = 10(362,880) = 3,628,800$

3. $26 \cdot 26 \cdot 9 \cdot 10 \cdot 10 = 608,400$ license plates

4. a. $6 \cdot 10 = 60$ **b.** $3 \cdot 6 \cdot 10 = 180$ **5. a.** $2 \cdot 4 = 8$ **b.** $2 + 4 = 6$

6. $10 \cdot 9 \cdot 8 = 720$

7. a. There are $9! = 362,880$ arrangements in all, so there are 362,879 other arrangements.

   b. $26 \cdot 25 \cdot 24 \cdot 23 \cdot 22 \cdot 21 \cdot 20 \cdot 19 \cdot 18 =$ 1,133,836,704,000

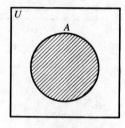

8. The number of elements in set A is the same as the number of elements in the universal set minus the number of elements not in set A. See the Venn diagram at the right. The shaded region is what is left of U when $\overline{A}$ is removed.

Ex. 8

Written Exercises 15-2 • pages 575–577

1. a. $5! = 5 \cdot 4 \cdot 3 \cdot 2 \cdot 1 = 120$ **b.** $6! = 6 \cdot 5 \cdot 4 \cdot 3 \cdot 2 \cdot 1 = 720$

   c. $7! = 7 \cdot 6 \cdot 5 \cdot 4 \cdot 3 \cdot 2 \cdot 1 = 5040$ **d.** $0! = 1$

2. a. $\dfrac{10!}{9!} = \dfrac{10 \cdot 9!}{9!} = 10$ **b.** $\dfrac{20!}{18!} = \dfrac{20 \cdot 19 \cdot 18!}{18!} = 20 \cdot 19 = 380$

   c. $\dfrac{n!}{(n-1)!} = \dfrac{n(n-1)!}{(n-1)!} = n$ **d.** $\dfrac{(n+1)!}{(n-1)!} = \dfrac{(n+1)(n)(n-1)!}{(n-1)!} = n^2 + n$

3. $5! = 5 \cdot 4 \cdot 3 \cdot 2 \cdot 1 = 120$ **4.** $9! = 9 \cdot 8 \cdot 7 \cdot 6 \cdot 5 \cdot 4 \cdot 3 \cdot 2 \cdot 1 = 362,880$

5. $2 \cdot 2 \cdot 2 \cdot 2 \cdot 2 \cdot 2 \cdot 2 \cdot 2 \cdot 2 \cdot 2 = 2^{10} = 1024$

6. There are 5 choices for each of the 10 questions: $5^{10} = 9,765,625$

7. a. $1 \cdot 26 \cdot 26 \cdot 26 = 17,576$ **b.** $1 \cdot 25 \cdot 24 \cdot 23 = 13,800$

8. a. $5 \cdot 5 \cdot 5 = 125$ **b.** $5 \cdot 4 \cdot 3 = 60$

9. $12 \cdot 11 \cdot 10 \cdot 9 = 11,880$ **10.** $10 \cdot 9 \cdot 8 \cdot 7 = 5040$

11. a.

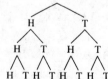

   b. 14 branches, 8 leaves

   c. number of branches $= 2 + 4 + 8 + \cdots + 2^{10} =$
   $\dfrac{2(1 - 2^{10})}{1 - 2} = 2046$; number of leaves $= 2^{10} = 1024$

12.

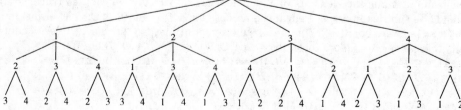

13. a. The coach can fill the first place in 9 ways, the second in 8 ways, and so on. Hence, there are $9! = 362,880$ different batting orders.

    b. Since the last place is fixed, the coach has 8 choices for the first place, 7 for the second, and so on. Hence, there are $8! = 40,320$ different batting orders.

    c. Since the last place and the third place are fixed, the coach has 7 choices for the first place, 6 for the second, and so on. Hence, there are $7! = 5040$ different batting orders.

14. For the first team: $6 \cdot 5 \cdot 4 \cdot 1 = 120$ ways and for the second: $3 \cdot 2 \cdot 1 \cdot 1 = 6$ ways; total number of teams: $120 \cdot 6 = 720$.

15. We can form 6 one-digit numbers, $6 \cdot 5 = 30$ two-digit numbers, and $6 \cdot 5 \cdot 4 = 120$ three-digit numbers, so we can form $6 + 30 + 120 = 156$ numbers in all.

16. You can form 5 one-flag messages, $5 \cdot 4 = 20$ two-flag messages, $5 \cdot 4 \cdot 3 = 60$ three-flag messages, $5 \cdot 4 \cdot 3 \cdot 2 = 120$ four-flag messages, and $5 \cdot 4 \cdot 3 \cdot 2 \cdot 1 = 120$ five-flag messages, so you can form $5 + 20 + 60 + 120 + 120 = 325$ messages in all.

17. For 3 letters followed by 2 digits, there are $26 \cdot 26 \cdot 26 \cdot 10 \cdot 10 = 1,757,600$ possibilities. For 3 letters followed by 3 digits, there are $26 \cdot 26 \cdot 26 \cdot 10 \cdot 10 \cdot 10 = 17,576,000$ possibilities. There are $1,757,600 + 17,576,000 = 19,333,600$ possibilities in all.

18. For 2 letters followed by 3 nonzero digits, there are $26 \cdot 26 \cdot 9 \cdot 9 \cdot 9 = 492,804$ possibilities; for 2 letters followed by 4 nonzero digits, there are $26 \cdot 26 \cdot 9 \cdot 9 \cdot 9 \cdot 9 = 4,435,236$ possibilities; there are $492,804 + 4,435,236 = 4,928,040$ possibilities in all.

19. a. There are 8 choices for the first digit (all but 0 and 7) and 9 choices for the other two digits: $8 \cdot 9 \cdot 9 = 648$ numbers.

    b. The first 3-digit number is 100, and the last is 999, so there are 900 3-digit numbers. From part (a), 648 of them contain no 7's, so $900 - 648 = 252$ contain at least one 7.

20. a. The first digit can be 1–7 and each of the others can be 0–7; there are $7 \cdot 8 \cdot 8 \cdot 8 = 3584$ possibilities.

    b. There are 9000 four-digit numbers (1000–9999) and 3584 with no 8's or 9's, so there are $9000 - 3584 = 5416$ four-digit numbers with at least one 8 or 9.

21. By the Complement Principle, n(at least one 3) $= 2000 - n$(no 3's), since there are 2000 numbers from 5000 to 6999. For numbers with no 3's, there are 2 choices for the first digit, a 5 or a 6, and 9 choices for each of the other digits; thus, there are $2 \cdot 9 \cdot 9 \cdot 9 = 1458$ numbers with no 3's, and $2000 - 1458 = 542$ possible numbers with at least one 3.

22. The total number of license plates is $26 \cdot 26 \cdot 26 \cdot 9 \cdot 10 \cdot 10 = 15,818,400$. The total number of plates with *no* vowels is $21 \cdot 21 \cdot 21 \cdot 9 \cdot 10 \cdot 10 = 8,334,900$. Thus, there are $15,818,400 - 8,334,900 = 7,483,500$ possible plates with at least one vowel.

23. a. There are $8 \cdot 2 \cdot 10 = 160$ possible area codes.

    b. There are $8 \cdot 8 \cdot 10 = 640$ possible exchanges within an area code.

    c. There are 10^4 four-digit numbers, so there are $10^4 - 1 = 9999$ that are not all zeros.

    d. There are $640 \cdot 9999 = 6,399,360$ possible phone numbers in the 312 area code.

    e. There are $160 \cdot 640 \cdot 9999 = 1,023,897,600$ possible phone numbers in all.

24. From Ex. 23(d), there are 6,399,360 possible phone numbers in each area code. Thus, there are $5 \cdot (6,399,360) = 31,996,800$ possible phone numbers without adding more area codes.

25. a. There are 9 different letters, so $9! = 362,880$ "words" can be formed.

   b. There are 3 choices for the first letter (I, E, A) and 2 choices for the last letter; once these have been chosen, 7 letters remain; $3 \cdot 7 \cdot 6 \cdot 5 \cdot 4 \cdot 3 \cdot 2 \cdot 1 \cdot 2 = 6 \cdot 7! = 30,240$ "words" begin and end with a vowel.

26. a. $7! = 5040$ "words" can be formed.

   b. $7 \cdot 6 \cdot 5 \cdot 4 \cdot 3 \cdot 2 = 5040$ 6-letter "words" can be formed; the answers are the same.

   c. There are 2 choices for the first letter (E, O) and 5 for the last (V, R, M, N, T). Five letters are left for the remaining choices: $2 \cdot 5 \cdot 4 \cdot 3 \cdot 5 = 600$ 5-letter "words" begin with a vowel and end with a consonant.

27. There are $26 \cdot 26 = 676$ different pairs of initials, so the 677th student's initials must match the initials of another student.

28. a. There are $26 \cdot 26 \cdot 26 = 17,576$ different sets of initials; since $17,576 < 40,000$, at least 2 students must have the same initials.

   b. $17,576 + 1$, or 17,577 students

29. Then at least one pigeonhole will contain more than one pigeon. In Exs. 27 and 28, the students correspond to pigeons and the initials correspond to pigeonholes. Because our alphabet has 26 letters, there are 26^2 sets of 2-letter initials and 26^3 sets of 3-letter initials. If a given number of people is greater than the number of sets of possible initials, then there will be people who have the same initials.

30. Answers may vary. A **bit** is a **bi**nary digi**t**. We can think of a bit as storing either a 0 or a 1. Since a byte is composed of 8 bits, $2^8 = 256$ different pieces of information in binary code can be stored in a byte. The ASCII code associates with each keyboard character a unique number between 0 and 255. Thus, when a key is pressed on the keyboard, the computer stores not the character itself but a number corresponding to that character. For example, the ASCII code for the character H is 72, or 01001000 in binary code.

31. $\log_{10} 9! \approx 5.56$; $\log_{10} 10! = \log_{10}(10 \cdot 9!) = \log_{10} 10 + \log_{10} 9! \approx 1 + 5.56 = 6.56$

32. $\log_{10} 100! - \log_{10} 99! = \log_{10} \dfrac{100!}{99!} = \log_{10} \dfrac{100 \cdot 99!}{99!} = \log_{10} 100 = 2$

33. a. $10 \cdot 9 \cdot 8 \cdot 7 = \dfrac{(10 \cdot 9 \cdot 8 \cdot 7)(6 \cdot 5 \cdot 4 \cdot 3 \cdot 2 \cdot 1)}{(6 \cdot 5 \cdot 4 \cdot 3 \cdot 2 \cdot 1)} = \dfrac{10!}{6!}$

   b. For $1 \le r \le n$, $n(n-1)(n-2) \cdots (n-r+1) =$
   $$n(n-1)(n-2) \cdots (n-r+1) \cdot \frac{(n-r)(n-r-1) \cdots 1}{(n-r)(n-r-1) \cdots 1} = \frac{n!}{(n-r)!}$$

34. Of the factors in 100!, 20 have factors of 5. Of those, 4 have 2 factors of 5 (25, 50, 75, and 100) and 16 have only 1 factor of 5; hence, there are $8 + 16$, or 24 factors of 5 among the factors of 100! and at least 24 factors of 2 since there are 50 even numbers. Thus $100! = 5^{24} \cdot 2^{24} \cdot n = (5 \cdot 2)^{24} \cdot n = 10^{24} \cdot n$ for some integer n. Hence, 100! ends in 24 zeros.

Class Exercises 15-3 · page 580

1. a. $_5P_2 = \dfrac{5!}{3!} = 5 \cdot 4 = 20$ **b.** $_5C_2 = \dfrac{5!}{3!2!} = \dfrac{5 \cdot 4}{1 \cdot 2} = 10$

2. a. $_6P_3 = \dfrac{6!}{3!} = 6 \cdot 5 \cdot 4 = 120$ **b.** $_6C_3 = \dfrac{6!}{3!3!} = \dfrac{6 \cdot 5 \cdot 4}{1 \cdot 2 \cdot 3} = 20$

3. a. $_{10}P_3 = \dfrac{10!}{7!} = 10 \cdot 9 \cdot 8 = 720$ **b.** $_{10}C_3 = \dfrac{10!}{7!3!} = \dfrac{10 \cdot 9 \cdot 8}{1 \cdot 2 \cdot 3} = 120$

4. a. $_4P_4 = \dfrac{4!}{0!} = 4! = 24$ **b.** $_4C_4 = \dfrac{4!}{4!0!} = 1$ **5.** $_{10}P_3 = \dfrac{10!}{7!} = 10 \cdot 9 \cdot 8 = 720$

6. $_{10}C_3 = \dfrac{10!}{7!3!} = \dfrac{10 \cdot 9 \cdot 8}{1 \cdot 2 \cdot 3} = 120$

7. The order of the three numbers is important. For example, $8 - 20 - 17$ is different from $20 - 8 - 17$ and $17 - 8 - 20$. Hence you use permutations to determine the number of "combinations"!

8. a. ABC   BAC   CAB   BCA   CBA   ACB
　　ABD   BAD   DAB   BDA   DBA   ADB
　　ACD   CAD   DAC   CDA   DCA   ADC
　　BCD   CBD   DBC   CDB   DCB   BDC

b. There are 24 entries and $24 = {}_4P_3 = 4 \cdot 3 \cdot 2$. Six entries involve A, B, and C and six involve A, C, and D. (Note that $3! = 6$.)

c. ABC, ABD, ACD, BCD. (Note that $_4C_3 = \dfrac{4!}{1!3!} = 4$.)

In this case, the order is not important.

Written Exercises 15-3 · pages 580–582

1. a. $_{20}P_2 = \dfrac{20!}{18!} = 20 \cdot 19 = 380$ **b.** $_{20}C_2 = \dfrac{20!}{18!2!} = \dfrac{20 \cdot 19}{1 \cdot 2} = 190$

2. a. $_{13}P_4 = \dfrac{13!}{9!} = 13 \cdot 12 \cdot 11 \cdot 10 = 17{,}160$ **b.** $_{13}C_4 = \dfrac{13!}{9!4!} = \dfrac{13 \cdot 12 \cdot 11 \cdot 10}{1 \cdot 2 \cdot 3 \cdot 4} = 715$

3. a. $_{10}C_4 = \dfrac{10!}{6!4!} = \dfrac{10 \cdot 9 \cdot 8 \cdot 7}{1 \cdot 2 \cdot 3 \cdot 4} = 210$ **b.** $_{10}P_4 = \dfrac{10!}{6!} = 10 \cdot 9 \cdot 8 \cdot 7 = 5040$

4. a. $_{20}P_5 = \dfrac{20!}{15!} = 20 \cdot 19 \cdot 18 \cdot 17 \cdot 16 = 1{,}860{,}480$

b. $_{20}C_5 = \dfrac{20!}{15!5!} = \dfrac{20 \cdot 19 \cdot 18 \cdot 17 \cdot 16}{1 \cdot 2 \cdot 3 \cdot 4 \cdot 5} = 15{,}504$

5. a. $_{200}C_3 = \dfrac{200!}{197!3!} = \dfrac{200 \cdot 199 \cdot 198}{1 \cdot 2 \cdot 3} = 1{,}313{,}400$

b. $_{200}P_3 = \dfrac{200!}{197!} = 200 \cdot 199 \cdot 198 = 7{,}880{,}400$

6. a. $_4P_4 = \dfrac{4!}{0!} = 4! = 24$ **b.** $_4C_4 = \dfrac{4!}{0!4!} = 1$

7. a. $_8P_3 = 8 \cdot 7 \cdot 6 = 336$ **b.** $_8C_3 = \dfrac{8!}{5!3!} = \dfrac{8 \cdot 7 \cdot 6}{1 \cdot 2 \cdot 3} = 56$

8. a. $_{52}P_5 = \dfrac{52!}{47!} = 52 \cdot 51 \cdot 50 \cdot 49 \cdot 48 = 311{,}875{,}200$

b. $_{52}C_5 = \dfrac{52!}{47!5!} = \dfrac{52 \cdot 51 \cdot 50 \cdot 49 \cdot 48}{1 \cdot 2 \cdot 3 \cdot 4 \cdot 5} = 2{,}598{,}960$

9. a. $_{12}P_6 = \dfrac{12!}{6!} = 12 \cdot 11 \cdot 10 \cdot 9 \cdot 8 \cdot 7 = 665{,}280$

  b. $_{12}C_6 = \dfrac{12!}{6!6!} = \dfrac{12 \cdot 11 \cdot 10 \cdot 9 \cdot 8 \cdot 7}{1 \cdot 2 \cdot 3 \cdot 4 \cdot 5 \cdot 6} = 924$

10. a. $_{12}P_5 = \dfrac{12!}{7!} = 12 \cdot 11 \cdot 10 \cdot 9 \cdot 8 = 95{,}040$

  b. $_{12}C_5 = \dfrac{12!}{7!5!} = \dfrac{12 \cdot 11 \cdot 10 \cdot 9 \cdot 8}{5 \cdot 4 \cdot 3 \cdot 2 \cdot 1} = 792$

11. a. $_7C_4 = \dfrac{7!}{3!4!} = \dfrac{7 \cdot 6 \cdot 5}{1 \cdot 2 \cdot 3} = 35$

  b. $_7C_3 = \dfrac{7!}{4!3!} = \dfrac{7 \cdot 6 \cdot 5}{1 \cdot 2 \cdot 3} = 35$; the answers are equal since choosing 3 out of 7 *to go*

  is the same as choosing 4 out of 7 *to stay*. $\left(\text{Note: } _7C_4 = \dfrac{7!}{4!3!} = {_7C_3}.\right)$

12. a. $_9C_3 = \dfrac{9!}{6!3!} = \dfrac{9 \cdot 8 \cdot 7}{1 \cdot 2 \cdot 3} = 84$ **b.** $_9C_6 = \dfrac{9!}{3!6!} = \dfrac{9 \cdot 8 \cdot 7}{1 \cdot 2 \cdot 3} = 84$; they are equal.

13. a. $_{100}C_2 = \dfrac{100!}{98!2!}$; $_{100}C_{98} = \dfrac{100!}{2!98!} = \dfrac{100!}{98!2!} = {_{100}C_2}$

  b. Answers will vary. In choosing 98 out of 100, we are finding the number of ways to choose 98 things and leave out 2 things; that is, we are finding the number of ways to choose 2 things out of 100.

14. a. $_{11}C_3 = \dfrac{11!}{8!3!}$; $_{11}C_8 = \dfrac{11!}{3!8!} = \dfrac{11!}{8!3!} = {_{11}C_3}$

  b. $_nC_r = {_nC_{n-r}}$; $_nC_r = \dfrac{n!}{(n-r)!\,r!}$; $_nC_{n-r} = \dfrac{n!}{[n-(n-r)]!\,(n-r)!} = \dfrac{n!}{r!\,(n-r)!} = \dfrac{n!}{(n-r)!\,r!} = {_nC_r}$, $0 \le r \le n$

15. a. $_5C_0 = \dfrac{5!}{5!0!} = 1$ **b.** $_5C_5 = \dfrac{5!}{0!5!} = 1$

16. $_nC_n = \dfrac{n!}{0!n!} = \dfrac{1}{0!}$. There is only one way to select n objects from a group of n; take them all. Thus, $\dfrac{1}{0!} = 1$ and so $0! = 1$.

17. There are 28 single-scoop cones, $28^2 = 784$ double-scoop cones, and $28^3 = 21{,}952$ triple-scoop cones, or 22,764 possible cones in all.

18. There are $_{28}C_2 = \dfrac{28!}{26!2!} = \dfrac{28 \cdot 27}{1 \cdot 2} = 378$ double-scoop cones with different flavors and 28 double-scoop cones with the same flavor in both scoops. Thus, there are $378 + 28$, or 406 cones in all.

19. Consider the six seats to be filled. There are 6 choices for the first seat, and 1 choice for the next seat; any of the remaining 4 can go next and there is only 1 choice for the next seat; finally either of the remaining 2 can go next and there is 1 choice for the last seat. Hence there are $6 \cdot 1 \cdot 4 - 1 \cdot 2 \cdot 1 = 48$ ways they may be seated.

20. The algebra books can be arranged in $4! = 24$ ways, the geometry books can be arranged in 2 ways, and the precalculus books can be arranged in $3! = 6$ ways. Additionally, there are $3! = 6$ ways in which the groups of algebra, geometry, and precalculus books can be arranged. Hence, there are $(4! \cdot 2! \cdot 3!)3! = (24 \cdot 2 \cdot 6) \cdot 6 = 1728$ arrangements.

21. a. There are 12 face cards in a standard deck, so $_{12}C_5 = \dfrac{12!}{7!5!} = \dfrac{12 \cdot 11 \cdot 10 \cdot 9 \cdot 8}{1 \cdot 2 \cdot 3 \cdot 4 \cdot 5} = $ 792 ways.

b. There are 40 cards that are not face cards, so $_{40}C_5 = \dfrac{40!}{35!5!} = \dfrac{40 \cdot 39 \cdot 38 \cdot 37 \cdot 36}{1 \cdot 2 \cdot 3 \cdot 4 \cdot 5} = $ 658,008 ways.

c. The number of hands with at least one face card is the total number of hands possible less the number with no face cards.
$$_{52}C_5 - {}_{40}C_5 = \frac{52!}{47!5!} - 658{,}008 = \frac{52 \cdot 51 \cdot 50 \cdot 49 \cdot 48}{1 \cdot 2 \cdot 3 \cdot 4 \cdot 5} - 658{,}008 = $$
$2{,}598{,}960 - 658{,}008 = 1{,}940{,}952$ ways.

22. $_{52}P_5 - {}_{40}P_5 = 5!(_{52}C_5 - {}_{40}C_5) = 120(1{,}940{,}942) = 232{,}914{,}240$ ways

23. a. Since order is unimportant, there are $_6C_2 = \dfrac{6!}{4!2!} = \dfrac{6 \cdot 5}{1 \cdot 2} = 15$ line segments.

b. In part (a) it is assumed that no set of 3 points is collinear; if 3 or more points are collinear, the number of line segments is less than 15.

24. Number of diagonals $= {}_nC_2 - n = \dfrac{n!}{(n-2)!2!} - n = \dfrac{n(n-1)}{1 \cdot 2} - n = $
$\dfrac{n(n-1) - 2n}{2} = \dfrac{n(n-1-2)}{2} = \dfrac{n(n-3)}{2}$

25. $_nC_2 = 45; \dfrac{n!}{(n-2)!2!} = 45; \dfrac{n(n-1)(n-2)!}{(n-2)!\,2!} = 45; \dfrac{n(n-1)}{2 \cdot 1} = 45; n(n-1) = 90;$
$n^2 - n - 90 = 0; (n-10)(n+9) = 0; n = 10$ or $n = -9$. Choose $n = 10$ since $n > 0$.

26. $_nC_2 = {}_{n-1}P_2; \dfrac{n!}{(n-2)!2!} = \dfrac{(n-1)!}{(n-1-2)!}; \dfrac{n!}{(n-2)!2!} = \dfrac{(n-1)!}{(n-3)!}; \dfrac{n(n-1)(n-2)!}{(n-2)!2!} = $
$\dfrac{(n-1)(n-2)(n-3)!}{(n-3)!}; \dfrac{n(n-1)}{2!} = (n-1)(n-2); n \geq 2,$ so $\dfrac{n}{2 \cdot 1} = n - 2; n = $
$2n - 4; n = 4$

27. By applying the formula $_nC_r = \dfrac{n!}{(n-r)!r!}$ to the given values, we obtain the array at the right. The pattern (Pascal's Triangle) has many properties and patterns. The most obvious is that the first and last entry of each row is a 1; every other entry is equal to the sum of the two entries just above it. Also, each row sum is equal to $2^{\text{row no.}}$.

				1		1			
			1		2		1		
		1		3		3		1	
	1		4		6		4		1
1		5		10		10		5		1

28. $\dfrac{n(n-1)(n-2)\cdots(n-r+1)}{1\cdot 2\cdot 3\cdot\cdots r} = \dfrac{n(n-1)(n-2)\cdots(n-r+1)}{1\cdot 2\cdot 3\cdot\cdots r}\cdot\dfrac{(n-r)!}{(n-r)!} =$

$\dfrac{n(n-1)(n-2)\cdots(n-r+1)\cdot(n-r)(n-r-1)\cdots 3\cdot 2\cdot 1}{1\cdot 2\cdot 3\cdot\cdots r\cdot(n-r)!} = \dfrac{n!}{r!(n-r)!} =$

$\dfrac{n!}{(n-r)!\,r!}$ for $1 \le r \le n$

29. $_{52}C_{13}\cdot {}_{39}C_{13}\cdot {}_{26}C_{13}\cdot {}_{13}C_{13} = \dfrac{52!}{39!13!}\cdot\dfrac{39!}{26!13!}\cdot\dfrac{26!}{13!13!}\cdot\dfrac{13!}{0!13!} = \dfrac{52!39!26!13!}{39!26!13!0!(13!)^4} =$

$\dfrac{52!}{(13!)^4}; \dfrac{52!}{(13!)^4} \approx 5.36 \times 10^{28}$

Computer Exercise · page 583

```
100 PRINT "ENTER N: ";
110 INPUT N
120 PRINT "ENTER R: ";
130 INPUT R
140 LET P = 1
150 LET C = 1
160 FOR I = 1 TO R
170    LET P = P*(N - I + 1)
180    LET C = C*(N - I + 1)/I
190 NEXT I
200 PRINT"PERMUTATIONS: ";P
210 PRINT"COMBINATIONS: ";C
220 END
```

Activity 15-4 · page 583

a. MOP, MPO, OMP, OPM, PMO, POM   **b.** MOM, MMO, OMM

c. The repetition of the letter M

Class Exercises 15-4 · page 585

1. a. $_4P_4 = 4! = 24$ **b.** $\dfrac{4!}{2!1!1!} = \dfrac{24}{2} = 12$ **2. a.** $_5P_5 = 5! = 120$ **b.** $\dfrac{5!}{2!1!1!1!} = \dfrac{120}{2} = 60$

3. a. $_6P_6 = 6! = 720$ **b.** $\dfrac{6!}{2!2!1!1!} = \dfrac{720}{4} = 180$

4. a. $_{11}P_{11} = 11! = 39{,}916{,}800$ **b.** $\dfrac{11!}{2!2!3!1!1!1!1!} = \dfrac{39{,}916{,}800}{2\cdot 2\cdot 6} = 1{,}663{,}200$

5. Answers will vary.   **6.** Answers will vary.

7. You must travel 6 blocks east and 3 blocks south. The number of routes is

$\dfrac{9!}{6!3!} = \dfrac{9\cdot 8\cdot 7}{1\cdot 2\cdot 3} = 84.$

8. Suppose five people (A, B, C, D, and E) are seated in that order around a circular table. Then each of the linear permutations except BCDAE corresponds to the circular permutation just described.

Written Exercises 15-4 · pages 585–587

1. $\dfrac{8!}{2!2!1!1!1!1!} = \dfrac{40{,}320}{2 \cdot 2} = 10{,}080$ 2. $\dfrac{12!}{3!2!1!1!1!1!1!1!1!} = \dfrac{479{,}001{,}600}{6 \cdot 2} = 39{,}916{,}800$

3. $\dfrac{11!}{4!4!2!1!} = \dfrac{39{,}916{,}800}{24 \cdot 24 \cdot 2} = 34{,}650$ 4. $\dfrac{11!}{3!2!2!1!1!1!1!} = \dfrac{39{,}916{,}800}{6 \cdot 2 \cdot 2} = 1{,}663{,}200$

5. $\dfrac{10!}{6!3!1!} = \dfrac{3{,}628{,}800}{720 \cdot 6} = 840$ ways   6. $\dfrac{12!}{2!7!3!} = \dfrac{479{,}001{,}600}{2(5040)(6)} = 7920$ ways

7. There are $\dfrac{28!}{4!3!3!5!1!4!2!1!1!1!2!1!} \approx 3.0632 \times 10^{22}$ arrangements. There are $(60 \cdot 60 \cdot 24 \cdot 365)$ seconds in a year. The printing would take about $\dfrac{3.0632 \times 10^{22}}{60 \cdot 60 \cdot 24 \cdot 365} \approx 9.71 \times 10^{14}$ years.

8. OHIO has $\dfrac{4!}{2!1!1!} = 12$ permutations.

9. **a.** 6 south and 8 east; $\dfrac{14!}{6!8!} = 3003$ routes

    b. 2 south and 5 east; $\dfrac{7!}{2!5!} = 21$ routes

    c. 4 south and 3 east; $\dfrac{7!}{4!3!} = 35$ routes

    d. From parts (b) and (c), there are 21 routes from X to Q and 35 routes from Q to Y; hence, there are $21 \cdot 35 = 735$ routes from X to Y via Q.

10. Every route from X to P covers 7 blocks of which 3 are south and 4 are east; there are $\dfrac{7!}{3!4!} = 35$ routes from X to P. Similarly, there are $\dfrac{7!}{3!4!} = 35$ routes from P to Y. Therefore, there are $35 \cdot 35 = 1225$ routes from X to Y via P. From Ex. 9(a), there are 3003 routes from X to Y; $\dfrac{1225}{3003} \approx 0.4079$; about 40.8% of the routes from X to Y pass through P.

11. There are $5! = 120$ linear permutations of 5 people. For each arrangement, each person can shift position one place clockwise 5 times without changing the relative positions of the people. There are $\dfrac{120}{5} = 24$ different circular permutations.

12. **a.** $n!$

    b. There are $n!$ linear arrangements of n people. For any given arrangement of the people, each person can shift position one place clockwise n times without changing the relative positions of the people. Thus, there are $\dfrac{n!}{n} = \dfrac{n(n-1)!}{n} = (n-1)!$ different seating arrangements.

13. This is an arrangement of 31 days in which 3 are alike, 12 are alike, and 16 are alike. There are $\dfrac{31!}{3!12!16!} \approx 1.37 \times 10^{11}$ ways in all.

14. $\dfrac{52!}{13!13!13!13!} \approx 5.36 \times 10^{28}$; the answer in Ex. 29 on p. 582 was $\dfrac{52!}{(13!)^4}$, the same as in this exercise.

15. a. Pair the position of a name from the grade-book list with the letter grade in the same position in the 25-letter "word." For example, the first letter of the "word" is the grade for the first person on the class list.

   b. $\dfrac{25!}{5!\,10!\,6!\,3!\,1!} \approx 8.25 \times 10^{12}$ ways in all.

16. Each "word" is an arrangement of 6 characters, 4 ×'s and 2 bars. There are $\dfrac{6!}{4!\,2!} = 15$ possible combinations of orders.

17. Form "words" as in Ex. 16. Each "word" is an arrangement of 17 characters, 12 ×'s and 5 separator bars. There are $\dfrac{17!}{12!\,5!} = 6188$ possible subgroup combinations.

18. Use the approach suggested by Ex. 16. Form "words" by using the consonants and bars to represent the vowels in alphabetical order. For example, a "word" could be SB ||| STM ||. There are $\dfrac{10!}{5!\,2!\,1!\,1!\,1!} = 15{,}120$ permutations with the vowels in alphabetical order.

Assorted Exercises 15-1 through 15-4 • pages 587–589

1. $_{100}C_3 = \dfrac{100!}{97!\,3!} = \dfrac{100 \cdot 99 \cdot 98}{3 \cdot 2 \cdot 1} = 161{,}700$ ways

2. $26 \cdot 26 \cdot 26 \cdot 9 \cdot 9 \cdot 9 = 12{,}812{,}904$ possible license plates

3. a. There are 9 choices for each digit; $9 \cdot 9 \cdot 9 \cdot 9 = 6561$ 4-digit integers.

   b. There are 8 choices for the first digit; there are 9 choices for the remaining 3 digits; $8 \cdot 9 \cdot 9 \cdot 9 = 5832$ 4-digit integers.

   c. There are 4 choices for the first digit $(2, 4, 6, 8)$, 10 choices for each of the next 2 digits, and 5 choices for the last digit; $4 \cdot 10 \cdot 10 \cdot 5 = 2000$ 4-digit integers.

4. There are 10 possible times to take exams, so the number of ways to schedule the 4 exams is $_{10}P_4 = \dfrac{10!}{6!} = 5040$ ways.

5. a. $_{30}P_2 = \dfrac{30!}{28!} = 30 \cdot 29 = 870$ different tickets

   b. $_{30}C_2 = \dfrac{30!}{28!\,2!} = \dfrac{30 \cdot 29}{2 \cdot 1} = 435$ kinds of tickets

6. a. All 6 letters are different, so $6! = 720$ arrangements are possible.

   b. Two letters, S's, are identical; therefore, $\dfrac{6!}{2!\,1!\,1!\,1!\,1!} = 360$ arrangements are possible.

   c. There are 2 T's and 2 O's; therefore, $\dfrac{6!}{2!\,2!\,1!\,1!} = 180$ arrangements are possible.

7. By the Multiplication Principle, there are $3 \cdot 5 \cdot 4 = 60$ such routes. There are $3 \cdot 5 \cdot 4 \cdot 4 \cdot 5 \cdot 3 = 3600$ different round trips.

8. $_{30}C_3 = \dfrac{30!}{27!\,3!} = \dfrac{30 \cdot 29 \cdot 28}{3 \cdot 2 \cdot 1} = 4060$ possible choices

9. Refer to the Venn diagram at the right. Since all students must take at least one of the three courses, $43 + 38 + 55 + (88 - x) + (107 - x) + (132 - x) + x = 297$; $463 - 2x = 297$; $x = 83$; 83 students intend to take all three courses.

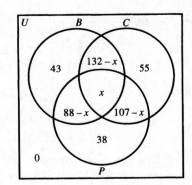

10. $_{70}P_{10} = \dfrac{70!}{60!} \approx \dfrac{\sqrt{140\pi}\left(\dfrac{70}{e}\right)^{70}}{\sqrt{120\pi}\left(\dfrac{60}{e}\right)^{60}} = \sqrt{\dfrac{7}{6}} \cdot \left(\dfrac{70}{60}\right)^{60} \cdot$

$70^{10} \cdot e^{-10} \approx 1.44 \times 10^{18}$

11. a. There are $3! = 6$ orders for the numbers 8, 15, and 17; 6×10 s $= 1$ min to try all 6.

b. Exactly 2 the same = total no. − all different − all 3 the same = $40 \cdot 40 \cdot 40 - 40 \cdot 39 \cdot 38 - 40 \cdot 1 \cdot 1 = 4680$ possibilities; 4680×10 s = $46{,}800$ s $= \dfrac{46{,}800}{3600} = 13$ h to try all the possibilities.

12. a. $_{8}P_{8} = 8! = 40{,}320$ ways   **b.** $\dfrac{1}{8} \cdot 8! = 5040$ ways

13. a. $_{8}C_{4} = \dfrac{8!}{4!4!} = \dfrac{8 \cdot 7 \cdot 6 \cdot 5}{4 \cdot 3 \cdot 2 \cdot 1} = 70$ different 4-member committees

b. If the mayor must be on the committee, we must choose 3 more members out of the remaining 7: $_{7}C_{3} = \dfrac{7!}{3!4!} = \dfrac{7 \cdot 6 \cdot 5}{3 \cdot 2 \cdot 1} = 35$ committees that include the mayor.

c. We must choose 4 members from the remaining 7: $_{7}C_{4} = \dfrac{7!}{3!4!} = \dfrac{7 \cdot 6 \cdot 5}{3 \cdot 2 \cdot 1} = 35$ committees that do not include the mayor.

d. $35 + 35 = 70$, as required

14. a. $_{9}C_{4} = \dfrac{9!}{5!4!} = \dfrac{9 \cdot 8 \cdot 7 \cdot 6}{4 \cdot 3 \cdot 2 \cdot 1} = 126$ different 4-member committees

b. If the mayor must be on the committee, we must choose 3 more members out of the remaining 8: $_{8}C_{3} = \dfrac{8!}{5!3!} = \dfrac{8 \cdot 7 \cdot 6}{3 \cdot 2 \cdot 1} = 56$ committees that include the mayor.

c. $_{8}C_{4} = \dfrac{8!}{4!4!} = \dfrac{8 \cdot 7 \cdot 6 \cdot 5}{4 \cdot 3 \cdot 2 \cdot 1} = 70$ committees that do not include the mayor

d. $56 + 70 = 126$, as required

15. With one bill, there are $_{4}C_{1} = \dfrac{4!}{3!1!} = 4$ different amounts; with two bills, there are $_{4}C_{2} = \dfrac{4!}{2!2!} = \dfrac{4 \cdot 3}{2 \cdot 1} = 6$ different amounts; with three bills, there are $_{4}C_{3} = \dfrac{4!}{3!1!} = 4$ different amounts; with four bills, there are $_{4}C_{4} = \dfrac{4!}{4!0!} = 1$ amount possible. In all, $4 + 6 + 4 + 1 = 15$ different amounts can be formed.

16. Each unit can be either a dot or a dash; there are 2 one-unit sequences, $2 \cdot 2 = 4$ two-unit sequences, $2 \cdot 2 \cdot 2 = 8$ three-unit sequences, $2 \cdot 2 \cdot 2 \cdot 2 = 16$ four-unit sequences, $2^5 = 32$ five-unit sequences, and $2^6 = 64$ six-unit sequences. Thus, there are $2 + 4 + 8 + 16 + 32 + 64 = 126$ possible sequences in all.

17. One can choose 0, 1, 2, 3, 4, 5, or 6 toppings: $_6C_0 + {_6C_1} + {_6C_2} + {_6C_3} + {_6C_4} +$

$_6C_5 + {_6C_6} = \dfrac{6!}{6!0!} + \dfrac{6!}{1!5!} + \dfrac{6!}{4!2!} + \dfrac{6!}{3!3!} + \dfrac{6!}{2!4!} + \dfrac{6!}{1!5!} + \dfrac{6!}{0!6!} = 1 + 6 + 15 + 20 + 15 +$

6 + 1 = 64 different pizzas can be made.

18. a. $_8P_4 = \dfrac{8!}{4!} = 8 \cdot 7 \cdot 6 \cdot 5 = 1680$ 4-letter "words"

   b. $5 \cdot 4 \cdot 3 \cdot 2 = 120$ 4-letter "words" with no vowels

   c. 1680 − 120 = 1560 4-letter "words" with at least one vowel

19. a. The group of boys can stand in 5! ways. The five girls can stand in 5! ways. There are 6 places for the group of boys to stand: at the beginning of the line, after the first girl, after the second girl, and so on, including after all five girls. Therefore, there are $5! \cdot 5! \cdot 6 = 86,400$ arrangements if all the boys stand in succession.

   b. If a boy is first, there are $5 \cdot 5 \cdot 4 \cdot 4 \cdot 3 \cdot 3 \cdot 2 \cdot 2 \cdot 1 \cdot 1 = 14,400$ arrangements. Similarly, if a girl is first, there are 14,400 arrangements. Thus, there are $14,400 + 14,400 = 28,800$ arrangements in all.

20. a. If 5 boys and 5 girls stand in a circle and if all the boys stand in succession, then all of the girls stand in succession, too. That is, the circle of boys and girls breaks into two semicircles, one consisting of boys and one consisting of girls. There are 5! ways of arranging the boys in their semicircle, and for each of these, there are 5! ways of arranging the girls in their semicircle. Thus, there are $5! \cdot 5! = 14,400$ ways of forming a circle of boys and girls.

   b. For every circular permutation, there are 10 linear permutations. From Ex. 19(b), there are 28,800 linear arrangements, so there are $\dfrac{1}{10} \cdot 28,800 = 2880$ circular arrangements.

21. The total number of 5-digit numbers is $9 \cdot 10 \cdot 10 \cdot 10 \cdot 10 = 90,000$. The number of 5-digit numbers that contain no 3's is $8 \cdot 9 \cdot 9 \cdot 9 \cdot 9 = 52,488$. There are $90,000 - 52,488 = 37,512$ 5-digit numbers that contain at least one 3.

22. a. The 3 represents the number of ways of arranging the letters of the "word" SEE: $\dfrac{3!}{2!1!} = 3$. The 4 represents the number of ways of arranging the letters of the "word" SEEE: $\dfrac{4!}{3!1!} = 4$.

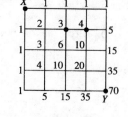

   b. To find the number of paths from X to another point, determine s, the number of steps south and e, the number of blocks east. Then the number of paths is $\dfrac{(s + e)!}{s!e!}$.

23. a. The team winning the series must win the last game of the series. In a 6-game series won by team A, A must win 3 of the remaining 5 games. Which 3 games is not important, so there are $_5C_3 = \dfrac{5!}{3!2!}$ different series.

b. Suppose team A wins the series by winning the last game of the series. If there are 4 games in the series, then there is $_3C_3 = 1$ way for A to win. If there are 5 games in the series, then there are $_4C_3 = 4$ ways for A to win. If there are 6 games in the series, then there are $_5C_3 = 10$ ways for A to win. If there are 7 games in the series, then there are $_6C_3 = 20$ ways for A to win. Thus, there are $1 + 4 + 10 + 20 = 35$ different sequences of games in which team A wins. A parallel analysis can be made for team B's winning the series. Thus, there are $2 \cdot 35 = 70$ different sequences of games possible.

24. Player A could win in 3, 4, or 5 sets. If 3 sets are needed, then there is one way for this to happen: player A wins all 3 sets. If 4 sets are needed, player A wins the last set and 2 of the remaining 3 sets in any order; this can occur in $_3C_2 = 3$ ways. If 5 sets are needed, then player A wins the last set and 2 of the remaining 4 in any order; this can occur in $_4C_2 = 6$ ways. Hence, there are $1 + 3 + 6$ different ways for player A to win the championship.

Class Exercises 15-5 · page 592

1. The numbers in Pascal's triangle have symmetry in a vertical line drawn through the center of the triangle.

2. $1, 7, 21, 35, 35, 21, 7, 1$

3. a. $1, 8, 28, 56$ **b.** $x^8, 8x^7y, 28x^6y^2, 56x^5y^3$ **c.** $x^8, -8x^7y, 28x^6y^2, -56x^5y^3$

4. $(a - b)^3 = a^3 - 3a^2b + 3ab^2 - b^3$

5. $(a + b)^4 = a^4 + 4a^3b + 6a^2b^2 + 4ab^3 + b^4$

6. $(a - b)^4 = a^4 - 4a^3b + 6a^2b^2 - 4ab^3 + b^4$

7. a. $_6C_2\, x^4y^2 = \dfrac{6 \cdot 5}{1 \cdot 2}x^4y^2 = 15x^4y^2$ **b.** $_9C_3\, x^6y^3 = \dfrac{9 \cdot 8 \cdot 7}{1 \cdot 2 \cdot 3}x^6y^3 = 84x^6y^3$

Written Exercises 15-5 · pages 592–594

1. a. $(a + b)^3 = a^3 + 3a^2b + 3ab^2 + b^3$

  b. $(20 + 1)^3 = 20^3 + 3(20)^2 + 3(20) + 1 = 8000 + 1200 + 60 + 1 = 9261$

  c. $(20 - 1)^3 = 20^3 - 3(20)^2 + 3(20) - 1 = 8000 - 1200 + 60 - 1 = 6859$

2. a. $(x + y)^4 = x^4 + 4x^3y + 6x^2y^2 + 4xy^3 + y^4$

  b. $(10 + 1)^4 = 10^4 + 4(10)^3 + 6(10)^2 + 4(10) + 1 = 14{,}641$

  c. $(10 - 1)^4 = 10^4 - 4(10)^3 + 6(10)^2 - 4(10) + 1 =$
  $10{,}000 - 4000 + 600 - 40 + 1 = 6561$

3. a. $(a + b)^5 = a^5 + 5a^4b + 10a^3b^2 + 10a^2b^3 + 5ab^4 + b^5$

  b. $(a - b)^5 = a^5 - 5a^4b + 10a^3b^2 - 10a^2b^3 + 5ab^4 - b^5$

  c. $(2a + 1)^5 = (2a)^5 + 5(2a)^4 + 10(2a)^3 + 10(2a)^2 + 5(2a) + 1 =$
  $32a^5 + 80a^4 + 80a^3 + 40a^2 + 10a + 1$

4. a. $p^6 + 6p^5q + 15p^4q^2 + 20p^3q^3 + 15p^2q^4 + 6pq^5 + q^6$

  b. $p^6 - 6p^5q + 15p^4q^2 - 20p^3q^3 + 15p^2q^4 - 6pq^5 + q^6$

  c. $(3p - 2)^6 = (3p)^6 - 6(3p)^5(2) + 15(3p)^4(2)^2 - 20(3p)^3(2)^3 + 15(3p)^2(2)^4 -$
  $6(3p)(2)^5 + 2^6 = 729p^6 - 2916p^5 + 4860p^4 - 4320p^3 + 2160p^2 - 576p + 64$

5. **a.** $x^7 + 7x^6y + 21x^5y^2 + 35x^4y^3 + 35x^3y^4 + 21x^2y^5 + 7xy^6 + y^7$

    b. $x^7 - 7x^6y + 21x^5y^2 - 35x^4y^3 + 35x^3y^4 - 21x^2y^5 + 7xy^6 - y^7$

    c. $(x^2 - 2y)^7 = (x^2)^7 - 7(x^2)^6(2y) + 21(x^2)^5(2y)^2 - 35(x^2)^4(2y)^3 + 35(x^2)^3(2y)^4 - $
        $21(x^2)^2(2y)^5 + 7(x^2)(2y)^6 - (2y)^7 = x^{14} - 14x^{12}y + 84x^{10}y^2 - 280x^8y^3 + $
        $560x^6y^4 - 672x^4y^5 + 448x^2y^6 - 128y^7$

6. **a.** $(a + b)^8 = a^8 + 8a^7b + 28a^6b^2 + 56a^5b^3 + 70a^4b^4 + 56a^3b^5 + 28a^2b^6 + 8ab^7 + b^8$

    b. $(a - b)^8 = a^8 - 8a^7b + 28a^6b^2 - 56a^5b^3 + 70a^4b^4 - 56a^3b^5 + 28a^2b^6 - 8ab^7 + b^8$

    c. $(2a - b^2)^8 = (2a)^8 - 8(2a)^7b^2 + 28(2a)^6(b^2)^2 - 56(2a)^5(b^2)^3 + 70(2a)^4(b^2)^4 - $
        $56(2a)^3(b^2)^5 + 28(2a)^2(b^2)^6 - 8(2a)(b^2)^7 + (b^2)^8 = 256a^8 - 1024a^7b^2 + $
        $1792a^6b^4 - 1792a^5b^6 + 1120a^4b^8 - 448a^3b^{10} + 112a^2b^{12} - 16ab^{14} + b^{16}$

7. $(x^2 - y^2)^3 = 1(x^2)^3 + 3(x^2)^2(-y^2)^2 + 3(x^2)(-y^2)^2 + 1(-y^2)^3 = x^6 - 3x^4y^2 + $
    $3x^2y^4 - y^6$

8. $(2x^2 - 1)^4 = (2x^2)^4 + 4(2x^2)^3(-1) + 6(2x^2)^2(-1)^2 + 4(2x^2)(-1)^3 + (-1)^4 = $
    $16x^8 - 32x^6 + 24x^4 - 8x^2 + 1$

9. $(x^2 + 1)^5 = (x^2)^5 + 5(x^2)^4(1) + 10(x^2)^3(1)^2 + 10(x^2)^2(1)^3 + 5(x^2)(1)^4 + (1)^5 = $
    $x^{10} + 5x^8 + 10x^6 + 10x^4 + 5x^2 + 1$

10. $\left(1 + \dfrac{x}{2}\right)^3 = 1 + 3\left(\dfrac{x}{2}\right) + 3\left(\dfrac{x}{2}\right)^2 + \left(\dfrac{x}{2}\right)^3 = 1 + \dfrac{3}{2}x + \dfrac{3}{4}x^2 + \dfrac{1}{8}x^3$

11. $\left(x + \dfrac{1}{x}\right)^6 = x^6 + 6x^5\left(\dfrac{1}{x}\right) + 15x^4\left(\dfrac{1}{x}\right)^2 + 20x^3\left(\dfrac{1}{x}\right)^3 + 15x^2\left(\dfrac{1}{x}\right)^4 + 6x\left(\dfrac{1}{x}\right)^5 + $

    $\left(\dfrac{1}{x}\right)^6 = x^6 + 6x^4 + 15x^2 + 20 + 15x^{-2} + 6x^{-4} + x^{-6}$

12. $\left(2x - \dfrac{1}{x}\right)^8 = (2x)^8 - 8(2x)^7\left(\dfrac{1}{x}\right) + 28(2x)^6\left(\dfrac{1}{x}\right)^2 - 56(2x)^5\left(\dfrac{1}{x}\right)^3 + 70(2x)^4\left(\dfrac{1}{x}\right)^4 - $

    $56(2x)^3\left(\dfrac{1}{x}\right)^5 + 28(2x)^2\left(\dfrac{1}{x}\right)^6 - 8(2x)\left(\dfrac{1}{x}\right)^7 + \left(\dfrac{1}{x}\right)^8 = 256x^8 - 1024x^6 + 1792x^4 - $

    $1792x^2 + 1120 - 448x^{-2} + 112x^{-4} - 16x^{-6} + x^{-8}$

13. $(a^2)^{100} + 100(a^2)^{99}(-b) + \dfrac{100 \cdot 99}{1 \cdot 2}(a^2)^{98}(-b)^2 + \dfrac{100 \cdot 99 \cdot 98}{1 \cdot 2 \cdot 3}(a^2)^{97}(-b)^3$

14. $(3p)^{20} + 20(3p)^{19}(2q) + \dfrac{20 \cdot 19}{1 \cdot 2}(3p)^{18}(2q)^2 + \dfrac{20 \cdot 19 \cdot 18}{1 \cdot 2 \cdot 3}(3p)^{17}(2q)^3$

15. $(\sin x)^{10} + 10(\sin x)^9(\sin y) + \dfrac{10 \cdot 9}{1 \cdot 2}(\sin x)^8(\sin y)^2 + \dfrac{10 \cdot 9 \cdot 8}{1 \cdot 2 \cdot 3}(\sin x)^7(\sin y)^3$

16. $(\sin x)^{30} + 30(\sin x)^{29}(-\cos y) + \dfrac{30 \cdot 29}{1 \cdot 2}(\sin x)^{28}(-\cos y)^2 + $

    $\dfrac{30 \cdot 29 \cdot 28}{1 \cdot 2 \cdot 3}(\sin x)^{27}(-\cos y)^3$

17. $(1.01)^5 = (1 + 0.01)^5 = 1^5 + 5 \cdot 1^4(0.01) + 10 \cdot 1^3(0.01)^2 + 10 \cdot 1^2(0.01)^3 + $
    $5 \cdot 1(0.01)^4 + (0.01)^5 = 1 + 0.05 + 0.0010 + 0.000010 + \cdots \approx 1.05$

18. $(0.99)^5 = (1 - 0.01)^5 = 1^5 - 5 \cdot 1^4(0.01) + 10 \cdot 1^3(0.01)^2 - 10 \cdot 1^2(0.01)^3 + $
    $5 \cdot 1(0.01)^4 - (0.01)^5 = 1 - 0.05 + 0.0010 - 0.000010 + \cdots \approx 0.95$

19. a. $_{12}C_4 = \dfrac{12!}{8!4!} = \dfrac{12 \cdot 11 \cdot 10 \cdot 9}{4 \cdot 3 \cdot 2 \cdot 1} = 495$ **b.** $_{12}C_8 = \dfrac{12!}{8!4!} = \dfrac{12 \cdot 11 \cdot 10 \cdot 9}{4 \cdot 3 \cdot 2 \cdot 1} = 495$

20. a. $_{20}C_3 = \dfrac{20!}{17!3!} = \dfrac{20 \cdot 19 \cdot 18}{3 \cdot 2 \cdot 1} = 1140$ **b.** $_{20}C_{17} = \dfrac{20!}{3!17!} = \dfrac{20 \cdot 19 \cdot 18}{3 \cdot 2 \cdot 1} = 1140$

21. Since the term is of the form $_{12}C_a \cdot (x^2)^{12-a}\left(\dfrac{2}{x}\right)^a$, we know that $(x^2)^{12-a}\left(\dfrac{2}{x}\right)^a = b \cdot x^0$

where b is some constant. So $x^{24-2a} \cdot 2^a \cdot x^{-a} = b \cdot x^0$, or $24 - 2a - a = 0$;

$24 - 3a = 0$; $a = 8$. Thus, the ninth term contains no x; $_{12}C_8 \cdot (x^2)^{12-8}\left(\dfrac{2}{x}\right)^8 =$

$\dfrac{12!}{4!8!} \cdot x^8 \cdot \dfrac{2^8}{x^8} = \dfrac{12 \cdot 11 \cdot 10 \cdot 9}{4 \cdot 3 \cdot 2 \cdot 1} \cdot 2^8 = 126{,}720$.

22. Since the term is of the form $_{10}C_b \cdot (a^3)^{10-b}(-2)^b$, we know that $(a^3)^{10-b}(-2)^b = k \cdot a^{18}$

where k is some constant. So $a^{30-3b} \cdot (-2)^b = k \cdot a^{18}$, or $30 - 3b = 18$; $12 = 3b$;

$b = 4$. The required term is the fifth term; $_{10}C_4 \cdot (a^3)^6(-2)^4 = \dfrac{10!}{6!4!}a^{18}(16) =$

$\dfrac{10 \cdot 9 \cdot 8 \cdot 7}{4 \cdot 3 \cdot 2 \cdot 1} \cdot 16a^{18} = 3360a^{18}$.

23. a. $_{n+1}C_k = \dfrac{(n+1)!}{k!(n-k+1)!}$

   b. If the mayor is on the committee, we must fill $k - 1$ places with n possible choices,

   so we have $_nC_{k-1} = \dfrac{n!}{(k-1)!(n-k+1)!}$ committees.

   c. If the mayor is not on the committee, we must fill k places with n possible choices,

   so we have $_nC_k = \dfrac{n!}{k!(n-k)!}$ committees.

   d. $_{n+1}C_k = {_nC_{k-1}} + {_nC_k}$

   e. $_nC_{k-1} + {_nC_k} = \dfrac{n!}{(n-k+1)!(k-1)!} + \dfrac{n!}{(n-k)!k!} = \dfrac{n!}{(n-k+1)!(k-1)!} \cdot \dfrac{k}{k} +$

   $\dfrac{n!}{(n-k)!k!} \cdot \dfrac{(n-k+1)}{(n-k+1)} = \dfrac{n!(k)}{(n-k+1)!(k-1)! \cdot k} + \dfrac{n!(n-k+1)}{(n-k)!(n-k+1) \cdot k!} =$

   $\dfrac{n!(k) + n!(n-k+1)}{(n-k+1)!k!} = \dfrac{n!(k+n-k+1)}{(n-k+1)!k!} = \dfrac{n!(n+1)}{(n-k+1)!k!} = \dfrac{(n+1)!}{(n-k+1)!k!} = {_{n+1}C_k}$

24. $(a + b)^n = \displaystyle\sum_{k=0}^{n} {_nC_k}a^{n-k}b^k$

25. The triangular numbers are $1, 3, 6, 10, 15, \ldots$; they occur in the third column. In the

   nth row, the third entry is $_nC_2 = \dfrac{n!}{(n-2)!2!} = \dfrac{n(n-1)}{2} = (n-1)$st triangular number.

26. a. The nonnegative integral powers of 2   **b.** The Fibonacci sequence

27. Pascal's triangle for which even numbers are highlighted shows a pattern of inverted
   equilateral triangles of various sizes. Specifically, in each row n, where $n = 2^k$ and
   $k = 1, 2, 3, \ldots$, an inverted triangle having $n - 1$ numbers on a side begins. Moreover,
   to the left and to the right of each inverted triangle is the same pattern of triangles that
   appears immediately above the inverted triangle.

28. Pascal's triangle for which numbers divisible by three are highlighted shows a pattern of inverted equilateral triangles of various sizes. Specifically, at each row $n = 3^k$ for $k = 1, 2, 3, \ldots$, there begins a trio of equilateral triangles having $n - 1$ numbers on a side. (The triangles in each trio are themselves arranged in a triangular pattern.) Moreover, in the regions between the triangles of each trio is the same pattern of triangles that appears immediately above the uppermost triangle in the trio.

29. a. $(\cos \theta + i \sin \theta)^3 = (\cos \theta)^3 + 3(\cos \theta)^2(i \sin \theta)^1 + 3(\cos \theta)^1(i \sin \theta)^2 + (i \sin \theta)^3 = \cos^3\theta + 3i \cos^2\theta \sin \theta + 3(i^2)\cos \theta \sin^2\theta + i^3\sin^3\theta = (\cos^3\theta - 3 \cos \theta \sin^2\theta) + i(3 \sin \theta \cos^2\theta - \sin^3\theta)$

b. $(\cos \theta + i \sin \theta)^3 = \cos 3\theta + i \sin 3\theta$

c. $a + bi = c + di$ if and only if $a = c$ and $b = d$; thus, if the values for $(\cos \theta + i \sin \theta)^3$ are equal, $\cos 3\theta = \cos^3\theta - 3 \cos \theta \sin^2\theta$ and $\sin 3\theta = 3 \sin \theta \cos^2\theta - \sin^3\theta$.

30. $(\cos \theta + i \sin \theta)^4 = (\cos \theta)^4 + 4(\cos \theta)^3(i \sin \theta)^1 + 6(\cos \theta)^2(i \sin \theta)^2 + 4(\cos \theta)(i \sin \theta)^3 + (i \sin \theta)^4 = \cos^4\theta + 4i \cos^3\theta \sin \theta + 6i^2 \cos^2\theta \sin^2\theta + 4i^3 \cos \theta \sin^3\theta + i^4 \sin^4\theta = (\cos^4\theta - 6 \cos^2\theta \sin^2\theta + \sin^4\theta) + i(4 \cos^3\theta \sin \theta - 4 \cos \theta \sin^3\theta)$; also, by De Moivre's theorem, $(\cos \theta + i \sin \theta)^4 = \cos 4\theta + i \sin 4\theta$; thus, $\cos 4\theta = \cos^4\theta - 6 \cos^2\theta \sin^2\theta + \sin^4\theta$ and $\sin 4\theta = 4 \cos^3\theta \sin \theta - 4 \cos \theta \sin^3\theta$.

31. $(1 + x)^{-1} = 1 + (-1)x + \dfrac{(-1)(-2)}{1 \cdot 2}x^2 + \dfrac{(-1)(-2)(-3)}{1 \cdot 2 \cdot 3}x^3 + \cdots +$ $\dfrac{(-1)(-2)(-3) \cdots (-1 - (r - 1))}{1 \cdot 2 \cdot 3 \cdot 4 \cdots r}x^4 + \cdots = 1 - x + x^2 - x^3 + \cdots + (-1)^r x^r + \cdots$

32. $(1 + x)^{-2} = 1 + (-2)x + \dfrac{(-2)(-3)}{1 \cdot 2}x^2 + \dfrac{(-2)(-3)(-4)}{1 \cdot 2 \cdot 3}x^3 + \cdots +$ $\dfrac{(-2)(-3)(-4) \cdots (-(r + 1))}{1 \cdot 2 \cdot 3 \cdots r}x^4 + \cdots = 1 - 2x + 3x^2 - 4x^3 + \cdots +$ $(-1)^r(r + 1)x^r + \cdots$

33. $(1 + x)^{1/2} = 1 + \dfrac{1}{2}x + \dfrac{\dfrac{1}{2}\left(-\dfrac{1}{2}\right)}{1 \cdot 2}x^2 + \dfrac{\dfrac{1}{2}\left(-\dfrac{1}{2}\right)\left(-\dfrac{3}{2}\right)}{1 \cdot 2 \cdot 3}x^3 + \cdots = 1 + \dfrac{1}{2}x - \dfrac{1}{8}x^2 +$ $\dfrac{1}{16}x^3 + \cdots$; when $|x|$ is small, this sum is approximately equal to $1 + \dfrac{1}{2}x$. $\sqrt{1.04} = (1.04)^{1/2} = (1 + 0.04)^{1/2} \approx 1 + \dfrac{1}{2}(0.04) = 1.02$; $\sqrt{0.98} = (0.98)^{1/2} = (1 - 0.02)^{1/2} \approx 1 + \dfrac{1}{2}(-0.02) = 0.99$.

34. $(1 + x)^{1/3} = 1 + \dfrac{1}{3}x + \dfrac{\dfrac{1}{3}\left(-\dfrac{2}{3}\right)}{1 \cdot 2}x^2 + \dfrac{\left(\dfrac{1}{3}\right)\left(-\dfrac{2}{3}\right)\left(-\dfrac{5}{3}\right)}{1 \cdot 2 \cdot 3}x^3 + \cdots = 1 + \dfrac{1}{3}x -$ $\dfrac{1}{9}x^2 + \dfrac{5}{81}x^3 + \cdots$; when $|x|$ is small, this sum is approximately equal to $1 + \dfrac{1}{3}x$.

$\sqrt[3]{1.12} = (1.12)^{1/3} = (1 + 0.12)^{1/3} \approx 1 + \dfrac{1}{3}(0.12) = 1.04$; $\sqrt[3]{67} = \sqrt[3]{64\left(1 + \dfrac{3}{64}\right)} =$ $4\left(1 + \dfrac{3}{64}\right)^{1/3} \approx 4\left(1 + \dfrac{1}{3} \cdot \dfrac{3}{64}\right) = 4\left(1 + \dfrac{1}{64}\right) = 4\dfrac{1}{16} \approx 4.06$

Chapter Test · page 595

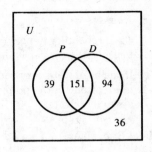

1. $n(P \cap \overline{D}) = n(P) - 151 = 190 - 151 = 39$;
$n(\overline{P} \cap D) = n(D) - 151 = 245 - 151 = 94$;
$n(\overline{P \cup D}) = n(U) - 39 - 151 - 94 = 320 - 284 = 36$;
36 children buy neither.

2. There are 4 choices for left guard, 3 choices for right guard, 5 choices for left forward, 4 choices for right forward, and 3 choices for center. By the Multiplication Principle, there are $4 \cdot 3 \cdot 5 \cdot 4 \cdot 3 = 720$ ways to form a team.

3. The order of the letters or digits is important. There are $26 \cdot 25 \cdot 24 \cdot 23 = 358{,}800$ 4-letter passwords and $10 \cdot 9 \cdot 8 \cdot 7 = 5040$ 4-digit passwords, so $358{,}800 + 5040 = 363{,}840$ different passwords are possible.

4. The order of the books is important; $_7P_5 = \dfrac{7!}{2!} = 7 \cdot 6 \cdot 5 \cdot 4 \cdot 3 = 2520$ arrangements are possible.

5. We need to find the total number of pairs of people; $_6C_2 = \dfrac{6!}{4!2!} = \dfrac{6 \cdot 5}{1 \cdot 2} = 15$ handshakes in all.

6. $\dfrac{8!}{3!3!2!} = \dfrac{8 \cdot 7 \cdot 6 \cdot 5 \cdot 4}{(3 \cdot 2 \cdot 1)(2 \cdot 1)} = 560$ sequences are possible.

7. For each circular arrangement of n objects, there corresponds n linear permutations. For example, if 3 football players are in a huddle, there are only $2! = 2$ ways in which they can be arranged. If the same players are seated on a bench, however, there are 3 linear arrangements for each circular arrangement (so that there are $3 \cdot 2! = 3! = 6$ linear permutations in all). In general, there are $n!$ linear arrangements of n objects and $(n - 1)!$ circular arrangements.

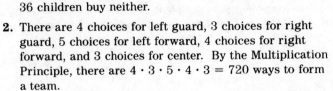

circular   linear

ACB
CBA
BAC

ABC
BCA
CAB

8. $(x - 2)^{15} = x^{15} + 15x^{14}(-2) + \dfrac{15 \cdot 14}{1 \cdot 2}x^{13}(-2)^2 + \dfrac{15 \cdot 14 \cdot 13}{1 \cdot 2 \cdot 3}x^{12}(-2)^3 + \cdots$;

the coefficient of x^{13} is $105 \cdot 4 = 420$; the coefficient of x^{12} is $455(-8) = -3640$.

Activity 16-1 · page 597

a–c. Answers will vary.

d. The probability in part (c) is likely to be a better predictor than the probability in part (b) because the former is based on a much larger number of observations.

e. If the head were very wide relative to the length of the point, the tack would land point up more often. If the point were very long, and the head small, the tack would land point down more often.

Class Exercises 16-1 · page 602

1. $S = \{1, 2, 3, 4, 5, 6\}$ **a.** 1 and 4 are perfect squares; $P(\text{perfect square}) = \dfrac{2}{6} = \dfrac{1}{3}$.

b. All elements of S are factors of 60; $P(\text{factor of 60}) = \dfrac{6}{6} = 1$.

c. No element of S is negative; $P(\text{negative}) = \dfrac{0}{6} = 0$.

2. a. $\dfrac{1}{52}$ **b.** $\dfrac{13}{52} = \dfrac{1}{4}$ **c.** $\dfrac{4}{52} = \dfrac{1}{13}$ **d.** $\dfrac{26}{52} = \dfrac{1}{2}$

e. Jack, queen, and king in each suit; $\dfrac{12}{52} = \dfrac{3}{13}$ **f.** Half the face cards are red; $\dfrac{6}{52} = \dfrac{3}{26}$.

3. Refer to the sample space of ordered pairs on p. 601; the sample points with the same number are $(1, 1)$, $(2, 2)$, $(3, 3)$, $(4, 4)$, $(5, 5)$, and $(6, 6)$; $P(\text{same number}) = \dfrac{6}{36} = \dfrac{1}{6}$.

4. The sample space, shown on p. 601, has 36 elements.

a. $P((1, 2) \text{ or } (2, 1)) = \dfrac{2}{36} = \dfrac{1}{18}$ **b.** $P((1, 3), (2, 2), \text{ or } (3, 1)) = \dfrac{3}{36} = \dfrac{1}{12}$

c. $P((1, 2), (2, 1), (1, 3), (2, 2), \text{ or } (3, 1)) = \dfrac{5}{36}$

5. $P(\text{no rain}) = 1 - P(\text{rain}) = 1 - 0.40 = 0.60$; 60% chance

6. $P(\text{at least one accident}) = 1 - P(\text{none}) = 1 - 0.82 = 0.18$

7. The possible sums are not equally likely. For example, there is a much greater probability of getting a sum of 7 than a sum of 2.

8. a. Yes **b.** The sample points are not equally likely.

9. S is not a sample space because some outcomes, such as queen of hearts, correspond to more than one element in S.

10. This is not valid reasoning because the populations of the 50 states are not equal.

Written Exercises 16-1 · pages 603–605

1–4. Refer to Example 2 on p. 600.

1. a. $\dfrac{26}{52} = \dfrac{1}{2}$ **b.** $\dfrac{13}{52} = \dfrac{1}{4}$ **c.** $1 - \dfrac{1}{4} = \dfrac{3}{4}$ **2. a.** $\dfrac{6}{52} = \dfrac{3}{26}$ **b.** $\dfrac{2}{52} = \dfrac{1}{26}$ **c.** $1 - \dfrac{1}{26} = \dfrac{25}{26}$

3. a. $\dfrac{13}{52} = \dfrac{1}{4}$ **b.** $\dfrac{0}{52} = 0$ **c.** $1 - 0 = 1$ **4. a.** $\dfrac{4}{52} = \dfrac{1}{13}$ **b.** $\dfrac{8}{52} = \dfrac{2}{13}$ **c.** $1 - \dfrac{2}{13} = \dfrac{11}{13}$

5. a. $\dfrac{10 + 10}{4280} = \dfrac{1}{214}$ **b.** $\dfrac{20 + 3(4)}{4280} = \dfrac{4}{535}$ **c.** $1 - \dfrac{4}{535} = \dfrac{531}{535}$

6. a. $P(12, 14, 16, 18, \text{ or } 20) = \dfrac{5}{10} = \dfrac{1}{2}$ **b.** $P(12, 15, \text{ or } 18) = \dfrac{3}{10}$

   c. $P(11, 13, 17, \text{ or } 19) = \dfrac{4}{10} = \dfrac{2}{5}$

7. a. There are not an equal number of telephone numbers in each borough.

   b. The number of telephone numbers in Manhattan and the number of telephone numbers in New York City

8. Each state has the same number of senators, but the number of representatives varies according to the population of the state.

9. a. $P((1,5), (2,4), (3,3), (4,2), \text{ or } (5,1)) = \dfrac{5}{36}$

   b. $P((1,6), (2,5), (3,4), (4,3), (5,2), \text{ or } (6,1)) = \dfrac{6}{36} = \dfrac{1}{6}$

   c. $P((2,6), (3,5), (4,4), (5,3), \text{ or } (6,2)) = \dfrac{5}{36}$

10. a. $\dfrac{18}{36} = \dfrac{1}{2}$ **b.** $P((6,6)) = \dfrac{1}{36}$ **c.** $P(\text{sum} < 12) = 1 - P(\text{sum} = 12) = 1 - \dfrac{1}{36} = \dfrac{35}{36}$

11. From Class Ex. 3, $P(\text{same number}) = \dfrac{1}{6}$; $P(\text{different nos.}) = 1 - P(\text{same number}) =$

   $1 - \dfrac{1}{6} = \dfrac{5}{6}$.

12. $r > w$ in the following: $(2,1), (3,1), (3,2), (4,1), (4,2), (4,3), (5,1), (5,2), (5,3), (5,4),$

   $(6,1), (6,2), (6,3), (6,4), \text{ and } (6,5); P(r > w) = \dfrac{15}{36} = \dfrac{5}{12}$.

13. a.

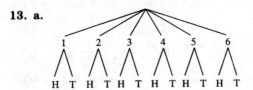

   b. $P(2H, 4H, \text{ or } 6H) = \dfrac{3}{12} = \dfrac{1}{4}$

14. a.

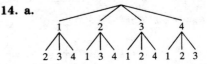

   b. $P((2,4), (3,4), (4,2), \text{ or } (4,3)) = \dfrac{4}{12} = \dfrac{1}{3}$

15. a. If odds are 4 to 3, then $P(\text{Nat'l League wins}) = \dfrac{4}{4+3} = \dfrac{4}{7}$.

   b. $P(\text{American League wins}) = 1 - P(\text{Nat'l League wins}) = 1 - \dfrac{4}{7} = \dfrac{3}{7}$

16. a. If odds in favor of winning are 2 to 3, then $P(\text{incumbent wins}) = \dfrac{2}{2+3} = \dfrac{2}{5}$.

   b. $P(\text{incumbent loses}) = 1 - P(\text{incumbent wins}) = 1 - \dfrac{2}{5} = \dfrac{3}{5}$

17. Each die can land 8 different ways, so there are $8 \cdot 8 = 64$ different outcomes.

   a. $P((1,1)) = \dfrac{1}{64}$ **b.** $P((1,2) \text{ or } (2,1)) = \dfrac{2}{64} = \dfrac{1}{32}$

   c. A sum of 9 can occur in 8 ways: $(1,8), (2,7), (3,6), (4,5), (5,4), (6,3), (7,2),$ and $(8,1)$. This sum is most likely since each number from 1 to 8 is used.

18. There are $12 \cdot 12$, or 144, different outcomes possible.   **a.** $P((12, 12)) = \dfrac{1}{144}$

   b. $P((11, 12)$ or $(12, 11)) = \dfrac{2}{144} = \dfrac{1}{72}$

   c. A sum of 13 can occur in 12 ways; since each number from 1 to 12 can be used to make a sum of 13, this is the sum most likely to appear.

19. a. {{Alvin, Bob}, {Alvin, Carol}, {Alvin, Donna}, {Bob, Carol}, {Bob, Donna}, {Carol, Donna}}

   b. $\dfrac{1}{6}$ **c.** $P(\text{not Carol}) = \dfrac{3}{6} = \dfrac{1}{2}$

20. a. {(black, black), (black, red), (red, black), (red, red)}

   b. After a red card is drawn, a black card is more likely to be drawn than a red card, because there are 26 black cards left and only 25 red cards.

21. The inclusion-exclusion principle states that "For any sets A and B, $n(A \cup B) = n(A) + n(B) - n(A \cap B)$." If A and B are events in a sample space of s equally likely outcomes, then dividing both sides of the above equation will turn each term into a probability: $\dfrac{n(A \cup B)}{s} = \dfrac{n(A)}{s} + \dfrac{n(B)}{s} - \dfrac{n(A \cap B)}{s}$ becomes $P(A$ or $B) = P(A) + P(B) - P(A$ and $B)$, which is the equation for the probability of either of two events.

22. $P(\text{dog or cat}) = P(\text{dog}) + P(\text{cat}) - P(\text{dog and cat}) = \dfrac{632}{1260} + \dfrac{568}{1260} - \dfrac{114}{1260} = \dfrac{1086}{1260} = \dfrac{181}{210}$

23. There are $5! = 120$ ways to arrange the letters of the word TEXAS. TAXES is one such arrangement, so $P(\text{TAXES}) = \dfrac{1}{120}$.

24. There are $6! = 720$ ways to arrange the letters of the word COSINE. SONICE is one such arrangement, so $P(\text{SONICE}) = \dfrac{1}{720}$.

25. If $P(x) = x^2 - 2x + k$, then $P(x) = 0$ when $x = \dfrac{2 \pm \sqrt{4 - 4k}}{2} = 1 \pm \sqrt{1 - k}$. Hence $P(x)$ has linear factors with integral coefficients when $1 - k$ is a perfect square. This is true when $k = -3$, 0, or 1. $P(\text{can be factored}) = \dfrac{3}{8}$.

26. If the discriminant is ≥ 0, then $y = x^2 - 4x + c$ will intersect the x-axis. $D = 16 - 4c \geq 0$ when $c \leq 4$; there are four elements of the set that are ≤ 4. P (intersect) $= \dfrac{4}{6} = \dfrac{2}{3}$.

27. There are $_5C_3 = \dfrac{5!}{2!3!} = 10$ ways of selecting 3 letters from the word OCEAN. Only one of these has all vowels. $P(\text{all vowels}) = \dfrac{1}{10}$.

28. There are $_7C_3 = \dfrac{7!}{4!3!} = 35$ ways of selecting 3 letters from the word PAINTED. Only one of these has all vowels. $P(\text{all vowels}) = \dfrac{1}{35}$.

29. There are 900 numbers from 100 to 999, inclusive.

    a. There are $9 \cdot 9 \cdot 9 = 729$ 3-digit numbers that contain no 0's. $P(\text{no 0's}) = \dfrac{729}{900} = \dfrac{81}{100}$.

    b. $P(\text{at least one 0}) = 1 - P(\text{no 0's}) = 1 - \dfrac{81}{100} = \dfrac{19}{100}$

30. There are 9000 numbers from 1000 to 9999, inclusive.

    a. There are 8 choices for the first digit (not 0 and not 9), and 9 choices for the other digits, so there are $8 \cdot 9 \cdot 9 \cdot 9 = 5832$ numbers that contain no 9's.

        $P(\text{no 9's}) = \dfrac{5832}{9000} = \dfrac{81}{125}$.

    b. $P(\text{at least one 9}) = 1 - P(\text{no 9's}) = 1 - \dfrac{81}{125} = \dfrac{44}{125}$

Activity 16-2 · page 605

    a. $\{HH, HT, TH, TT\}$ **b.** $P(TT) = \dfrac{1}{4}$ **c.** $P(\text{T on first toss}) = \dfrac{1}{2}$;

    $P(\text{T on second toss}) = \dfrac{1}{2}$

    d. Multiply: $\dfrac{1}{2} \cdot \dfrac{1}{2} = \dfrac{1}{4}$. **e.** $P(TTT) = \dfrac{1}{2} \cdot \dfrac{1}{2} \cdot \dfrac{1}{2} = \dfrac{1}{8}$

Class Exercises 16-2 · page 609

1. Answers may vary. Rule 1 is used in Example 2 when finding $P(YY)$ and $P(GG)$. Rule 2 is used in Example 1 when finding $P(YY)$ and $P(GG)$.

2. No; $P(\text{jack}) = \dfrac{4}{52} = \dfrac{1}{13}$; $P(\text{jack} \,|\, \text{face card}) = \dfrac{4}{12} = \dfrac{1}{3}$; since $P(\text{jack}) \neq P(\text{jack} \,|\, \text{face card})$, the events are not independent.

3. a. $P(S) = \dfrac{16}{30} = \dfrac{8}{15}$ **b.** $P(B \,|\, S) = \dfrac{9}{16}$

    c. $P(B) = \dfrac{12}{30} = \dfrac{6}{15}$ and $P(B \,|\, S) = \dfrac{9}{16}$; since $P(B) \neq P(B \,|\, S)$, the events are not independent.

Written Exercises 16-2 · pages 609–613

1. a. See the figure at the right.

    b. No; $P(\text{second is R} \,|\, \text{first is R}) \neq P(\text{second is R})$

    c. $P(RR) + P(GG) = \dfrac{3}{8} \cdot \dfrac{2}{7} + \dfrac{5}{8} \cdot \dfrac{4}{7} = \dfrac{26}{56} = \dfrac{13}{28}$

    d. $P(\text{not same color}) = 1 - P(\text{same color}) = 1 - \dfrac{13}{28} = \dfrac{15}{28}$

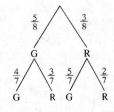

2. a. See below. **b.** No; $P(\text{second is R} \mid \text{first is R}) \neq P(\text{second is R})$

   c. $P(\text{GG}) + P(\text{RR}) = \dfrac{3}{4} \cdot \dfrac{2}{3} + \dfrac{1}{4} \cdot 0 = \dfrac{1}{2}$

   d. $P(\text{not same color}) = 1 - P(\text{same color}) = 1 - \dfrac{1}{2} = \dfrac{1}{2}$

Ex. 2

Ex. 3

3. a. See above. **b.** Yes; $P(\text{second is R} \mid \text{first is R}) = P(\text{second is R})$

   c. $P(\text{GG}) + P(\text{RR}) = \dfrac{5}{8} \cdot \dfrac{5}{8} + \dfrac{3}{8} \cdot \dfrac{3}{8} = \dfrac{34}{64} = \dfrac{17}{32}$

   d. $P(\text{not same color}) = 1 - P(\text{same color}) = 1 - \dfrac{17}{32} = \dfrac{15}{32}$

4. a. See below. **b.** Yes; $P(\text{second is R} \mid \text{first is R}) = P(\text{second is R})$

   c. $P(\text{GG}) + P(\text{RR}) = \dfrac{3}{4} \cdot \dfrac{3}{4} + \dfrac{1}{4} \cdot \dfrac{1}{4} = \dfrac{10}{16} = \dfrac{5}{8}$

   d. $P(\text{not same color}) = 1 - P(\text{same color}) = 1 - \dfrac{5}{8} = \dfrac{3}{8}$

Ex. 4

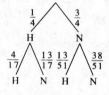

Ex. 5

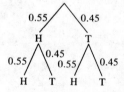

Ex. 6

5. a. See above.

   b. $P(\text{HH}) = \dfrac{1}{4} \cdot \dfrac{4}{17} = \dfrac{1}{17}$; $P(\text{HN}) = \dfrac{1}{4} \cdot \dfrac{13}{17} = \dfrac{13}{68}$; $P(\text{NH}) = \dfrac{3}{4} \cdot \dfrac{13}{51} = \dfrac{13}{68}$;

   $P(\text{NN}) = \dfrac{3}{4} \cdot \dfrac{38}{51} = \dfrac{19}{34}$

6. a. See above.

   b. $P(\text{HH}) = (0.55)(0.55) = 0.3025$; $P(\text{HT}) = (0.55)(0.45) = 0.2475$;
   $P(\text{TH}) = (0.45)(0.55) = 0.2475$; $P(\text{TT}) = (0.45)(0.45) = 0.2025$

7. a. $P(\text{F}) = \dfrac{12}{52} = \dfrac{3}{13}$; $P(\text{F} \mid \text{D}) = \dfrac{3}{13}$; since $P(\text{F}) = P(\text{F} \mid \text{D})$, the events are independent.

   b. $P(\text{R}) = \dfrac{26}{52} = \dfrac{1}{2}$; $P(\text{R} \mid \text{D}) = \dfrac{13}{13} = 1$; since $P(\text{R}) \neq P(\text{R} \mid \text{D})$, the events are not
   independent.

8. a. $P(\text{A}) = \dfrac{3}{6} = \dfrac{1}{2}$; $P(\text{A} \mid \text{H}) = \dfrac{1}{3}$; not independent

   b. $P(\text{A}) = \dfrac{3}{6} = \dfrac{1}{2}$; $P(\text{A} \mid \text{C}) = \dfrac{1}{2}$; independent

9. a.

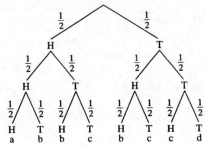

10. a.

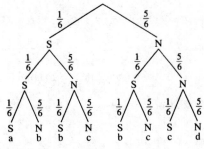

b. $P(\text{HHH}) = \dfrac{1}{8}$; $P(2 \text{ heads}) = 3 \cdot \dfrac{1}{8} =$

$\dfrac{3}{8}$ (labeled (b) in diagram);

$P(1 \text{ head}) = \dfrac{3}{8}$ (labeled (c) in

diagram); $P(0 \text{ heads}) = P(\text{TTT}) = \dfrac{1}{8}$

b. $P(3 \text{ 6's}) = \left(\dfrac{1}{6}\right)^3 = \dfrac{1}{216}$; $P(2 \text{ 6's}) =$

$3\left(\dfrac{1}{6}\right)^2\left(\dfrac{5}{6}\right) = \dfrac{15}{216} = \dfrac{5}{72}$ (labeled

(b) in diagram); $P(1 \text{ 6}) =$

$3\left(\dfrac{1}{6}\right)\left(\dfrac{5}{6}\right)^2 = \dfrac{75}{216} = \dfrac{25}{72}$ (labeled

(c) in diagram); $P(0 \text{ 6's}) =$

$P(\text{NNN}) = \left(\dfrac{5}{6}\right)^3 = \dfrac{125}{216}$

11. a. $P(3 \text{ vowels}) = (\text{no. of 3-letter words with 3 vowels}) \div (\text{no. of 3-letter words}) = \dfrac{{}_3P_3}{{}_8P_3} =$

$\dfrac{3 \cdot 2 \cdot 1}{8 \cdot 7 \cdot 6} = \dfrac{1}{56}$; $P(\text{first is vowel}) = \dfrac{3}{8}$; $P(\text{second is vowel} \mid \text{first is vowel}) = \dfrac{2}{7}$;

$P(\text{third is vowel} \mid \text{1st two are vowels}) = \dfrac{1}{6}$; $P(3 \text{ vowels}) = \dfrac{3}{8} \cdot \dfrac{2}{7} \cdot \dfrac{1}{6} = \dfrac{1}{56}$

b. $P(3 \text{ consonants}) = \dfrac{5}{8} \cdot \dfrac{4}{7} \cdot \dfrac{3}{6} = \dfrac{5}{28}$ or $P(3 \text{ consonants}) = \dfrac{{}_5P_3}{{}_8P_3} = \dfrac{5 \cdot 4 \cdot 3}{8 \cdot 7 \cdot 6} = \dfrac{5}{28}$

12. $P(\text{GGGG}) = P(G) \cdot P(G \mid G) \cdot P(G \mid G \text{ and } G) \cdot P(G \mid G \text{ and } G \text{ and } G) = \dfrac{13}{24} \cdot \dfrac{12}{23} \cdot \dfrac{11}{22} \cdot \dfrac{10}{21} =$

$\dfrac{65}{966}$ or $P(\text{GGGG}) = \dfrac{{}_{13}P_4}{{}_{24}P_4} = \dfrac{13 \cdot 12 \cdot 11 \cdot 10}{24 \cdot 23 \cdot 22 \cdot 21} = \dfrac{65}{966}$

13. a. $\dfrac{147}{400}$ **b.** $\dfrac{20}{400} = \dfrac{1}{20}$ **14.** $\dfrac{40}{100} = \dfrac{2}{5}$ **15. a.** $\dfrac{29}{90}$ **b.** $\dfrac{29}{147}$ **16. a.** $\dfrac{39}{100}$ **b.** $\dfrac{39}{156} = \dfrac{1}{4}$

17. $P(\text{junior}) = \dfrac{100}{400} = \dfrac{1}{4}$; $P(\text{junior} \mid \text{humanities}) = \dfrac{39}{156} = \dfrac{1}{4}$; events are independent.

18. $P(\text{junior}) = \dfrac{100}{400} = \dfrac{1}{4}$; $P(\text{junior} \mid \text{nat. sciences}) = \dfrac{33}{147}$; since $\dfrac{1}{4} \neq \dfrac{33}{147}$, the events are not

independent.

19. $P(B \mid A) = \dfrac{n(A \cap B)}{n(A)} = \dfrac{n(A \cap B)}{n(A)} \cdot \dfrac{\dfrac{1}{n(S)}}{\dfrac{1}{n(S)}} = \dfrac{\dfrac{n(A \cap B)}{n(S)}}{\dfrac{n(A)}{n(S)}} = \dfrac{P(A \text{ and } B)}{P(A)}$

20. Rule 1 says $P(A \text{ and } B) = P(A) \cdot P(B \mid A)$. If A and B are independent, then
$P(B \mid A) = P(B)$. Substituting $P(B)$ for $P(B \mid A)$ yields Rule 2.

21. a. $P(3 \text{ clubs}) = \dfrac{13}{52} \cdot \dfrac{12}{51} \cdot \dfrac{11}{50} = \dfrac{11}{850}$ **b.** $P(3 \text{ red}) = \dfrac{26}{52} \cdot \dfrac{25}{51} \cdot \dfrac{24}{50} = \dfrac{2}{17}$

c. $P(3 \text{ aces}) = \dfrac{4}{52} \cdot \dfrac{3}{51} \cdot \dfrac{2}{50} = \dfrac{1}{5525}$ **d.** $P(\text{no ace}) = \dfrac{48}{52} \cdot \dfrac{47}{51} \cdot \dfrac{46}{50} = \dfrac{4324}{5525}$

e. $P(\text{at least 1 ace}) = 1 - P(\text{no ace}) = 1 - \dfrac{4324}{5525} = \dfrac{1201}{5525}$

f. $P(1 \text{ or } 2 \text{ aces}) = 1 - [P(3 \text{ aces}) + P(\text{no ace})] = 1 - \left(\dfrac{1}{5525} + \dfrac{4324}{5525}\right) = \dfrac{48}{221}$

22. a. $P(5 \text{ spades}) = \dfrac{13}{52} \cdot \dfrac{12}{51} \cdot \dfrac{11}{50} \cdot \dfrac{10}{49} \cdot \dfrac{9}{48} = \dfrac{33}{66,640}$

b. The probability is $\dfrac{33}{66,640}$ for 5 of any one suit; $P(5 \text{ hearts}) + P(5 \text{ diamonds}) +$

$P(5 \text{ clubs}) + P(5 \text{ spades}) = 4 \cdot \dfrac{33}{66,640} = \dfrac{33}{16,660}$

c. $P(\text{no ace}) = \dfrac{48}{52} \cdot \dfrac{47}{51} \cdot \dfrac{46}{50} \cdot \dfrac{45}{49} \cdot \dfrac{44}{48} = \dfrac{35,673}{54,145}$

d. $P(\text{at least 1 ace}) = 1 - P(\text{no ace}) = 1 - \dfrac{35,673}{54,145} = \dfrac{18,472}{54,145}$

23. a. $P(3 \text{ clubs}) = \left(\dfrac{13}{52}\right)^3 = \left(\dfrac{1}{4}\right)^3 = \dfrac{1}{64}$ **b.** $P(3 \text{ red}) = \left(\dfrac{26}{52}\right)^3 = \left(\dfrac{1}{2}\right)^3 = \dfrac{1}{8}$

c. $P(3 \text{ aces}) = \left(\dfrac{4}{52}\right)^3 = \left(\dfrac{1}{13}\right)^3 = \dfrac{1}{2197}$ **d.** $P(\text{no ace}) = \left(\dfrac{48}{52}\right)^3 = \left(\dfrac{12}{13}\right)^3 = \dfrac{1728}{2197}$

e. $P(\text{at least 1 ace}) = 1 - P(\text{no ace}) = 1 - \dfrac{1728}{2197} = \dfrac{469}{2197}$

f. $P(1 \text{ or } 2 \text{ aces}) = 1 - [P(\text{no ace}) + P(3 \text{ aces})] = 1 - \left(\dfrac{1728}{2197} + \dfrac{1}{2197}\right) = \dfrac{36}{169}$

24. a. $P(5 \text{ spades}) = \left(\dfrac{13}{52}\right)^5 = \left(\dfrac{1}{4}\right)^5 = \dfrac{1}{1024}$

b. $P(5 \text{ hearts}) + P(5 \text{ diamonds}) + P(5 \text{ clubs}) + P(5 \text{ spades}) = 4 \cdot \dfrac{1}{1024} = \dfrac{1}{256}$

c. $P(\text{no ace}) = \left(\dfrac{48}{52}\right)^5 = \left(\dfrac{12}{13}\right)^5 = \dfrac{248,832}{371,293}$

d. $P(\text{at least one ace}) = 1 - P(\text{no ace}) = 1 - \dfrac{248,832}{371,293} = \dfrac{122,461}{371,293}$

25. a. $P(\text{different days}) = \dfrac{7}{7} \cdot \dfrac{6}{7} \cdot \dfrac{5}{7} = \dfrac{30}{49}$

b. $P(\text{at least 2 on same day}) = 1 - P(\text{all different}) = 1 - \dfrac{30}{49} = \dfrac{19}{49}$

26. a. $P(\text{all different}) = \dfrac{6}{6} \cdot \dfrac{5}{6} \cdot \dfrac{4}{6} = \dfrac{5}{9}$

b. $P(\text{at least 2 the same}) = 1 - P(\text{all different}) = 1 - \dfrac{5}{9} = \dfrac{4}{9}$

27. The following draws constitute a win for person A: B and RRB. $P(B) = \dfrac{2}{5}$ and

$P(\text{RRB}) = \dfrac{3}{5} \cdot \dfrac{2}{4} \cdot \dfrac{2}{3} = \dfrac{1}{5}$; $P(\text{A wins}) = \dfrac{2}{5} + \dfrac{1}{5} = \dfrac{3}{5}$.

28. The following draws constitute a win for person A: B, RRB, and RRRRB. P(A wins) $=$
P(B) $+ P$(RRB) $+ P$(RRRRB) $= \dfrac{1}{5} + \dfrac{4}{5} \cdot \dfrac{3}{4} \cdot \dfrac{1}{3} + \dfrac{4}{5} \cdot \dfrac{3}{4} \cdot \dfrac{2}{3} \cdot \dfrac{1}{2} \cdot 1 = \dfrac{1}{5} + \dfrac{1}{5} + \dfrac{1}{5} = \dfrac{3}{5}$.

29. a. For Y to win from point A, Y must win the next 3 games; P(YYY) $= (0.4)^3 = 0.064$.

    b. From point $(1,3)$, Y can win the championship in 3 ways: Y, XY, and XXY. P(Y wins from $(1,3)$) $= P$(Y) $+ P$(XY) $+ P$(XXY) $= 0.4 + (0.6)(0.4) + (0.6)^2(0.4) = 0.784$.

    c. From point $(2,1)$, P(X wins the championship) $= P$(XX) $+ P$(XYX) $+ P$(YXX) $+$ P(XYYX) $+ P$(YXYX) $+ P$(YYXX) $= (0.6)^2 + 2(0.4)(0.6)^2 + 3(0.4)^2(0.6)^2 = 0.8208$.

30. a. In a 4-game series, there is 1 order of wins: XXXX. P(XXXX) $= p^2(1-p)^2$ since team X is the home team for games 1 and 2. In a 5-game series with X winning the last game, the orders are found by arranging the other 4 games' outcomes, YXXX. There are $\dfrac{4!}{3!1!} = 4$ ways of arranging these. List all 4 possibilities and find the probability of each; this yields $2p(1-p)^4 + 2p^3(1-p)^2$. In a 6-game series with X winning the last game, the orders are found by arranging the other 5 games' outcomes: YYXXX. There are $\dfrac{5!}{3!2!} = 10$ ways of arranging these. List all 10 possibilities and find the probability of each; this yields $p(1-p)^5 + 6p^3(1-p)^3 + 3p^5(1-p)$. In a 7-game series, the orders are found by arranging the first 6 outcomes: YYYXXX. There are $\dfrac{6!}{3!3!} = 20$ ways of arranging these. List all 20 possibilities and find the probability of each; this yields $p(1-p)^6 + 9p^3(1-p)^4 + 9p^5(1-p)^2 + p^7$. Hence P(X wins) $= p^2(1-p)^2 + 2p(1-p)^4 + 2p^3(1-p)^2 + p(1-p)^5 + 6p^3(1-p)^3 + 3p^5(1-p) + p(1-p)^6 + 9p^3(1-p)^4 + 9p^5(1-p)^2 + p^7$; simplifying, $f(p) = 20p^7 - 70p^6 + 108p^5 - 95p^4 + 52p^3 - 18p^2 + 4p$.

    b. One way to solve this is to graph $y = f(p)$ and $y = 0.5$ on the same axes using a computer or graphing calculator. When $p = 0.5$, $f(p) = 0.5$ and so team X's probability of winning is 0.5.

Class Exercises 16-3 · page 615

1. $4\left(\dfrac{1}{2}\right)^3\left(\dfrac{1}{2}\right)$ represents P(3 heads) in 4 tosses: HHHT, HHTH, HTHH, THHH;

$6\left(\dfrac{1}{2}\right)^2\left(\dfrac{1}{2}\right)^2$ represents P(2 heads) in 4 tosses: HHTT, HTHT, HTTH, THHT, THTH,

TTHH; $4\left(\dfrac{1}{2}\right)\left(\dfrac{1}{2}\right)^3$ represents P(1 head) in 4 tosses: HTTT, THTT, TTHT, TTTH; $\left(\dfrac{1}{2}\right)^4$

represents P(0 heads) in 4 tosses: TTTT.

2. Find the second term in $\left(\dfrac{1}{6} + \dfrac{5}{6}\right)^4$, that is, $_4C_3\left(\dfrac{1}{6}\right)^3\left(\dfrac{5}{6}\right)^1$.

3. $\dfrac{1}{2}$, because the outcomes are independent.

4. a. P(RR) $= \dfrac{50}{100} \cdot \dfrac{49}{99} = \dfrac{49}{198}$

    b. The first and second drawings are not independent trials.   **c.** P(RR) $\approx \dfrac{1}{2} \cdot \dfrac{1}{2} = \dfrac{1}{4}$.

Written Exercises 16-3 · pages 616–618

1.

	3 heads	2 heads	1 head	0 heads
Outcome	HHH	HHT, HTH, THH	HTT, THT, TTH	TTT
Probability	$\left(\dfrac{1}{2}\right)^3$	$3\left(\dfrac{1}{2}\right)^3$	$3\left(\dfrac{1}{2}\right)^3$	$\left(\dfrac{1}{2}\right)^3$

a. $\left(\dfrac{1}{2}\right)^3 = \dfrac{1}{8}$ **b.** $_3C_2 \cdot \left(\dfrac{1}{2}\right)^2 \cdot \dfrac{1}{2} = 3 \cdot \dfrac{1}{4} \cdot \dfrac{1}{2} = \dfrac{3}{8}$

c. $_3C_1 \cdot \left(\dfrac{1}{2}\right)\left(\dfrac{1}{2}\right)^2 = 3 \cdot \dfrac{1}{2} \cdot \dfrac{1}{4} = \dfrac{3}{8}$ **d.** $\left(\dfrac{1}{2}\right)^3 = \dfrac{1}{8}$

2.

	4 sixes	3 sixes	2 sixes	1 six	0 sixes
Outcome	SSSS	SSSN, SSNS, SNSS, NSSS	SSNN, SNSN, NSSN, SNNS, NNSS, NSNS	SNNN, NSNN, NNSN, NNNS	NNNN
Prob.	$\left(\dfrac{1}{6}\right)^4$	$4\left(\dfrac{1}{6}\right)^3\left(\dfrac{5}{6}\right)$	$6\left(\dfrac{1}{6}\right)^2\left(\dfrac{5}{6}\right)^2$	$4\left(\dfrac{1}{6}\right)\left(\dfrac{5}{6}\right)^3$	$\left(\dfrac{5}{6}\right)^4$

a. $\left(\dfrac{1}{6}\right)^4 = \dfrac{1}{1296}$ **b.** $_4C_3\left(\dfrac{1}{6}\right)^3\left(\dfrac{5}{6}\right) = 4 \cdot \dfrac{1}{216} \cdot \dfrac{5}{6} = \dfrac{5}{324}$

c. $_4C_2\left(\dfrac{1}{6}\right)^2\left(\dfrac{5}{6}\right)^2 = 6 \cdot \dfrac{1}{36} \cdot \dfrac{25}{36} = \dfrac{25}{216}$ **d.** $_4C_1\left(\dfrac{1}{6}\right)\left(\dfrac{5}{6}\right)^3 = 4 \cdot \dfrac{1}{6} \cdot \dfrac{125}{216} = \dfrac{125}{324}$

e. $\left(\dfrac{5}{6}\right)^4 = \dfrac{625}{1296}$

3. The possible outcomes are BBBB, BBBG, BBGB, BGBB, GBBB, BBGG, BGBG, BGGB, GBBG, GGBB, GBGB, BGGG, GBGG, GGBG, GGGB, and GGGG.

a. $\left(\dfrac{1}{2}\right)^4 = \dfrac{1}{16}$ **b.** $_4C_3\left(\dfrac{1}{2}\right)^3\left(\dfrac{1}{2}\right) = 4 \cdot \dfrac{1}{8} \cdot \dfrac{1}{2} = \dfrac{1}{4}$

c. $_4C_2\left(\dfrac{1}{2}\right)^2\left(\dfrac{1}{2}\right)^2 = 6 \cdot \dfrac{1}{4} \cdot \dfrac{1}{4} = \dfrac{3}{8}$ **d.** $_4C_1\left(\dfrac{1}{2}\right)\left(\dfrac{1}{2}\right)^3 = 4 \cdot \dfrac{1}{2} \cdot \dfrac{1}{8} = \dfrac{1}{4}$

e. $\left(\dfrac{1}{2}\right)^4 = \dfrac{1}{16}$

4. a. $\left(\dfrac{2}{5}\right)^3 = \dfrac{8}{125}$ **b.** $_3C_2 \cdot \left(\dfrac{2}{5}\right)^2 \cdot \dfrac{3}{5} = 3 \cdot \dfrac{4}{25} \cdot \dfrac{3}{5} = \dfrac{36}{125}$

c. $_3C_1 \cdot \dfrac{2}{5} \cdot \left(\dfrac{3}{5}\right)^2 = 3 \cdot \dfrac{2}{5} \cdot \dfrac{9}{25} = \dfrac{54}{125}$ **d.** $\left(\dfrac{3}{5}\right)^3 = \dfrac{27}{125}$

5. $P(2\ 5\text{'s}) = {}_4C_2 \cdot \left(\dfrac{1}{6}\right)^2\left(\dfrac{5}{6}\right)^2 = \dfrac{4 \cdot 3}{2 \cdot 1} \cdot \dfrac{1}{36} \cdot \dfrac{25}{36} = \dfrac{25}{216} \approx 0.116$

6. $P(1\ 3) = {}_7C_1 \cdot \left(\dfrac{1}{6}\right)\left(\dfrac{5}{6}\right)^6 = 7 \cdot \dfrac{1}{6} \cdot \dfrac{15{,}625}{46{,}656} = \dfrac{109{,}375}{279{,}936} \approx 0.391$

7. a. $\dfrac{13}{52} = \dfrac{1}{4}$ **b.** $\dfrac{3}{4} \cdot \dfrac{3}{4} = \dfrac{9}{16}$; $_2C_1 \cdot \dfrac{1}{4} \cdot \dfrac{3}{4} = 2 \cdot \dfrac{3}{16} = \dfrac{3}{8}$; $\dfrac{1}{4} \cdot \dfrac{1}{4} = \dfrac{1}{16}$

8. On each draw, $P(\text{spades}) = \dfrac{13}{52} = \dfrac{1}{4}$. **a.** $\dfrac{1}{4} \cdot \dfrac{1}{4} \cdot \dfrac{1}{4} = \dfrac{1}{64}$

b. $_3C_2 \cdot \left(\dfrac{1}{4}\right)^2 \cdot \dfrac{3}{4} = 3 \cdot \dfrac{1}{16} \cdot \dfrac{3}{4} = \dfrac{9}{64}$ **c.** $_3C_1 \cdot \dfrac{1}{4} \cdot \left(\dfrac{3}{4}\right)^2 = 3 \cdot \dfrac{1}{4} \cdot \dfrac{9}{16} = \dfrac{27}{64}$

d. $\dfrac{3}{4} \cdot \dfrac{3}{4} \cdot \dfrac{3}{4} = \dfrac{27}{64}$

9. Assume that for each question $P(\text{correct answer}) = \dfrac{1}{4}$.

a. $P(6 \text{ correct}) = \left(\dfrac{1}{4}\right)^6 = \dfrac{1}{4096}$

b. $P(5 \text{ correct}) = {}_6C_5 \cdot \left(\dfrac{1}{4}\right)^5 \cdot \dfrac{3}{4} = 6 \cdot \dfrac{3}{4096} = \dfrac{9}{2048}$

10. On each draw, $P(\text{red}) = \dfrac{4}{7}$ and $P(\text{white}) = \dfrac{3}{7}$.

a. $P(0 \text{ red}) = \left(\dfrac{3}{7}\right)^3 = \dfrac{27}{343}$

b. $P(1 \text{ red}) = {}_3C_1 \cdot \dfrac{4}{7} \cdot \left(\dfrac{3}{7}\right)^2 = 3 \cdot \dfrac{4}{7} \cdot \dfrac{9}{49} = \dfrac{108}{343}$

c. $P(2 \text{ red}) = {}_3C_2 \cdot \left(\dfrac{4}{7}\right)^2 \cdot \dfrac{3}{7} = 3 \cdot \dfrac{16}{49} \cdot \dfrac{3}{7} = \dfrac{144}{343}$ **d.** $P(3 \text{ red}) = \left(\dfrac{4}{7}\right)^3 = \dfrac{64}{343}$

11. $P(\text{recommend}) = \dfrac{8}{10} = \dfrac{4}{5}$ and $P(\text{not recommend}) = 1 - \dfrac{4}{5} = \dfrac{1}{5}$

a. $P(3 \text{ recommend}) = \left(\dfrac{4}{5}\right)^3 = \dfrac{64}{125}$ **b.** $P(\text{none recommends}) = \left(\dfrac{1}{5}\right)^3 = \dfrac{1}{125}$

c. $P(\text{at least 1 recommends}) = 1 - P(\text{none recommends}) = 1 - \dfrac{1}{125} = \dfrac{124}{125}$

12. $P(\text{at least 6 ft}) = \dfrac{1}{3}$, so $P(\text{under 6 ft}) = \dfrac{2}{3}$.

a. $P(7 \text{ under 6 ft}) = \left(\dfrac{2}{3}\right)^7 = \dfrac{128}{2187}$

b. $P(7 \text{ at least 6 ft}) = \left(\dfrac{1}{3}\right)^7 = \dfrac{1}{2187}$

c. $P(\text{all but 1 under 6 ft}) = {}_7C_1 \cdot \dfrac{1}{3} \cdot \left(\dfrac{2}{3}\right)^6 = 7 \cdot \dfrac{1}{3} \cdot \dfrac{64}{729} = \dfrac{448}{2187}$

13. a. The empirical probability (0.3125) is greater than the theoretical probability (0.125).

    b. $0.1\overline{6}$

    c. $250 \div 3 = 83.\overline{3}$; 83 simulations can be carried out; about 0.13253

    d. Answers may vary. While often an approximation of an event's theoretical probability will be improved by increasing the number of simulations, this will not always be the case. Given the nature of random-number tables and of probabilities, it is certainly possible for a large number of simulations to produce a very inaccurate approximation of an event's theoretical probability.

14. a. Answers may vary. Examples:

    Exercise 2: Assign the digits 1–6 to the outcomes "1"–"6," respectively. Cross off all occurrences of the digits 0, 7, 8, and 9.

    Exercise 3: Assign the digits 0–4 to the outcome "boy" and assign the digits 5–9 to the outcome "girl."

    Exercise 4: Assign the digits 0–3 to the outcome "heads" and assign the digits 4–9 to the outcome "tails."

    Exercise 5: See Exercise 2.

    Exercise 6: See Exercise 2.

    Exercise 7: Assign the digits 0 and 1 to the outcome "spade" and assign the digits 2–7 to the outcome "not a spade." Cross off all occurrences of the digits 8 and 9.

    Exercise 8: See Exercise 7.

    Exercise 9: Assign the digits 0 and 1 to the outcome "right" and assign the digits 2–7 to the outcome "wrong." Cross off all occurrences of the digits 8 and 9.

    Exercise 10: Assign the digits 0–3 to the outcome "red" and assign the digits 4–6 to the outcome "white." Cross off all occurrences of the digits 7, 8, and 9.

    Exercise 11: Assign the digits 0–7 to the outcome "recommend Brand X" and assign the digits 8 and 9 to the outcome "do not recommend Brand X."

    Exercise 12: Assign the digits 1–3 to the outcome "at least 6 ft tall" and assign the digits 4–9 to the outcome "under 6 ft tall." Cross off all occurrences of the digit 0.

    b. Answers may vary, depending upon the string of random numbers used, and upon the assignment of digits to outcomes. Answers are given to three significant digits and are based on **(1)** conducting simulations by starting with the first number in the random-number table given in Exercise 13 and continuing as far as possible through the table, and **(2)** assigning digits to outcomes as specified in part (a). Decimal approximations of the theoretical probabilities calculated in the original exercise are given in parentheses.

Exercise 2: a. 0 (0.001) **b.** 0.026 (0.015) **c.** 0.105 (0.116) **d.** 0.342 (0.386) **e.** 0.526 (0.482)

Exercise 7: a. 0.255 (0.25) **b.** 0.541 (0.5625); 0.408 (0.375); 0.051 (0.0625)

15. a. $P(4 \text{ of } 5) = {}_5C_4 \cdot (0.75)^4(0.25) = \dfrac{405}{1024} \approx 0.396$

    b. No; Answers may vary. A player's free-throw percentage changes slightly each time he or she takes a free throw because, for example, the player may be tired or may have spent a great deal of time practicing. Also, the free-throw percent changes after each free throw attempted.

16. a.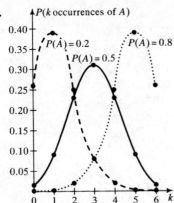

k	0	1	2	3	4	5	6
$P(k)$	$\dfrac{1}{64}$	$\dfrac{6}{64}$	$\dfrac{15}{64}$	$\dfrac{20}{64}$	$\dfrac{15}{64}$	$\dfrac{6}{64}$	$\dfrac{1}{64}$

b. A bell-shaped curve

c. When $P(A) \neq \dfrac{1}{2}$, the resulting curve is not

symmetric. For $P(A) > \dfrac{1}{2}$, the peak is shifted to

the right; for $P(A) < \dfrac{1}{2}$, the peak is shifted to

the left.

17. $P(\text{sprout}) = \dfrac{9}{10} = 0.9$, so $P(\text{not sprout}) = 0.1$; $P(1 \text{ does not sprout}) =$

$_{10}C_1 \cdot (0.1)(0.9)^9 = (0.9)^9 \approx 0.387$.

18. $P(\text{at least 2 rings}) = 1 - [P(\text{no rings}) + P(1 \text{ ring})]$

$$= 1 - \left[\left(\frac{4}{5}\right)^5 + {}_5C_1 \cdot \frac{1}{5} \cdot \left(\frac{4}{5}\right)^4 \right]$$

$$= 1 - \left(\frac{1024}{3125} + \frac{5 \cdot 4^4}{3125} \right)$$

$$= \frac{821}{3125} \approx 0.263$$

19. a. Ignoring leap year, if there are 366 people, then there must be at least one common birthday.

b. $P(\text{at least 1 match}) = 1 - P(\text{no match}) = 1 - \dfrac{{}_{364}P_9}{365^9} =$

$1 - \dfrac{364 \cdot 363 \cdot 362 \cdot 361 \cdot 360 \cdot 359 \cdot 358 \cdot 357 \cdot 356}{365^9} \approx 1 - 0.8830518 \approx 0.117$

c. 23 people, since $P(\text{at least one match}) = 1 - \dfrac{{}_{364}P_{22}}{365^{22}} \approx 1 - 0.4927028 \approx 0.507 \approx 50\%$.

20. a. Use the method of Ex. 19. $P(\text{at least one match}) = 1 - P(\text{no match})$. Since your birthday is one specific day out of 365, $P(\text{not your birthday}) = \dfrac{364}{365}$.

$P(\text{no match in } n \text{ tries}) = \left(\dfrac{364}{365}\right)^n$. $P(\text{at least one match}) = 1 - \left(\dfrac{364}{365}\right)^n$.

b. If $n = 253$, $P(\text{at least one match}) \approx 0.5005 \approx 50\%$.

Computer Exercise · page 618

```
10 PRINT "WHAT IS THE NUMBER OF TRIALS"
20 PRINT "IN A SIMULATION";
30 INPUT N
40 PRINT "WHAT IS THE PROBABILITY THAT THE";
50 PRINT "DESIRED OUTCOME OCCURS ON EACH TRIAL";
60 INPUT P
70 PRINT "HOW MANY SIMULATIONS DO YOU";
80 PRINT "WANT PERFORMED";
90 INPUT S
100 DIM A(N)
110 FOR I = 0 TO N
120 LET A(I) = 0
130 NEXT I
140 FOR I = 1 TO S
150 LET T = 0
160 FOR J = 1 TO N
170 IF RND (1) < = P THEN LET T = T + 1
180 NEXT J
190 LET A(T) = A(T) + 1
200 NEXT I
210 PRINT "T", "T/"S
220 FOR I = 0 TO N
230 PRINT I,A(I)/S
240 NEXT I
250 END
```

Class Exercises 16-4 · pages 620–621

1. a. $\dfrac{{}_4C_3 \cdot {}_5C_0}{{}_9C_3}$ **b.** $\dfrac{{}_4C_2 \cdot {}_5C_1}{{}_9C_3}$ **c.** $\dfrac{{}_4C_1 \cdot {}_5C_2}{{}_9C_3}$ **d.** $\dfrac{{}_4C_0 \cdot {}_5C_3}{{}_9C_3}$

2. a. $\dfrac{{}_4C_4 \cdot {}_{48}C_1}{{}_{52}C_5}$ **b.** $\dfrac{{}_4C_0 \cdot {}_{48}C_5}{{}_{52}C_5}$ **c.** $\dfrac{{}_{13}C_4 \cdot {}_{39}C_1}{{}_{52}C_5}$ **d.** $\dfrac{{}_4C_4 \cdot {}_4C_1}{{}_{52}C_5}$

3. a. no aces

   b. $P(\text{at least 1 ace}) = 1 - P(\text{no aces}) = 1 - \dfrac{{}_4C_0 \cdot {}_{48}C_5}{{}_{52}C_5} = 1 - \dfrac{1 \cdot 1{,}712{,}304}{2{,}598{,}960} \approx$
   $1 - 0.6588 = 0.3412$

Written Exercises 16-4 · pages 621–623

1. a. $\dfrac{{}_{13}C_5 \cdot {}_{39}C_0}{{}_{52}C_5} = \dfrac{\dfrac{13!}{8!5!} \cdot 1}{\dfrac{52!}{47!5!}} = \dfrac{13!\,47!}{8!\,52!}$ **b.** $\dfrac{{}_{13}C_0 \cdot {}_{39}C_5}{{}_{52}C_5} = \dfrac{1 \cdot \dfrac{39!}{34!5!}}{\dfrac{52!}{47!5!}} = \dfrac{39!\,47!}{34!\,52!}$

   c. $P(\text{at least 1 heart}) = 1 - P(\text{no hearts}) = 1 - \dfrac{39!\,47!}{34!\,52!}$

2. a. $\dfrac{_{13}C_{13}}{_{52}C_{13}} = \dfrac{1}{\dfrac{52!}{39!\,13!}} = \dfrac{39!\,13!}{52!}$ **b.** $\dfrac{_{39}C_{13}}{_{52}C_{13}} = \dfrac{\dfrac{39!}{26!\,13!}}{\dfrac{52!}{39!\,13!}} = \dfrac{39!\,39!\,13!}{26!\,13!\,52!} = \dfrac{39!\,39!}{26!\,52!}$

c. $P(\text{at least 1 club}) = 1 - P(\text{no clubs}) = 1 - \dfrac{39!\,39!}{26!\,52!}$

3. There are $_8C_2 = 28$ ways to draw 2 marbles; $P(0\ \text{red}) = \dfrac{_5C_0 \cdot {}_3C_2}{_8C_2} = \dfrac{1 \cdot 3}{28} = \dfrac{3}{28}$;

$P(1\ \text{red}) = \dfrac{_5C_1 \cdot {}_3C_1}{_8C_2} = \dfrac{5 \cdot 3}{28} = \dfrac{15}{28}$; $P(2\ \text{red}) = \dfrac{_5C_2 \cdot {}_3C_0}{_8C_2} = \dfrac{10 \cdot 1}{28} = \dfrac{5}{14}$;

$\dfrac{3}{28} + \dfrac{15}{28} + \dfrac{10}{28} = \dfrac{28}{28} = 1.$

4. There are $_8C_2 = 28$ ways to draw 2 marbles; $P(0\ \text{red}) = \dfrac{_6C_0 \cdot {}_2C_2}{_8C_2} = \dfrac{1 \cdot 1}{28} = \dfrac{1}{28}$;

$P(1\ \text{red}) = \dfrac{_6C_1 \cdot {}_2C_1}{_8C_2} = \dfrac{6 \cdot 2}{28} = \dfrac{3}{7}$; $P(2\ \text{red}) = \dfrac{_6C_2 \cdot {}_2C_0}{_8C_2} = \dfrac{15 \cdot 1}{28} = \dfrac{15}{28}$;

$\dfrac{1}{28} + \dfrac{12}{28} + \dfrac{15}{28} = \dfrac{28}{28} = 1.$

5. a. $\dfrac{_{12}C_4 \cdot {}_8C_0}{_{20}C_4} = \dfrac{\dfrac{12!}{8!\,4!} \cdot 1}{\dfrac{20!}{16!\,4!}} = \dfrac{12!\,16!\,4!}{8!\,4!\,20!} = \dfrac{12!\,16!}{8!\,20!} = \dfrac{33}{323} \approx 0.102$

b. $\dfrac{_{12}C_3 \cdot {}_8C_1}{_{20}C_4} = \dfrac{\dfrac{12!}{9!\,3!} \cdot 8}{\dfrac{20!}{16!\,4!}} = \dfrac{12!\,16!\,4! \cdot 8}{9!\,3!\,20!} = \dfrac{352}{969} \approx 0.363$

c. $\dfrac{_{12}C_2 \cdot {}_8C_2}{_{20}C_4} = \dfrac{\dfrac{12!}{10!\,2!} \cdot \dfrac{8!}{6!\,2!}}{\dfrac{20!}{16!\,4!}} = \dfrac{12!\,8!\,16!\,4!}{10!\,2!\,6!\,2!\,20!} = \dfrac{616}{1615} \approx 0.381$

d. $\dfrac{_{12}C_1 \cdot {}_8C_3}{_{20}C_4} = \dfrac{12 \cdot \dfrac{8!}{5!\,3!}}{\dfrac{20!}{16!\,4!}} = \dfrac{12 \cdot 8!\,16!\,4!}{5!\,3!\,20!} = \dfrac{224}{1615} \approx 0.139$

e. $\dfrac{_{12}C_0 \cdot {}_8C_4}{_{20}C_4} = \dfrac{1 \cdot \dfrac{8!}{4!\,4!}}{\dfrac{20!}{16!\,4!}} = \dfrac{8!\,16!\,4!}{4!\,4!\,20!} = \dfrac{14}{969} \approx 0.014$

6. a. $\dfrac{_8C_2 \cdot {}_7C_1}{_{20}C_3} = \dfrac{\dfrac{8!}{6!\,2!} \cdot \dfrac{7!}{6!\,1!}}{\dfrac{20!}{17!\,3!}} = \dfrac{28 \cdot 7}{1140} = \dfrac{49}{285} \approx 0.172$

b. $\dfrac{_5C_3}{_{20}C_3} = \dfrac{\dfrac{5!}{2!\,3!}}{\dfrac{20!}{17!\,3!}} = \dfrac{10}{1140} = \dfrac{1}{114} \approx 0.009$ **c.** $\dfrac{_{15}C_3}{_{20}C_3} = \dfrac{\dfrac{15!}{12!\,3!}}{\dfrac{20!}{17!\,3!}} = \dfrac{455}{1140} = \dfrac{91}{228} \approx 0.399$

d. $\dfrac{_8C_1 \cdot {}_7C_1 \cdot {}_5C_1}{_{20}C_3} = \dfrac{8 \cdot 7 \cdot 5}{1140} = \dfrac{14}{57} \approx 0.246$

7. (1) $P(E) = \dfrac{10}{20} = \dfrac{1}{2}$; (2) $P(E) = \dfrac{{}_1C_1 \cdot {}_5C_2}{{}_6C_3} = \dfrac{1 \cdot \frac{5!}{2!3!}}{\frac{6!}{3!3!}} = \dfrac{10}{20} = \dfrac{1}{2}$

8. (1) $P(\text{E and F}) = \dfrac{4}{20} = \dfrac{1}{5}$; (2) $P(\text{E and F}) = \dfrac{{}_2C_2 \cdot {}_4C_1}{{}_6C_3} = \dfrac{1 \cdot 4}{20} = \dfrac{1}{5}$

9. (1) $P(\text{E or F}) = \dfrac{16}{20} = \dfrac{4}{5}$; (2) $P(\text{E or F}) = \dfrac{{}_2C_2 \cdot {}_4C_1}{{}_6C_3} + \dfrac{{}_2C_1 \cdot {}_4C_2}{{}_6C_3} = \dfrac{1 \cdot 4}{20} + \dfrac{2 \cdot 6}{20} = \dfrac{16}{20} = \dfrac{4}{5}$

10. (1) $P(\text{not E and not F}) = \dfrac{4}{20} = \dfrac{1}{5}$; (2) $P(\text{not E and not F}) = \dfrac{{}_4C_3}{{}_6C_3} = \dfrac{4}{20} = \dfrac{1}{5}$

11. (1) $P(\text{A and not B}) = \dfrac{6}{20} = \dfrac{3}{10}$; (2) $P(\text{A and not B}) = \dfrac{{}_1C_1 \cdot {}_1C_0 \cdot {}_4C_2}{{}_6C_3} = \dfrac{1 \cdot 1 \cdot 6}{20} = \dfrac{3}{10}$

12. (1) $P(\text{A and B and not C}) = \dfrac{3}{20}$; (2) $P(\text{A and B and not C}) = \dfrac{{}_2C_2 \cdot {}_1C_0 \cdot {}_3C_1}{{}_6C_3} =$

$\dfrac{1 \cdot 1 \cdot 3}{20} = \dfrac{3}{20}$

13. (1) $P(\text{D}\,|\,\text{A}) = \dfrac{4}{10} = \dfrac{2}{5}$; (2) $P(\text{D}\,|\,\text{A}) = \dfrac{P(\text{D and A})}{P(\text{A})} = \dfrac{\frac{{}_2C_2 \cdot {}_4C_1}{{}_6C_3}}{\frac{{}_1C_1 \cdot {}_5C_2}{{}_6C_3}} = \dfrac{\frac{4}{20}}{\frac{10}{20}} = \dfrac{2}{5}$

14. (1) $P(\text{D}\,|\,\text{not A and not B}) = \dfrac{3}{4}$; (2) $P(\text{D}\,|\,\text{not A and not B}) = \dfrac{P(\text{D and not A and not B})}{P(\text{not A and not B})} =$

$\dfrac{{}_1C_1 \cdot {}_3C_2}{{}_6C_3} \div \dfrac{{}_4C_3}{{}_6C_3} = \dfrac{1 \cdot 3}{20} \div \dfrac{4}{20} = \dfrac{3}{4}$

15. a. $\dfrac{{}_{26}C_{13}}{{}_{52}C_{13}} = \dfrac{26!}{13!13!} \div \dfrac{52!}{39!13!} = \dfrac{26!39!13!}{13!13!52!} = \dfrac{26!39!}{13!52!} =$

$\dfrac{26 \cdot 25 \cdot 24 \cdot 23 \cdot 22 \cdot 21 \cdot 20 \cdot 19 \cdot 18 \cdot 17 \cdot 16 \cdot 15 \cdot 14}{52 \cdot 51 \cdot 50 \cdot 49 \cdot 48 \cdot 47 \cdot 46 \cdot 45 \cdot 44 \cdot 43 \cdot 42 \cdot 41 \cdot 40} = \dfrac{23 \cdot 19}{7 \cdot 47 \cdot 46 \cdot 43 \cdot 41} =$

$\dfrac{437}{26,681,242} \approx 1.64 \times 10^{-5}$

b. $\dfrac{{}_{13}C_7 \cdot {}_{13}C_6}{{}_{52}C_{13}} = \dfrac{13!}{6!7!} \cdot \dfrac{13!}{7!6!} \div \dfrac{52!}{39!13!} = \left(\dfrac{13 \cdot 12 \cdot 11 \cdot 10 \cdot 9 \cdot 8}{6 \cdot 5 \cdot 4 \cdot 3 \cdot 2 \cdot 1}\right)^2 \div$

$\left(\dfrac{52 \cdot 51 \cdot 50 \cdot 49 \cdot 48 \cdot 47 \cdot 46 \cdot 45 \cdot 44 \cdot 43 \cdot 42 \cdot 41 \cdot 40}{13 \cdot 12 \cdot 11 \cdot 10 \cdot 9 \cdot 8 \cdot 7 \cdot 6 \cdot 5 \cdot 4 \cdot 3 \cdot 2 \cdot 1}\right) =$

$(13 \cdot 12 \cdot 11)(13 \cdot 12 \cdot 11) \div (50 \cdot 49 \cdot 47 \cdot 46 \cdot 43 \cdot 41 \cdot 17 \cdot 4) \approx \dfrac{2,944,656}{6.350 \times 10^{11}} \approx$

4.64×10^{-6}

c. $1 - \dfrac{{}_{40}C_{13}}{{}_{52}C_{13}} = 1 - \left(\dfrac{40!}{27!13!} \div \dfrac{52!}{39!13!}\right) = 1 - \dfrac{40!39!}{27!52!} = 1 - \dfrac{19 \cdot 37 \cdot 2 \cdot 31 \cdot 29}{7 \cdot 47 \cdot 23 \cdot 5 \cdot 43 \cdot 41} =$

$1 - \dfrac{1,263,994}{66,703,105} \approx 1 - 0.0189 = 0.981$

d. 0, since there are only 12 face cards

16. a. $P(\text{same suit}) = 4 \cdot P(\text{all spades}) = 4 \cdot \dfrac{{}_{13}C_{13}}{{}_{52}C_{13}} = \dfrac{4 \cdot 1}{\dfrac{52!}{13!\,39!}} = \dfrac{4 \cdot 13!\,39!}{52!} \approx 6.30 \times 10^{-12}$

b. $\dfrac{{}_{13}C_7 \cdot {}_{13}C_3 \cdot {}_{13}C_3}{{}_{52}C_{13}} = \dfrac{13!}{7!\,6!} \cdot \dfrac{13!}{10!\,3!} \cdot \dfrac{13}{10!\,3!} \div \dfrac{52!}{39!\,13!} \approx 2.21 \times 10^{-4}$

c. $\dfrac{{}_{12}C_{12} \cdot {}_{40}C_1}{{}_{52}C_{13}} = \dfrac{1 \cdot 40}{\dfrac{52!}{39!\,13!}} \approx 6.30 \times 10^{-11}$

d. $P(\text{at least 1 diamond}) = 1 - P(\text{no diamonds}) = 1 - \dfrac{{}_{39}C_{13}}{{}_{52}C_{13}} = 1 - \dfrac{\dfrac{39!}{13!\,26!}}{\dfrac{52!}{13!\,39!}} =$

$1 - \dfrac{39!\,39!}{26!\,52!} \approx 1 - 0.0128 \approx 0.987$

17. There are ${}_5C_4$ ways to choose 4 couples and ${}_2C_1$ ways to choose one person from each

couple. $\dfrac{{}_5C_4 \cdot {}_2C_1 \cdot {}_2C_1 \cdot {}_2C_1 \cdot {}_2C_1}{{}_{10}C_4} = \dfrac{5 \cdot 2 \cdot 2 \cdot 2 \cdot 2}{210} = \dfrac{8}{21}$

18. a. $\dfrac{{}_5C_5}{{}_{11}C_5} = 1 \div \dfrac{11!}{5!\,6!} = \dfrac{5!\,6!}{11!} = \dfrac{5 \cdot 4 \cdot 3 \cdot 2 \cdot 1}{11 \cdot 10 \cdot 9 \cdot 8 \cdot 7} = \dfrac{1}{462}$

b. $\dfrac{{}_2C_2 \cdot {}_9C_3}{{}_{11}C_5} = 1 \cdot \dfrac{9!}{6!\,3!} \div \dfrac{11!}{5!\,6!} = \dfrac{9 \cdot 8 \cdot 7}{3 \cdot 2 \cdot 1} \cdot \dfrac{5 \cdot 4 \cdot 3 \cdot 2 \cdot 1}{11 \cdot 10 \cdot 9 \cdot 8 \cdot 7} = \dfrac{2}{11}$

c. $1 - P(\text{no B's}) = 1 - \dfrac{{}_9C_5}{{}_{11}C_5} = 1 - \left(\dfrac{9 \cdot 8 \cdot 7 \cdot 6}{4 \cdot 3 \cdot 2 \cdot 1} \div 462\right)$ from part (a) $= 1 - \dfrac{126}{462} = \dfrac{8}{11}$

d. There are ${}_5C_5 = 1$ way to choose 5 different letters from the word. There are 5 A's,

2 B's, 2 R's, 1 C, and 1 D, so $P(\text{one of each letter}) = \dfrac{{}_5C_1 \cdot {}_2C_1 \cdot {}_2C_1 \cdot {}_1C_1 \cdot {}_1C_1}{{}_{11}C_5} =$

$\dfrac{5 \cdot 2 \cdot 2 \cdot 1 \cdot 1}{462}$ from part (a) $= \dfrac{10}{231}$

19. $P(\text{not accepted}) = P(\text{at least 1 defective}) = 1 - P(\text{none defective}) = 1 - \dfrac{{}_{90}C_4}{{}_{100}C_4} =$

$1 - \left(\dfrac{90 \cdot 89 \cdot 88 \cdot 87}{4 \cdot 3 \cdot 2 \cdot 1} \div \dfrac{100 \cdot 99 \cdot 98 \cdot 97}{4 \cdot 3 \cdot 2 \cdot 1}\right) = 1 - \dfrac{15{,}486}{23{,}765} = \dfrac{8279}{23{,}765} \approx 0.348$

20. a. $\dfrac{{}_6C_3}{{}_{20}C_3} = \dfrac{6 \cdot 5 \cdot 4}{3 \cdot 2 \cdot 1} \div \dfrac{20 \cdot 19 \cdot 18}{3 \cdot 2 \cdot 1} = \dfrac{1}{57}$

b. $1 - P(\text{no defectives}) = 1 - \dfrac{{}_{14}C_3}{{}_{20}C_3} = 1 - \left(\dfrac{14 \cdot 13 \cdot 12}{3 \cdot 2 \cdot 1} \div \dfrac{20 \cdot 19 \cdot 18}{3 \cdot 2 \cdot 1}\right) =$

$1 - \dfrac{364}{1140} = \dfrac{194}{285}$

21. a. $P(\text{first not ace}) \cdot P(\text{second not ace} \mid \text{first not ace}) \cdot P(\text{third is ace} \mid \text{first 2 not aces}) =$

$\dfrac{48}{52} \cdot \dfrac{47}{51} \cdot \dfrac{4}{50} = \dfrac{376}{5525} \approx 0.0681$

b. $P(\text{first ace on or before 9th card}) = P(\text{ace on 1st}) + P(\text{ace on 2nd}) +$

$P(\text{ace on 3d}) + \cdots + P(\text{ace on 9th}) = \dfrac{4}{52} + \dfrac{48}{52} \cdot \dfrac{4}{51} + \dfrac{48}{52} \cdot \dfrac{47}{51} \cdot \dfrac{4}{50} + \cdots \approx 0.0769 +$

$0.0724 + 0.0681 + 0.0639 + 0.0599 + 0.0561 + 0.0524 + 0.0489 + 0.0456 =$

$0.5442 > 50\%$ (or $P(\text{first ace on or before 9th}) = 1 - P(\text{first ace after 9th}) =$

$1 - \dfrac{48}{52} \cdot \dfrac{47}{51} \cdot \dfrac{46}{50} \cdot \dfrac{45}{49} \cdot \dfrac{44}{48} \cdot \dfrac{43}{47} \cdot \dfrac{42}{46} \cdot \dfrac{41}{45} \cdot \dfrac{40}{44} \cdot 1 \cdot 1 \cdot \cdots \cdot 1 = 1 - 0.45585 \approx$

$0.544 > 50\%)$

COMMUNICATION: Reading · page 623

1. Answers may vary. In probability, an *experiment* is an action having various outcomes that occur unpredictably, such as tossing a coin or rolling a die. Usually, such an experiment has many different outcomes, or events, and probability reasoning is used to predict the likelihood of these various outcomes. In science, the term *experiment* refers more broadly to a test made to demonstrate a known truth, or to examine the validity of a hypothesis. The most valuable scientific experiments are those that can be repeated to demonstrate than an outcome occurs predictably.

2. Answers may vary, depending on the dictionary definitions students select. For example, consider the word *degree*.

   (1) A unit of angular measure equal in magnitude to the central angle subtended by 1/360 of the circumference of a circle. "Each angle of an equilateral triangle measures 60 degrees." (2) The greatest sum of the exponents of the variables in a term of a polynomial or polynomial equation. "A quadratic equation involves second-degree polynomials." (3) A unit division of a temperature scale. "Bake the cookies at 375 degrees." (4) Relative intensity or amount of a quality, attribute, or the like. "The firefighter suffered first-degree burns." (5) A division or classification of a specific crime according to its seriousness. "He was convicted of second-degree murder."
   (6) An academic title given by a college or university to a student who has completed a course of study. "She received a B.A. degree in history."

Class Exercises 16-5 · page 625

1. a.
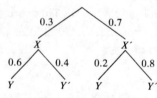

   b. $P(Y|X')$; $P(X')$

   c. $P(X \text{ and } Y) = P(X) \cdot P(Y|X) = (0.3)(0.6) = 0.18$

   d. $P(X' \text{ and } Y) = P(X') \cdot P(Y|X') = (0.7)(0.2) = 0.14$

   e. $P(Y) = P(X \text{ and } Y) + P(X' \text{ and } Y) = 0.18 + 0.14 = 0.32$

   f. $P(X|Y) = \dfrac{P(X \text{ and } Y)}{P(Y)} = \dfrac{0.18}{0.32} = 0.5625$

2. a. $P(A|R) = 0.2$ b. $P(A|S) = 0$ c. $P(R \text{ and } C) = P(R) \cdot P(C|R) = (0.1)(0.5) = 0.05$

   d. $P(S \text{ and } C) = P(S) \cdot P(C|S) = (0.9)(0.4) = 0.36$

   e. $P(C) = P(R \text{ and } C) + P(S \text{ and } C) = 0.05 + 0.36 = 0.41$

   f. $P(R|C) = \dfrac{P(R \text{ and } C)}{P(C)} = \dfrac{0.05}{0.41} \approx 0.122$

3. $P(F'|T) = 1 - P(F|T) \approx 1 - 0.802 = 0.198;\ P(F'|T) = \dfrac{P(F' \text{ and } T)}{P(T)} =$

   $\dfrac{P(F') \cdot P(T|F')}{0.393} = \dfrac{(0.65)(0.12)}{0.393} \approx 0.198$

Written Exercises 16-5 · pages 626–629

1. **a.** $P(X') = 1 - P(X) = 1 - 0.4 = 0.6$ **b.** $P(Y'|X) = 1 - P(Y|X) = 1 - 0.3 = 0.7$

   c. $P(Y'|X') = 1 - P(Y|X') = 1 - 0.5 = 0.5$

   d. $P(X \text{ and } Y) = P(X) \cdot P(Y|X) = (0.4)(0.3) = 0.12$

   e. $P(X' \text{ and } Y) = P(X') \cdot P(Y|X') = (0.6)(0.5) = 0.3$

   f. $P(Y) = P(X \text{ and } Y) + P(X' \text{ and } Y) = 0.12 + 0.30 = 0.42$

   g. $P(X|Y) = \dfrac{P(X \text{ and } Y)}{P(Y)} = \dfrac{0.12}{0.42} \approx 0.286$

   h. $P(X'|Y) = 1 - P(X|Y) \approx 1 - 0.286 = 0.714$

2. **a.** $P(S) = 1 - P(R) = 1 - 0.7 = 0.3$

   b. $P(L|R) = 1 - (P(J|R) + P(K|R)) = 1 - 0.3 - 0.2 = 0.5$

   c. $P(L|S) = 1 - (P(J|S) + P(K|S)) = 1 - 0.5 - 0.4 = 0.1$

   d. $P(J \text{ and } R) = P(R) \cdot P(J|R) = (0.7)(0.3) = 0.21$

   e. $P(J \text{ and } S) = P(S) \cdot P(J|S) = (0.3)(0.5) = 0.15$

   f. $P(J) = P(R \text{ and } J) + P(S \text{ and } J) = 0.21 + 0.15 = 0.36$

   g. $P(R|J) = \dfrac{P(R \text{ and } J)}{P(J)} = \dfrac{0.21}{0.36} \approx 0.583$ **h.** $P(S|J) = \dfrac{P(S \text{ and } J)}{P(J)} = \dfrac{0.15}{0.36} \approx 0.417$

3. **a.** $P(A \text{ and } D) + P(A \text{ and } E) = P(A) \cdot P(D|A) + P(A) \cdot P(E|A) = (0.5)(0.1) +$
   $(0.5)(0.9) = 0.5 = P(A)$

   b. $P(B \text{ and } D) + P(B \text{ and } E) = (0.4)(0.7) + (0.4)(0.3) = 0.4 = P(B)$

   c. $P(C \text{ and } D) + P(C \text{ and } E) = P(C) = 1 - P(A) - P(E) = 1 - 0.5 - 0.4 = 0.1$

4. $P(A|D) = \dfrac{P(A \text{ and } D)}{P(D)} = \dfrac{P(A \text{ and } D)}{P(A \text{ and } D) + P(B \text{ and } D) + P(C \text{ and } D)} =$

   $\dfrac{(0.5)(0.1)}{(0.5)(0.1) + (0.4)(0.7) + 0} = \dfrac{0.05}{0.05 + 0.28} = \dfrac{0.05}{0.33} \approx 0.152; P(B|D) = \dfrac{P(B \text{ and } D)}{P(D)} =$

   $\dfrac{P(B) \cdot P(D|B)}{P(D)} = \dfrac{(0.4)(0.7)}{0.33} \approx 0.848; P(C|D) = \dfrac{P(C \text{ and } D)}{P(D)} = \dfrac{0}{P(D)} = 0$

5. **a.**

   b. $P(R) = P(A \text{ and } R) + P(B \text{ and } R) = \dfrac{1}{2} \cdot \dfrac{2}{5} + \dfrac{1}{2} \cdot \dfrac{4}{5} = \dfrac{1}{5} + \dfrac{2}{5} = \dfrac{3}{5} = 0.6$

   c. $P(A|R) = \dfrac{P(A \text{ and } R)}{P(R)} = \dfrac{\dfrac{1}{2} \cdot \dfrac{2}{5}}{\dfrac{3}{5}} = \dfrac{1}{3} \approx 0.333$

6. a.

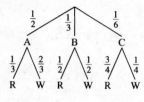

b. $P(R) = P(A \text{ and } R) + P(B \text{ and } R) + P(C \text{ and } R) =$
$$\frac{1}{2} \cdot \frac{1}{3} + \frac{1}{3} \cdot \frac{1}{2} + \frac{1}{6} \cdot \frac{3}{4} = \frac{11}{24}$$

c. $P(A|R) = \dfrac{P(A \text{ and } R)}{P(R)} = \dfrac{\frac{1}{6}}{\frac{11}{24}} = \dfrac{4}{11} \approx 0.364;$

$P(B|R) = \dfrac{P(B \text{ and } R)}{P(R)} = \dfrac{\frac{1}{6}}{\frac{11}{24}} = \dfrac{4}{11} \approx 0.364;$

$P(C|R) = 1 - \dfrac{4}{11} - \dfrac{4}{11} = \dfrac{3}{11} \approx 0.273$

7. a.

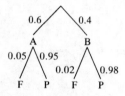

b. $P(F) = (0.6)(0.05) + (0.4)(0.02) =$
0.038; 3.8%

c. $P(A|F) = \dfrac{P(A \text{ and } F)}{P(F)} = \dfrac{(0.6)(0.05)}{0.038} \approx$
0.789

8. a.

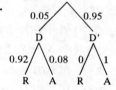

b. $P(A) = (0.05)(0.08) + (0.95)(1) =$
0.954; 95.4%

c. $P(D|A) = \dfrac{P(D \text{ and } A)}{P(A)} = \dfrac{(0.05)(0.08)}{0.954} \approx$
0.004

9. a. $P(\text{accident}) = (0.22)(0.11) + (0.43)(0.03) + (0.35)(0.02) = 0.0441; 4.41\%$

b. $P(A|\text{accident}) = \dfrac{P(A \text{ and accident})}{P(\text{accident})} = \dfrac{(0.22)(0.11)}{0.0441} \approx 0.549$

10. $P(D'|\text{accident}) = \dfrac{P(D' \text{ and accident})}{P(\text{accident})} = \dfrac{(0.75)(0.13)}{(0.25)(0.05) + (0.75)(0.13)} \approx 0.886$

11. a. Answers may vary. A false negative can be serious because someone with the disease will not receive treatment. On the other hand, a false positive can be serious because someone who is healthy might receive expensive, potentially damaging treatment that isn't needed.

b. See right.

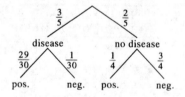

c. $P(\text{disease}|\text{pos.}) = \dfrac{P(\text{disease and pos.})}{P(\text{pos.})} = \dfrac{(0.6)\left(\frac{29}{30}\right)}{(0.6)\left(\frac{29}{30}\right) + (0.4)\left(\frac{1}{4}\right)} \approx 0.853$

d. $P(\text{no disease}|\text{neg.}) = \dfrac{P(\text{no disease and neg.})}{P(\text{neg.})} = \dfrac{(0.4)\left(\frac{3}{4}\right)}{(0.6)\left(\frac{1}{30}\right) + (0.4)\left(\frac{3}{4}\right)} \approx 0.938$

12. a. (1) $(0.85)(0.01) = 0.0085$; (2) $(0.15)(0.015) = 0.00225$

b. (1) $(6.85)(0.99) + (0.15)(0.985) \approx 0.989$;

(2) $P(\text{disease} \mid \text{pos.}) = \dfrac{P(\text{disease and pos.})}{P(\text{pos.})} = \dfrac{(0.85)(0.99)}{(0.85)(0.99) + (0.15)(0.015)} \approx 0.997$

(3) $P(\text{no disease} \mid \text{neg.}) = \dfrac{P(\text{no disease and neg.})}{P(\text{neg.})} = \dfrac{(0.15)(0.985)}{(0.85)(0.01) + (0.15)(0.985)} \approx 0.946$

13. $P(\text{game 1}) = \dfrac{1}{2} = P(\text{game 2})$; $P(\text{sum of 2 in game 1}) = \dfrac{1}{36}$; $P(\text{2 in game 2}) = \dfrac{1}{6}$;

$$P(\text{game 1} \mid 2) = \frac{P(\text{game 1 and sum of 2})}{P(2)} = \frac{\dfrac{1}{2} \cdot \dfrac{1}{36}}{\dfrac{1}{2} \cdot \dfrac{1}{36} + \dfrac{1}{2} \cdot \dfrac{1}{6}} = \frac{1}{1 + 6} = \frac{1}{7}$$

14. a. $P(4 \mid \text{game 1}) = \dfrac{1}{12}$; $P(4 \mid \text{game 2}) = \dfrac{1}{6}$; $P(\text{game 1} \mid 4) = \dfrac{P(\text{game 1 and 4})}{P(4)} =$

$$\frac{\dfrac{1}{2} \cdot \dfrac{1}{12}}{\dfrac{1}{2} \cdot \dfrac{1}{12} + \dfrac{1}{2} \cdot \dfrac{1}{6}} = \frac{1}{1 + 2} = \frac{1}{3}$$

b. $P(\text{game 1} \mid 7) = \dfrac{P(\text{game 1 and 7})}{P(7)} = \dfrac{\dfrac{1}{2} \cdot \dfrac{1}{6}}{\dfrac{1}{2} \cdot \dfrac{1}{6} + \dfrac{1}{2} \cdot 0} = 1$

c. Since it isn't possible to get a sum of 1 in the two-dice game, the probability is 0.

15. $P(\text{weighted dice} \mid \text{2 sixes}) = \dfrac{P(\text{weighted dice and 2 sixes})}{P(\text{2 sixes})} = \dfrac{\dfrac{1}{2} \cdot \left(\dfrac{1}{4} \cdot \dfrac{1}{4}\right)}{\dfrac{1}{2} \cdot \left(\dfrac{1}{6} \cdot \dfrac{1}{6}\right) + \dfrac{1}{2} \cdot \left(\dfrac{1}{4} \cdot \dfrac{1}{4}\right)} =$

$\dfrac{9}{13} \approx 0.692$

16. a. $P(\text{2-headed} \mid \text{all heads}) = \dfrac{P(\text{2-headed and all heads})}{P(\text{all heads})}$; if $n = 1$, $\dfrac{\dfrac{1}{2} \cdot 1}{\dfrac{1}{2} \cdot 1 + \dfrac{1}{2} \cdot \dfrac{1}{2}} = \dfrac{2}{3}$;

if $n = 2$, $\dfrac{\dfrac{1}{2} \cdot 1}{\dfrac{1}{2} \cdot 1 + \dfrac{1}{2} \left(\dfrac{1}{2} \cdot \dfrac{1}{2}\right)} = \dfrac{4}{5}$; if $n = 10$, $\dfrac{\dfrac{1}{2} \cdot 1}{\dfrac{1}{2} \cdot 1 + \dfrac{1}{2} \left(\dfrac{1}{2}\right)^{10}} = \dfrac{1024}{1025}$

b. $P(\text{2-headed} \mid \text{all heads}) = \dfrac{P(\text{2-headed and all heads})}{P(\text{all heads})} = \dfrac{\dfrac{1}{2} \cdot 1}{\dfrac{1}{2} \cdot 1 + \dfrac{1}{2} \left(\dfrac{1}{2}\right)^n} =$

$$\frac{\dfrac{1}{2}}{\dfrac{1}{2} + \left(\dfrac{1}{2}\right)^{n+1}} \cdot \frac{2^{n+1}}{2^{n+1}} = \frac{2^n}{2^n + 1}$$

17. a. $P(2R|B) = \frac{4}{5} \cdot \frac{4}{5} = \frac{16}{25};\ P(2W|B) =$

$\frac{1}{5} \cdot \frac{1}{5} = \frac{1}{25};\ P(1R, 1W|B) = 1 - \frac{16}{25} -$

$\frac{1}{25} = \frac{8}{25}$

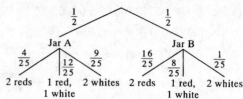

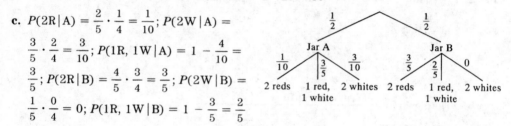

b. $P(2W \text{ and } A) = P(A) \cdot P(2W|A) =$

$\frac{1}{2} \cdot \frac{9}{25} = \frac{9}{50};\ P(2W \text{ and } B) = \frac{1}{2} \cdot \frac{1}{25} =$

$\frac{1}{50}$; choose Jar A. $P(1R, 1W \text{ and } A) = \frac{1}{2} \cdot \frac{12}{25} = \frac{6}{25};\ P(1R, 1W \text{ and } B) = \frac{1}{2} \cdot \frac{8}{25} = \frac{4}{25}$;

choose Jar A. $P(\text{correct}) = P(B) \cdot P(2R|B) + P(A) \cdot P(2W|A) +$

$P(A) \cdot P(1R, 1W|A) = \frac{1}{2} \cdot \frac{16}{25} + \frac{1}{2} \cdot \frac{9}{25} + \frac{1}{2} \cdot \frac{12}{25} = \frac{37}{50} = 0.74.$

c. $P(2R|A) = \frac{2}{5} \cdot \frac{1}{4} = \frac{1}{10};\ P(2W|A) =$

$\frac{3}{5} \cdot \frac{2}{4} = \frac{3}{10};\ P(1R, 1W|A) = 1 - \frac{4}{10} =$

$\frac{3}{5};\ P(2R|B) = \frac{4}{5} \cdot \frac{3}{4} = \frac{3}{5};\ P(2W|B) =$

$\frac{1}{5} \cdot \frac{0}{4} = 0;\ P(1R, 1W|B) = 1 - \frac{3}{5} = \frac{2}{5}$

d. $P(2R \text{ and } A) = P(A) \cdot P(2R|A) = \frac{1}{2} \cdot \frac{1}{10} = \frac{1}{20};\ P(2R \text{ and } B) = \frac{1}{2} \cdot \frac{3}{5} = \frac{3}{10}$; choose

Jar B. $P(1R, 1W \text{ and } A) = \frac{1}{2} \cdot \frac{3}{5} = \frac{3}{10};\ P(1R, 1W \text{ and } B) = \frac{1}{2} \cdot \frac{2}{5} = \frac{2}{10}$; choose

Jar A. $P(2W \text{ and } A) = \frac{1}{2} \cdot \frac{3}{10} = \frac{3}{20};\ P(2W \text{ and } B) = \frac{1}{2} \cdot 0 = 0$; choose Jar A.

$P(\text{correct}) = P(B) \cdot P(2R|B) + P(A) \cdot P(1R, 1W|A) + P(A) \cdot P(2W|A) =$

$\frac{1}{2} \cdot \frac{3}{5} + \frac{1}{2} \cdot \frac{3}{5} + \frac{1}{2} \cdot \frac{3}{10} = \frac{3}{4} = 0.75$

e. Experiment 2 since $0.75 > 0.74$

f. $P(R \text{ and } A) = \frac{1}{2} \cdot \frac{2}{5} = \frac{1}{5};\ P(R \text{ and } B) = \frac{1}{2} \cdot \frac{4}{5} = \frac{2}{5}$; choose Jar B. $P(W \text{ and } A) =$

$\frac{1}{2} \cdot \frac{3}{5} = \frac{3}{10};\ P(W \text{ and } B) = \frac{1}{2} \cdot \frac{1}{5} = \frac{1}{10}$; choose Jar A. $P(\text{correct}) = P(B) \cdot$

$P(R|B) + P(A) \cdot P(W|A) = \frac{1}{2} \cdot \frac{4}{5} + \frac{1}{2} \cdot \frac{3}{5} = \frac{7}{10} = 0.7.$ **The probability of being**

correct when choosing just one ball is no more than 0.05 from the probability of
being correct when choosing two balls.

Class Exercises 16-6 • pages 632–633

1. Expected payoff $= 5(0.6) + 10(0.4) = 7$

2. Expected payoff $= 3(0.5) + 1(0.1) + (-1)(0.4) = 1.2$

3. Expected value of game $= 2\left(\frac{2}{6}\right) + (-1)\left(\frac{4}{6}\right) = 0$; yes

4. Answers may vary. For example, if the sum is greater than 7, win \$1; if the sum
equals 7, no money is exchanged; if the sum is less than 7, lose \$1.

5. $P(3 \text{ heads}) = \left(\dfrac{1}{2}\right)^3 = \dfrac{1}{8}$; $P(2 \text{ heads}) = {}_3C_2 \cdot \left(\dfrac{1}{2}\right)^2 \cdot \dfrac{1}{2} = \dfrac{3}{8}$; $P(1 \text{ head}) =$

${}_3C_1 \cdot \dfrac{1}{2} \cdot \left(\dfrac{1}{2}\right)^2 = \dfrac{3}{8}$; $P(0 \text{ heads}) = \left(\dfrac{1}{2}\right)^3 = \dfrac{1}{8}$. Expected payoff $= 5 \cdot \dfrac{1}{8} + 3 \cdot \dfrac{3}{8} +$

$1 \cdot \dfrac{3}{8} + (-9) \cdot \dfrac{1}{8} = 1$; $\$1$.

6. The student may not have $\$1500$ available to replace the car. If the student has $\$160$ for insurance, he or she would be able to replace the car.

Written Exercises 16-6 · pages 633–635

1. Expected payoff $= 9(0.1) + 7(0.3) + (-5)(0.6) = 0$

2. Expected payoff $= 6(0.2) + 3(0.1) + (-5)(0.7) = -2$

3. Expected payoff $= 60(0.4) + 52(0.5) + 50(0.1) = 55$

4. Expected payoff $= 13(0.4) + (-7)(0.2) + (-12)(0.4) = -1$

5.

Event	Die shows 1, 3, 5	Die shows 2 or 4	Die shows 6
A's Payoff	$\$1$	$-\$3$	$\$2$
Probability	$\dfrac{1}{2}$	$\dfrac{1}{3}$	$\dfrac{1}{6}$

Expected payoff $= 1 \cdot \dfrac{1}{2} + (-3)\dfrac{1}{3} + 2 \cdot \dfrac{1}{6} = -\dfrac{1}{6}$; not fair; B has the advantage.

6.

Event	2R	1R, 1W	2W
A's Payoff	$\$5$	$-\$2$	$-\$2$
Probability	$\dfrac{1}{3}$	$\dfrac{2}{3}$	0

Expected payoff $= 5 \cdot \dfrac{1}{3} + (-2)\dfrac{2}{3} = \dfrac{1}{3}$; not fair; A has the advantage.

7.

Sum	6, 7, 8	Other
A's Payoff	$\$5$	$-\$4$
Probability	$\dfrac{4}{9}$	$\dfrac{5}{9}$

(See p. 601 for the sample space.)
Expected payoff $= 5 \cdot \dfrac{4}{9} + (-4)\dfrac{5}{9} = 0$; fair.

8.

Event	Odd sum	Same no.	Other
A's Payoff	$\$1$	$\$3$	$-\$3$
Probability	$\dfrac{1}{2}$	$\dfrac{1}{6}$	$\dfrac{1}{3}$

(See p. 601 for the sample space.)
Expected payoff $= 1 \cdot \dfrac{1}{2} + 3 \cdot \dfrac{1}{6} + (-3)\dfrac{1}{3} = 0$; fair.

9. $P(\text{joker}) = \dfrac{2}{54} = \dfrac{1}{27}$; $P(\text{face card}) = \dfrac{12}{54} = \dfrac{6}{27}$; expected payoff $= 5 \cdot \dfrac{1}{27} + 2 \cdot \dfrac{6}{27} +$

$(-1)\left(\dfrac{20}{27}\right) = -\dfrac{1}{9}$; expected loss of $\dfrac{1}{9}$ of a dollar, or about 11¢.

10. In 10 one-dollar games you would expect to lose about $10 \cdot 11¢$, or 1.10. In a \$10 game the expected payoff is $5 \cdot \dfrac{1}{27} + 2 \cdot \dfrac{6}{27} + (-10)\left(\dfrac{20}{27}\right) = -\dfrac{183}{27}$, so you would expect to lose about \$6.78.

11. Expected value $= 1 \cdot \dfrac{1}{5} + \left(-\dfrac{1}{4}\right)\left(\dfrac{4}{5}\right) = 0$; no gain or loss.

12. $P(\text{correct}) = \dfrac{1}{3}$; expected value $= 1 \cdot \dfrac{1}{3} + \left(-\dfrac{1}{4}\right)\left(\dfrac{2}{3}\right) = \dfrac{1}{6}$; expected gain of $\dfrac{1}{6}$ point.

13.

Event	2G	1G, 1R	2R
Payoff	\$2	\$1	−\$1
Probability	$\dfrac{1}{10}$	$\dfrac{3}{5}$	$\dfrac{3}{10}$

$P(2G) = \dfrac{2}{5} \cdot \dfrac{1}{4} = \dfrac{1}{10}$; $P(2R) = \dfrac{3}{5} \cdot \dfrac{2}{4} = \dfrac{3}{10}$;

$P(1G, 1R) = 1 - \dfrac{4}{10} = \dfrac{3}{5}$. Expected payoff $=$

$2 \cdot \dfrac{1}{10} + 1 \cdot \dfrac{3}{5} + (-1)\dfrac{3}{10} = \dfrac{1}{2}$; 50¢ expected gain.

14.

No. of matches	3	2	1	0
Payoff	\$3	\$2	\$1	−\$1
Probability	$\dfrac{1}{216}$	$\dfrac{15}{216}$	$\dfrac{75}{216}$	$\dfrac{125}{216}$

$P(\text{each guess}) = \dfrac{1}{6}$; $P(3 \text{ wins}) =$

$\left(\dfrac{1}{6}\right)^3 = \dfrac{1}{216}$; $P(2 \text{ wins}) = 3 \cdot \left(\dfrac{1}{6}\right)^2 \cdot$

$\dfrac{5}{6} = \dfrac{15}{216}$; $P(1 \text{ win}) = 3 \cdot \dfrac{1}{6} \cdot \left(\dfrac{5}{6}\right)^2 =$

$\dfrac{75}{216}$; $P(0 \text{ wins}) = \left(\dfrac{5}{6}\right)^3 = \dfrac{125}{216}$.

Expected payoff $= \dfrac{3}{216} + \dfrac{30}{216} + \dfrac{75}{216} -$

$\dfrac{125}{216} = -\dfrac{17}{216}$; expected loss of about 8¢.

15. Expected value $= \displaystyle\sum_{i=1}^{n} x_i P(x_i)$

16. Let $\mathbf{u} = (x_1, x_2, \ldots, x_n)$ and $\mathbf{v} = (P(x_1), P(x_2), \ldots, P(x_n))$; then the expected value equals $\mathbf{u} \cdot \mathbf{v}$.

17. Expected profit per gallon $= 1.10(0.3) + 0.90(0.38) + 0.70(0.2) + 0.40(0.06) + (-0.10)(0.02) = 0.834$; for 25,000 gallons the expected profit is $(\$.834)(25,000) = \$20,850$.

18. Expected gain per policy $= 1 \cdot \dfrac{10^6 - 0.45}{10^6} + (-100,000)\left(\dfrac{0.45}{10^6}\right) = 0.99999955 - 0.045 \approx 0.955$, or about 95.5¢; on 100,000 policies the expected gain is about $100,000(\$.955) = \$95,500$.

19. Answers may vary. Expected payoff $= 26,000(0.2) + 0(0.8) = 5200$; expected value $= 5200 - 4000 = 1200$. The expected value is positive, but the company stands an 80% chance of losing \$4000. Perhaps the company should bid on other jobs with lower costs of bidding and higher probabilities of being accepted.

20. Answers may vary. Expected payoff $= 140(0.12) = \$16.80$ and expected value $= 16.80 - 35 < 0$. Since the expected value is negative your family should not buy the contract.

21. Expected payoff $= 1000\left(\dfrac{1}{2000}\right) + 100\left(\dfrac{5}{2000}\right) + 10\left(\dfrac{20}{2000}\right) + (0)\left(\dfrac{1974}{2000}\right) = 0.85$; expected value $= 0.85 - 1.00 = -0.15$; expected loss of 15¢.

22. $P(\text{pick } 5) = \dfrac{_5C_5}{_{30}C_5} = \dfrac{1}{142{,}506}$; $P(\text{pick } 4) = \dfrac{_5C_4 \cdot {_{25}}C_1}{_{30}C_5} = \dfrac{5 \cdot 25}{142{,}506}$; expected payoff =

$100{,}000 \cdot \dfrac{1}{142{,}506} + 100 \cdot \dfrac{125}{142{,}506} = \dfrac{6250}{7917} \approx 0.789$; expected value $\approx 0.79 - 1.00 =$

-0.21; expected loss of 21¢.

23. The only way B can win is for 2 consecutive tails to appear.

$P(\text{B wins}) = \dfrac{1}{2} \cdot \dfrac{1}{2} = \dfrac{1}{4}$; $P(\text{A wins}) = 1 - \dfrac{1}{4} = \dfrac{3}{4}$. A's

expected gain $= (100)\dfrac{3}{4} + (0)\dfrac{1}{4} = 75$; \$75.

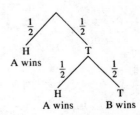

24. Since you double your bet each time, the actual winnings each game is \$1. For example, if you win on the third try, the payoff is \$4 but you paid \$3 in the first 2 tries, so you win \$1. This is true throughout the game. $P(\text{win in 1 game}) = \dfrac{1}{2}$, $P(\text{win in 2 games}) =$

$\dfrac{1}{2} \cdot \dfrac{1}{2} = \dfrac{1}{4}$, $P(\text{win in 3 games}) = \dfrac{1}{2} \cdot \dfrac{1}{2} \cdot \dfrac{1}{2} = \dfrac{1}{8}$, and so on. You will have lost \$15

after 4 losses (\$1, \$2, \$4, and \$8). Expected payoff $= 1 \cdot \dfrac{1}{2} + 1 \cdot \dfrac{1}{4} \cdot 1 \cdot \dfrac{1}{8} + 1 \cdot \dfrac{1}{16} +$

$(-15)\dfrac{1}{16} = 0$; no expected gain or loss.

25. a. $\left(\dfrac{5}{6}\right)^3 \cdot \dfrac{1}{6}$; $\left(\dfrac{5}{6}\right)^4 \cdot \dfrac{1}{6}$; $\left(\dfrac{5}{6}\right)^{n-1} \cdot \dfrac{1}{6}$ **b.** $\displaystyle\sum_{n=1}^{\infty} n\left(\dfrac{5}{6}\right)^{n-1} \cdot \dfrac{1}{6} = \dfrac{1}{6}\displaystyle\sum_{n=1}^{\infty} n\left(\dfrac{5}{6}\right)^{n-1}$

c. $\dfrac{1}{6}\displaystyle\sum_{n=1}^{\infty} n\left(\dfrac{5}{6}\right)^{n-1} = \dfrac{1}{6} \cdot \dfrac{1}{\left(1 - \dfrac{5}{6}\right)^2} = \dfrac{1}{6} \cdot \dfrac{1}{\dfrac{1}{36}} = \dfrac{36}{3} = 6$

Chapter Test · pages 636–637

1. Empirical probability is determined by observing the outcomes of experiments. For example, you could find the empirical probability of getting heads when tossing a coin by tossing a coin many times and recording the outcomes. Theoretical probability is determined by reasoning about the events. For example, you could reason that the probability of getting heads is $\dfrac{1}{2}$, since a coin has 2 sides and each is as likely as the other to land up.

2. a. $P(\text{spade and 5}) = \dfrac{13}{52} \cdot \dfrac{1}{6} = \dfrac{1}{24}$ **b.** $P(\text{not spade and 5}) = \dfrac{39}{52} \cdot \dfrac{1}{6} = \dfrac{1}{8}$

c. $P(\text{not spade and not 5}) = \dfrac{39}{52} \cdot \dfrac{5}{6} = \dfrac{5}{8}$

3. a. $\dfrac{5}{9} \cdot \dfrac{4}{8} = \dfrac{5}{18}$ **b.** $P(6) = \dfrac{1}{9}$ **c.** $\dfrac{3}{9} \cdot \dfrac{3}{8} = \dfrac{1}{8}$

4. a. $(0.9)^5 \approx 0.59$

b. $P(\text{at least 3 sprout}) = P(3) + P(4) + P(5) = {_5}C_3(0.9)^3(0.1)^2 + {_5}C_4(0.9)^4(0.1) +$
$(0.9)^5 = 0.0729 + 0.32805 + 0.59049 \approx 0.991$

5. a. $\dfrac{_{12}C_3 \cdot {}_{99}C_3}{_{24}C_3 \cdot {}_{145}C_3} = \dfrac{\dfrac{12 \cdot 11 \cdot 10}{3 \cdot 2 \cdot 1} \cdot \dfrac{99 \cdot 98 \cdot 97}{3 \cdot 2 \cdot 1}}{\dfrac{24 \cdot 23 \cdot 22}{3 \cdot 2 \cdot 1} \cdot \dfrac{145 \cdot 144 \cdot 143}{3 \cdot 2 \cdot 1}} = \dfrac{12 \cdot 11 \cdot 10 \cdot 99 \cdot 98 \cdot 97}{24 \cdot 23 \cdot 22 \cdot 145 \cdot 144 \cdot 143} \approx 0.034$

b. $\dfrac{_{12}C_1 \cdot {}_{12}C_2 \cdot {}_{99}C_1 \cdot {}_{46}C_2}{_{24}C_3 \cdot {}_{145}C_3} = \dfrac{12 \cdot \dfrac{12 \cdot 11}{2 \cdot 1} \cdot 99 \cdot \dfrac{46 \cdot 45}{2 \cdot 1}}{\dfrac{24 \cdot 23 \cdot 22}{3 \cdot 2 \cdot 1} \cdot \dfrac{145 \cdot 144 \cdot 143}{3 \cdot 2 \cdot 1}} \approx 0.081$

6. a.

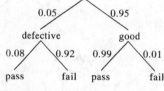

b. $P(\text{pass}) = P(\text{def.}) \cdot P(\text{pass} \mid \text{def.}) + P(\text{good}) \cdot P(\text{pass} \mid \text{good}) = 0.05(0.08) + 0.95(0.99) = 0.9445$; about 94.5%

c. $P(\text{def.} \mid \text{pass}) = \dfrac{P(\text{def. and pass})}{P(\text{pass})} = \dfrac{0.05(0.08)}{0.9445} \approx 0.004$

7. a.

Number of "heads"	0	1	2	3	4	5
Payoff	−$20	$1	$2	$3	$4	−$20
Probability	$\dfrac{1}{32}$	$\dfrac{5}{32}$	$\dfrac{10}{32}$	$\dfrac{10}{32}$	$\dfrac{5}{32}$	$\dfrac{1}{32}$

$P(1) = {}_5C_1\left(\dfrac{1}{2}\right)^5 = \dfrac{5}{32};$

$P(2) = {}_5C_2\left(\dfrac{1}{2}\right)^5 = \dfrac{10}{32};$

$P(3) = {}_5C_3\left(\dfrac{1}{2}\right)^5 = \dfrac{10}{32};$

$P(4) = {}_5C_4\left(\dfrac{1}{2}\right)^5 = \dfrac{5}{32};$

$P(5) = {}_5C_5\left(\dfrac{1}{2}\right)^5 = \dfrac{1}{32};$

b. Expected payoff $= -20\left(\dfrac{1}{32}\right) + 1\left(\dfrac{5}{32}\right) + 2\left(\dfrac{10}{32}\right) + 3\left(\dfrac{10}{32}\right) + 4\left(\dfrac{5}{32}\right) - 20\left(\dfrac{1}{32}\right) = \dfrac{35}{32} \approx \$1.09$

Class Exercises 17-1 · page 643

1. a. $\bar{x} = \dfrac{1 + 2 + 3 + 5 + 5}{5} = \dfrac{16}{5} = 3.2$; median $= 3$; mode: 5

b. $\bar{x} = \dfrac{3 + 4 + 4 + 7 + 8 + 8 + 10 + 11}{8} = \dfrac{55}{8} = 6.875$; median $= \dfrac{7 + 8}{2} = 7.5$;

modes: 4 and 8

c. $\bar{x} = \dfrac{1 + 2 + 3}{3} = \dfrac{6}{3} = 2$; median $= 2$; no mode

d. $\bar{x} = \dfrac{0 \cdot 3 + 1 \cdot 2 + 2 \cdot 5 + 3 \cdot 8 + 4 \cdot 2}{3 + 2 + 5 + 8 + 2} = \dfrac{44}{20} = 2.2$; median $= \dfrac{2 + 3}{2} = 2.5$; mode: 3

2. Answers will vary.

3. a. Yes; if the sum of the integers is not a multiple of the number of integers, the mean is not an integer.

b. Yes; if the number of measurements is even, the median is not always an integer.

c. No

4. The median and mode remain the same and the mean decreases.

Written Exercises 17-1 · pages 643–648

1. Numbers of mice: 3, 4, 4, 5, 6, 6, 6, 7, 8; $\bar{x} = \dfrac{3 + 2(4) + 5 + 3(6) + 7 + 8}{9} = \dfrac{49}{9} \approx 5.4$;

median $= 6$; mode: 6

2. Test scores: 74, 78, 84, 85, 87, 90, 92; $\bar{x} = \dfrac{74 + 78 + 84 + 85 + 87 + 90 + 92}{7} = \dfrac{590}{7} \approx$

84.3; median $= 85$; no mode

3. $\bar{x} = \dfrac{0 + 1 + 3 + 3(5) + 7 + 2(9) + 11 + 15 + 99}{12} = \dfrac{169}{12} \approx 14.1$; median $= \dfrac{5 + 7}{2} = 6$;

mode: 5

4. $\bar{x} = \dfrac{185 + 189 + 191 + 2(193) + 195 + 196 + 2(198) + 200}{10} = \dfrac{1938}{10} = 193.8$ cm;

median $= \dfrac{193 + 195}{2} = 194$ cm; modes: 193 cm and 198 cm

5. a. 4, 7, 8, 9, 9, 10, 11, 12, 13, 13, 14, 15, 16, 16, 18, 19, 21, 22, 25, 28. Since there are 20 measurements, the median is the mean of the 10th and 11th. Median $= \dfrac{13 + 14}{2} = 13.5$; the modes are 9, 13, and 16.

b.

0	4
·	7 8 9 9
1	0 1 2 3 3 4
·	5 6 6 8 9
2	1 2
·	5 8

6. a.

```
5 | 2 3 3 5 5 5 6 7 8 9
6 | 0 0 1 2 2 3 4 7 8
7 | 0 3 4 4 5 6 8
8 | 0 1 2 3
```

5 | 2 represents 52°.

b.

```
5 | 2 3 3
· | 5 5 5 6 7 8 9
6 | 0 0 1 2 2 3 4
· | 7 8
7 | 0 3 4 4
· | 5 6 8
8 | 0 1 2 3
·
```

5 ı 2 represents 52°.

c. median $= \dfrac{62 + 63}{2} = 62.5$;

mode $= 55$

7. a. Use stems of 24, 25, 26, 27.

c. median $= \dfrac{.260 + .261}{2} = .2605$

b.

```
24 | 2 5
25 | 2 6 7 9
26 | 0 1 1 3 9
27 | 1 6 7
```

24 ı 2 represents .242

8. a. Use stems of 9, 10, 11, 12, 13, 14.

c. median $= \dfrac{126 + 127}{2} = 126.5$

b.

```
 9 | 4
10 | 1 8
11 | 6 7
12 | 2 6 7 7 9
13 | 0 4
14 | 2 5
```

9 ı 4 represents 94.

9. a. Find 7 on the horizontal axis and read up to the graph. Then read left to the vertical axis to find the value 60%; 60% of the students had scores less than or equal to 7. At 8 the corresponding vertical value is 80%; $100 - 80 = 20\%$; 20% of the students had scores greater than 8.

b. Find 50% on the vertical axis and take the corresponding value of 6 on the horizontal axis. The median score is 6.

Exs. 10–12. Note that the frequencies used in the cumulative frequency polygons for these exercises are expressed in percents. Integers can be used instead of percents.

10.

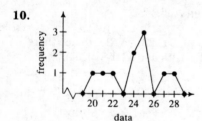
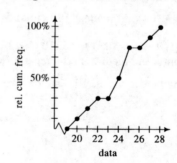

11. a.

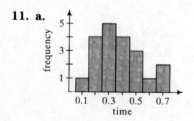

b. $\bar{x} = \dfrac{1(0.1) + 4(0.2) + 5(0.3) + 4(0.4) + 3(0.5) + 1(0.6) + 2(0.7)}{20} = \dfrac{7.5}{20} = 0.375;$

median $= \dfrac{0.3 + 0.4}{2} = 0.35;$ mode $= 0.3$

c.

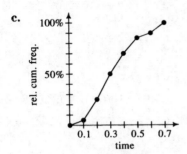

12. a.

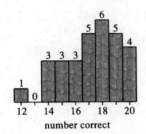

b. $\bar{x} = \dfrac{12 + 3(14) + 3(15) + 3(16) + 5(17) + 6(18) + 5(19) + 4(20)}{30} = \dfrac{515}{30} \approx 17.2;$

median $= \dfrac{17 + 18}{2} = 17.5;$ mode $= 18$

c.

13. a. $\bar{x} = \dfrac{8 + 8 + 8.50 + 10.50 + 12}{5} = \dfrac{47}{5} =$ $9.40;$ median $= \$8.50;$ mode $= \$8.00$

b. The mean is raised to $11.00 per hour. The middle wage and most frequent wage do not change, hence the median and mode do not change.

14. Answers may vary. For example: 45, 50, 55, 60, 65, 70, 75, 84, 88, 88; median $= 67.5$

15. The x- and y-coordinates of each vertex of a frequency polygon are the middle value and the frequency of each class, respectively. Thus, we can approximate the mean of the data by summing the products of the x- and y-coordinates of each vertex and then dividing by the sum of the y-coordinates. For the frequency polygon on page 641, the sum of the products of the x- and y-coordinates is $2(350) + 13(450) + 19(550) + 11(650) + 5(750) = 27{,}900;$ the sum of the y-coordinates is $2 + 13 + 19 + 11 + 5 = 50;$ so the mean of the data is about $\dfrac{27{,}900}{50} = 558.$

16. In architecture, an ogive is the rib of a Gothic vault. Because a cumulative frequency polygon tends to rise sharply and then level off, its shape suggests an ogive.

17. a.

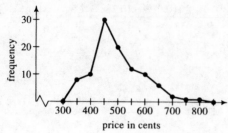

b. $\bar{x} = (8(350) + 10(400) + 30(450) + 20(500) + 12(550) + 10(600) + 6(650) + 2(700) + 750 + 800)/100 = \dfrac{49{,}750}{100} \approx \$4.98;$

median = \$5.00; mode = \$4.50

18. a.

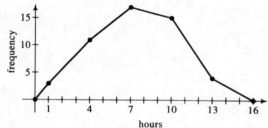

b. $\bar{x} = \dfrac{3(1) + 11(4) + 17(7) + 15(10) + 4(13)}{50} = \dfrac{368}{50} = 7.36$ h; median = 7 h; mode = 7 h

19. Answers will vary.

20. a. Yes; multiply the average by 12 to get the total earnings. **b.** No **c.** No

21. The mean is incorrect since $100\bar{x}$ is the sum of the number of children and therefore must be an integer. However, $100(2.038)$ is not an integer. The median will be an integer or end in .5. The mode must be an integer.

22. a.

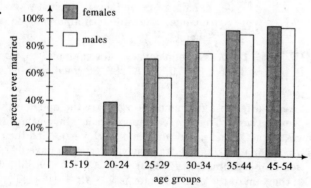

b. Females marry earlier than males. By age 54, they have married at about the same rate.

23. a.

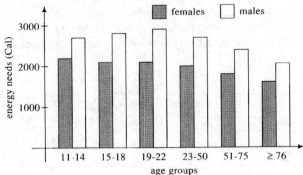

energy needs (Cal)

females ☐ males

age groups

11-14 15-18 19-22 23-50 51-75 ≥ 76

b. Data show that men require more calories than women and the requirements build for men and then decrease, whereas they never increase for women after the age of eleven.

24. To determine the mean, divide the new sum by 11. $\bar{x} = \dfrac{10(17.1) + 21}{11} = \dfrac{192}{11} \approx 17.5$. To have a median of 16.5, the 5th and 6th data points could be 16 and 17, 15 and 18, or 14 and 19, respectively; however, the mode is 16 so 16 must be one of the data points. Thus, the 6th data point is 17. When 21 is added to the list, 17 is the median. Since all the members of the group are teenagers, adding 21 to the list does not change the mode. Thus, the mean = 17.5, median = 17, and mode = 16.

25. a. $\bar{x} = 17{,}800$; total salary = 100(17,800); new mean $= \dfrac{100(17{,}800) + 15(1000)}{100} =$ $17,950. Since the salaries of the 50th and 51st employees are unchanged, the median stays at $14,500.

b. New mean $= \dfrac{100(17{,}800) + 80(1000)}{100} = \$18{,}600$. Since the salaries of the 50th and 51st employees have been raised $1000, the new median is $15,500.

Class Exercises 17-2 · page 651

1. Extremes: 137, 225; range = 225 − 137 = 88

2. 170, 190; 190 − 170 = 20

3. Median = 178

4. An outlier is more than 1.5 × interquartile range from the nearer quartile. In this example, 1.5 × 20 = 30, so any data point greater than 220 or less than 140 is an outlier. Thus, 225 is an outlier. The other outliers are 137 and 222.

Written Exercises 17-2 · pages 651–653

1. a.

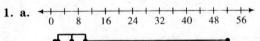

0   8   16   24   32   40   48   56

b. No; 1 is less than 1.5(8) = 12 from 2. Yes; 52 is more than 12 from 8.

2. a.

0   5   10   15

b. Yes; 15 is more than 1.5(3) = 4.5 from 5.

3. In general, the students have better mathematics scores than verbal scores. The median mathematics score is higher than the median verbal score. Although the upper quartiles are the same, the distance to the median of the mathematics scores is much smaller, indicating more mathematics than verbal scores clustered in the upper 500 range. The verbal scores extended over a wider range than the mathematics scores.

4. In general, the students grew. The median height increased about 5 cm. The interquartile range and the range are greater, indicating that the heights are less clustered. There is an outlier now, indicating that one student grew considerably taller than the others.

5.

1	1
2	3 6
3	1
4	1 4 9
5	2 6

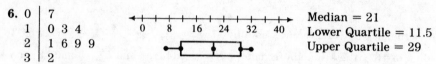

Median = 410
Lower Quartile = 245
Upper Quartile = 505

$5 \mid 2 = 520$ calories

6.

0	7
1	0 3 4
2	1 6 9 9
3	2

Median = 21
Lower Quartile = 11.5
Upper Quartile = 29

7.

Protein		Carb.
7	0	6 8
9 5 4 2	1	
9 5 3 1	2	
	3	1 1 4 5 8
	4	0 3

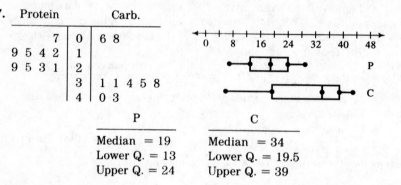

P	C
Median = 19	Median = 34
Lower Q. = 13	Lower Q. = 19.5
Upper Q. = 24	Upper Q. = 39

8.

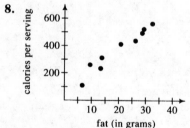

It appears that as the amount of fat increases, the number of calories also increases.

Exs. 9–15. Answers will vary.

Class Exercises 17-3 · pages 657–658

1. a. Team A's heights are more closely clustered around 192 cm than Team B's.

b. Team B has one or two very tall players who are likely to get the rebounds.

2.

x_i	1	7	9	15	32
$x_i{}^2$	1	49	81	225	356

$\overline{x} = \dfrac{32}{4} = 8;\ s^2 = \dfrac{356}{4} - 8^2 = 25;\ s = \sqrt{25} = 5$

3. The mean would be decreased by one and the standard deviation would be unchanged.

4. a. $\dfrac{14 - 10}{4} = 1$ **b.** $\dfrac{18 - 10}{4} = 2$ **c.** $\dfrac{20 - 10}{4} = 2.5$ **d.** $\dfrac{8 - 10}{4} = -0.5$ **e.** $\dfrac{10 - 10}{4} = 0$

5. Formula (1a) would be easier to use since you don't take the difference between each data point and the mean.

6. Yes; $s^2 = 0$ if $\displaystyle\sum_{i=1}^{n} (x_i - \overline{x})^2 = 0$; this occurs when $x_i = \overline{x}$ for all i; thus, if every data entry in a set is the same, the variance is 0.

Written Exercises 17-3 · pages 658–660

1. $\overline{x} = \dfrac{10 + 8 + 7 + 5 + 5}{5} = \dfrac{35}{5} = 7$; $s^2 = \dfrac{(10 - 7)^2 + (8 - 7)^2 - (7 - 7)^2 + 2(5 - 7)^2}{5} = \dfrac{18}{5} =$

3.6; $s = \sqrt{3.6} \approx 1.90$

2. $\overline{x} = \dfrac{3 + 8 + 7 + 0 + 4 + 7 + 13}{7} = \dfrac{42}{7} = 6$;

$s^2 = \dfrac{(3 - 6)^2 + (8 - 6)^2 + (7 - 6)^2 + (0 - 6)^2 + (4 - 6)^2 + (7 - 6)^2 + (13 - 6)^2}{7} =$

$\dfrac{104}{7} \approx 14.857$; $s \approx 3.85$

3. a. Row A: $\overline{x} = \dfrac{1 + 2 + 3 + 4 + 5}{5} = \dfrac{15}{5} = 3$; $s^2 = \dfrac{4 + 1 + 0 + 1 + 4}{5} = \dfrac{10}{5} = 2$;

$s = \sqrt{2} \approx 1.41$. Row B: $\overline{x} = \dfrac{11 + 12 + 13 + 14 + 15}{5} = \dfrac{65}{5} = 13$;

$s^2 = \dfrac{4 + 1 + 0 + 1 + 4}{5} = \dfrac{10}{5} = 2$; $s = \sqrt{2} \approx 1.41$. The means differ by 10 and the

standard deviations are equal.

b. mean, 23; standard deviation, $\sqrt{2} \approx 1.41$

4. a. Row A: $\overline{x} = \dfrac{2(5) + 2(7) + 8 + 10}{6} = \dfrac{42}{6} = 7$; $s^2 = \dfrac{4 + 4 + 0 + 0 + 1 + 9}{6} = \dfrac{18}{6} = 3$;

$s = \sqrt{3} \approx 1.73$. Row B: $\overline{x} = \dfrac{2(20) + 2(28) + 32 + 40}{6} = \dfrac{168}{6} = 28$;

$s^2 = \dfrac{2(64) + 2(0) + 16 + 144}{6} = \dfrac{288}{6} = 48$; $s = \sqrt{48} = 4\sqrt{3} \approx 6.93$. Both the mean

and the standard deviation of B are four times as large as those of A.

b. $\overline{x} = 7 \cdot 10 = 70$; $s = 10\sqrt{3} \approx 17.3$

5. $\overline{x} = 68 + 10 = 78$; no change in the standard deviation, $s = 14$

6. $\overline{x} = 66(2.5) = 165$ cm; $s = 4(2.5) = 10$ cm

7. a. $\dfrac{90 - 78}{6} = 2$ **b.** $\dfrac{75 - 78}{6} = -0.5$ **c.** $\dfrac{78 - 78}{6} = 0$ **d.** $\dfrac{70 - 78}{6} = -\dfrac{4}{3} \approx -1.3$

8. a. $\dfrac{720 - 760}{80} = -0.5$ **b.** $\dfrac{860 - 760}{80} = 1.25$ **c.** $\dfrac{600 - 760}{80} = -2$ **d.** $\dfrac{776 - 760}{80} = 0.2$

9. Math: $\dfrac{610 - 500}{100} = 1.1$; Verbal: $\dfrac{580 - 504}{98} = 0.8$

10. Test 1: $\dfrac{82 - 72}{5} = 2$; Test 2: $\dfrac{88 - 80}{6} = 1.3$; she did better on Test 1.

11. Since the physics students are likely to have higher scores than the students selected at random, the group of physics students will have a greater mean score. The physics students' scores are likely to be clustered near the top while randomly selected scores will vary more greatly; thus, the group selected at random will have the greater standard deviation.

12. The randomly selected people will have great variability in their typing rates, so this group will have the greater standard deviation.

13. a. Zero; if each person thinks of the same number, then $s = 0$.

b. 4; if 2 people choose the number 1 and the other 2 choose the number 9, then $s = \sqrt{16} = 4$.

14. $\bar{x} = \dfrac{3(1) + 4(3) + 5(6) + 6(9) + 7(7) + 8(4)}{1 + 3 + 6 + 9 + 7 + 4} = \dfrac{180}{30} = 6;\ s^2 = \dfrac{\sum\limits_{i=1}^{r} x_i^2 f_i}{n} - \bar{x}^2 =$

$\dfrac{9 \cdot 1 + 16 \cdot 3 + 25 \cdot 6 + 36 \cdot 9 + 49 \cdot 7 + 64 \cdot 4}{1 + 3 + 6 + 9 + 7 + 4} - 6^2 = \dfrac{1130}{30} - 36 = \dfrac{5}{3};\ s = \sqrt{\dfrac{5}{3}} \approx 1.29$

15. $\bar{x} = \dfrac{7(1) + 8(2) + 9(5) + 10(4) + 11(4) + 12(4)}{1 + 2 + 5 + 4 + 4 + 4} = \dfrac{200}{20} = 10;\ s^2 = \dfrac{\sum\limits_{i=1}^{r} x_i^2 f_i}{n} - \bar{x}^2 =$

$\dfrac{49 \cdot 1 + 64 \cdot 2 + 81 \cdot 5 + 100 \cdot 4 + 121 \cdot 4 + 144 \cdot 4}{20} - 10^2 = \dfrac{2042}{20} - 100 = 2.1;$

$s = \sqrt{2.1} \approx 1.45$

16. Let $S = $ the original sum of the data entries and let $n = $ the number of entries. Then $\dfrac{S}{n}$ is the original mean. $\bar{x} = \dfrac{S + nc}{n} = \dfrac{S}{n} + c$, so the mean value increases by c. If x_i is any data entry, then $(x_i + c) - \left(\dfrac{S}{n} + c\right) = x_i - \dfrac{S}{n}$; thus $s = \sqrt{\dfrac{\sum\limits_{i=1}^{n} (x_i - \bar{x})^2}{n}}$ is unchanged.

17. Let $S = $ the original sum of the data entries and let $n = $ number of entries. Then $\bar{x} = \dfrac{S}{n}$ is the original mean. The new mean $= \dfrac{cS}{n} = c\left(\dfrac{S}{n}\right)$, so the new mean is c times as large as the original. $cx_i - c \cdot \bar{x} = c(x_i - \bar{x})$, so the new standard deviation $=$

$\sqrt{\dfrac{\sum\limits_{i=1}^{n} (c(x_i - \bar{x}))^2}{n}} = c\sqrt{\dfrac{\sum\limits_{i=1}^{n} (x_i - \bar{x})^2}{n}} = cS.$ Thus, the new standard deviation is c times as large as the original.

18. $s^2 = \dfrac{\sum\limits_{i=1}^{n} (x_i - \bar{x})^2}{n} = \dfrac{1}{n} \sum\limits_{i=1}^{n} (x_i^2 - 2x_i\bar{x} + (\bar{x})^2) = \dfrac{1}{n}\left[\sum\limits_{i=1}^{n} x_i^2 - 2\bar{x} \sum\limits_{i=1}^{n} x_i + n(\bar{x})^2\right] =$

$\dfrac{1}{n}\left[\sum\limits_{i=1}^{n} x_i^2 - 2\bar{x}(n \cdot \bar{x}) + n(\bar{x})^2\right] = \dfrac{1}{n}\left[\sum\limits_{i=1}^{n} x_i^2 - n(\bar{x})^2\right] = \dfrac{\sum\limits_{i=1}^{n} x_i^2}{n} - \bar{x}^2$

Activity 17-4 · page 662

a. 80 seconds, the middle of the distribution

b. $\frac{4.9}{2} + 6.1 + 6.6 + 7.6 + 7.9 + 8.0 + 7.8 + 7.4 + 6.7 + 6.0 + \frac{5.1}{2} = 69.1$; about 69%

Class Exercises 17-4 · page 666

1. Answers will vary.　**2.** $P(1.2) = 0.8849$; 88.49%　**3.** $P(-1.8) = 0.0359$; 3.59%

4. $P(0.6) = 0.7257$; $1 - 0.7257 = 0.2743$; 27.43%

5. $P(-1.3) = 0.0968$; $1 - 0.0968 = 0.9032$; 90.32%

6. $P(1) - P(0) = 0.8413 - 0.5 = 0.3413$; 34.13%

7. $P(1) - P(-1) = 0.8413 - 0.1587 = 0.6826$; 68.26%

8. $P(2) - P(-2) = 0.9772 - 0.0228 = 0.9544 \approx 95\%$

9. a.

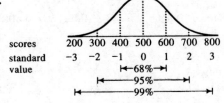

b. (1) $z = \dfrac{300 - 500}{100} = -2$; $P(-2) = 0.0228$; 2.28%

(2) $z = \dfrac{700 - 500}{100} = 2$; $P(2) = 0.9772$; $1 - 0.9772 = 0.0228$; 2.28%

(3) $\dfrac{600 - 500}{100} = 1$; $\dfrac{400 - 500}{100} = -1$; $P(1) - P(-1) = 0.8413 - 0.1587 = 0.6826$; 68.26%

c. $P(z) = 0.9000$; $z = 1.3$; $1.3 = \dfrac{x - 500}{100}$; $130 = x - 500$; $x = 630$

d. $P(z) = 0.2500$; $z = -0.7$; $-0.7 = \dfrac{x - 500}{100}$; $-70 = x - 500$; $x = 430$

10. a.

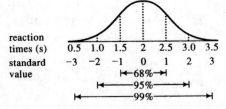

b. (1) $z = \dfrac{0.5 - 2}{0.5} = -3$; $P(-3) = 0.0013$; 0.13%

(2) $z = \dfrac{3 - 2}{0.5} = 2$; $P(2) = 0.9772$; $1 - 0.9772 = 0.0228$; 2.28%

(3) $\dfrac{2.5 - 2}{0.5} = 1$; $\dfrac{1.5 - 2}{0.5} = -1$; $P(1) - P(-1) = 0.8413 - 0.1587 = 0.6826$; 68.26%

(Continued)

c. $P(z) = 0.2000$; $z = -0.8$; $-0.8 = \dfrac{x - 2}{0.5}$; $-0.4 = x - 2$; $x = 1.6$ s

d. $P(z) = 0.7500$; $z = 0.7$; $0.7 = \dfrac{x - 2}{0.5}$; $0.35 = x - 2$; $x \approx 2.4$ s

Written Exercises 17-4 · pages 667–669

1. a. $\dfrac{95 - 100}{10} = -0.5$; $P(-0.5) = 0.3085$; 30.85%

b. $\dfrac{115 - 100}{10} = 1.5$; $1 - P(1.5) = 1 - 0.9332 = 0.0668$; 6.68%

c. $\dfrac{90 - 100}{10} = -1.0$; $\dfrac{110 - 100}{10} = 1.0$; $P(1.0) - P(-1.0) = 0.8413 - 0.1587 = 0.6826$;

68.26%

d. $\dfrac{105 - 100}{10} = 0.5$, $\dfrac{125 - 100}{10} = 2.5$; $P(2.5) - P(0.5) = 0.9938 - 0.6915 = 0.3023$;

30.23%

2. a. $\dfrac{150 - 200}{20} = -2.5$; $P(-2.5) = 0.0062$; 0.62%

b. $\dfrac{220 - 200}{20} = 1$; $P(1) = 0.8413$; $1 - 0.8413 = 0.1587$; 15.87%

c. $\dfrac{170 - 200}{20} = -1.5$; $\dfrac{230 - 200}{20} = 1.5$; $P(1.5) - P(-1.5) = 0.9332 - 0.0668 = 0.8664$;

86.64%

d. $\dfrac{160 - 200}{20} = -2$, $\dfrac{190 - 200}{20} = -0.5$; $P(-0.5) - P(-2) = 0.3085 - 0.0228 = 0.2857$;

28.57%

3. a. $\dfrac{500 - 600}{80} = -1.25$; $P(-1.3) = 0.0968$; 9.68%

b. $\dfrac{640 - 600}{80} = 0.5$; $P(0.5) = 0.6915$; $1 - 0.6915 = 0.3085$; 30.85%

c. $\dfrac{600 - 600}{80} = 0$; $\dfrac{720 - 600}{80} = 1.5$; $P(1.5) - P(0) = 0.9332 - 0.5000 = 0.4332$;

43.32%

d. $\dfrac{760 - 600}{80} = 2$; $\dfrac{800 - 600}{80} = 2.5$; $P(2.5) - P(2) = 0.9938 - 0.9772 = 0.0166$; 1.66%

4. a. $\dfrac{20{,}200 - 20{,}000}{100} = 2$; $P(2) = 0.9772$; $1 - 0.9772 = 0.0228$; 2.28%

b. $\dfrac{19{,}900 - 20{,}000}{100} = -1$; $P(-1) = 0.1587$; 15.87%

c. $\dfrac{19{,}950 - 20{,}000}{100} = -0.5; \dfrac{20{,}050 - 20{,}000}{100} = 0.5; P(0.5) - P(-0.5) = 0.6915 -$

0.3085 = 0.3830; 38.3%

d. $\dfrac{20{,}100 - 20{,}000}{100} = 1; \dfrac{20{,}150 - 20{,}000}{100} = 1.5; P(1.5) - P(1) = 0.9332 - 0.8413 =$

0.0919; 9.19%

5. a. $\dfrac{8 \text{ min}}{4 \text{ min}} = 2; P(2) = 0.9772; 97.72\%$ arrive on time, so 2.28% are late.

b. $\dfrac{7}{4} = 1.75; \dfrac{8}{4} = 2; P(2) - P(1.75) \approx 0.9772 - 0.9641 = 0.0131; 1.31\%$

c. $\dfrac{-7}{4} = -1.75; P(-1.75) \approx 0.0359; 3.59\%$

6. a. $\dfrac{610 - 750}{70} = -2.0; P(-2.0) = 0.0228; 2.28\%$

b. $\dfrac{940 - 750}{70} \approx 2.7; P(2.7) = 0.9965; 1 - 0.9965 = 0.0035; 0.35\%$

c. $\dfrac{700 - 750}{70} \approx -0.7; \dfrac{800 - 750}{70} \approx 0.7; P(0.7) - P(-0.7) = 0.7580 - 0.2420 = 0.516;$

51.6%

d. $\dfrac{730 - 750}{70} \approx -0.3; \dfrac{770 - 750}{70} \approx 0.3; P(0.3) - P(-0.3) = 0.6179 - 0.3821 = 0.2358;$

23.58%

7. $\dfrac{17 - 17.6}{0.3} = -2; \dfrac{18 - 17.6}{0.3} \approx 1.3; P(1.3) - P(-2) = 0.9032 - 0.0228 = 0.8804,$ so

88.04% pass. Hence 100% − 88.04% = 11.96% fail.

8. 290 − 262 = 28; if there are fewer than 28 "no-shows", then the airline will be unable

to seat all the passengers who show up; $\dfrac{28 - 45}{10} = -1.7; P(-1.7) = 0.0446; 4.46\%.$

9. a. $P(0.8) \approx 80\%; 0.8 = \dfrac{x - 500}{100}; x = 580$ **b.** $P(-0.7) \approx 25\%; -0.7 = \dfrac{x - 500}{100};$

$x = 430$

10. a. $P(1.3) \approx 90\%; 1.3 = \dfrac{x - 1250}{70}; x = 1341$

b. $P(0.7) \approx 75\%; 0.7 = \dfrac{x - 1250}{70}; x = 1299$

11. $P(1.3) \approx 90\%; 1.3 = \dfrac{x - 72}{9}; x = 83.7 \approx 84$ is the minimum for an A. $P(0.3) \approx 60\%;$

$0.3 = \dfrac{x - 72}{9}; x = 74.7 \approx 75$ is the minimum for a B.

12. a. $P(-1.6) \approx 5\%; -1.6 = \dfrac{x - 72}{9}; x = 57.6 \approx 58$ is the minimum passing score.

b. $P(-1.0) \approx 15\%; -1.0 = \dfrac{x - 72}{9}; x = 63$ is the minimum for a C.

13. a. $\dfrac{2 - 2.01}{0.002} = \dfrac{-0.01}{0.002} = -5; P(-5) \approx 0; 100\% - 0\% = 100\%.$

b. $P(-1.3) \approx 10\%; -1.3 = \dfrac{x - 2.01}{0.002}; x \approx 2.0074$ L.

c. This is the same as the amount that 20% have less than. $P(-0.8) \approx 20\%;$

$-0.8 = \dfrac{x - 2.01}{0.002}; x \approx 2.0084$ L

14. a. $P(-0.7) \approx 25\%; P(0.7) \approx 75\%; \pm 0.7 = \dfrac{x - 65}{5}; x = 65 \pm 3.5; 61.5$ kg to 68.5 kg

b. $P(-1.3) \approx 10\%; -1.3 = \dfrac{x - 65}{5}; x = 58.5$ kg

c. $P(1.7) \approx 95\%; 1.7 = \dfrac{x - 65}{5}; x = 73.5$ kg

15. $\dfrac{1.2 - 0.92}{0.24} \approx 1.2; P(1.2) = 0.8849; 1 - 0.8849 = 0.1151; 11.51\%$

16. $\dfrac{600}{3750} = 0.16; P(-1) \approx 0.16; -1 = \dfrac{x - (198 \text{ min } 36 \text{ s})}{23 \text{ min } 14 \text{ s}}; x = 175$ min 22 s

17. a. The curve is symmetric about the y-axis. Since the area between the curve and the x-axis is 1, the area to the left of the y-axis is $\frac{1}{2} \cdot 1 = \frac{1}{2}$.

b. Blue area $= (0.4) \left(\dfrac{1}{\sqrt{2\pi}} e^{-(0.4)^2/2} \right) = (0.4)(0.3989 e^{-0.08}) \approx 0.1473$

c. $P(0.4) \approx 0.5 + 0.1473 = 0.6473$

d. Too small; there is unshaded area under the curve. $P(0.4) = 0.6554$

e.

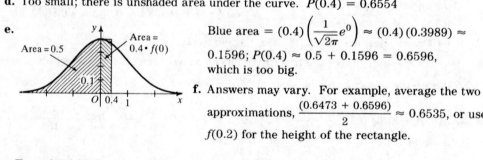

Blue area $= (0.4) \left(\dfrac{1}{\sqrt{2\pi}} e^0 \right) \approx (0.4)(0.3989) \approx$ 0.1596; $P(0.4) \approx 0.5 + 0.1596 = 0.6596$, which is too big.

f. Answers may vary. For example, average the two approximations, $\dfrac{(0.6473 + 0.6596)}{2} \approx 0.6535$, or use $f(0.2)$ for the height of the rectangle.

Class Exercises 17-5 · page 672

1. (c) sampling by questionnaire. The subscribers of the television magazine may have opinions that are different from those of the general public. In fact, the magazine may have influenced the subscribers' opinions.

2. (b) judgment sampling. The consultant must use his or her own judgment to define "successful" and then choose 50 successful managers. Another consultant might define and choose differently.

3. (a) convenience sampling. The location of the market or the time of day may make people's answers unrepresentative of those of the general public.

4. (d) simple random sampling. There are no apparent limitations.

5. (c) sampling by questionnaire. Each question is biased. The first gives a favorable impression of the store, while the second gives an unfavorable impression.

6. (c) sampling by questionnaire. People who still use their cars may be more likely to respond than those who do not.

7. $0.2\left(\dfrac{30}{50}\right) + 0.8\left(\dfrac{20}{50}\right) = 0.12 + 0.32 = 0.44;\ 44\%$

Written Exercises 17-5 · pages 673–674

1–4. Answers may vary. Examples are given.

1. One special interest group may have been responsible for most of the favorable mail.

2. People who do not want their taxes raised to support the zoo may have been more likely to respond in favor of an increased admission charge in order to put the financing burden on those who use the zoo.

3. It is likely that people who call will have strong feelings about the highway. These feelings may not reflect those of the general population.

4. People may not respond truthfully. Busy people may not take the time to answer. The people who shop at this store may be more or less affluent than the general population.

5. $\dfrac{12}{40} \cdot \dfrac{360}{360 + 320} + \dfrac{24}{40} \cdot \dfrac{320}{360 + 320} = \dfrac{3}{10} \cdot \dfrac{9}{17} + \dfrac{3}{5} \cdot \dfrac{8}{17} = \dfrac{75}{170} = \dfrac{15}{34} \approx 44\%$

6. $0.25\left(\dfrac{8}{10}\right) + 0.75\left(\dfrac{4}{10}\right) = 0.2 + 0.3 = 0.5;\ 50\%$

7. $\bar{x} = \dfrac{15 + 8 + 20 + 5 + 12}{5} = \dfrac{60}{5} = 12;\ x_{est.} = n \cdot \bar{x} = 200 \cdot 12 = 2400$ years of experience

8. $\bar{x} = \dfrac{1 + 2 + 2(3) + 4 + 2(5) + 9}{8} = \dfrac{32}{8} = 4$ defective bulbs per box; $x_{est.} = 4 \cdot 150 =$ 600 defective bulbs

9. $\bar{x} = \dfrac{12}{156} = \dfrac{1}{13};\ 80 = n \cdot \dfrac{1}{13};\ n = 1040$ deer

10. $\bar{x} = \dfrac{m}{d};\ c = x \cdot \dfrac{m}{d};\ x = \dfrac{cd}{m}$

11. $\dfrac{300}{1000} \cdot \dfrac{15}{25} + \dfrac{250}{1000} \cdot \dfrac{12}{25} + \dfrac{250}{1000} \cdot \dfrac{10}{25} + \dfrac{200}{1000} \cdot \dfrac{9}{25} = \dfrac{3}{10} \cdot \dfrac{3}{5} + \dfrac{1}{4} \cdot \dfrac{12}{25} + \dfrac{1}{4} \cdot \dfrac{2}{5} + \dfrac{1}{5} \cdot \dfrac{9}{25} =$
$0.18 + 0.12 + 0.10 + 0.072 = 0.472;\ 47\%$

12. $\dfrac{300}{1000} \cdot \dfrac{20}{25} + \dfrac{250}{1000} \cdot \dfrac{14}{25} + \dfrac{250}{1000} \cdot \dfrac{18}{25} + \dfrac{200}{1000} \cdot \dfrac{16}{25} = \dfrac{3}{10} \cdot \dfrac{4}{5} + \dfrac{1}{4} \cdot \dfrac{14}{25} + \dfrac{1}{4} \cdot \dfrac{18}{25} + \dfrac{1}{5} \cdot \dfrac{16}{25} =$
$0.24 + 0.14 + 0.18 + 0.128 = 0.688;\ 69\%$

13. During the 1948 presidential campaign, most public opinion polls predicted that the challenger, Thomas E. Dewey, would win a landslide victory over the incumbent, Harry S. Truman. Truman, however, won the election. The polls were incorrect for two basic reasons: (1) they were conducted too far in advance of the election (so that people had time to change their minds, and they did), and (2) they were based on quota samples that did not accurately reflect the people who voted.

Computer Exercise · page 674

```
10 DIM S(50)
20 FOR I = 1 TO 50
30 READ S(I)
40 NEXT I
50 PRINT "SAMPLE SIZE";
60 INPUT SS
70 PRINT "NUMBER OF SAMPLES";
80 INPUT NS
90 PRINT "INTERVAL WIDTH FOR HISTOGRAM";
100 INPUT W
110 DIM M(20 * W + 1)
120 FOR I = 0 TO 20 * W + 1
130 LET M(I) = 0
140 NEXT I
150 FOR I = 1 TO NS
160 LET T = 0
165 LET NA = 50
170 FOR J = 1 TO SS
180 LET D = INT(NA * RND(1)) + 1
190 LET T = T + S(D)
192 LET TEMP = S(NA)
194 LET S(NA) = S(D)
196 LET S(D) = TEMP
198 LET NA = NA - 1
200 NEXT J
210 LET SM = INT(T / SS) - 549 + 10 * W
220 IF SM < = 0 THEN LET M(0) = M(0) + 1
230 IF SM > 0 AND SM < 20 * W + 1 THEN LET M(SM) = M(SM) + 1
240 IF SM > = 20 * W + 1 THEN LET M(20 * W + 1) = M(20 * W + 1) + 1
250 NEXT I
260 HOME
270 PRINT " <";550 - 10 * W;" ";
280 LET N = M(0): GOSUB 500
290 FOR I = 1 TO 20 * W STEP W
300 PRINT 549 - 10 * W + I;"-";549 - 9 * W + I;" ";
310 LET N = 0
320 FOR J = 0 TO W - 1
330 LET N = N + M(I + J)
340 NEXT J
350 GOSUB 500
360 NEXT I
370 PRINT " >";550 + 10 * W;" ";
380 LET N = M(20 * W + 1): GOSUB 500
390 END
500 IF N = 0 THEN GOTO 540
510 FOR J = 1 TO N
520 PRINT "*";
```

```
530 NEXT J
540 PRINT
550 RETURN
1000 DATA 480,520,670,580,500,570,700,540,690,500
1010 DATA 520,450,570,540,620,450,460,790,580,610
1020 DATA 550,630,460,470,510,630,510,720,490,540
1030 DATA 690,510,500,710,520,480,750,450,670,610
1040 DATA 320,680,530,510,350,400,400,640,430,480
```

Based on four runs of sample size = 5, 10, 15, and 20, it appears that the distribution of sample means is normally distributed with a mean that tends towards the population mean and a standard deviation that decreases as the sample size increases.

Class Exercises 17-6 · pages 677–678

1. a. $\sqrt{\dfrac{\overline{p}(1-\overline{p})}{n}} = \sqrt{\dfrac{(0.4)(0.6)}{24}} = \sqrt{0.01} = 0.1$

  b. $0.4 - 2(0.1) < p < 0.4 + 2(0.1)$; $0.2 < p < 0.6$; $0.4 - 3(0.1) < p < 0.4 + 3(0.1)$; $0.1 < p < 0.7$

2. a. $\sqrt{\dfrac{\overline{p}(1-\overline{p})}{n}} = \sqrt{\dfrac{0.5(0.5)}{25}} = \sqrt{0.01} = 0.1$

  b. $0.5 - 2(0.1) < p < 0.5 + 2(0.1)$; $0.3 < p < 0.7$; $0.5 - 3(0.1) < p < 0.5 + 3(0.1)$; $0.2 < p < 0.8$

3. a. $\overline{p} = \dfrac{50}{100} = 0.5$; $s = \sqrt{\dfrac{\overline{p}(1-\overline{p})}{n}} = \sqrt{\dfrac{0.5(0.5)}{100}} = \sqrt{0.0025} = 0.05$

  b. $0.5 - 2(0.05) < p < 0.5 + 2(0.05)$; $0.4 < p < 0.6$

4. the sample size

5. No; however, the students from Central High School are probably a more homogeneous group than the students selected from the entire state.

Written Exercises 17-6 · pages 678–680

1. $\overline{p} = \dfrac{7}{10} = 0.7$; $s = \sqrt{\dfrac{0.7(0.3)}{10}} = \sqrt{0.021} \approx 0.145$; $0.7 - 2(0.145) < p < 0.7 + 2(0.145)$; $0.41 < p < 0.99$

2. $\overline{p} = \dfrac{6}{10} = 0.6$; $s = \sqrt{\dfrac{0.6(0.4)}{10}} = \sqrt{0.024} \approx 0.155$; $0.6 - 2(0.155) < p < 0.6 + 2(0.155)$; $0.29 < p < 0.91$

3. $\overline{p} = \dfrac{36}{100} = 0.36$; $s = \sqrt{\dfrac{0.36(0.64)}{100}} = 0.048$; $0.36 - 2(0.048) < p < 0.36 + 2(0.048)$; $0.264 < p < 0.456$

4. $\overline{p} = \dfrac{16}{25} = 0.64$; $s = \sqrt{\dfrac{0.64(0.36)}{25}} = 0.096$; $0.64 - 2(0.096) < p < 0.64 + 2(0.096)$; $0.448 < p < 0.832$

5. $\overline{p} = \dfrac{25}{400} = 0.0625$; $s = \sqrt{\dfrac{0.0625(0.9375)}{400}} \approx 0.012$; $0.0625 - 2(0.012) < p < 0.0625 + 2(0.012)$; $0.0385 < p < 0.0865$

6. $\bar{p} = \dfrac{80}{100} = 0.8$; $s = \sqrt{\dfrac{0.8(0.2)}{100}} = 0.04$; $0.8 - 2(0.04) < p < 0.8 + 2(0.04)$;

$0.72 < p < 0.88$

7. $\bar{p} = \dfrac{30}{1000} = 0.03$; $s = \sqrt{\dfrac{0.03(0.97)}{1000}} \approx 0.005$; $0.03 - 3(0.005) < p < 0.03 + 3(0.005)$;

$0.015 < p < 0.045$

8. $\bar{p} = \dfrac{1280}{1600} = 0.8$; $s = \sqrt{\dfrac{0.8(0.2)}{1600}} = 0.01$; $0.8 - 3(0.01) < p < 0.8 + 3(0.01)$;

$0.77 < p < 0.83$

9. $\bar{p} = \dfrac{9}{100} = 0.09$; $s = \sqrt{\dfrac{0.09(0.91)}{100}} \approx 0.0286$; $0.09 - 3(0.0286) < p < 0.09 + 3(0.0286)$;

$0.004 < p < 0.176$

10. $\bar{p} = \dfrac{10}{100} = 0.1$; $s = \sqrt{\dfrac{0.1(0.9)}{100}} = 0.03$; $0.1 - 3(0.03) < p < 0.1 + 3(0.03)$;

$0.01 < p < 0.19$

11. $\bar{p} = \dfrac{200}{250} = 0.8$; $s = \sqrt{\dfrac{0.8(0.2)}{250}} \approx 0.0253$; $0.8 - 3(0.0253) < p < 0.8 + 3(0.0253)$;

$0.724 < p < 0.876$

12. $\bar{p} = \dfrac{150}{200} = 0.75$; $s = \sqrt{\dfrac{0.75(0.25)}{200}} \approx 0.0306$; $0.75 - 3(0.0306) < p < 0.75 + 3(0.0306)$;

$0.658 < p < 0.842$

13. 4   **14.** 100

15. $\sqrt{\dfrac{0.36(0.64)}{900}} = 0.016$; $0.016(2) = 0.032$ or about 3.2%, so the margin of error is
associated with a 95% confidence interval.

16. $\sqrt{\dfrac{0.55(0.45)}{380}} \approx 0.026$; $0.026(2) = 0.052$ or about 5%, so the margin of error is associated
with a 95% confidence interval.

17. $\bar{p} = \dfrac{144}{240} = 0.6$; $s = \sqrt{\dfrac{0.6(0.4)}{240}} \approx 0.0316$; $0.6 - 2(0.0316) < p < 0.6 + 2(0.0316)$;

$0.537 < p < 0.663$; $0.6 - 3(0.0316) < p < 0.6 + 3(0.0316)$; $0.505 < p < 0.695$

18. $\bar{p} = 0.3$; $s = \sqrt{\dfrac{0.3(0.7)}{840}} \approx 0.0158$; $0.3 - 2(0.0158) < p < 0.3 + 2(0.0158)$;

$0.268 < p < 0.332$. $0.3 - 3(0.0158) < p < 0.3 + 3(0.0158)$; $0.253 < p < 0.347$

19. a.

b. $f\left(\dfrac{1}{2}\right) = \dfrac{1}{4}$

c. $s = \sqrt{\dfrac{f(\bar{p})}{n}} \le \sqrt{\dfrac{\frac{1}{4}}{n}} = \sqrt{\dfrac{1}{4n}}$

20. A 95% confidence interval for p is $\bar{p} - 2s < p < \bar{p} + 2s$. The maximum value for

$2s$ is $2\sqrt{\dfrac{1}{4n}} = \sqrt{\dfrac{1}{n}}$. Thus, the widest possible 95% confidence interval is

$$\bar{p} - \sqrt{\dfrac{1}{n}} < p < \bar{p} + \sqrt{\dfrac{1}{n}}.$$

21. $\sqrt{\dfrac{1}{n}} < 0.01; \dfrac{1}{n} < 0.0001; n > 10{,}000$

22. $\sqrt{\dfrac{1}{n}} < 0.03; \dfrac{1}{n} < 0.0009; n > 1111.1; n \geq 1112$

23. $\sqrt{\dfrac{1}{n}} < 0.05; \dfrac{1}{n} < 0.0025; n > 400$

24. $\bar{p} = \dfrac{18}{50} = 0.36; s = \sqrt{\dfrac{0.36(0.64)}{50}} \approx 0.0679$; from the table on p. 664, about

80% of the time, p is within 1.3 standard deviations of $\bar{p}$; $\bar{p} - 1.3s < p < \bar{p} + 1.3s$;
$0.36 - 1.3(0.0679) < p < 0.36 + 1.3(0.0679); 0.272 < p < 0.448$

Chapter Test 17 · page 681

1. Descriptive statistics involves collecting, organizing, and summarizing data. Inferential statistics involves drawing conclusions about a population based on a sample.

2. a.

b. $\bar{x} = \dfrac{(4(80) + 9(100) + 11(120) + 6(140))1000}{30} = \$112{,}667$;

median $= \$120{,}000$; mode $= \$120{,}000$

3. a. median $= \dfrac{25 + 27}{2} = 26$; lower quartile $= 22$; upper quartile $=$

$\dfrac{29 + 31}{2} = 30$

b. range $= 36 - 18 = 18$; interquartile range $= 30 - 22 = 8$

c.

4. $\bar{x}_E = 70, s_E = 5, z_E = \dfrac{80 - 70}{5} = 2; \bar{x}_M = 65, s_M = 10, z_M = \dfrac{80 - 65}{10} = 1.5.$

Zina did better on the English test since her score was two standard deviations above the mean as compared to 1.5 standard deviations above the mean on the math test.

5. $\bar{x} = (4 + 7 + 10 + 11 + 12 + 14 + 15 + 21 + 32)/9 = 14,$
$s^2 = (10^2 + 7^2 + 4^2 + 3^2 + 2^2 + 0^2 + 1^2 + 7^2 + 18^2)/9 \approx 61.33, s = \sqrt{61.33} \approx 7.83$

6. a. $z = \dfrac{1300 - 1200}{100} = 1; 1 - P(1) = 1 - 0.8413 = 0.1587 = 15.87\%$

b. $z = \dfrac{1000 - 1200}{100} = -2; P(-2) = 0.0228 = 2.28\%$

c. $P(z) = 90\%$ when $z \approx 1.3; 1.3 = \dfrac{x - 1200}{100}; x = 1330$

d. $P(z) = 75\%$ when $z \approx 0.7; 0.7 = \dfrac{x - 1200}{100}; x = 1270$

7. $\bar{p} = 0.62\left(\dfrac{27}{50}\right) + 0.38\left(\dfrac{14}{50}\right) = 0.4412; 44\%$

8. $\bar{p} = \dfrac{486}{600} = 0.81; s = \sqrt{\dfrac{(0.81)(0.19)}{600}} = 0.016; 0.81 - 2(0.016) < p < 0.81 + 2(0.016);$
$0.778 < p < 0.842$ is the 95% confidence interval. The 99% interval is
$0.81 + 3(0.016) < p < 0.81 + 3(0.016); 0.762 < p < 0.858$

Class Exercises 18-1 · pages 687–688

1. Answers may vary. Examples are given. **a.** $\dfrac{1}{2}$ **b.** 2 **c.** $y = \dfrac{1}{2}x + 2$

2. **a.** $\bar{x} = 11; \bar{y} = 7; \overline{xy} = 83; s_x = \sqrt{\dfrac{\sum\limits_{i=1}^{n} x_i^2 - n(\bar{x})^2}{n}} = \sqrt{\dfrac{1070 - 8(11)^2}{8}} = \sqrt{12.75} \approx$

$3.571; s_y = \sqrt{\dfrac{\sum\limits_{i=1}^{n} y_i^2 - n(\bar{y})^2}{n}} = \sqrt{\dfrac{424 - 8(7)^2}{8}} = \sqrt{4} = 2$

b. The least squares line contains $(\bar{x}, \bar{y}) = (11, 7)$.

c. $m = \dfrac{\overline{xy} - \bar{x} \cdot \bar{y}}{s_x^2} = \dfrac{83 - (11)(7)}{12.75} \approx 0.471$

d. $y - 7 = 0.471(x - 11); y = 0.471x + 1.82$

e. $r = \dfrac{\overline{xy} - \bar{x} \cdot \bar{y}}{s_x s_y} = \dfrac{83 - (11)(7)}{(3.571)(2)} \approx 0.84$

3. The least-squares line is the line for which the sum of the squares of the vertical distances from the data points to the line is minimum or least.

4. **a.** 0.6 **b.** 0.8 **c.** −0.6 **d.** −0.8 **e.** 0 **f.** 0.2

5. **a.** Positive; in general, as a person's height increases, so does his or her weight.

   b. Positive; a person's height is to a large extent genetically determined, and therefore is related to the parents' average height.

   c. Negative; as an automobile's age increases, its value decreases.

   d. Negative; the incidence of flu is greater during the winter months, when temperatures are lower.

   e. Zero; among the adult population there is no relationship between a person's height and the person's number of years of formal education.

6. Answers may vary. Examples are given. **a.** Approximately 1.75

   b. Thousands of dollars per year. **c.** About $1750

7. No, both the number of firefighters and the number of students are related to the total population of Chesterfield, rather than to each other.

Written Exercises 18-1 · pages 688–691

1. Positive; generally, the more time spent studying, the better one knows the subject, and the better one does on a test.

2. Negative; the more absences in the course, the more poorly the class is likely to perform on examinations, due to the increase in missed information.

3. Negative; the lighter a car is, the better its fuel economy is likely to be.

4. Positive; obesity can strain the heart, resulting in increased blood pressure.

5. Positive; blood pressure generally increases with age.

6. Approximately zero; in general, ability in one area is not associated with ability in the other area.

7. Negative; demand for an item is generally high when its supply is low.

8. Positive; the better rested one is, the better one can concentrate and think clearly.

9. 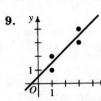 $\bar{x} = 2; \bar{y} = 2.5; \overline{xy} = 6; s_x = \sqrt{\dfrac{20 - 4(2)^2}{4}} = \sqrt{1} = 1;$

$s_y = \sqrt{\dfrac{30 - 4(2.5)^2}{4}} = \sqrt{1.25} \approx 1.118; r = \dfrac{6 - (2)(2.5)}{(1)(1.118)} \approx 0.89.$ The

slope of the least-squares line is $m = \dfrac{6 - (2)(2.5)}{1} = 1;$ the equation of

the least-squares line is $y - 2.5 = 1(x - 2)$, or $y = x + 0.5$.

10. $\bar{x} = 2; \bar{y} = 2; \overline{xy} = 3; s_x = \sqrt{\dfrac{18 - 4(2)^2}{4}} = \sqrt{0.5} \approx 0.7071;$

$s_y = \sqrt{\dfrac{26 - 4(2)^2}{4}} = \sqrt{2.5} \approx 1.5811; r = \dfrac{3 - (2)(2)}{(0.7071)(1.5811)} \approx -0.89.$

The slope of the least-squares line is $m = \dfrac{3 - (2)(2)}{0.5} = -2;$ the

equation of the least squares lines is $y - 2 = -2(x - 2)$, or $y = -2x + 6$.

11. $\bar{x} = 3; \bar{y} = 4; \overline{xy} = 9; s_x = \sqrt{\dfrac{55 - 5(3)^2}{5}} = \sqrt{2} \approx 1.4142;$

$s_y = \sqrt{\dfrac{106 - 5(4)^2}{5}} = \sqrt{5.2} \approx 2.2804; r = \dfrac{9 - (3)(4)}{(1.4142)(2.2804)} \approx$

$-0.93.$ The slope of the least-squares line is $m = \dfrac{9 - 3(4)}{2} =$

$-1.5;$ the equation of the least squares line is $y - 4 =$

$-1.5(x - 3),$ or $y = -1.5x + 8.5.$

12. 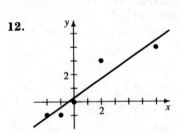 $\bar{x} = 1; \bar{y} = 1; \overline{xy} = 6.6; s_x = \sqrt{\dfrac{45 - 5(1)^2}{5}} = \sqrt{8} \approx 2.8284;$

$s_y = \sqrt{\dfrac{27 - 5(1)^2}{5}} = \sqrt{4.4} \approx 2.0976; r = \dfrac{6.6 - (1)(1)}{(2.8284)(2.0976)} \approx$

$0.94.$ The slope of the least-squares line is $m =$

$\dfrac{66 - (1)(1)}{8} = 0.7;$ the equation of the least-squares line is

$y - 1 = 0.7(x - 1),$ or $y = 0.7x + 0.3.$

13. $y = 3.5 - 0.5x$ **a.** $3.5 - 0.5(1) = 3.0;$ B **b.** $3.5 - 0.5(3) = 2.0;$ C

   c. $3.5 - 0.5(8) = -0.5;$ F

14. $y = 14x + 0.7$ **a.** $14(10) + 0.7 = 140.7$ **b.** $14(5) + 0.7 = 70.7$

   c. $14(7) + 0.7 = 98.7$

15. $\bar{x} = 226.4; \bar{y} = 4; \overline{xy} = 1018.8; s_x = \sqrt{\dfrac{288,376 - 5(226.4)^2}{5}} = \sqrt{6418.24} \approx 80.114;$

$m = \dfrac{1018.8 - (226.4)(4)}{6418.24} \approx 0.017637;$ the equation of the least-squares line is $y - 4 =$

$0.018(x - 226.4),$ or $y = 0.018x + 0.007;$ when $y = 7, x = \dfrac{7 - 0.007}{0.018} = 388.5.$

16. $\overline{x} = 583.33$; $\overline{y} = 74.17$; $\overline{xy} = 45{,}166.67$; $s_x = \sqrt{\dfrac{2{,}150{,}000 - 6(583.33)^2}{6}} =$

$\sqrt{18{,}059.44443} \approx 134.385$; $m = \dfrac{45{,}166.67 - (583.33)(74.17)}{18{,}059.44443} \approx 0.1053$; the equation of

the least-squares line is $y - 74.17 = 0.1053(x - 583.33)$, or $y = 0.105x + 12.7$;

when $x = 550$, $y = 0.105(550) + 12.7 \approx 70$.

17. Answers may vary. The correlation coefficient in Ex. 15 is $r = \dfrac{1018.8 - (226.4)(4)}{(80.114)(1.414)} \approx$

0.9993. The correlation coefficient in Ex. 16 is $r = \dfrac{45{,}166.67 - (583.33)(74.17)}{(134.385)(14.554)} \approx 0.972$.

Since the correlation in Ex. 15 is higher, that prediction has a better chance of being
correct. Also, in Ex. 16, there were very different results for the two values of 500 near
the 550 value used for prediction.

18.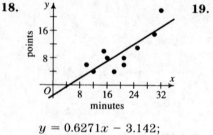

$y = 0.6271x - 3.142$;
$r = 0.8234$

19.

$y = 0.17x - 1.075$;
$r = 0.8246$

20.

$y = 2.279x + 4.159$;
$r = 0.6123$

21. No, the correlation coefficients in Exs. 18, 19, and 20
show that both fouls committed and points scored
seem to be related to minutes played, rather than to
each other. Committing more fouls is not a way to
increase points scored.

22. a. $d_i = |y_i - y| = |y_i - (\overline{y} + m(x_i - \overline{x}))| = |(y_i - \overline{y}) - m(x_i - \overline{x})|$

b. $D = \displaystyle\sum_{i=1}^{n} d_i^2 = \sum_{i=1}^{n} [(y_i - \overline{y}) - m(x_i - \overline{x})]^2 = \sum_{i=1}^{n} [(y_i - \overline{y})^2 - 2m(y_i - \overline{y})(x_i - \overline{x}) +$

$m^2(x_i - \overline{x})^2] = \displaystyle\sum_{i=1}^{n} (y_i - \overline{y})^2 - 2m \sum_{i=1}^{n} (y_i - \overline{y})(x_i - \overline{x}) + m^2 \sum_{i=1}^{n} (x_i - \overline{x})^2 =$

$m^2 \displaystyle\sum_{i=1}^{n} (x_i - \overline{x})^2 - 2m \sum_{i=1}^{n} (y_i - \overline{y})(x_i - \overline{x}) + \sum_{i=1}^{n} (y_i - \overline{y})^2$

c. D is a quadratic in m and is minimized when $m = \dfrac{-(\text{coefficient of linear term})}{2(\text{coefficient of quadratic term})} =$

$\dfrac{-(-2)\displaystyle\sum_{i=1}^{n}(y_i - \overline{y})(x_i - \overline{x})}{2\displaystyle\sum_{i=1}^{n}(x_i - \overline{x})^2} = \dfrac{\displaystyle\sum_{i=1}^{n}(y_i - \overline{y})(x_i - \overline{x})}{\displaystyle\sum_{i=1}^{n}(x_i - \overline{x})^2}.$

$$\textbf{d. } m = \frac{\displaystyle\sum_{i=1}^{n}(y_i - \bar{y})(x_i - \bar{x})}{\displaystyle\sum_{i=1}^{n}(x_i - \bar{x})^2} \cdot \frac{\dfrac{1}{n}}{\dfrac{1}{n}} = \frac{\dfrac{1}{n}\displaystyle\sum_{i=1}^{n}(x_i y_i - x_i \bar{y} - \bar{x} y_i + \bar{x}\cdot\bar{y})}{\dfrac{1}{n}\displaystyle\sum_{i=1}^{n}(x_i - \bar{x})^2} =$$

$$\frac{\dfrac{1}{n}\displaystyle\sum_{i=1}^{n}x_i y_i - \dfrac{1}{n}\displaystyle\sum_{i=1}^{n}x_i - \bar{x}\cdot\dfrac{1}{n}\displaystyle\sum_{i=1}^{n}y_i + \dfrac{1}{n}\displaystyle\sum_{i=1}^{n}\bar{x}\cdot\bar{y}}{\dfrac{1}{n}\displaystyle\sum_{i=1}^{n}(x_i - \bar{x})^2} =$$

$$\frac{\dfrac{1}{n}\displaystyle\sum_{i=1}^{n}x_i y_i - \bar{y}\cdot\dfrac{1}{n}\displaystyle\sum_{i=1}^{n}x_i - \bar{x}\cdot\dfrac{1}{n}\displaystyle\sum_{i=1}^{n}y_i + \dfrac{1}{n}\displaystyle\sum_{i=1}^{n}\bar{x}\cdot\bar{y}}{\dfrac{1}{n}\displaystyle\sum_{i=1}^{n}(x_i - \bar{x})^2} =$$

$$\frac{\overline{xy} - \bar{y}\cdot\bar{x} - \bar{x}\cdot\bar{y} + \dfrac{1}{n}\cdot n\bar{x}\cdot\bar{y}}{\dfrac{1}{n}\displaystyle\sum_{i=1}^{n}(x_i - \bar{x})^2} = \frac{\overline{xy} - \bar{x}\cdot\bar{y}}{s_x^{\,2}}$$

Class Exercises 18-2 · page 694

1. a. $y = 2 \cdot 3^x$; $\log y = \log(2 \cdot 3^x) = \log 2 + \log 3^x = \log 2 + x \log 3$;
$\log y = (\log 3)x + \log 2$

b. $\log 3$ **c.** The vertical-axis intercept is $\log 2$.

2. $\log y = 1.3x + 0.8$; $y = 10^{1.3x + 0.8} = (10^{1.3})^x \cdot 10^{0.8}$; $y \approx (6.31)19.95^x$

3. $\log y = \dfrac{1}{2}x + 2$; $y = 10^{(1/2)x + 2} = (10^{1/2})^x \cdot 10^2$; $y \approx (100)3.162^x$

4. $\log y = -1.4x + 2.5$; $y = 10^{-1.4x + 2.5} = (10^{-1.4})^x \cdot 10^{2.5}$; $y \approx (316.2)0.0398^x$

5. $\ln y = 1.5x + 4$; $y = e^{1.5x + 4} = (e^{1.5})^x \cdot e^4$; $y \approx (54.6)4.482^x$

Written Exercises 18-2 · pages 694–697

For exercises in section 18-2, equations of least-squares lines are found using a calculator or computer software.

1. $\log y = 1.6x + 1.8$; $y = 10^{1.6x + 1.8} = (10^{1.6})^x \cdot 10^{1.8}$; $y \approx (63.1)39.81^x$

2. $\log y = 2.41x + 0.75$; $y = 10^{2.41x + 0.75} = (10^{2.41})^x \cdot 10^{0.75}$; $y \approx (5.623)257.0^x$

3. $\log y = -0.4x + 2.3$; $y = 10^{-0.4x + 2.3} = (10^{-0.4})^x \cdot 10^{2.3}$; $y \approx (199.5)0.3981^x$

4. $\log y = -1.05x - 1.92$; $y = 10^{-1.05x - 1.92} = (10^{-1.05})^x \cdot 10^{-1.92}$; $y \approx (0.012)0.0891^x$

5. $\ln y = 0.5x + 3.2$; $y = e^{0.5x + 3.2} = (e^{0.5})^x \cdot e^{3.2}$; $y \approx (24.53)1.649^x$

6. $\ln y = 4.5 - 2x$; $y = e^{4.5 - 2x} = e^{4.5} \cdot (e^{-2})^x$; $y \approx (90.02)0.1353^x$

7.

x	1	3	4	5
$\log y$	0.4771	0.8293	1.0056	1.1816

The equation of the least-squares line is $\log y = 0.1761x + 0.301$;
$y = 10^{0.1761x + 0.301} = 10^{0.301} \cdot (10^{0.1761})^x$; $y = (2)1.5^x$.

8.

x	2	2.5	4	5.5
log y	0.5599	0.5809	0.6425	0.7050

The equation of the least-squares line is log $y = 0.0414x + 0.4772$;
$y = 10^{0.0414x + 0.4772} = 10^{0.4772} \cdot (10^{0.0414})^x$; $y = (3)1.1^x$.

9. a.

x	1	2	3	4	5	6
log y	4.0792	3.9823	3.8865	3.7924	3.6902	3.5911

The equation of the least-squares line is log $y = -0.0975x + 4.1780$;
$y = 10^{-0.0975x + 4.1780} = 10^{4.1780} \cdot (10^{-0.0975})^x$; $y = (15,070)0.7989^x$.

b. $(15,070)0.7989^{10} \approx \$1596$ **c.** $(15,070)0.7989^0 \approx \$15,000$

10. a.

x = no. of years from 1800	-10	0	10	20	30	40	50
log P	0.5911	0.7243	0.8573	0.9823	1.1106	1.2330	1.3655

60	70	80	90	100	110	120
1.4969	1.5999	1.7007	1.7987	1.8808	1.9638	2.0241

130	140	150	160	170	180	190
2.0892	2.1196	2.1781	2.2536	2.3081	2.3551	2.3957

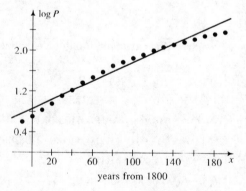

years from 1800

b. The equation of the least-squares line is log $P = 0.009x + 0.8563$.

c. $P = 10^{0.009x + 0.8563} = 10^{0.8563} \cdot (10^{0.009})^x$; $P = (7.183)1.021^x$; when $x = 200$ (the year 2000), the population will be $(7.183)1.021^{200} \approx 458.6$ million.

11. a. Between 1840 and 1850, Florida, Iowa, Wisconsin, and Texas became states. Between 1950 and 1960, Alaska and Hawaii became states.

b. Yes, given that $D = P/A$ and that A is likely to remain constant, exponential growth in P leads to exponential growth in D.

12. a.

d	0	0.5	0.75	1	1.5	2
$\log r$	0	0.2041	0.3010	0.3979	0.6021	0.7993

3	4	5	6	10	15
1.2041	1.6021	2	2.3997	4	6

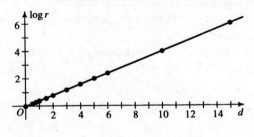

The line passes through the points $(0,0)$ and $(10,4)$, so the slope is $\frac{2}{5}$.

The equation of the line is $\log r = \frac{2}{5}d$, so $d = \frac{5}{2}\log r$ or $d = 2.5\log r$.

b. $\log r = 0.4d$; $r = 10^{0.4d} = (10^{0.4})^d$; $r = 2.512^d$

c. When $d = 14$, $r = 2.512^{14} \approx 400{,}000$.

13. a.

t	0	0.5	1.0	1.5	2.0	2.5	3.0	3.5
$\log V$	0.1818	0.0969	0.0043	−0.0862	−0.1739	−0.2676	−0.3468	−0.4318

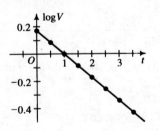

The equation of the least-squares line is $\log V = -0.1766t + 0.1811$.

b. $V = 10^{-0.1766t + 0.1811} = 10^{0.1811} \cdot (10^{-0.1766})^t$; $V = (1.517)0.6659^t$

14. a.

x	0	1	2	4	7	10	20
$\log (y - 6)$	1.1461	1.0492	0.9542	0.7559	0.4771	0.1761	−0.699

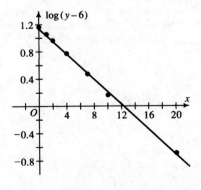

The equation of the least-squares line is $\log (y - 6) = -0.0926x + 1.1331$.

b. $y - 6 = 10^{-0.0926x + 1.1331} = 10^{1.1331} \cdot (10^{-0.0926})^x = (13.59)0.808^x$; $y = (13.59)0.808^x + 6$

15. a. 74°F **b.** $T = 74$

c.

t	0	10	30	50	70
$\log (T - 74)$	1.9243	1.7694	1.5024	1.2625	1.0086

90	110	120	125	130
0.7160	0.3222	0	−0.1549	−0.3010

The equation of the least-squares line is $\log (T - 74) = -0.0165t + 2.0142$;
$T - 74 = 10^{-0.0165t + 2.0142} = 10^{2.0142} \cdot (10^{-0.0165})^t = (103.3)0.9627^t$;
$T = (103.3)0.9627^t + 74$.

16. a.

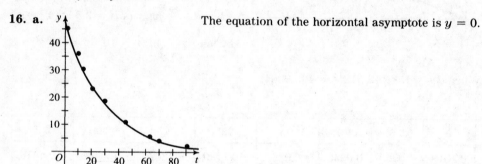

The equation of the horizontal asymptote is $y = 0$.

b.

t	2	10	14	21	30	45	63	70	91
$\log y$	1.6551	1.5575	1.48	1.3636	1.2718	1.0414	0.7482	0.6335	0.3802

The equation of the least-squares line is $\log y = -0.0147t + 1.689$;
$y = 10^{-0.0147t + 1.6892} = 10^{1.6892} \cdot (10^{-0.0147})^t$; $y = (48.89)0.9667^t$.

17. a. In some year, a person will run 800 m in 0 seconds.

b. Answers may vary. An example is given. $\log (y - 100) = -0.0075x + 1.225$;
$y - 100 = 10^{-0.0075x + 1.225} = 10^{1.225} \cdot (10^{-0.0075})^x = (16.79)0.9829^x$;
$y = (16.79)0.9829^x + 100$

18. If $y = ab^x$, then $\log y = \log (ab^x) = \log a + x \log b$; thus $\log y = (\log b)x + \log a$.
Therefore, the points $(x, \log y)$ lie on a line with slope $\log b$. If the line through the
points $(x, \log y)$ has negative slope, then $\log b < 0$, and therefore $0 < b < 1$.

Activity · page 700

1. No **2.** Answers may vary. **3.** $t = 2\pi \sqrt{\dfrac{l}{g}}$, where $g = 980$ cm/s^2

Class Exercises 18-3 · page 701

1. a. $y = 4x^3$; $\log y = \log 4x^3 = \log 4 + \log x^3$; $\log y = \log 4 + 3 \log x$ **b.** 3
2. $\log y = 5 \log x + 2$; $y = 10^{5 \log x + 2} = 10^{\log x^5} \cdot 10^2$; $y = 100x^5$

3. $\log y = 2.5 \log x + 3.5$; $y = 10^{2.5 \log x + 3.5} = 10^{\log x^{2.5}} \cdot 10^{3.5}$; $y = 3162x^{2.5}$

4. $\ln y = -3 \ln x + 1$; $y = e^{-3 \ln x + 1} = e^{\ln x^{-3}} \cdot e^1$; $y = ex^{-3}$

5. $\ln y = 2 \ln x + 0.5$; $y = e^{2 \ln x + 0.5} = e^{\ln x^2} \cdot e^{0.5}$; $y = 1.649x^2$

6. 2; the cost of a pizza depends on its area, which varies as the square of the diameter.

7. 3; the weight of a whale depends on its volume, which varies as the cube of its length.

Written Exercises 18-3 · pages 701–704

For exercises in section 18-3, equations of least-squares lines are found using a calculator or computer software.

1. $\log y = 4 \log x + 1$; $y = 10^{4 \log x + 1} = 10^{\log x^4} \cdot 10^1$; $y = 10x^4$

2. $\log y = 2.5 \log x + 1.3$; $y = 10^{2.5 \log x + 1.3} = 10^{\log x^{2.5}} \cdot 10^{1.3}$; $y = 19.95x^{2.5}$

3. $\log y = -\log x + 0.7$; $y = 10^{-\log x + 0.7} = 10^{\log x^{-1}} \cdot 10^{0.7} = 5.012x^{-1}$; $y = \dfrac{5.012}{x}$

4. $\log y = -2.6 \log x - 0.7$; $y = 10^{-2.6 \log x - 0.7} = 10^{\log x^{-2.6}} \cdot 10^{-0.7}$; $y = 0.1995x^{-2.6}$

5. $\ln y = 1.6 \ln x + 3.2$; $y = e^{1.6 \ln x + 3.2} = e^{\ln x^{1.6}} \cdot e^{3.2}$; $y = 24.53x^{1.6}$

6. $\ln y = -0.4 \ln x + 4.2$; $y = e^{-0.4 \ln x + 4.2} = e^{\ln x^{-0.4}} \cdot e^{4.2}$; $y = 66.69x^{-0.4}$

7.

$\log x$	0.3010	0.4771	0.6990	0.7782
$\log y$	0.6021	1.1303	1.7959	2.0334

The equation of the least-squares line is $\log y = 2.999 \log x - 0.3007$;
$y = 10^{2.999 \log x - 0.3007} = 10^{\log x^{2.999}} \cdot 10^{-0.3007} \approx 0.5004x^{2.999}$; $y = \dfrac{1}{2}x^3$.

8.

$\log x$	−0.0458	0.1761	0.3802	0.4771
$\log y$	0.6937	0.2504	−0.1549	−0.3565

The equation of the least-squares line is $\log y = -2.004x + 0.6029$;
$y = 10^{-2.004 \log x + 0.6029} = 10^{\log x^{-2.004}} \cdot 10^{0.6029} \approx 4.0077x^{-2.004}$; $y = 4x^{-2}$.

9. a. 2; area varies as the square of a length.

   b. 3; weight depends on volume; volume varies as the cube of a length.

10. a. 3; capacity depends on volume; volume varies as the cube of a length.

   b. 2; the amount of paint depends on the surface area; surface area varies as the square of a length.

11. 2; area varies as the square of a length.

12. $\dfrac{1}{3}$; the weight of a cube depends upon the volume; volume varies as the cube of the diagonal, so $x = ky^3$; $y = ax^{1/3}$.

13. $\dfrac{1}{2}$; the cost of a circular rug depends on its area; area varies as the square of the diameter, so $x = ky^2$; $y = ax^{1/2}$.

14. $\dfrac{3}{2}$; the surface area is $x = 4\pi r^2$, so $r = \dfrac{1}{2\sqrt{\pi}}x^{1/2}$; the volume is

$$y = \frac{4}{3}\pi r^3 = \frac{4}{3}\pi\left(\frac{1}{2\sqrt{\pi}}x^{1/2}\right)^3 = ax^{3/2}, \text{ where } a = \frac{1}{6\sqrt{\pi}}.$$

15. $\dfrac{2}{3}$; the weight, x, depends on z^3, the cube of the length of the elephant's ear, so

$z = kx^{1/3}$; the surface area of the elephant's ear, y, depends on z^2, the square of the length of the elephant's ear, so $y = tz^2 = t(kx^{1/3})^2 = ax^{2/3}$.

16. a. Using the points $(\log x, \log y)$ for Mercury and Pluto we find $m = \dfrac{4.9581 - 1.9445}{3.7718 - 1.7627} \approx$

1.5, and the equation of the line is $Y - 4.9581 = 1.5(X - 3.7718)$, or $Y = 1.5X - 0.7$.

b. $\log y = 1.5 \log x - 0.7$; $y = 10^{1.5 \log x - 0.7} = 10^{\log x^{1.5}} \cdot 10^{-0.7}$; $y = 0.1995x^{1.5}$

c. $y = 0.1995(7000)^{1.5} \approx 116{,}800$ days

17.

$\log a$	5.6253	5.8267	6.0294	6.2749
$\log T$	1.6284	1.9304	2.2348	2.6026

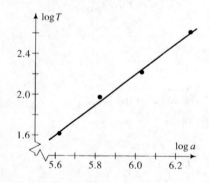

The equation of the least-squares line is $\log T = 1.4999 \log a - 6.8087$; $T = 10^{1.4999 \log a - 6.8087} = 10^{\log a^{1.4999}} \cdot 10^{-6.8087} \approx (1.553 \times 10^{-7})a^{1.5}$;
$T = (1.553 \times 10^{-7})a^{3/2}$.

18. a.

$\log x$	2	2.3010	2.6021	2.9031	3.1761	3.6990	4
$\log y$	0.9965	1.2956	1.6422	2.0147	2.3344	2.8986	3.2152

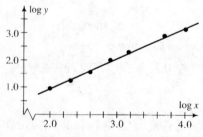

The equation of the least-squares line is $\log y = 1.1253 \log x - 1.2677$.

b. $y = 10^{1.1253 \log x - 1.2677} = 10^{\log x^{1.1253}} \cdot 10^{-1.2677}$; $y = 0.054x^{1.125}$

c. time $= 0.054(1000)^{1.125} \approx 128$ s

19. a.

log d	1.1761	1.2041	1.2304	1.2553	1.2788	1.3010	1.3222	1.3424	1.3617
log B	1.2714	1.3589	1.4411	1.5185	1.5917	1.6612	1.7274	1.7904	1.8506

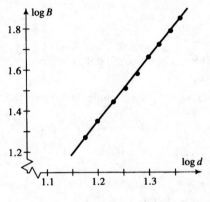

Using $(\log d, \log B)$ for the points $(16, 22.85)$ and $(22, 61.72)$, $m = \dfrac{1.7904 - 1.3589}{1.3424 - 1.2041} \approx 3$.

b. Since $\log B = 3 \log d + k$,
$B = 10^{3 \log d + k} = 10^{\log d^3} \cdot 10^k = a d^3$.

Thus, the number of board feet, a measure of the volume of wood, varies as the cube of the diameter of the tree.

c. If height is proportional to the radius, then $h = kr$ and $V = \pi r^2 h = k\pi r^3$. Thus the volume is proportional to the radius cubed, which is proportional to the diameter cubed.

20. a.

log D	0.4771	0.9031	1.1139	1.2553	1.3617	1.4472	1.5185
log P	1.8451	1.6335	1.4771	1.3617	1.2788	1.2304	1.1761

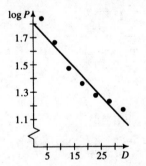
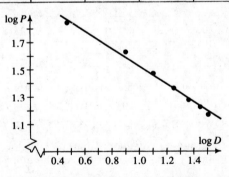

b. $\log P$ versus $\log D$; the equation of the least-squares line is $\log P = -0.6583 \log D + 2.1885$; $P = 10^{-0.6583 \log D + 2.1885} = 10^{\log D^{-0.6583}} \cdot 10^{2.1885}$; $P = (154.3)D^{-0.6583}$.

c. $P = (154.3)45^{-0.6583} = 12.6$; about 13% of such shots succeed.

d. No; very few shots are attempted at such distances; the model is based on data that are mostly at or below 30 ft. It would be impossible to tell whether the model would apply beyond the experimental range.

21.

log x	2.1367	2.1492	2.1673	2.1959
log y	1.7924	1.8129	1.8195	1.8325

The equation of the least-squares line is $\log y = 0.6141 \log x + 0.4865$; $y = 10^{0.6141 \log x + 0.4865} = 10^{\log x^{0.6141}} \cdot 10^{0.4865}$; $y = (3.065)x^{0.6141}$.

22. a. The cost of a pizza depends on its area. Since area varies as the square of the diameter, cost varies as the square of the diameter.

b. $4 = a(100) + k$; $7 = a(225) + k$; $k = 4 - 100a$; $k = 7 - 225a$; $4 - 100a = 7 - 225a$; $125a = 3$; $a = 0.024$; $k = 4 - 100(0.024) = 1.6$; $y = 0.024x^2 + 1.6$

c. cost $= 0.024(144) + 1.6 \approx \$5.06$

23. Answers will vary.

Mixed Exercises 18-4 · pages 705–709

For exercises in section 18-4, equations of least-squares lines are found using a calculator or computer software.

1. a. A power function is used to model weight as a function of length. (See p. 698.)

$\log x$	0.7404	1.0253	1.1761	1.2304	1.2923	1.3424	1.3979	1.4472
$\log y$	-1	-0.3979	0	0.2041	0.3979	0.5441	0.7324	0.8692

The equation of the least-squares line is $\log y = 2.6898 \log x - 3.0765$;
$y = 10^{2.6898 \log x - 3.0765} = 10^{\log x^{2.6898}} \cdot 10^{-3.0765}$; $y = (8.385 \times 10^{-4})x^{2.69}$.

b. weight $= (8.385 \times 10^{-4})12^{2.69} \approx 0.6707$; about 0.7 lb

2. a. Increases in salaries are partially due to inflation; therefore an exponential function model is chosen.

$x = $ no. of years after 1967	0	9	12	15	17	19	22
$\log y$	1.2788	1.7118	2.0569	2.3820	2.5172	2.6160	2.6964

The equation of the least-squares line is $\log y = 0.0706x + 1.2312$;
$y = 10^{0.0706x + 1.2312} = 10^{1.2312}(10^{0.0706})^x$; $y = (17.03)1.177^x$.

b. In 1970, $x = 3$ and $y = 17.03(1.177)^3 \approx 27.77$; about \$27,770.

3. a. Using a graphing calculator to graph (P, T), $(P, \log T)$, and $(\log P, \log T)$ shows that $(\log P, \log T)$ looks the most linear; therefore a power function model is chosen.

$\log P$	2.1732	2.3692	2.5502	2.7210	2.8808	3.0314
$\log T$	1.7782	1.8451	1.9031	1.9542	2	2.0414

The equation of the least-squares line is $\log T = 0.3061 \log P + 1.1182$;
$T = 10^{0.3061 \log P + 1.1182} = 10^{\log P^{0.3061}} \cdot 10^{1.1182}$; $T = 13.13P^{0.3061}$.

b. $T = (13.13)(620)^{0.3061} \approx 93.98$; about 94°C

4. a. Graphing the data shows that the points (d, P) appear to lie on a line. The equation of the least-squares line is $P = 1.47d$.

b. $P = 1.47(13) = 19.11$; 19.11 lb/in.^{2}

5. a. $W = aC^3$; weight depends on volume; the volume varies as the cube of the radius and the radius varies with the circumference.

b. (671 lb, 11.9375 ft); $671 = a(11.9375)^3$; $a \approx 0.394$; $W = 0.394C^3$

c. 18 in. $= 1.5$ ft; $C = \pi(1.5) \approx 4.71$ ft; $W = 0.394(4.71)^3 \approx 41$ lb

d. The equation used was based on only one piece of data.

6. a. Exponential growth is the best model for increasing costs due to inflation.

b.

x = no. of years from 1945	0	10	20	30	40
log C	1.2553	1.4281	1.4983	1.7308	2.0318

The equation of the least-squares line is log $C = 0.0186x + 1.2177$;

$C = 10^{0.0186x + 1.2177} = 10^{1.2177} \cdot (10^{0.0186})^x$; $C = (16.51)1.044^x$.

c. In the year 2000, $x = 55$ and $C = (16.51)1.044^{55} \approx 176.3$. Accuracy of the prediction might be affected by unexpected occurrences such as war, the discovery of a new inexpensive fuel, changes in political alliances, or changes in demographics.

7. a. The Great Depression and World War II

b. The GNP is directly influenced by inflation; therefore an exponential model is chosen.

x = no. of years from 1900	0	20	40	60	80	88
log G	1.2788	1.9638	2	2.7050	3.4365	3.6885

The equation of the least-squares line is log $G = 0.0266x + 1.2343$;

$G = 10^{0.0266x + 1.2343} = 10^{1.2343}(10^{0.0266})^x$; $G = (17.15)1.063^x$.

c. In the year 2000, $x = 100$ and $G = (17.15)1.063^{100} \approx 7719.52$; about 7720

8. The asymptotic behavior of the function suggests the use of an exponential model.

t	0	2	4	6	8	10	12
log $(r - 70)$	1.9031	1.6021	1.3010	0.9031	0.4771	0.3010	0

The equation of the least-squares line is log $(r - 70) = -0.1631t + 1.9056$;

$r - 70 = 10^{-0.1631t + 1.9056} = 10^{1.9056} \cdot (10^{-0.1631})^t$; $r = (80.46)0.6869^t + 70$.

9. a. About 19°C

b.

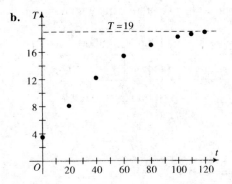

The equation of the horizontal asymptote is $T = 19$.

(Continued on next page)

c.

t	0	20	40	60	80	100	110	120
log $(19 - T)$	1.1903	1.0374	0.8325	0.5563	0.3010	−0.0969	−0.3979	−1

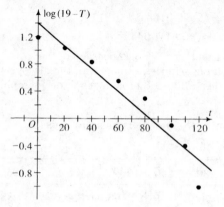

The equation of the least-squares line is
log $(19 - T) = -0.0166t + 1.4037$.

d. $19 - T = 10^{-0.0166t + 1.4037} =$
$10^{1.4037} \cdot (10^{-0.0166})^t = (25.33)0.9625^t;$
$T = 19 - (25.33)0.9625^t$

10. a. \$150

b. The asymptotic behavior of the function suggests the use of an exponential model.

x	50	300	800	2000	3000
log $(y - 150)$	2.5647	2.3747	2	1.0792	0.3010

The equation of the least-squares line is log $(y - 150) = -0.000767x + 2.6074$;
$y - 150 = 10^{-0.000767x + 2.6074} = 10^{2.6074} \cdot (10^{-0.000767})^x = (404.9)0.9982^x;$
$y = (404.9)0.9982^x + 150.$

c. $y = 404.9(0.9982)^{500} + 150 \approx 314.48$; about \$314

11. The asymptotic behavior of the function suggests the use of an exponential model.

x	100	200	400	800	1500	3000
log $(y - 5)$	0.5682	0.5563	0.4314	0.2041	0.0414	−0.1549

The equation of the least-squares line is log $(y - 5) = -0.0002x + 0.5277$;
$y - 5 = 10^{-0.0002x + 0.5277} = 10^{0.5277} \cdot (10^{-0.0002})^x = (3.371)0.9995^x;$
$y = (3.371)0.9995^x + 5.$

12. Answers will vary.

13. a. 5; $(3(9) + 8 + 5) \div 5 = 8$, which is a grade of B

b. x = perceived grade; y = actual grade; the equation of the least-squares line is
$y = 0.9054x + 0.4932$. If the perceived grade is B, then $x = 8$, and the actual
grade = $0.9054(8) + 0.4932 = 7.736$, or about 7.7.

14. Answers will vary. Examples: weight vs. daily calorie intake, income vs. taxes.

Chapter Test · page 710

1. a. $r = \dfrac{\overline{xy} - \overline{x} \cdot \overline{y}}{s_x s_y} = \dfrac{2240 - (55)(40)}{(7)(8)} \approx 0.71$

b. $m = \dfrac{\overline{xy} - \overline{x} \cdot \overline{y}}{s_x^2} = \dfrac{2240 - (55)(40)}{7^2} = 0.8163;\ y - 40 = 0.8163(x - 55);$

$y = 0.82x - 4.9$

c. Since the correlation coefficient is only 0.71, the model probably has limited accuracy in predicting precipitation.

2. No, both scores are probably related to some other factor, perhaps I.Q., rather than to each other.

3. $\log y = 3x + 2;\ y = 10^{3x+2} = (10^3)^x \cdot 10^2;\ y = (100)1000^x$

4. a.

x = no. of years after 1900	35	45	55	65	75	85
$\log y$	2.6758	3.0874	3.2744	3.4429	3.7672	4.1429

The equation of the least-squares line is $\log y = 0.02727x + 1.7624$;

$y = 10^{0.0273x + 1.7624} = 10^{1.7624} \cdot (10^{0.02727})^x;\ y = (57.86)1.065^x.$

b. In 1995, $x = 95$ and income $= 57.86(1.065)^{95} \approx \$22{,}940$.

5. $\log y = 1.7 \log x - 2;\ y = 10^{1.7 \log x - 2} = 10^{\log x^{1.7}} \cdot 10^{-2};\ y = (0.01)x^{1.7}$

6. $m = 2$; using the Pythagorean Theorem, the length of a side of the square rug is $\dfrac{x}{\sqrt{2}}$,

and therefore the area of the rug is $\dfrac{x^2}{2}$. The cost of the rug is likely to vary as its area; since the area varies as the square of the length of the diagonal, $m = 2$ is a reasonable conjecture.

7. a. A power function is used to model the periods of orbiting planets.

$\log d$	7.5563	7.8277	8.1514
$\log y$	1.9445	2.3522	2.837

The equation of the least-squares line is $\log y = 1.4997 \log d - 9.3873$;

$y = 10^{1.4997 \log d - 9.3873} = 10^{\log d^{1.4997}} \cdot 10^{-9.3873};\ y = (4.099 \times 10^{-10})d^{1.5}.$

b. $365 = (4.099 \times 10^{-10})d^{1.5};\ d = \left(\dfrac{365}{4.099 \times 10^{-10}}\right)^{2/3} \approx 92{,}560{,}000$ mi

Cumulative Review · pages 711–712

1. Let F be the set of students who answered the first question correctly, and let S be the set of students who answered the second question correctly. $n(F \cap \overline{S}) = 62 - 37 = 25$; $n(F \cup S) = n(S) + n(F \cap \overline{S}) = 48 + 25 = 73;\ n(\overline{F} \cap \overline{S}) = 100 - 73 = 27.$

2. There are 21 choices for the first letter, 5 choices for the second letter, and 20 choices for the third letter. There are $21 \cdot 5 \cdot 20 = 2100$ different sequences.

3. $_9C_4 = \dfrac{9!}{4!\,5!} = 126$ **4.** $_9P_3 = 9 \cdot 8 \cdot 7 = 504$ **5.** $\dfrac{6!}{2!\,3!} = 60$

6. $\left(2x - \dfrac{1}{2}\right)^4 = (2x)^4 + 4(2x)^3\left(-\dfrac{1}{2}\right) + 6(2x)^2\left(-\dfrac{1}{2}\right)^2 + 4(2x)\left(-\dfrac{1}{2}\right)^3 + \left(-\dfrac{1}{2}\right)^4 =$

$16x^4 - 16x^3 + 6x^2 - x + \dfrac{1}{16}$

7. a. $P(\text{diamond}) = \dfrac{13}{52} = \dfrac{1}{4}$ **b.** $P(\text{not a diamond}) = 1 - \dfrac{1}{4} = \dfrac{3}{4}$ **c.** $P(\text{red club}) = 0$

   d. $P(\text{not a red club}) = 1$ **e.** $P(10) = \dfrac{4}{52} = \dfrac{1}{13}$ **f.** $P(\text{ace or king}) = \dfrac{8}{52} = \dfrac{2}{13}$

8. a. $P(3 \text{ spades}) = \dfrac{{}_{13}C_3}{{}_{52}C_3} = \dfrac{13 \cdot 12 \cdot 11}{52 \cdot 51 \cdot 50} = \dfrac{11}{850} \approx 0.013$

   b. Define the events as follows: A = at least one ace is drawn, B = all cards are black, and C = none of the cards is the ace of spades.

$P(A) = 1 - \dfrac{{}_{48}C_3}{{}_{52}C_3} \approx 1 - 0.783 = 0.217; \; P(B) = \dfrac{{}_{26}C_3}{{}_{52}C_3} \approx 0.118; \; P(C) = \dfrac{{}_{51}C_3}{{}_{52}C_3} \approx 0.942.$

$P(A\,|\,B) = 1 - \dfrac{{}_{24}C_3}{{}_{26}C_3} \approx 1 - 0.778 = 0.222 \neq P(A); \; \therefore A \text{ and } B \text{ are dependent events.}$

$P(A\,|\,C) = 1 - \dfrac{{}_{48}C_3}{{}_{51}C_3} \approx 1 - 0.831 = 0.169 \neq P(A); \; \therefore A \text{ and } C \text{ are dependent events.}$

$P(C\,|\,B) = \dfrac{{}_{25}C_3}{{}_{26}C_3} \approx 0.885 \neq P(C); \; \therefore B \text{ and } C \text{ are dependent events.}$

9. $P(\text{three}) = \dfrac{1}{6}, \; P(\text{not three}) = \dfrac{5}{6}; \; P(\text{exactly 2 threes}) = {}_5C_2 \cdot \left(\dfrac{1}{6}\right)^2\left(\dfrac{5}{6}\right)^3;$

$= 10 \cdot \dfrac{5^3}{6^5} = \dfrac{625}{3888}$

10. $P(\text{diamond}\,|\,\text{diamond}) = \dfrac{12}{51} = \dfrac{4}{17}; \; P(\text{spade}\,|\,\text{diamond}) = \dfrac{13}{51}$

11. Let D stand for a defective output.

   a. $P(D) = 0.60 \times 0.04 + 0.40 \times 0.07 = 0.024 + 0.028 = 0.052; \; 5.2\%$

   b. $P(A\,|\,D) = \dfrac{P(A \cap D)}{P(D)} = \dfrac{\frac{6}{10} \cdot \frac{4}{100}}{\frac{52}{1000}} = \dfrac{24}{1000} \cdot \dfrac{1000}{52} = \dfrac{6}{13}$

12. Expected payoff $= \dfrac{4}{10}(50) + \dfrac{3}{10}(100) + \dfrac{2}{10}(500) + \dfrac{1}{10}(1000) = \$250$

13.

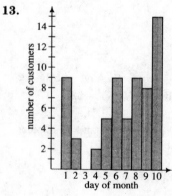

$\text{mean} = \dfrac{9 + 6 + 0 + 8 + 25 + 54 + 35 + 72 + 72 + 150}{9 + 3 + 0 + 2 + 5 + 9 + 5 + 9 + 8 + 15}$

≈ 6.63

$\text{median} = 7$

$\text{mode} = 10$

14. a.

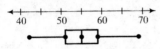

4	2 3
·	6 7 8 9 9
5	0 1 1 1 1 1 2 2 4 4 4 4
·	5 5 5 5 6 6 6 7 7 7 7 8
6	0 1 1 1 2 4 4
·	5 8 9

4|2 represents an age of 42.

b.

$$15.\ \overline{x} = \frac{385}{10} = 38.5; \quad s^2 = \frac{\sum\limits_{i=1}^{n} x_i^2}{n} - \overline{x}^2; \quad s^2 = \frac{25{,}333}{10} - (38.5)^2; \quad s^2 = 1051.05; \quad s \approx 32.4$$

16. a. $z = \dfrac{x - \overline{x}}{s} = \dfrac{18 - 25}{4} \approx -1.8; \quad P(-1.8) = 0.0359;$ about 4%

b. $z = \dfrac{x - \overline{x}}{s} = \dfrac{20 - 25}{4} \approx -1.3; \quad P(-1.3) = 0.0986; \quad 1 - 0.0986 = 0.9014;$ about 90%

c. $z_1 = \dfrac{x - \overline{x}}{s} = \dfrac{22 - 25}{4} \approx -0.8; \quad z_2 = \dfrac{x - \overline{x}}{s} = \dfrac{27 - 25}{4} = 0.5;$

$P(0.5) - P(-0.8) = 0.6915 - 0.2119 = 0.4796;$ about 48%

17. People who are at home during business hours are not representative of the general population. They may be at home with children, retired, or unemployed, conditions that may influence their feelings about taxes. Opinions of those who work during business hours would not be included.

18. $\overline{p} = \dfrac{430}{1000} = 0.43; \quad 3\sqrt{\dfrac{\overline{p}(1 - \overline{p})}{n}} = 3\sqrt{\dfrac{(0.43)(0.57)}{1000}} \approx 0.047;$

$0.430 - 0.047 < p < 0.430 + 0.047; \quad 0.383 < p < 0.477$

19. a. $r = \dfrac{\overline{xy} - \overline{x} \cdot \overline{y}}{s_x s_y} = \dfrac{9.6 - (2)(7.2)}{(1.414)(3.487)} \approx -0.973$

b. $m = \dfrac{\overline{xy} - \overline{x} \cdot \overline{y}}{s_x^2} = \dfrac{9.6 - (2)(7.2)}{2} = -2.4;$ the equation of the

least-squares line is $y - 7.2 = -2.4(x - 2)$ or $y = -2.4x + 12.$

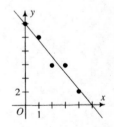

Ex. 19

20. a. Power; the volume of a pipe varies as the square of the diameter.

b. Linear; sales tax is usually a percentage of cost.

c. Asymptotic; the temperature would not exceed the boiling point of water.

21.

$\log x$	0.1761	0.3010	0.3979	0.4771
$\log y$	0.8261	1.2014	1.4942	1.7340

The equation of the least-squares line is $\log y = 3.0163 \log x + 0.2944;$

$y = 10^{3.0163 \log x + 0.2944} = 10^{\log x^{3.0163}} \cdot 10^{0.2944}; \quad y = 1.97x^{3.016}.$

22. a. An exponential model is best for population growth because populations tend to double at regular intervals.

b.

t	0	1	2	3	4	5	6	7
log P	0	0.3010	0.6021	0.9031	1.0792	1.3222	1.5682	1.7853

The equation of the least-squares line is log $P = 0.252t + 0.063$;
$P = 10^{0.252t + 0.063} = 10^{0.063} \cdot (10^{0.252})^t$; $P = (1.156)1.786^t$.

c. $(1.156)1.786^{10} \approx 382$

Class Exercises 19-1 · page 723

1. a. The limit of $\dfrac{2x^2}{x^2 + 1}$ as x approaches infinity is 2. This means that for large values of x, the value of $\dfrac{2x^2}{x^2 + 1}$ is close to 2.

b. The limit of $\dfrac{3x + 1}{2x - 5}$ as x approaches negative infinity is $\dfrac{3}{2}$. This means that for negative values of x where $|x|$ is large, the value of $\dfrac{3x + 1}{2x - 5}$ is close to $\dfrac{3}{2}$.

c. The limit of $f(x)$ as x approaches 2 from the right is 3. This means that for values of x near 2, but greater than 2, the value of $f(x)$ is close to 3.

d. The limit of $g(x)$ as x approaches 2 from the left is infinity. This means that for values of x near 2, but less than 2, the value of $g(x)$ is arbitrarily large.

2. $0, -1, 4, 1$

3. The function is discontinuous at $x = 2$ because $\lim\limits_{x \to 2} f(x) = 4$ and $f(2) = 1$, so $\lim\limits_{x \to 2} f(x) \neq f(2)$.

4. a. $0; -3$ **b.** Because the right-hand and left-hand limits are different.

5. The limit of a quotient of two functions equals the quotient of the limits of the functions.

6. a. $\lim\limits_{x \to \infty} \dfrac{5 - \dfrac{3}{x} + \dfrac{1}{x^2}}{7 + \dfrac{9}{x^2}} = \dfrac{5}{7}$

b. $f(1.1) = 10; f(1.01) = 100; f(1.001) = 1000; \lim\limits_{x \to 1^+} \dfrac{1}{x - 1} = \infty$

c. $f(0.9) = -10; f(0.99) = -100; f(0.999) = -1000; \lim\limits_{x \to 1^-} \dfrac{1}{x - 1} = -\infty$

d. $\lim\limits_{x \to 4} (x + 4) = 8$

Written Exercises 19-1 · pages 724–726

1. $\lim\limits_{x \to \infty} \dfrac{3x - 5}{4x + 9} = \lim\limits_{x \to \infty} \dfrac{3 - \dfrac{5}{x}}{4 + \dfrac{9}{x}} = \dfrac{3 - 0}{4 + 0} = \dfrac{3}{4}$ **2.** $\lim\limits_{x \to \infty} \dfrac{2x^2 - 7x}{3x^2 + 5} = \lim\limits_{x \to \infty} \dfrac{2 - \dfrac{7}{x}}{3 + \dfrac{5}{x^2}} = \dfrac{2 - 0}{3 + 0} = \dfrac{2}{3}$

3. $\lim\limits_{x \to -\infty} \dfrac{8x^2 - 7x + 5}{4x^2 + 9} = \lim\limits_{x \to -\infty} \dfrac{8 - \dfrac{7}{x} + \dfrac{5}{x^2}}{4 + \dfrac{9}{x^2}} = \dfrac{8 - 0 + 0}{4 + 0} = \dfrac{8}{4} = 2$

4. $\lim\limits_{x \to -\infty} \dfrac{5x^3}{7x^3 + 8x^2} = \lim\limits_{x \to -\infty} \dfrac{5}{7 + \dfrac{8}{x}} = \dfrac{5}{7 + 0} = \dfrac{5}{7}$

5. $\lim\limits_{x \to \infty} \dfrac{(x^2 + 1)(x^2 - 1)}{2x^4} = \lim\limits_{x \to \infty} \dfrac{x^4 - 1}{2x^4} = \lim\limits_{x \to \infty} \dfrac{1 - \dfrac{1}{x^4}}{2} = \dfrac{1 - 0}{2} = \dfrac{1}{2}$

6. $\displaystyle\lim_{x\to\infty}\frac{x^2\cos\dfrac{1}{x}}{2x^2-1}=\lim_{x\to\infty}\frac{\cos\dfrac{1}{x}}{2-\dfrac{1}{x^2}}=\frac{\displaystyle\lim_{x\to\infty}\cos\dfrac{1}{x}}{\displaystyle\lim_{x\to\infty}\left(2-\dfrac{1}{x^2}\right)}=\frac{\cos 0}{2-0}=\frac{1}{2}$

7. $\displaystyle\lim_{x\to1}\frac{(x+1)(x-1)}{x-1}=\lim_{x\to1}(x+1)=1+1=2$

8. $\displaystyle\lim_{x\to3}\frac{2x^2-6x}{x-3}=\lim_{x\to3}\frac{2x(x-3)}{x-3}=\lim_{x\to3}2x=2\cdot3=6$

9. $\displaystyle\lim_{x\to2}\frac{x^2+x-6}{2x-4}=\lim_{x\to2}\frac{(x+3)(x-2)}{2(x-2)}=\lim_{x\to2}\frac{x+3}{2}=\frac{2+3}{2}=\frac{5}{2}$

10. $\displaystyle\lim_{x\to4}\frac{x-4}{x^2-x-12}=\lim_{x\to4}\frac{x-4}{(x-4)(x+3)}=\lim_{x\to4}\frac{1}{x+3}=\frac{1}{4+3}=\frac{1}{7}$

11. If $f(x)=\dfrac{x+1}{x}=1+\dfrac{1}{x}$, then $f(0.1)=11, f(0.01)=101,$ and $f(0.001)=1001$;

$\displaystyle\lim_{x\to0^+}\frac{x+1}{x}=\infty.$

12. If $g(x)=\dfrac{x}{x+1}$, then $g(-0.9)=-9, g(-0.99)=-99,$ and $g(-0.999)=-999$;

$\displaystyle\lim_{x\to-1^+}\frac{x}{x+1}=-\infty.$

13. If $h(x)=\dfrac{x-4}{x-3}$, then $h(3.1)=-9, h(3.01)=-99,$ and $h(3.001)=-999$;

$\displaystyle\lim_{x\to3^+}\frac{x-4}{x-3}=-\infty.$

14. $\displaystyle\lim_{x\to3^-}\frac{2x-6}{x^2-3x}=\lim_{x\to3^-}\frac{2(x-3)}{x(x-3)}=\lim_{x\to3^-}\frac{2}{x}=\frac{2}{3}$

15. $\displaystyle\lim_{x\to-2^-}\frac{x^2+4x+4}{x^2+3x+1}=\frac{4-8+4}{4-6+1}=\frac{0}{-1}=0$

16. $\displaystyle\lim_{x\to-2^+}\frac{3x+6}{2x+4}=\lim_{x\to-2^+}\frac{3(x+2)}{2(x+2)}=\lim_{x\to-2^+}\frac{3}{2}=\frac{3}{2}$

17. $\dfrac{x}{1-\sqrt{1-x}}\cdot\dfrac{1+\sqrt{1-x}}{1+\sqrt{1-x}}=\dfrac{x(1+\sqrt{1-x})}{1-(1-x)}=\dfrac{x(1+\sqrt{1-x})}{x}=1+\sqrt{1-x}\,;$

$\displaystyle\lim_{x\to0}\frac{x}{1-\sqrt{1-x}}=\lim_{x\to0}(1+\sqrt{1-x})=1+\sqrt{1}=1+1=2$

18. $\dfrac{2-\sqrt{4-x}}{x}\cdot\dfrac{2+\sqrt{4-x}}{2+\sqrt{4-x}}=\dfrac{4-(4-x)}{x(2+\sqrt{4-x})}=\dfrac{x}{x(2+\sqrt{4-x})}=\dfrac{1}{2+\sqrt{4-x}}\,;$

$\displaystyle\lim_{x\to0}\frac{2-\sqrt{4-x}}{x}=\lim_{x\to0}\frac{1}{2+\sqrt{4-x}}=\frac{1}{2+\sqrt{4}}=\frac{1}{2+2}=\frac{1}{4}$

19. a. When $x>0$, $|x|=x$, so $\displaystyle\lim_{x\to0^+}\frac{|x|}{x}=\lim_{x\to0^+}\frac{x}{x}=1.$

b. When $x<0$, $|x|=-x$, so $\displaystyle\lim_{x\to0^-}\frac{|x|}{x}=\lim_{x\to0^-}\frac{-x}{x}=-1.$

c. Since the right-hand and left-hand limits are different, the limit does not exist.

20. a. If $f(x) = \dfrac{x-3}{x^2-1}$, then $f(1.1) \approx -9$, $f(1.01) \approx -99$, and $f(1.001) \approx -999$;

$$\lim_{x \to 1^+} \frac{x-3}{x^2-1} = -\infty.$$

b. If $f(x) = \dfrac{x-3}{x^2-1}$, then $f(0.9) \approx 11$, $f(0.99) \approx 101$, and $f(0.999) \approx 1001$;

$$\lim_{x \to 1^-} \frac{x-3}{x^2-1} = \infty.$$

c. Since the right-hand and left-hand limits are different, the limit does not exist.

21. a. When $x > 2$, $\dfrac{x-2}{\sqrt{x^2-4}} = \dfrac{x-2}{\sqrt{(x-2)(x+2)}} = \dfrac{\sqrt{x-2}}{\sqrt{x+2}}$; $\displaystyle\lim_{x \to 2^+} \dfrac{x-2}{\sqrt{x^2-4}} = \lim_{x \to 2^+} \dfrac{\sqrt{x-2}}{\sqrt{x+2}} = \dfrac{\sqrt{0}}{\sqrt{4}} = 0.$

b. When $-2 < x < 2$, $\sqrt{x^2-4}$ is not defined, so $\displaystyle\lim_{x \to 2^-} \dfrac{x-2}{\sqrt{x^2-4}}$ does not exist.

c. Since the limit as x approaches 2 from the left does not exist, the limit does not exist.

22. a. When $3 < x < 4$, $\lfloor x \rfloor = 3$; $\displaystyle\lim_{x \to 3^+} \lfloor x \rfloor = 3.$

b. When $2 < x < 3$, $\lfloor x \rfloor = 2$; $\displaystyle\lim_{x \to 3^-} \lfloor x \rfloor = 2.$

c. Since the right-hand and left-hand limits are different, the limit does not exist.

23. a. 1 **b.** 0 **c.** ∞ **d.** $-\infty$ **24. a.** ∞ **b.** 0 **c.** $-\infty$ **d.** ∞

25. $\dfrac{1 - \sqrt{x^2+1}}{x^2} \cdot \dfrac{1 + \sqrt{x^2+1}}{1 + \sqrt{x^2+1}} = \dfrac{1 - (x^2+1)}{x^2(1 + \sqrt{x^2+1})} = \dfrac{-x^2}{x^2(1 + \sqrt{x^2+1})} = -\dfrac{1}{1 + \sqrt{x^2+1}}$;

$$\lim_{x \to 0} \frac{1 - \sqrt{x^2+1}}{x^2} = \lim_{x \to 0} -\frac{1}{1 + \sqrt{x^2+1}} = -\frac{1}{1 + \sqrt{1}} = -\frac{1}{2}.$$

26. $\sqrt{4x^2 - 4x} - 2x = \sqrt{4x^2}\sqrt{1 - \dfrac{1}{x}} - 2x = -2x\sqrt{1 - \dfrac{1}{x}} - 2x$ (when $x < 0$) $=$

$-2x\left(\sqrt{1 - \dfrac{1}{x}} + 1\right)$; as $x \to -\infty$, $\sqrt{1 - \dfrac{1}{x}} \to 1$, $-2\left(\sqrt{1 - \dfrac{1}{x}} + 1\right) \to -4$, and

$-2x\left(\sqrt{1 - \dfrac{1}{x}} + 1\right) \to \infty$; thus, $\displaystyle\lim_{x \to -\infty}(\sqrt{4x^2 - 4x} - 2x) = \infty.$

27. $(\sqrt{x+1} - \sqrt{x}) \cdot \dfrac{\sqrt{x+1} + \sqrt{x}}{\sqrt{x+1} + \sqrt{x}} = \dfrac{x+1-x}{\sqrt{x+1} + \sqrt{x}} = \dfrac{1}{\sqrt{x+1} + \sqrt{x}}$; as $x \to \infty$,

$\sqrt{x+1} \to \infty$, $\sqrt{x} \to \infty$, $\sqrt{x+1} + \sqrt{x} \to \infty$, and $\dfrac{1}{\sqrt{x+1} + \sqrt{x}} \to 0$; thus,

$\displaystyle\lim_{x \to \infty}(\sqrt{x+1} - \sqrt{x}) = 0.$

28. $(\sqrt{x^2 + 2x} - x) \cdot \dfrac{\sqrt{x^2 + 2x} + x}{\sqrt{x^2 + 2x} + x} = \dfrac{x^2 + 2x - x^2}{\sqrt{x^2 + 2x} + x} = \dfrac{2x}{\sqrt{x^2 + 2x} + x} =$

$\dfrac{2x}{\sqrt{x^2}\sqrt{1 + \dfrac{2}{x}} + x} = \dfrac{2x}{x\sqrt{1 + \dfrac{2}{x}} + x}$ (when $x > 0$) $= \dfrac{2x}{x\left(\sqrt{1 + \dfrac{2}{x}} + 1\right)} = \dfrac{2}{\sqrt{1 + \dfrac{2}{x}} + 1}$;

thus, $\displaystyle\lim_{x \to \infty}(\sqrt{x^2 + 2x} - x) = \lim_{x \to \infty} \dfrac{2}{\sqrt{1 + \dfrac{2}{x}} + 1} = \dfrac{2}{\sqrt{1+0} + 1} = \dfrac{2}{2} = 1.$

29. a. $\dfrac{(1 + h)^2 - 1}{h} = \dfrac{1 + 2h + h^2 - 1}{h} = \dfrac{h^2 + 2h}{h} = h + 2;\ \lim\limits_{h \to 0} \dfrac{(1 + h)^2 - 1}{h} = \lim\limits_{h \to 0} (h + 2) =$
$0 + 2 = 2$

b. $\dfrac{(2 + h)^2 - 4}{h} = \dfrac{4 + 4h + h^2 - 4}{h} = \dfrac{h^2 + 4h}{h} = h + 4;\ \lim\limits_{h \to 0} \dfrac{(2 + h)^2 - 4}{h} = \lim\limits_{h \to 0} (h + 4) =$
$0 + 4 = 4$

c. $\dfrac{(x + h)^2 - x^2}{h} = \dfrac{2xh + h^2}{h} = 2x + h;\ \lim\limits_{h \to 0} \dfrac{(x + h)^2 - x^2}{h} = \lim\limits_{h \to 0} (2x + h) = 2x$

30. a. $\lim\limits_{h \to 0} \dfrac{(1 + h)^3 - 1}{h} = \lim\limits_{h \to 0} \dfrac{1 + 3h + 3h^2 + h^3 - 1}{h} = \lim\limits_{h \to 0} (3 + 3h + h^2) = 3 + 0 + 0 = 3$

b. $\lim\limits_{h \to 0} \dfrac{(2 + h)^3 - 8}{h} = \lim\limits_{h \to 0} \dfrac{8 + 12h + 6h^2 + h^3 - 8}{h} = \lim\limits_{h \to 0} (12 + 6h + h^2) =$
$12 + 0 + 0 = 12$

c. $\lim\limits_{h \to 0} \dfrac{(x + h)^3 - x^3}{h} = \lim\limits_{h \to 0} \dfrac{x^3 + 3x^2h + 3xh^2 + h^3 - x^3}{h} = \lim\limits_{h \to 0} (3x^2 + 3xh + h^2) =$
$3x^2 + 0 + 0 = 3x^2$

31. $\lim\limits_{x \to 1^-} f(x) = \lim\limits_{x \to 1^-} (2 - x^2) = 1;\ \lim\limits_{x \to 1^+} f(x) = \lim\limits_{x \to 1^+} x = 1;$ so $\lim\limits_{x \to 1} f(x) = 1 = f(1);$ thus,
f is continuous at $x = 1$. Since f is defined by polynomials elsewhere, f is continuous for all x.

32. Since f is defined by polynomials, it is continuous for all x except possibly $x = 0$.
$\lim\limits_{x \to 0^-} f(x) = \lim\limits_{x \to 0^-} (x^2 + 1) = 1;\ \lim\limits_{x \to 0^+} f(x) = \lim\limits_{x \to 0^+} x^2 = 0;$ so $\lim\limits_{x \to 0} f(x)$ does not exist;
thus, f is discontinuous at $x = 0$.

33. The only possible discontinuity is at $x = 2$ (see Ex. 32). $\lim\limits_{x \to 2} f(x) = \lim\limits_{x \to 2} \dfrac{x^2 - 4}{x - 2} =$
$\lim\limits_{x \to 2} (x + 2) = 4;\ f(2) = 4;$ so f is continuous at $x = 4$. f is continuous for all x.

34. Since f is not defined at $x = 1$, f is discontinuous at $x = 1$.

35. $f(-1) = 1;\ \lim\limits_{x \to -1^-} f(x) = \lim\limits_{x \to -1^-} x^2 = 1 = \lim\limits_{x \to -1^+} (ax + b) = -a + b;\ -a + b = 1.$
$f(1) = -1;\ \lim\limits_{x \to 1^-} f(x) = \lim\limits_{x \to 1^-} (ax + b) = a + b = \lim\limits_{x \to 1^+} -x^2 = -1;\ a + b = -1.$
$2b = 0;\ b = 0,\ a = -1.$

36. $f(-2) = 5;\ \lim\limits_{x \to -2^-} f(x) = \lim\limits_{x \to -2^-} (1 - 2x) = 5 = \lim\limits_{x \to -2^+} (ax + b) = -2a + b;$
$-2a + b = 5.\ f(1) = 5;\ \lim\limits_{x \to 1^-} f(x) = \lim\limits_{x \to 1^-} (ax + b) = a + b = \lim\limits_{x \to 1^+} (3x + 2) = 5;$
$a + b = 5.\ 3a = 0;\ a = 0,\ b = 5.$

37. As $x \to \infty,\ \dfrac{1}{x} \to 0,$ so $\lim\limits_{x \to \infty} (43{,}987)^{1/x} = 43{,}987^0 = 1.$

38. As $x \to 0^+,\ \dfrac{1}{x} \to \infty$ and so $2^{1/x} \to \infty;$ as $x \to 0^-,\ \dfrac{1}{x} \to -\infty$ and so $2^{1/x} \to 0;$ since
$\lim\limits_{x \to 0^+} 2^{1/x} \neq \lim\limits_{x \to 0^-} 2^{1/x},\ \lim\limits_{x \to 0} 2^{1/x}$ does not exist.

39. $\lim\limits_{x \to 0} \dfrac{3^x - 3^{-x}}{3^x + 3^{-x}} = \dfrac{\lim\limits_{x \to 0} (3^x - 3^{-x})}{\lim\limits_{x \to 0} (3^x + 3^{-x})} = \dfrac{3^0 - 3^0}{3^0 + 3^0} = \dfrac{1 - 1}{1 + 1} = \dfrac{0}{2} = 0$

40. $\lim\limits_{x \to -\infty} \dfrac{3^x - 3^{-x}}{3^x + 3^{-x}} = \lim\limits_{x \to -\infty} \left(\dfrac{3^x - 3^{-x}}{3^x + 3^{-x}} \cdot \dfrac{3^x}{3^x}\right) = \lim\limits_{x \to -\infty} \dfrac{3^{2x} - 1}{3^{2x} + 1} = \dfrac{0 - 1}{0 + 1} = -1$

41. As $x \longrightarrow 0^+$, $\dfrac{1}{x} \longrightarrow \infty$, and $\sin \dfrac{1}{x}$ oscillates between -1 and 1; since $\sin \dfrac{1}{x}$ has no single

limiting value as $x \longrightarrow 0^+$, the limit does not exist.

42. $\displaystyle\lim_{x \to \infty} \sin \dfrac{1}{x} = \sin 0 = 0$

43. a. The radian measure θ of a central angle is given by $\theta = \dfrac{s}{r}$, so $s = r\theta$; length of

$\overparen{PQ} = 1 \cdot \theta = \theta$.

b. As $\theta \longrightarrow 0$, the length of $\overline{PN}$ approaches the length of $\overparen{PQ}$, so the ratio of their

lengths approaches 1; $PN = \dfrac{\sin \theta}{1} = \sin \theta$ and the length of $\overparen{PQ} = \theta$ (from part (a)),

so $\displaystyle\lim_{\theta \to 0} \dfrac{\sin \theta}{\theta} = 1$.

44. $\displaystyle\lim_{\theta \to 0} \dfrac{\tan \theta}{\theta} = \lim_{\theta \to 0} \left(\dfrac{\sin \theta}{\cos \theta} \cdot \dfrac{1}{\theta} \right) = \lim_{\theta \to 0} \left(\dfrac{\sin \theta}{\theta} \cdot \dfrac{1}{\cos \theta} \right) = 1 \cdot \dfrac{1}{\cos 0} = 1 \cdot 1 = 1$

45. $\displaystyle\lim_{\theta \to 0} \dfrac{\sin 2\theta}{\theta} = \lim_{\theta \to 0} \dfrac{2 \sin \theta \cos \theta}{\theta} = \lim_{\theta \to 0} \left(2 \cdot \dfrac{\sin \theta}{\theta} \cdot \cos \theta \right) = 2 \cdot 1 \cdot \cos 0 = 2 \cdot 1 = 2$

46. $\displaystyle\lim_{t \to 0} \dfrac{1 - \cos t}{t^2} = \lim_{t \to 0} \dfrac{1 - \cos t}{t^2} \cdot \dfrac{1 + \cos t}{1 + \cos t} = \lim_{t \to 0} \dfrac{1 - \cos^2 t}{t^2(1 + \cos t)} = \lim_{t \to 0} \left(\dfrac{\sin^2 t}{t^2} \cdot \dfrac{1}{1 + \cos t} \right) =$

$\displaystyle\lim_{t \to 0} \left[\left(\dfrac{\sin t}{t} \right)^2 \cdot \dfrac{1}{1 + \cos t} \right] = 1^2 \cdot \dfrac{1}{1 + 1} = \dfrac{1}{2}$.

47. Let $t = \dfrac{1}{x}$. As $x \longrightarrow \infty$, $t \longrightarrow 0$; thus, $\displaystyle\lim_{x \to \infty} x \sin \dfrac{1}{x} = \lim_{t \to 0} \left(\dfrac{1}{t} \cdot \sin t \right) = \lim_{t \to 0} \dfrac{\sin t}{t} = 1$.

Activity · page 726

a.

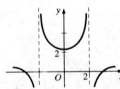

b. x-intercepts ± 3; y-intercept $\dfrac{9}{4}$.

c. To find the x-intercepts algebraically, solve $x^2 - 9 = 0$.
To find the y-intercepts, evaluate the expression for
$x = 0$.

d. $x = \pm 2$ are vertical asymptotes; $y = 1$ is a horizontal asymptote.

e. To find equations for the vertical asymptotes, solve $x^2 - 4 = 0$. To find the
horizontal asymptotes, evaluate $\displaystyle\lim_{x \to \pm\infty} \left(\dfrac{x^2 - 9}{x^2 - 4} \right)$.

Class Exercises 19-2 · page 728

1. x-intercept at 2; vertical asymptotes at $x = 1$ and $x = 3$; horizontal asymptote at $y = 0$

2. x-intercept at 0; vertical asymptotes at $x = \pm 2$; horizontal asymptote at $y = 0$

3. x-intercepts at 2 and -2; vertical asymptote at $x = 0$; no horizontal asymptotes

4. x-intercepts at 3 and -3; no vertical asymptotes; horizontal asymptote at $y = 1$

5. x-intercept at 0; vertical asymptotes at $x = \pm\sqrt{6}$; horizontal asymptote at $y = 2$

6. x-intercept at 0; vertical asymptotes at $x = 1$ and $x = -1$; no horizontal asymptotes

7. If c is a zero of greater multiplicity for g than for f, then there is a vertical asymptote at
$x = c$; otherwise, there is a "hole" in the graph at $x = c$.

Written Exercises 19-2 · pages 728–729

1.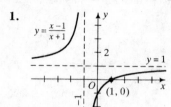
$y = \dfrac{x-1}{x+1}$

2.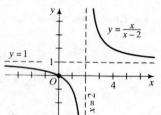
$y = \dfrac{x}{x-2}$

3.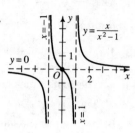
$y = \dfrac{x}{x^2-1}$

4.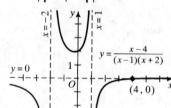
$y = \dfrac{2x^2}{x^2-9}$

5.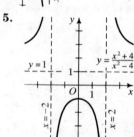
$y = \dfrac{x^2+4}{x^2-4}$

6.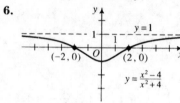
$y = \dfrac{x^2-4}{x^2+4}$

7.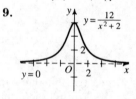
$y = \dfrac{x-4}{(x-1)(x+2)}$

8.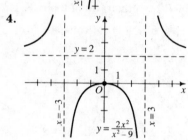
$y = \dfrac{x+3}{(x+1)(x-3)}$

9.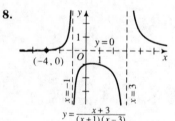
$y = \dfrac{12}{x^2+2}$

10.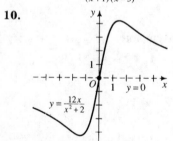
$y = \dfrac{12x}{x^2+2}$

11.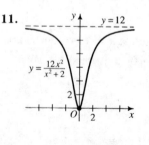
$y = \dfrac{12x^2}{x^2+2}$

12.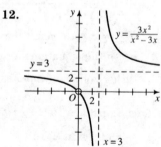
$y = \dfrac{3x^2}{x^2-3x}$

13.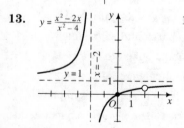
$y = \dfrac{x^2-2x}{x^2-4}$

14.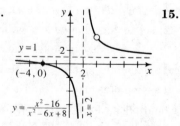
$y = \dfrac{x^2-16}{x^2-6x+8}$

15.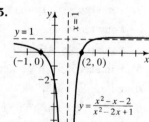
$y = \dfrac{x^2-x-2}{x^2-2x+1}$

16. As $x \longrightarrow \pm\infty$, $\dfrac{x^3}{x^2+1} \longrightarrow \dfrac{x^3}{x^2} = x$; thus, $\displaystyle\lim_{x \to \pm\infty} \dfrac{x^3}{x^2+1} = x$ and $y = x$ is an asymptote of

the graph.

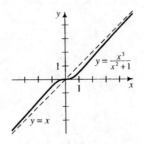

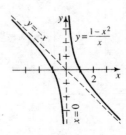

Ex. 16   **Ex. 17**

17. As $x \longrightarrow \pm\infty$, $\dfrac{1-x^2}{x} \longrightarrow \dfrac{-x^2}{x} = -x$; thus, $\displaystyle\lim_{x \to \pm\infty} \dfrac{1-x^2}{x} = -x$, and $y = -x$ is an asymptote

of the graph.

18. a. the nonzero real numbers

   b. zeros at $x = n\pi$ with n a nonzero
integer

   c. As $x \longrightarrow \infty$, $\sin x$ oscillates between
-1 and 1, so $\displaystyle\lim_{x \to \infty} \dfrac{\sin x}{x} = 0$. From
Exercise 43, page 725, we know that
$\displaystyle\lim_{x \to 0} \dfrac{\sin x}{x} = 1$.

d.

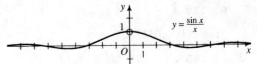

Activity 19-3 · page 730

Change lines 20 and 40:

```
20 FOR X = .001 TO 1.570 STEP .001
40 LET A = A + .001 * Y
```

Area ≈ 0.999499996

Class Exercises 19-3 · page 731

 1. Change lines 20 and 30:

```
20 FOR X = .1 TO 3.1 STEP .1
30 LET Y = SIN(X)
```

 2. Use line 20 in Ex. 1, but change line 30:

```
30 LET Y = 2 * SIN(X)
```

3. a.

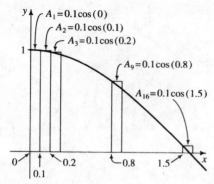

b. Approximation from part (a) is greater than actual area, because each rectangle contains area not under the curve.

Written Exercises 19-3 · pages 731–733

For Exs. 1–4, change lines 20, 30, and 40 as indicated.

1. 20 FOR X = .01 TO .79 STEP .01
30 LET Y = TAN(X)
40 LET A = A + Y * .01

Area ≈ 0.346158926

2. 20 FOR X = .01 TO 3.14 STEP .01
30 LET Y = SIN(X)
40 LET A = A + Y * .01

Area ≈ 1.99999006

3. 20 FOR X = .01 TO 1 STEP .01
30 LET Y = X ∧ 2
40 LET A = A + Y * .01

Area ≈ 0.33835001

4. 20 FOR X = .01 TO 1 STEP .01
30 LET Y = X ∧ 3
40 LET A = A + Y * .01

Area ≈ 0.255025001

5. a. $\frac{1}{2}$, because we are finding the area of a right triangle with $b = h = 1$:
$$A = \frac{1}{2}(1)(1) = \frac{1}{2}.$$

b. For $n = 4$, area $= \frac{1}{5}$. Using a computer and a step size of 0.01, we find that the area is approximately 0.205033334.

6. a. Using a computer and a step size of 0.01, we get: area of $A \approx 0.690653435$, area of $B \approx 1.09528637$, and area of $C \approx 1.78760092$. These areas are good approximations of ln 2, ln 3, and ln 6, respectively.

b. area of A + area of B = area of C; the area under the curve $y = \frac{1}{x}$ from $x = 1$ to $x = a$ is ln a.

7. 10 LET A = 0
20 LET X = 1.01
30 LET Y = 1/X
40 LET A = A + Y * .01
50 IF ABS(A – 1) < .005 THEN 80
60 LET X = X + .01
70 GOTO 30
80 PRINT "X = "; X
90 END

Running the program above gives $x \approx 2.72$. (The exact value of x is e.)

8. Using a computer and a step size of 0.001, we find that the area is approximately 3.1405925. The exact area is π.

9. 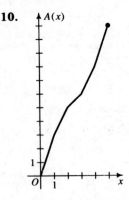   10.

11. 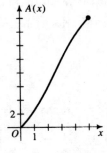   12.

13. $A = 2$; the graph of $y = \cos x$ is symmetric in the y-axis, so the area from $x = -\dfrac{\pi}{2}$ to $x = 0$ equals the area from $x = 0$ to $x = \dfrac{\pi}{2}$.

14. $A = 4$; the graphs of $y = \cos x$ and $y = -\cos x$ are symmetric with respect to each other in the x-axis, so the area between the x-axis and the graph of $y = -\cos x$ equals the area between the graph of $y = \cos x$ and the x-axis.

15. $A = 2$; the horizontal values remain the same while the vertical values are doubled, so the area is doubled.

16. $A = 2$; the sine curve is a horizontal translation of the cosine curve, so the area under $y = \sin x$ from $x = 0$ to $x = \pi$ is the same as the area under $y = \cos x$ from $x = -\dfrac{\pi}{2}$ to $x = \dfrac{\pi}{2}$ (see Ex. 13).

17. $A = 2$; the graph of $y = |\cos x|$ is symmetric in the vertical line $x = \dfrac{\pi}{2}$, so the area from $x = 0$ to $x = \dfrac{\pi}{2}$ equals the area from $x = \dfrac{\pi}{2}$ to $x = \pi$.

18. a. Students will predict either that $A = 0$ or that $A = 2$.

    b. Students should change line 20 to:

        `20 FOR X = .1 TO 3.1 STEP .1`

    If students do not change line 40, the computer gives an approximate area of 0. If students change line 40 to:

        `40 LET A = A + ABS(Y) * .1`

    the computer gives an approximate area of 2.

    c. Students should recognize that the "heights" of the rectangles used to approximate the area between the x-axis and $y = \cos x$ from $x = \dfrac{\pi}{2}$ to $x = \pi$ are negative. Thus, the "area" from $x = \dfrac{\pi}{2}$ to $x = \pi$ is the opposite of the area from $x = 0$ to $x = \pi$, so that the sum of the areas is 0. (This is the result that integral calculus would give.) However, if students deliberately make the area from $x = \dfrac{\pi}{2}$ to $x = \pi$ positive, then the sum of the areas is 2.

Class Exercises 19-4 · page 735

1. 6 terms, out to $\dfrac{x^{11}}{11!}$; the power series and sine graphs nearly coincide at $x = 4$.

2. $-\sin x$; in other words, sine is an odd function.

Written Exercises 19-4 · pages 735–736

1. $\cos \dfrac{\pi}{4} = 1 - \dfrac{\pi^2}{4^2 \cdot 2!} + \dfrac{\pi^4}{4^4 \cdot 4!} - \dfrac{\pi^6}{4^6 \cdot 6!} + \cdots$; $\sin\left(-\dfrac{\pi}{4}\right) = \dfrac{-\pi}{4 \cdot 1!} - \dfrac{(-\pi)^3}{4^3 \cdot 3!} + \dfrac{(-\pi)^5}{4^5 \cdot 5!} -$ $\dfrac{(-\pi)^7}{4^7 \cdot 7!} + \cdots = -\dfrac{\pi}{4 \cdot 1!} + \dfrac{\pi^3}{4^3 \cdot 3!} - \dfrac{\pi^5}{4^5 \cdot 5!} + \dfrac{\pi^7}{4^7 \cdot 7!} - \cdots$

2. $e = 1 + \dfrac{1}{1!} + \dfrac{1}{2!} + \dfrac{1}{3!} + \cdots$; $e^{-1} = 1 - \dfrac{1}{1!} + \dfrac{1}{2!} - \dfrac{1}{3!} + \cdots$

3. a. $\ln 2 = \ln(1 + 1) = 1 - \dfrac{1}{2} + \dfrac{1}{3} - \dfrac{1}{4} + \cdots$

    b. The series is defined only when $-1 < x \le 1$.

4. $\dfrac{\pi}{4} = \operatorname{Tan}^{-1} 1 = 1 - \dfrac{1}{3} + \dfrac{1}{5} - \dfrac{1}{7} + \cdots$

5. $e^{-x} = 1 + \dfrac{(-x)}{1!} + \dfrac{(-x)^2}{2!} + \dfrac{(-x)^3}{3!} + \cdots = 1 - \dfrac{x}{1!} + \dfrac{x^2}{2!} - \dfrac{x^3}{3!} + \cdots$, all real x

6. $e^{-x^2} = 1 + \dfrac{(-x^2)}{1!} + \dfrac{(-x^2)^2}{2!} + \dfrac{(-x^2)^3}{3!} + \cdots = 1 - \dfrac{x^2}{1!} + \dfrac{x^4}{2!} - \dfrac{x^6}{3!} + \cdots$, all real x

7. $\operatorname{Tan}^{-1} 2x = 2x - \dfrac{(2x)^3}{3} + \dfrac{(2x)^5}{5} - \dfrac{(2x)^7}{7} + \cdots$, $-\dfrac{1}{2} \le x \le \dfrac{1}{2}$

8. $\ln(1 - x) = \ln[1 + (-x)] = -x - \dfrac{(-x)^2}{2} + \dfrac{(-x)^3}{3} - \dfrac{(-x)^4}{4} + \cdots =$ $-x - \dfrac{x^2}{2} - \dfrac{x^3}{3} - \dfrac{x^4}{4} - \cdots$, $-1 \le x < 1$

9. $\sin x^2 = x^2 - \dfrac{(x^2)^3}{3!} + \dfrac{(x^2)^5}{5!} - \dfrac{(x^2)^7}{7!} + \cdots$, all real x

10. $\cos 2x = 1 - \dfrac{(2x)^2}{2!} + \dfrac{(2x)^4}{4!} - \dfrac{(2x)^6}{6!} + \cdots$, all real x

11. $\displaystyle\lim_{x \to 0} \frac{\sin x}{x} = \lim_{x \to 0} \frac{1}{x}\left(x - \frac{x^3}{3!} + \frac{x^5}{5!} - \frac{x^7}{7!} + \cdots\right) = \lim_{x \to 0}\left(1 - \frac{x^2}{3!} + \frac{x^4}{5!} - \frac{x^6}{7!} + \cdots\right) =$

$1 - 0 + 0 - 0 + \cdots = 1$

12. $\displaystyle\lim_{x \to 0} \frac{1 - \cos x}{x^2} = \lim_{x \to 0} \frac{1}{x^2}\left[1 - \left(1 - \frac{x^2}{2!} + \frac{x^4}{4!} - \frac{x^6}{6!} + \cdots\right)\right] =$

$\displaystyle\lim_{x \to 0} \frac{1}{x^2}\left[\frac{x^2}{2!} - \frac{x^4}{4!} + \frac{x^6}{6!} - \cdots\right] = \lim_{x \to 0}\left(\frac{1}{2!} - \frac{x^2}{4!} + \frac{x^4}{6!} - \cdots\right) = \frac{1}{2 \cdot 1} - 0 + 0 - \cdots = \frac{1}{2}$

13. $\displaystyle\lim_{x \to 0} \frac{\ln(1 + x)}{x} = \lim_{x \to 0} \frac{1}{x}\left(x - \frac{x^2}{2} + \frac{x^3}{3} - \frac{x^4}{4} + \cdots\right) = \lim_{x \to 0}\left(1 - \frac{x}{2} + \frac{x^2}{3} - \frac{x^3}{4} + \cdots\right) =$

$1 - 0 + 0 - 0 + \cdots = 1$

14. $e^{i\theta} = 1 + \dfrac{i\theta}{1!} + \dfrac{(i\theta)^2}{2!} + \dfrac{(i\theta)^3}{3!} + \cdots = 1 + \dfrac{i\theta}{1!} - \dfrac{\theta^2}{2!} - \dfrac{i\theta^3}{3!} + \dfrac{\theta^4}{4!} + \dfrac{i\theta^5}{5!} - \cdots =$

$\left(1 - \dfrac{\theta^2}{2!} + \dfrac{\theta^4}{4!} - \dfrac{\theta^6}{6!} + \cdots\right) + i\left(\dfrac{\theta}{1!} - \dfrac{\theta^3}{3!} + \dfrac{\theta^5}{5!} - \dfrac{\theta^7}{7!} + \cdots\right) = \cos \theta + i \sin \theta$

15. $e^{i\theta} = \cos \theta + i \sin \theta$; $e^{i\pi} = \cos \pi + i \sin \pi = -1 + i \cdot 0 = -1$;
$e^{2i\pi} = \cos 2\pi + i \sin 2\pi = 1 + i \cdot 0 = 1$

16. $e^{i\theta} = \cos \theta + i \sin \theta$; $e^{-i\theta} = \cos(-\theta) + i \sin(-\theta) = \cos \theta - i \sin \theta$;

thus, $\dfrac{e^{i\theta} + e^{-i\theta}}{2} = \dfrac{1}{2}[(\cos \theta + i \sin \theta) + (\cos \theta - i \sin \theta)] = \dfrac{1}{2}(2 \cos \theta) = \cos \theta$

17. $e^{i\theta} = \cos \theta + i \sin \theta$; $e^{-i\theta} = \cos(-\theta) + i \sin(-\theta) = \cos \theta - i \sin \theta$;

thus, $\dfrac{e^{i\theta} - e^{-i\theta}}{2i} = \dfrac{1}{2i}[(\cos \theta + i \sin \theta) - (\cos \theta - i \sin \theta)] = \dfrac{1}{2i}(2i \sin \theta) = \sin \theta$

18. Substituting, $(\sin \theta)^2 + (\cos \theta)^2 = \left[\dfrac{e^{i\theta} - e^{-i\theta}}{2i}\right]^2 + \left[\dfrac{e^{i\theta} + e^{-i\theta}}{2}\right]^2 =$

$\dfrac{1}{4i^2}(e^{2i\theta} - 2e^0 + e^{-2i\theta}) + \dfrac{1}{4}(e^{2i\theta} + 2e^0 + e^{-2i\theta}) = -\dfrac{1}{4}e^{2i\theta} + \dfrac{1}{2} - \dfrac{1}{4}e^{-2i\theta} + \dfrac{1}{4}e^{2i\theta} +$

$\dfrac{1}{2} + \dfrac{1}{4}e^{-2i\theta} = \dfrac{1}{2} + \dfrac{1}{2} = 1$

19. $1 + \dfrac{1}{2} + \dfrac{1}{3} + \cdots + \dfrac{1}{n} + \cdots = 1 + \left(\dfrac{1}{2}\right) + \left(\dfrac{1}{3} + \dfrac{1}{4}\right) + \left(\dfrac{1}{5} + \dfrac{1}{6} + \dfrac{1}{7} + \dfrac{1}{8}\right) +$

$\left(\dfrac{1}{9} + \cdots + \dfrac{1}{16}\right) + \cdots + \left(\dfrac{1}{2^{n-1} + 1} + \cdots + \dfrac{1}{2^n}\right) + \cdots$; within the nth pair of parentheses,

there are 2^{n-1} terms; $\dfrac{1}{2^{n-1} + 1} > \dfrac{1}{2^n}$, $\dfrac{1}{2^{n-1} + 2} > \dfrac{1}{2^n}$, and so on; thus, $\dfrac{1}{2^{n-1} + 1} + \dfrac{1}{2^{n-1} + 2} +$

$\cdots + \dfrac{1}{2^n} > \dfrac{1}{2^n} + \dfrac{1}{2^n} + \cdots + \dfrac{1}{2^n}$ with 2^{n-1} addends $= 2^{n-1} \cdot \dfrac{1}{2^n} = \dfrac{1}{2}$. Since each group of

terms is $\dfrac{1}{2}$ or more, $\displaystyle\sum_{n=1}^{\infty} \frac{1}{n} > \sum_{n=1}^{\infty} \frac{1}{2}$; since the latter series becomes arbitrarily large as

n becomes large, the series diverges; the harmonic series has partial sums that exceed those of the divergent series, so it, too, is divergent.

20. $\dfrac{1}{1^p} + \dfrac{1}{2^p} + \dfrac{1}{3^p} + \cdots + \dfrac{1}{n^p} + \cdots = \dfrac{1}{1^p} + \left(\dfrac{1}{2^p} + \dfrac{1}{3^p}\right) + \left(\dfrac{1}{4^p} + \dfrac{1}{5^p} + \dfrac{1}{6^p} + \dfrac{1}{7^p}\right) + \cdots < \dfrac{1}{1^p} +$

$\left(\dfrac{1}{2^p} + \dfrac{1}{2^p}\right) + \left(\dfrac{1}{4^p} + \dfrac{1}{4^p} + \dfrac{1}{4^p} + \dfrac{1}{4^p}\right) + \left(\dfrac{1}{8^p} + \cdots + \dfrac{1}{8^p}\right) + \cdots = 1 + \dfrac{2}{2^p} + \dfrac{4}{4^p} + \dfrac{8}{8^p} + \cdots =$

$1 + 2^{1-p} + 4^{1-p} + 8^{1-p} + \cdots = \displaystyle\sum_{n=0}^{\infty} (2^{1-p})^n$, which is a geometric series with common

ratio $r = 2^{1-p}$; $|r| < 1$ if $p > 1$; thus $\displaystyle\sum_{n=0}^{\infty} (2^{1-p})^n$ converges, and therefore

$\displaystyle\sum_{n=1}^{\infty} \dfrac{1}{n^p} < \sum_{n=0}^{\infty} (2^{1-p})^n$ converges, if $p > 1$.

21. If $p < 1$, then $\dfrac{1}{1^p} + \dfrac{1}{2^p} + \dfrac{1}{3^p} + \dfrac{1}{4^p} + \cdots > \dfrac{1}{1} + \dfrac{1}{2} + \dfrac{1}{3} + \dfrac{1}{4} + \cdots$, which diverges (see
   Ex. 19).

Activity 1   19-5 · page 737

a. Enter 10 and repeatedly press the square root key. The limit of the orbit is 1.

b. Enter 0.1 and repeatedly press the square root key. The limit of the orbit is 1.

Activity 2   19-5 · page 739

The web diagram at the bottom of page 739 shows that $\lim_{n \to \infty} f^n(x_0) = 1$ for a seed x_0 where
$0 < x_0 < 1$. The choice of seed does not affect the analysis.

Class Exercises 19-5 · page 741

1. $x_0 = 5$, $x_1 = 5 + 2 = 7$, $x_2 = 7 + 2 = 9$, $x_3 = 9 + 2 = 11$

2. $x_0 = 0$, $x_1 = 1 - 0 = 1$, $x_2 = 1 - 1 = 0$, $x_3 = 1 - 0 = 1$

3. $x_0 = -1$, $x_1 = 3(-1) + 2 = -1$, $x_2 = 3(-1) + 2 = -1$, $x_3 = 3(-1) + 2 = -1$

4. $x_0 = 1$, $x_1 = \dfrac{1}{10} + 1 = 1.1$, $x_2 = \dfrac{1.1}{10} + 1 = 1.11$, $x_3 = \dfrac{1.11}{10} + 1 = 1.111$

5. Answers may vary. For example, let $f(x) = x$. Then $f^2(x_0) = x_0$, and $(f(x_0))^2 = x_0^2$.

6. a. 0   **b.** slowly

7. The fixed points occur at the intersection of the graph of $y = f(x)$ and the line $y = x$.

8. $1 - \sqrt{3}$ is attracting; $1 + \sqrt{3}$ is repelling.

9. a. $x^2 = x$; $x^2 - x = 0$; $x(x - 1) = 0$; $x = 0$ or $x = 1$; the fixed points are 0, 1.

   b. -1 is an eventually fixed point because $f(-1) = 1$, which is a fixed point.

   c. $\lim_{n \to \infty} f^n(x_0) = \begin{cases} \infty & \text{if } |x_0| > 1 \\ 0 & \text{if } |x_0| < 1 \end{cases}$

   d. 0 is attracting; 1 is repelling.

10. a. $x^3 = x$; $x^3 - x = 0$; $x(x - 1)(x + 1) = 0$; $x = 0$, $x = 1$, or $x = -1$; the fixed points are $-1, 0, 1$.

b. There are no eventually fixed points.

c. $\displaystyle\lim_{x \to \infty} f^n(x_0) = \begin{cases} -\infty & \text{if } x_0 < -1 \\ 0 & \text{if } -1 < x_0 < 1 \\ \infty & \text{if } x_0 > 1 \end{cases}$

d. 0 is attracting; $-1, 1$ are repelling.

Written Exercises 19-5 · pages 741–743

1. $x_0 = 5$, $x_1 = 2(5) - 1 = 9$, $x_2 = 2(9) - 1 = 17$, $x_3 = 2(17) - 1 = 33$

2. $x_0 = 0$, $x_1 = \sqrt{0 + 1} = 1$, $x_2 = \sqrt{1 + 1} = \sqrt{2} \approx 1.41$, $x_3 = \sqrt{\sqrt{2} + 1} \approx 1.55$

3. $x_0 = 2.5$, $x_1 = \dfrac{1}{2.5} = 0.4$, $x_2 = \dfrac{1}{0.4} = 2.5$, $x_3 = \dfrac{1}{2.5} = 0.4$

4. $x_0 = 20$, $x_1 = \dfrac{20}{2} - 2 = 8$, $x_2 = \dfrac{8}{2} - 2 = 2$, $x_3 = \dfrac{2}{2} - 2 = -1$

5. $5x - 16 = x$; $4x = 16$; $x = 4$ **6.** $3 - \dfrac{1}{2}x = x$; $3 = \dfrac{3}{2}x$; $x = 2$

7. $x^2 + \dfrac{1}{4} = x$; $x^2 - x + \dfrac{1}{4} = 0$; $\left(x - \dfrac{1}{2}\right)^2 = 0$; $x = \dfrac{1}{2}$

8. $x^2 - 1 = x$; $x^2 - x - 1 = 0$; $x = \dfrac{1 \pm \sqrt{5}}{2}$

9. $2 + \sqrt{x} = x$; $\sqrt{x} = x - 2$; $x = x^2 - 4x + 4$; $x^2 - 5x + 4 = 0$; $(x - 1)(x - 4) = 0$; $x = 1$ (reject) or $x = 4$

10. $\dfrac{1}{2}\left(x + \dfrac{4}{x}\right) = x$; $x + \dfrac{4}{x} = 2x$; $\dfrac{4}{x} = x$; $x^2 = 4$; $x = \pm 2$

11. $x_0 = 235$, $x_1 \approx 2.371$, $x_2 \approx 0.375$, $x_3 \approx -0.426$; x_4 is not defined because $x_3 < 0$

12. $x_0 = 101$, $x_1 = 10$, $x_2 = 3$, $x_3 \approx 1.414$, $x_4 \approx 0.644$; x_5 is not defined because $x_4 < 1$

13. a. See web diagram below. $\displaystyle\lim_{n \to \infty} f^n(x_0) = 2$ **b.** There is no change.

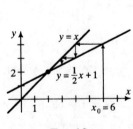

Ex. 13

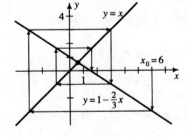

Ex. 14

14. a. See web diagram above. $\displaystyle\lim_{n \to \infty} f^n(x_0) = \dfrac{3}{5}$ **b.** There is no change.

15. a. $\lim_{n \to \infty} f^n(x_0) \approx 0.739$

b. The web diagram below is for the seed $x_0 = 1$, but web diagrams for other seeds would be similar.

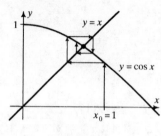

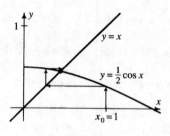

Ex. 15   **Ex. 16**

16. a. $\lim_{n \to \infty} f^n(x_0) \approx 0.450$

b. The web diagram above is for the seed $x_0 = 1$, but web diagrams for other seeds would be similar.

17. a. If $x_0 > 0$, $\lim_{n \to \infty} f^n(x_0) = 0$. **b.** If $-1 < x_0 < 0$, $\lim_{n \to \infty} f^n(x_0)$ is not defined.

18. a. If $0 < x_0 < 1$, $\lim_{n \to \infty} f^n(x_0) = 0$. **b.** If $x_0 < 0$ or $x_0 > 1$, $\lim_{n \to \infty} f^n(x_0) = -\infty$.

19. $2x - x^2 = x$; $x^2 - x = 0$; $x(x - 1) = 0$; $x = 0$ or $x = 1$; 0 and 1 are fixed points.

Since $f(2) = 0$, 2 is an eventually fixed point. $\lim_{n \to \infty} f^n(x_0) = \begin{cases} 1 & \text{if } 0 < x_0 < 2 \\ -\infty & \text{if } x_0 < 0 \text{ or } x_0 > 2 \end{cases}$

20. $2 \sin x = x$; $x = 0$ or $x \approx \pm 1.895$; 0, 1.895, and -1.895 are fixed points. Since $f(\pi) = 0$, $f(1.245) \approx 1.895$, and $f(-1.245) \approx -1.895$, $\pm n\pi$ and $\pm 1.245 \pm n\pi$, where n is a natural number, are eventually fixed points.

$\lim_{n \to \infty} f^n(x_0) = \begin{cases} 1.895 & \text{if } 2k\pi < x_0 < (2k + 1)\pi, k \text{ an integer} \\ -1.895 & \text{if } (2k - 1)\pi < x_0 < 2k\pi, k \text{ an integer} \end{cases}$

21. There will be a finite limit if $|m| < 1$ (that is, if the slope of the graph of f is less than the slope of the line $y = x$) or if $m = 1$ and $k = 0$.

22. a. $f(x) = x$; $x^2 + c = x$; $x^2 - x + c = 0$; fixed points exist when the discriminant of $x^2 - x + c = 0$ is positive: $1 - 4c \geq 0$; $c \leq \dfrac{1}{4}$.

b. For $c > \dfrac{1}{4}$ in $f(x) = x^2 + c$, $\lim_{n \to \infty} f^n(x_0) = \infty$ for any x_0.

23. a. $x_0 = 1, x_1 = 2, x_2 = 1.5, x_3 = 1.\overline{6}, x_4 = 1.6, x_5 = 1.625, x_6 \approx 1.615, x_7 \approx 1.619,$ $x_8 \approx 1.618, x_9 \approx 1.618$; $\lim_{n \to \infty} f^n(1) \approx 1.62$

b. As long as $f^n(x_0)$ is defined for all natural numbers n, the choice of x_0 does not matter. In a web diagram, $\dfrac{1 + \sqrt{5}}{2} \approx 1.62$ is an attracting fixed point.

24. $x_0 = 2, x_1 = 1.5, x_2 = 1.\overline{3}, x_3 = 1.25, \ldots$; this sequence converges very slowly to 1.

25. a. $x_0 = 3$, $x_1 = \dfrac{3 + \frac{10}{3}}{2} \approx 3.17$, $x_2 \approx 3.16$, $x_3 \approx 3.16$, $x_4 \approx 3.16$

b. $\dfrac{x + \frac{N}{x}}{2} = x$; $2x = x + \dfrac{N}{x}$; $x = \dfrac{N}{x}$;

$x^2 = N$; $x = \pm\sqrt{N}$, so $\sqrt{N}$ is a fixed point. The web diagram is for $N = 10$ and $x_0 = 3$; web diagrams for other positive N and x_0 are similar.

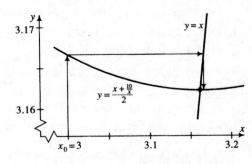

26. Use a program like the following, which gives the first 200 terms in the orbit of a seed:

```
10 INPUT "SEED: "; X
20 PRINT "ORBIT: ";
30 FOR N = 1 TO 200
40 PRINT X; " ";
50 IF INT(X/2) = X/2 THEN LET X = X/2: GOTO 70
60 IF INT(X/2) < > X/2 THEN LET X = 3 * X + 1
70 NEXT N
80 END
```

The seeds $x_0 = 1$, $x_0 = 2$, and $x_0 = 4$ all have periodic orbits. All other integral seeds less than 27 have orbits that are eventually periodic, settling down to the same cycle of $4, 2, 1, 4, 2, 1, \dots$ within the first twenty iterations. The seed $x_0 = 27$, however, requires over 100 iterations before the orbit becomes periodic. One might conjecture that all seeds other than 1, 2, and 4 have orbits that are eventually periodic.

Class Exercises 19-6 · page 746

1. a. $x_0 = 100$; $x_n = 1.01(x_{n-1}) - 20.6$

b. The orbit of $x_0 = 100$ for iterations of $f(x) = 1.01x - 20.6$

2. $x_0 = 0.2$, $x_1 = 0.32$, $x_2 \approx 0.435$, $x_3 \approx 0.492$, $x_4 \approx 0.500$; 0.5 is an equilibrium point.

3. $x_0 = 0.2$, $x_1 = 0.08$, $x_2 \approx 0.037$, $x_3 \approx 0.018$, $x_4 \approx 0.009$; 0 is an equilibrium point.

Written Exercises 19-6 · pages 746–748

1. a. $x_0 = 100$; $x_n = 1.06(x_{n-1})$ **b.** The orbit of $x_0 = 100$ for iterations of $f(x) = 1.06x$

2. a. $x_0 = 5000$; $x_n = 1.08(x_{n-1}) + 3000$

b. The orbit of $x_0 = 5000$ for iterations of $f(x) = 1.08x + 3000$

3. a. $x_0 = 400{,}000$; $x_n = 1.065(x_{n-1}) - 30{,}000$

b. The orbit of $x_0 = 400{,}000$ for iterations of $f(x) = 1.065x - 30{,}000$

4. a. $x_0 = 400$; $x_n = 1.015(x_{n-1}) - 23.3$

b. The orbit of $x_0 = 400$ for iterations of $f(x) = 1.015x - 23.3$

For Exs. 5–9, students should recognize that the choice of seed does not affect the result of iterating the logistic function for the given value of c.

5. Equilibrium point at 0.333   **6.** Period-2 cycle: 0.513, 0.799

7. Period-4 cycle: 0.501, 0.875, 0.383, 0.827

8. Period-8 cycle: 0.506, 0.887, 0.355, 0.813, 0.540, 0.882, 0.370, 0.828

9. Period-3 cycle: 0.505, 0.957, 0.156

10. For $x_0 = 0.5$, the function has an equilibrium point at 0. For $x_0 = 0.50001$ or $x_0 = 0.49999$, the function fluctuates chaotically.

11. a. Iterate the function $f(x) = 1.1x - 50{,}000$ using the seed x_0. After 20 iterations, you

have $x_{20} = (1.1)^{20}x_0 - \left[\sum_{n=1}^{20} (1.1)^{n-1}\right]50{,}000$. Since $x_{20} = 0$,

$$x_0 = \frac{\left[\sum_{n=1}^{20}(1.1)^{n-1}\right]50{,}000}{(1.1)^{20}} \approx 425{,}678.19.$$ This result can also be obtained using

trial-and-error and the following program:

```
10 INPUT "SEED: "; X
20 FOR N = 1 TO 20
30 LET X = 1.1 * X - 50000
40 NEXT N
50 PRINT "FINAL BALANCE: "; X
60 END
```

b. You would be earning interest on the money that has not been withdrawn. Twenty iterations of the function $f(x) = 1.1x - 50{,}000$ for the seed $x_0 = 1{,}000{,}000$ result in 20 annual withdrawals of $50,000 and an account balance of $3,863,749.98.

12. $f^n(x_0) = (1 + r)^n x_0$; the growth is exponential

13. a. If $0 < x_0 < 1$, $\lim_{n \to \infty} f^n(x_0) = 1$. The population has an equilibrium point at 1, that is, 100% of capacity.

b. Answers will vary. If $c = 2.5$ and $x_0 = 0.6$, a period-2 cycle (0.6 and 1.2) is produced. If $c = 2.7$ and $x_0 = 0.5$, the function behaves chaotically.

14. a. If $0 < x_0 \le 0.146$, $\lim_{n \to \infty} f^n(x_0) = -\infty$. If $0.147 \le x_0 < 1$, there is an equilibrium point at 0.854.

b. Answers will vary. If $c = 2.5$, $h = -0.1$, and $x_0 = 0.5$, a period-2 cycle (0.676 and 1.124) is produced. If $c = 3$, $h = -0.1$, and $x_0 = 0.5$, the function behaves chaotically.

15. Iterate the function $f(x) = (1 + c(1 - x)x) + hx$

a. For any seed between 0 and 1, there is an equilibrium point at 0.875.

b. Answers will vary. If $c = 1.95$, $h = 0.1$, and $x_0 = 0.5$, a period-2 cycle (0.923 and 1.154) is produced. If $c = 2.8$, $h = -0.1$, and $x_0 = 0.5$, the function behaves chaotically.

Chapter Test · page 749

1. a. $\lim\limits_{x \to \infty} \dfrac{x^2 - x}{x^2 - 1} = \lim\limits_{x \to \infty} \dfrac{1 - \frac{1}{x}}{1 - \frac{1}{x^2}} = \dfrac{1 - 0}{1 - 0} = 1$

b. $\lim\limits_{x \to 1} \dfrac{x^2 - x}{x^2 - 1} = \lim\limits_{x \to 1} \dfrac{x(x - 1)}{(x + 1)(x - 1)} = \lim\limits_{x \to 1} \dfrac{x}{x + 1} = \dfrac{1}{1 + 1} = \dfrac{1}{2}$

c. $\lim\limits_{x \to 9} \dfrac{\sqrt{x} - 3}{x - 9} = \lim\limits_{x \to 9} \dfrac{\sqrt{x} - 3}{(\sqrt{x} - 3)(\sqrt{x} + 3)} = \lim\limits_{x \to 9} \dfrac{1}{\sqrt{x} + 3} = \dfrac{1}{\sqrt{9} + 3} = \dfrac{1}{6}$

d. $\lim\limits_{x \to 0^-} \dfrac{|x|}{x} = -1$, but $\lim\limits_{x \to 0^+} \dfrac{|x|}{x} = 1$; thus, $\lim\limits_{x \to 0} \dfrac{|x|}{x}$ does not exist.

2. a.

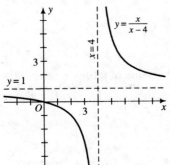

b.

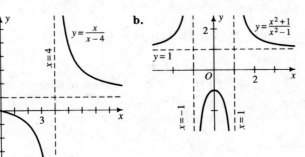

3. a. Change lines 20, 30, and 40:

```
20 FOR X = .01 TO 2 STEP .01
30 LET Y = 4 - X ^ 2
40 LET A = A + .01 * Y
```

Area ≈ 5.31330004

b. Because the graph of $y = 4 - x^2$ is symmetric in the y-axis, the area is doubled. Area ≈ 10.62660008

4. a. $\cos 1 = 1 - \dfrac{1}{2!} + \dfrac{1}{4!} - \dfrac{1}{6!} + \cdots$

b. $\ln(1 + x^2) = x^2 - \dfrac{(x^2)^2}{2} + \dfrac{(x^2)^3}{3} - \dfrac{(x^2)^4}{4} + \cdots = x^2 - \dfrac{x^4}{2} + \dfrac{x^6}{3} - \dfrac{x^8}{4} + \cdots,$

$-1 \le x \le 1$

5. To the nearest thousandth, the orbit is $1, 0.718, 0.051, -0.948, -1.612, -1.801, -1.835,$ $-1.840, -1.841, -1.841, \ldots$; thus, $\lim\limits_{n \to \infty} f^n(1) \approx -1.841$.

6. 0 is a repelling fixed point; $\lim\limits_{n \to \infty} f^n(x_0) = \begin{cases} \infty & \text{if } x_0 > 0 \\ -\infty & \text{if } x_0 < 0 \end{cases}$

7. $\lim\limits_{n \to \infty} f^n(x_0) = 0$. The population will not survive regardless of the initial population.

Class Exercises 20-1 · page 761

1. $f'(x) = 8x^7$　**2.** $f'(x) = 12x^{11}$　**3.** $f'(x) = -4x^{-5} = -\dfrac{4}{x^5}$　**4.** $g'(x) = 3(7x^6) = 21x^6$

5. $h'(x) = \dfrac{3}{2}x^{1/2} = \dfrac{3}{2}\sqrt{x}$　**6.** $f'(x) = 4 \cdot \left(-\dfrac{3}{2}\right)x^{-5/2} = -\dfrac{6}{x^2\sqrt{x}} = -\dfrac{6\sqrt{x}}{x^3}$

7. $g'(x) = \dfrac{1}{3} \cdot x^{-2/3} = \dfrac{1}{3\sqrt[3]{x^2}} = \dfrac{\sqrt[3]{x}}{3x}$

8. $h(x) = x^{-1/3};\ h'(x) = -\dfrac{1}{3} \cdot x^{-4/3} = \dfrac{-1}{3\sqrt[3]{x^4}} = -\dfrac{1}{3x\sqrt[3]{x}} = -\dfrac{\sqrt[3]{x^2}}{3x^2}$

9. $f'(x) = 8(3x^2) - 7(2x) + 3 \cdot x^0 + 0 = 24x^2 - 14x + 3$

10. $g'(x) = 4(-5)x^{-6} - 2(-3)x^{-4} + 9(-1)x^{-2} = -\dfrac{20}{x^6} + \dfrac{6}{x^4} - \dfrac{9}{x^2}$

11. a. $f'(x) = 8x; f'(-1) = 8(-1) = -8$　**b.** Slope at $(-1, 4)$ is $f'(-1) = -8$.

12. a. $f'(x) = m$, so $f'(1) = f'(2) = f'(3) = m$.　**b.** The slope of a line is constant.

13. For $x > 0$, the slope of $f(x) = |x|$ is 1, and for $x < 0$ the slope is -1. The slope at $x = 0$ is undefined since the slope abruptly changes sign at $x = 0$.

Written Exercises 20-1 · pages 762–763

1. a. $\dfrac{8-1}{2-1} = 7$　**b.** $\dfrac{3.375-1}{1.5-1} = \dfrac{2.375}{0.5} = 4.75$　**c.** $\dfrac{1.331-1}{1.1-1} = \dfrac{0.331}{0.1} = 3.31$

　d. $f'(x) = 3x^2; f'(1) = 3(1)^2 = 3$

2. a. $\dfrac{\frac{1}{3} - \frac{1}{2}}{3-2} = -\dfrac{1}{6}$　**b.** $\dfrac{\frac{1}{2.5} - \frac{1}{2}}{2.5-2} = \dfrac{\frac{2-2.5}{5.0}}{0.5} = \dfrac{-0.1}{0.5} = -\dfrac{1}{5}$

　c. $\dfrac{\frac{1}{2.1} - \frac{1}{2}}{2.1-2} = \dfrac{\frac{2-2.1}{4.2}}{0.1} = \dfrac{\frac{-0.1}{4.2}}{0.1} = -\dfrac{0.1}{4.2} \cdot \dfrac{1}{0.1} = -\dfrac{1}{4.2} = -\dfrac{10}{42} = -\dfrac{5}{21}$

　d. $f(x) = x^{-1}; f'(x) = -1(x)^{-2} = -\dfrac{1}{x^2}; f'(2) = -\dfrac{1}{(2)^2} = -\dfrac{1}{4}$

3. $f'(x) = 2 \cdot 5x^4 = 10x^4$　**4.** $f'(x) = 3 \cdot 6x^5 = 18x^5$　**5.** $f'(x) = -2(3)x^2 = -6x^2$

6. $f'(x) = 8 \cdot \dfrac{3}{4}x^{-1/4} = \dfrac{6x^{3/4}}{x} = \dfrac{6\sqrt[4]{x^3}}{x}$　**7.** $f'(x) = 3(-2)x^{-3} = -\dfrac{6}{x^3}$

8. $g'(x) = 4\left(-\dfrac{1}{2}\right)x^{-3/2} = \dfrac{-2}{x^{3/2}} = -\dfrac{2}{x\sqrt{x}} = -\dfrac{2\sqrt{x}}{x^2}$

9. $f(x) = \sqrt{5} \cdot x^{1/2}; f'(x) = \sqrt{5}\left(\dfrac{1}{2}\right)x^{-1/2} = \dfrac{\sqrt{5}}{2} \cdot \dfrac{1}{\sqrt{x}} \cdot \dfrac{\sqrt{x}}{\sqrt{x}} = \dfrac{\sqrt{5x}}{2x}$

10. $f(x) = \dfrac{1}{4}x^{-4}; f'(x) = \dfrac{1}{4}(-4)x^{-5} = -\dfrac{1}{x^5}$

11. $g'(x) = \dfrac{7}{2}(2)x - 5 + 0 - 1(-1)x^{-2} = 7x - 5 + \dfrac{1}{x^2}$

12. $f'(x) = \dfrac{4}{3}(3)x^2 - 2(-1)x^{-2} - 5(-2)x^{-3} + 0 = 4x^2 + \dfrac{2}{x^2} + \dfrac{10}{x^3}$

13. $f'(x) = 3x^2$; the slope at $P(-1, -1)$ is $f'(-1) = 3(-1)^2 = 3$.

14. $f'(x) = \frac{1}{2}(4)x^3 = 2x^3$; the slope at $P\left(-1, \frac{1}{2}\right)$ is $f'(-1) = 2(-1)^3 = 2(-1) = -2$.

15. $f'(x) = 3(2)x - 2 = 6x - 2$; the slope at $P(0, 1)$ is $f'(0) = 6(0) - 2 = -2$.

16. $f'(x) = 2\left(\frac{1}{2}\right)x^{-1/2} = x^{-1/2} = \frac{1}{\sqrt{x}}$; the slope at $P(9, 6)$ is $f'(9) = \frac{1}{\sqrt{9}} = \frac{1}{3}$.

17. $f'(x) = 8(4)x^3 - 7(2)x + 5 = 32x^3 - 14x + 5$; the slope at $P(-1, 2)$ is $f'(-1) = 32(-1)^3 - 14(-1) + 5 = -13$.

18. $f'(x) = 3x^2 - 5(2)x + 4 = 3x^2 - 10x + 4$; the slope at $P(2, -2)$ is $f'(2) = 3(2)^2 - 10(2) + 4 = -4$.

19. $f'(x) = 2(-1)x^{-2} - 4(-2)x^{-3} = -\frac{2}{x^2} + \frac{8}{x^3}$; the slope at $P(2, 0)$ is $f'(2) =$

$-\frac{2}{(2)^2} + \frac{8}{(2)^3} = -\frac{2}{4} + \frac{8}{8} = \frac{1}{2}$.

20. $f'(x) = \frac{1}{3}x^{-2/3} = \frac{1}{3\sqrt[3]{x^2}}$; the slope at $P(-1, -1)$ is $f'(-1) = \frac{1}{3\sqrt[3]{(-1)^2}} = \frac{1}{3 \cdot 1} = \frac{1}{3}$.

21. a. $f'(x) = a \cdot 2x + b = 2ax + b$ **b.** $f'\left(-\frac{b}{2a}\right) = 2a\left(-\frac{b}{2a}\right) + b = -b + b = 0$; since

the slope is 0 at $x = -\frac{b}{2a}$, there is a horizontal tangent to the parabola at the vertex.

22. If $f(x)$ and $g(x)$ are two functions such that $g(x) - f(x) = c$ for some constant c, then $g(x) = f(x) + c$. The graph of $g(x)$ is therefore a vertical translation of the graph of $f(x)$. Since the vertical translation does not affect the slope of the tangent at any point, $g'(x) = f'(x)$ for all x.

23. $f'(x) = 3x^2$; the slope of the curve at $P(-2, -8)$ is $f'(-2) = 3(-2)^2 = 12$ and the line contains $(-2, -8)$, so its equation is $\frac{y - (-8)}{x - (-2)} = 12$; $y + 8 = 12x + 24$; $y = 12x + 16$.

24. $f'(x) = 6x - 5$; the slope of the curve at $P(2, 2)$ is $f'(2) = 6(2) - 5 = 7$ and the line contains $(2, 2)$, so its equation is $\frac{y - 2}{x - 2} = 7$; $y - 2 = 7x - 14$; $y = 7x - 12$.

25. $f'(x) = -2x^{-3} = \frac{-2}{x^3}$; the slope of the curve at $P(-1, 1)$ is $f'(-1) = \frac{-2}{(-1)^3} = \frac{-2}{-1} = 2$ and the line contains $(-1, 1)$, so its equation is $\frac{y - 1}{x - (-1)} = 2$; $y - 1 = 2x + 2$; $y = 2x + 3$.

26. $f'(x) = -\frac{1}{3}x^{-4/3} = \frac{-1}{3\sqrt[3]{x^4}} = -\frac{1}{3x\sqrt[3]{x}}$; the slope of the curve at $P\left(8, \frac{1}{2}\right)$ is

$f'(8) = -\frac{1}{3(8)\sqrt[3]{8}} = -\frac{1}{24 \cdot 2} = -\frac{1}{48}$ and the line contains $\left(8, \frac{1}{2}\right)$, so its equation is

$\frac{y - \frac{1}{2}}{x - 8} = -\frac{1}{48}$; $48\left(y - \frac{1}{2}\right) = -(x - 8)$; $48y - 24 = -x + 8$; $y = -\frac{1}{48}x + \frac{2}{3}$.

27. a. See below.

    b. $f'(x) = 6 \cdot 2x - 3x^2 = 12x - 3x^2; f'(x) = 0$ when $3x(4 - x) = 0; x = 0$ or $x = 4$

    c. See below.

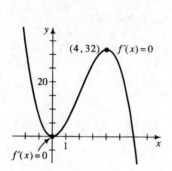

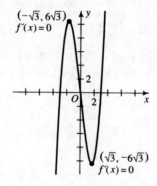

     Exs. 27(a), 27(c)   **Exs. 28(a), 28(c)**

28. a. See above.   **b.** $f'(x) = 3x^2 - 9; f'(x) = 0$ when $3x^2 = 9; x^2 = 3; x = \pm\sqrt{3} \approx \pm 1.73$

    c. See above.

29. a. $f'(x) = \dfrac{2}{3}x^{-1/3} = \dfrac{2}{3\sqrt[3]{x}}$ **b.** $x = 0$

    c. See graph below. The tangent to the graph of $f(x) = x^{2/3}$ is vertical at $x = 0$. Since the tangent has no slope there, $f'(0)$ is undefined.

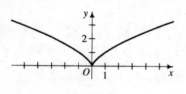

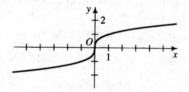

        Ex. 29(c)   **Ex. 30(c)**

30. a. $f'(x) = \dfrac{1}{3}x^{-2/3} = \dfrac{1}{3\sqrt[3]{x^2}}$ **b.** $x = 0$

    c. See graph above. The tangent to the graph of $f(x)$ is vertical at $x = 0$. Since the tangent has no slope there, $f'(0)$ is undefined.

31–36. Let the given function be ax^n; its derivative is anx^{n-1}. Answers may vary by a constant.

31. $an = 4$ and $n - 1 = 3; n = 4; a \cdot 4 = 4; a = 1; f(x) = ax^n = 1 \cdot x^4 = x^4$

32. $an = 5$ and $n - 1 = 4; n = 5; a \cdot 5 = 5; a = 1; f(x) = ax^n = 1 \cdot x^5 = x^5$

33. $an = 3$ and $n - 1 = 5; n = 6; a \cdot 6 = 3; a = \dfrac{1}{2}; g(x) = ax^n = \dfrac{1}{2}x^6$

34. $an = 1$ and $n - 1 = 7; n = 8; a \cdot 8 = 1; a = \dfrac{1}{8}; g(x) = ax^n = \dfrac{1}{8}x^8$

35. $h'(x) = 3x^{1/2}; an = 3$ and $n - 1 = \dfrac{1}{2}; n = \dfrac{3}{2}; a \cdot \dfrac{3}{2} = 3; a = 2; h(x) = ax^n = 2x^{3/2}$,

    or $2x\sqrt{x}$

36. $h'(x) = x^{1/3}$; $an = 1$ and $n - 1 = \dfrac{1}{3}$; $n = \dfrac{4}{3}$; $a \cdot \dfrac{4}{3} = 1$; $a = \dfrac{3}{4}$; $h(x) = ax^n = \dfrac{3}{4}x^{4/3}$

37. a. $x^n + nx^{n-1}h + \dfrac{n(n-1)}{2}x^{n-2}h^2 + \cdots + h^n$

    b. $f'(x) = \lim\limits_{h \to 0} \dfrac{f(x+h) - f(x)}{h} = \lim\limits_{h \to 0} \dfrac{(x+h)^n - x^n}{h} =$

$$\lim_{h \to 0} \dfrac{x^n + nx^{n-1}h + \dfrac{n(n-1)}{2}x^{n-2}h^2 + \cdots + h^n - x^n}{h} =$$

$$\lim_{h \to 0} \dfrac{nx^{n-1}h + \dfrac{n(n-1)}{2}x^{n-2}h^2 + \cdots + h^n}{h} =$$

$$\lim_{h \to 0} \left(nx^{n-1} + \dfrac{n(n-1)}{2}x^{n-2}h + \cdots + h^{n-1} \right) = nx^{n-1}$$

38. If $f(x) = cx^n$, then $f'(x) = \lim\limits_{h \to 0} \dfrac{f(x+h) - f(x)}{h} = \lim\limits_{h \to 0} \dfrac{c(x+h)^n - cx^n}{h} =$

$\lim\limits_{h \to 0} c\left[\dfrac{(x+h)^n - x^n}{h} \right] = c \cdot \lim\limits_{h \to 0} \dfrac{(x+h)^n - x^n}{h} = c \cdot nx^{n-1}$, from Ex. 37. Thus

$f'(x) = cnx^{n-1}$.

39. If $f(x) = p(x) + q(x)$, then $f'(x) = \lim\limits_{h \to 0} \dfrac{f(x+h) - f(x)}{h} =$

$\lim\limits_{h \to 0} \dfrac{[p(x+h) + q(x+h)] - [p(x) + q(x)]}{h} = \lim\limits_{h \to 0} \dfrac{[p(x+h) - p(x)] + [q(x+h) - q(x)]}{h} =$

$\lim\limits_{h \to 0} \left[\dfrac{p(x+h) - p(x)}{h} + \dfrac{q(x+h) - q(x)}{h} \right] = \lim\limits_{h \to 0} \dfrac{p(x+h) - p(x)}{h} + \lim\limits_{h \to 0} \dfrac{q(x+h) - q(x)}{h} =$

$p'(x) + q'(x)$ by the definition of derivative.

Communication: Visual Thinking · page 763

1. The large square has an area of $(a + b)^2$. It contains four blue rectangles, each with an area of ab, and one red square, with an area of $(a - b)^2$. Thus, $(a + b)^2 = (a - b)^2 + 4ab$.

2. Since the four blue rectangles are completely contained in the large square (with some room to spare), $(a + b)^2 > 4ab$.

3. $(a + b)^2 > 4ab$; $\sqrt{(a + b)^2} > \sqrt{4ab}$; $a + b > 2\sqrt{ab}$; $\dfrac{a + b}{2} > \sqrt{ab}$, so the arithmetic mean of a and b is greater than the geometric mean of a and b.

Class Exercises 20-2 · page 767

1. a. $x = -5$, $x = -1$, $x = 2$, $x = 6$

    b. $-5 < x < -1$, $2 < x < 6$

    c. $x < -5$, $-1 < x < 2$, $x > 6$

2. a. $x = -4$, $x = 1$, $x = 4$

    b. $x < -4$, $-4 < x < 1$, $x > 4$ **c.** $1 < x < 4$

3. Yes; see the figure at the right.

4. Since $f'(x)$ changes from positive to zero to negative in the interval, the graph is a parabola that opens downward; there is a local maximum at $x = 2$.

5. $x = 0$ does not give a maximum or minimum since $f'(x) < 0$ on both sides of $x = 0$; there is a local maximum at $x = -1$ since $f'(-1) = 0$ and the curve changes from rising to falling at $x = -1$; similarly, there is a local minimum at $x = 2$.

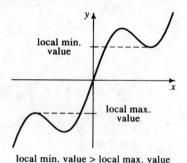

local min. value > local max. value

Ex.3

6. $f'(x) = \frac{1}{3} \cdot 3x^2 - 4 = x^2 - 4$; $f'(x) = 0$ when $x = \pm 2$; $f'(x) > 0$ when $x < -2$ or $x > 2$ and $f'(x) < 0$ when $-2 < x < 2$; thus there is a local maximum at $x = -2$ and a local minimum at $x = 2$.

Written Exercises 20-2 · pages 768–769

1. $f(x) = x(x^2 - 12)$; $f'(x) = 3x^2 - 12$; if $f'(x) = 0$, $3(x^2 - 4) = 0$; $x = \pm 2$; $f'(x) > 0$ when $x < -2$ or $x > 2$; $f'(x) < 0$ when $-2 < x < 2$; local maximum at $(-2, 16)$ and local minimum at $(2, -16)$

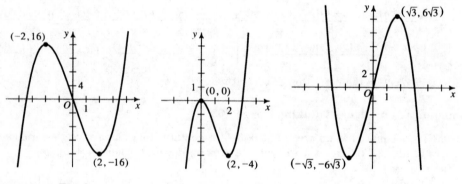

Ex. 1   Ex. 2   Ex. 3

2. $f(x) = x^2(x - 3)$; $f'(x) = 3x^2 - 6x = 3x(x - 2)$; $f'(x) = 0$ when $x = 0$ or $x = 2$; $f'(x) > 0$ when $x < 0$ or $x > 2$; $f'(x) < 0$ when $0 < x < 2$; local maximum at $(0, 0)$ and local minimum at $(2, -4)$

3. $f(x) = x(3 + x)(3 - x)$; $f'(x) = 9 - 3x^2 = 3(3 - x^2)$; $f'(x) = 0$ when $x = \pm \sqrt{3}$; $f'(x) < 0$ when $x < -\sqrt{3}$ or $x > \sqrt{3}$; $f'(x) > 0$ when $-\sqrt{3} < x < \sqrt{3}$; local minimum at $(-\sqrt{3}, -6\sqrt{3})$ and local maximum at $(\sqrt{3}, 6\sqrt{3})$

4. $f(x) = (x^2 - 3)(x + 4)$; $f'(x) = 3x^2 + 8x - 3 = (3x - 1)(x + 3)$; $f'(x) = 0$ when $x = \frac{1}{3}$ or $x = -3$; $f'(x) > 0$ when $x < -3$ or $x > \frac{1}{3}$; $f'(x) < 0$ when $-3 < x < \frac{1}{3}$; local maximum at $(-3, 6)$ and local minimum at $\left(\frac{1}{3}, -\frac{338}{27}\right)$

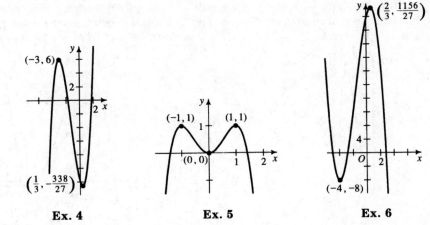

Ex. 4   Ex. 5   Ex. 6

5. $g(x) = x^2(2 - x^2)$; $g'(x) = 4x - 4x^3 = 4x(1 - x^2)$; $g'(x) = 0$ when $x = 0$ or $x = \pm 1$; $g'(x)$ changes sign at each zero, so $(0, 0)$ is a local minimum and $(-1, 1)$ and $(1, 1)$ are local maximums.

6. $h(x) = (8 - x^2)(5 + x)$; $h'(x) = -3x^2 - 10x + 8 = -(3x - 2)(x + 4)$; $h'(x) = 0$ when $x = \dfrac{2}{3}$ or $x = -4$; $h'(x)$ changes sign at each zero, so $(-4, -8)$ is a local minimum and $\left(\dfrac{2}{3}, \dfrac{1156}{27}\right)$ is a local maximum.

7. $g(x) = (x + 2)(x - 2)(x^2 - 2)$; $g'(x) = 4x^3 - 12x = 4x(x^2 - 3)$; $g'(x) = 0$ when $x = 0$ or $x = \pm\sqrt{3}$; $g'(x)$ changes sign at each zero, so $(-\sqrt{3}, -1)$ and $(\sqrt{3}, -1)$ are local minimums, and $(0, 8)$ is a local maximum.

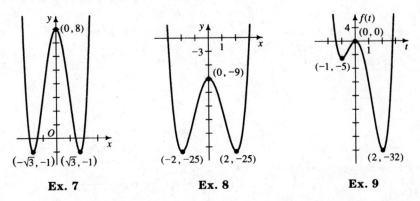

Ex. 7   Ex. 8   Ex. 9

8. $f(x) = (x + 3)(x - 3)(x^2 + 1)$; $f'(x) = 4x^3 - 16x = 4x(x^2 - 4)$; $f'(x) = 0$ when $x = 0$ or $x = \pm 2$; $f'(x)$ changes sign at each zero, so $(-2, -25)$ and $(2, -25)$ are local minimums and $(0, -9)$ is a local maximum.

9. $f(t) = t^2(3t^2 - 4t - 12)$; $f'(t) = 12t^3 - 12t^2 - 24t = 12t(t^2 - t - 2) = 12t(t + 1)(t - 2)$; $f'(t) = 0$ when $t = -1$, $t = 0$, or $t = 2$; $f'(t)$ changes sign at each zero, so $(-1, -5)$ and $(2, -32)$ are local minimums and $(0, 0)$ is a local maximum.

10. $g(t) = t^3(8 - 3t); g'(t) = 24t^2 - 12t^3 = -12t^2(t - 2); g'(t) = 0$ when $t = 0$ or $t = 2$;
$t = 0$ is a double root of $g'(t)$, so $g(t)$ does not have a local maximum or minimum at
$t = 0; g'(t) > 0$ when $0 < t < 2$ and $g'(t) < 0$ when $t > 2$, so $(2, 16)$ is a local maximum.

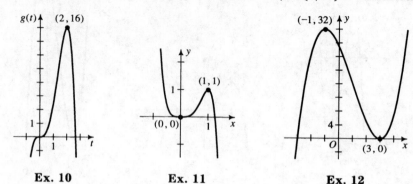

Ex. 10   Ex. 11   Ex. 12

11. $h(x) = x^4(5 - 4x); h'(x) = 20x^3 - 20x^4 = 20x^3(1 - x); h'(x) = 0$ when $x = 0$ or
$x = 1; h'(x)$ changes sign at each zero, so $(0, 0)$ is a local minimum and $(1, 1)$ is a local
maximum.

12. $f(x) = (x + 3)(x - 3)^2; f'(x) = 3x^2 - 6x - 9 = 3(x + 1)(x - 3); f'(x) = 0$ when
$x = -1$ or $x = 3; f'(x)$ changes sign at each zero, so $(-1, 32)$ is a local maximum and
$(3, 0)$ is a local minimum.

13. If $f(x) = x^2 - 6x + 4, f'(x) = 2x - 6$; the minimum occurs when $2x - 6 = 0; x = 3$;
$f(3) = -5; V(3, -5)$.

14. If $g(x) = 5 + 8x - 4x^2, g'(x) = 8 - 8x$; the maximum occurs when $8 - 8x = 0; x = 1$;
$g(1) = 9; V(1, 9)$.

15. If $h(x) = (x - a)^2 = x^2 - 2ax + a^2, h'(x) = 2x - 2a$; the vertex occurs when
$2x - 2a = 0; x = a; h(a) = (a - a)^2 = 0; V(a, 0)$.

16. If $f(x) = ax^2 + bx + c, f'(x) = 2ax + b$; the vertex occurs when $2ax + b = 0$;
$$x = -\frac{b}{2a}; f\left(-\frac{b}{2a}\right) = a\left(-\frac{b}{2a}\right)^2 + b\left(-\frac{b}{2a}\right) + c = \frac{ab^2 - 2ab^2 + 4a^2c}{4a^2} = \frac{4ac - b^2}{4a};$$
$$V\left(-\frac{b}{2a}, \frac{4ac - b^2}{4a}\right).$$

17.   **18.**

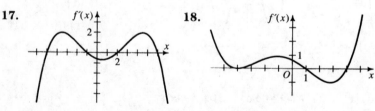

19. $f'(x) = 3x^2 - 6x + 3 = 3(x^2 - 2x + 1) = 3(x - 1)^2$; if $f'(x) = 0$, then $x = 1$; however,
$f'(x) > 0$ if $x \neq 1$; since $f'(x) > 0$ on both sides of $x = 1, f(x)$ has neither a local
maximum nor a local minimum.

20. a. $f(x) = x^3 - 6x^2 + 9x - 5$; $x = -\dfrac{b}{3a} = -\dfrac{-6}{3(1)} = 2$; $f(2) = -3$; $S(2, -3)$

b. $f'(x) = 3x^2 - 12x + 9 = 3(x - 3)(x - 1)$; $f'(x) = 0$ when $x = 1$ or $x = 3$; $f(x)$ has local maximum point $P(1, -1)$ and local minimum point $Q(3, -5)$.

c. Methods may vary. Slope of $\overline{PS} = \dfrac{-3 - (-1)}{2 - 1} = -2$; slope of $\overline{SQ} = \dfrac{-5 - (-3)}{3 - 2} = -2$; since the slopes are equal, P, Q, and S are collinear.

d. Midpoint of $\overline{PQ} = \left(\dfrac{1 + 3}{2}, \dfrac{-1 - 5}{2} \right) = (2, -3) = S$

21. a. $\displaystyle\lim_{h \to 0} \dfrac{\sin (0 + h) - \sin 0}{h} =$

$\displaystyle\lim_{h \to 0} \left(\dfrac{\sin h}{h} - 0 \right) = 1$ by Ex. 43 on p. 725.

b. From the diagram at the right, when $x = 0$,

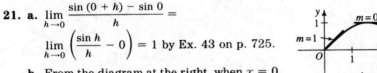

slope $= 1$; when $x = \dfrac{\pi}{2}$, slope $= 0$; when

$x = \pi$, slope $= -1$; when $x = \dfrac{3\pi}{2}$, slope $= 0$; when $x = 2\pi$, slope $= 1$.

c. Note that $\cos 0 = 1 = \cos 2\pi$, $\cos \dfrac{\pi}{2} = 0 = \cos \dfrac{3\pi}{2}$, and $\cos \pi = -1$; the derivative of $\sin x$ is $\cos x$.

22. $g(x) = x + 2 \cos x$; $g'(x) = 1 + 2(-\sin x) = 1 - 2 \sin x$; if $g'(x) = 0$, then $\sin x = \dfrac{1}{2}$;

$x = \dfrac{\pi}{6} + 2n\pi$ or $x = \dfrac{5\pi}{6} + 2n\pi$, where n is an integer. Since the sign of $g'(x)$ changes from positive to negative at each $x = \dfrac{\pi}{6} + 2n\pi$, $g(x)$ has infinitely many local maximum values. Since the sign of $g'(x)$ changes from negative to positive at each $x = \dfrac{5\pi}{6} + 2n\pi$, $g(x)$ has infinitely many local minimum values.

23. $f(x) = x + 2x^{-1}$; $f'(x) = 1 - 2x^{-2} = 1 - \dfrac{2}{x^2}$; neither $f(x)$ nor $f'(x)$ is defined at $x = 0$;

$f'(x) = 0$ when $1 = \dfrac{2}{x^2}$, $x^2 = 2$, and $x = \pm\sqrt{2}$; local maximum at $(-\sqrt{2}, -2\sqrt{2})$ and local minimum at $(\sqrt{2}, 2\sqrt{2})$; for large $|x|$, $f(x) \approx x$, so $y = x$ is an asymptote; $\displaystyle\lim_{x \to 0^+} (x + 2x^{-1}) = \infty$ and $\displaystyle\lim_{x \to 0^-} (x + 2x^{-1}) = -\infty$, so the y-axis is an asymptote.

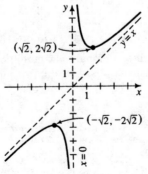

Ex. 23

24. $g(x) = x^2 + x^{-2}$; $g'(x) = 2x - 2x^{-3} = 2x - \dfrac{2}{x^3}$; neither $g(x)$ nor $g'(x)$ is defined at

$x = 0$; $g'(x) = 0$ when $2x = \dfrac{2}{x^3}$, $x^4 = 1$, and $x = \pm 1$; local minimums at $(1, 2)$ and

$(-1, 2)$; for large $|x|$, $g(x) \approx x^2$, so $y = x^2$ is an asymptote; $\lim\limits_{x \to 0} g(x) = \infty$, so the y-axis
is an asymptote.

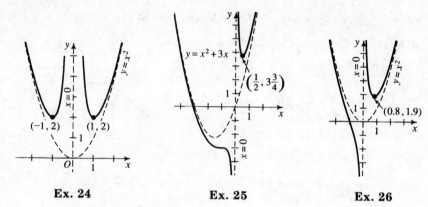

Ex. 24   Ex. 25   Ex. 26

25. $h(x) = x^2 + 3x + \dfrac{1}{x}$; $h'(x) = 2x + 3 - x^{-2} = 2x + 3 - \dfrac{1}{x^2}$; neither $h(x)$ nor $h'(x)$ is

defined at $x = 0$; $h'(x) = 0$ when $2x + 3 - \dfrac{1}{x^2} = 0$; $2x^3 + 3x^2 - 1 = 0$;

$(x + 1)^2(2x - 1) = 0$; $x = -1$ or $x = \dfrac{1}{2}$; since -1 is a double root of $h'(x)$, there is

no maximum or minimum at $x = -1$; local minimum at $\left(\dfrac{1}{2}, 3\dfrac{3}{4}\right)$; for large $|x|$, $h(x) \approx$

$x^2 + 3x$, so $y = x^2 + 3x$ is an asymptote; $\lim\limits_{x \to 0^+} h(x) = \infty$ and $\lim\limits_{x \to 0^-} h(x) = -\infty$, so the
y-axis is an asymptote.

26. $f(x) = x^2 + \dfrac{1}{x}$; $f'(x) = 2x - x^{-2} = 2x - \dfrac{1}{x^2}$; neither $f(x)$ nor $f'(x)$ is defined at $x = 0$;

$f'(x) = 0$ when $2x = \dfrac{1}{x^2}$; $x^3 = \dfrac{1}{2}$; $x = \sqrt[3]{\dfrac{1}{2}} = \dfrac{1}{\sqrt[3]{2}} = \dfrac{\sqrt[3]{4}}{2}$; $\left(\dfrac{1}{2}\sqrt[3]{4}, \dfrac{3}{2}\sqrt[3]{2}\right) \approx (0.8, 1.9)$ is

a local minimum; for large $|x|$, $f(x) \approx x^2$, so $y = x^2$ is an asymptote; $\lim\limits_{x \to 0^+} f(x) = \infty$ and
$\lim\limits_{x \to 0^-} f(x) = -\infty$, so the y-axis is an asymptote.

Activity 20-3 • page 769

a. On $-2 \le x \le 0$: 16 and 9; on $0 \le x \le 4$: 11 and -16; on $0 \le x \le 1$: 11 and 0

b. A global minimum, -16, exists; a global maximum does not exist because $f(6)$ is not
included in the interval, so it isn't possible to name a maximum value of $f(x)$ in the
interval.

c. $g(6) = 65$; $g(1) = 0$

d. Since $g(x)$ has a "corner" at each point for which $g(x) = 0$, $g'(x)$ is undefined at those
x-values.

Class Exercises 20-3 · page 772

1. $f'(x) = 2 \neq 0$, so the extreme values occur at the endpts.; $f(0) = -1; f(3) = 5$; max. value of 5 at $x = 3$ and min. value of -1 at $x = 0$.

2. $f'(x) = -1 \neq 0; f(1) = 2; f(4) = -1$; max. value of 2 at $x = 1$ and min. value of -1 at $x = 4$.

3. $g'(x) = 2x; g'(x) = 0$ if $x = 0; g(0) = 2; g(-1) = 3; g(2) = 6$; max. value of 6 at $x = 2$ and min. value of 2 at $x = 0$.

4. $g'(x) = -2x; g'(x) = 0$ if $x = 0; g(0) = 2; g(-2) = -2; g(1) = 1$; max. value of 2 at $x = 0$ and min. value of -2 at $x = -2$.

5. $h'(x) = 3x^2 - 3 = 3(x + 1)(x - 1); h'(x) = 0$ if $x = \pm 1; x = -1$ is not in the interval; $h(1) = -2; h(0) = 0; h(2) = 2$; max. value of 2 at $x = 2$ and min. value of -2 at $x = 1$.

6. $h'(x) = 3x^2 + 3 > 0$ for all $x; h(-2) = -14; h(2) = 14$; max. value of 14 at $x = 2$ and min. value of -14 at $x = -2$.

7. $f'(x) = 2 \cos x$ (see Ex. 21, p. 768); $f'(x) = 0$ when $x = \dfrac{\pi}{2}; f\left(\dfrac{\pi}{2}\right) = 2 \sin \dfrac{\pi}{2} = 2$;

   $f(0) = 2 \sin 0 = 0; f(\pi) = 2 \sin \pi = 0$; max. value of 2 at $x = \dfrac{\pi}{2}$ and min. value of 0

   at $x = 0$ and $x = \pi$.

8. Note that in the interval $0 \leq x \leq \pi, f(x) = \cos \dfrac{x}{2}$ is decreasing steadily; thus, the

   maximum value, 1, occurs at $x = 0$ and the minimum value, 0, occurs at $x = \pi$.

Written Exercises 20-3 · pages 772–774

1. **a.** Let the coordinates of the vertices of the lower side be $(x, 0)$ and $(-x, 0)$. Since the upper vertices are on the parabola $y = 3 - x^2$, their coordinates are $(x, 3 - x^2)$ and $(-x, 3 - x^2)$. The length of the rectangle is $|x - (-x)| = |2x|$ and its width is $|3 - x^2 - 0| = |3 - x^2|$. If $x \geq 0$, the length is $2x$ and the width is $3 - x^2$, so the area is $A(x) = (2x)(3 - x^2) = 6x - 2x^3$.

   b. Since $A(0) = 0 = A(\sqrt{3})$, $x = 0$ and $x = \sqrt{3}$ can be eliminated. The maximum area occurs when $A'(x) = 0; A'(x) = 6 - 6x^2 = 0$ when $6 = 6x^2; 1 = x^2; x = 1$ or -1. Since $x \geq 0$, choose $x = 1$.

2. **a.** The coordinates of the vertices of the lower base are $(3, 0)$ and $(-3, 0)$. Since the upper vertices are on the parabola $y = 9 - x^2$, their coordinates are $(x, 9 - x^2)$ and $(-x, 9 - x^2)$. The height of the trapezoid is $9 - x^2$, and its area is

   $A(x) = \dfrac{1}{2}h(b_1 + b_2); b_1 = |3 - (-3)| = 6$ and $b_2 = |x - (-x)| = |2x|$. If $x \geq 0$,

   $b_2 = 2x$, so $A(x) = \dfrac{1}{2}(9 - x^2)(6 + 2x)$.

   b. $A(0) = 27; A(3) = 0; A(x) = -x^3 - 3x^2 + 9x + 27; A'(x) = -3x^2 - 6x + 9 = 0$ when $-3(x^2 + 2x - 3) = 0; (x + 3)(x - 1) = 0; x = -3$ or $x = 1$. Since $x \geq 0$, choose $x = 1$. $A(1) = 32 > A(0)$, so the greatest area occurs when $x = 1$.

3. Let the coordinates of the vertex on the positive x-axis be $(x, 0)$; then the coordinates of the vertex on the line $y = 8 - 2x$ are $(x, 8 - 2x)$, and the coordinates of the vertex on the positive y-axis are $(0, 8 - 2x)$. The length of the rectangle is $|x - 0| = |x| = x$ when $x > 0$ and the width is $|8 - 2x - 0| = |8 - 2x|$. Since $(4, 0)$ is the x-intercept of $y = 8 - 2x$, $x < 4$. When $0 < x < 4$, the width becomes $8 - 2x$. The area $A(x) = x(8 - 2x) = 8x - 2x^2$. $A'(x) = 8 - 4x$; $A'(x) = 0$ when $8 - 4x = 0$; $x = 2$. The maximum area occurs when $x = 2$; $A(2) = 8(2) - 2(2)^2 = 8$ square units.

4. a. Since area $=$ length $\cdot$ width, $A = x \cdot$ width; width $= \dfrac{A}{x}$.

b. perimeter $= 2 \cdot$ length $+ 2 \cdot$ width $= 2x + \dfrac{2A}{x}$

c. $P(x) = 2x + \dfrac{2A}{x}$; $P'(x) = 2 + 2A(-1)x^{-2} = 2 - \dfrac{2A}{x^2}$. $P'(x) = 0$ when $2 - \dfrac{2A}{x^2} = 0$; $2x^2 - 2A = 0$; $2x^2 = 2A$; $x^2 = A$; $x = \pm\sqrt{A}$. Since x is a length, choose $x = \sqrt{A}$. The minimum perimeter occurs when the length is $\sqrt{A}$ and the width is $\dfrac{A}{\sqrt{A}} = \dfrac{A\sqrt{A}}{A} = \sqrt{A}$.

d. the square with the given area

5. a. The length of the box is $15 - 2x$, the width is $8 - 2x$, and the height is x; $V(x) = x(15 - 2x)(8 - 2x)$, $0 \le x \le 4$.

b. $V(x) = 120x - 46x^2 + 4x^3$; $V'(x) = 120 - 92x + 12x^2$; $V'(x) = 0$ when $4(30 - 23x + 3x^2) = 0$; $(3x - 5)(x - 6) = 0$; $x = \dfrac{5}{3}$ or $x = 6$. Reject $x = 6$ since it results in a negative width. $V(x)$ is maximum when $x = \dfrac{5}{3}$ cm.

6. a. $V(x) = lwh = (1 - x)(1 - 2x)x = 2x^3 - 3x^2 + x$, $0 \le x \le \dfrac{1}{2}$

b. $V'(x) = 6x^2 - 6x + 1$; $V'(x) = 0$ when $x = \dfrac{6 \pm \sqrt{36 - 4(6)(1)}}{2(6)} = \dfrac{3 \pm \sqrt{3}}{6}$; reject $x = \dfrac{3 + \sqrt{3}}{6}$, since $\dfrac{3 + \sqrt{3}}{6} > \dfrac{1}{2}$; thus, $x = \dfrac{3 - \sqrt{3}}{6} \approx 0.21$ m.

7. a. $x = x^3$; $x^3 - x = 0$; $x(x + 1)(x - 1) = 0$; $x = 0, \pm 1$; $(0, 0)$, $(1, 1)$, $(-1, -1)$

b. At $x = a$, $a > 0$, the vertical distance between the graphs is $D(a) = a - a^3$ since $a > a^3$ when $0 < a < 1$. $D'(a) = 1 - 3a^2$; $D'(a) = 0$ when $1 = 3a^2$; $a^2 = \dfrac{1}{3}$; $a = \pm\sqrt{\dfrac{1}{3}} = \pm\dfrac{\sqrt{3}}{3}$; $a > 0$, so reject $-\dfrac{\sqrt{3}}{3}$; $a = \dfrac{\sqrt{3}}{3}$.

8. a. $V = lwh$; $4000 = x \cdot x \cdot h$; $h = \dfrac{4000}{x^2}$

b. $A(x) = x^2 + 4xh = x^2 + 4x\left(\dfrac{4000}{x^2}\right) = x^2 + \dfrac{16,000}{x} = x^2 + 16,000x^{-1}$, $x > 0$

c. $A'(x) = 2x - 16,000x^{-2} = 2x - \dfrac{16,000}{x^2}$; $A'(x) = 0$ when $2x = \dfrac{16,000}{x^2}$; $x^3 = 8000$; $x = 20$ cm and $h = \dfrac{4000}{20^2} = 10$ cm.

9. profit = revenue − costs; $P(x) = 18x − (300 + 12x + 0.2x^{3/2}) =$

$-0.2x^{3/2} + 6x − 300; P'(x) = -0.2\left(\dfrac{3}{2}x^{1/2}\right) + 6 = -0.3x^{1/2} + 6; P'(x) = 0$ when

$0.3x^{1/2} = 6; x^{1/2} = 20; \sqrt{x} = 20; x = 400;$ 400 sets a day

10. a. After t hours, the truck has traveled $60t$ kilometers and the car, $80t$ kilometers.
$d^2 = (60t)^2 + (100 − 80t)^2; d(t) = \sqrt{(60t)^2 + (10 − 80t)^2}, t \ge 0$

b. In simplified form, $d(t) = 20\sqrt{25t^2 − 40t + 25}$. Note that $d(0) = 100$ km. For
$t > 0$, $\sqrt{t}$ is at its minimum when t is at its minimum. Thus, $\sqrt{25t^2 − 40t + 25}$ is
at its minimum when $25t^2 − 40t + 25$ is minimum; taking the derivative,
$50t − 40 = 0; t = 0.8$ hours. $d(0.8) < d(0)$, so the distance is minimum 0.8 h after
noon, or at 12:48 P.M.

c. $d(0.8) = 20\sqrt{25(0.8)^2 − 40(0.8) + 25} = 20(3) = 60$ km

11. Let the base edges be x ft long and the height be h ft; $V = lwh; 8 = x \cdot x \cdot h; h = \dfrac{8}{x^2}$.

$C(x)$ = total cost = cost of base + cost of sides = cost of base + 4(cost of each side) =

$6x^2 + 4(3xh) = 6x^2 + 12x\left(\dfrac{8}{x^2}\right) = 6x^2 + \dfrac{96}{x}$ where $x > 0$; $C'(x) = 12x − 96x^{-2} =$

$12x − \dfrac{96}{x^2}; C'(x) = 0$ when $12x = \dfrac{96}{x^2}; x^3 = 8; x = 2; h = \dfrac{8}{2^2} = 2;$ a cube 2 ft on each

side will minimize the cost.

12. a. Use the cross-sectional diagram shown. The right triangles are
similar by AA similarity, so $\dfrac{12 − h}{r} = \dfrac{12 − h + h}{6}; \dfrac{12 − h}{r} = \dfrac{12}{6};$
$12 − h = 2r; h = 12 − 2r$ and $0 \le r \le 6$ since the cylinder is
inscribed in the cone.

b. $V(r) = \pi r^2 h = \pi r^2(12 − 2r) = 12\pi r^2 − 2\pi r^3;$ note that
$V(0) = V(6) = 0; V'(r) = 24\pi r − 6\pi r^2; V'(r) = 0$ when $6\pi r(4 − r) = 0;$
$r = 0$ (reject) or $r = 4; V(4) = \pi \cdot 4^2(12 − 2 \cdot 4) = 64\pi$ cubic units.

13. a. The cone has radius $\sqrt{9 − x^2}$ and height $3 + x; V = \dfrac{1}{3}\pi r^2 h; V(x) =$

$\dfrac{1}{3}\pi(9 − x^2)(3 + x)$ and $0 \le x \le r$ since the cone is inscribed in the sphere.

b. Since $V(3) = 0$, reject $x = 3; V(x) = \dfrac{1}{3}\pi(27 + 9x − 3x^2 − x^3) = -\dfrac{1}{3}\pi x^3 −$

$\pi x^2 + 3\pi x + 9\pi; V'(x) = -\pi x^2 − 2\pi x + 3\pi;$ if $V'(x) = 0, -\pi(x^2 + 2x − 3) = 0;$
$(x + 3)(x − 1) = 0; x = -3$ (reject since $x \ge 0$) or $x = 1; V(1) =$

$\dfrac{1}{3}\pi(9 − 1)(3 + 1) = \dfrac{32\pi}{3};$ since $V(0) = 9\pi < \dfrac{32\pi}{3}$, the maximum volume is

$\dfrac{32\pi}{3}$ cubic units.

14. The cylinder shown has radius $\sqrt{100 − x^2}$ and height $2x;$
$V = \pi r^2 h; V(x) = \pi(100 − x^2)(2x) = 200\pi x − 2\pi x^3;$ since the
cylinder is inscribed in the sphere, $0 \le x \le 10; V(0) = 0 = V(10),$
so reject $x = 0$ and $x = 10$. $V'(x) = 200\pi − 6\pi x^2; V'(x) = 0$

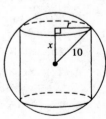

when $6\pi x^2 = 200\pi; x^2 = \dfrac{100}{3}; x \ge 0,$ so $x = \dfrac{10\sqrt{3}}{3}; V\left(\dfrac{10\sqrt{3}}{3}\right) =$

$\pi\left(100 − \dfrac{100}{3}\right)\left(\dfrac{20\sqrt{3}}{3}\right) = \dfrac{4000\pi\sqrt{3}}{9}$ cubic units.

15. a. The total cost per mile at 30 mi/h $= 6 + 0.004(30) + \dfrac{14.40}{30} = 6.12 + 0.48 = \$6.60$;

total cost per mile at 40 mi/h $= 6 + (0.004)(40) + \dfrac{14.40}{40} = 6.16 + 0.36 = \$6.52$

b. $C(x) = 6 + 0.004x + \dfrac{14.40}{x}$; $C'(x) = 0.004 - 14.40x^{-2} = 0.004 - \dfrac{14.40}{x^2}$; $C'(x) = 0$

when $0.004 = \dfrac{14.40}{x^2}$; $x^2 = 3600$; $x = 60$ since $x > 0$; 60 mi/h

c. Answers may vary. For example, a speed of 60 mi/h may be over the local speed limit or it may be an unsafe speed given the truck design, the load being carried, or the road conditions.

Class Exercises 20-4 • page 778

1. a. velocity at time $t = h'(t) = (48 - 32t)$ ft/s; $v(1) = h'(1) = 48 - 32(1) = 16$ ft/s

b. acceleration at time $t = h''(t) = v'(t) = -32$; the acceleration is always -32 ft/s^2.

2. a. 2 m **b.** average velocity $= \dfrac{\Delta s}{\Delta t} = \dfrac{2 - 0}{2 - 0} = 1$ m/s

c. The car moves to the right when $0 \le t < 2$ and moves to the left when $t > 2$.

d. at $t = 4$ seconds **e.** about 1 m/s; about 0 m/s; about -1 m/s

f. about $t = 0$ seconds; about $t = 4$ seconds **g.** 2 m

3. a. When $0 < t < 5$, $s(t) < 0$; left from the origin

b. $s(1) = 1^2 - 5(1) = -4$; 4 m to the left of the origin

c. $s(2) = 2^2 - 5(2) = -6$; 6 m **d.** average velocity $= \dfrac{\Delta s}{\Delta t} = \dfrac{-6 - 0}{2 - 0} = -3$ m/s

e. $s(t) = 0$ when $t^2 - 5t = 0$; $t(t - 5) = 0$; $t = 0$ (reject) or $t = 5$; at $t = 5$ s

f. $v(t) = s'(t) = 2t - 5$; $a(t) = v'(t) = 2$; the acceleration is 2 m/s^2 at $t = 1$ s and at $t = 2$ s.

4. a. to the left of the origin **b.** $v(t) > 0$; moving to the right **c.** $a(t) < 0$; decelerating

Written Exercises 20-4 • pages 778–781

1. a. average velocity $= \dfrac{\Delta s}{\Delta t} = \dfrac{h(2) - h(0)}{2 - 0} = \dfrac{144 - 80}{2} = 32$ ft/s

b. $h'(t) = (64 - 32t)$ ft/s; $h'(1) = 64 - 32(1) = 32$ ft/s

c. $h'(t) = 0$ when $64 = 32t$; $t = 2$; at $t = 2$ s

d. The maximum height is reached when $h'(t) = 0$, or $t = 2$; $h(2) = 144$; 144 ft.

e. $h(t) = 0$ when $80 + 64t - 16t^2 = 0$; $t^2 - 4t - 5 = 0$; $(t - 5)(t + 1) = 0$; $t = 5$ or $t = -1$ (reject); at $t = 5$ s

f. The ball is falling when $2 < t < 5$.

g. $h''(t) = -32$ since $h'(t) = 64 - 32t$; the acceleration is a constant -32 ft/s^2.

h. See right.

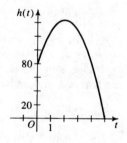

2. a. average velocity $= \dfrac{\Delta s}{\Delta t} = \dfrac{h(2) - h(0)}{2 - 0} = \dfrac{176.4 - 98}{2} = 39.2$ m/s

  b. $h'(t) = (49 - 9.8t)$ m/s; $h'(1) = 49 - 9.8(1) = 39.2$ m/s

  c. $h'(t) = 0$ when $49 = 9.8t$; $t = 5$; at $t = 5$ s

  d. The maximum height is reached when $h'(t) = 0$, or $t = 5$; $h(5) = 220.5$; 220.5 m

  e. $h(t) = 0$ when $98 + 49t - 4.9t^2 = 0$; $t^2 - 10t - 20$

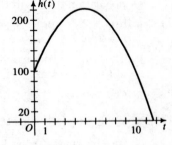

$$= 0; t = \dfrac{10 \pm \sqrt{100 - 4(1)(-20)}}{2 \cdot 1} = \dfrac{10 \pm 6\sqrt{5}}{2} =$$

    $5 \pm 3\sqrt{5}$; reject $5 - 3\sqrt{5}$ since it is negative; the helicopter lands at $t = 5 + 3\sqrt{5} \approx 11.7$ s.

  f. The helicopter is descending when $5 < t < 5 + 3\sqrt{5}$.

  g. $h'(t) = 49 - 9.8t$, so $h''(t) = -9.8$; -9.8 m/s^2

  h. See right.

3. a. about 2 m **b.** $\dfrac{2 \text{ m}}{2 \text{ s}} = 1$ m/s **c.** $0 \le t < 3.5$; $t > 3.5$ **d.** at $t = 6$ s

  e. The instantaneous velocity at $t = 1$ is about 1 m/s, at $t = 2$ is about 1 m/s, and at $t = 3$ is about 0.5 m/s. **f.** $t \approx 1.5$ s; $t \approx 5$ s **g.** 3 m

4. a. 1 m (to the left) **b.** $\dfrac{-1 \text{ m}}{2 \text{ s}} = -0.5$ m/s **c.** $t > 2$; $0 \le t < 2$ **d.** at $t = 3$ s

  e. The instantaneous velocity at $t = 1$ is about -0.5 m/s, at $t = 2$ is about 0 m/s, and at $t = 3$ is about 2 m/s. **f.** $t \approx 3$ s; $t \approx 1$ s **g.** 3 m right of the starting point

5.

7.

8.

6.

9.

10.

11. a. At the present time t, $P'(t) > 0$.

  b. At the present time t, $P''(t) < 0$, or $P'(t)$ is decreasing.

  c. At some future time t, $P'(t)$ will equal 0.

  d. See right.

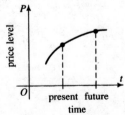

12. a. At time t_1, $P'(t_1) > 0$ and $P''(t_1) > 0$, while at time t_2, $P'(t_2) > 0$ and $P''(t_2) < 0$.

  b. Biologists who are studying the endangered species should conclude that at time t_1 both the population and rate of growth in the population are increasing; that is, the population is growing at an increasing rate. (Note that this is indicative of exponential growth.) At time t_2, the population is still growing, but at a decreasing rate. (This indicates that the population may be reaching a steady state, the population the environment will support.)

13. a. $d(t) = 2t^3 - 6t^2 + 8 = 2(t - 2)^2(t + 1)$; $d(t) = 0$ when $t = -1$ (reject) and when $t = 2$; at $t = 2$ s.

  b. The particle isn't moving when its velocity is 0; $v(t) = d'(t) = 6t^2 - 12t = 6t(t - 2)$; at $t = 0$ s and at $t = 2$ s.

  c. $a(t) = d''(t) = 12t - 12 = 0$; $t = 1$; at $t = 1$ s

14. a. $s'(t) = 10\left(\dfrac{3}{2}t^{1/2}\right) - 15 = 15\sqrt{t} - 15$; $s'(0.25) = 15\sqrt{0.25} - 15 = -7.5$; the car is traveling eastward at 7.5 mi/h.

  b. $s'(t) = 0$ when $15\sqrt{t} - 15 = 0$; $\sqrt{t} = 1$; $t > 0$ so $t = 1$; $s(1) = 10 \cdot 1^{3/2} - 15 \cdot 1 + 20 = 15$; when the velocity is 0, the time is 1:00 P.M. and the car is 15 mi west of Rockford.

15. a.

  b. $d \approx \sqrt{(160 - 80)^2 + (80 - 128)^2} = \sqrt{8704} \approx$ 93.3 ft

  c. $s(1.9) = (152, 86.24)$; $d \approx$
  $\sqrt{(152 - 160)^2 + (86.24 - 80)^2} =$
  $\sqrt{102.9376} \approx 10.1458$ ft; average
  speed $\approx \dfrac{10.1458}{0.1 \text{ s}} \approx 101.5$ ft/s

  d. $v(t) = (80, -32t)$; $v(2) = (80, -64)$; $|v(2)| =$
  $\sqrt{80^2 + (-64)^2} \approx 102.4$ ft/s; the ball hits the
  ground at time $t = 3$ s with a speed of $|v(3)| =$
  $|(80, -96)| = \sqrt{80^2 + (-96)^2} \approx 125.0$ ft/s.

Chapter Test · page 782

1. a. $f'(x) = 15x^2 - 6x + 1$

  b. $f'(x) = 3(-2x^{-3}) - 2(-x^{-2}) = -\dfrac{6}{x^3} + \dfrac{2}{x^2}$

  c. $f(x) = \sqrt[3]{3}x^{2/3} - x^{-1/3}$; $f'(x) = \sqrt[3]{3} \cdot \dfrac{2}{3}x^{-1/3} + \dfrac{1}{3}x^{-4/3} = \dfrac{2\sqrt[3]{3} \cdot x^{2/3}}{3x} + \dfrac{x^{2/3}}{3x^2} =$
  $\dfrac{2\sqrt[3]{3x^2}}{3x} + \dfrac{\sqrt[3]{x^2}}{3x^2}$

  d. $f'(x) = 2(-6x^{-7}) + 3(6x^5) = -\dfrac{12}{x^7} + 18x^5$

2. To sketch the graph of a function $f(x)$ using its derivative, first find the derivative and determine where it is zero or not defined. On an interval where the derivative is positive, the graph of $f(x)$ is rising; on an interval where the derivative is negative, the graph of $f(x)$ is falling. At a point where the derivative is zero, the graph is "flat," possibly a local maximum or minimum. If the derivative changes sign from positive to negative, the "flat" point is a local maximum; if the derivative changes sign from negative to positive, the "flat" point is a local minimum. If the derivative does not change sign at a point where the derivative is 0, the graph of $f(x)$ "flattens out," but is neither a maximum nor a minimum.

3. $f(x) = 12x - x^3; f'(x) = 12 - 3x^2 = -3(x + 2)(x - 2);$
$f'(x) = 0$ when $x = \pm 2; f'(x) < 0$ when $x < -2$ or $x > 2;$
$f'(x) > 0$ when $-2 < x < 2;$ thus, $(-2, -16)$ is a local
minimum point and $(2, 16)$ is a local maximum point.

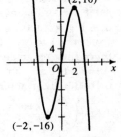

4. $f(2)$ is the maximum value of $f(x)$.

5. a. $A(x, h) = 2x^2 + 4xh$ **b.** $x^2 h = 1000; h = \dfrac{1000}{x^2}$

c. $A(x) = 2x^2 + 4x\left(\dfrac{1000}{x^2}\right) = 2x^2 + \dfrac{4000}{x}, x > 0$

d. $A'(x) = 4x - \dfrac{4000}{x^2}; A'(x) = 0$ when $4x = \dfrac{4000}{x^2}; x^3 = 1000; x = 10; h = \dfrac{1000}{10^2} = 10;$
the surface area is minimized when $x = h = 10$ units.

6. $s(5) = 3 \cdot 5^2 - 5 \cdot 5 + 4 = 54$, so the position is 54 units to the right of the origin.
$v(t) = s'(t) = 6t - 5; v(5) = 25;$ at $t = 5$, the velocity is 25 units/s. $a(5) = v'(5) = 6;$
at $t = 5$, the acceleration is 6 units/s^2.

7. a. The arrow passes the edge of the cliff when $160t - 16t^2 = 336;$
$-16(t^2 - 10t + 21) = 0; -16(t - 7)(t - 3) = 0; t = 3$ or $t = 7;$ the arrow passes
the cliff in 3 s on its upward trip.

b. $v(t) = 160 - 32t; v(t) = 0$ when $t = 5; s(5) = 160 \cdot 5 - 16 \cdot 5^2 = 400; 400$ ft

c. From part (a), the arrow reaches the cliff's edge on its downward trip when $t = 7;$
7 s.

d. $v(7) = 160 - 32 \cdot 7 = -64; -64$ ft/s

Cumulative Review, Chapters 19–20 · page 783

1. a. $\displaystyle\lim_{x \to 1} \dfrac{2x^2 - x - 1}{2x^2 - 3x + 1} = \lim_{x \to 1} \dfrac{(2x + 1)(x - 1)}{(2x - 1)(x - 1)} = \dfrac{2 \cdot 1 + 1}{2 \cdot 1 - 1} = 3$

b. $\displaystyle\lim_{x \to \infty} \dfrac{3x^2 + 2}{2x^2 + x} = \lim_{x \to \infty} \dfrac{3 + \frac{2}{x^2}}{2 + \frac{1}{x}} = \dfrac{3 + 0}{2 + 0} = \dfrac{3}{2}$ **c.** $\displaystyle\lim_{x \to 2^+} \dfrac{x + 1}{2 - x} = -\infty$

2. a.

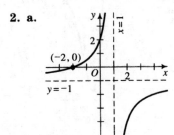

b.

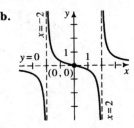

c.

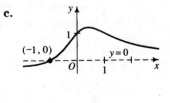

3. a.
```
10 LET A = 0
20 FOR X = 1.01 TO 2 STEP .01
30    LET Y = LOG(X)/LOG(10)
40    LET A = A + .01 * Y
50 NEXT X
60 PRINT "THE AREA IS APPROXIMATELY ";A
70 END
```

LOG(X) is the *natural* logarithmic function. To get the *common* logarithmic function, use the change of base formula:

$$\log_{10} x = \frac{\log_e x}{\log_e 10}$$

THE AREA IS APPROXIMATELY .1692687

b. Each point on the graph of $y = 1 + \log x$ is 1 unit higher than the point with the same x-coordinate on the graph of $y = \log x$. Thus, the area under the curve $y = 1 + \log x$ from $x = 1$ to $x = 2$ is equal to the area under the curve $y = \log x$ from $x = 1$ to $x = 2$ plus the area of a rectangle that is 1 unit high and 1 unit wide. Since the area under the curve $y = \log x$ from $x = 1$ to $x = 2$ is approximately 0.1693, the area under the curve $y = 1 + \log x$ from $x = 1$ to $x = 2$ is approximately 0.1693 + 1, or about 1.169.

4. a. $\text{Tan}^{-1}\left(\dfrac{1}{x}\right) = \dfrac{1}{x} - \dfrac{1}{3x^3} + \dfrac{1}{5x^5} - \dfrac{1}{7x^7} + \cdots$ **b.** $-1 \le \dfrac{1}{x} \le 1; x \ge 1$ or $x \le -1$

5. a. $10, 1, -0.8, -1.16$ **b.** Use a method from Section 19-5. $\lim\limits_{n \to \infty} f^n(10) = -1.25$

6. Solve $x = x^2 + \dfrac{1}{4}$; $4x^2 - 4x + 1 = 0$; $(2x - 1)^2 = 0$; $x = \dfrac{1}{2}$; thus, $x_0 = \dfrac{1}{2}$ is a fixed point; since $f(-x) = f(x)$ for all x, $x_0 = -\dfrac{1}{2}$ is an eventually fixed point; if $|x_0| < \dfrac{1}{2}$, then the orbit of x_0 tends to $\dfrac{1}{2}$. If $|x_0| > \dfrac{1}{2}$, then the orbit of x_0 tends to infinity.

7. Change line 80 to: 80 IF G > 90 THEN PRINT G, X
Input 2.5 for C and 0.3 for X. The population has an equilibrium point of 0.6 (that is, 60% of capacity).

8. a. $f'(x) = 3(2x) - 1 + 4(-x^{-2}) - \dfrac{1}{2}(-2x^{-3}) = 6x - 1 - \dfrac{4}{x^2} + \dfrac{1}{x^3}$

b. $f(x) = \sqrt{2} \cdot x^{1/2} - 2x^{-1/2}; f'(x) = \sqrt{2} \cdot \dfrac{1}{2}x^{-1/2} - 2\left(-\dfrac{1}{2}x^{-3/2}\right) = \dfrac{\sqrt{2}}{2\sqrt{x}} + \dfrac{1}{\sqrt{x^3}} = \dfrac{\sqrt{2x}}{2x} + \dfrac{\sqrt{x}}{x^2}$

9. $f(x) = 1 - 2x^2; f'(x) = -4x; f'(-1) = -4(-1) = 4$; slope is 4.

10. $f(x) = 3x^4 + 4x^3 - 12x^2 + 6; f'(x) = 12x^3 + 12x^2 - 24x = 12x(x + 2)(x - 1)$. If $f'(x) = 0$, then $x = -2$, or $x = 0$, or $x = 1; f'(x) < 0$ for $x < -2$ and for $0 < x < 1; f'(x) > 0$ for $-2 < x < 0$ and for $x > 1$. Thus, $f(x)$ has a local maximum at $(0, 6)$ and local minimums at $(-2, -26)$ and $(1, 1)$.

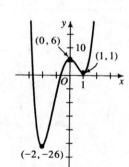

11. The length and width of the box are $10 - 2x$, and the height is x. $V(x) = x(10 - 2x)^2 = 4x^3 - 40x^2 + 100x$, $0 \le x \le 5$. Since $V(0) = V(5) = 0$, reject $x = 0$ and $x = 5$. $V'(x) = 12x^2 - 80x + 100 = 4(3x - 5)(x - 5)$; if $V'(x) = 0$, then $x = \dfrac{5}{3}$ or $x = 5$ (reject). The volume is maximum when $x = \dfrac{5}{3}$, that is, when the box is $6\frac{2}{3}$ in. by $6\frac{2}{3}$ in. by $1\frac{2}{3}$ in.

12. $s(t) = 8t - t^2$; $v(t) = s'(t) = 8 - 2t$; $v(t) > 0$ for $0 \le t < 4$ and $v(t) < 0$ for $t > 4$.

    a. $0 \le t < 4$ **b.** $t > 4$

Appendix 1

Exercises • pages 823–824

1. 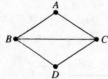 Vertex valences: A: 2; B: 3; C: 3, D: 2. No Euler circuit is possible.

2. 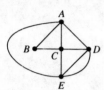 Vertex valences: A: 4; B: 2; C: 4; D: 4; E: 4. An Euler circuit is possible. Answers may vary. For example, *ABCDECAEDA.*

3. Answers may vary. For example, *BACBDC.*

4.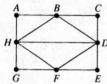
a. No; vertices D and H have valence 5.

b. Yes; for example, *HABCDEFGHBDFHD.*

5.
a. No, since vertices U and O (respresenting the utility room and the outdoors) have valence 3.

b. Yes; for example, *OHLDKHNKUOU.*

6. a.

V	In	Out
B	1	1
C	1	3
D	1	1
E	2	2
F	1	1
G	2	2
H	1	1
I	3	1

b. No Euler circuit is possible because the invalence and outvalence of vertex C are not equal. A similar problem occurs at vertex I.

c. Reverse the arrow from C to I so that it points from I to C. Euler circuits may vary. For example, *ABCAICDEIGFEGHA.*

7. a. The Bears and the Tigers have the most wins so far (2 each).

b.

508

c. The Bears should be ranked first, with three wins and no losses. The Eagles should be ranked last, with three losses and no wins. To rank the teams, subtract the invalence (losses) from the outvalence (wins) for each vertex (team) and order the results from greatest to least as shown at the right. The last game, between the Bears and the Eagles, does not need to be played. Even if the Eagles beat the Bears, leaving the Bears and Lions both with a 3–1 win-loss record, the Bears rank first because they beat the Lions. Likewise, the Panthers and Eagles both would have a 1–3 win-loss record, but the Eagles still rank last because the Panthers beat them.

V	Out	In
B	3	0
L	3	1
T	2	2
P	1	3
E	0	3

Exercises · pages 826–827

1. No

2. Yes. For example, *DQFXYHPEZVLD*.

3. a. Yes

b. No

c. At least one of the values x or y must be odd.

4.

Circuit	Distance
FEHGF	470
FEGHF	625
FHEGF	515
FHGEF	625
FGHEF	470
FGEHF	515

The shortest distance is 470 mi for circuit *FEHGF* or its reverse circuit *FGHEF*.

5. Yes, the nearest-neighbor algorithm gives the shortest circuit, *FGHEF*.

6. a. $14! \approx 8.7178 \times 10^{10}$

b. $14 \times 14! \times 10^{-9} \approx 1220.5$ s; it will take the computer about 20 min 20.5 s to compute the distances for all circuits visiting 14 cities.

7. a. 20! **b.** 20 **c.** $20 \times 20!$ **d.** $20 \times 20! \times 10^{-9} \approx 4.8658 \times 10^{10}$ s ≈ 1542 yr

8. a. 25! **b.** 25 **c.** $25 \times 25!$ **d.** $25 \times 25! \times 10^{-9} \approx 3.8778 \times 10^{17}$ s $\approx 1.229 \times 10^{10}$ yr

9. a. The graph is not drawn exactly to scale.

b. *ADCEBA*, the circuit given by the nearest-neighbor algorithm, costs $850. (A less expensive circuit exists: *ADBCEA* costs $700.)

10. Answers may vary. For example, school bus routes to pick up students, or bicycle courier routes to drop off packages.

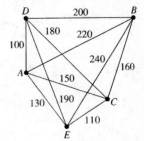

Exercises · pages 828–829

1. 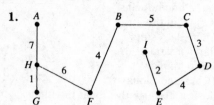   **2.**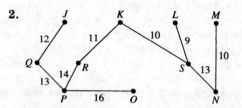

3. $n - 1$

4. $FE + EA + AB + ED + DC = 150 + 140 +$
$250 + 100 + 200 = \$840$

5. a. Answers may vary. For example, agriculturalists might design a system of irrigation ditches to carry a maximum flow of water.

b.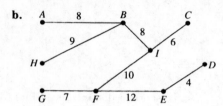

6.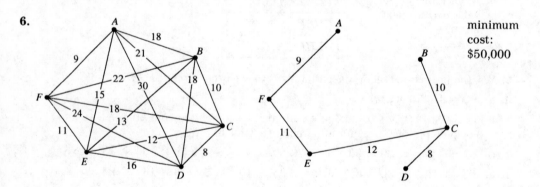

minimum
cost:
$50,000

7. $9900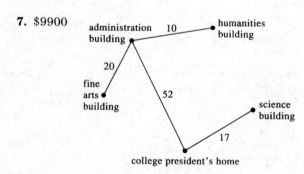

Appendix 2

Exercises · pages 832–834

1. x: $1, 2, 3, 5$; median is $\dfrac{2+3}{2} = 2.5$; y: $1, 2, 4, 5$; median is $\dfrac{2+4}{2} = 3$;

   summary point $= (2.5, 3)$

2. x: $1, 2, 3, 4, 5, 6, 8$; median is 4; y: $1, 2, 2, 3, 4, 4, 5$; median is 3; summary point $= (4, 3)$

3. **a.** $8, 8, 8$ **b.** $6, 5, 6$ **c.** $20, 21, 20$

4. **a.** $m_1 = \dfrac{6-3}{8-2} = \dfrac{1}{2}$;

   l_1: $y = \dfrac{1}{2}x + 2$;

   l_2: $y = \dfrac{1}{2}x + \dfrac{7}{2}$

   b. $k = \dfrac{2}{3}(2) + \dfrac{1}{3}\left(\dfrac{7}{2}\right) = \dfrac{5}{2}$;

   l_m: $y = \dfrac{1}{2}x + \dfrac{5}{2}$

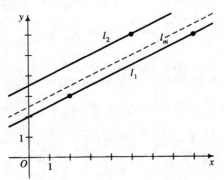

5. summary points: $(1, 2)$, $(4, 3)$, $(7, 5)$;

   $m = \dfrac{5-2}{7-1} = \dfrac{1}{2}$;

   l_1: $y = \dfrac{1}{2}x + \dfrac{3}{2}$; l_2: $y = \dfrac{1}{2}x + 1$;

   $k = \dfrac{2}{3}\left(\dfrac{3}{2}\right) + \dfrac{1}{3}(1) = \dfrac{4}{3}$; l_m: $y = \dfrac{1}{2}x + \dfrac{4}{3}$

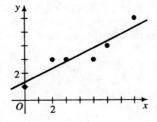

6. summary points: $(4, 8)$, $(8, 6)$, $(12, 2)$;

   $m = \dfrac{2-8}{12-4} = -\dfrac{6}{8} = -\dfrac{3}{4}$;

   l_1: $y = -\dfrac{3}{4}x + 11$; l_2: $y = -\dfrac{3}{4}x + 12$;

   $k = \dfrac{2}{3}(11) + \dfrac{1}{3}(12) = \dfrac{34}{3}$; l_m: $y = -\dfrac{3}{4}x + \dfrac{34}{3}$

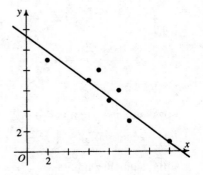

7. summary points: $(-1, 4)$, $\left(\dfrac{7}{2}, \dfrac{5}{2}\right)$, $(9, -1)$;

   $m = \dfrac{-1-4}{9+1} = -\dfrac{1}{2}$;

   l_1: $y = -\dfrac{1}{2}x + \dfrac{7}{2}$; l_2: $y = -\dfrac{1}{2}x + \dfrac{17}{4}$;

   $k = \dfrac{2}{3}\left(\dfrac{7}{2}\right) + \dfrac{1}{3}\left(\dfrac{17}{4}\right) = \dfrac{15}{4}$; l_m: $y = -\dfrac{1}{2}x + \dfrac{15}{4}$

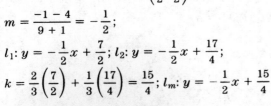
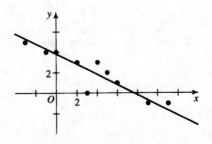

8. summary points: $(2,4)$, $(5,5)$, $(8,7)$;

$$m = \frac{7-4}{8-2} = \frac{1}{2};$$

$$l_1: y = \frac{1}{2}x + 3; \; l_2: y = \frac{1}{2}x + \frac{5}{2};$$

$$k = \frac{2}{3}(3) + \frac{1}{3}\left(\frac{5}{2}\right) = \frac{17}{6}; \; l_m: y = \frac{1}{2}x + \frac{17}{6}$$

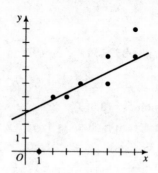

9. a. No change

b. No change

c. The point $(\overline{x}, \overline{y})$ is changed and so is the least-squares line. Before replacement, $(\overline{x}, \overline{y}) = (3.6, 2)$. After replacement, $(\overline{x}, \overline{y}) = (3.6, 4)$.

10. summary points: $(7,5)$, $(11,7)$, $(15,8)$; $m = \dfrac{8-5}{15-7} = \dfrac{3}{8}; \; l_1: y = \dfrac{3}{8}x + \dfrac{19}{8};$

$l_2: y = \dfrac{3}{8}x + \dfrac{23}{8}; \; k = \dfrac{2}{3}\left(\dfrac{19}{8}\right) + \dfrac{1}{3}\left(\dfrac{23}{8}\right) = \dfrac{61}{24};$ the median-median line has

equation $y = \dfrac{3}{8}x + \dfrac{61}{24}$ or $y = 0.38x + 2.54$; the least-squares line has equation

$y = 0.47x + 1.82$. The median-median line has a larger y-intercept and a smaller slope. The median-median line predicts a 20-year-old will wear a size 10 shoe and the least-squares line predicts a 20-year-old will wear a shoe size greater than 11.

11. a.

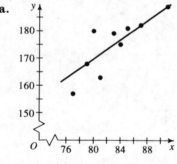

Boys

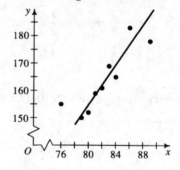

Girls

Boys: summary points are $(79, 168)$, $(83, 175)$, $(87, 182)$. l_1, l_2, l_m all have the same

equation, $y = \dfrac{7}{4}x + \dfrac{119}{4}$ or $y = 1.75x + 29.75$.

Girls: summary points are $(79, 152)$, $(82, 161)$, $(86, 178)$; $l_1: y = \dfrac{26}{7}x - \dfrac{990}{7};$

$l_2: y = \dfrac{26}{7}x - \dfrac{1005}{7}; \; l_m: y = \dfrac{26}{7}x - \dfrac{995}{7} \approx 3.71x - 142.14$

b. boy at 18: ≈ 166 cm; girl at 18: ≈ 147 cm

12. a. summary points: $(472, 97.625)$,
$(1305, 130.25)$, $(2396.5, 204.25)$;

$$m = \frac{204.25 - 97.625}{2396.5 - 472} = \frac{106.625}{1924.5} \approx 0.0554;$$

$l_1: y = 0.0554x + 71.4743;$

$l_2: y = 0.0554x + 57.9478;$

$l_m: y = 0.0554x + 66.9655$

b. Cost $\approx \$136.50$

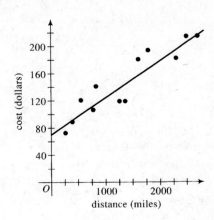

13. a. summary points: $(5.95, 10.15)$,
$(46, 40.1)$, $(77, 39.55)$;

$$m = \frac{39.55 - 10.15}{77.0 - 5.95} \approx 0.4138;$$

$l_1: y = 0.4138x + 7.6879;$

$l_2: y = 0.4138x + 21.0655;$

$k = \dfrac{2}{3}(7.6879) + \dfrac{1}{3}(21.0655);$

$k = 12.1471;$

$l_m: y = 0.4138x + 12.1471$

b. U.S.A. (prediction = 43.6,
reality = 81.1)

c. predicted number =
$(0.4138)(89.0) + 12.1471 \approx$
49.0 televisions per 100 people

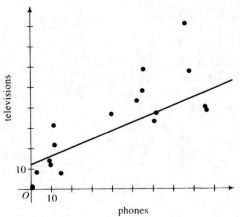

14. a.

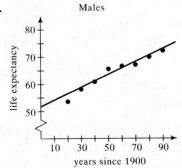

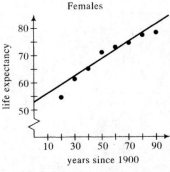

Males: summary points = $(30, 58.1)$, $(55, 66.1)$, $(80, 70)$; $m = \dfrac{70 - 58.1}{80 - 30} = 0.238;$

$l_1: y = 0.238x + 50.96;\ l_2: y = 0.238x + 53.01;$

$k = \dfrac{2}{3}(50.96) + \dfrac{1}{3}(53.01) \approx 51.64;$

$l_m: y = 0.238x + 51.64$

(Continued on next page)

       Females: summary points $= (30, 61.6), (55, 72.1), (80, 77.4)$;

$$m = \frac{77.4 - 61.6}{80 - 30} = 0.316;$$

       $l_1: y = 0.316x + 52.12$; $l_2: y = 0.316x + 54.72$;

$$k = \frac{2}{3}(52.12) + \frac{1}{3}(54.72) \approx 52.99;$$

       $l_m: y = 0.316x + 52.99$

  b. The life expectancy for females is greater than for males and is improving at a greater rate than for males.

  c. Boy: $0.238(100) + 51.64 = 75.44 \approx 75.4$ yr;
     Girl: $0.316(100) + 52.99 = 84.59 \approx 84.6$ yr

15. Answers may vary.

Appendix 3

Exercises · page 837

	Sign Changes in $P(x)$	Sign Changes in $P(-x)$	Number of Positive Roots	Number of Negative Roots	Number of Imaginary Roots
1.	1	1	1	1	0
2.	2	0	2	0	0
			0	0	2
3.	2	1	2	1	0
			0	1	2
4.	1	2	1	2	0
			1	0	2
5.	2	2	2	2	0
			2	0	2
			0	2	2
			0	0	4
6.	2	0	2	0	2
			0	0	4
7.	3	1	3	1	0
			1	1	2
8.	3	1	3	1	0
			1	1	2

9–16. M is the least nonnegative integral upper bound for the real roots of $P(x) = 0$ and L is the greatest nonpositive integral lower bound.

9. a.

x	coefficients of quotient and remainder			
1	1	−1	2	7
2	1	0	3	11
−1	1	−3	0	5

all ≥ 0; $M = 2$

alternately ≥ 0 and ≤ 0; $L = -1$

b. $-1 < x < 2$

10. a.

x	coefficients of quotient and remainder			
3	1	−1	5	10
4	1	0	8	27
−1	1	−5	13	−18

all ≥ 0; $M = 4$

alternately ≥ 0 and ≤ 0; $L = -1$

b. $-1 < x < 4$

11. a.

x	coefficients of quotient and remainder			
2	2	1	−1	1
3	2	3	6	21
−1	2	−5	2	1
−2	2	−7	11	−19

all ≥ 0; $M = 3$

alternately ≥ 0 and ≤ 0; $L = -2$

b. $-2 < x < 3$

12. a.

x	coefficients of quotient and remainder			
1	2	0	1	9
−1	2	−4	5	3
−2	2	−6	13	−18

all ≥ 0; $M = 1$

alternately ≥ 0 and ≤ 0; $L = -2$

b. $-2 < x < 1$

13. a.

y	coefficients of quotient and remainder			
0	1	1	0	0	16
−1	1	0	0	0	16

all ≥ 0; $M = 0$

alternately ≥ 0 and ≤ 0; $L = -1$

b. $-1 < y < 0$

14. a.

y	coefficients of quotient and remainder				
1	1	1	1	−1	4
2	1	2	4	6	9
−1	1	−1	1	−3	0

all ≥ 0; $M = 2$

alternately ≥ 0 and ≤ 0; $L = -1$

b. $-1 < y < 2$

15. a.

y	coefficients of quotient and remainder				
2	1	0	−3	−2	4
3	1	1	0	4	20
−1	1	−3	0	4	4
−2	1	−4	5	−6	20

all ≥ 0; $M = 3$

alternately ≥ 0 and ≤ 0; $L = -2$

b. $-2 < y < 3$

16. a.

y	coefficients of quotient and remainder				
1	1	3	2	9	4
−3	1	−1	2	1	−8
−4	1	−2	7	−21	79

all ≥ 0; $M = 1$

alternately ≥ 0 and ≤ 0; $L = -4$

b. $-4 < y < 1$

Appendix 4

Exercises · pages 840–841

1. $\dfrac{6x}{(x-1)(x+5)} = \dfrac{A}{x-1} + \dfrac{B}{x+5}$; $6x = A(x+5) + B(x-1)$; If $x = 1$, $A = 1$; If $x = -5$,

$B = 5$; $\dfrac{6x}{(x-1)(x+5)} = \dfrac{1}{x-1} + \dfrac{1}{x+5}$

2. $\dfrac{2x+1}{(x-4)(x-1)} = \dfrac{A}{x-4} + \dfrac{B}{x-1}$; $2x+1 = A(x-1) + B(x-4)$; If $x = 1$, $B = -1$;

If $x = 4$, $A = 3$; $\dfrac{2x+1}{(x-4)(x-1)} = \dfrac{3}{x-4} - \dfrac{1}{x-1}$

3. $7x + 6 = A(4x - 3) + B(x + 3)$; If $x = -3$, $A = 1$; If $x = \dfrac{3}{4}$, $B = 3$;

$\dfrac{7x+6}{4x^2 + 9x - 9} = \dfrac{3}{4x-3} + \dfrac{1}{x+3}$

4. $\dfrac{4x + 1}{6x^2 - 7x + 2} = \dfrac{A}{3x - 2} + \dfrac{B}{2x - 1}$; $4x + 1 = A(2x - 1) + B(3x - 2)$; If $x = \dfrac{2}{3}$, $A = 11$;

If $x = \dfrac{1}{2}$, $B = -6$; $\dfrac{4x + 1}{6x^2 - 7x + 2} = \dfrac{11}{3x - 2} - \dfrac{6}{2x - 1}$

5. $\dfrac{x^2 + 8}{x^3 + 3x^2 + 2x} = \dfrac{A}{x} + \dfrac{B}{x + 1} + \dfrac{C}{x + 2}$; $x^2 + 8 = A(x + 1)(x + 2) + Bx(x + 2) + Cx(x + 1)$;

If $x = 0$, $A = 4$; If $x = -1$, $B = -9$; If $x = -2$; $C = 6$; $\dfrac{x^2 + 8}{x^3 + 3x^2 + 2x} = \dfrac{4}{x} - \dfrac{9}{x + 1} + \dfrac{6}{x + 2}$

6. $\dfrac{x - 6}{x^3 + x^2 - 6x} = \dfrac{A}{x} + \dfrac{B}{x - 2} + \dfrac{C}{x + 3}$; $x - 6 = A(x - 2)(x + 3) + Bx(x + 3) + Cx(x - 2)$;

$A + B + C = 0$

$A + 3B - 2C = 1$

$-6A = -6$

$A = 1$, $B = -\dfrac{2}{5}$, $C = -\dfrac{3}{5}$; $\dfrac{x - 6}{x^3 + x^2 - 6x} = \dfrac{1}{x} - \dfrac{\frac{2}{5}}{x - 2} - \dfrac{\frac{3}{5}}{x + 3}$

7. $\dfrac{x + 6}{x^2 - 4} = \dfrac{A}{x - 2} + \dfrac{B}{x + 2}$; $x + 6 = A(x + 2) + B(x - 2)$; If $x = 2$, $A = 2$;

If $x = -2$, $B = -1$; $\dfrac{x + 6}{x^2 - 4} = \dfrac{2}{x - 2} - \dfrac{1}{x + 2}$

8. $\dfrac{x^2 + x + 6}{x^3 + 2x^2 - 5x - 6} = \dfrac{A}{x + 3} + \dfrac{B}{x + 1} + \dfrac{C}{x - 2}$;

$x^2 + x + 6 = A(x + 1)(x - 2) + B(x + 3)(x - 2) + C(x + 3)(x + 1)$;

$A + B + C = 1$

$-A + B + 4C = 1$

$-2A + B + 3C = 6$

$A = \dfrac{6}{5}$; $B = -1$, $C = \dfrac{4}{5}$; $\dfrac{x^2 + x + 6}{x^3 + 2x^2 - 5x - 6} = \dfrac{\frac{6}{5}}{x + 3} - \dfrac{1}{x + 1} + \dfrac{\frac{4}{5}}{x - 2}$

9. $\dfrac{3x - 1}{(x + 4)^2} = \dfrac{A}{x + 4} + \dfrac{B}{(x + 4)^2}$; $3x - 1 = A(x + 4) + B$; If $x = -4$, $B = -13$; If $x = 0$ and

$B = -13$, $A = 3$; $\dfrac{3x - 1}{(x + 4)^2} = \dfrac{3}{x + 4} - \dfrac{13}{(x + 4)^2}$

10. $\dfrac{x^2 + x + 2}{(x - 2)^3} = \dfrac{A}{x - 2} + \dfrac{B}{(x - 2)^2} + \dfrac{C}{(x - 2)^3}$; $x^2 + x + 2 = A(x - 2)^2 + B(x - 2) + C$;

If $x = 2$, $C = 8$; $x^2 + x + 2 = Ax^2 + (-4A + B)x + 8$; $A = 1$;

$1 = -4 + B$; $B = 5$; $\dfrac{x^2 + x + 2}{(x - 2)^3} = \dfrac{1}{x - 2} + \dfrac{5}{(x - 2)^2} + \dfrac{8}{(x - 2)^3}$

11. $\dfrac{x^2 + 1}{x^3 + 3x^2 + 3x + 1} = \dfrac{A}{x + 1} + \dfrac{B}{(x + 1)^2} + \dfrac{C}{(x + 1)^3}$; $x^2 + 1 = A(x + 1)^2 + B(x + 1) + C$;

If $x = -1$, $C = 2$; $x^2 + 1 = Ax^2 + (2A + B)x + 2$; $A = 1$; $0 = 2A + B$; $B = -2$;

$\dfrac{x^2 + 1}{x^3 + 3x^2 + 3x + 1} = \dfrac{1}{x + 1} - \dfrac{2}{(x + 1)^2} + \dfrac{2}{(x + 1)^3}$

12. $\dfrac{x + 2}{x^2 - 2x + 1} = \dfrac{A}{x - 1} + \dfrac{B}{(x - 1)^2}$; $x + 2 = A(x - 1) + B$; If $x = 1$, $B = 3$;

$x + 2 = Ax - A + 3$; $A = 1$; $\dfrac{x + 2}{x^2 - 2x + 1} = \dfrac{1}{x - 1} + \dfrac{3}{(x - 1)^2}$

13. $\dfrac{2x^2}{x^3 + x^2 - x - 1} = \dfrac{A}{x - 1} + \dfrac{B}{x + 1} + \dfrac{C}{(x + 1)^2}$;

$2x^2 = A(x + 1)^2 + B(x + 1)(x - 1) + C(x - 1)$;

$2x^2 = (A + B)x^2 + (2A + C)x + (A - B - C)$;

$A + B = 2$

$2A + C = 0$

$A - B - C = 0$

$A = \dfrac{1}{2}; B = \dfrac{3}{2}, C = -1; \dfrac{2x^2}{x^3 + x^2 - x - 1} = \dfrac{\frac{1}{2}}{x - 1} + \dfrac{\frac{3}{2}}{x + 1} - \dfrac{1}{(x + 1)^2}$

14. $\dfrac{6}{x^4 - x^2} = \dfrac{A}{x - 1} + \dfrac{B}{x + 1} + \dfrac{C}{x^2}; 6 = A(x + 1)x^2 + B(x - 1)x^2 + C(x - 1)(x + 1)$;

If $x = 0, C = -6$; If $x = 1, A = 3$, If $x = -1, B = -3$; $\dfrac{6}{x^4 - x^2} = \dfrac{3}{x - 1} - \dfrac{3}{x + 1} - \dfrac{6}{x^2}$

15. $\dfrac{x^3 - x^2}{(x + 3)^4} = \dfrac{A}{x + 3} + \dfrac{B}{(x + 3)^2} + \dfrac{C}{(x + 3)^3} + \dfrac{D}{(x + 3)^4}$;

$x^3 - x^2 = A(x + 3)^3 + B(x + 3)^2 + C(x + 3) + D; A = 1$; If $x = -3, D = -36$;

$x^3 - x^2 = x^3 + (3A + B)x^2 + (3A + 2B + C)x + 9B + 3C - 9; B = -10; C = 33$;

$\dfrac{x^3 - x^2}{(x + 3)^4} = \dfrac{1}{x + 3} - \dfrac{10}{(x + 3)^2} + \dfrac{33}{(x + 3)^3} - \dfrac{36}{(x + 3)^4}$

16. $5x^2 - 27x + 21 = A(x - 4)(x + 3) + B(x + 3) + C(x - 4)^2$;

$5x^2 - 27x + 21 = (A + C)x^2 + (-A + B - 8C)x + (-12A + 3B + 16C)$;

$5 = A + C$

$-27 = -A + B - 8$

$21 = -12A + 3B + 16C$

$A = 2, B = -1$, and $C = 3$

17. a. For all values of A and B, the expression $A(x + 2) + B(x - 8)$ is linear and thus will not produce the quadratic term x^2.

b. $x^2 - 6x - 16 = \overset{\displaystyle 2}{\overline{)2x^2 + 0x - 28}}$

$ \underline{2x^2 - 12x - 32}$

$ 12x + 4$

Thus, $\dfrac{2x^2 - 28}{x^2 - 6x - 16} = 2 + \dfrac{12x + 4}{x^2 - 6x - 16}$

c. $\dfrac{10}{x - 8} + \dfrac{2}{x + 2} = \dfrac{10(x + 2) + 2(x - 8)}{(x - 8)(x + 2)} = \dfrac{12x + 4}{x^2 - 6x - 16}$

18. $\dfrac{4x^2 - x}{2x^2 + x - 3} = 2 + \dfrac{-3x + 6}{2x^2 + x - 3} = 2 + \dfrac{-3x + 6}{(x - 1)(2x + 3)}; \dfrac{-3x + 6}{(x - 1)(2x + 3)} = \dfrac{A}{x - 1} + \dfrac{B}{2x + 3}$;

$-3x + 6 = A(2x + 3) + B(x - 1)$; If $x = 1, A = \dfrac{3}{5}$; If $x = -\dfrac{3}{2}; B = -\dfrac{21}{5}$;

$\dfrac{4x^2 - x}{2x^2 + x - 3} = 2 + \dfrac{\frac{3}{5}}{x - 1} - \dfrac{\frac{21}{5}}{(2x + 3)}$

19. $\dfrac{4x^3 - 3x^2 + 2x + 1}{x^2 + x} = 4x - 7 + \dfrac{9x + 1}{x^2 + x}; \dfrac{9x + 1}{x^2 + x} = \dfrac{9x + 1}{x(x + 1)}$;

$9x + 1 = A(x + 1) + Bx$; If $x = -1, B = 8$; If $x = 0, A = 1$;

$\dfrac{4x^3 - 3x^2 + 2x + 1}{x^2 + x} = 4x - 7 + \dfrac{1}{x} + \dfrac{8}{x + 1}$

20. $\dfrac{4x^3 - x + 2}{4x^2 + 4x + 1} = x - 1 + \dfrac{2x + 3}{4x^2 + 4x + 1} = x - 1 + \dfrac{2x + 3}{(2x + 1)^2}$;

$\dfrac{2x + 3}{(2x + 1)^2} = \dfrac{A}{2x + 1} + \dfrac{B}{(2x + 1)^2}$

$2x + 3 = A(2x + 1) + B$; If $x = -\dfrac{1}{2}, B = 2$; $2x + 3 = 2Ax + A + 2$;

If $x = 0, A = 1$; $\dfrac{4x^3 - x + 2}{4x^2 + 4x + 1} = x - 1 + \dfrac{1}{2x + 1} + \dfrac{2}{(2x + 1)^2}$

21. $\dfrac{5x^2 - 2x - 1}{(x - 2)(x^2 + x - 1)} = \dfrac{A}{x - 2} + \dfrac{Bx + C}{x^2 + x - 1}$;

$5x^2 - 2x - 1 = A(x^2 + x - 1) + (Bx + C)(x - 2)$; If $x = 2, A = 3$;

$5x^2 - 2x - 1 = (3 + B)x^2 + (3 - 2B + C)x + (-3 - 2C)$; $B = 2$; $C = -1$;

$\dfrac{5x^2 - 2x - 1}{(x - 2)(x^2 + x - 1)} = \dfrac{3}{x - 2} + \dfrac{2x - 1}{x^2 + x - 1}$

22. $\dfrac{x^4 + 8}{x^3 - 15x - 4} = x + \dfrac{15x^2 + 4x + 8}{x^3 - 15x - 4}$; $\dfrac{15x^2 + 4x + 8}{x^3 - 15x - 4} = \dfrac{15x^2 + 4x + 8}{(x + 4)(x^2 + 4x + 1)}$;

$\dfrac{15x^2 + 4x + 8}{(x + 4)(x^2 + 4x + 1)} = \dfrac{A}{x + 4} + \dfrac{Bx + C}{x^2 + 4x + 1}$

$15x^2 + 4x + 8 = A(x^2 + 4x + 1) + (Bx + C)(x + 4)$;

$15 = A + B$

$4 = 4A + 4B + C$

$8 = A + 4C$;

$A = 8, B = 7, C = 0$; $\dfrac{x^4 + 8}{x^3 - 15x - 4} = x + \dfrac{8}{x + 4} + \dfrac{7x}{x^2 + 4x + 1}$

Appendix 5

Exercises · page 845

1. a. Let m be the slope of the line; $y - 5 = m(x - 0)$; $y = mx + 5$

   b. Let k be the y-intercept; $y = -2x + k$

2. $y = 3x + k$ and $x^2 + y^2 = 10$; $x^2 + (3x + k)^2 = 10$; $10x^2 + 6xk + k^2 - 10 = 0$;
$(6k)^2 - 4(10)(k^2 - 10) = -4k^2 + 400$;

   a. $-4k^2 + 400 > 0$ when $-10 < k < 10$

   b. $-4k^2 + 400 = 0$ when $k = \pm 10$

   c. $-4k^2 + 400 < 0$ when $k < -10$ or $k > 10$

3. $y = mx$ and $x^2 + (y - 4)^2 = 8$; $x^2 + (mx - 4)^2 = 8$; $(1 + m^2)x^2 - 8mx + 8 = 0$;
$(-8m)^2 - 4(1 + m^2)(8) = 0$; $32m^2 - 32 = 0$; $m = \pm 1$; $y = \pm x$

4. $y = mx$ and $y = (x + 3)^2 - 5$; $x^2 + (6 - m)x + 4$; $(6 - m)^2 - 4(1)(4) = 0$;
$m^2 - 12m + 20 = 0$; $(m - 2)(m - 10) = 0$; $m = 2$ or 10; $y = 2x, y = 10x$

5. $y = -x + k$ and $xy = 1$; $x(-x + k) - 1 = 0$; $x^2 - kx - 1 = 0$; $k^2 - 4(1)(-1) = 0$;
$k = \pm 2$; $y = -x + 2, y = -x - 2$

6. $y = -3x + k$ and $9x^2 + 4y^2 = 36$;
$9x^2 + 4(-3x + k)^2 - 36 = 0$;
$45x^2 - 24kx + (4k^2 - 36) = 0$;
$(-24k)^2 - 4(45)(4k^2 - 36) = 0$;
$-144k^2 + 6480 = 0$;
$k^2 = 45$; $k = \pm 3\sqrt{5}$;
$y = -3x + 3\sqrt{5}$; $y = -3x - 3\sqrt{5}$

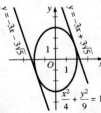

7. $y = mx - 1$ and $(x + 1)^2 + 4y^2 = 4$;
$(x + 1)^2 + 4(mx - 1)^2 - 4 = 0$;
$(1 + 4m^2)x^2 + (2 - 8m)x + 1 = 0$;
$(2 - 8m)^2 - 4(1 + 4m^2)(1) = 0$; $48m^2 - 32m = 0$;
$m = 0$ or $\dfrac{2}{3}$; $y = -1$, $y = \dfrac{2}{3}x - 1$

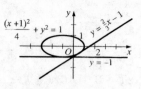

8. $y = m(x + 2)$ and $y = 6x - x^2$; $x^2 + (m - 6)x + 2m$; $(m - 6)^2 - 4(1)(2m) < 0$;
$m^2 - 20m + 36 < 0$; $(m - 18)(m - 2) < 0$; $2 < m < 18$

9. a. $y = -2x + k$ and $y = x^2 + 4x$; $x^2 + 6x - k = 0$; $k = -9$; $y = -2x - 9$

   b. $-2x - 9 = x^2 + 4x$; $x^2 + 6x + 9 = 0$; $x = -3$, $y = -3$; point of tangency:
     $(-3, -3)$

   c. The normal has slope $\dfrac{1}{2}$ and contains $(-3, -3)$; $y = \dfrac{1}{2}x - \dfrac{3}{2}$

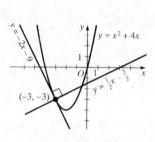
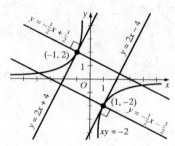

Ex. 9   Ex. 10

10. a. $y = 2x + k$ and $xy = -2$; $x(2x + k) = -2$; $2x^2 + kx + 2 = 0$; $k^2 - 4(2)(2) = 0$;
    $k = \pm 4$; $y = 2x \pm 4$

   b. If $y = 2x + 4$, $x(2x + 4) + 2 = 0$; $2x^2 + 4x + 2 = 0$; $x = -1$; If $x = -1$, $y = 2$.
    If $y = 2x - 4$, $x(2x - 4) + 2 = 0$; $2x^2 - 4x + 2 = 0$; $x = 1$; If $x = 1$, $y = -2$.

   c. The normal at $(-1, 2)$ has slope $-\dfrac{1}{2}$; $y = -\dfrac{1}{2}x + \dfrac{3}{2}$

    The normal at $(1, -2)$ has slope $-\dfrac{1}{2}$; $y = -\dfrac{1}{2}x - \dfrac{3}{2}$

11. a. $y = mx + 4$ and $x^2 + y^2 = 8$;
    $x^2 + (mx + 4)^2 = 8$;
    $(1 + m^2)x^2 + 8mx + 8 = 0$;
    $(8m)^2 - 4(1 + m^2)(8) = 0$; $32m^2 - 32 = 0$;
    $m = \pm 1$; $y = x + 4$, $y = -x + 4$

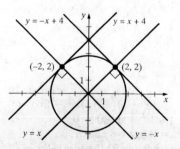

   b. If $y = x + 4$, $2x^2 + 8x + 8 = 0$; $x = -2$;
    If $x = -2$, $y = 2$.
    If $y = -x + 4$, $2x^2 - 8x + 8 = 0$; $x = 2$;
    If $x = 2$, $y = 2$.

   c. The normal at $(-2, 2)$ has slope -1, $y = -x$;
    The normal at $(2, 2)$ has slope 1, $y = x$

12. a. $y = mx$ and $y = x^2 + 4$; $x^2 - mx + 4 = 0$;
$(-m)^2 - 4(1)(4) = 0$; $m = \pm 4$; $y = \pm 4x$

b. If $y = 4x$, $x^2 - 4x + 4 = 0$; $x = 2$;
If $x = 2$, $y = 8$
If $y = -4x$, $x^2 + 4x + 4 = 0$; $x = -2$;
If $x = -2$, $y = 8$

c. The normal at $(2, 8)$ has slope $-\dfrac{1}{4}$; $y = -\dfrac{1}{4}x + \dfrac{17}{2}$

The normal at $(-2, 8)$ has slope $\dfrac{1}{4}$; $y = \dfrac{1}{4}x + \dfrac{17}{2}$

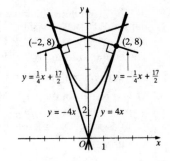

Appendix 6

Exercises • page 847

1. The equation $X_{1T} = 350$ locates the wall 350 ft from the batter. The equation $Y_{1T} = 5T$ makes the wall extend from $y = 5(0) = 0$ ft (the ground) to $y = 5(7.4) = 37$ ft (the height of the Green Monster).

2. No. The ball hits the wall 17 ft above the ground, too high for an outfielder.

3. a. $Y_{2T} = 3 + 65T - 16T^2$

b. No. The ball hits the wall 34.5 ft above the ground, or 2.5 ft below the wall's top.

4. Yes. The equations for the ball's path are $X_{2T} = 100T$ and $Y_{2T} = 3 + 66T - 16T^2$. Trace the graph of these equations to find that $Y_{2T} = 38$ when $X_{2T} = 350$. Since the height of the wall is 37 ft, the ball goes 1 ft over the top of the wall.

5. $v = \sqrt{100^2 + 60^2} = \sqrt{13{,}600} \approx 117$ ft/s; $\theta = \mathrm{Tan}^{-1}\left(\dfrac{60}{100}\right) \approx 31.0°$

6. a, b. A home run occurs when $36° \le \theta \le 59°$.

Exercises • pages 849–850

1. a. Enter the initial concentration, 115, and press the ENTER key. Press the multiplication key and enter 0.3. Press the ENTER key twice to display the concentration after 2 days. This concentration is about 10.4 ng/mL.

b. Set the graphing calculator to sequence mode. Enter the equation $U_n = 0.3U_{n-1}$. Set the WINDOW variables as follows: U_nStart = 115, nStart = 0, nMin = 0, and nMax = 7. Use the TABLE feature to display the value of U_n for $n = 0, 1, 2, \ldots, 7$. The concentration after 7 days is $U_7 \approx 0.0252$ ng/mL.

2. a. Enter the initial concentration, 230, and press the ENTER key. Press the multiplication key and enter 0.3. Press the ENTER key twice to display the concentration after 2 days, 20.7 ng/mL. The information entered here differs from that for Exercise 1(a) in that the initial concentration is 230 rather than 115.

b. Set the graphing calculator to sequence mode. Enter the equation $U_n = 0.3U_{n-1}$. Set the WINDOW variables as follows: U_nStart = 230, nStart = 0, nMin = 0, and nMax = 7. Use the TABLE feature to display the value of U_n for $n = 0, 1, 2, \ldots, 7$. The concentration after 7 days is $U_7 \approx 0.0503$ ng/mL. The information entered here differs from that for Exercise 1(b) in that the value of U_nStart is 230 rather than 115.

3. Enter the initial concentration, 40, and press the ENTER key. Press the multiplication key, enter 0.5, press the addition key, and enter 40. Press the ENTER key 4 times to display the concentration after 4 days. This concentration is 77.5 ng/mL.

4. Set the graphing calculator to sequence mode. Enter the equation $U_n = 0.5U_{n-1} + 40$. Set the WINDOW variables as follows: $U_n\text{Start} = 40$, $n\text{Start} = 0$, $n\text{Min} = 0$, and $n\text{Max} = 8$. Use the TABLE feature to display the value of U_n for $n = 0, 1, 2, \ldots, 8$. The concentration after 8 days is $U_8 \approx 79.8$ ng/mL.

5. The concentration increases, approaching a limiting value of 80 ng/mL.

6. **a.** Use either Method 1 or Method 2, changing 40 to 120 in the steps described on page 849. Using either method, you obtain a concentration of 180 ng/mL after 1 day and 239.9 ng/mL after 10 days.

   b. 240 ng/mL

7. The results from Exercises 3–6 suggest that the initial concentration is half the limiting concentration. So to obtain a limiting concentration of 400 ng/mL, the size of the daily dose should be such that the initial concentration is 200 ng/mL.

Exercises • pages 851–852

1. Answers will vary. One good viewing window is defined by Xmin = 55, Xmax = 100, Xscl = 10, Ymin = 0, Ymax = 250, and Yscl = 50.

2. Yes, the scatter plot appears to be linear, since the plotted points lie close to a line.

3. Answers will vary. One example follows: The population of the United States has increased during the period 1960–1995, and a greater number of people generate a greater amount of trash per year.

4. The values in list L_4 (by year) are 0.49 (in 1960), 0.53 (in 1965), 0.60 (in 1970), 0.60 (in 1975), 0.67 (in 1980), 0.69 (in 1985), 0.79 (in 1990), and 0.79 (in 1995).

5. **a.** 0.53 tons   **b.** 0.67 tons   **c.** It increased from 0.49 tons to 0.79 tons.

6. The statement is false. The statement implies that the average amount of trash generated per person per year remained constant, when in fact it increased.

7. The values in list L_5 (by year) are 2.68 (in 1960), 2.91 (in 1965), 3.28 (in 1970), 3.26 (in 1975), 3.64 (in 1980), 3.78 (in 1985), 4.31 (in 1990), and 4.33 (in 1995).

8. Each value k in list L_4 is in tons per year. Therefore, the corresponding value

   in list L_5 is $\dfrac{k \text{ tons}}{1 \text{ year}} \times \dfrac{2000 \text{ lb}}{1 \text{ ton}} \times \dfrac{1 \text{ year}}{365 \text{ days}} \approx 5.48k$ lb/day. So the formula

   for L_5 converts tons per year to pounds per day.

9. **a.** Multiply the average amount of trash each person generated in 1995, 4.33 lb/day, by a percentage (expressed as a decimal) from the table to obtain one of these values: food wastes, 0.65 lb/day; paper, 1.65 lb/day; cardboard, 0.22 lb/day; plastics, 0.22 lb/day; leather and textiles, 0.087 lb/day; yard wastes, 0.43 lb/day; wood, 0.13 lb/day; metals and glass, 0.78 lb/day; dirt, ashes, etc., 0.17 lb/day.

   b. Answers will vary.

Exercises • page 853

1. In this exercise, the starting term is $U_{30} = 14{,}487$ rather than $U_{21} = 1000$, and the equation for U_n no longer has the term "+ 1000."

2. **a.** In the table, look at the row in which $n = 31$, and note that $U_{31} \approx 15{,}646$ and $V_{31} = 1000$. So Tim has \$15,646 and Tom has \$1000.

   b. In the table, look at the row in which $n = 32$, and note that $U_{32} \approx 16{,}898$ and $V_{32} = 2080$. So Tim has \$16,898 and Tom has \$2080.

3. In the table, scroll down to the row in which $n = 65$, and note that $U_{65} \approx 214{,}195$ and $V_{65} \approx 172{,}317$. So Tim has \$214,195 and Tom has \$172,317.

4. Tim's money has compounded for 10 more years than Tom's money.

5. Change the equation for U_n on page 852 to $U_n = 1.06U_{n-1} + 1000$. In the table, scroll down to the row in which $n = 30$, and note that $U_{30} \approx 13{,}181$. So Tim has \$13,181.

6. Use the procedure described in Exercise 2, but change the equations for U_n and V_n to $U_n = 1.06U_{n-1}$ and $V_n = 1.06V_{n-1} + 1000$, and change U_nStart to 13,181. Scroll down the table to the row in which $n = 65$, and note that $U_{65} \approx 101{,}310$ and $V_{65} \approx 111{,}435$. So Tim has \$101,310 and Tom has \$111,435.

Refresher Exercises • page 854

1. II
2. No; $2(4) - 5(-1) = 13 \neq 7$
3. **a.** $3\sqrt{63} = 3\sqrt{9}\sqrt{7} = 9\sqrt{7}$ **b.** $\sqrt{10}(\sqrt{2} + 2) = \sqrt{20} + 2\sqrt{10} = 2\sqrt{5} + 2\sqrt{10}$
4. $\sqrt{3^2 + 6^2} = \sqrt{45} = 3\sqrt{5}$ cm
5. $3(-2) + 4y = -9;\ 4y = -3;\ y = -\dfrac{3}{4}$
6. **a.** $5x - 8y = 16;\ 8y = 5x - 16;\ y = \dfrac{5}{8}x - 2$

   b. $\dfrac{y - 4}{x + 3} = \dfrac{3}{4};\ 4(y - 4) = 3(x + 3);\ 4y = 3x + 25;\ y = \dfrac{3}{4}x + \dfrac{25}{4}$
7. **a.** $(2x - 5)^2 = (2x)^2 - 2(2x)(-5) + (-5)^2 = 4x^2 - 20x + 25$
   b. $(\sqrt{3} + 2)(\sqrt{3} - 2) = (\sqrt{3})^2 - 2^2 = -1$
8. **a.** $x^2 - 5x + 6 = (x - 3)(x - 2)$ **b.** $x^2 - 25 = x^2 - 5^2 = (x - 5)(x + 5)$
9. $0 = x^2 + 2x - 3;\ (x + 3)(x - 1);\ x = -3, 1$
10. $(x - 2)^2 = 9;\ x - 2 = \pm 3;\ x = -1, 5$

Refresher Exercises • page 854

1. **a.** linear **b.** quadratic **c.** linear
2. $f(0) = 3(0)^2 - 0 - 1 = -1;\ f(-2) = 3(-2)^2 - (-2) - 1 = 13;$
   $f(i) = 3(i)^2 - i - 1 = 3(-1) - i - 1 = -4 - i$
3. $(-1)^3 - 3(-1)^2 - (-1) + 3 = 0;\ -1$ is a zero of $P(x) = x^3 - 3x^2 - x + 3$.
4. **a.** $x = \dfrac{2 \pm \sqrt{(-2)^2 - 4(1)(6)}}{2(1)} = 1 \pm i\sqrt{5}$ **b.** $2x(3 - 2x) = 0;\ x = 0$ or $\dfrac{3}{2}$;
   c. $(x - 1)(x^2 + 8) = 0;\ x = 1$ or $x = \pm 2i\sqrt{2}$
5. $5x^2 - 20x = 5x(x^2 - 4) = 5x(x + 2)(x - 2)$
6. The vertex is $(-1, -8)$; $y + 8 = k(x + 1)^2$; Since $(1, 0)$ is on the graph, $8 = 4k$; $k = 2$; $y + 8 = 2(x + 1)^2$; $y = 2x^2 + 4x - 6$

7. **a.**   **b.** maximum value 0

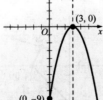

8. a. $A(x) = x(3 - 2x)$ **b.** $P(x) = 2x + 2(3 - 2x) = 6 - 2x$

   c. If $x = 2$, then $3 - 2x < -1$. A rectangle cannot have a side of negative length.

9. Height = diameter = $2r$; $V(r) = \pi r^2 h = 2\pi r^3$

10. $1 - 3i + 1 + 3i = 2$; $(1 - 3i)(1 + 3i) = 1 + 3i - 3i - 9i^2 = 1 + 3i - 3i + 9 = 10$

Refresher Exercises • page 855

1.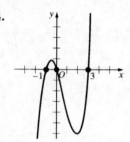

2. $5 - 4x \le -7$; $-4x \le -12$; $x \ge 3$

3. a. $5, -5$ **b.** $-5 < x < 5$ **c.** $-5 < x$ or $x > 5$

4.   **a.** below   **b.** above

                                          c. below   **d.** above

5. $y = (2y - 6)^2 - 4(2y - 6) + 3$

                          $0 = 4y^2 - 32y + 15$

                          $y = \dfrac{32 \pm \sqrt{32^2 - 4(4)(15)}}{2(4)} = 3 \text{ or } \dfrac{21}{4}; \ x = 0 \text{ or } \dfrac{9}{2};$

                          $(0,3), \left(\dfrac{9}{2}, \dfrac{21}{4}\right)$

6. a. $P(0,0) = 10(0) + 8(0) = 0$ **b.** $P(0,5) = 10(0) + 8(5) = 40$

   c. $P(3,3) = 10(3) + 8(3) = 54$ **d.** $P(5,2) = 10(5) + 8(2) = 66$

7. $a^3 - 2a^2 - a + 2 = a^2(a - 2) - (a - 2) = (a^2 - 1)(a + 2) = (a - 1)(a + 1)(a + 2)$

8. $4x^4 - 21x^2 + 27 = (4x^2 - 9)(x^2 - 3); \ 4x^2 - 9 = 0 \text{ or } x^2 - 3 = 0; \ x = \pm\dfrac{3}{2} \text{ or } \pm\sqrt{3}$

Refresher Exercises • page 855

1. The function is undefined when $x + 7 = 0$, that is, when $x = -7$.

2. Since $(x - 2)^2 \ge 0$, and $(x - 2)^2 - 1 \ge -1$, $y \ge -1$.

3. $k(a) = 3a + 2$; $k(2a) = 3(2a) + 2 = 6a + 2$; $k(a + 1) = 3(a + 1) + 2 = 3a + 5$

4.   **5.**   **6.**

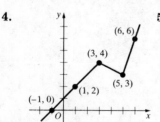
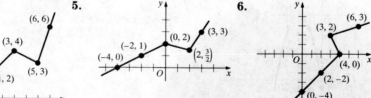
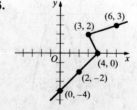
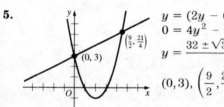

7. If $x = y^3 - 4$, $y^3 = x + 4$; $y = \sqrt[3]{x + 4}$

8. $y = 2x^2 + 8x + 5 = 2(x^2 + 4x + 4) - 3 = 2(x + 2)^2 - 3$; vertex $(-2, -3)$; since the parabola opens upward, the axis of symmetry is $x = -2$.

9. For example, 2 cm $\times$ 3 cm, 6 cm $\times$ 1 cm, $\sqrt{2}$ cm $\times$ $3\sqrt{2}$ cm

10. If each leg is twice as long as the base and the base has length b, then
perimeter $= b + 2b + 2b = 5b$

Refresher Exercises • page 856

1. a. $x^3 x^2 (4x^3)^2 = 4^2 (x^{3+2+3(2)}) = 16x^{11}$ **b.** $\dfrac{(3x)^3 y^4}{(6xy^2)^2} = \dfrac{27x^3 y^4}{36x^2 y^4} = \dfrac{3x}{4}$

2. a. 3.722 **b.** 2.718

3. $3, 10, 4$; $\sqrt{x}$, $x \geq 0$

4. $2^x = 64 = 2^6$; $x = 6$

5. $10^x = 1000 = 10^3$; $x = 3$

6. $\left(\dfrac{x}{2}\right)^3 = 125 = 5^3$; $\dfrac{x}{2} = 5$; $x = 10$

7. $x^5 = 0.00001 = 10^{-5}$; $x = \dfrac{1}{10} = 0.1$

8. $89(1.07) = 95.23$; $95.23

9.

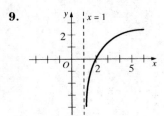

10. If $f(2) = 5$ and $f(6) = -1$, then $y - 5 = \dfrac{5 - (-1)}{2 - 6}(x - 2)$; $f(x) = y = -\dfrac{3}{2}x + 8$

Refresher Exercises • page 856

1. a. $f(x) = 2x^2 + 8x + 5 = 2(x^2 + 4x + 4) - 3 = 2(x + 2)^2 - 3$ **b.** $(-2, -3)$

2.

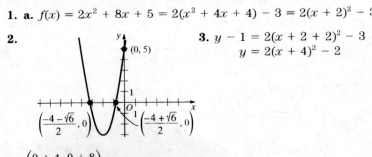

3. $y - 1 = 2(x + 2 + 2)^2 - 3$
$y = 2(x + 4)^2 - 2$

4. $\left(\dfrac{0 + 4}{2}, \dfrac{0 + 8}{2}\right) = (2, 4)$

5. Slope of $\overline{BC} = \dfrac{8 - 0}{4 - 10} = -\dfrac{4}{3}$; $y - 4 = -\dfrac{4}{3}(x - 2)$; $4x + 3y = 20$

6. a. Slope of $\overline{CM} = \dfrac{0 - 4}{10 - 2} = -\dfrac{1}{2}$; $y - 4 = -\dfrac{1}{2}(x - 2)$; $y = -\dfrac{1}{2}x + 5$

b. Slope of $\overline{AB} = \dfrac{8 - 0}{4 - 0} = 2$; since $2\left(-\dfrac{1}{2}\right) = -1$, $\overline{AB}$ and $\overline{CM}$ are perpendicular and since $\overline{AB}$ and $\overline{CM}$ contain M, $\overline{CM}$ is the perpendicular bisector of $\overline{AB}$.

7. $AC = \sqrt{(10 - 0)^2 + (0 - 0)^2} = 10$; $BC = \sqrt{(10 - 4)^2 + (0 - 8)^2} = 10$; the triangle is isosceles

8. a. $9 - x^2 \geq 0$; $x^2 \leq 9$; $-3 \leq x \leq 3$; $0 \leq y \leq 3$

b. $x - 3 \geq 0$, so $x \geq 3$; $y \leq 0$

9. $5 - 5x - 3x^2 = 2x - 1$; $3x^2 + 7x - 6 = 0$; $x = \dfrac{-7 \pm \sqrt{7^2 - 4(3)(-6)}}{2(3)} = -3$ or $\dfrac{2}{3}$;

If $x = -3$, $y = -7$; If $x = \dfrac{2}{3}$, $y = \dfrac{1}{3}$; $(-3, -7)$ or $\left(\dfrac{2}{3}, \dfrac{1}{3}\right)$

10. a. $x^2 + 4(-y)^2 = x^2 + 4y^2 = 1$; yes   **b.** $(-x)^2 + 4y^2 = x^2 + 4y^2 = 1$; yes

c. $y^2 + 4x^2 \neq x^2 + 4y^2$; no   **d.** $(-x)^2 + 4(-y)^2 = x^2 + 4y^2$; yes

Refresher Exercises • page 857

1. a.

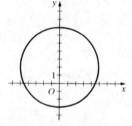

b. $r = 5$; circumference $= 2\pi(5) \approx 31.4$; area $= \pi(5)^2 \approx 78.5$

2. $\sqrt{6^2 - 4^2} = 2\sqrt{5}$; since P is in the second quadrant, $x = -2\sqrt{5}$

3. height $= \sqrt{10^2 - 5^2} = 5\sqrt{3}$

4. $\sqrt{5^2 + 5^2} = 5\sqrt{2}$

5. a. $x = -2$ **b.** $y = -3$ **c.** $x = 0, y = 0$ **d.** $\dfrac{x^2}{36} - \dfrac{y^2}{4} = 1$; $y = \pm\dfrac{1}{3}x$

6. Since $f(-1) = f(1) = 3$, the function is not one-to-one and hence has no inverse.

7. If the function is restricted to all nonnegative real numbers, the function has an inverse given by $f^{-1}(x) = \sqrt{x - 2}$ where $x \geq 2$.

8. $f(f^{-1}(x)) = (\sqrt{x - 2})^2 + 2 = x$; $f^{-1}(f(x)) = \sqrt{(x^2 + 2) - 2} = x$

Refresher Exercises • page 857

1.

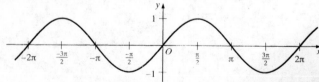

2. The hypotenuse of the associated right triangle is $\sqrt{29}$.

    a. Since θ is a third-quadrant angle, $\csc \theta < 0$; $\csc \theta = -\dfrac{\sqrt{29}}{5}$

    b. Since θ is a third-quadrant angle, $\sec \theta < 0$; $\sec \theta = -\dfrac{\sqrt{29}}{2}$

    c. $\cot \theta = \dfrac{\frac{1}{5}}{\frac{2}{5}} = \dfrac{2}{5}$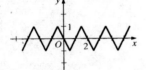

3. a. $\dfrac{\pi}{2}$ **b.** $-\dfrac{\pi}{6}$ **c.** $\dfrac{\pi}{4}$

4. 4; 1

5. a. 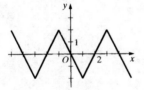   **b.** 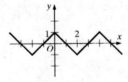   **c.**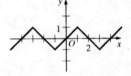

    d. 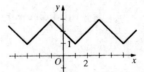   **e.** 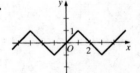   **f.**

6. $2y = 2y(y - 1)$; $2y^2 - 4y = 0$; $2y(y - 2) = 0$; $y = 0$ or 2

7. $2t^2 - t = 2 - \dfrac{1}{t}$; $2t^3 - t^2 - 2t + 1 = 0$; $t^2(2t - 1) - (2t - 1) = 0$;

    $(t^2 - 1)(2t - 1) = 0$; $t = \pm 1, \dfrac{1}{2}$

8. $2\sqrt{a} = a - 3$; $4a = a^2 - 6a + 9$; $a^2 - 10a + 9 = 0$; $(a - 9)(a - 1) = 0$; $a = 9$ or 1;
reject 1; $a = 9$

9. $\dfrac{\frac{a}{b} - \frac{b}{a}}{a - b} = \dfrac{a^2 - b^2}{ab(a - b)} = \dfrac{(a - b)(a + b)}{ab(a - b)} = \dfrac{a + b}{ab}$

10. $1 - x^2 = x^2 - 3x + 2$; $2x^2 - 3x + 1 = 0$; $(2x - 1)(x - 1) = 0$; $x = 1$ or $\dfrac{1}{2}$;

    If $x = 1$, $y = 0$; If $x = \dfrac{1}{2}$, $y = \pm \dfrac{\sqrt{3}}{2}$.

Refresher Exercises • page 858

  1. $x = 10 \sin 43° \approx 6.43$ **2.** $x = 3 \cot 27° \approx 5.89$

  3. Area $= 0.5(10)(8) \sin 30° = 20$ **4.**

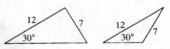

  5. $9 = 16 + 36 - 2(4)(6) \cos \theta$; $\cos \theta = 0.8957$; $\theta = 26.4°$

6. $\dfrac{\sin \theta}{4} = \dfrac{\sin 27.2°}{5}$; $\sin \theta = 0.3649$; $\theta = 21.4°$

7. $230° - 180° = 50°$; $120° - 50° = 70°$

8.

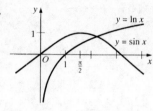

between 2 and 3

Refresher Exercises • page 858

1. $\sqrt{17^2 - 15^2} = 8$; Since the angle is a third-quadrant angle, $\cos x = -\dfrac{8}{17}$, $\tan x = \dfrac{15}{8}$,

$\csc x = -\dfrac{17}{15}$, $\sec x = -\dfrac{17}{8}$, and $\cot x = \dfrac{8}{15}$

2. Let $\theta = \mathrm{Cos}^{-1}\left(-\dfrac{1}{3}\right)$; $\sin \theta = \dfrac{2\sqrt{2}}{3}$

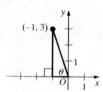

3. -1

4. $\dfrac{\tan 210° + \tan 120°}{1 - (\tan 210°)(\tan 120°)} = \tan(210° + 120°) = \tan 330° = -\dfrac{\sqrt{3}}{3}$

5. $\tan(-\theta) = \dfrac{\sin(-\theta)}{\cos(-\theta)} = \dfrac{-\sin \theta}{\cos \theta} = -\tan \theta$

6. $\tan \theta = \dfrac{3}{5}$; $\theta = 31°$

7. $\sin \theta \cdot \sec^2(90° - \theta) = \dfrac{\sin \theta}{\cos^2(90° - \theta)} = \dfrac{\sin \theta}{\sin^2 \theta} = \dfrac{1}{\sin \theta} = \csc \theta$

8. $2 + \cos^2 \theta = 3\sin^2 \theta$; $4\sin^2 \theta - 3 = 0$; $\sin \theta = \pm\dfrac{\sqrt{3}}{2}$; $\theta = 60°, 120°, 240°, 300°$

9.

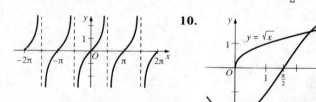

10.

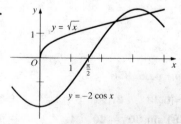

domain: all real numbers except the odd multiples of $\dfrac{\pi}{2}$; range: all real numbers

2 solutions

Refresher Exercises • page 859

1. $\cos 216° = -\cos 36°$ **2.** $x = 8 \cos 135° = -4\sqrt{2}$, $y = 8 \sin 135° = 4\sqrt{2}$

3. a. $AC = \sqrt{(7-0)^2 + (4-0)^2} = 8.1$ **b.** $\tan \angle BAC = \dfrac{7}{4}$; $\angle BAC = 60.3°$

4. $(1 + i)^3 = (1 + i)(1 + i)^2 = (1 + i)(2i) = -2 + 2i$

5. $(4 - i\sqrt{3})(2 + i\sqrt{3}) = 8 + (4\sqrt{3} - 2\sqrt{3})i - 3i^2 = 11 + 2\sqrt{3}i$

6. $(4 - i\sqrt{3}) + (2 + i\sqrt{3}) = 6$

7. $\dfrac{2 - 4i}{1 + i} = \dfrac{2 - 4i}{1 + i} \cdot \dfrac{1 - i}{1 - i} = \dfrac{-2 - 6i}{2} = -1 - 3i$

8. Since $(-\sqrt{3} + i)^2 = 3 - 2i\sqrt{3} + i^2 = 2 - 2i\sqrt{3}$, $-\sqrt{3} + i$ is a square root of $2 - 2i\sqrt{3}$.

9. $\cos 4\theta = 2 \cos^2 2\theta - 1 = 2(\cos^2 \theta - \sin^2 \theta)^2 - 1 = 8 \cos^4 \theta - 8 \cos^2 \theta + 1$

10. $2x^2 + 2x + 1 = 0$; $x = \dfrac{-2 \pm \sqrt{2^2 - 4(2)(1)}}{2(2)}$; $x = \dfrac{-1 \pm i}{2}$

Refresher Exercises • page 859

1. a. $AB = 7 - 3 = 4$; $OA = \sqrt{(3-0)^2 + (4-0)^2} = 5$; $OB = \sqrt{(7-0)^2 + (4-0)^2} = \sqrt{65}$

b. $5 = 4^2 + (\sqrt{65})^2 - 2(4)(\sqrt{65}) \cos \angle OBA$; $\cos \angle OBA = -0.6018$; $\angle OBA = 127°$

2. Let x be the required distance; $x = \sqrt{200^2 + 250^2 - 2(200)(250) \cos 70°} = 261.74$; $x \approx 261$ km

3. $y - 4 = \dfrac{4 - (-5)}{-2 - 1}(x + 2)$; $y - 4 = -3(x + 2)$; $y = -3x - 2$

4.

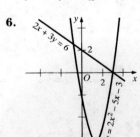

5. The center is $\left(\dfrac{2 + 8}{2}, \dfrac{-3 - 11}{2}\right) = (5, -7)$; the radius is $\sqrt{(8 - 5)^2 + (-11 - (-7))^2} = 5$; $(x - 5)^2 + (y + 7)^2 = 25$

6.

$2x^2 - 5x - 3 = 2 - \dfrac{2}{3}x$; $6x^2 - 13x - 15 = 0$

$x = \dfrac{13 \pm \sqrt{(-13)^2 - 4(6)(-15)}}{2(6)} = \dfrac{13 \pm 23}{12}$; $x = 3$ or $-\dfrac{5}{6}$;

If $x = 3$, $y = 0$; If $x = -\dfrac{5}{6}$, $y = \dfrac{23}{9}$.

7. $(x - 1)^2 + y^2 = (2 \cos^2 \theta - 1)^2 + (\sin 2\theta)^2 = \cos^2 \theta + \sin^2 \theta = 1$

8. $8 \operatorname{cis} 240° = 8(\cos 240° + i \sin 240°) = 8\left(-\dfrac{1}{2} - i\dfrac{\sqrt{3}}{2}\right) = -4 - 4i\sqrt{3}$

9. a. $2(0) + 3(0) - 4z = 10; z = -\dfrac{5}{2}$

b. Let $y = z = 0$, then $x = 5$; Let $x = 0$ and $y = 2$, then $z = -1$; so $(5, 0, 0)$ and $(0, 2, -1)$ are also solutions.

10. None, since both lines have the same slope, 1.5, but different y-intercepts.

Refresher Exercises • page 860

1. a.

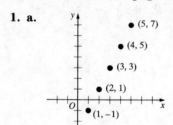

b.

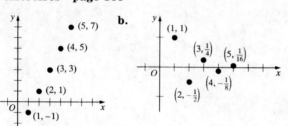

2. a. Area triangle $1 = \dfrac{1}{2}(8)(8) = 32$; Area triangle $2 = \dfrac{1}{2}(4)(4) = 8$;

Area triangle $3 = \dfrac{1}{2}(2)(2) = 2$;

b. Area of triangle $n = \dfrac{1}{2}(4^{4-n})$; If $n = 10$, area of triangle $10 = \dfrac{1}{2}(4^{-6}) = 2^{-7} = \dfrac{1}{128}$

3. $4 + 3(n - 1) = 511; 3n = 510; n = 170$

4. $5 \cdot 2^{n-1} = 20{,}480; 2^{n-1} = 4096 = 2^{12}; n - 1 = 12; n = 13$

5. $\dfrac{n(5 + 3n)}{2} = 175; n(5 + 3n) = 350; 3n^2 + 5n - 350 = 0;$

$n = \dfrac{-5 \pm \sqrt{5^2 - 4(3)(-350)}}{2(3)} = \dfrac{-5 \pm 65}{6} = 10 \text{ or } -\dfrac{35}{3}$

6. $\dfrac{1}{1 - 4n} = 2; 1 = 2 - 8n; n = \dfrac{1}{8}$

7. a. $\dfrac{4(1000)^2 + 3(1000) - 5}{2(1000)^2 - 1000} = 2.002$ **b.** $\dfrac{2 \sin 1000}{1000} = 0.002$

8. $k^2 + k + 2(k + 1) = k(k + 1) + 2(k + 1) = (k + 1)(k + 2)$

Refresher Exercises • page 860

1. a. $2x + 5y = -11$
$3x - 7y = 27$
$14x + 35y = -77$
$15x - 35y = 135$
$29x = 58;$
$x = 2; y = -3$

b. $x + 2y - z = -3$
$\qquad 2x - y + 2z = 7$
$\qquad\qquad 3y + 4z = 5$

Multiply the first equation by -2 and add to the second equation.

$-5y + 2z = 6$
$3y + 4z = 5$

Solve the system above to obtain $y = -1$ and $z = 2$.
Substitute these values into the first or second original equation to obtain $x = 1$.
The solution is $(1, -1, 2)$.

2. a. $(3)(-5) - (1)(2) = -17$

b. $\begin{vmatrix} 1 & 2 & -3 \\ 4 & -4 & 1 \\ 2 & 1 & 0 \end{vmatrix} = 1\begin{vmatrix} -4 & 1 \\ 1 & 0 \end{vmatrix} - 2\begin{vmatrix} 4 & 1 \\ 2 & 0 \end{vmatrix} + (-3)\begin{vmatrix} 4 & -4 \\ 2 & 1 \end{vmatrix}$

$\qquad\qquad = 1(-1) - 2(-2) + (-3)(12) = -33$

3. $\mathbf{u} - \mathbf{v} = (2, -7, 0) - (-3, 1, -6) = (5, -8, 6);$
$\qquad 2\mathbf{u} + \mathbf{v} = 2(2, -7, 0) + (-3, 1, -6) = (1, -13, -6);$
$\qquad \mathbf{u} \cdot \mathbf{v} = 2(-3) + (-7)(1) + 0(-6) = -13$

4. $\mathbf{u} \times \mathbf{v} = \begin{vmatrix} \mathbf{i} & \mathbf{j} & \mathbf{k} \\ 2 & -7 & 0 \\ -3 & 1 & -6 \end{vmatrix} = 42\mathbf{i} + 12\mathbf{j} - 19\mathbf{k};$

$\qquad \mathbf{v} \times \mathbf{u} = \begin{vmatrix} \mathbf{i} & \mathbf{j} & \mathbf{k} \\ -3 & 1 & -6 \\ 2 & -7 & 0 \end{vmatrix} = -42\mathbf{i} - 12\mathbf{j} + 19\mathbf{k}; \ \mathbf{u} \times \mathbf{v} \neq \mathbf{v} \times \mathbf{u}$

5. a. $(1, 0) \to (1, 0), (0, 1) \to (0, -1)$ **b.** $y = x - 1$

6. a. $(1, 0) \to (-1, 0), (0, 1) \to (0, 1)$ **b.** $y = x + 1$

7. a. $(1, 0) \to (0, 1), (0, 1) \to (1, 0)$ **b.** $y = -x + 1$

8. a. $(1, 0) \to (-1, 1), (0, 1) \to (-2, 2)$ **b.** $y = -x$

Refresher Exercises • page 861

1. By the Fundamental Counting Principle, there are $3 \times 2 = 6$ paths.

2. The first digit can be $1, 2, 3,$ or 4. The second digit can be one of the three digits not already used, and so on. There are $4 \times 3 \times 2 \times 1 = 24$ four-digit numbers using $1, 2, 3,$ and 4 only once.

3. $(1, 2), (1, 3), (1, 4), (2, 3), (2, 4), (3, 4)$; 6 different ways

4. a. 4 **b.** 5 **c.** 3 **d.** 4

5. $(x - y)^3 = (x^2 - 2xy + y^2)(x - y) = x^3 - 3x^2y + 3xy^2 - y^3$

6. $(2x + 1)^3 = (4x^2 + 4xy + 1)(2x + 1) = 8x^3 + 12x^2 + 6x + 1$

7. a. $2 + 2i = 2\sqrt{2}\left(\dfrac{\sqrt{2}}{2} + \dfrac{\sqrt{2}}{2}i\right) = 2\sqrt{2}(\cos 45° + i \sin 45°) = 2\sqrt{2} \text{ cis } 45°$

$\qquad$ **b.** $(2 + 2i)^4 = (2\sqrt{2})^4 \text{ cis }(4 \cdot 45°) = 64(-1) = -64$

8. a. $\mathbf{u} \cdot \mathbf{v} = 1(0.5) + 4(0.25) + (-6)(0.25) = 0$ **b.** The vectors are perpendicular.

Refresher Exercises • page 861

1. a. $\dfrac{13}{52} = \dfrac{1}{4}$ **b.** $\dfrac{12}{52} = \dfrac{3}{13}$ **c.** $\dfrac{4}{52} = \dfrac{1}{13}$

2. a. $4! = 24$ **b.** $\dfrac{4!}{1!\,1!\,2!} = 12$

3.

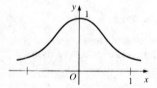

4. $_{39}C_5 = \dfrac{39!}{5!(39 - 5)!} = 575{,}757$

5. $_{52}C_5 - {_{39}C_5} = \dfrac{52!}{5!(52 - 5)!} - \dfrac{39!}{5!(39 - 5)!} = 2{,}023{,}203$

6. $_{13}C_5 = \dfrac{13!}{5!(13 - 5)!} = 1287$

7. $\dfrac{_4C_1 \cdot {_5C_2}}{_9C_3} = \dfrac{4 \cdot 10}{84} = \dfrac{10}{21}$

8. The middle term is the sixth term: $_{10}C_5(3x)^5(-2y)^5 = -1{,}959{,}552x^5y^5$

Refresher Exercises • page 862

1. 72% of the 25 students scored 70 or above; $0.72 \times 25 = 18$

2. $\dfrac{21}{32} = 0.656$, about 66%

3. a. $\dfrac{6 + 8 + 9 + 13}{4} = 9$ **b.** $\dfrac{(-3)^2 + (-1)^2 + 0^2 + 4^2}{4} = 6.5$

4.

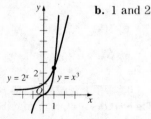

5. $(f \circ g)(x) = (\sqrt{x + 3})^2 - 4 = x + 3 - 4 = x - 1;\ x + 3 \geq 0$, that is, $x \geq -3$

6. a.   **b.** 1 and 2

$y = 2^x$ $y = x^3$

Refresher Exercises • page 862

1. Using the points $(85.2, 56.4)$ and $(325.8, 215)$, which are on the line, $y - 215 = \dfrac{215 - 56.4}{325.8 - 85.2}(x - 325.8)$; $y = 0.659x + 0.237$

2. The mean is 7.24; $(2.8 - 7.24)^2 + (3.7 - 7.24)^2 + (8.4 - 7.24)^2 + (9.6 - 7.24)^2 + (12.5 - 7.24)^2 = 66.828$; standard deviation $= \sqrt{\dfrac{66.828}{5}} = 3.65$

3. **a.**　　　　　　　　**b.**　　　　　　　　**c.**

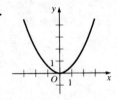　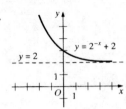　

4. **a.** $10 - 0.31(6) = 8.14$　**b.** $50(1.06)^6 = 70.9$　**c.** $25.8(6)^{0.4} = 52.8$

5. **a.** $\log x = 0.72$, $10^{\log x} = 10^{0.72}$; $x = 5.248$

   b. $4^x = 3$; $\log 4^x = \log 3$; $x = \log 3 = 0.792$

   c. $\ln x = 1.09$; $e^{\ln x} = e^{1.09}$; $x = 2.974$

6. **a.** $\log y = x + 2$; $10^{\log y} = 10^{x+2}$; $y = 10^{x+2}$; $y = (100)10^x$

   b. $\ln y = 2 \ln x$; $\ln y = \ln x^2$; $y = x^2$

   c. $\log y = 4 \log x + 3$; $\log y = \log x^4 + \log 1000$; $\log y = \log (1000x^4)$; $y = 1000x^4$

Refresher Exercises • page 863

1. **a.** As n increases without bound, $\sqrt{n}$ increases without bound; $\lim\limits_{x \to \infty} \sqrt{n} = +\infty$

   b. $\lim\limits_{x \to \infty} \dfrac{2n^3 - n}{7n^3 - n^2} = \lim\limits_{n \to \infty} \dfrac{2 - \dfrac{1}{n^2}}{7 - \dfrac{1}{n}} = \dfrac{2}{7}$

2. $\dfrac{x^2 - 9}{x^2 + 4x + 3} = \dfrac{x^2 - 9}{(x + 1)(x + 3)} = \dfrac{x - 3}{x + 1}$

3. $\dfrac{x - 4}{\sqrt{x - 4}} = \dfrac{x - 4}{\sqrt{x - 4}} \cdot \dfrac{\sqrt{x - 4}}{\sqrt{x - 4}} = \dfrac{(x - 4)(\sqrt{x - 4})}{x - 4} = \sqrt{x - 4}$

4. $(x - 4)(x + 2)(x - 1)^2 \geq 0$ when $(x - 4)(x + 2) \geq 0$ and when $x = 1$; $x - 4 \geq 0$ and $x + 2 \geq 0$, or $x - 4 \leq 0$ and $x + 2 \leq 0$, or $x = 1$; $x \leq -2$, or $x = 1$, or $x \geq 4$.

5. $(x - 5)(x + 3) > 0$ and $x^2 - 1 < 0$ or $(x - 5)(x + 3) < 0$ and $x^2 - 1 > 0$; $[(x < -3 \text{ or } x > 5) \text{ and } (-1 < x < 1)]$ or $[(-3 < x < 5) \text{ and } (x < -1 \text{ or } x > 1)]$; $-3 < x < -1$ or $1 < x < 5$

6. Area $= \dfrac{1}{2}(1)(1) + \dfrac{1}{2}(1)(1 + 4) = 3$

7. $1 + 1 + 0.5 + 0.1666 + 0.4166 + 0.0083 = 2.716$; e

8. $f(1) = \sqrt{2} + 1 = 2$; $f(f(1)) = f(2) = \sqrt{2} + 1 = 2.414$; $f(f(f(1))) = f(2.414) = \sqrt{2.414} + 1 = 2.554$

Refresher Exercises • page 863

1. $y + 4 = -\dfrac{2}{3}(x + 1)$; $3y + 12 = -2x - 2$; $2x + 3y = -14$

2. a. $\dfrac{2^x - 1}{x - 0} = \dfrac{2^x - 1}{x}$

   b. $\dfrac{2^2 - 1}{2} = 1.5$; $\dfrac{2^1 - 1}{1} = 1$; $\dfrac{2^{0.5} - 1}{0.5} = 0.828$; $\dfrac{2^{0.1} - 1}{0.1} = 0.718$

3. $\dfrac{2(x + h)^2 + 1 - (2x^2 + 1)}{h} = \dfrac{2x^2 + 4xh + 2h^2 + 1 - 2x^2 - 1}{h} = \dfrac{4xh + h^2}{h} = 4x + 2h$

4. $y = x^2 - 6x + 2 = (x - 3)^2 - 7$; vertex $(3, -7)$

5.

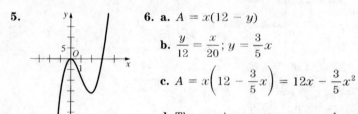

6. a. $A = x(12 - y)$

   b. $\dfrac{y}{12} = \dfrac{x}{20}$; $y = \dfrac{3}{5}x$

   c. $A = x\left(12 - \dfrac{3}{5}x\right) = 12x - \dfrac{3}{5}x^2$

   d. The maximum area occurs when $x = \dfrac{-12}{2\left(-\dfrac{3}{5}\right)} = 10$;

         maximum area $= 12(10) - \dfrac{3}{5}(10^2) = 60$